Skeletal Variation and Adaptation in Europeans

Skeletal Variation and Adaptation in Europeans

Upper Paleolithic to the Twentieth Century

Edited by Christopher B. Ruff

Center for Functional Anatomy and Evolution
Johns Hopkins University School of Medicine
Baltimore, MD, USA

WILEY Blackwell

This edition first published 2018

Registered Offices
John Wiley & Sons, Inc., 111 River Street, Hoboken, NJ 07030, USA

Editorial Office
111 River Street, Hoboken, NJ 07030, USA

For details of our global editorial offices, customer services, and more information about Wiley products visit us at www.wiley.com.

Wiley also publishes its books in a variety of electronic formats and by print-on-demand. Some content that appears in standard print versions of this book may not be available in other formats.

Library of Congress Cataloging-in-Publication Data

Names: Ruff, Christopher, editor.
Title: Skeletal variation and adaptation in Europeans : upper Paleolithic to the Twentieth Century / edited by Christopher B. Ruff.
Description: Hoboken, NJ : John Wiley & Sons, 2018. | Includes bibliographical references and index. |
Identifiers: LCCN 2017052568 (print) | LCCN 2017058923 (ebook) | ISBN 9781118628034 (pdf) | ISBN 9781118628027 (epub) | ISBN 9781118627969 (cloth)
Subjects: LCSH: Human skeleton–Variation–Europe. | Human skeleton–Analysis. | Paleoanthropology–Europe.
Classification: LCC QM101 (ebook) | LCC QM101 .S548 2018 (print) | DDC 611/.71–dc23
LC record available at https://lccn.loc.gov/2017052568

Cover Design: Wiley
Cover Image: Simon Bening. September: plowing and sowing seed. Book of Hours ("Da Costa Hours"), Bruges, c. 1515. MS. M.399, f.10v.
Image courtesy of Akademische Druck- u. Verlagsanstalt, Graz/Austria.
The Morgan Library & Museum, New York, NY, USA
Photo Credit: The Morgan Library & Museum / Art Resource, NY

Set in 10/12pt Warnock by SPi Global, Pondicherry, India
Printed in Singapore by C.O.S. Printers Pte Ltd

10 9 8 7 6 5 4 3 2 1

Errata: Skeletal Variation and Adaptation in Europeans: Upper Paleolithic to the Twentieth Century

In Chapter 2, p. 20, the formula for estimation of stature from skeletal height is given incorrectly. The correct formula is: Stature = $1.009 \times SKH - 0.0426 \times age + 12.1$.

In Chapter 8, p. 230, lines 3-7, several indices derived as ratios are mistakenly given in cm units. These include relative bi-iliac breadth (17.4), relative sitting height (39.9), right and left brachial indices (72.6 and 74.6), and clavicle/humerus length (18.4).

In Chapter 9, p. 243, the last three columns of Table 9.1, with sample sizes, were left off. The complete table is given below.

Table 9.1 Dates and sample sizes for each time period.

Time period	Date range*	Males (N)	Females (N)	Total
Modern	1600 AD to ≥ 1900 AD	46	22	68
Medieval	500–1599 AD	39	32	71
Iron Age/Roman	2250–1650 BC	31	33	64
Bronze Age	4350–2950 BC	17	17	34
Neolithic	7000–4050 BC	20	23	43
Mesolithic	10500–5880 BC	20	9	29
Upper Paleolithic	33388–11367 BP	18	14	32

*Dates indicated as BP are calibrated.

Contents

List of Contributors

Margit Berner
Department of Anthropology
Natural History Museum
Vienna
Austria

Evan Garofalo
Center for Functional Anatomy and Evolution
Johns Hopkins University School of Medicine
Baltimore, MD
USA

Heather Garvin
Center for Functional Anatomy and Evolution
Johns Hopkins University School of Medicine
Baltimore, MD
USA

Brigitte Holt
Department of Anthropology
University of Massachusetts
Amherst, MA
USA

Martin Hora
Department of Anthropology and Human Genetics
Faculty of Science, Charles University
Prague
Czech Republic

Juho-Antti Junno
Department of Archaeology
University of Oulu
Oulu
Finland

Anna Kjellström
Osteological Research Laboratory
Stockholm University
Stockholm
Sweden

Heli Maijanen
Department of Archaeology
University of Oulu
Oulu
Finland

Eliška Makajevová
Department of Anthropology and Human Genetics
Faculty of Science, Charles University
Prague
Czech Republic

Petra Molnar
Osteological Research Laboratory
Stockholm University
Stockholm
Sweden

Sirpa Niinimäki
Department of Archaeology
University of Oulu
Oulu
Finland

Markku Niskanen
Department of Archaeology
University of Oulu
Oulu
Finland

Christopher B. Ruff
Center for Functional Anatomy and Evolution
Johns Hopkins University School of Medicine
Baltimore, MD
USA

Anna-Kaisa Salmi
Department of Archaeology
University of Oulu
Oulu
Finland

Kati Salo
Department of Archaeology
University of Helsinki
Helsinki
Finland

Vladimir Sládek
Department of Anthropology and Human Genetics
Faculty of Science, Charles University
Prague
Czech Republic

Dannielle Tompkins
Department of Anthropology
University of Massachusetts
Amherst, MA
USA

Tiina Väre
Department of Archaeology
University of Oulu
Oulu
Finland

Petr Velemínský
Department of Anthropology
National Museum
Prague
Czech Republic

Rosa Vilkama
Department of Archaeology
University of Oulu
Oulu
Finland

Erin Whittey
Department of Anthropology
University of Massachusetts
Amherst, MA
USA

Preface

Human body form is subject to a complex array of environmental influences, both natural and cultural in origin, acting throughout the lifespan. One way to approach this issue is through comparative studies of different populations exposed to varying environments. Adding temporal depth to such comparisons allows one to assess how populations respond through time to changing environmental conditions. In anthropology, this usually involves the use of skeletal material from historical, archaeological, or paleontological contexts. Many such studies have concentrated on fairly broad characteristics based on standard skeletal metrics, or have focused on specific areas of inquiry such as paleopathological indicators. More can be learned from skeletal remains, and the people represented by them, by incorporating other methods of analysis, including those based on engineering principles. The information obtained can shed new light on both the behavior and biology of past populations, and provide context for understanding modern human variation. This is what we have attempted to do here, for a large, representative sampling of European populations spanning the past 30,000 years and several major environmental transitions.

The investigations reported in this volume are the result of a collaboration that began in 2007 between Christopher Ruff, Brigitte Holt, Markku Niskanen, Vladimir Sládek, and Margit Berner. Subsequent data collection was primarily funded by the National Science Foundation (BCS-0642297 and BCS-0642710), with additional support from the Grant Agency of the Czech Republic (206/09/0589) and the Academy of Finland and Finnish Cultural Foundation. A symposium reporting some preliminary results was held at the annual meeting of the American Association of Physical Anthropologists in Portland, Oregon, in 2012.

The project expanded to include many co-authors from a number of different countries, who are represented among the chapters of the present volume. Many other people also helped with various aspects of the project. Those who assisted in collecting or processing of data include: Trang Diem Vu, Sarah Reedy, Quan Tran, Andrew Merriweather, Juho-Antti Junno, Anna-Kaisa Salmi, Tiina Väre, Rosa Vilkama, Jaroslav Roman, and Petra Spevackova. For access to skeletal collections and otherwise facilitating data acquisition, we thank Andrew Chamberlain, Rob Kruszynski, Jay Stock, Mercedes Okumura, Jane Ellis-Schön, Jacqueline McKinley, Lisa Webb, Jillian Greenaway, Alison Brookes, Jo Buckberry, Chris Knüsel, Horst Bruchhaus, Ronny Bindl, Hugo Cardoso, Sylvia Jiménez-Brobeil, Maria Dolores Garralda, Michèle Morgan, Clive Bonsall, Adina Boroneant, Alexandru Vulpe, Monica Zavattaro, Elsa Pacciani, Fulvia Lo Schiavo, Maria Giovanna Belcastro, Alessandro Riga, Nico Radi, Giorgio Manzi, Maryanne Tafuri, Pascal Murail, Patrice Courtaud, Dominique Castex, Fredérik Léterlé, Emilie Thomas, Aurore Schmitt, Aurore Lambert, Sandy Parmentier, Alessandro Canci, Gino Fornaciari, Davide Caramella, Jan Storå, Anna Kjellström, Petra Molnar, Niels Lynnerup, Pia Bennike, Leena Drenzel, Torbjörn Ahlström, Per Karsten, Bernd Gerlach, Lars Larsson, Petr Velemínsky,

Maria Teschler-Nicola, and Anna Pankowska. We also thank Erik Trinkaus, Steven Churchill, and Trent Holliday for generously making available data for Upper Paleolithic and Mesolithic specimens included in the study. Erin Whittey produced all of the maps included throughout the volume.

Christopher B. Ruff
Baltimore, Maryland

1

Introduction

Christopher B. Ruff

Center for Functional Anatomy and Evolution, Johns Hopkins University School of Medicine, Baltimore, MD, USA

The modern human body is the product of a long evolutionary history stretching back millions of years (Aiello and Dean, 1990). A basically modern body plan is usually considered to have been achieved with *Homo erectus* about 1.5 million years ago (Walker and Leakey, 1993), although with some significant subsequent modifications and regional variation en route to *Homo sapiens* (Ruff, 1995; Weaver, 2003; Larson *et al.*, 2007). Early 'anatomically modern humans' (EAM), that is, *H. s. sapiens*, appeared in the late Middle Pleistocene and were the only form to survive until the end of the Pleistocene (Klein, 2009). With the appearance of EAM humans, it is often assumed that "...fundamental evolutionary change in body form ceased" (ibid.: 615). However, some systematic changes in body size and shape continued through the terminal Pleistocene and early Holocene (Frayer, 1980, 1984; Jacobs, 1985; Ruff *et al.*, 1997; Meiklejohn and Babb, 2011). In addition, changes in skeletal robusticity, or bone strength relative to body size, occurred during this time period, again with regional variation (Holt, 2003; Ruff, 2005; Ruff *et al.*, 2006b; Shackelford, 2007; Marchi *et al.*, 2011).

Studies of such trends within EAM humans, however, have generally been carried out on a relatively limited number of samples and/or within narrow geographic areas or time periods. There has also been a tendency to divide work between academic disciplines, with paleoanthropological studies more focused on the Pleistocene, and bioarchaeological studies on more recent variation within the Holocene. However, as demonstrated in some of the studies cited above, morphological changes in the human skeleton form a continuum across the Pleistocene–Holocene boundary, as populations adapted to various environmental changes – climatic, technological, and ecological – that also bridged this time period. Behavioral changes within the Holocene can also be viewed as extensions of those initiated earlier in the terminal Pleistocene, for example, increasing sedentism. Characterization of long-term trends extending from the Late Pleistocene through very recent populations should shed light on both the full adaptability of the modern human skeleton, as well as the effects of a number of major transitions occurring over this time range, including from foraging to early food production, the intensification of agriculture, and increasing urbanization, mechanization, and social complexity.

The present volume is an attempt to do this, for one broad but well-defined geographic region: Europe. In terms of available late Pleistocene and Holocene skeletal material, Europe is very well sampled (e.g., Jaeger *et al.*, 1998; Roberts and Cox, 2003). It also has a rich archaeological and expanding genetic record (Milisauskas, 2002, Brandt *et al.*, 2013), as well as

Skeletal Variation and Adaptation in Europeans: Upper Paleolithic to the Twentieth Century,
First Edition. Edited by Christopher B. Ruff.

much historic evidence for the latest periods. Thus, there is both abundant primary material as well as considerable context within which to interpret temporal and regional variation in skeletal morphology.

The overall intent of the study was to sample adult skeletal material from as broad a representation of European populations as possible, beginning at about 30,000 years ago through the 20th century. Although ontogenetic analyses would have provided additional potentially quite interesting data, it was felt that this initial sampling should be limited to adults, to ensure feasibility of the study within a reasonable time frame. Also, for the same general reason, the study samples were concentrated primarily in Western, Central, and northern Europe – that is, not including Greece and the southern Balkans or much of Eastern Europe (see Figure 1.1).

The primary focus of the study is behavioral reconstruction within these past populations – that is, to document changes in behavior as reflected in skeletal morphology. There are many possible ways to do this (Larsen, 2015), but our approach here has been through the assessment of strength characteristics of the major limb bones, derived from diaphyseal cross sections. As discussed in more detail in Chapter 3, there is much theoretical and empirical

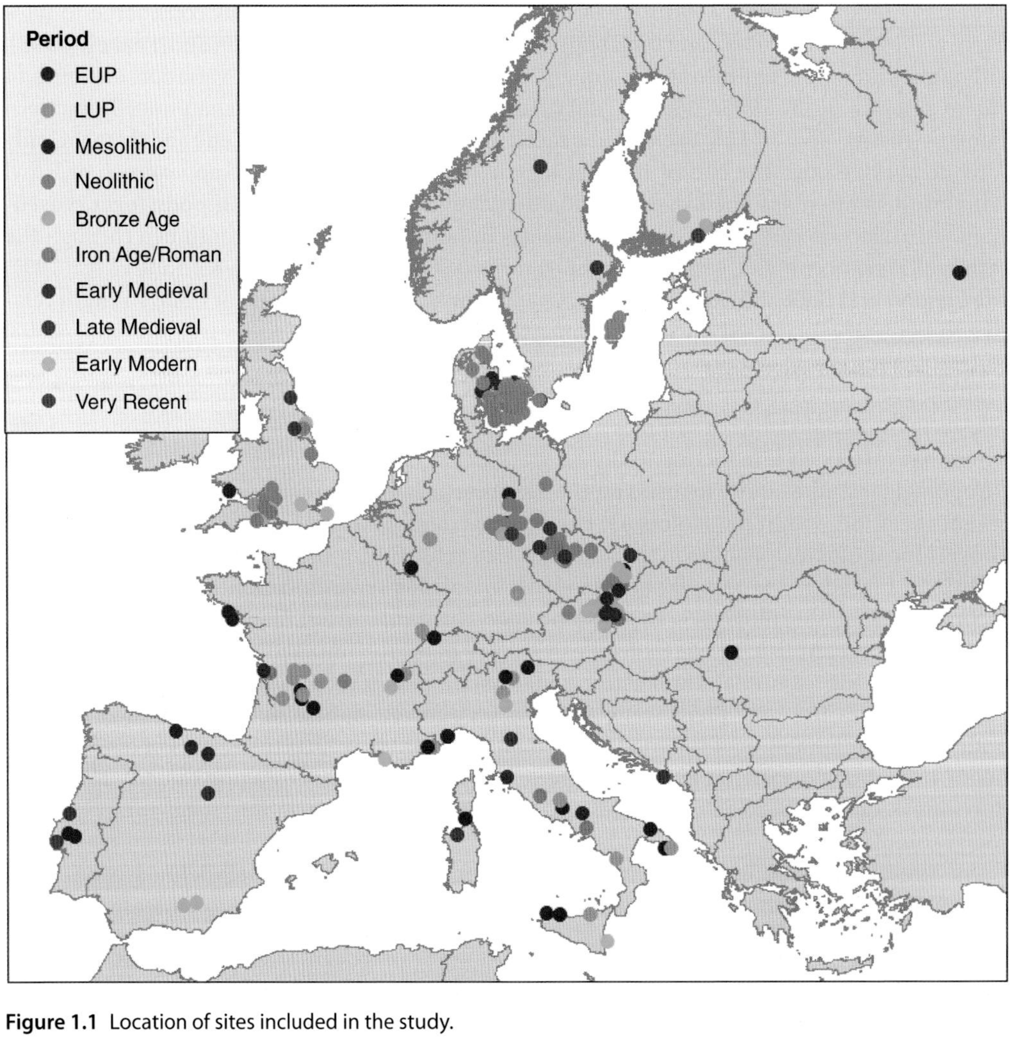

Figure 1.1 Location of sites included in the study.

evidence tying variation in long bone diaphyseal structure to variation in mechanical loadings on the limbs, and thus to behavioral use during life (Ruff *et al.*, 2006a; Ruff, 2008). By sampling both the upper and lower limbs, we can assess both locomotor behavior (e.g., mobility) and manipulative behavior (e.g., tool use). A number of studies have applied this approach to European skeletal samples (Holt, 2003; Marchi *et al.*, 2006; Ruff *et al.*, 2006b; Sládek *et al.*, 2006a,b, 2007; Trinkaus and Svoboda, 2006; Marchi, 2008; Sparacello and Marchi, 2008; Marchi *et al.*, 2011; Sparacello *et al.*, 2011; Macintosh *et al.*, 2014). However, as noted earlier, these have all been limited in scope in various ways. The present study is the first to apply this method to a broad temporal and geographic sampling of populations across Late Pleistocene and Holocene Europe.

The other major focus of the study is the reconstruction of body size and shape in these populations. Both body size and shape must be considered when assessing long-bone strength, because each influences the mechanical loadings on the limbs (Ruff, 2000; Shaw and Stock, 2011; also see Chapter 3). Thus, these factors must be controlled in order to infer behavior from long-bone structure. Body size and shape are also informative characteristics in themselves, as they are related to other systemic environmental effects such as climate and diet (see Chapter 4). In this regard, this study overlaps with a number of other studies of Holocene populations that examined temporal trends in body size, usually stature (Cohen and Armelagos, 1984; Steckel and Rose, 2002; Roberts and Cox, 2003; Cohen and Crane-Kramer, 2007). However, the primary emphasis of those studies was health status – that is, using variation in stature to assess variations in health between populations. Although we consider implications of our findings in the context of changes in health and nutrition, we did not assess other skeletal indicators of health and disease in our samples, although we did eliminate individuals showing clear pathologies that may have affected bone structure (see below). How overall body form was assessed from skeletal dimensions is described in Chapter 2.

Another purpose of gathering the present study data was to create a general reference source within which to evaluate other samples, including those of living populations. For example, it has been suggested that archaeological skeletal samples may represent good 'baselines' to address clinical issues such as the increasing incidence of osteoporosis in very recent Western populations (Eaton and Nelson, 1991; Lees *et al.*, 1993; Ekenman *et al.*, 1995; Mays, 1999; Brickley, 2002; Sievanen *et al.*, 2007). However, we actually know relatively little about broad patterns of variation in skeletal structure across time and space within the Holocene. The lifestyle and behavior of specific past populations or individual specimens are also more easily interpreted within a general temporal and geographic context (e.g., Ruff *et al.*, 2006b; Marchi *et al.*, 2011). The large database generated in this study should serve as a good starting point for comparisons of this kind.

Finally, several techniques of body size reconstruction and structural analysis were refined or developed specifically for this study. New formulae for reconstructing stature and body mass from skeletal dimensions, developed using the present study data set, have been published (Ruff *et al.*, 2012), and further extensions of these techniques are described in Chapter 2. Methods for both noninvasively extracting and analyzing cross-sectional diaphyseal properties were also advanced during the course of the study (see Chapter 3). It is hoped that these experiences and new results will help future researchers in designing and carrying out similar investigations.

1.1 Study Sample

Material from a total of 2179 individual skeletons was included in the study. All individuals were adult and possessed enough pelvic or cranial material to sex the individual (see below for sexing and aging methods). The other criteria for inclusion were possession of a relatively well-preserved femur, tibia, or humerus, and an intact femoral head for body mass estimation. In a few cases without a measurable femoral head, body mass could be estimated using pelvic bi-iliac breadth and stature, as described in Chapter 2. Some 95% of the individuals in the sample fulfilled these criteria. The relatively small number of individuals without body mass estimates were included in analyses of bilateral asymmetry, stature, or other body proportions not requiring body mass.

Given the emphasis of the study on diaphyseal structural properties, an important consideration for inclusion in the sample was the state of preservation of the cortex of the major long bones. Because of the sensitivity of cross-sectional properties to small changes in the outer contour of the section (see Chapter 3), no diaphyseal cross sections with significant wear on the periosteal surface were included in the study. This was a major constraint on the selection of specimens. Cross-sectional data were obtained on one or more long bones of 1955 individuals (90% of the total sample), with the femur the best represented (1830; 84%), followed by the tibia (1652; 76%) and humerus (1578; 72% with at least one humerus). The other 10% of the sample was used in various body size and shape analyses; the majority of these (80%) were from Medieval Scandinavia or Bronze Age Central Europe.

As noted earlier, we did not include bones showing clear signs of pathology that could affect bone structure or mechanical loading, for example, healed fractures, rickets, osteomyelitis or periostitis. However, we did not exclude individuals with arthritis, first, because some degree of arthritis was a near-ubiquitous feature of older individuals in most of the samples, and thus could be considered a 'normal' condition in this age range, and second, because arthritis does not itself directly affect bone structure in the diaphysis. (There may be indirect effects from changes in mechanical loading of the limb; however, again it could be argued that this is a 'normal' feature of aging, i.e., a characteristic of the individual and population.)

The geographic distribution of study sample sites is shown in Figure 1.1, subdivided by temporal periods (see below). For regional analyses, the samples were divided into seven groups: 1) British Isles; 2) Scandinavia and Finland; 3) North-Central Europe (including Germany, Czech Republic, Austria, and Switzerland); 4) France; 5) Italy; 6) Iberian peninsula (Spain and Portugal); and 7) Balkans (Bosnia-Herzegovina and Romania [Iron Gates region]). Samples from Sardinia and Corsica were assigned to Italy and France, respectively. Sunghir 1, from Russia (Early Upper Paleolithic), was assigned to Scandinavia/Finland, although a good case could be made for grouping it with North-Central Europe as well. All regions are represented by at least 142 individuals, except for the Balkans, with 71 (divided between only two sites – the Late Medieval Mistihalj and Mesolithic Schela Cladovei). The best represented regions are North-Central Europe and Scandinavia/Finland.

The distribution of individuals by region within 10 archaeologically/historically defined temporal periods is shown in Table 1.1. Note that dates for the Mesolithic, Neolithic, and Bronze Age overlap slightly due to variable transition times in different regions of Europe (Milisauskas, 2002). Some of the analyses in this volume use slightly modified temporal groupings, for example, separating the Early Neolithic Copper Age from other Neolithic samples, and Iron Age from Roman samples, or combining the two Upper Paleolithic or two most recent samples. Where this is done it is explicitly noted. So-called Neolithic Foragers from Scandinavia (the Pitted Ware culture; Linderholm, 2011) are also distinguished from other Neolithic cultures for certain analyses. Sample sizes vary temporally, with predictably more individuals in the Holocene time periods. The best-represented periods are the Neolithic through Late Medieval. All periods include data from several regions, although regional representation is understandably sparse in the earliest periods.

A complete listing of the individual sites included in the study is given in Appendix 1. A data file with all individual measurements and derived variables is given in a on-line file that can be accessed at: http://www.hopkinsmedicine.org/fae/CBR.html.

Table 1.1 Study sample sizes by temporal period and region.

Period	Date range[a]	Britain	Scand.	North-Central	France	Italy	Iberia	Balkans	Total
Very Recent	≥1900 AD	0	70	26	0	10	51	0	**157**
Early Modern	1500–1850 AD	40	53	0	35	23	0	0	**151**
Late Medieval	1000–1450 AD	121	177	131	0	14	48	54	**545**
Early Medieval	600–950 AD	0	24	175	17	40	63	0	**319**
Iron/Roman	300 BC–300 AD	118	65	52	19	45	0	0	**299**
Bronze	2400–1000 BC	17	0	180	0	34	34	0	**265**
Neolithic	5350–2050 BC	19	109	130	34	9	0	0	**301**
Mesolithic	8550–4000 BC	1	31	3	23	7	11	17	**93**
Late Upper Paleolithic	22,000–11,000 BP	0	0	4	8	12	0	0	**24**
Early Upper Paleolithic	33,000–26,000 BP	1	1	10	6	7	0	0	**25**
Total		**317**	**530**	**711**	**142**	**201**	**207**	**71**	**2179**

[a] Date of death, calibrated ^{14}C.

1.2 Osteological Measurements

Skeletal elements included in the study are indicated in Figure 1.2. One femur and one tibia were selected from each individual, choosing the best-preserved side or, in cases where both sides were equally well preserved, a right or left side at random. Bilateral asymmetry in the lower limb bones is quite small (Auerbach and Ruff, 2006). Both right and left humeri and radii, when available, were included. Many individuals possessed bones from both upper limbs, allowing direct assessments of bilateral asymmetry. The other bones measured in the study included other elements used to reconstruct anatomical stature (Fully, 1956; Raxter *et al.*, 2006; see Chapter 2) – the talus and calcaneus (either right or left), sacrum, vertebrae, and cranium; and elements used in analyses of body mass and body proportions – the articulated pelvis and both clavicles.

A list of all of the linear dimensions included in the study is given in Table 1.2. All of these are standard dimensions (Martin, 1928) or have previously been defined and illustrated in Ruff (2002) or Raxter *et al.* (2006). Abbreviations used throughout the volume are also listed. Body size and shape variables derived from these dimensions are defined in Chapter 2, and in individual chapters. Structural variables derived from bone cross sections are defined in Chapter 3.

Maximum lengths of the femur, tibia, humerus, radius, and clavicle were taken; in addition, femoral bicondylar and tibial lateral condylar lengths were measured for use in anatomical stature reconstruction. The length′ ('biomechanical length') dimensions of the femur, tibia, and humerus were taken as distances parallel to the long axis of the diaphysis, as described by Ruff (2002): for the femur, from the average distal projection of the condyles to the superior surface of the femoral neck; for the tibia, from the average proximal projection of the midpoint of the tibial plateaus to the midpoint of the talar articular surface; and for the humerus, from the superior surface of the head to the lateral lip of the trochlea. Bones were first 'leveled' and positioned as illustrated by Ruff (2002), using a special osteometric board and small wedges positioned at the relevant landmarks (see Fig. 3 in Ruff and Hayes, 1983). Length′ was used to define locations for taking anteroposterior (A-P) and mediolateral (M-L) diaphyseal breadths and cross-sectional measurements: at 50% of length′ in the femur and tibia, and at 35% of

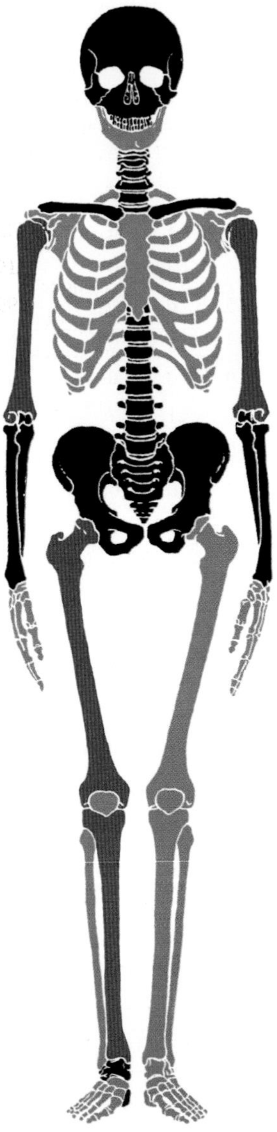

Figure 1.2 Skeleton showing elements measured in the study. Red: elements included in biomechanical analyses (note that only the right femur and tibia are highlighted, to indicate that only one side was included in these analyses; however, as explained in the text, either a right or a left side was chosen for measurement). Black: elements only included in stature, body mass, and body proportions analyses (including all presacral vertebrae inferior to C1).

length′ from the distal end of the humerus (to avoid the deltoid tuberosity). Diaphyseal breadths at 50% of radius length were also taken for a subset (about one-third) of the sample. Cross-sectional dimensions at these sites were determined from radiographically or computed tomography (CT)-derived images. Chapter 3 describes these methods and the cross-sectional parameters in more detail. Other linear measurements of long bones included superoinferior (S-I) breadths of the femoral and humeral heads, and M-L articular breadths of the distal femur, proximal tibia, and distal humerus, the last three as defined by Ruff (2002).

Bi-iliac (maximum M-L) breadth of the pelvis was taken when available. The two coxal bones and sacrum were articulated and held together by several large rubber bands. No additional material was placed between the bones. Bi-iliac breadth could be measured or estimated on about 55% of the total sample. About two-thirds of these were taken on complete pelves; the

Table 1.2 Linear osteological measurements.

Element	Dimension	Abbrev.	Reference[a]	Instrument
Femur	Maximum ln.	FML	M1	Osteometric board
	Bicondylar ln.	FBICL	M2	Osteometric board
	Length′	FBIOL	Ruff, 2002	See text
	Head S-I bd.	FHSI	M18	Digital calipers
	Distal M-L artic. bd.	FDML	Ruff, 2002	Digital calipers
	F50% A-P bd.	F50AP	Ruff, 2002	Digital calipers
	F50% M-L bd.	F50ML	Ruff, 2002	Digital calipers
Tibia	Maximum ln.	TML	M1a	Osteometric board
	Lat. Condylar ln.	TLCL	M1	Osteometric board
	Length′	TBIOL	Ruff, 2002	See text
	Proximal M-L artic. bd.	TPML	Ruff, 2002	Digital calipers
	T50% A-P bd.	T50AP	Ruff, 2002	Digital calipers
	T50% M-L bd.	T50ML	Ruff, 2002	Digital calipers
Humerus	Maximum ln.	HML	M1	Osteometric board
	Length′	HBIOL	Ruff, 2002	See text
	Head S-I bd.	HHSI	M10	Digital calipers
	Distal M-L artic. bd.	HDML	Ruff, 2002	Digital calipers
	H35% A-P bd.	H35AP	Ruff, 2002	Digital calipers
	H35% M-L bd.	H35ML	Ruff, 2002	Digital calipers
Radius	Maximum ln.	RML	M1	Osteometric board
	R50% A-P bd.	R50AP	Ruff, 2002	Digital calipers
	R50% M-L bd.	R50ML	Ruff, 2002	Digital calipers
Clavicle	Maximum ln.	CML	M1	Osteometric board
Calcaneus and Talus	Articulated ht.	TCH	Raxter *et al.*, 2006	Osteometric board
Sacrum	S1 ht.	–	Raxter *et al.*, 2006	Digital calipers
L5–C2 vertebrae	Ant. height of body	–	Raxter *et al.*, 2006	Digital calipers
Cranium	Basion-Bregma ht.	BBH	M17	Spreading calipers
Pelvis	Bi-iliac bd.	BIB	M2	Osteometric board

[a] M: Martin, R. (1928) Lehrbuch der Anthropologie. Fischer, Jena. Ruff, C.B. (2002) Long bone articular and diaphyseal structure in Old World monkeys and apes, I: Locomotor effects. *Am. J. Phys. Anthropol.*, **119**, 305–342. Raxter, M.H., Auerbach, B.M., and Ruff, C.B. (2006) A revision of the Fully technique for estimating statures. *Am. J. Phys. Anthropol.*, **130**, 374–384.

remainder included either some estimation of missing portions or reconstruction from hemi-pelves. The means, standard deviations and ranges of values of bi-iliac breadth for complete and incomplete pelves were almost identical, indicating that no systematic bias was introduced through the estimation process.

All of the other dimensions used for 'anatomical' reconstruction of stature were measured as described and illustrated by Raxter *et al.* (2006). The height of the articulated talus and calcaneus was measured in the physiological position, with space under the anterior calcaneus. Missing talocrural heights (about 20% of the sample) were estimated from femoral and tibial lengths as described in Ruff *et al.* (2012). With these estimations, about 87% of the total sample had this dimension. Vertebral body heights were taken as the maximum S–I height of the body anterior to the pedicles (or in the case of the first sacral vertebra, the sacral alae). Heights of missing vertebrae were estimated as described by Auerbach (2011) and Ruff *et al.* (2012). With these estimations, about 55% of the sample had vertebral column (S1 through C2) lengths. In about 40% of these the columns were complete, 20% had single nonadjacent missing vertebrae, and the remainder had multiple adjacent missing vertebrae in the cervical and/or thoracic regions, requiring estimation of total column length using either lumbar + thoracic or lumbar region lengths. Means and distributions of column lengths in the different categories of completeness were very similar (means within 2%). First sacral body height was estimated from combined presacral column height in a small percentage (14%) of the sample.

Basion-bregma height was measurable in only 795 individuals (36%) of the study sample. In our previous methodological study (Ruff *et al.*, 2012), following Auerbach (2011), we did not attempt to estimate missing cranial heights. However, for the present study we developed and tested a new method for doing this from total skeletal height inferior to the cranium, which was then applied to 527 additional individuals, increasing the sample with skeletal heights to 1322 individuals, or 61% of the sample. The technique is described in more detail in Chapter 2. Missing long bone lengths were not estimated for any individuals, because this presupposes specific linear length proportions, which vary between populations and were a subject of investigation here.

All data for Neolithic through Very Recent samples were collected by the contributors to this volume. Some Mesolithic data and much of the Upper Paleolithic data used in the study were made available from other colleagues, in particular Erik Trinkaus, Steven Churchill, and Trent Holliday, or were obtained from the literature (Matiegka, 1938). Details are given in the on-line data set.

1.3 Other Variables

The primary features used in assigning sex were pelvic, including ischio/pubic length proportion, shape of the sciatic notch, subpubic features, and the position of the auricular surface relative to the sciatic notch ('composite arch') (Phenice, 1969; Buikstra and Ubelaker, 1994; Brůžek, 2002; Walker, 2005). Cranial features such as development of supraorbital ridges, mastoid processes, and other characteristics (Buikstra and Ubelaker, 1994) were used as secondary indicators when diagnostic pelvic regions were not available. Because these features also show populational variation (e.g., Garvin and Ruff, 2012), they were always first assessed in combination with pelvic morphology, when available, within the same population, and then applied to individuals without pelvic indicators. The sample as a whole was slightly male-biased (56%).

Adult age status was defined as epiphyseal fusion of the long bones, except that almost complete fusion of the proximal humerus was allowed. Beyond this, the main purpose of estimating age was to factor that variable into the calculation of anatomical stature (see Chapter 2). Individuals were placed into age ranges based on a number of indicators, including fusion of epiphyses and pseudoepiphyses of the coxal bones, vertebrae, and medial clavicle (Krogman, 1962; Buikstra and Ubelaker, 1994); pubic symphyseal and auricular surface changes (Brooks and Suchey, 1990; Buckberry and Chamberlain, 2002; Falys *et al.*, 2006); and dental wear (Brothwell, 1972), the latter 'calibrated' within populations using skeletal indicators. Depending on the age and number of indicators present, age ranges could be as narrow as a year or two (e.g., 20–21 years) or as wide as a decade, or for the oldest individuals, simply 50+ or 60+ years. Age range midpoints were calculated (using 55 and 65 years for 50+ and 60+ categories), and used in the stature estimation procedure. Almost all of the Very Recent (≥1900 AD) individuals had known ages at death, along with 63 of the Early Modern (1600–1850 AD) individuals. The mean age for the entire age-able sample was 38 years, with mean ages for all temporal periods varying within ±5 years from this. About 5% of the sample could not be assigned even a broad age because of a lack of diagnostic elements (most were from the Upper Paleolithic or Mesolithic); these were assigned the mean age of 38 years.

Table 1.3 Rural and urban samples.

Period	Rural	Urban
Very Recent	13	144
Early Modern	46	105
Late Medieval	202	343
Early Medieval	241	78
Iron/Roman	214	85
Bronze	265	0
Neolithic	301	0
Mesolithic	93	0
Late Upper Paleolithic	24	0
Early Upper Paleolithic	25	0
Total	**1424**	**755**

In addition to temporal period and region, sites were further categorized in two other ways: by local terrain, and as either rural or urban. Both of these factors were predicted to have potential effects on body size and mechanical loading of the skeleton. As described in detail in Chapter 5, terrain was subdivided into three categories ranging from 'flat' to 'hilly' to 'mountainous,' based on topographic relief within a radius of 10 km from the site. 'Urban' sites were those characterized by large agglomerated settlements with evidence for economic specialization (i.e., towns or cities), while 'rural' sites did not show these characteristics. A site was considered 'urban' even if some of its occupants engaged in non-urban occupations, for example they walked out to fields. It is recognized that these are very broad categories with much overlap, and that to some extent the definitions of what constitutes each is relative to the time periods considered; for example, an Iron Age city may not be equivalent to a 19th or 20th century city. However, as a first approximation this categorization was still felt to be useful, particularly for making comparisons within similar temporal ranges. The distribution of rural and urban sites by time period is shown in Table 1.3. The first urban sites in our sample appear in the later Iron Age/Roman period (the Romano-British site of Poundbury and two Roman sites in Italy); there is only one Very Recent rural site (Sassari in Sardinia). More details can be found in the on-line data set.

1.4 Organization of the Book

This book is organized into three major sections. The first section presents the methodologies employed in the rest of the studies (an exception is the terrain variable, which is described in Chapter 5). As noted earlier, the development of new or modified methodologies for both body size/shape and bone structural analyses was one of the major outcomes of the study. Because body size and shape are important in mechanical analyses, that topic is presented first in Chapter 2, followed by the techniques used to derive and statistically evaluate diaphyseal structural properties in Chapter 3.

The second section of the book includes four chapters on general pan-European variation in body size and shape (Chapter 4), long-bone robusticity (Chapter 5), sexual dimorphism in body size/shape and robusticity (Chapter 6), and upper-limb bone bilateral asymmetry (Chapter 7). Temporal variation across Europe as a whole is the main focus of these chapters, although the effects of terrain and rural/urban distinctions are noted when appropriate, and some cross-region comparisons are also carried out.

The third section considers changes in body size/shape and bone structural properties within six specific regions (see Table 1.1), including more local archeological and historical context. Comparisons of region-specific trends with those for Europe as a whole are also carried out. Most of these chapters include some anthropometric data for recent living populations from the regions, to allow further assessment of very recent secular changes in body size. Methods summaries are given in each chapter, although detailed methods are presented in Chapters 1–3. Finally, a conclusions chapter draws together findings from earlier in the book and highlights some of the major themes identified in both general and regional analyses.

As noted in the Preface this study was the result of collaborations and the joint efforts of a large group of researchers. Authorship of the general chapters in Section 2 is limited to the primary collaborators and other personnel involved in developing specific techniques utilized in those chapters, even though many other people contributed data incorporated in those analyses. Authorship of the regional chapters includes all people who worked on the collection and/or analysis of data for that region.

References

Aiello, L. and Dean, C. (1990) *An Introduction to Human Evolutionary Anatomy*. Academic Press, London, 596 p.

Auerbach, B.M. (2011) Methods for estimating missing human skeletal element osteometric dimensions employed in the revised fully technique for estimating stature. *Am. J. Phys. Anthropol.*, **145**, 67–80.

Auerbach, B.M. and Ruff, C.B. (2006) Limb bone bilateral asymmetry: variability and commonality among modern humans. *J. Hum. Evol.*, **50**, 203–218.

Brandt, G., Haak, W., Adler, C.J., *et al.* (2013) Ancient DNA reveals key stages in the formation of central European mitochondrial genetic diversity. *Science*, **342**, 257–261.

Brickley, M. (2002) An investigation of historical and archaeological evidence for age-related bone loss and osteoporosis. *Int. J. Osteoarch.*, **12**, 364–371.

Brooks, S. and Suchey, J.M. (1990) Skeletal age determination based on the os pubis: a comparison of the Ascádi-Nemeskéri and Suchey-Brooks methods. *Hum. Evol.*, **5**, 227–238.

Brothwell, D.R. (1972) *Digging Up Bones*. British Museum of Natural History, London.

Brůžek, J. (2002) A method for visual determination of sex, using the human hip bone. *Am. J. Phys. Anthropol.*, **117**, 157–168.

Buckberry, J.L. and Chamberlain, A.T. (2002) Age estimation from the auricular surface of the ilium: a revised method. *Am. J. Phys. Anthropol.*, **119**, 231–239.

Buikstra, J.E. and Ubelaker, D.H. (1994) *Standards for Data Collection from Human Skeletal Remains* (ed. J. Haas), Arkansas Archaeological Survey, Fayetteville, Arkansas.

Cohen, M.N. and Armelagos, G.J. (1984) *Paleopathology at the Origins of Agriculture*. Academic Press, New York.

Cohen, M.N. and Crane-Kramer, G.M.M. (eds) (2007) *Ancient Health: Skeletal Indicators of Agricultural and Economic Intensification*. University Press of Florida, Gainesville.

Eaton, S.B. and Nelson, D.A. (1991) Calcium in evolutionary perspective. *Am. J. Clin. Nutr.*, **54** (Suppl. 1), 281S–287S.

Ekenman, I., Eriksson, S.A., and Lindgren, J.U. (1995) Bone density in medieval skeletons. *Calcif. Tissue Int.*, **56**, 355–358.

Falys, C.G., Schutkowski, H., and Weston, D.A. (2006) Auricular surface aging: worse than expected? A test of the revised method on a documented historic skeletal assemblage. *Am. J. Phys. Anthropol.*, **130**, 508–513.

Frayer, D.W. (1980) Sexual dimorphism and cultural evolution in the late Pleistocene and Holocene of Europe. *J. Hum. Evol.*, **9**, 399–415.

Frayer, D.W. (1984) Biological and cultural change in the European Late Pleistocene and Early Holocene. In: *The Origins of Modern Humans: A World Survey of the Fossil Evidence* (eds F.H. Smith and F. Spencer), Wiley-Liss, New York, pp. 211–250.

Fully, G. (1956) Une nouvelle méthode de détermination de la taille. *Ann. Med. Legale*, **35**, 266–273.

Garvin, H.M. and Ruff, C.B. (2012) Sexual dimorphism in skeletal browridge and chin morphologies determined using a new quantitative method. *Am. J. Phys. Anthropol.*, **147**, 661–670.

Holt, B.M. (2003) Mobility in Upper Paleolithic and Mesolithic Europe: evidence from the lower limb. *Am. J. Phys. Anthropol.*, **122**, 200–215.

Jacobs, K.H. (1985). Evolution in the postcranial skeleton of Late Glacial and early Postglacial European hominids. *Z. Morph. Anthropol.*, **75**, 307–326.

Jaeger, U., Bruchhaus, H., Finke, L., Kromeyer-Hauschild, K., and Zellner, K. (1998) Säkularer trend bei der körperhöhe seit dem Neolithikum. *Anthropol. Anz.*, **56**, 117–130.

Klein, R. (2009) *The Human Career*, 3rd edn, University of Chicago Press. Chicago.

Krogman, W.M. (1962). *The Human Skeleton in Forensic Medicine.* C.C. Thomas, Springfield, Illinois.

Larsen, C.S. (2015) *Bioarchaeology: Interpreting Behavior from the Human Skeleton*, 2nd edn. Cambridge University Press, Cambridge.

Larson, S.G., Jungers, W.L., Morwood, M.J., Sutikna, T., Jatmiko, Saptomo, E.W., Due, R.A., and Djubiantono, T. (2007) *Homo floresiensis* and the evolution of the hominin shoulder. *J. Hum. Evol.*, **53**, 718–731.

Lees, B., Molleson, T., Arnett, T.R., and Stevenson, J.C. (1993) Differences in proximal femur bone density over two centuries. *Lancet*, **341**, 673–675.

Linderholm, A. (2011) The genetics of the Neolithic transition: New light on differences between hunter-gatherers and farmers in southern Sweden. In: *Human Bioarchaeology of the Transition to Agriculture* (eds R. Pinhasi and J.T. Stock). Wiley-Blackwell, New York, pp. 385–402.

Macintosh, A.A., Pinhasi, R., and Stock, J.T. (2014) Femoral and tibial cross-sectional morphology reflects complex change in sex roles, mobility, and division of labor across similar to 6200 years of agriculture in Central Europe. *Am. J. Phys. Anthropol.*, **153**, 173–173.

Marchi, D. (2008) Relationships between lower limb cross-sectional geometry and mobility: the case of a Neolithic sample from Italy. *Am. J. Phys. Anthropol.*, **137**, 188–200.

Marchi, D., Sparacello, V., and Shaw, C. (2011) Mobility and lower limb robusticity of a pastoralist Neolithic population from north-western Italy. In: *Human Bioarchaeology of the Transition to Agriculture* (eds R. Pinhasi and J.T. Stock). Wiley-Blackwell, New York, pp. 317–346.

Marchi, D., Sparacello, V.S., Holt, B.M., and Formicola, V. (2006) Biomechanical approach to the reconstruction of activity patterns in Neolithic Western Liguria, Italy. *Am. J. Phys. Anthropol.*, **131**, 447–455.

Martin, R. (1928) *Lehrbuch der Anthropologie.* Fischer, Jena.

Matiegka, J. (1938) *Homo Predmostensis: Fosilní Clovek z Predmostí na Morave II. Ostatní Cásti Kostrové.* Nákaladem Ceské Akademie Ved e Umení, Prague.

Mays, S.A. (1999) Osteoporosis in earlier human populations. *J. Clin. Densitom.*, **2**, 71–78.

Meiklejohn, C. and Babb, J. (2011) Long bone length, stature and time in the European Late Pleistocene and Early Holocene. In: *Human Bioarchaeology of the Transition to Agriculture*. (eds R. Pinhasi and J.T. Stock). Wiley-Blackwell, Chichester, pp. 153–175.
Milisauskas, S. (ed.) (2002) *European Prehistory: A Survey*. Kluwer Academic/Plenum, New York.
Phenice, T.W. (1969) A newly developed visual method of sexing the os pubis. *Am. J. Phys. Anthropol.*, **30**, 297–301.
Raxter, M.H., Auerbach, B.M., and Ruff, C.B. (2006) A revision of the Fully technique for estimating statures. *Am. J. Phys. Anthropol.*, **130**, 374–384.
Roberts, C. and Cox, M. (2003) *Health and Disease in Britain: From Prehistory to the Present Day*. Sutton Publishing, Stroud, UK.
Ruff, C.B. (1995) Biomechanics of the hip and birth in early *Homo*. *Am. J. Phys. Anthropol.*, **98**, 527–574.
Ruff, C.B. (2000) Body size, body shape, and long bone strength in modern humans. *J. Hum. Evol.*, **38**, 269–290.
Ruff, C.B. (2002) Long bone articular and diaphyseal structure in Old World monkeys and apes, I: Locomotor effects. *Am. J. Phys. Anthropol.*, **119**, 305–342.
Ruff, C.B. (2005) Mechanical determinants of bone form: Insights from skeletal remains. *J. Musculoskelet. Neuronal. Interact.*, **5**, 202–212.
Ruff, C.B. (2008) Biomechanical analyses of archaeological human skeletal samples. In:. *Biological Anthropology of the Human Skeleton*, 2nd edn (eds M.A. Katzenburg and S.R. Saunders). John Wiley & Sons, Inc., New York, pp. 183–206.
Ruff, C.B. and Hayes, W.C. (1983) Cross-sectional geometry of Pecos Pueblo femora and tibiae – a biomechanical investigation. I. Method and general patterns of variation. *Am. J. Phys. Anthropol.*, **60**, 359–381.
Ruff, C.B., Holt, B.H., and Trinkaus, E. (2006a) Who's afraid of the big bad Wolff? Wolff's Law and bone functional adaptation. *Am. J. Phys. Anthropol.*, **129**, 484–498.
Ruff, C.B., Holt, B.M., Niskanen, M., *et al.* (2012) Stature and body mass estimation from skeletal remains in the European Holocene. *Am. J. Phys. Anthropol.*, **148**, 601–617.
Ruff, C.B., Holt, B.M., Sládek, V., *et al.* (2006b) Body size, body proportions, and mobility in the Tyrolean 'Iceman'. *J. Hum. Evol.*, **51**, 91–101.
Ruff, C.B., Trinkaus, E., and Holliday, T.W. (1997) Body mass and encephalization in Pleistocene *Homo*. *Nature*, **387**, 173–176.
Shackelford, L.L. (2007) Regional variation in the postcranial robusticity of Late Upper Paleolithic humans. *Am. J. Phys. Anthropol.*, **133**, 655–668.
Shaw, C.N. and Stock, J.T. (2011) The influence of body proportions on femoral and tibial midshaft shape in hunter-gatherers. *Am. J. Phys. Anthropol.*, **144**, 22–29.
Sievanen, H., Jozsa, L., Pap, I., *et al.* (2007) Fragile external phenotype of modern human proximal femur in comparison with medieval bone. *J. Bone Miner. Res.*, **22**, 537–543.
Sládek, V., Berner, M., and Sailer, R. (2006a) Mobility in Central European Late Eneolithic and Early Bronze Age: Femoral cross-sectional geometry. *Am. J. Phys. Anthropol.*, **130**, 320–332.
Sládek, V., Berner, M., and Sailer, R. (2006b) Mobility in Central European Late Eneolithic and Early Bronze Age: Tibial cross-sectional geometry. *J. Archaeol. Sci.*, **33**, 470–482.
Sládek, V., Berner, M., Sosna, D., and Sailer, R. (2007) Human manipulative behavior in the Central European Late Eneolithic and Early Bronze Age: humeral bilateral asymmetry. *Am. J. Phys. Anthropol.*, **133**, 669–681.
Sparacello, V. and Marchi, D. (2008) Mobility and subsistence economy: a diachronic comparison between two groups settled in the same geographical area (Liguria, Italy). *Am. J. Phys. Anthropol.*, **136**, 485–495.

Sparacello, V.S., Pearson, O.M., Coppa, A., and Marchi, D. (2011) Changes in skeletal robusticity in an iron age agropastoral group: the Samnites from the Alfedena necropolis (Abruzzo, Central Italy). *Am. J. Phys. Anthropol.*, **144**, 119–130.

Steckel, R.H. and Rose, J.C. (eds) (2002) *The Backbone of History: Health and Nutrition in the Western Hemisphere.* Cambridge University Press, Cambridge.

Trinkaus, E.T. and Svoboda, J. (2006) *Early Modern Human Evolution in Central Europe: The People in Dolni Vestonice and Pavlov.* Oxford University Press, Oxford.

Walker, A. and Leakey, R. (eds) (1993) *The Nariokotome* Homo erectus *Skeleton.* Harvard University Press, Cambridge.

Walker, P.L. (2005) Greater sciatic notch morphology: sex, age, and population differences. *Am. J. Phys. Anthropol.*, **127**, 385–391.

Weaver, T.D. (2003) The shape of the Neandertal femur is primarily the consequence of a hyperpolar body form. *Proc. Natl Acad. Sci. USA*, **100**, 6926–6929.

2

Body Size and Shape Reconstruction

Markku Niskanen[1] *and Christopher B. Ruff*[2]

[1] *Department of Archaeology, University of Oulu, Oulu, Finland*
[2] *Center for Functional Anatomy and Evolution, Johns Hopkins University School of Medicine, Baltimore, MD, USA*

2.1 Introduction

The reconstruction of body size and shape of past European populations is one of the major focuses of this study. The considerable importance of body size and shape reconstruction of past populations is addressed in Chapters 1 and 4.

When this study commenced it utilized the most recent methods to estimate body mass (Ruff *et al.*, 2005) and stature (Raxter *et al.*, 2006) available at that time. Based on the new data collected during the study, new regression equations to predict stature from long bone lengths and body mass from femoral head breadth were developed and published (Ruff *et al.*, 2012). The development of additional methods for body size and shape estimation has continued. In this chapter, we review, describe, and evaluate previously published and new methods used in reconstructing body size and body shape in this study.

2.2 Body Size and Shape Estimation

Most linear (vertical) and lateral dimensions of the body are largely determined by underlying skeletal dimensions, and can thus be estimated from these dimensions with different levels of precision. Linear and lateral dimensions combined (e.g., stature and trunk breadth) determine the skeletal frame size, which is an important determinant of body mass (e.g., body mass estimation from stature and bi-iliac breadth; Ruff *et al.*, 2005). Proportional relationships of different linear (e.g., trunk length, limb length) and lateral (e.g., biacromial breadth, bi-iliac breadth) dimensions represent body shape. For example, stature is often used as a denominator in determining body proportions, such as relative body segment lengths (e.g., relative sitting height) and relative body breadths (e.g., biacromial and bi-iliac breadths relative to stature).

Stature estimation has a long history. Estimation methods are divided into 'anatomical' and 'mathematical' methods (Lundy, 1985). The anatomical method was pioneered by Dwight (1894), simplified by Fully (1956), and most recently modified by Raxter *et al.* (2006) and Niskanen *et al.* (2013). This method takes into account differences in body proportions because it is based on skeletal components of stature from the skull to the heel. It also provides more accurate stature estimates than the mathematical method. In the most recent versions of this

Skeletal Variation and Adaptation in Europeans: Upper Paleolithic to the Twentieth Century, First Edition. Edited by Christopher B. Ruff.

method (Raxter *et al.*, 2006; Niskanen *et al.*, 2013), stature is estimated from the sum of contributing skeletal elements – from the so-called skeletal height (here abbreviated as SKH) – with a regression equation. In this study, we have used Equation 1 of Raxter *et al.* (2006), which includes an age term.

The mathematical method refers to a method of estimating stature from a particular skeletal dimension (most often long bone lengths) generally using a sex- and population-specific regression equation. The most often and widely used equations to estimate stature of past and present Europeans are those of Trotter and Gleser (1952, 1958), developed for Euroamericans. These equations may provide reasonably accurate stature estimates for the most recent Europeans, but not necessarily for earlier Europeans (Formicola and Franceschi, 1996; Maijanen and Niskanen, 2006; Vercellotti *et al.*, 2009). Partly due to this reason, population-specific equations have been developed more recently for both European and non-European skeletal samples utilizing a 'hybrid' approach in which long bone lengths are regressed against stature estimates provided by the anatomical method (e.g., Sciulli *et al.*, 1990; Formicola and Franceschi, 1996; Sciulli and Hetland, 2007; Raxter *et al.*, 2008; Vercellotti *et al.*, 2009; Auerbach and Ruff, 2010; Maijanen and Niskanen, 2010). This 'hybrid' approach was also used to develop the equations used in this study (Ruff *et al.*, 2012).

The most precise regression equations to estimate stature from long bone lengths are naturally those that are designed for specific samples. Unfortunately, this requires larger sample sizes than are generally available. A sample size of $N > 40$ is needed to be confident that distributions for x (e.g. bone length) and y (stature) are reasonably representative of the population, and that the regression line plotted through the data is neither too steep nor shallow (on sample sizes, see Maijanen, 2011). For this reason, Ruff *et al*.'s (2012) equations provide the most viable option for stature estimation for European skeletal samples for which developing sample-specific equations is not feasible. Some temporal and regional effects on stature estimation precision are naturally expected when applying long bone regression equations to skeletal samples representing different temporal and regional groups.

Body mass estimation from skeletal dimensions has a much shorter history than stature estimation (Ruff *et al.*, 2012 and references therein). These estimations can also never rival stature estimations from skeletal dimensions in accuracy due to considerable variation in the skeletal size–soft tissue mass relationship.

The two main categories of body mass estimation are the 'morphometric' approach and the 'mechanical' approach (Auerbach and Ruff, 2004). Morphometric body mass estimation is analogous to anatomical stature estimation. The body mass estimate is based on estimated or even measured body dimensions, that is, stature and bi-iliac breadth as in the stature/bi-iliac breadth method (Ruff 1994, 2000; Ruff *et al.*, 1997, 2005). These estimates exhibit little or no size-related estimation error (M. Niskanen, unpublished), but biacromial shoulder breadth relative to bi-iliac breadth (Ruff, 2000; Ruff *et al.*, 2005) and relative sitting height (Niinimäki and Niskanen, 2015) affect estimation precision.

This 'morphometric' stature/bi-iliac breadth method has been applied in many studies to estimate body mass of archaeological specimens (Ruff, 1994; Arsuaga *et al.*, 1999; Rosenberg *et al.*, 2006; Ruff *et al.*, 2006; Kurki *et al.*, 2010). We use the most recent version of this method (Ruff *et al.*, 2005) in this study.

Mechanical body mass estimation is analogous to mathematical stature estimation. Body mass is estimated from skeletal dimensions (e.g., joint surface size) that mechanically support body mass. Femoral head breadth is often used because it is so often preserved, and demonstrably correlates positively with body mass (Ruff *et al.*, 2012 and references therein). Four different studies have provided regression equations to estimate body mass of recent human samples from femoral head breadth (Ruff *et al.*, 1991, 2012; McHenry, 1992; Grine *et al.*, 1995). We use sex-specific equations introduced in Ruff *et al.* (2012) generated by a 'hybrid' approach in which morphometric body mass estimates provided by the stature/bi-iliac breadth method (Ruff *et al.*, 2005) are regressed against femoral head breadth. We assess if our body mass estimates from femoral head breadth provided by Ruff *et al.*'s (2012) equation exhibit any directional prediction error and its possible effects on results (on size-related error; see Auerbach and Ruff, 2004).

Only a few published attempts have been made to estimate body dimensions other than stature from skeletal dimensions. Olivier and Pineau (1957) used a cadaveric sample to generate regression equations to estimate upper arm length from humeral length, forearm length from radial length, thigh length from femoral length, and lower leg length from tibial length. Olivier and Tissier (1975) used the same cadaveric material and generated equations to estimate long bone lengths from upper limb and lower limb segment lengths. Piontek (1979) used measured biacromial breadths and X-ray images of living subjects to generate regression equations to estimate biacromial shoulder breadth from clavicular lengths. Ruff *et al.* (1997) introduced a regression equation to estimate living bi-iliac breadth from skeletal bi-iliac breadth, based on a few population means of living bi-iliac breadth and skeletal bi-iliac breadth.

We are not aware of any published methods to estimate sitting height from skeletal dimensions, but skeletal trunk height represented by the summed dorsal body heights of the thoracic and lumbar vertebrae plus sacral anterior length has been used to represent trunk length (Franciscus and Holliday, 1992; Holliday, 1995, 1997). Crural and brachial indices have been used for decades to represent intra-limb ratios, and the clavicular length–humeral length ratio has been used to indicate relative shoulder breadth for many decades (Martin, 1928; Martin and Saller, 1957; Trinkaus, 1981; Holliday, 1995, 1997).

In this chapter, all methods used in the study to reconstruct body size and shape are described and their precision evaluated. New, previously unpublished, methods are discussed and tested more thoroughly.

2.3 Materials and Methods

Skeletal samples included in this study are discussed in Chapter 1. More detailed information on regional and temporal samples is provided in regional chapters presented in Section III of this book.

In addition to applying previously published methods to estimate stature (Raxter *et al.*, 2006), body mass (Ruff *et al.*, 2006, 2012), and living bi-iliac breadth (Ruff *et al.*, 1997), we have estimated sitting height, subischial lower limb length, and biacromial breadth) and their commonly used ratios (e.g., relative sitting height, relative shoulder breadth, etc.) from skeletal dimensions to compare body size and shape of past Europeans with living Europeans. Procedures used to generate these new estimations are discussed in detail.

Linear osteological measurements taken on skeletal specimens are defined in Table 1.2 in Chapter 1, which also gives abbreviations. Dimensions and ratios computed using these measurements and estimated body dimensions are provided in Table 2.1. Some additional dimensions used in comparisons with living populations in Chapters 4 and 12 were also derived and are listed in the table; these are defined further below.

2.3.1 Estimation of Missing Elements

Some estimation of missing skeletal dimensions was done to increase sample sizes. In a few cases, particular long bone length dimensions were missing and were estimated from other length dimensions of the same bones, using equations generated from the study sample

Table 2.1 Variables computed from skeletal dimensions.

Variable	Abbrev.	Description†
Vertebral column height	VCH	Maximum heights anterior to pedicles from C2 to S1
Skeletal height	SKH	BBH + VCL + FBICL + TLCL + TCH
Partial skeletal height	PSKH	VCL + FBICL + TLCL + TCH
Stature	STA	Estimated from SKH, PSKH, or lower limb long bone lengths.
Anatomical stature	ASTA	Estimated from SKH or PSKH using equations: ASTA = 1.009 × SKH − 0.0426 × age + 12.1 ASTA = 1.045 × PSKH + 18.911
Sitting height	SHT	BBH + VCH
Relative sitting height	RSHT	(SHT/ASTA estimated from SKH) × 100
Subischial lower limb length*	LlimbL	See text
Living sitting height*	LSHT	ASTA − LlimbL
Living relative sitting height*	LRSHT	LSHT/ASTA × 100
Lower leg length*	LEGL	1.01 × TLCL
Body mass	BM	Estimation equations are provided in Table 2.5
Body mass index	BMI	BM/STA in meters^2
Relative clavicular breadth	RCB	Left and right CML summed or preserved length doubled/Skeletal BIB × 100
Biacromial breadth*	BAB	See text on estimating BAB from CML
Relative (biacromial) shoulder breadth*	RBAB	BAB/STA × 100
Living bi-iliac breadth*	BIB	1.17 × Skeletal bi-iliac breadth in cm − 3.0 cm
Relative bi-iliac breadth*	RBIB	(BIB/STA) × 100
Living biacromial breadth/bi-iliac breadth index*	BAB / BIB	(BAB/BIB) × 100
Crural index	CI	TLCL/FBICL × 100
Relative lower leg length*	RLEGL	(LEGL/STA) × 100
Brachial index	BI	(RML/HML) × 100 (Means of left and right index are used here)

* Anthropometric-equivalent dimension or index.
† Basic linear measurements defined in Chapter 1; Table 1.2.
BBH: basion-bregma height; FBICL: femoral bicondylar length; TLCL: tibial lateral condylar length; TCH: Talar-calcaneal height; CML: clavicle maximum length; RML: radius maximum length; HML: humerus maximum length.

(Table 2.2). As correlation coefficients and standard error of estimation values presented in this table demonstrate, estimation errors are very small in these missing value estimations.

In the case of missing vertebral heights, the total vertebral column length was estimated from existing vertebral heights by procedures described in Auerbach (2011) and Ruff *et al.* (2012). In the case of 53 Upper Paleolithic, Mesolithic, and Neolithic specimens, the total vertebral column length was estimated with sex-specific regression equations from existing posterior midline heights of vertebra, anterior midline heights of vertebra and/or from maximum heights of

Table 2.2 Equations to estimate missing femoral and tibial lengths from available lengths.

Length predicted	Predictor length	Slope	Intercept	r	SEE	N
FMAXL	FBIOL	1.036	12.528	0.983	5.38	1905
FBICL	FMAXL	0.998	−2.40	0.977	2.32	1922
FBIOL	FMAXL	0.933	1.998	0.983	5.10	1905
TMAXL	TLCL	1.009	2.850	0.998	1.71	812
TMAXL	TBIOL	1.052	6.308	0.995	2.58	812
TLCL	TMAXL	0.987	−1.300	0.998	1.69	812
TLCL	TBIOL	1.041	4.110	0.995	2.50	812
TBIOL	TLCL	0.942	−2.648	0.995	2.44	812

vertebra anterior to pedicles (see Appendix 2(a)). In the case of two Upper Paleolithic and seven Mesolithic specimens, maximum midline vertebral heights provided by Vincenzo Formicola were converted to corresponding maximum heights anterior to pedicles using mean ratios of these measurements (see Appendix 2(b)). Missing S1 heights were estimated from the total presacral vertebral column height with a pooled-sex regression equation (S1 height = 0.0458 × presacral height + 10.2) provided in Ruff *et al.* (2012). In all of the above cases, estimation error is so small that it has little or no effect on results.

Talocrural height (TCH) was estimated from femoral bicondylar and tibial lateral condylar lengths, abbreviated as FBICL and TLCL, respectively, (see Table 1.2) for 172 individuals using sex-specific equations from Ruff *et al.* (2012):

$$\text{male TCH} = 0.094 \times \text{FBICL} + 0.036 \times \text{TLCL} + 15.9$$

$$\text{female TCH} = 0.096 \times \text{FBICL} + 0.012 \times \text{TLCL} + 20.6$$

This skeletal ankle height has such a small absolute contribution to the total skeletal height and stature that estimation error affects stature estimation by only a few millimeters in the vast majority of cases.

2.3.2 Statistical Procedures

Reduced major axis (RMA) regression is used widely in this study because least squares (LS) equations tend to underestimate stature and body mass at the high end of variation and underestimate them at the low end of variation (Ruff *et al.*, 2012 and references therein). We converted LS slopes to RMA slopes by dividing LS slopes by the correlation coefficient value. After that, we recalculated y-intercepts by subtracting from the mean of y variable (e.g., anatomical stature) the mean of x variable (e.g., partial skeletal height) multiplied by the RMA slope value (on this procedure, see Hofman, 1988; Hens *et al.*, 1998).

In those cases where the actual living dimension is unknown (e.g., sitting height, subischial lower limb length), correction factors based on anatomical data (e.g., fresh bone length including cartilage versus dry bone length without cartilage) are used instead of regression equations to convert skeletal dimensions to corresponding living dimensions. Estimation precisions of these estimates cannot be evaluated by comparing observed and predicted values of matched individuals (e.g., observed value minus predicted value) because there are no observed values available for these comparisons. Instead, we have compared values estimated from skeletal dimensions with observed anthropometric mean values of the same-sex samples representing the same population (e.g. the Euroamerican males).

Residual values (observed value minus predicted value) or the percentage prediction error (PPE or %PE), computed as [(observed – predicted)/predicted] × 100 (Smith, 1984), were used in assessing estimation error and its direction. The stature estimated using the anatomical method and the body mass estimated using the morphometric stature/bi-iliac breadth method represent 'observed' stature and body mass values, respectively. These assessments of estimation error were performed to determine possible effects of temporal period and geography, as well as sex and age. Analyses were done separately for both sexes and for pooled-sex samples, depending on the case. Results are not compared with those obtained by applying previous stature and body mass estimation equations because this was already done in Ruff *et al.* (2012).

All statistical analyses were performed using SPSS (Version 22). Stature is expressed in centimeters and body mass in kilograms. All dimensions used in stature estimation equations are expressed in centimeters, and also in applying the morphometric stature/bi-iliac breadth method in body mass estimation. Femoral head breadth is expressed in millimeters in body mass estimation.

2.4 Estimating Body Size and Shape from Skeletal Dimensions

2.4.1 Stature

Stature was estimated anatomically from skeletal height (SKH) using Equation 1 (Stature = 1.009 × SKH + 12.1; r = 0.956; SEE = 2.22; N = 119) of Raxter *et al.* (2006), based on the Terry Collection reference sample, whenever possible. A partial skeletal height (PSKH) was used if an individual had all other essential skeletal elements of the SKH except the basion-bregma height. We generated an RMA-equation (Stature = 1.045 × PSKH + 18.911; r = 0.996; SEE = 0.707; N = 537) by regressing stature estimates provided by Equation 1 of Raxter *et al.* (2006) against PSKH, so that statures derived from SKH and PSKH would be as directly comparable as possible. (Note that the correlation is extremely high here because stature is based on total skeletal height.) Stature can be estimated with about equal precision from SKH and PSKH in the Terry Collection reference data used in Niskanen *et al.* (2013), but temporal and geographic differences in basion-bregma height naturally have some effect in our European data set. This directional bias is assessed in this study.

If neither SKH nor PSKH was available, stature was estimated from lower limb long bone lengths using the region- and sex-specific equations in Ruff *et al.* (2012) (see Table 2.3 here). The one exception was that statures of all Early Upper Paleolithic individuals, including those from northern Europe, were estimated using equations designed for South Europeans because these individuals have elongated limbs, especially distal segments, relative to stature and there are no obvious latitudinal differences in this small sample (see Chapter 4). Figure 2.1 demonstrates a lack of latitudinal differences in relative lower leg length in the Early Upper Paleolithic sample, some suggestion of a trend in the Late Upper Paleolithic sample (with the exception of one high latitude outlier [Oberkassel 1]), and clear latitudinal differences in the Mesolithic sample.

In all cases, regardless of sample or period, combined maximum femoral and maximum tibial length (FMAXL + TMAXL) was used whenever possible. For the remaining individuals, stature was estimated from either maximum femoral or maximum tibial length. Long bone regression equations from Ruff *et al.* (2012) used here are presented in Table 2.3.

Stature was estimated for a total of 2132 individuals (1204 males, 928 females). Of these, anatomical stature (ASTA) was estimated from SKH for 536 individuals (290 males, 246 females)

Table 2.3 Equations to estimate stature from lower limb long bone lengths reproduced from Ruff *et al.* (2012; their Table 3).

Bone	Region	Sex	Slope	Intercept	r	SEE	N
Femur	All	Males	2.72	42.85	0.907	3.21	268
		Females	2.69	43.56	0.875	2.92	233
Tibia	North	Males	3.09	52.04	0.881	3.58	154
		Females	2.92	56.94	0.832	3.20	146
	South	Males	2.78	60.76	0.913	3.05	114
		Females	3.05	49.68	0.888	2.90	87
Fem. + Tib.	North	Males	1.49	43.55	0.918	2.93	154
		Females	1.42	48.59	0.889	2.60	146
	South	Males	1.40	49.68	0.930	2.74	114
		Females	1.47	42.96	0.894	2.82	87

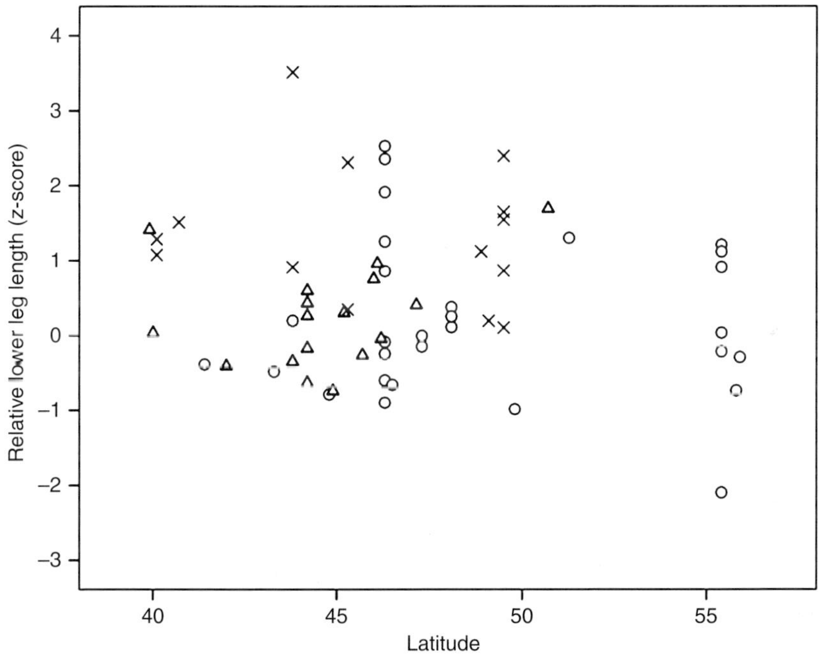

Figure 2.1 Relative lower leg length (z-score) plotted against latitude in Early Upper Paleolithic (crosses), Late Upper Paleolithic (triangles), and Mesolithic (circles) individuals.

and from PSKH for 527 individuals (317 males, 210 females), providing a total sample of 1063 individuals (607 males, 456 females) whose stature is estimated using at least a partial anatomical method. Stature was estimated from lower limb long bone lengths for a total of 1069 individuals (597 males, 472 females). Combined femoral and tibial length was possible to use for this estimation for 750 individuals (429 males, 321 females). Only femoral length was used in 239 individuals (126 males, 113 females), and only tibial length in 80 individuals (42 males, 38 females). Table 2.4 provides numerical information for how many individuals in each temporal period each stature estimation method was used.

Table 2.4 Numbers of individuals and stature estimation method applied.

Temporal sample	Total	SKH	PSKH	FTL	FL	TL
REC	157	58	0	68	27	4
EMO	150	75	30	32	9	4
LME	531	157	162	132	67	13
EME	312	55	79	136	31	11
IRO	298	80	87	88	29	14
BRO	254	23	60	129	26	16
NEO (farmers)	216	38	49	93	25	11
NEO (foragers)	76	17	33	23	3	0
MES	88	14	15	37	16	6
LUP	25	9	8	5	2	1
EUP	25	10	5	7	4	0
Total	2132	536	527	750	239	80

2.4.2 Sitting Height and Total Lower Limb Length

The basion-bregma height of the braincase (BBH) plus vertebral column height (VCH, computed as the sum of vertebral bodies from C2 through S1) represents skeletal trunk height (STH) and is used as a proxy for sitting height (SHT). This dimension, divided by anatomically estimated stature, is used as a proxy for relative sitting height (RSHT) in several chapters of this volume. Estimations of living sitting height (LSHT), relative sitting height (LRSHT), subischial lower limb length (LlimbL), and lower leg length (LEGL) from skeletal dimensions are used in Chapters 4 and 12, mainly so that direct comparisons to living populations can be carried out. Procedures to estimate these anthropometric values are described below.

We estimated living sitting height by subtracting estimated subischial lower limb length from stature estimated from either complete or partial skeletal height. Because this lower limb length is largely determined by skeletal dimensions, it can be more reliably estimated than living sitting height, which includes a large non-skeletal (e.g., intervertebral disks, vertebral column curvature) component. The estimation procedures described below provided living subischial lower limb length and sitting height estimates for 607 males and 456 females.

The total lower limb length from the femoral head to the heel can be estimated by multiplying the sum of femoral bicondylar length, tibial biomechanical length, and talocrural height by 1.007 (correction factor based on data in Ingalls, 1927) to convert dry-bone dimensions to fresh-bone dimensions, and then adding sex-specific soft-tissue additions (19.30 mm in males and 15.98 mm in females). These additions are derived by adding together average summed cartilage thicknesses of knee and ankle (males 8.3 mm; females 6.98 mm) and heel pad thicknesses under pressure (males 12 mm; females 10 mm; Uzel *et al.*, 2006), and subtracting 1 mm due to overlap between slightly convex distal tibial and slightly concave talar articular surfaces. These cartilage thicknesses were computed from data in Hall and Wyshak (1980) and Shepherd and Seedhom (1999). The femoral head cartilage was not included because it was not included in measuring the distance from a line drawn between the superior surfaces of both femoral heads to a line drawn between both ischial tuberosities (see below).

Because both cartilage and heel pad thicknesses are positively correlated with stature, and thus long bone lengths, within a sex (Hall and Wyshak, 1980; Shepherd and Seedhom, 1999;

Connolly *et al.*, 2008, Uzel *et al.*, 2006), we used simple sex-specific multiplications. Based on the Euroamerican reference sample, the sum of femoral bicondylar length, tibial lateral condylar length and talocrural height should be multiplied by 1.008 in males and by 1.005 in females.

The height of the femoral head above the ischial tuberosity including the soft-tissue thickness at this tuberosity (FH to Isch.Tub.) must be subtracted from the total lower limb length estimated as above to convert it to the living subischial lower limb length. We estimated this dimension from femoral head superoinferior breadth (FHSI) using a regression equation (FH to Isch.Tub. = $1.357 \times \text{FHSI} + 29.676$; $r = 0.714$; $\text{SEE} = 4.57$; $N = 38$) based on measurements taken from 30 male and eight female Finnish pelvises and femora articulated together with the pelvis oriented so that the pubic symphysis and the anterior superior iliac spines are at the same level. Because the tissue thickness between the ischial tuberosity and the seat is between 10–15 mm (Brodeur *et al.*, 1995: their Appendix C), 12.5 mm was added to the intercept (FH to Isch.Tub. = $1.357 \times \text{FHSI} + 42.176$). Complete sex-specific equations to estimate subischial lower limb length (LlimbL) are as follows:

$$\text{LlimbL}(\text{males}) = \left[1.008 \times (\text{FBICL} + \text{TLCL} + \text{TCH})\right] - \left[(1.357 \times \text{FHSI}) + 42.176\right]$$

$$\text{LlimbL}(\text{females}) = \left[1.005 \times (\text{FBICL} + \text{TLCL} + \text{TCH})\right] - \left[(1.357 \times \text{FHSI}) + 42.176\right]$$

For 10 males and 13 females that have anatomical statures but no femoral head breadth values, we estimated subischial lower limb length (LlimbL) directly from the sum of femoral bicondylar length (FBICL), tibial lateral condylar length (TLCL), and talocrural height (TCH) with the following sex-specific equations based on our European data:

$$\text{LlimbL}(\text{males}) = 0.957 \times (\text{FBICL} + \text{TLCL} + \text{TCH}) - 61.387$$
$$(r = 0.998; \text{SEE} = 3.15; N = 1005)$$
$$\text{LlimbL}(\text{females}) = 0.953 \times (\text{FBICL} + \text{TLCL} + \text{TCH}) - 57.351$$
$$(r = 0.998; \text{SEE} = 2.69; N = 755)$$

Our estimated living sitting height is predictably highly correlated with skeletal trunk height ($\text{STH} = \text{BBH} + \text{VCL}$) (males, $r = 0.982$, $N = 290$; females, $r = 0.982$, $N = 246$) because neither of these dimensions here are true living dimensions, but are estimated from skeletal dimensions. Estimated relative sitting height (estimated sitting height/ anatomical stature × 100) is for the same reason also highly correlated with a proxy of relative sitting height computed by dividing skeletal trunk length by anatomical stature (males, $r = 0.931$, $N = 290$; females, $r = 0.937$, $N = 246$). Thus, the two types of measurements and indices (skeletal and anthropometric) should provide similar results when applied across skeletal samples, but the anthropometric equivalents allow direct comparisons to living population data.

2.4.3 Limb Segment Lengths

Brachial index (RMAXL/HMAXL × 100) is computed to represent the forearm length relative to upper arm length. Crural index (TLCL/FBICL × 100) is computed to represent the leg length relative to thigh length. In addition, living relative lower leg length (RLEGL = LEGL/ASTA × 100) is computed to facilitate comparisons with anthropometric samples in Chapters 4 and 12. The procedure to estimate living lower leg length (center of knee to center of tibiotalar joint) from tibial length (TLCL) was based on information given in

Olivier and Pineau (1957) and Olivier and Tissier (1975) on the relationship between TMCl and lower leg length, and Sládek *et al.* (2000) on the relationship between TMCL and TLCL. The resulting conversion equation is:

$$\text{LEGL} = 1.01 \times \text{TLCL}$$

2.4.4 Body Breadths

The sum of left and right maximum clavicular length – or the preserved length, either right or left, doubled – is used as a proxy for biacromial breadth of the shoulders in many chapters in this volume. This measure, divided by stature, gives the relative clavicular breadth (RCB). Biacromial breadth (BAB), estimated from maximum clavicular length, is used in Chapters 4 and 12 to facilitate comparing skeletal samples with anthropometric samples. Dividing this by stature gives the relative biacromial breadth (RBAB), and dividing it with estimated living bi-iliac breadth provides the biacromial breadth/bi-iliac breadth index.

Our equations to estimate biacromial breadth from clavicular length were modified from those reported by Piontek (1979), based on anthropometric biacromial breadths and clavicular lengths measured radiographically. Biacromial breadth estimates provided by Piontek's (1979) LS equations exhibit clear directional estimation error because the difference between the estimated biacromial breadths and summed clavicular lengths decreases with increasing clavicular length. This difference should slightly increase because the jugular notch and acromion process breadths correlate positively with clavicular lengths (based on our unpublished data). This problem may be a result of the afore-mentioned issue with LS equations when applied near the extremes of the data distribution.

We therefore converted Pionteks's (1979) LS regression slopes to RMA slopes and recomputed intercepts using averaged mean biacromial breadth and clavicular length values from Piontek (1979), supplemented with large data sets of biacromial breadths of modern Euroamericans (NHANES III; McDowell *et al.*, 2009; their Tables 54 and 55), and clavicular lengths of US Forensic Anthropology Data Bank Euroamericans born on or after 1930 (Spradley and Jantz, 2011; their Table 8). Averaging these means with those provided in Piontek (1979) provides the following values: clavicular length 151.965 mm for males and 136.52 mm for females; biacromial breadths 400.935 mm for males and 355.64 mm for females. Dividing Piontek's LS slopes by his correlation coefficients yields the RMA slopes (Hofman, 1988), which together with these mean x and y values provides the intercepts. The resulting RMA equation for males is satisfactory, with no apparent bias, but that of females still provides biased estimates. The male equation is thus calibrated for females using female means of biacromial breadth and clavicular length to generate an equation that produces more accurate estimates for females at the two extremes of size range. The resultant equations are as follows:

$$\text{BAB}(\text{males}) = 2.353 \times \text{CML} + 43.361$$
$$\text{BAB}(\text{females}) = 2.353 \times \text{CML} + 34.408$$

We recommend that additional studies be carried out to further test these equations in living samples of various body sizes and proportions.

For use in comparisons with living samples in Chapters 4 and 12, skeletal bi-iliac breadth was converted to living bi-iliac breadth (BIB) using Ruff *et al.*'s (1997) equation (Living bi-iliac breadth = 1.17 × Skeletal bi-iliac breadth – 3; all dimensions in cm). Living bi-iliac breadth divided by stature gives living relative bi-iliac breadth (RBIB).

2.4.5 Body Mass

Body mass was estimated from stature and living bi-iliac breadth for 1202 individuals (664 males and 538 females) using the sex-specific equations of Ruff *et al.* (2005). For 863 individuals (505 males and 358 females) without a bi-iliac breadth value, body mass was estimated from femoral head supero-inferior breadth (FHSI) using sex-specific equations of Ruff *et al.* (2012) generated by applying a 'hybrid' approach – that is, body masses estimated by Ruff *et al.*'s (2005) equations were regressed against femoral head breadths. Table 2.5 presents these two sets of sex-specific equations and Table 2.6 presents numerical information for how many individuals in each temporal period each method was used.

Body mass index (BMI) was computed to represent relative laterality or linearity of the body. As in living individuals, BMI = body mass in kg/ (stature in m)^2. Individuals that have a broad bi-iliac breadth or large femoral head breadth relative to stature have high BMI values, whereas those that have narrow bi-iliac breadth or small femoral head breadth relative to stature have low BMI values.

Table 2.5 Equations to estimate body mass (BM) from stature (STA) and living bi-iliac breath (LBIB) or from femoral head breadth (FHSI).

Sex	Equation	r	SEE	N	Source
Males	BM = 0.422 × STA + 3.126 × BIB − 92.9	0.913	3.7	32	Ruff *et al.* (2005)
Females	BM = 0.504 × STA + 1.804 × BIB − 92.9	0.819	4.0	26	
Males	BM = 2.80 × FHSI − 66.70	0.636	6.84	624	Ruff *et al.* (2012)
Females	BM = 2.18 × FHSI − 3581	0.671	4.44	521	

Table 2.6 Number of individuals and body mass estimation method.

Temporal sample	Total	STA and LBIB	FHSI
REC	154	102	52
EMO	144	91	53
LME	522	396	126
EME	307	165	142
IRO	287	145	142
BRO	244	93	151
NEO (farmers)	218	107	111
NEO (foragers)	75	49	26
MES	67	30	37
LUP	24	12	12
EUP	23	12	11
Total	2065	1202	863

2.5 Evaluation of Errors in Estimating Body Size and Shape from Skeletal Dimensions

The actual results of applying methods reviewed and described above to the study samples as a whole are presented in Chapter 4. In the remainder of this chapter, we evaluate the estimation precision (reliability) of the methods used and determine possible systematic effects of temporal and regional variation, as well as of sex and age of individuals on estimations.

2.5.1 Stature Estimation

Subtracting statures estimated from partial skeletal heights from statures estimated from complete skeletal heights reveals that male stature is underestimated by an average of 0.05 cm and female stature overestimated by 0.10 cm when partial skeletal height (not including the cranium) is used to estimate skeletal height. Figure 2.2 reveals that the sex difference is relatively small. The total estimation error is less than 1 cm in 77.4% of all cases (415/536), and over 2 cm in only 2.4% of all cases (13/536).

There are some temporal differences in this estimation error. Statures estimated from partial skeletal height tend to be underestimated in the case of Upper Paleolithic, Mesolithic, Neolithic, and Bronze Age Europeans, whereas there is a slight tendency of overestimation in later samples (see Table 2.7 and Fig. 2.3). This pattern is due to the earlier Europeans having taller braincases relative to postcranial length than the more recent Europeans. However, except for Neolithic farmers, the mean difference between estimates within any period is less than 0.5 cm. Thus, using partial skeletal height to estimate stature in some individuals does not appear to introduce any significant bias.

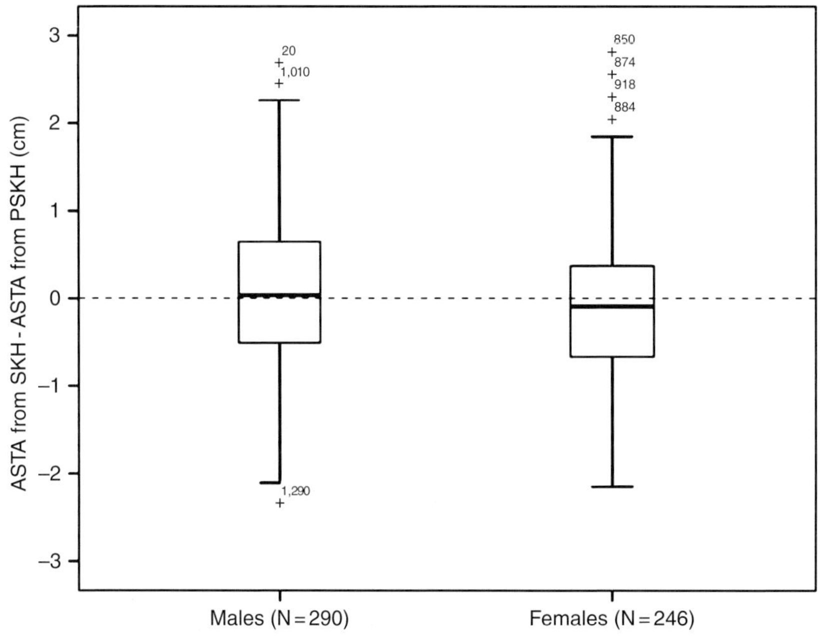

Figure 2.2 Stature estimated from complete skeletal height (ASTA from SKH) minus stature estimated from partial skeletal height (ASTA from PSKH). Differences between sexes.

Table 2.7 Directional estimation error (cm) of estimating stature from partial skeletal height. Computed as stature estimated from SKH minus stature estimated from PSKH.

Temporal sample	N	Mean	Median	Minimum	Maximum
REC	58	– 0.11	– 0.12	– 1.94	1.36
EMO	75	– 0.34	– 0.23	– 2.10	1.85
LME	157	– 0.23	– 0.24	– 2.34	2.26
EME	55	– 0.09	– 0.36	– 1.75	2.35
IRO	80	– 0.01	– 0.01	– 1.95	1.56
BRO	23	0.52	0.18	– 0.48	2.72
NEO (farmers)	38	1.02	1.13	– 0.67	2.83
NEO (foragers)	17	0.10	0.17	– 1.21	1.05
MES	14	0.44	0.57	– 2.15	2.45
LUP	9	0.47	0.52	– 0.76	1.46
EUP	10	0.19	0.33	– 0.68	0.78
Total	536	– 0.02	– 0.04	– 2.34	2.83

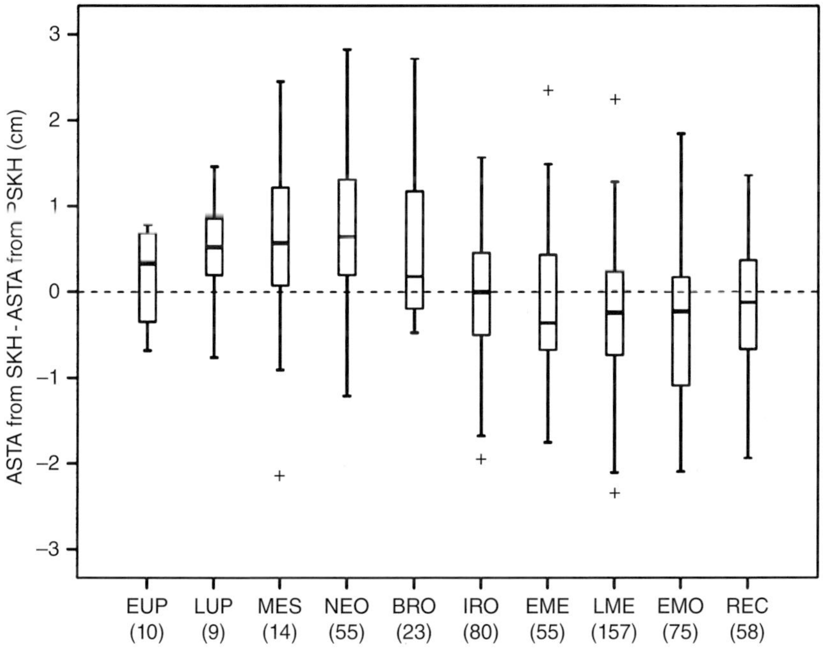

Figure 2.3 Stature estimated from complete skeletal height (ASTA from SKH) minus stature estimated from partial skeletal height (ASTA from PSKH). Differences between temporal samples. Number in parentheses below period abbreviation is sample size.

Ruff *et al.* (2012) demonstrated that their long bone regression equations provide stature estimates that exhibit little or no bias within European Holocene samples, and that geography and temporal period have relatively little effect on estimation error. Since the present study sample includes pre-Holocene time periods, some additional individuals from more recent periods, and subdivision of the Neolithic sample into farmers and foragers, we re-evaluated those results here (Table 2.8), by subtracting stature estimates provided by these long bone regression equations from anatomical stature estimates provided by Equation 1 of Raxter *et al.* (2006). In the case of the recent Finnish autopsy sample, stature estimates provided by long bone regression equations are subtracted from cadaveric statures converted to living stature by multiplying them by 0.9851 (correction factor from Niskanen *et al.*, 2013).

As shown previously (Ruff *et al.*, 2012), for most Holocene comparisons the new mathematical stature equations are relatively unbiased, with average residuals against anatomical stature of about 1 cm or less. The most noticeable directional bias is in estimating statures of the Neolithic period foragers from Sweden associated with the Pitted Ware Culture (i.e., Neolithic foragers), whose anatomical statures average about 2 cm taller than statures predicted from their limb bone lengths (Table 2.8; also see Chapter 12). The Mesolithic and Late Upper Paleolithic people exhibit the second-most estimation bias. Their anatomical statures are over 2 cm taller than statures predicted from their femoral lengths, within 1–2 cm taller than those predicted from combined femoral and tibial lengths, and within 1 cm of those estimated from tibial lengths. Statures of these Late Pleistocene and Early Holocene individuals are underestimated from long bone lengths, partly for the same reason that their statures are underestimated from partial skeletal heights: their braincases are tall.

Our equations provide reasonably accurate estimates for the Finnish autopsy sample, composed of individuals born between 1840 and 1914 (see Table 2.8; also see Chapter 12). Statures estimated from long bone lengths are very similar to cadaveric statures converted to living statures. Stature estimates for the more recent German autopsy sample cannot be similarly evaluated because we do not have statures, either cadaveric or living, for these individuals.

2.5.2 Sitting Height, Lower Limb Subischial Length, and Lower Leg Length

As mentioned earlier, Equation 1 of Raxter *et al.* (2006) may slightly underestimate statures of tall individuals, and overestimate statures of short individuals. This slight directional error may also have some effect on absolute and relative sitting height estimates.

Because we do not have any sample that has living sitting height and subischial lower limb lengths as well as skeletal measures, we tested our estimation precision indirectly using a sample of 34 Euroamerican males from the William Bass Collection in Maijanen (2009; see their Table 3). The mean subischial lower limb length estimate for this sample, calculated as described above, is 83.96 cm. We subtracted this value from mean stature provided by Equation 1 of Raxter *et al.* (2006), using 25 as the age (176.52 cm), which produced a mean living sitting height estimate of 92.01 cm and a mean living relative sitting height estimate of 52.44%. These estimates are similar to expected values. For example, 86 Euroamerican males under 30 years of age in Friedlaender *et al.* (1977: their Table 2) had mean stature and sitting height of 177.8 cm and 93.3 cm, respectively. Mean relative sitting height and subischial lower limb length are thus 52.47% and 84.50 cm, respectively, quite close to our values.

Another source of potential differences between skeletally estimated and true anthropometric dimensions is error in measuring the anthropometric dimensions themselves. Diurnal changes (Krishnan and Vij, 2007), age-related changes (Niskanen *et al.*, 2013 and references therein), fatness (Brodeur *et al.*, 1995: their Appendix C; Bogin and Varela-Silva, 2008), and different techniques to measure stature (Damon, 1964; Niskanen *et al.*, 2013) and sitting height (Carr *et al.*, 1989) all have their cumulative effects. In this light, our estimations of living sitting heights, subischial lower limb lengths, and relative sitting heights are certainly precise enough to allow comparing past (skeletal) and present (anthropometric) populations.

Table 2.8 Residuals of estimating anatomical stature from long bone lengths. Stature is estimated from SKH using Equation 1 of Raxter *et al.* (2006), except in the recent autopsy sample where it is cadaveric stature × 0.9851 (a correction factor from Niskanen *et al.*, 2013). Bold values indicate mean residuals of ≥1 cm. Positive values indicate underestimation and negative values overestimation.

	N		FTL	FL	TL
REC (autopsy)	48	Mean	0.21	– 0.02	0.51
		Median	– 0.21	0.03	0.19
		Minimum	– 7.61	– 7.16	– 9.27
		Maximum	11.28	10.21	12.56
REC	58	Mean	0.97	0.54	0.98
		Median	1.01	– 0.03	0.82
		Minimum	– 5.32	– 6.74	– 6.67
		Maximum	8.75	10.38	8.01
EMO	75	Mean	– 0.89	– 0.75	**– 1.16**
		Median	– 0.81	– 0.41	**– 1.47**
		Minimum	– 7.30	– 10.14	– 7.63
		Maximum	5.42	6.04	6.52
LME	157	Mean	– 0.11	– 0.31	0.26
		Median	– 0.06	– 0.40	0.54
		Minimum	– 9.60	– 10.24	– 7.93
		Maximum	7.66	6.92	9.30
EME	55	Mean	– 0.46	0.32	**– 1.32**
		Median	– 0.58	– 0.02	**– 1.56**
		Minimum	– 5.53	– 5.21	– 6.66
		Maximum	5.21	5.59	6.57
IRO	80	Mean	0.25	0.44	0.14
		Median	0.27	0.22	– 0.02
		Minimum	– 4.86	– 5.61	– 6.94
		Maximum	6.89	9.16	11.63
BRO	23	Mean	– 0.65	– 0.80	– 0.52
		Median	0.06	– 0.37	0.05
		Minimum	– 6.37	[illegible]	[illegible]
		Maximum	5.01	5.00	6.08
NEO (farmers)	38	Mean	0.63	0.50	0.56
		Median	0.22	0.36	0.50
		Minimum	– 5.15	– 7.05	– 4.79
		Maximum	7.73	7.53	9.72
NEO (foragers)	17	Mean	**2.20**	**2.10**	**2.18**
		Median	**2.35**	**1.86**	**1.60**
		Minimum	– 3.49	– 3.49	– 4.12
		Maximum	7.10	9.18	6.95
MES	14	Mean	0.91	**2.34**	– 0.64
		Median	0.75	**2.55**	– 1.45
		Minimum	– 2.61	– 0.33	– 6.51
		Maximum	6.53	6.09	7.38
LUP	9	Mean	**1.63**	**2.72**	0.13
		Median	**0.61**	**2.46**	– 0.08
		Minimum	– 2.79	– 1.03	– 6.03
		Maximum	5.07	6.33	4.80
EUP	10	Mean	0.19	0.87	– 0.58
		Median	0.72	0.93	– 1.07
		Minimum	– 2.88	– 2.83	– 4.10
		Maximum	1.75	5.08	4.34
Total	584	Mean	0.09	0.14	0.00
		Median	0.01	– 0.03	– 0.12
		Minimum	– 9.60	– 10.24	– 9.27
		Maximum	11.28	10.38	12.56

We cannot evaluate the precision of lower leg length estimation due to a lack of appropriate data. Some estimation error must exist, but it cannot have much effect because the total soft-tissue component is only a few millimeters.

2.5.3 Trunk Breadths

Error in estimating biacromial breadth from clavicular length cannot be properly evaluated here because, as also in the case of sitting height, we do not have 'true' biacromial breadths for comparison. Correlation coefficients between clavicular lengths and biacromial breadths (males r = 0.816; females r = 0.758, from Piontek, 1979, their Table 2) indicate that this estimation precision is about equal to that in estimating stature from long bone lengths.

Measurement technique matters in measuring biacromial breadth of living individuals and could, therefore, affect comparisons of biacromial breadths estimated from clavicular lengths with measured biacromial breadths of anthropometric samples. Instructing the subjects to keep "...shoulders relaxed to the point of slumping forward" (Tanner, 1964:25) provides greater biacromial breadths than instructing them to maintain good posture and not to allow shoulders to slump forward (Martin and Saller, 1957:331). Instructing the subjects to stand at attention and to square shoulders backward naturally provides the narrowest biacromial breadths. These different techniques can easily produce biacromial breadths that differ by 2 cm or even 3 cm (M. Niskanen, unpublished experiment).

There are far fewer technical problems and thus considerably less measurement error in bi-iliac breadth than in bi-acromial breadth (Friedlaender *et al.*, 1977; Bennett and Osborne, 1986). Living bi-iliac breadth is essentially skeletal bi-iliac breadth plus an absolutely and relatively small soft-tissue addition. Inter-individual differences in soft-tissue thickness exist, but these differences probably amount to a few millimeters only.

Our estimates of living bi-iliac breadth are very likely reasonably accurate at the middle of the size range, but there is some estimation bias. Some very-small-bodied humans (skeletal bi-iliac breadth less than 17.64 cm) can receive living bi-iliac breadth estimates that are actually narrower than their skeletal bi-iliac breadths (see Stock, 2013; their Fig. 8), but these individuals are outside the normal range of variation. The smallest skeletal bi-iliac breadth value in our data set (21.4 cm; Ruff *et al.*, 2012, their Table 2) produces a living bi-iliac breadth estimate of 22.04 cm.

2.5.4 Body Mass Estimation

The two methods used in body mass estimation in this study provide similar, but not identical, body mass estimates for individuals in our sample. In addition to individual variation in the relationship between joint size and skeletal frame size, any directional bias in equations to estimate body mass as well as in those used to estimate stature and living bi-iliac breadth introduces additional variation (see Auerbach and Ruff, 2004).

Table 2.9 shows differences in these estimates as residual values (body mass estimated from femoral head breadth is subtracted from that estimated from stature and bi-iliac breadth) and percentage prediction errors ('morphometric' body mass estimated from stature and bi-iliac breadth representing 'true' body mass) for each temporal period. In most periods, average differences are quite small, with the largest differences (>2 kg) in the Bronze Age sample and among Neolithic period foragers (the Pitted Ware sample) from Scandinavia, where femoral head estimates are smaller than those estimated from stature and bi-iliac breadth. These are still within about 6% of each other, however.

We examine if differences in body size and shape are behind these differences in body mass estimates in our sample. Correlations between percentage prediction errors and body size and

Table 2.9 Difference in body mass estimation. Body mass estimated from femoral head breadth subtracted from body mass estimated using stature/bi-iliac method (kg) is above and percentage prediction error (%) is below. Bold values indicate mean residuals ≥2 kg. Positive values indicate underestimation and negative values overestimation.

	N	Mean	Median	Minimum	Maximum
REC	97	1.51	1.54	– 15.22	18.69
		3.13	2.50	– 22.11	31.08
EMO	90	– 0.33	0.27	– 19.00	12.52
		0.31	0.53	– 27.24	24.92
LME	394	– 0.45	0.25	– 24.86	19.20
		– 0.15	0.34	– 29.42	39.92
EME	163	– 0.87	– 0.80	– 20.55	20.60
		– 0.93	– 1.35	– 25.95	0.30
IRO	141	0.33	– 0.16	– 15.78	15.90
		0.54	– 0.23	– 20.14	25.37
BRO	93	**2.30**	**2.45**	– 10.97	14.67
		4.15	**4.02**	– 17.76	23.57
NEO (farmers)	101	1.02	0.53	– 10.80	18.12
		1.75	1.01	– 20.50	32.55
NEO (foragers)	49	**3.59**	**2.38**	– 7.00	17.69
		6.02	**3.97**	– 10.65	30.11
MES	28	– 1.00	– 1.32	– 8.72	11.94
		– 1.09	– 2.10	– 12.04	25.57
LUP	12	0.66	1.50	– 11.11	7.34
		1.26	2.70	– 17.41	11.06
FLUP	12	– 1.24	– 0.78	– 9.81	8.38
		– 1.71	– 1.11	– 14.62	12.91
Total	1180	0.26	0.47	– 24.86	20.60
		0.86	0.76	– 29.42	39.92

shape variables are presented in Table 2.10, separately for males and females. These correlations reveal that, in both sexes, the percentage prediction error has significant positive correlations with morphometric body mass, stature, absolute bi-iliac breadth, and bi-iliac breadth relative to stature, whereas correlations with biacromial breadth/bi-iliac breadth are negative. Body masses estimated from femoral head breadth thus tend to be larger than morphometric body masses if stature is short and bi-iliac breadth absolutely and relatively narrow, while the reverse is the case if stature is tall and bi-iliac breadth absolutely and relatively broad. These results are generally consistent with earlier studies comparing femoral head and stature/bi-iliac estimations of body mass in very small and large modern humans (Auerbach and Ruff, 2004; Kurki *et al.*, 2010).

The above patterns partly explain why body mass estimated from femoral head breadth tends to be smaller than body mass estimated from stature and bi-iliac breadth in the Bronze Age and Pitted Ware samples, which are characterized by absolutely and relatively broad bi-iliac breadths, although not necessarily tall statures (see Chapter 4). The earlier-mentioned directional bias to estimate living bi-iliac breadth from skeletal bi-iliac breadth does not explain this pattern because its effect is very small here.

Table 2.10 Directional body mass prediction error when body mass is estimated from femoral head breadth. Directional estimation error is computed as body mass estimated from stature and bi-iliac breadth minus body mass estimated from femoral head breadth. Bold values indicate correlations that are significant at $p < 0.01$.

		Males	Females
BM from STA and LBIB	Correlation	**0.400**	**0.402**
	Sig.	**0.000**	**0.000**
	N	**652**	**528**
ASTA	Correlation	0.080	**0.236**
	Sig.	0.090	**0.000**
	N	455	**349**
STA	Correlation	**0.141**	**0.241**
	Sig.	**0.000**	**0.000**
	N	**652**	**528**
BAB	Correlation	0.115	0.066
	Sig.	0.013	0.218
	N.	474	347
LBIB	Correlation	**0.456**	**0.421**
	Sig.	**0.000**	**0.000**
	N	**652**	**528**
BAB / STA	Correlation	0.054	– 0.115
	Sig.	0.241	0.032
	N	474	347
LBIB / STA	Correlation	**0.423**	**0.320**
	Sig.	**0.000**	**0.000**
	N	**652**	**528**
BAB / LBIB	Correlation	**– 0.306**	**– 0.416**
	Sig.	**0.000**	**0.000**
	N	**474**	**347**
LRSHT	Correlation	0.059	– 0.121
	Sig.	0.210	0.024
	N	455	349

2.5.5 Sex and Age Effects

Age certainly has some effect on both body size and shape. Age-related changes in skeletal and soft-tissue components of stature affect living stature, which starts to decline during the fourth or fifth decade of life. Male stature starts to decline earlier than female stature, but female stature loss accelerates more with age, resulting in a greater stature loss over the adult lifespan (see Sorkin *et al.*, 1999 and references therein). Sex differences in stature loss with age have little effect on our results, mainly because most individuals in our sample are too young to have lost much stature with age. We have further discussed this issue in Ruff *et al.* (2012).

There is some age-related increase in bi-iliac breadth in our sample (Fig. 2.4). Increase in living bi-iliac breadth through adulthood is due to about equal amounts of skeletal and soft-tissue changes (based on comparing data in Friedlaender *et al.*, 1977 with data in Völgyi *et al.*, 2010 and Berger *et al.*, 2011). This, combined with the stature decline with age, affects estimations of body mass using the stature/bi-iliac breadth method, and also results in broader bi-iliac breadth relative to stature with age. In addition, it affects bi-iliac breadth relative to biacromial breadth because, while bi-iliac breadth increases through adult life, biacromial breadth increases to about 45 years of age and declines slowly thereafter (see Borkan *et al.*, 1983 on

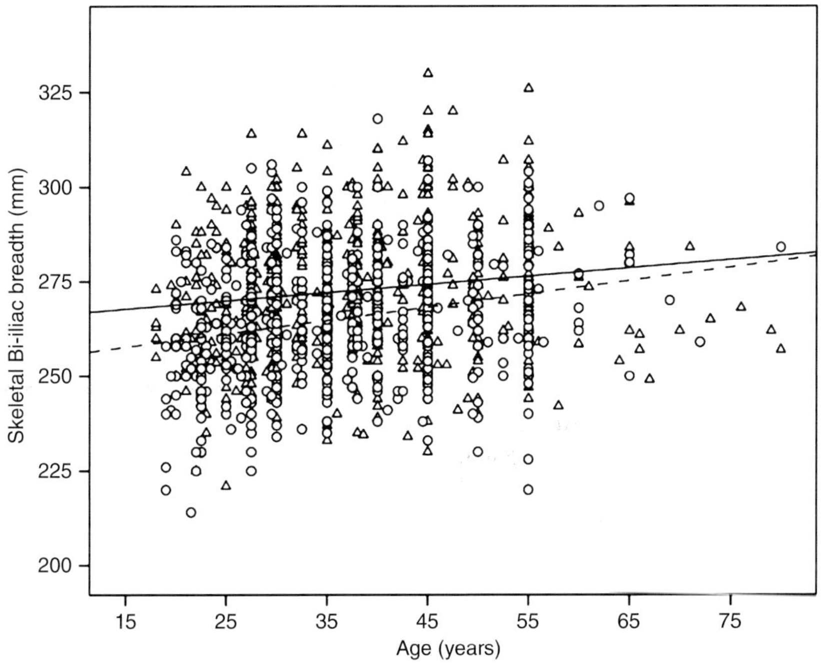

Figure 2.4 Skeletal bi-iliac breadth regressed against either known or estimated age. Males: triangles and solid lines (y = 0.217x + 264.517, r = 0.140, N = 667); females: dashed lines and circles (y = 351x + 252.476, r = 0.236, N = 540).

changes in biacromial breadth), reflecting an acromioclavicular joint space reduction (Petersson and Redlund-Johnell, 1983) and changes in muscle (and even fat) mass with age.

Estimated body mass, and especially body mass index, increase in with age in our subsample in which body mass is estimated from bi-iliac breadth and complete skeletal height (body mass and age r = 0.145, p = 0.001, N = 430; body mass index and age r = 0.240, p <0.001, N = 430). This trend is due to a decreasing stature estimate and increasing bi-iliac breadth with age. In a subsample where body mass is estimated from femoral head breadth and stature from summed femoral and tibial lengths correlations are much lower (body mass and age r = 0.065, p = 0.090, N = 432; body mass index and age r = 0.112, p = 0.010, N = 342). These low and positive correlations are due to age not being taken into account in estimating stature from long bone lengths in this study, and a possible slight increase in femoral head breadth with age, which has been noted by Berger *et al.* (2011) in their cross-sectional study of living individuals.

It should be noted here that the above pattern applies to skeletal samples. Applying the stature/bi-iliac breadth method on living samples results in an overestimation of body mass at both extremes of age range and underestimations of body mass of 'prime age' adults due to muscle mass increase in early adulthood and its decline in old age (Junno and Niskanen, 2015).

2.6 Discussion and Conclusions

There is not much room for improvement in estimating stature from complete skeletal height unless individual skeletons are articulated, dry bone dimensions are converted to wet-bone dimensions, and soft-tissue corrections are made. Because this reconstruction of stature through articulation is impractical and possible in only a few cases, we have to content ourselves to "...stature estimates that are accurate to within 4.5 cm in 95% of the individuals" (on this estimation precision, see Raxter *et al.*, 2006). However, equations that provide estimates which exhibit minimum directional estimation bias related to stature, age, and sex could be generated using larger reference samples of individuals whose stature, age, and sex are known.

Stature estimates from long bone lengths are the most imprecise for the Late Upper Paleolithic and Mesolithic samples, although even here average bias is 1.6 cm or less (i.e., ≤1%) using femur + tibia length. Due to small samples sizes for these periods, increased sample sizes of Paleolithic and Mesolithic individuals are required to fully assess stature estimation error in these samples, as well as to possibly generate period-specific equations, if warranted. The Neolithic foraging subsample also showed relatively higher estimation error, and this needs to be further investigated.

Errors in estimating stature and living bi-iliac breadth naturally affect precision in estimating body mass using the morphometric method (Auerbach and Ruff, 2004). Our estimations of sitting height, subischial lower limb length, leg length, and bi-acromial breadth from skeletal dimensions probably also include some error, which is more difficult to evaluate due to a lack of appropriate data (i.e., anthropometric dimensions and underlying skeletal dimensions from the same individuals).

However, these possible errors are less significant if they have a noticeable effect mainly at extremes of the size range (e.g., stature range, bi-iliac breadth range) and thus have little or no effect on sample means. Also, estimation errors are generally quite small compared to variation between periods, sexes, and so on. Taken together, these techniques provide a robust method for examining temporal and geographic trends in body size and shape. The refinement of these estimation methods is, however, a never-ending process due to the accumulation of new information and data.

Acknowledgments

We thank Tiina Väre and Laura Paiges Sotos for taking measurements from articulated femora and pelvises. It would have been impossible to determine the height of the femoral head above the ischial tuberosity without these measurements.

References

Arsuaga, J.-L., Lorenzo, C., Carretero, J.-M., Gracia, A., Martinez, I., Garcia, N., Bermudez de Castro, J.-M., and Carbonell, E. (1999) A complete human pelvis from the Middle Pleistocene of Spain. *Nature*, **399**, 255–258.

Auerbach, B.M. (2011) Methods for estimating missing human skeletal element osteometric dimensions employed in the revised Fully technique for estimating stature. *Am. J. Phys. Anthropol.*, **145**, 67–80.

Auerbach, B.M. and Ruff, C.B. (2004) Human body mass estimation: a comparison of "morphometric" and "mechanical" methods. *Am. J. Phys. Anthropol.*, **125**, 331–342.

Auerbach, B.M. and Ruff, C.B. (2010) Stature estimation formulae for indigenous North American populations. *Am. J. Phys. Anthropol.*, **141**, 190–207.

Bennett, K.A. and Osborne, R.H. (1986) Interobserver measurement reliability in anthropometry. *Hum. Biol.*, **58**, 751–759.

Berger, A.A., May, R., Renner, J.B., Viradia, N., and Dahners, L.E. (2011) Surprising evidence of pelvic growth (widening) after skeletal maturity. *J. Orthop. Res.*, **29**, 1719–1723.

Bogin, B. and Varela-Silva, M.I. (2008) Fatness biases the use of estimated leg length as an epidemiological marker in adults in the NHANES III sample. *Int. J. Epidemiol.*, **37**, 201–209.

Borkan, G.A., Hults, D.E., and Glynn, R.J. (1983) Role of longitudinal change and secular trend in age differences in male body dimensions. *Hum. Biol.*, **55**, 629–641.

Brodeur, R.R., Cui, Y., and Reynolds, H.M. (1995) The initial position and postural attitude of vehicle operators. Ergonomics Research Laboratory ERL-TR-95-003. MSU, East Lansing, MI. www.oandplibrary.org/cpo/pdf/1985_04_015.pdf.

Carr, R.V., Rempel, R.D., and Ross, W.D. (1989) Sitting height: an analysis of five measurement techniques. *Am. J. Phys. Anthropol.*, **79**, 339–344.

Connolly, A., FitzPatric, D., Moulton, J., Lee, J., and Lerner, A. (2008) Tibiofemoral cartilage thickness distribution and its correlation with anthropometric variables. *Proc. Inst. Mech. Eng. H*, **222**, 29–39.

Damon, A. (1964) Notes on anthropometric technique I. Stature against a wall and standing free. *Am. J. Phys. Anthropol.*, **22**, 73–78.

Dwight, T. (1894) Methods of estimating the height from parts of the skeleton. *Med. Rec. N.Y.*, **46**, 293–296.

Formicola, V. and Franceschi, M. (1996) Regression equations for estimating stature from long bones of early Holocene European samples. *Am. J. Phys. Anthropol.*, **100**, 83–88.

Franciscus, R.G. and Holliday, T.W. (1992) Hindlimb skeletal allometry in Plio-Pleistocene hominids with special reference to AL-288-1 ("Lucy"). *Bull. Mém. Soc. d'Anthropol. Paris*, n.s. 4, série 1-2:5–20.

Friedlaender, J.S., Costa, P.T., Jr, Bosse, R., Ellis, E., Rhoads, J.G., and Stoudt, H.W. (1977) Longitudinal physique changes among healthy white veterans at Boston. *Hum. Biol.*, **49**, 541–558.

Fully, G. (1956) Une nouvelle méthode de détermination de la taille. *Ann. Méd. Legale*, **35**, 266–273

Grine, F.E., Jungers, W.L., Tobias, P.V., and Pearson, O.M. (1995) Fossil *Homo* femur from Berg Aukas, northern Namibia. *Am. J. Phys. Anthropol.*, **97**, 151–185.

Hall, F.M. and Wyshak, G. (1980) Thickness of articular cartilage in the normal knee. *J. Bone Joint Surg. Am.*, **62**, 408–413.

Hens, S.M., Konigsberg, L.W., and Jungers, W.L. (1998) Estimation of African ape body length from femur length. *J. Hum. Evol.*, **34**, 401–411.

Holliday, T. (1995) Body Size and Proportions in the Late Pleistocene Western Old World and the Origins of Modern Humans. Ph.D. dissertation. The University of New Mexico, Albuquerque, New Mexico.

Holliday, T.W. (1997) Body proportions in Late Pleistocene Europe and modern human origins. *J. Hum. Evol.*, **32**, 423–447.

Hofman, M.A. (1988) Allometric scaling in paleontology: a critical survey. *Hum. Evol.*, **3**, 177–188.

Ingalls, N.W. (1927) Studies of the femur. III. The effects of maceration and drying in the white and negro. *Am. J. Phys. Anthropol.*, **10**, 297–321.

Junno, J.-A. and Niskanen, M. (2015) The effects of age on body mass estimation using the stature/bi-iliac method (abstract). *Am. J. Phys. Anthropol. Suppl.*, **60**, 181–182.

Krishan, K. and Vij, K. (2007) Diurnal variation of stature in three adults and one child. *Anthropologist*, **9**, 113–117.

Kurki, H.K., Ginter, J.K., Stock, J.T., and Pfeiffer, S. (2010) Body size estimation of small-bodied humans: applicability of current methods. *Am. J. Phys. Anthropol.*, **141**, 169–180.

Lundy, J.K. (1985) The mathematical versus anatomical methods of stature estimates from long bones. *Am. J. Forensic Med. Pathol.*, **6**, 73–75.

Maijanen, H. (2009) Testing anatomical methods for stature estimation on individuals from the W. M. Bass Donated Skeletal Collection. *J. Forensic Sci.*, **54**, 746–752.

Maijanen, H. (2011) Stature estimation from skeletal elements: general problems and small solutions. Ph.D. dissertation. University of Oulu, Oulu, Finland.

Maijanen, H. and Niskanen, M. (2006) Comparing stature-estimation methods on Medieval inhabitants of Westerhus, Sweden. *Fennoscandia Archaeol.*, **23**, 37–46.

Maijanen, H. and Niskanen, M. (2010) New regression equations for stature estimation for medieval Scandinavians. *Int. J. Osteoarchaeol.*, **20**, 472–480.

Martin, R. (1928) *Lehrbuch der Anthropologie.* Gustav Fischer Verlag, Jena.

Martin, R. and Saller, K. (1957) *Lehrbuch der Anthropologie.* Gustav Fischer Verlag, Stuttgart.

McDowell, M.A., Fryar, C.D., and Ogden, C.L. (2009) Anthropometric reference data for children and adults: United States, 1988–1994. National Center for Health Statistics. *Vital Health Stat.*, **11**(249), 1–68. Available at: www.cdc.gov/nchs/data/series/sr_11/sr11_249.pdf.

McHenry, H.M. (1992) Body size and proportions in early hominids. *Am. J. Phys. Anthropol.*, **87**, 407–431.

Niinimäki, S. and Niskanen, M. (2015) Testing the cylindrical model for weight reconstruction – the effects of fat mass, lean mass, and body proportions (abstract). *Am. J. Phys. Anthropol. Suppl.*, **60**, 236.

Niskanen, M., Maijanen, H., McCarthy, D., and Junno, J.-A. (2013) Application of the anatomical method to estimate the maximum adult stature and the age-at-death stature. *Am. J. Phys. Anthropol.*, **152**, 96–106.

Olivier, G. and Pineau, H. (1957) Comparisons entre les mensurations sur le squelette et sur le vivant. *Revue Anthropologique*, **3**, 1–16.

Olivier, G. and Tissier, H. (1975) Estimation de la stature feminine d'après les os longs des membres. *Bulletin de la Société d'anthropologie de Paris*, **2**, 1–11.

Petersson, C.J. and Redlund-Johnell, I. (1983) Radiographic joint space in normal acromioclavicular joints. *Acta Orthop. Scand.*, **54**, 431–433.

Piontek, J. (1979) Reconstruction of individual physical build features in the investigation of praehistoric population. *Coll. Anthropol.*, **2**, 251–253.

Raxter, M.H., Auerbach, B.M., and Ruff, C.B. (2006) Revision of the Fully technique for estimating statures. *Am. J. Phys. Anthropol.*, **130**, 374–484.

Raxter, M.H., Ruff, C.B., Azab, A., Erfan, M., Soliman, M., and El-Sawaf, A. (2008) Stature estimation in ancient Egyptians: a new technique based on anatomical reconstruction of stature. *Am. J. Phys. Anthropol.*, **136**, 147–155.

Rosenberg, K.R., Lü, Z., and Ruff, C.B. (2006) Body size, body proportions and encephalization in a Middle Pleistocene archaic human from northern China. *Proc. Natl Acad. Sci. USA*, **103**, 3552–3556.

Ruff, C.B. (1994) Morphological adaptation to climate in modern and fossil hominids. *Yrbk Phys. Anthropol.*, **37**, 65–107.

Ruff, C.B. (2000) Body mass prediction from skeletal frame size in elite athletes. *Am. J. Phys. Anthropol.*, **113**, 507–517.

Ruff, C.B., Holt, B.M., Sladék, V., Berner, M., Garofalo, E., Garvin, H.M., Hora, M., Maijanen, H., Niinimäki, S., Salo, K., Schuplerová, E., and Tompkins, D. (2012) Stature and body mass estimation from skeletal remains in the European Holocene. *Am. J. Phys. Anthropol.*, **148**, 601–617.

Ruff, C.B., Holt, B.M., Sládek, V., Berner, M., Murphy, W.A., Nedden, Dz., Seidler, H., and Reicheis, W. (2006) Body size, body shape, and long bone strength of the Tyrolean ice man. *J. Hum. Evol.*, **51**, 91–178.

Ruff, C.B., Niskanen, M., Junno, J.-A., and Jamison, P. (2005) Body mass prediction from stature and bi-iliac breadth in two high latitude populations, with application to earlier higher latitude humans. *J. Hum. Evol.*, **48**, 381–392.

Ruff, C.B., Trinkaus, E., and Holliday, T.W. (1997) Body mass and encephalization in Pleistocene Homo. *Nature*, **387**, 173–176.

Ruff, C.B., Scott, W.W., and Liu, A.Y.-C. (1991) Articular and diaphyseal remodeling of the proximal femur with changes in body mass in adults. *Am. J. Phys. Anthropol.*, **86**, 397–413.

Sciulli, P.W. and Hetland, B.M. (2007) Stature estimation for prehistoric Ohio Valley Native American populations based on revision to the Fully Technique. *Archaeol. East N. Am.*, **35**, 105–113.

Sciulli, P.W., Schneider, K.N., and Mahaney, M.C. (1990) Stature estimation in prehistoric Native Americans of Ohio. *Am. J. Phys. Anthropol.*, **83**, 275–280.

Shepherd, D.E.T. and Seedhom, B.B. (1999) Thickness of human articular cartilage in joints of the lower limb. *Ann. Rheum. Dis.*, **58**, 27–34.

Sládek, V., Trinkaus, E., Hillson, S.W., and Holliday, T.W. (2000) *The People of the Pavlovian: Skeletal Catalogue and Osteometrics of the Gravettian Fossil Hominids from Dolni Věstonice and Pavlov. The Dolni Věstonice Studies*. Volume 5/2000. The Academy of Sciences of the Czech Republic, Brno.

Smith, R.J. (1984) Allometric scaling in comparative biology: problems of concept and method. *Am. J. Physiol.*, **246**, R152–R160.

Sorkin, J.D., Muller, D.C., and Andres, R. (1999) Longitudinal changes in the heights of men and women: consequential effects on body mass index. *Epidemiol. Rev.*, **21**, 247–260.

Spradley, M.K. and Jantz, R.L. (2011) Sex estimation in forensic anthropology: skull versus postcranial elements. *J. Forensic Sci.*, **56**, 289–296.

Stock, J.T. (2013) The skeletal phenotype of "Negritos" from the Andaman Islands and Philippines relative to global variation among hunter-gatherers. *Hum. Biol.*, **85**, 67–94.

Tanner, J.M. (1964) *The physique of the Olympic athlete*. London, George Allen and Unwin.

Trinkaus, E. (1981) Neandertal limb proportions and cold adaptation. In: *Aspects of Human Evolution* (ed. C.B. Stringer), Taylor and Francis, London, pp. 187–224.

Trotter, M. and Gleser, G.C. (1952) Estimation of stature from long bones of American whites and Negroes. *Am. J. Phys. Anthropol.*, **16**, 79–123.

Trotter, M. and Gleser, G.C. (1958) A re-evaluation of estimation of stature based on measurements of stature taken during life and of long bones after death. *Am. J. Phys. Anthropol.*, **16**, 79–124.

Uzel, M., Cetinus, E., Ekerbicer, H.C., and Karaoguz, A. (2006) Heel pad thickness and athletic ability in young adults: a sonographic study. *J. Clin. Ultrasound*, **34**, 231–236.

Vercellotti, G., Agnew, A.M., Justus, H.M., and Sciulli, P.W. (2009) Stature estimation in an early medieval (XI–XII c.) Polish population: testing the accuracy of regression equations in bioarchaeological sample. *Am. J. Phys. Anthropol.*, **140**, 135–142.

Völgyi, E., Tylavsky, F.A., Lu, J., Wang, Q., Alén, M., and Cheng, S. (2010) Bone and body segment lengthening and widening: a 7-year follow-up study in pubertal girls. *Bone*, **47**, 773–782.

3

Quantifying Skeletal Robusticity

Christopher B. Ruff

Center for Functional Anatomy and Evolution, Johns Hopkins University School of Medicine, Baltimore, MD, USA

3.1 Cross-Sectional Properties

Adult skeletal form is the product of a complex interaction between genetic and environmental influences acting during growth and development (Pearson and Lieberman, 2004; Ruff *et al.*, 2006a). Although genetics plays a role, some skeletal features – including cross-sectional dimensions of long bone diaphyses – appear to be very developmentally plastic and responsive to applied mechanical loadings during life (Trinkaus *et al.*, 1994; Ruff *et al.*, 2006a). For example, bilateral asymmetry in humeral diaphyseal strength in modern athletes engaging in unimanual activities, as well as European Neandertal and Upper Paleolithic samples, can reach 30% or more (Trinkaus *et al.*, 1994; Churchill and Formicola, 1997; Ruff, 2008a; Warden *et al.*, 2014; and see Chapter 7 of this volume). Both overall bone strength and strength in particular planes have been shown to correlate with specific behaviors (Macdonald *et al.*, 2009; Shaw and Stock, 2009a,b).

The developmental plasticity of diaphyseal structure has been exploited by a number of researchers to reconstruct behavior in past human populations, with more than 100 published studies taking this approach over the past 40 years (for a recent review, see Ruff and Larsen, 2014). Lower limb bone structure, in particular anteroposterior versus mediolateral bending strength, can give insights into mobility, with more mobile populations or subpopulations exhibiting relatively greater A-P bending strength in mid-diaphyseal regions (Ruff, 1987; Stock and Pfeiffer, 2001; Stock, 2006). Other factors, including body shape (Ruff, 1995; Ruff *et al.*, 2006b; Shaw and Stock, 2011) and terrain (Ruff, 1999; and see Chapter 5 of this volume) also influence lower limb bone strength, so these must also be considered when interpreting results. Variation in upper limb bone strength may be affected by patterns of mobility as well, for example water transport (Weiss, 2003; Stock and Pfeiffer, 2004), but in most cases is more directly related to manipulative behaviors, including tool and weapon use, and food procurement and processing techniques (Bridges, 1989; Sparacello *et al.*, 2015; and see Chapter 7 of this volume).

Cross-sectional geometric analyses of long bone diaphyses are based on modeling the bone as an engineering beam (Gere and Timoshenko, 1990), and calculating properties that reflect strength and rigidity of the beam under mechanical loading. A number of different properties may be determined (see Ruff, 2008a for a review). In this study we focus primarily on a few parameters that are most critical for assessing mechanical performance, and thus reconstructing behavior: bending strength in A-P and M-L planes, and average bending/torsional strength.

Skeletal Variation and Adaptation in Europeans: Upper Paleolithic to the Twentieth Century, First Edition. Edited by Christopher B. Ruff.

Table 3.1 Cross-sectional geometric properties.

Property	Abbrev.	Units	Mechanical interpretation
Section modulus about M-L (x) axis	Z_x	mm^3	A-P bending strength
Section modulus about A-P (y) axis	Z_y	mm^3	M-L bending strength
Polar section modulus	Z_p	mm^3	Torsional and average bending strength
Section modulus ratio	Z_x/Z_y	–	A-P/M-L bending strength
Cortical area	CA	mm^2	Compressive/tensile strength
Total subperiosteal area	TA	mm^2	–
Percentage cortical area ((CA/TA) × 100)	%CA	–	–

These are represented by the section moduli Z_x, Z_y, and Z_p (polar section modulus), respectively (see Table 3.1 for a list of all properties). The ratio Z_x/Z_y is used as a measure of relative A-P/M-L bending strength, and as such is one mechanical representation of cross-sectional 'shape.' We also include some measures of bone areal dimensions, including total subperiosteal (TA) and cortical areas (CA), and %CA (CA/TA). Cortical area measures axial compressive and tensile strength (which are less significant mechanically in most long bones), but also the total amount of bone. Both TA and CA, as well as %CA, are useful in describing relative changes occurring on the subperiosteal and endosteal surfaces, which may be significant in terms of diet and other systemic influences (e.g., Garn *et al.*, 1969).

Many studies of long bone cross-sectional geometry have examined second moments of area (I and J) rather than section moduli (e.g., Ruff and Hayes, 1983; Holt, 2003; Stock and Pfeiffer, 2004; Sládek *et al.*, 2006a,b; Macintosh *et al.*, 2014). The two properties are closely related (Ruff, 2008a), but section moduli measure strengths, while second moments of area (SMAs) measure rigidities. We use section moduli here because it seems more likely that long bone diaphyses are adapted to avoid maximum bending/torsional stresses that would exceed material strength, rather than to maintain an appropriate rigidity under loading (although both are important mechanical properties). Calculation of section moduli is slightly more involved, and has been done in different ways in the past, so we include some discussion of this issue and present new techniques for converting SMAs to section moduli.

3.2 Cross Section Reconstruction

Cross-sectional geometric properties were determined at 50% of length' of the femur and tibia, and at 35% of length' from the distal end in the humerus (see Chapter 1 for definitions of length'). All cross sections were taken perpendicular to the long axis of the diaphysis with the bones positioned in standardized anatomical position (see Chapter 1 and Ruff, 2002). Cross-sectional properties for the majority of the study sample were determined using a technique that combined external latex molding and biplanar radiography, as described below. Specimens from Central Europe, comprising one-third of the total sample with cross-sectional data (655/1955 individuals), were analyzed using computed tomography (CT), carried out at either the University of Veterinary Medicine in Vienna, or the Jena University Hospital in Jena, Germany. Cross-sectional properties were calculated from CT images using previously described custom-written software (CT-i: Sládek *et al.*, 2006a; MomentMacro: Ruff, n.d.). In addition, the Medieval sample from Mistihalj (see Chapter 13) was analyzed using a peripheral quantitative CT (pQCT) scanner, together with software provided with the system (Ferretti *et al.*, 1996; Ruff *et al.*, 2013).

For the remainder of the sample, cross-sectional outlines were obtained using a method first described by Trinkaus and Ruff (1989), and later referred to as the 'latex cast method' (Stock, 2002; O'Neill and Ruff, 2004). With the bone in standardized anatomical position, a ring of latex molding material (Coltène/Whaledent President Putty) was placed around the section of interest, with one edge of the mold flattened to conform to the section location, and allowed to harden. Medial and lateral marks equidistant above the support surface for the bone were made and labeled, along with the specimen number (anteriorly). The mold was then cut off (using a V-shaped cut to aid in realignment), reconnected, and the inner surface traced with a sharp pencil onto graph paper, which was scanned using a flatbed scanner. (Molds may also be scanned in directly, but care must be taken to avoid artifacts resulting from unevenness in the mold surface.)

A-P and M-L radiographs of the section locations were taken using a hand-held portable X-ray machine (Aribex Nomad Dental System) and a small dental digital sensor (AFP Digital Size 2 EVA Intraoral Sensor), with the bone oriented in the standard anatomical position. A small L-shaped balsa wood frame was constructed to hold the sensor in proper position for both A-P and M-L radiographs; the frame also included a built-in scale bar (thin aluminum sheet with notches at 1-cm intervals) placed at the midpoint of the bone and used later to correct for magnification. Anterior, posterior, medial, and lateral cortical breadths were measured from radiographs within the digital sensor software. These were entered along with external contour tracings into an R program that combines the dimensions to reconstruct the complete section image, including the medullary cavity (modeled as an ellipse), and calculates section properties (Sylvester *et al.*, 2010; for updates see http://www.hopkinsmedicine.org/fae/CBR.htm). A-P and M-L external breadths, measured using calipers at each section location (see Chapter 1), are also input to the program and used to correct for any remaining scaling and distortion problems.

This method, and the latex cast method in general, have been shown to be quite accurate (within 5% average error) for estimating cross-sectional properties of human long bones, when compared to more exact CT or direct sectioning methods (O'Neill and Ruff, 2004; Sylvester *et al.*, 2010). The prime advantage of the molding and biplanar radiographic method is that it can be applied in many situations in which CT is impractical or impossible, including museums not near to CT facilities or where specimens cannot be removed. The method also avoids expenses and other logistical (e.g., scheduling) issues often associated with CT. The major disadvantage is the extra time necessary to perform both the molding and reconstruction of sections. However, post-processing of sections can be done later at the home institution, minimizing initial data collection time, and the R program greatly helps in automating the final combination of outlines and radiographic breadths, as well as calculation of properties. Some additional software was developed for extracting properties from the comma-delimited files generated by the R program (Sylvester *et al.*, 2010) and placing them in correct order into an Excel file, further automating the process. About 4000 sections were analyzed for this study using the latex cast method, with an additional approximately 2100 analyzed using CT.

Data for the Upper Paleolithic sample and about half the Mesolithic sample were obtained from earlier studies that had used the latex cast method (Churchill, 1996; Holt, 2003; Trinkaus *et al.*, 2003; Trinkaus and Ruff, 2012; S. Churchill, pers. commun.; T.W. Holliday, pers. commun.; E. Trinkaus, pers. commun.). These earlier studies had not used the more automated R program, and had reconstructed endosteal contours manually. Examples of femoral midshaft cross sections for two specimens – Parabita 1 (Early Upper Paleolithic) and Roselle 1198 (Early Medieval), both males – are shown in Figure 3.1. The former was reconstructed manually

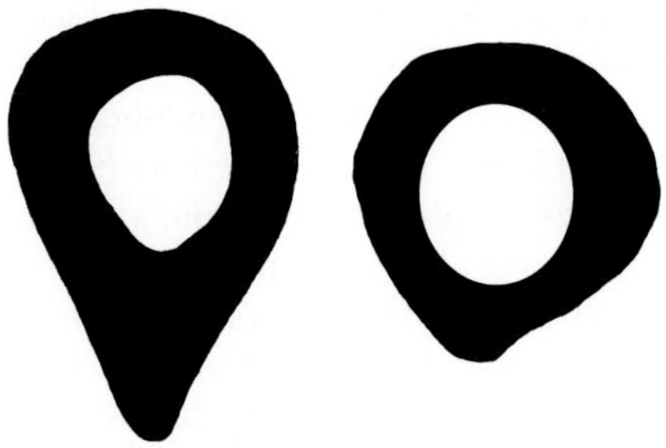

Figure 3.1 Femoral midshaft cross sections of Parabita 1, from the Early Upper Paleolithic, and Roselle 1198, from the Early Medieval. See Table 3.2 for cross-sectional properties.

Table 3.2 Properties of Parabita 1 and Roselle 1198 femoral 50% sections (see Figure 3.1).

Specimen	Z_x	Z_y	Z_p	Z_x/Z_y	CA	TA	%CA
Parabita 1	3030	2212	4927	1.37	617	768	87.4
Roselle 1198	1283	1425	2559	0.90	353	506	69.8

See Table 3.1 for definitions of properties and units

(Holt, 2003), while the latter used the automated R program. Both produce realistic endosteal contours and, as noted above, accurate cross-sectional properties (O'Neill and Ruff, 2004; Sylvester *et al.*, 2010). Section properties for each of these specimens are given in Table 3.2. The extreme difference in cross-sectional 'shape,' which is obvious in Figure 3.1, is reflected in very different values for Z_x/Z_y – much higher in Parabita 1, which also has relatively thicker cortices (higher %CA).

3.3 Section Moduli

As noted above, section moduli have been calculated in different ways in past studies, so some discussion of how that was done here is necessary. True section moduli are calculated as second moments of area divided by the maximum distance from the neutral axis in bending (e.g., for A-P bending, or Z_x, the A-P distance from the neutral axis to the outermost fiber of the section), or from the section centroid for torsion (Z_p) (Gere and Timoshenko, 1990). However, earlier software (e.g., Nagurka and Hayes, 1980) often only calculated second moments of area (I and J). In these cases, section moduli were derived by either dividing SMAs by half the relevant bone external diameter, obtained from caliper measurements (Ruff, 2002, 2008b) or, when breadths were not available, by taking SMAs to the 0.73 power, based on theoretical predictions and empirical observations (Ruff, 1995, 2000b, 2002). While internally consistent, these section moduli overestimate true section moduli on average, because the maximum distance from the neutral axis or centroid to the outermost fiber will always be equal to or greater than half of a diameter.

Both the R program used for the latex mold reconstructed sections, as well as the software used for analyzing sections derived from CT, calculate true section moduli. However, most of the section data used here that were obtained from previous studies had been analyzed using software that only calculated SMAs. To convert these to section moduli, we derived equations for calculating true section moduli from SMAs using the present study sample data, which

Table 3.3 Equations for converting second moments of area to section moduli.

	Equations using Breadths			Equations using Powers		
	Equation	%SEE	%PE	Equation	%SEE	%PE
Femur						
Z_x =	$I_x/(AP/2) \times 0.906 + 44$	4.6	0.0	$I_x^{0.73}$	6.7	−1.2
Z_y =	$I_y/(ML/2) \times 0.929 + 47$	4.5	0.0	$I_y^{0.73}$	3.7	5.4
Z_p =	$J/((AP + ML)/4) \times 0.842 + 115$	6.0	−0.1	$J^{0.74}$	6.7	2.7
Tibia						
Z_x =	$I_x/(AP/2) \times 0.908 + 58$	8.2	0.0	$I_x^{0.73}$	7.2	−4.2
Z_y =	$I_y/(ML/2) \times 0.836 + 59$	7.2	−0.1	$I_y^{0.73}$	7.2	0.4
Z_p =	$J/((AP + ML)/4) \times 0.756 + 50$	5.6	−0.1	$J^{0.73}$	6.2	−0.5
Right Humerus						
Z_x =	$I_x/(AP/2)\ 0.974 + 4$	4.5	−0.2	$I_x^{0.73}$	6.1	1.7
Z_y =	$I_y/(ML/2) \times 0.937 - 6$	7.2	−0.1	$I_y^{0.73}$	6.8	0.8
Z_p =	$J/((AP + ML)/4) \times 0.906 - 18$	5.6	−0.1	$J^{0.74}$	6.8	3.8
Left Humerus						
Z_x =	$I_x/(AP/2) \times 0.959 + 11$	4.4	0.0	$I_x^{0.73}$	6.1	1.7
Z_y =	$I_y/(ML/2) \times 0.942 - 2$	6.8	−0.2	$I_y^{0.73}$	6.8	0.8
Z_p =	$J/((AP + ML)/4) \times 0.889 - 2$	7.0	0.5	$J^{0.74}$	7.0	2.9

I_x: second moment of area in A-P plane; I_y: second moment of area in M-L plane; J: polar second moment of area; AP: anteroposterior external breadth, ML: mediolateral external breadth.

included SMAs as well as true section moduli (the number of sections ranged from about 1300 to 1800, depending on the section location). Two types of equations were derived: one type that used external breadths; and one type based on a power transformation. The latter were applied primarily to humeri, which did not always have caliper breadth measurements at the 35% location. However, equations for all sections are given here for future reference.

Table 3.3 lists the conversion equations for deriving section moduli from SMAs using external A-P and M-L breadths or powers of SMAs. For the breadth analyses, SMAs were first divided by the appropriate radius (or average radius), and then used in ordinary least squares (LS) regressions to predict section moduli. For the power analyses, regressions of log-transformed section moduli on SMAs, without a constant, were carried out. The slope indicates the correct power for the transformation. In all cases, this varied between 0.73 and 0.74, matching previous results (Ruff, 1995, 2000a, 2002). Percentage standard errors of estimate (%SEEs) and percentage prediction errors (%PEs) were used to assess the precision and accuracy (directional bias), respectively, of the equations. The %PEs for the regression equations using breadths are predictably very small; they are larger for the less exact power transformations, but still within about 5% for all properties. The %SEEs are reasonable, falling between about 4% and 8%. Given the very large and variable samples used to generate these equations (e.g., see Fig. 3.1), they should provide useful methods for converting data to section moduli when only SMAs are available. The regression equations for the femur and tibia were presented previously (Trinkaus and Ruff, 2012).

The polar section modulus, Z_p, is used here as a measure of torsional and average bending strength. Torsional strength (or rigidity) is less accurately estimated this way in sections that depart significantly from circularity (Piziali *et al.*, 1976; Daegling, 2002). Using a ratio of maximum to minimum bending rigidity of 1.5 as a threshold (Daegling, 2002), this applies mainly to the tibial midshaft section (almost all sections), and rarely to the femoral and humeral sections (15% or less of all sections). In such cases, however, Z_p is still an appropriate index of (twice) average bending strength, that is, bending strength averaged over different orientations. In our sample, correlations between Z_p and $Z_x + Z_y$ range between 0.98 and 0.99, with %SEEs of 4–5%. Z_p is always smaller than $Z_x + Z_y$ (because the maximum radial distance between the section centroid and the outermost fiber, used in calculating Z_p, is always greater than or equal to the A-P or M-L distances used in calculating Z_x and Z_y) – this averages 7% in the femur, 15% in the tibia, and 6–7% in humeri. The difference is not affected by asymmetry in A-P and M-L planes – note, for example, that the very differently shaped Parabita 1 and Roselle 1198 sections (see Fig. 3.1) both show a difference between Z_p and $Z_x + Z_y$ of about 6% (see Table 3.2). It is more affected by orientation of the major and minor axes of a section (maximum and minimum rigidity) relative to the A-P and M-L axes; thus, the tibial midshaft, in which the major axis lies at an angle to anatomical axes (Ruff and Hayes, 1983: their Fig. 6), has the greatest difference. However, because such orientations for particular sections tend to be fairly constant within recent humans (Ruff and Hayes, 1983: their Fig. 8), this factor should not bias comparisons within sections. Thus, Z_p can be used as a valid index of average bending strength in A-P and M-L planes for such comparisons.

We do not examine maximum and minimum bending strengths here. These parameters are difficult to calculate; maximum and minimum bending rigidities (I_{max} and I_{min}) have been used instead in many previous studies. However, the mechanical interpretation of these properties is unclear without further information – that is, the orientation of maximum and minimum rigidity (or strength), which adds another layer of complexity to analyses. Bending strengths in A-P and M-L planes are easier to relate to specific mechanical hypotheses, for example, lower limb bone structure in relation to mobility (see Chapter 5 and other chapters in this volume), and are thus the primary focus here. The polar section modulus, Z_p, is related to the average of maximum and minimum bending strengths (J is equal to the sum of I_{max} and I_{min}, as well as I_x and I_y), so Z_p can also be taken to reflect these properties.

3.4 Standardizing for Body Size

Because body size (body mass and bone length) contributes to limb bone mechanical loading, long bone diaphyseal strength and body size are closely related (Ruff, 2003). This is especially true in the weight-bearing lower limb, but also applies to the upper limb, through more indirect effects of body size such as muscle mass (Ruff, 2000b, 2003). Thus, in order to reconstruct behavioral differences between past populations and individuals using cross-sectional parameters, it is important to control for body size effects. This is in fact the basis for the definition of 'skeletal robusticity' as "...the strength or rigidity of a structure relative to the mechanically relevant measure of body size" (Ruff *et al.*, 1993:25).

The most 'mechanically relevant' body size parameter for long bone diaphyses is body mass, since that not only constitutes (through gravity) a mechanical load in itself but is also correlated with other characteristics such as muscle mass that contribute to mechanical loading of the limbs. For bending and torsion, which act via a moment arm around the bone, some linear dimension related to moment arm length is also required. Based on theoretical and empirical evidence, this can be taken as proportional to bone length (Polk *et al.*, 2000; Ruff, 2000b),

Table 3.4 Size-standardized section moduli of Parabita 1 and Roselle 1198 femoral 50% sections.

Specimen	Body mass (kg)	Femoral length' (mm)	$Z_x$std	$Z_y$std	$Z_p$std
Parabita 1	72.6	453	921	672	1498
Roselle 1198	66.3	420	461	512	919

See Table 3.2 for cross-sectional properties.

at least for sections near to midshaft. So, for section moduli, the appropriate 'size' parameter is body mass × bone length (we used length' here), with body mass in kilograms and length' in millimeters. This is multiplied by 10^4 to avoid using too many decimal places. Section moduli divided by this parameter are referred to as 'size-standardized.' Bone areas (TA and CA) can be size-standardized through dividing by body mass alone (and multiplied by 10^2). The estimation of body mass from skeletal dimensions is discussed in Chapter 2.

Table 3.4 presents body size data and size-standardized section moduli for the Parabita 1 and Roselle 1198 sections shown in Figure 3.1, based on cross-sectional properties given in Table 3.2. Body mass for Parabita 1 was determined from stature and bi-iliac breadth, and for Roselle 1198 from femoral head breadth (see Chapter 2). Parabita 1 is larger in both body mass and femoral length' than Roselle 1198. When standardized by body mass × length', Parabita 1 is still relatively stronger than Roselle 1198. However, while its relative M-L bending strength is 31% larger than the Medieval specimen, its relative A-P bending strength is fully twice that of Roselle 1198. This reflects the much larger decrease in relative A-P than M-L bending strength characteristic of both the femur and tibia during transition from the Upper Paleolithic to later temporal periods in Europe, a result of declining mobility (see Chapter 5). This example also illustrates the power of considering both bone strength and body size in such analyses, in terms of pinpointing specific mechanical effects and fine-tuning behavioral inferences.

References

Bridges, P.S. (1989) Changes in activities with the shift to agriculture in the southeastern United States. *Curr. Anthropol.*, **30**, 385–394.

Churchill, S.E. (1996) Particulate versus integrated evolution of the upper body in Late Pleistocene humans: a test of two models. *Am. J. Phys. Anthropol.*, **100**, 559–583.

Churchill, S.E. and Formicola, V. (1997) A case of marked bilateral asymmetry in the upper limbs of an Upper Palaeolithic male from Barma Grande (Liguria), Italy. *Int. J. Osteoarch.*, **7**, 18–38.

Daegling, D.J. (2002) Estimation of torsional rigidity in primate long bones. *J. Hum. Evol.*, **43**, 229–239.

Ferretti, J.L., Capozza, R.F., and Zanchetta, J.R. (1996) Mechanical validation of a tomographic (pQCT) index for noninvasive estimation of rat femur bending strength. *Bone*, **18**, 97–102.

Garn, S.M., Guzman, M.A., and Wagner, B. (1969) Subperiosteal gain and endosteal loss in protein-calorie malnutrition. *Am. J. Phys. Anthropol.*, **30**, 153–155.

Gere, J.M. and Timoshenko, S.P. (1990) *Mechanics of Materials*, 4th edn. PWS Publishing, Boston.

Holt, B.M. (2003) Mobility in Upper Paleolithic and Mesolithic Europe: evidence from the lower limb. *Am. J. Phys. Anthropol.*, **122**, 200–215.

Macdonald, H.M., Cooper, D.M., and McKay, H.A. (2009) Anterior-posterior bending strength at the tibial shaft increases with physical activity in boys: evidence for non-uniform geometric adaptation. *Osteoporosis Int.*, **20**, 61–70.

Macintosh, A.A., Pinhasi, R., and Stock, J.T. (2014) Lower limb skeletal biomechanics track long-term decline in mobility across ~6150 years of agriculture in Central Europe. *J. Arch. Sci.*, **52**, 376–390.

Nagurka, M.L. and Hayes, W.C. (1980) An interactive graphics package for calculating cross-sectional properties of complex shapes. *J. Biomech.*, **13**, 59–64.

O'Neill, M.C. and Ruff, C.B. (2004) Estimating human long bone cross-sectional geometric properties: a comparison of noninvasive methods. *J. Hum. Evol.*, **47**, 221–235.

Pearson, O.M. and Lieberman, D.E. (2004) The aging of Wolff's 'law': Ontogeny and responses to mechanical loading in cortical bone. *Yrbk Phys. Anthropol.*, **47**, 63–99.

Piziali, R.L., Hight, T.K., and Nagel, D.A. (1976) An extended structural analysis of long bones – Application to the human tibia. *J. Biomech.*, **9**, 695–701.

Polk, J.D., Demes, B., Jungers, W.L., Biknevicius, A.R., Heinrich, R.E., and Runestad, J.A. (2000) A comparison of primate, carnivoran and rodent limb bone cross-sectional properties: are primates really unique? *J. Hum. Evol.*, **39**, 297–325.

Ruff, C.B. (1987) Sexual dimorphism in human lower limb bone structure: Relationship to subsistence strategy and sexual division of labor. *J. Hum. Evol.*, **16**, 391–416.

Ruff, C.B. (1995) Biomechanics of the hip and birth in early *Homo*. *Am. J. Phys. Anthropol.*, **98**, 527–574.

Ruff, C.B. (1999) Skeletal structure and behavioral patterns of prehistoric Great Basin populations. In: *Prehistoric Lifeways in the Great Basin Wetlands: Bioarchaeological Reconstruction and Interpretation* (eds B.E. Hemphill and C.S. Larsen), University of Utah Press, Salt Lake City, pp. 290–320.

Ruff, C.B. (2000a) Biomechanical analyses of archaeological human skeletal samples. In: *Biological Anthropology of the Human Skeleton* (eds M.A. Katzenburg and S.R. Saunders), Alan R. Liss, New York, pp. 71–102.

Ruff, C.B. (2000b) Body size, body shape, and long bone strength in modern humans. *J. Hum. Evol.*, **38**, 269–290.

Ruff, C.B. (2002) Long bone articular and diaphyseal structure in Old World monkeys and apes, I: Locomotor effects. *Am. J. Phys. Anthropol.*, **119**, 305–342.

Ruff, C.B. (2003) Growth in bone strength, body size, and muscle size in a juvenile longitudinal sample. *Bone*, **33**, 317–329.

Ruff, C.B. (2008a) Biomechanical analyses of archaeological human skeletal samples. In: *Biological Anthropology of the Human Skeleton*, 2nd edn (eds M.A. Katzenburg and S.R. Saunders), John Wiley & Sons, Inc., New York, pp. 183–206.

Ruff, C.B. (2008b) Femoral/humeral strength in early African *Homo erectus*. *J. Hum. Evol.*, **54**, 383–390.

Ruff, C.B. (n.d.) MomentMacro: http://www.hopkinsmedicine.org/fae/mmacro.htm.

Ruff, C.B., Burgess, M.L., Bromage, T.G., Mudakikwa, A., and McFarlin, S.C. (2013) Ontogenetic changes in limb bone structural proportions in mountain gorillas (*Gorilla beringei beringei*). *J. Hum. Evol.*, **65**, 693–703.

Ruff, C.B. and Hayes, W.C. (1983) Cross-sectional geometry of Pecos Pueblo femora and tibiae – a biomechanical investigation. I. Method and general patterns of variation. *Am. J. Phys. Anthropol.*, **60**, 359–381.

Ruff, C.B., Holt, B.H., and Trinkaus, E. (2006a) Who's afraid of the big bad Wolff? Wolff's Law and bone functional adaptation. *Am. J. Phys. Anthropol.*, **129**, 484–498.

Ruff, C.B., Holt, B.M., Sladek, V., Berner, M., Murphy, W.A., Jr, zur Nedden, D., Seidler, H., and Recheis, W. (2006b) Body size, body proportions, and mobility in the Tyrolean 'Iceman'. *J. Hum. Evol.*, **51**, 91–101.

Ruff, C.B. and Larsen, C.S. (2014) Long bone structural analyses and reconstruction of past mobility: A historical review. In: *Mobility: Interpreting Behavior from Skeletal Adaptations and Environmental Interactions* (eds K. Carlson and D. Marchi), Springer, New York, pp. 13–29.

Ruff, C.B., Trinkaus, E., Walker, A., and Larsen, C.S. (1993) Postcranial robusticity in *Homo*, I: Temporal trends and mechanical interpretation. *Am. J. Phys. Anthropol.*, **91**, 21–53.
Shaw, C.N. and Stock, J.T. (2009a) Habitual throwing and swimming correspond with upper limb diaphyseal strength and shape in modern human athletes. *Am. J. Phys. Anthropol.*, **140**, 160–172.
Shaw, C.N. and Stock, J.T. (2009b) Intensity, repetitiveness, and directionality of habitual adolescent mobility patterns influence the tibial diaphysis morphology of athletes. *Am. J. Phys. Anthropol.*, **140**, 149–159.
Shaw, C.N. and Stock, J.T. (2011) The influence of body proportions on femoral and tibial midshaft shape in hunter-gatherers. *Am. J. Phys. Anthropol.*, **144**, 22–29.
Sládek, V., Berner, M., and Sailer, R. (2006a) Mobility in Central European Late Eneolithic and Early Bronze Age: Femoral cross-sectional geometry. *Am. J. Phys. Anthropol.*, **130**, 320–332.
Sládek, V., Berner, M., and Sailer, R. (2006b) Mobility in Central European Late Eneolithic and Early Bronze Age: Tibial cross-sectional geometry. J. Archaeol. Sci., **33**, 470–482.
Sparacello, V.S., d'Ercole, V., and Coppa, A. (2015) A bioarchaeological approach to the reconstruction of changes in military organization among Iron Age Samnites (Vestini) From Abruzzo, Central Italy. *Am. J. Phys. Anthropol.*, **156**, 305–316.
Stock, J. and Pfeiffer, S. (2001) Linking structural variability in long bone diaphyses to habitual behaviors: Foragers from the southern African Later Stone Age and the Andaman Islands. *Am. J. Phys. Anthropol.*, **115**, 337–348.
Stock, J.T. (2002) A test of two methods of radiographically deriving long bone cross-sectional properties compared to direct sectioning of the diaphysis. *Int. J. Osteoarch.*, **12**, 335–342.
Stock, J.T. (2006) Hunter-gatherer postcranial robusticity relative to patterns of mobility, climatic adaptation, and selection for tissue economy. *Am. J. Phys. Anthropol.*, **131**, 194–204.
Stock, J.T. and Pfeiffer, S.K. (2004) Long bone robusticity and subsistence behavior among Later Stone Age foragers of the forest and fynbos biomes of South Africa. *J. Arch. Sci.*, **31**, 999–1013.
Sylvester, A.D., Garofalo, E., and Ruff, C.B. (2010) An R program for automating bone cross section reconstruction. *Am. J. Phys. Anthropol.*, **142**, 665–669.
Trinkaus, E., Churchill, S.E., Holt, B., and Ruff, C.B. (2003) Patterns of diaphyseal cross sectional geometry between central and western European Early/Middle Upper Palaeolithic humans. In: *Changements Biologiques et Culturels en Europe de la Fin du Paléolithique Moyen au Néolithique. Talence: Laboratoire d'Anthropologie des Populations du Passé* (eds J. Bruzek, B. Vandermeersch, and M.D. Garralda), Université de Bordeaux, pp. 75–86.
Trinkaus, E., Churchill, S.E., and Ruff, C.B. (1994) Postcranial robusticity in *Homo*, II: Humeral bilateral asymmetry and bone plasticity. *Am. J. Phys. Anthropol.*, **93**, 1–34.
Trinkaus, E. and Ruff, C.B. (1989) Diaphyseal cross-sectional morphology and biomechanics of the Fond-de-Forêt 1 femur and the Spy 2 femur and tibia. *Bull. Soc. Roy. Bel. Anthropol. Préhist.*, **100**, 33–42.
Trinkaus, E. and Ruff, C.B. (2012) Femoral and tibial diaphyseal cross-sectional geometry in Pleistocene *Homo*. *PaleoAnthropology*, **2012**, 13–62.
Warden, S.J., Mantila Roosa, S.M., Kersh, M.E., Hurd, A.L., Fleisig, G.S., Pandy, M.G., and Fuchs, R.K. (2014) Physical activity when young provides lifelong benefits to cortical bone size and strength in men. *Proc. Natl Acad. Sci. USA*, **111**, 5337–5342.
Weiss, E. (2003) The effects of rowing on humeral strength. *Am. J. Phys. Anthropol.*, **121**, 293–302.

4

Temporal and Geographic Variation in Body Size and Shape of Europeans from the Late Pleistocene to Recent Times

Markku Niskanen[1], *Christopher B. Ruff*[2], *Brigitte Holt*[3], *Vladimir Sládek*[4], *and Margit Berner*[5]

[1] *Department of Archaeology, University of Oulu, Oulu, Finland*
[2] *Center for Functional Anatomy and Evolution, Johns Hopkins University School of Medicine, Baltimore, MD, USA*
[3] *Department of Anthropology, University of Massachusetts, Amherst, MA, USA*
[4] *Department of Anthropology and Human Genetics, Faculty of Science, Charles University, Prague, Czech Republic*
[5] *Department of Anthropology, Natural History Museum, Vienna, Austria*

In this chapter, we assess temporal and geographic variation in body size and body shape in Europe from the Early Upper Paleolithic to recent times. In order to establish the context for our interpretations of these trends, we first review some more general principles underlying variation in body form, as well as providing a brief summary of the population history of Late Pleistocene/Holocene Europe.

4.1 Environmental Adaptation

Humans, as a species, exhibit a considerable amount of temporal and geographic variation in body size (body mass) and body shape (proportional relationships of body segments). At the global level, geographic variation in body size, and especially in body shape, is in part due to genetic differences acquired via long-term natural selection to produce adaptation to different thermal environments (Ruff 2002 and references therein). However, especially within the same general region (e.g., Europe), body size and shape differences between generations and neighboring populations can also reflect differences in general nutritional status and health, and even physical activity because body size and shape exhibit plasticity during the years of growth. Because both genetic factors (long-term adaptation) and non-genetic factors (nutrition, health and physical activity) play a role in human body size and shape, it can be difficult to determine which one of these influences is the most significant in explaining observed differences between populations (Bogin and Rios, 2003 and references therein).

The human species is spread over different climatic zones and clearly adheres to Bergmann's rule, which states that the body size within a species increases with decreasing temperature (Bergmann, 1847), and to Allen's rule (Allen, 1877), which states that the relative size of extremities decreases with decrease of mean temperature. There is a general trend among humans for larger body size (e.g., mean body weight or mass), broader bodies and relatively shorter limbs

Skeletal Variation and Adaptation in Europeans: Upper Paleolithic to the Twentieth Century,
First Edition. Edited by Christopher B. Ruff.

(e.g., greater relative sitting height) in cold climates, and smaller body size, narrower bodies and relatively longer limbs in hot climates (Roberts, 1953, 1978; Ruff, 1991, 1994, 2002; Katzmarzyk and Leonard, 1998).

The 'cylindrical model' of the human body based on stature and bi-iliac breadth combines variation in body size and body width (Ruff, 1991, 1994). This model demonstrates that body breadth must change in order for the surface area/body mass ratio to change. It predicts that body breadth should remain constant regardless of stature changes within the same temperature zone, and that bodies should be broad in cold climates and narrow in warm climates regardless of stature, which is highly variable within climatic zones and across generations. These predictions are supported by anthropometric data on extant human populations (Ruff, 1994).

Interestingly, biacromial (shoulder) breadth does not differentiate cold- and heat-adapted populations from each other as well as bi-iliac (pelvic) breadth, as indicated by relatively small differences in biacromial breadth between Europeans and sub-Saharan Africans (Hiernaux, 1985: their Table 1). Instead, shoulder breadth has an apparently similar relationship with stature in populations adapted to different thermal environments. For example, 'white' and 'black' athletes have very similar mean statures and biacromial breadths, but 'white' athletes have over 2 cm broader bi-iliac breadths (based on data in Tanner, 1964). Biacromial breadth also tends to be more developmentally labile than bi-iliac breadth, which exhibits considerable stability as stature changes across generations (Bowles, 1932; Froehlich, 1970).

Comparisons of relative sitting height (sitting height/stature × 100) demonstrate that high-latitude populations have relatively longer trunks and shorter lower limbs than low-latitude populations (Cowgill *et al.*, 2012). The mean relative sitting heights of Eskimos (Inupiats and Inuits) average ca. 53.6% (Jamison and Zegura, 1970; Auger *et al.*, 1980; Ruff *et al.*, 2005), Europeans ca. 52.4%, and sub-Saharan Africans (excluding Pygmies) ca. 50.9% (Eveleth and Tanner, 1976). As will be pointed out ahead, relative sitting height correlates with both stature and environmental conditions.

Although about 80% of height variation within a given population at a given time can be explained by genetic differences, non-genetic factors (e.g., health, nutrition) can have considerable influence on stature, as indicated by a significant increase in the mean stature in many countries over the past few generations (Weedon and Frayling, 2008 and references therein). Male stature apparently exhibits more phenotypic plasticity than female stature. For example, according to de Beer (2004), the mean stature of young adult Dutch males increased by 6.0 cm (from 178.0 to 184.0 cm) and that of young adult females by 4.3 cm (from 166.3 to 170.6 cm) between birth cohorts 1944 and 1976, and the total stature increase in Netherlands has been ca. 17 cm among males and ca. 13 cm among females since the early 19th century. Sexual dimorphism in stature is thus expected to increase as the mean stature increases, and to decrease as the mean stature decreases (also see Stini, 1976; Stinson, 1985).

The mean stature can increase until most individuals in a population enjoy a sufficiently adequate growing environment to reach their genetic potential in growth. The increase in stature has apparently reached a plateau in recent years in northern Europe and Italy, which implies that these populations have reached this genetic potential (Larnkjær *et al.*, 2006). The most recent Dutch growth study of 2009 also indicates that the Dutch are no longer getting taller, and that male height has actually slightly reduced from the growth study conducted in 1997 (Schönbeck *et al.*, 2013).

Body proportions are apparently under stronger genetic control than body size because differences in relative limb length between sub-Saharan Africans and Europeans are already visible during early fetal development (Cowgill *et al.*, 2012). These differences are thus primarily caused by genetic differences acquired via long-term natural selection (Schultz 1923, 1926a,b; Warren *et al.*, 2002). Because of this, significant temporal changes are sometimes interpreted as indicating population replacement or immigration (e.g., Gallagher *et al.*, 2009).

Changes in body proportions within a population do not always result from genetic changes, however. For example, Paterson's (1996) study on Japanese macaques (*Macaca fuscata*) transferred from Mihara, Japan, to Oregon and Texas in the USA in the 1960s demonstrates that climate change can affect both body size and body proportions without genetic changes due to phenotypic plasticity.

In addition, stature changes across generations are generally associated with a change in body proportions because stature changes are more due to lower limb length changes than trunk length changes. For example, 74.2% of the stature increase of Japanese medical students between the early 1960s and early 1980s was due to an increase in lower limb length, and only 25.5% to a trunk length increase (based on data in Ohyama *et al.*, 1987; also see Tanner *et al.*, 1982). Some 80.3% of the stature increase of Norwegian conscripts between 1921 and 1962 was due to a lower limb length increase, and only 19.7% to a trunk length increase (based on data in Udjus, 1964). Stature increase of Greek students between the early 1970s and the late 1980s was also largely due to lower limb length increase. Some 89.7% of male stature and 81.8% of female stature increase were due to a lower limb length increase (computed from data in Manolis *et al.*, 1995: their Tables 1 and 2).

Although a better growing environment has resulted in taller statures and relatively longer lower limbs in East Asia, recent East Asians (e.g., Japanese) still have relatively shorter limbs than Europeans, indicating genetic differences. This is revealed by a comparison of the mean stature and sitting height of recent young adult Japanese males (Csukás *et al.*, 2006) with those of nine male European samples (from Eveleth and Tanner, 1976) selected due to similar statures with the recent Japanese males. The mean relative sitting height of the recent Japanese is 53.61 (sitting height/stature: 91.33/170.37), higher than that of the Europeans: 52.47 (89.58/170.71).

Leitch (1951) proposed decades ago that relative lower limb length reflects nutritional status and health. Improved environmental circumstances indeed produce relatively longer legs in all ethnic groups (Eveleth and Tanner, 1990:186), and leg length is an even better indicator of growing environment than stature (Fredriks *et al.*, 2005). For example, stature differences between the generally undernourished Maya children in Guatemala and better-nourished Maya children in the USA are largely due to longer lower limbs of the USA Maya children (Bogin *et al.*, 2002; Smith *et al.*, 2003; Bogin and Rios, 2003).

There is enough variation in overall nutritional status and health in recent developed Western European countries to result in differences in stature and relative lower limb length. Wadsworth *et al.*'s (2002) study of a 1946 British national birth cohort (N = 2879) and Li *et al.*'s (2007) study of a 1958 British birth cohort (N ~ 5900) found that lower limb length reflects early childhood environment, and that deficits in adult stature are mostly due to shorter lower limbs. This observation that limb length growth is more sensitive than trunk length growth (Wadsworth *et al.*, 2002) explains why recent Europeans enjoying adequate nutrition are taller and have relatively longer lower limbs than Europeans of past generations (for example, changes in Norwegian men in the 20th century reported by Udjus, 1964), who received less than adequate nutrition.

Physical activity during the years of somatic growth may also have some effect on stature and body proportions. This is demonstrated by observations that working children are shorter and lighter than the same-age non-working children living under similar conditions (Duyar and Özener, 2005). It should be borne in mind here, however, that it is often difficult to disentangle general health effects from physical activity effects.

Marital patterns (endogamy versus exogamy) and marital network sizes have very likely influenced temporal and geographic variation in stature due to their effect on genetic distances between parents. As discussed by Koziel *et al.* (2011), a large genetic distance may promote stature growth via increased heterozygosity promoting heterosis.

The present study of body size and body proportions of Europeans from the Early Upper Paleolithic to recent times suggests that observed temporal changes have been largely the result of phenotypic changes due to changes in environment, nutrition, and health. However, genetic adaptations must also have played some role, as indicated for example by the fact that North/Central Europeans of all time periods (after the Upper Paleolithic) tend to have relatively shorter tibias than Mediterranean Europeans. Genetic variation may also have been introduced by partial population replacement across the Last Glacial Maximum (the LGM), as well as across the Mesolithic–Neolithic transition. Therefore, it is also appropriate here to briefly review relevant evidence regarding the population history of Europe.

4.2 European Population History

Archaeological evidence suggests that anatomically modern *Homo sapiens* (AMHS) expansion into Europe started as early as 48,000 calBP (Hublin, 2015 and references therein). The earliest secure direct date of the European AMHS is 39,610 calBP from a mandible (Oase 1) from Romania (Fu *et al.*, 2016: their Extended Data Table 2). The early European AMHS replaced the Neandertals by about 40,000–39,000 calBP (Higham *et al.*, 2014), but not without genetic admixture. This admixing continued until the Neandertals disappeared, based on evidence of recent Neandertal admixture in Oase 1 from Romania (Fu *et al.*, 2015). A temporal decrease in Neandertal DNA commenced after the Neandertals had disappeared (Fu *et al.*, 2016).

The AMHS predating ca. 37,000 calBP contributed little or none to the later European gene pool, and individuals predating 32,000 calBP (individuals associated with Aurignacian) do not form a clear genetic cluster. The Gravettian (34,000–26,000 cal BP) individuals form a single cluster, which may have spread via population movements (Fu *et al.*, 2016) from the middle Danubian basin (Kozlowski, 2015).

Before the LGM commenced at ca. 26,000 calBP (Clark *et al.*, 2009; Hughes and Gibbard, 2015), the Early Upper Paleolithic (ca. 45,000–22,100 calBP) human settlement reached to about 53°N in Western and Central Europe, to about 58°N on the plains of Eastern Europe, and to about 62°N on the slopes of the Urals in the easternmost Europe. The northern limit of human settlement shifted considerably southward during the LGM when the northernmost continuously occupied regions were Southern France (the Franco-Cantabria) up to about 48/49°N, the Ukrainian/Russian Plain up to about 53°N, and the Middle/South Urals up to about 59°N (Verpoorte, 2009; Straus, 2016). These LGM refugiums were major source areas of the post-LGM colonization of more northern regions.

The early post-LGM dispersals from the Franco-Cantabria were associated with the Magdalenian culture and colonized Western and Central Europe (Otte, 2012; Straus, 2013, 2016) reaching the British Isles about 14,500 calBP (Verpoorte, 2009). These foragers descend partly from the 'Aurignacians' of Western Europe (Fu *et al.*, 2016).

There was a population turnover at least in Western and Central Europe ca. 14,500–14,000 calBP coinciding with the first significant warming period after the LGM (Posth *et al.*, 2016; Fu *et al.*, 2016). Dispersals from Southeastern Europe and/or Western Asia replaced the 'Magdalenians,' formed the Mesolithic 'Western European hunter-gatherer' (WHG) cluster of Lazaridis *et al.* (2014), and reached the northernmost Norway ca. 11,000 calBP (Bang-Andersen, 2012).

The post-LGM dispersals from the Ukrainian/Russian Plain colonized Eastern Europe (Dolukhanov, 1996) and encountered the WHG dispersals in northernmost Fennoscandia ca. 10,200 calBP (Rankama and Kankaanpää, 2011). These dispersals had partial Ancient North Eurasian (ANH) ancestry, and formed the 'Eastern European hunter-gatherer' (EHG) cluster. A 'Scandinavian hunter-gatherer' (SHG) cluster emerged as the WHG cluster intermixed with the EHG cluster in Scandinavia (Lazaridis *et al.*, 2014; Haak *et al.*, 2015).

This relatively recent permanent human colonization of northern parts of Central Europe and the entire Northern Europe region has implications for the relative cold adaptation of European populations. The Upper Paleolithic ancestral populations of the Mesolithic (10,000–5500 calBP) inhabitants of Europe north of about 45°N were exposed to rather similar mean temperatures for about 20,000 years between the initial colonization of Europe by modern humans and the post LGM colonization of northern regions. Considerable differences in cold adaptation are thus not expected between North and Central European populations due to the recent arrival of North Europeans from Central Europe.

The expansion of farming and herding over Europe during the Neolithic, which commences ca. 9000 calBP in the Balkans and no earlier than 6200 calBP in Britain and South Scandinavia (see Price, 2000 and references therein), had considerable effects on the European gene pool, as indicated by recent ancient DNA studies. The Mesolithic foragers of all regions of Europe appear to have closer genetic affinities with the extant North Europeans, whereas even Scandinavian Neolithic farmers exhibit genetic affinities with the extant South Europeans and Southwest Asians (Bramanti *et al.*, 2009; Malmström *et al.*, 2009; Haak *et al.*, 2010; Fu *et al.*, 2012; Skoglund *et al.*, 2012; Sánchez-Quinto *et al.*, 2012; Brandt *et al.*, 2013).

There were population movements in Europe at the end of Neolithic and during the so-called Copper Age (Brand *et al.*, 2013; Haak *et al.*, 2015). About three-quarters of the ancestry of the Corded Ware people of Germany was derived from people of the Yamnaya culture and, therefore, indicates a massive migration from the East European steppe into Central Europe (Haak *et al.*, 2015). Gene pools of the South Scandinavian Corded Ware and Late Neolithic people and the Bronze Age Montenegrians also partly derive from the steppe (Allentoft *et al.*, 2015). Domestic horses probably provided the steppe people an advantage, which allowed population expansions (Anthony and Brown, 2011).

Population relocations organized by the Roman Empire, migrations of the Germanic-speaking people (200 BC–500 AD) and Slavic-speaking people (500–700 AD) (see Heather, 2009 on these population movements) must have reshuffled the European gene pool.

Finally, climatic fluctuations during the Holocene itself should not be discounted as potential influences on body form. For example, the Neolithic agricultural expansion was probably aided by warm temperatures ca. 7800–7300 calBP during the Holocene Thermal Maximum (Paus *et al.*, 2011). It was quite warm during the so-called Medieval Warm Period (especially ca. 900–1100 AD based on Humlum *et al.*, 2011: their Fig. 8), but very cold during the so-called Little Ice Age, which began quite abruptly ca. 1275–1300 AD (Miller *et al.*, 2012) and ended ca. 1870 AD (Hendy *et al.*, 2002). Cold winters and rainy summers affected agricultural production, resulting in famine. The Great Famine of 1315–1317 affected practically all of Europe and resulted in millions of deaths (Jordan, 1996). There were severe famines in different parts of Europe during the 1690s. For example, 5–15% of the Scottish population starved during the so-called 'Ill Years' (Cullen, 2010), and ca. 28% of the Finnish population perished in just two years during the Great Famine of Finland (Muroma, 1991).

The Industrial Revolution, which began in 1760 in England and spread elsewhere in Europe soon after, was contemporary with the Little Ice Age. Considerable technological, economic, demographic (e.g., accelerated population growth rate, increased urbanization), and societal changes (e.g., the French Revolution) associated with this revolution affected the biological standard of living all over Europe (Komlos and Küchenhoff, 2012 and references therein).

4.3 Materials and Methods

4.3.1 Skeletal Samples and Variables

This study utilizes skeletal dimensions of more than 2000 individuals (see Chapter 1) dated from the Early Upper Paleolithic to recent times. Complete listing of sites and numbers of individuals, by sex, are given in Appendix 1 and the on-line data file. Sexing and aging of these skeletal individuals is discussed Ruff *et al.* (2012) and in Chapter 1 of this volume.

This material was subdivided into seven broad geographic regions and 10 general time periods (Table 4.1). Regions include: 1) The British Isles (all from England); 2) Scandinavia (Sweden and Denmark) and Finland; 3) North-Central Europe (Germany, Czech Republic, Austria); 4) France; 5) Italy; 6) The Iberian Peninsula (Spain and Portugal); and 7) the Balkans (Bosnia-Herzegovina and Romania). For some analyses (especially in Chapter 12), the above geographic regions were partly combined to form the following regional grouping: 1) Scandinavia and Finland (as above); 2) The British Isles and North-Central Europe including French individuals from north of 46°N; 3) Southwest Europe (the Iberian Peninsula, Italy, and France south of 46°N); and 4) the Balkans (as above). The geographic patterning of European populations revealed by a geographic map of European's genetic variation (see Novembre *et al.*, 2008 on this patterning) provides justification for the above geographic divisions of skeletal samples included in this study.

Time periods used in examining general trends include: 1) very recent (>1900 AD); 2) early modern (1600–1899 AD); 3) late Medieval (1000–1599 AD); 4) early Medieval (500–999 AD); 5) Iron Age/Roman (800 calBC–499 AD); 6) Bronze Age (2400–1000 calBC); 7) Neolithic (5400–2000 calBC); 8) Mesolithic (9000–4000/3900 calBC); 9) Late Upper Paleolithic (20,000–9000 calBC); and 10) Early Upper Paleolithic (32,000–20,000 calBC). The two Upper Paleolithic

Table 4.1 Temporal groupings.

Period	Abbreviation	Date (BC or AD)
Living anthropometric samples	LIV	ca. 2000 AD
Recent	REC	≥1900 AD
Early modern	EMO	1600–1899 AD
Late medieval	LME	1000–1599 AD
Early medieval	EME	500–999 AD
Iron / Roman Age	IRO	800 BC–499 AD
Bronze Age	BRO	2400–1000 BC
Neolithic	NEO	5400–2000 BC
Mesolithic	MES	9000–4000/3900 BC
Late Upper Paleolithic	LUP	20,000–9000 BC
Early Upper Paleolithic	EUP	32,000–20,000 BC

samples are separated by peak glacial conditions of the LGM dated to 22.1 ky calBP in northern Hemisphere (Shakun and Carlson, 2010), whereas the Late Upper Paleolithic and Mesolithic samples are separated by the Pleistocene–Holocene boundary dated to 11.7 ky calBP (Walker *et al.*, 2009), and thus the end of the Younger Dryas cold period (Alley and Clark, 1999). These periods, their abbreviations, and dates are listed in Table 4.1.

For some analyses, the Bronze Age, Neolithic, and Mesolithic samples are further subdivided to examine temporal changes across the Mesolithic–Neolithic transition and during the Neolithic. For these analyses, the Mesolithic sample is subdivided into Early Mesolithic (9000 calBC–6600 calBC) and Late Mesolithic (6200–4000/3900 calBC) samples. The Early Mesolithic individuals largely predate Neolithic farmers, whereas the Late Mesolithic ones are largely contemporary with them. There is an approximate 400-year time gap between these two Mesolithic subsamples. The Neolithic and Bronze Age individuals are subdivided into the following subsamples: Early Neolithic (ca. 5400–4200 calBC); Middle Neolithic (ca. 4200–2850 calBC); Copper Age (ca. 2850–1900 calBC); and Bronze Age (ca. 1900–1000 calBC). The Copper Age sample includes the Corded Ware, Bell Beaker, Late Neolithic Scandinavian, and Early Bronze Age individuals, whereas the Bronze Age sample includes all Bronze Age individuals younger than the end Neolithic anywhere in Europe.

Examinations of temporal trends in body size and body proportions focus on the overall patterns across all regions because many regions lack a complete temporal coverage. Examinations of geographic trends in body size and body proportions focus on comparing and contrasting individuals and samples above and below 46°N. This combining of Central and North Europeans is justified because Northern Europe was colonized relatively recently from Central Europe. Considerable differences in cold adaptation within this region are thus not expected.

4.3.2 Estimating Body Size and Shape from Skeletal Dimensions

Chapter 2 provides descriptions and evaluations of skeletal dimensions, body dimensions based on these skeletal dimensions, as well as indices computed from various dimensions used in this volume. Variables used in this chapter are listed in Table 4.2 and described below.

'Stature' in this volume is either 'anatomical stature' or stature estimated from long bone length. 'Anatomical stature' is estimated from either complete skeletal height (SKH) using Equation 1 of Raxter *et al.* (2006), or partial skeletal height (PSKH) using an equation presented in Chapter 2 of this volume. If neither skeletal height is available, stature is estimated from lower limb long bone lengths (preferably from combined femoral and tibial lengths) using region- and sex-specific equations in Ruff *et al.* (2012). Sitting height and lower limb length, and thus also relative sitting height are estimated from skeletal components of anatomical stature (see Chapter 2 for additional information).

Biacromial breadth of shoulders is estimated from clavicular length using sex-specific equations presented in Chapter 2, whereas living bi-iliac breadth is estimated from skeletal bi-iliac breadth using an equation provided by Ruff *et al.* (1997) to allow comparisons of skeletal and anthropometric samples. Biacromial breadth/stature ratio reflects relative upper trunk (shoulder) breadth, and bi-iliac breadth/stature ratio reflects relative lower trunk breadth.

Body mass is estimated from either stature and bi-iliac breadth (Ruff *et al.*, 2005) or femoral head breadth (Ruff *et al.*, 2012), depending on whether bi-iliac breadth is available. Body mass index (BMI) is computed from stature and body mass to reflect relative laterality of body because the true body mass is unknown. This index is high if bi-iliac breadth or femoral head breadth (depending on which method is used in body mass estimation) is large relative to stature.

Table 4.2 Variables used in this chapter, their abbreviations, and descriptions.

Variable	Abbreviation	Description
Stature	STA	Estimated from complete or partial skeletal height, when possible, and from lower limb long bone length, when not possible.
Anatomical stature	ASTA	Estimated from complete skeletal height, when possible, and from partial skeletal height when not possible
Sitting height	SHT	See Chapter 2
Lower limb length	LlimbL	See Chapter 2
Relative sitting height	RSHT	SHT/ASTA × 100
Body mass	BM	Estimated from stature and bi-iliac breadth or from femoral head breadth
Body mass index	BMI	BM / (Stature in meters)^2
Biacromial breadth	BAB	See Chapter 2 on estimating this dimension from clavicular lengths
Bi-iliac breadth	BIB	1.17 × Skeletal bi-iliac breadth in cm − 3.0 cm
Relative (biacromial) shoulder breadth (Biacromial breadth/Stature)	BAB/STA	BAB/STA × 100
Relative bi-iliac breadth (Bi-iliac breadth/Stature)	BIB/STA	BIB/STA × 100
Crural index	CI	Lateral condyle-medial malleolus tibial length/ Bicondylar femoral length × 100
Relative lower leg length	RLEGL	Lower leg length/Stature (see Chapter 2 for estimating lower leg length from tibial length)
Brachial index	BI	Radial maximum length/Humeral maximum length × 100 (Means of left and right index are used here)

Crural index reflects relative elongation of tibial relative to femoral length, whereas relative lower leg length reflects relative elongation of tibia relative to anatomical stature. Brachial index reflects relative elongation of radius relative to humerus.

For some analyses, the above dimensions were converted to z-scores (to mean of 0 and standard deviation of 1) using sex-specific means and standard deviations. This transformation eliminates male–female differences in size and shape, allowing analyses using pooled-sex samples and thus large sample sizes. Examining differences between temporally overlapping groups (e.g., foragers versus farmers, northern versus southern samples) is also made easier.

4.3.3 Anthropometric Samples and Variables

Anthropometric data on Europeans published in the 1960s or later were also obtained from the available literature. Means of selected anthropometric variables (stature, sitting height, subischial lower limb length, biacromial shoulder breadth, bi-iliac breadth, body mass, body mass index, as well as various indices based on these variables) are presented in Appendix 3(a,b). The most recent of these anthropometric data (i.e., representing individuals born in the 1960s or more recently) are also presented in various figures to represent living Europeans. For instance, stature means of 29 national level samples presented in Appendix 3(a) represent statures of living Europeans is stature figures. Other anthropometric variables presented in figures are from Appendix 3(b). Chapter 2 provides information on converting skeletal dimensions to anthropometric dimensions of living individuals.

4.3.4 Statistical and Graphical Analyses

All statistical analyses in this chapter are performed using IBM SPSS Statistics 22. Z scores are used in some analyses to allow combining of males and females. Sex-specific means of untransformed measurements, as well as combined sexes mean z-scores for each period and for both main geographic regions, are presented in Appendices 4 and 5. ANOVA is used (with z-scores as dependent variables) to determine if differences in body size and shape between temporal groups and various more-or-less contemporary groups (i.e., regional groups, foragers versus farmers, rural versus urban samples) are significant and thus real differences. F-ratios provide by ANOVA allow ranking variables according to their ability to differentiate the groups compared.

Many of the variables (both untransformed and their z-scores) are plotted against time periods or dates. Interpolation lines connect temporal sample means when variables are plotted against time periods, but regression lines are used when variables are plotted against dates. The non-parametric Loess procedure for fitting locally varying curves is used in some scatterplots. Loess is very useful for exploring complex temporal trends because it does not assume any particular (e.g., linear, quadratic) relationship (Jacoby, 2000).

The general patterns across all time periods are examined first using untransformed variables. Living Europeans are represented by population means or just their extremes (i.e., largest and smallest bi-iliac breadth) depending on the variable. This is followed by a more detailed consideration of body size and shape changes occurring across and/or during major environmental or cultural events using z-scores of variables. These events include: a) the Last Glacial Maximum; b) the Mesolithic–Neolithic transition; c) the Little Ice Age and partly contemporary Industrial Revolution; and d) very recent changes in body size and shape.

Geographic variation in body form is considered by comparing North/Central Europeans and South Europeans. Z-scores of variables are also used in these comparisons.

We will also utilize a subset of Iron Age, Early Medieval period, Late Medieval period, and Early Modern period individuals to examine possible urban–rural differences. This subsample is selected to equally represent these three periods as well as North/Central and Southern Europe. The Iron Age samples include Romano-British samples from England (rural from Poundbury Farm, urban from Poundbury town cemetery). The Early Medieval period rural and urban samples are from Central Europe only. Rural people are from the Avar culture sites of Mödling, Zwentendorf, Brucknendorf, and Laxenburger Str., whereas the urban sample is from Mikulcice. The Late Medieval samples are from Sweden (rural from Westerhus, urban from Sigtuna), Central Europe (rural from Dresden Briesnitz,urban from Opava-Vivirav), and Iberia (rural from San Baudelio de Berlanga, urban from Leiria). The Early Modern samples are from Finland (rural from Renko, urban from Porvoo) and France (rural from Moirans, urban from Observance). Comparisons are carried out as pooled rural versus urban samples, as well as within specific periods and regions to minimize the effects of temporal and geographic differences in body size and shape.

The effects of terrain on body size and shape mainly apply to Southern Europe because there is far less altitudinal variation in Northern Europe. There is thus no section on general terrain effects across Europe in this chapter.

4.4 Results

4.4.1 General Temporal Trends in Body Size and Body Proportions

Applying ANOVA using period number as an independent variable indicates that brachial index, crural index, lower limb length, and stature are the most variable over time; relative sitting height, relative lower leg length, body mass, and sitting height are intermediate in this respect; and absolute and relative trunk breadths and body mass index are the least variable. All of these differences are significant (see Table 4.3). We next examine specific temporal trends in these variables.

Mean stature has fluctuated considerably over millennia (Fig. 4.1a,b). Living Europeans are taller than all earlier Europeans, but only a few centimeters taller than the EUP Europeans. Early Modern and Recent Europeans were generally short-statured compared to all other groups. Stature thus declined considerably across the LGM from EUP to LUP, and at the end of the Medieval period.

A comparison of sitting heights and lower limb lengths in the same absolute scale demonstrates that temporal fluctuations in stature are due more to lower limb length changes than to trunk length changes (Fig. 4.2). This is especially true for the EUP–LUP transition but also, to a lesser extent, for the Medieval–Early Modern and Recent–Living transitions. Relative sitting height thus tends to decrease (and relative lower limb length to increase) as stature increases (Fig. 4.3).

Temporal fluctuations in body mass generally parallel those in stature (Fig. 4.4). The tallest mean statures are associated with the heaviest body masses, and the shortest mean statures with the lightest body masses. Thus, the EUP and living Europeans are the heaviest and the Early Modern Europeans are the lightest. Due to the above, BMI exhibits little temporal variation (Fig. 4.5).

Trunk breadths generally exhibit less temporal variation than stature (Fig. 4.6). There is more between- and within-period variation in biacromial breadth than in bi-iliac breadth. Biacromial

Table 4.3 ANOVA of 10 temporal groups. All variables are z-scores.

	F-ratio	Sig.
Stature	11.586	0.000
Anatomical stature	13.092	0.000
Sitting height	6.305	0.000
Lower limb length	14.359	0.000
Relative sitting height	9.352	0.000
Body mass	8.004	0.000
Body mass index	3.019	0.001
Biacromial breadth	4.849	0.000
Bi-iliac breadth	5.760	0.000
Relative (biacromial) shoulder breadth	3.236	0.001
Relative bi-iliac breadth	3.236	0.001
Crural index	22.311	0.000
Relative lower leg length	8.719	0.000
Brachial index	39.633	0.000

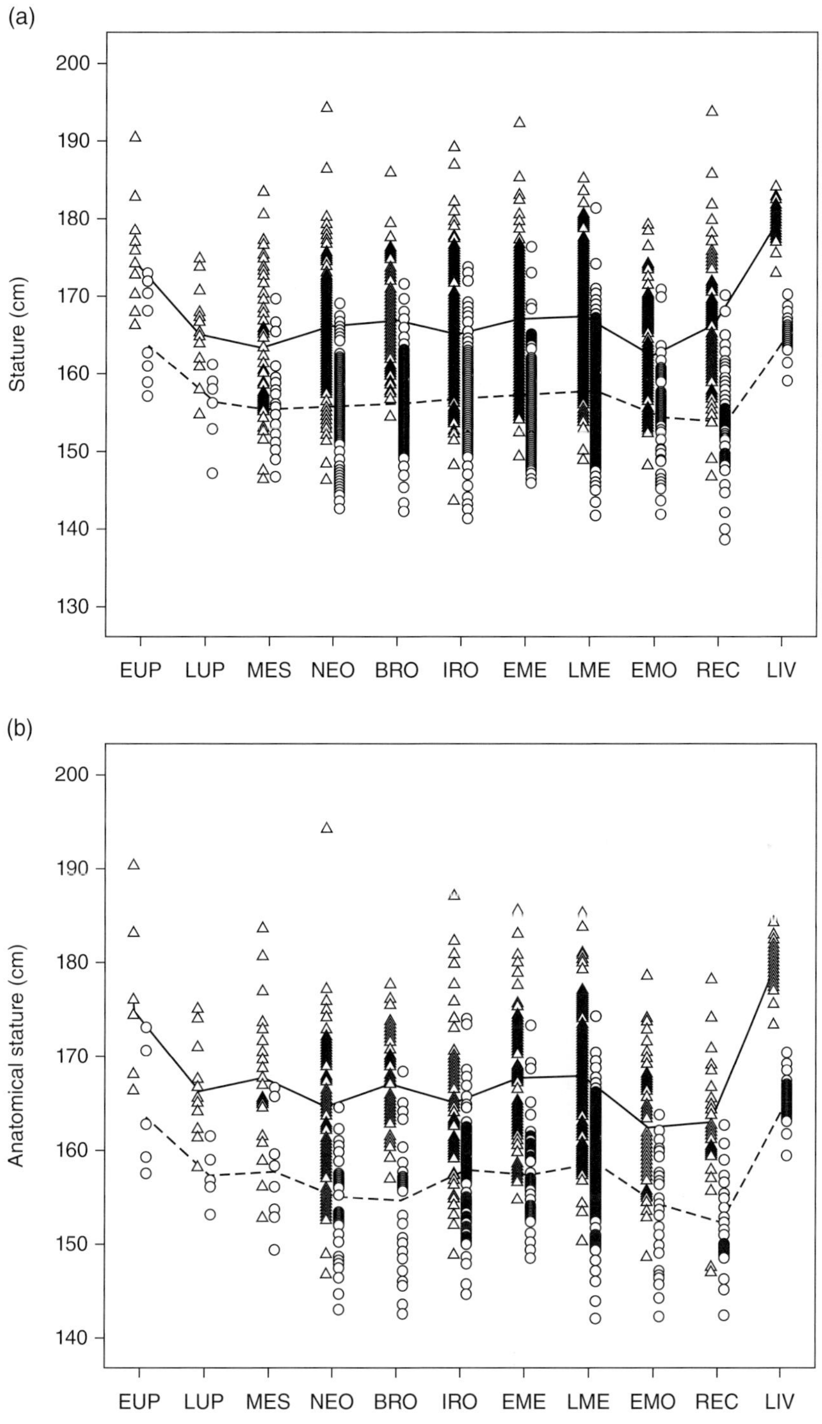

Figure 4.1 Temporal variation in stature. (a) Stature is estimated from skeletal heights or long bone lengths. (b) Anatomical stature is estimated from skeletal heights only. Males: triangles and solid lines; females: circles and dashed lines. Living Europeans are represented by means of 29 national level population samples from Appendix 3(a).

(a)

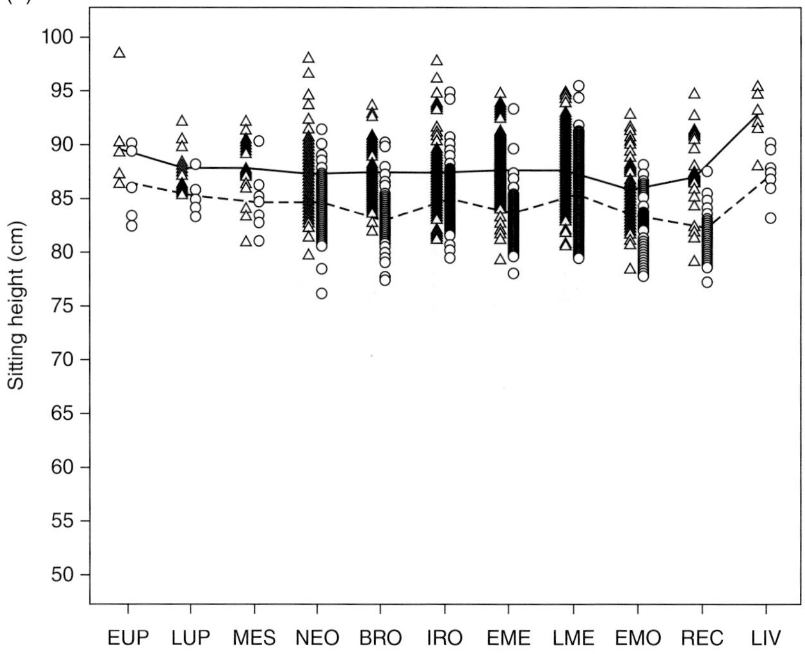

(b)

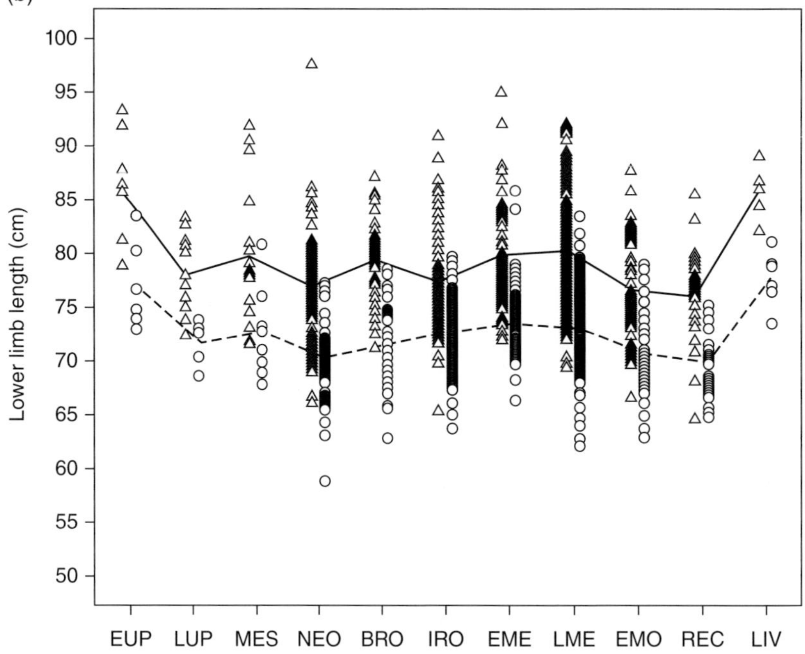

Figure 4.2 Temporal variation in sitting height (a) and lower limb length (b) in the same scale. Males: triangles and solid lines; females: circles and dashed lines. Living Europeans are represented by sex-specific means of seven samples (the Norwegians, Finns, Dutch, Czechs, Basques, Sardinians, and Greeks) from Appendix 3(b).

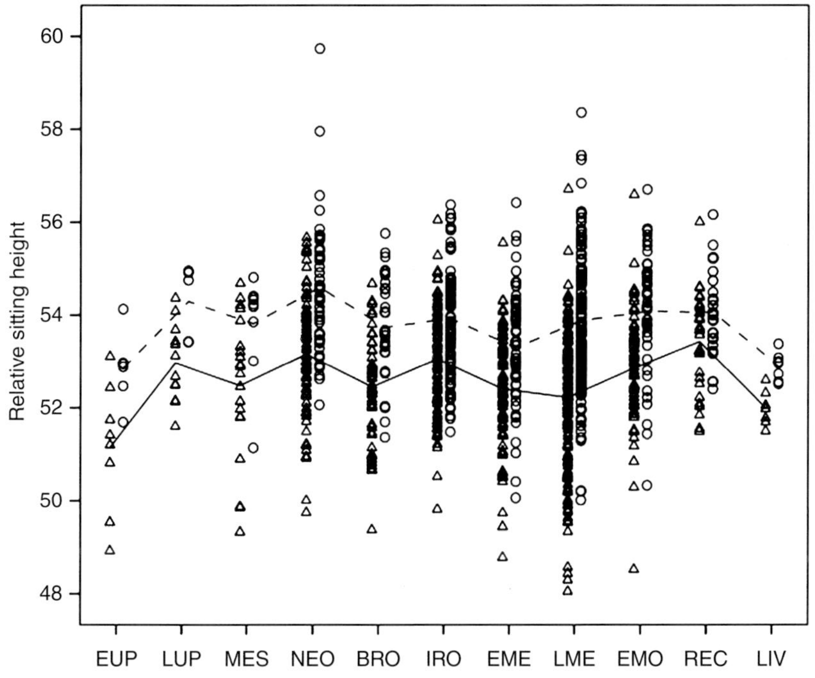

Figure 4.3 Temporal variation in relative sitting height. Males: triangles and solid lines; females: circles and dashed lines. Living Europeans are represented by sex-specific means of seven samples (the Norwegians, Finns, Dutch, Czechs, Basques, Sardinians, and Greeks) from Appendix 3(b).

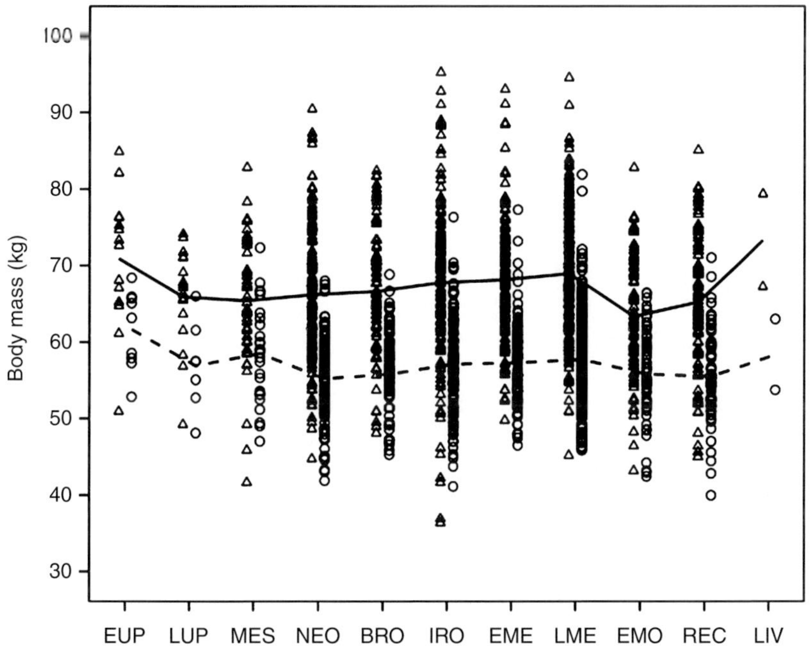

Figure 4.4 Temporal variation in body mass. Males: triangles and solid lines; females: circles and dashed lines. Living Europeans are represented by maximum and minimum sex-specific population mean values from Appendix 3(b).

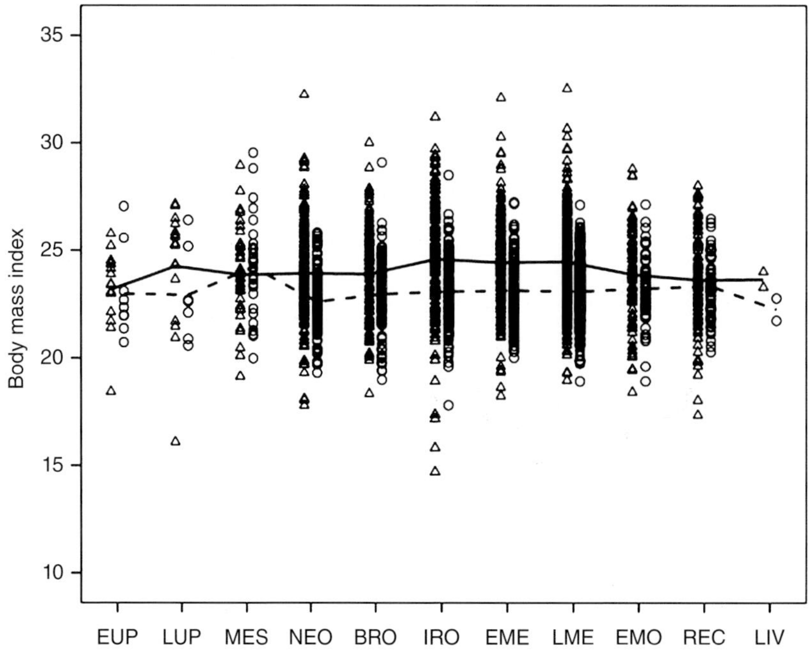

Figure 4.5 Temporal variation in body mass index. Males: triangles and solid lines; females: circles and dashed lines. Living Europeans are represented by maximum and minimum sex-specific population mean values from Appendix 3(b).

shoulder breadth was by far the largest during the EUP. It reduced considerably (by several centimeters) across the LGM to the LUP but thereafter was quite stable, changing less than 1 cm between periods except between the Recent sample and living Europeans when it increased about 1 cm. Bi-iliac breadth reduced slightly from the EUP through the Mesolithic and Neolithic, but increased thereafter, reaching maximum mean values during the Bronze and Iron Ages. It declined during the Early Modern period and has remained below peak values since then. However, total variation in mean bi-iliac breadth across all periods (within sex) was less than 2 cm.

Because bi-iliac breadth exhibits relatively little temporal variation in comparison to stature, relative bi-iliac breadth tends to show temporal trends that are inversely related to those for stature (Fig. 4.7b). Thus, the tall EUP and living samples have relatively narrow lower trunks. This is also true for biacromial breadth in the living sample (Fig. 4.7a). However, because of their extremely high values for biacromial breadth, relative shoulder breadth in the EUP is still quite high despite their high statures. There is an abrupt decline in relative biacromial breadth in the LUP and Mesolithic, especially among males, followed by an increase in the Neolithic, again more marked in males.

Crural index was the highest during the two Paleolithic periods and declined fairly continuously thereafter through the Iron Age, where it reached the recent European level (Fig. 4.8). Brachial index was also high during the Paleolithic, declined from the LUP to the Mesolithic, and increased from the Mesolithic to Bronze Age, after which it exhibited a slight overall trend of decline through recent Europeans (Fig. 4.9).

(a)

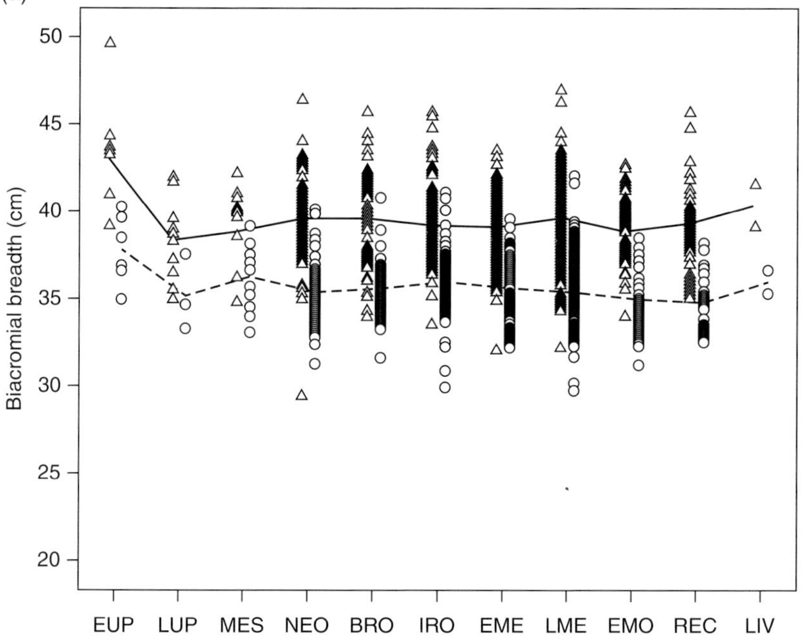

(b)

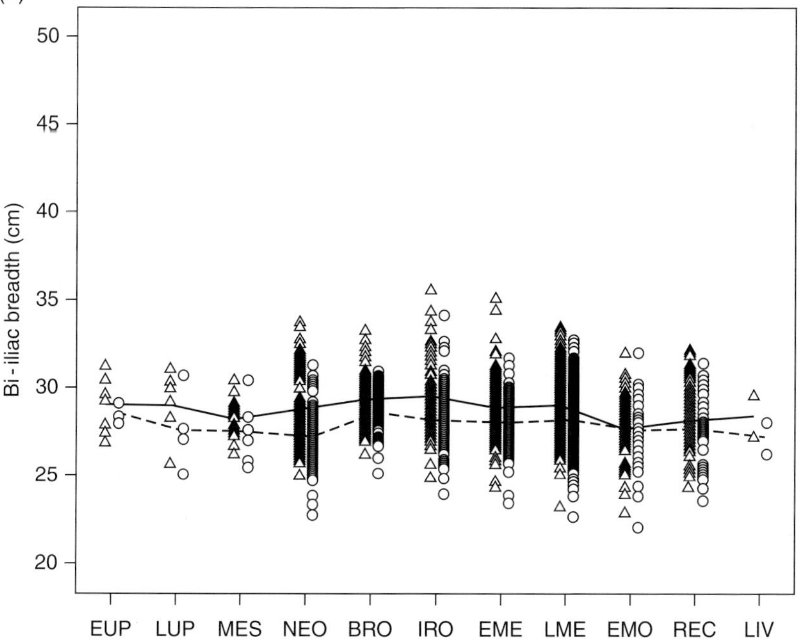

Figure 4.6 Temporal variation in trunk breadths. Biacromial breadth (a) and bi-iliac breadth (b). Males: triangles and solid lines; females: circles and dashed lines. Living Europeans are represented by maximum and minimum sex-specific population mean values from Appendix 3(b) because only extremes of European size-variation are represented by these absolute values.

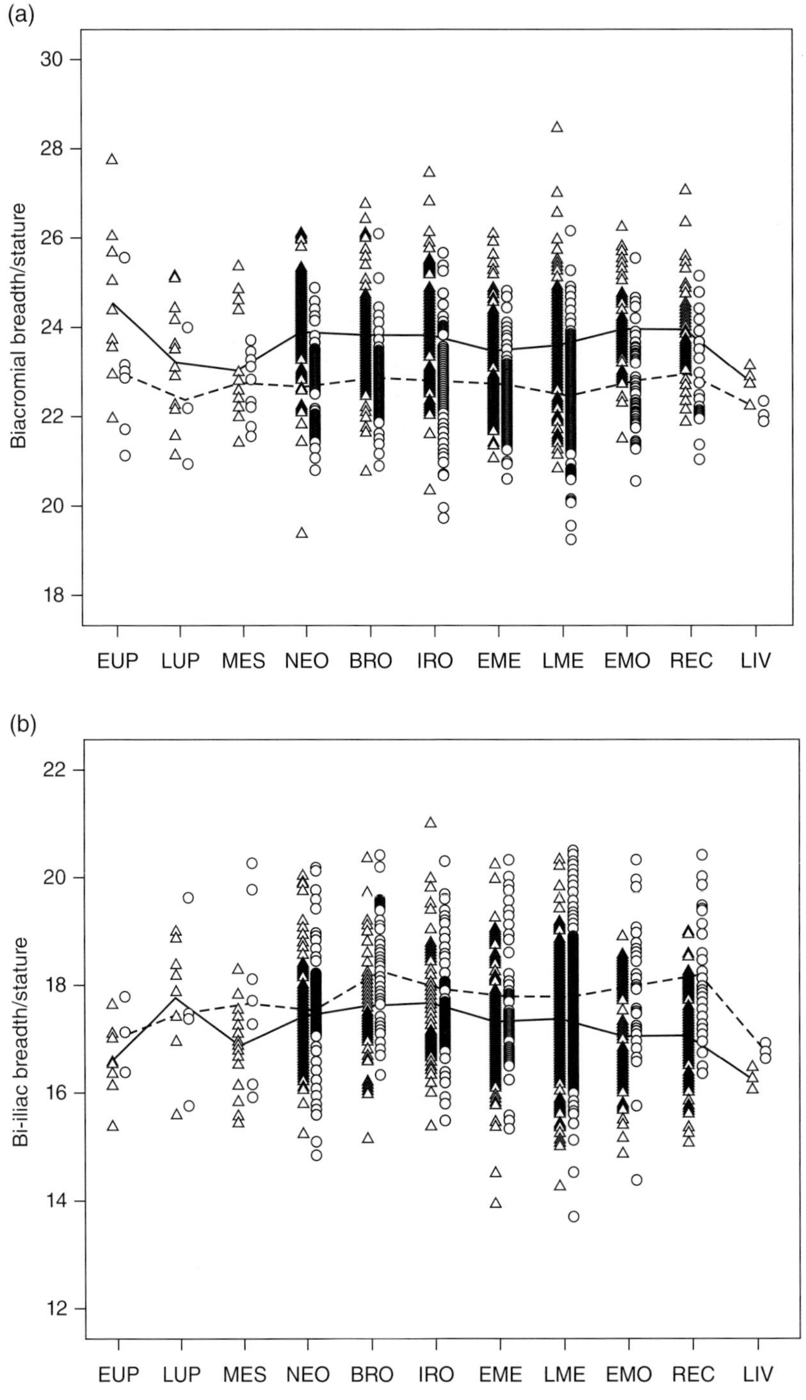

Figure 4.7 Temporal variation in trunk breadths relative to stature. Relative biacromial breadth (a) and relative bi-iliac breadth (b). Males: triangles and solid lines; females: circles and dashed lines. Living Europeans are represented by a few available mean values because these relative values cluster closely.

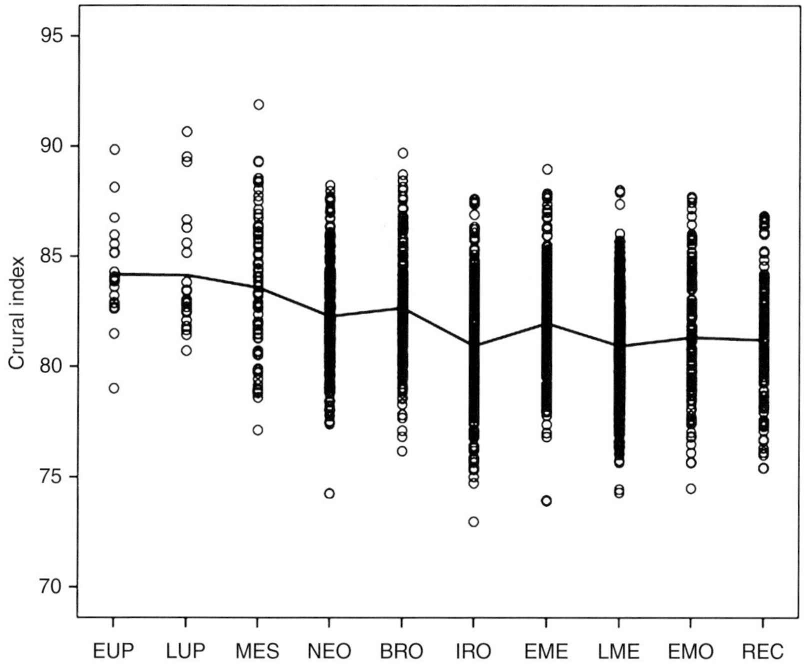

Figure 4.8 Temporal variation in crural index. Sexes are combined because there is no sexual dimorphism in this index.

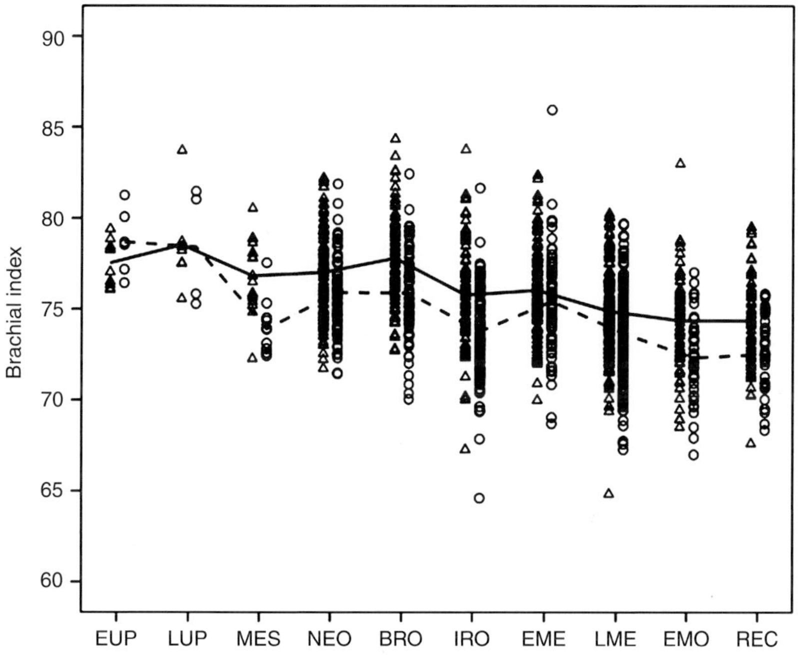

Figure 4.9 Temporal variation in brachial index. Males: triangles and solid lines; females: circles and dashed lines.

4.4.2 Body Size and Shape of Foragers Before and After the Last Glacial Maximum

The figures presented earlier demonstrate that there were considerable changes in both body size and shape across the LGM, and thus across the EUP–LUP transition. The LUP Europeans were much shorter, lighter, and shorter-limbed, had absolutely and relatively narrower biacromial breadths, and relatively broader bi-iliac breadths than their EUP predecessors. There were minor changes in absolute sitting heights, bi-iliac breadths, body mass index, and crural and brachial indices. Most, if not all, of the above changes were due to stature decline. Changes in body size and shape from the LUP to the Mesolithic were much smaller. We examine the pattern of Upper Paleolithic and Mesolithic body size changes in greater detail below.

The Loess curve through z-scores of stature plotted against calibrated dates reveals a curvilinear pattern of stature change during the Upper Paleolithic and Mesolithic periods (Fig. 4.10). This curvilinearity results from little or no stature change during the EUP, a considerable decline across the LGM, and a lesser decline within the LUP and Mesolithic periods. A quadratic regression curve thus provides a slightly better fit for this stature decline than a linear one (Fig. 4.10).

As revealed by the Loess curve, body mass decline across the LGM was less marked than stature decline in z-score units (Fig. 4.11). Body mass also remained quite stable during the LUP and Mesolithic periods. A linear regression line provides an adequate fit for this pattern (Fig. 4.11). This lesser change in body mass reflects the relative stability of absolute bi-iliac breadth (see Fig. 4.6(b)).

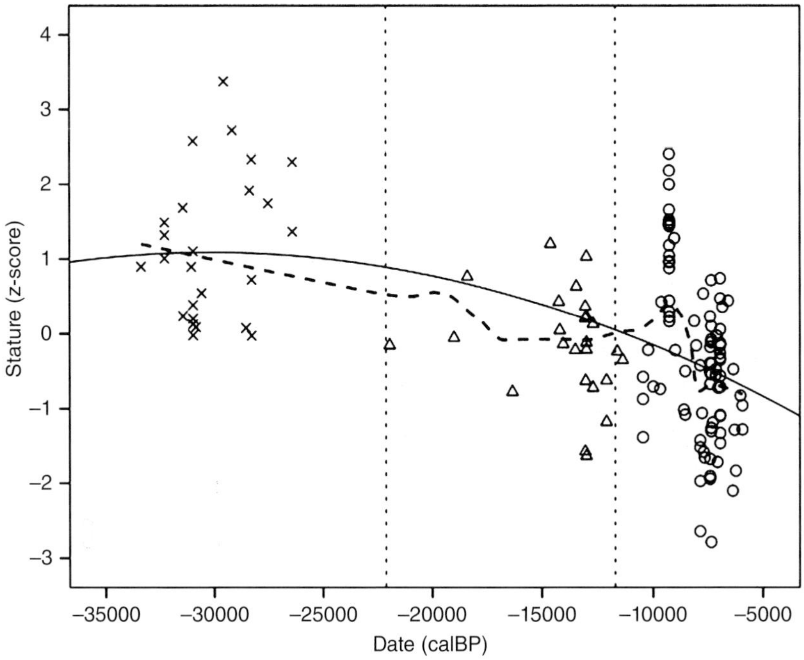

Figure 4.10 Stature (z-score) plotted against date (calBP) during Upper Paleolithic and Mesolithic. Dashed vertical lines mark the Holocene–Pleistocene boundary dated to 11 700 calBP (Walker *et al.*, 2009) and peak glacial condition of the LGM dated to 22 100 calBP (Shakun and Carlson, 2010). Solid line is quadratic regression curve (Stature z-score = −1.675 + 0.000 × calBP − 3.038E-9 × calBP^2; r = −0.510; N = 138) and dashed line is Loess curve through all data points. EUP are represented by crosses; LUP by triangles; and MES by circles.

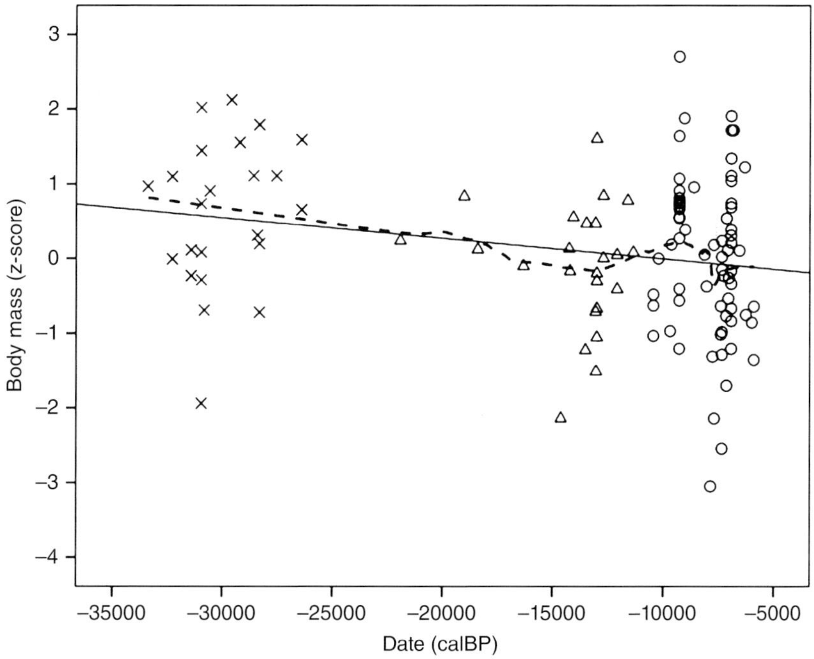

Figure 4.11 Body mass (z-score) plotted against date (calBP) during Upper Paleolithic and Mesolithic. Dashed vertical lines mark the Holocene–Pleistocene boundary dated to 11 700 calBP (Walker *et al.*, 2009) and peak glacial condition of the LGM dated to 22 100 calBP (Shakun and Carlson, 2010). Solid line is least-squares regression line (Body mass z-score = 0.00002818 × calBP − 0.291; r = 0.0236, SEE = 1.03, N = 112) and dashed line is Loess curve through all data points. EUP are represented by crosses; LUP by triangles; and MES by circles.

4.4.3 Changes Across the Mesolithic–Neolithic Transition and During the Neolithic

There were changes in body size and shape across the Mesolithic–Neolithic transition as well as during Neolithic. We focus here on changes in stature and relative body mass.

As shown in Figure 4.12, stature declined during the Mesolithic and the Late Mesolithic, when Europeans were almost as short-statured as their Neolithic farmer contemporaries. Stature of farmers increased during the Neolithic, resulting in tall statures during the Copper Age. The Pitted Ware-using Swedish foragers were shorter-statured than their Copper Age farmer contemporaries (Fig. 4.12). Body mass changes exhibit a very similar pattern and are thus not presented here.

Temporal changes in body mass/stature reflect changes in relative laterality of the body. The farmers were lighter for stature than their forager predecessors and/or contemporaries (Fig. 4.13a). In the case of the Early Neolithic farmers, this was due to their very narrow bi-iliac breadths (Fig. 4.13b). Although the bi-iliac breadth of farmers increased over time, the tall-statured Copper Age farmers had relatively narrower trunks than their shorter-statured, but about equally broad-bodied Pitted Ware-using foragers, which resulted in lower body mass indices of the farmers.

4.4.4 The Little Ice Age, the Industrial Revolution, and Very Recent Changes

There were considerably fewer changes in body size and proportions between the Bronze Age and Late Medieval periods than there had been during the Neolithic transition. Observed

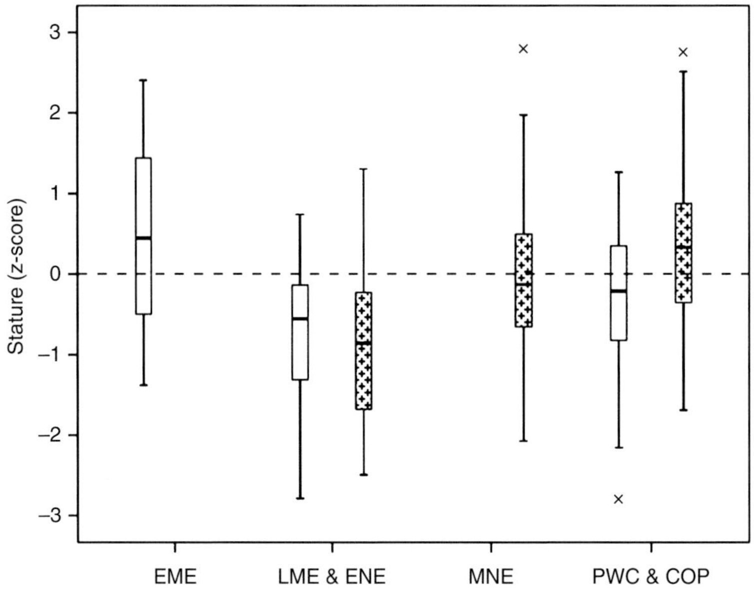

Figure 4.12 Boxplot of stature during Early Mesolithic (EME), Late Mesolithic (LME), Early Neolithic (ENE), and Copper Age (COP = Copper Age farmers, PWC = foragers associated with the Pitted Ware Culture). Foragers shown as unhatched boxes, farmers shown as hatched boxes.

changes may be due more to sampling than actual temporal changes. This relative stability ended with the onset of the Little Ice Age, resulting in considerable reduction in body size and associated changes in body shape from the Late Medieval to Early Modern period. The Industrial Revolution overlaps temporarily with the Little Ice Age.

Stature (Fig. 4.1) and body mass (Fig. 4.4) reduced significantly from the Late Medieval to Early Modern period. The Early Modern Europeans were as small-bodied as the Early Neolithic farmers. Changes in body proportional changes were non-significant, but relative sitting heights (Fig. 4.3) and relative trunk breadths increased somewhat (Fig. 4.7), reflecting a large reduction in lower limb length and less reduction in body breadth than in stature, respectively.

Stature and body mass increased significantly from the Early Modern to Recent period. Some European countries (e.g., Scandinavian countries, Netherlands, and Italy) reached a plateau in stature increase ca. 1990, but stature has increased in many other countries (e.g., Belgium, Spain, and Portugal) since this time (see Larnkjær *et al.*, 2006). At least in Portugal this continuing stature increase is due to its late commencement (see Schmidt *et al.*, 1995: their Fig. 1). This recent stature increase, as well as the above-discussed earlier stature changes, has been more due to limb length changes than to trunk length changes, resulting in lower relative sitting heights (see Udjus, 1964 and Manolis *et al.*, 1995 on recent changes in Norway and Greece, respectively). Biacromial breadth has increased some with stature increase, but bi-iliac breadth has been remarkably stable, resulting in narrower trunks (especially the lower trunk) relative to stature than in most earlier Europeans (see Fig. 4.7).

4.4.5 Geographic Differences in Body Size and Proportions

This section on regional/geographic differences only presents general patterns across Europe in light of skeletal data, except in the case of current Europeans, who are represented by anthropometric data. Regional chapters of this volume provide more detailed discussions of this topic.

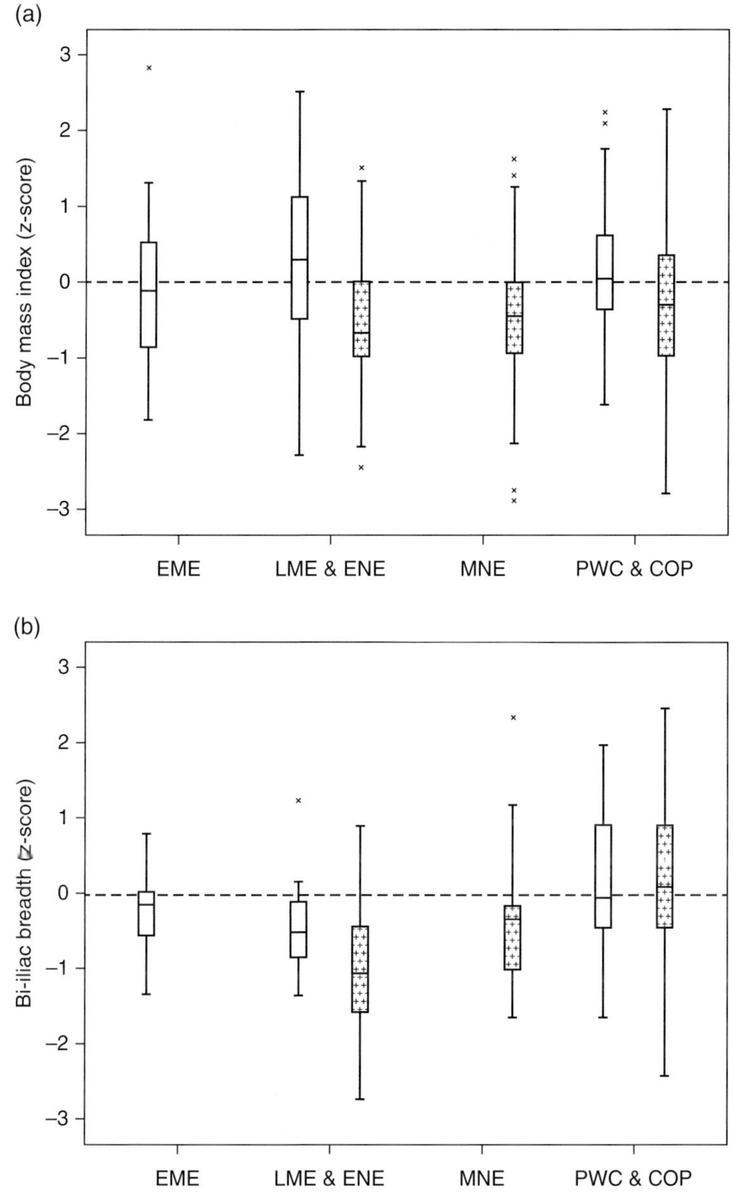

Figure 4.13 Boxplots of body mass index (a) and bi-iliac breadth (b) during Early Mesolithic (EME), Late Mesolithic (LME), Early Neolithic (ENE), and Copper Age (COP = Copper Age farmers, PWC = foragers associated with the Pitted Ware Culture). Foragers shown as unhatched boxes, farmers shown as hatched boxes.

Inter-regional variability in dimensions (Table 4.4) follows a pattern that is generally similar to that shown earlier for temporal comparisons. Stature, sitting height, body mass, lower leg length relative to both stature and femoral length, and body mass exhibit the most regional variation; lower limb length, relative sitting height, and brachial index are intermediate; and BMI and absolute and relative trunk breadths are the least variable between regions. All differences except bi-iliac breadth relative to stature are significant.

Table 4.4 ANOVA of seven regional groups. All variables are z-scores.

	F-ratio	Sig.
Stature	18.740	0.000
Anatomical stature	10.254	0.000
Sitting height	11.244	0.000
Lower limb length	7.843	0.000
Relative sitting height	9.155	0.000
Body mass	14.289	0.000
Body mass index	2.492	0.015
Biacromial breadth	7.632	0.000
Bi-iliac breadth	4.230	0.000
Relative (biacromial) shoulder breadth	2.627	0.011
Relative bi-iliac breadth	1.168	0.321
Crural index	16.482	0.000
Relative lower leg length	11.539	0.000
Brachial index	9.895	0.000

Applying ANOVA to broad latitudinal groups (North/Central versus South) demonstrates that there are clear differences between North/Central and South Europeans in body size and shape (Table 4.5). The North/Central Europeans differ from South Europeans most noticeably in their taller stature, greater sitting height, heavier body mass, shorter lower leg length relative to both stature and femoral length, higher relative sitting height, and absolutely broader shoulders. Their lower limbs are only barely significantly longer and body mass indices higher. Differences in brachial index, relative trunk breadths, and absolute bi-iliac breadth are non-significant.

Correlations of body size and shape variables with latitude and longitude provide largely similar results to the above (Table 4.6). This is not surprising because latitude and longitude are positively correlated ($r = 0.403$, $N = 2184$) in our data, and there is a general southwest to northeast gradient of European winter temperatures (see Huijzer and Vandenberghe, 1998 on these gradients). Therefore, we discuss only correlations with latitude below.

Absolute trunk breadths are relatively weakly correlated with latitude. The trunk breadths relative to stature are either stable or actually reduce with latitude and stature, because absolute body breadths are quite constant whereas stature can vary considerably. These findings also apply to anthropometric samples of recent and living Europeans included in Appendix 3(b). Relative sitting height has a significant positive correlation with latitude – that is, North/Central Europeans have relatively longer trunks and shorter lower limbs. This finding also applies to living Europeans. Tibial length relative to both stature and femoral length (crural index), and brachial index have significant negative correlations with latitude.

We examine pooled sex z-scores subdivided by latitude and plotted against periods to examine the temporal dimension of north-to-south differences in body size and shape. Living Europeans are included in some of these comparisons (e.g., stature), but there are too few anthropometric samples to represent current geographic variation in body shape adequately. We will not go through results period-by-period, but summarize how different general time

Table 4.5 ANOVA of North to South differences in total material. All variables are z-scores.

	Region	N	Mean	F-ratio	Sig.
Stature	North	1562	0.1047	65.971	0.000
	South	570	−0.2868		
Anatomical stature	North	701	0.1120	26.476	0.000
	South	362	−0.2169		
Sitting height	North	701	0.1646	58.847	0.000
	South	362	−0.3187		
Lower limb length	North	701	0.0472	4.613	0.032
	South	362	−0.0915		
Relative sitting height	North	701	0.0921	17.752	0.000
	South	362	−0.1783		
Body mass	North	1526	0.0855	43.645	0.000
	South	539	−0.2421		
Body mass index	North	1507	0.0272	4.253	0.039
	South	537	−0.0763		
Bi-acromial breadth	North	874	0.0661	12.696	0.000
	South	381	−0.1516		
Bi-iliac breadth	North	893	0.0227	1.779	0.183
	South	314	−0.0647		
Relative (biacromial) shoulder breadth	North	862	−0.0194	1.075	0.300
	South	377	0.0445		
Relative bi-iliac breadth	North	890	−0.0293	2.947	0.086
	South	312	0.0835		
Crural index	North	1314	−0.0485	11.296	0.001
	South	499	0.1277		
Relative lower leg length	North	701	−0.1241	32.674	0.000
	South	362	0.2403		
Brachial index	North	1028	−0.0037	0.049	0.825
	South	405	0.0093		

ranges follow – or do not follow – the overall geographic trends described above. It should also be borne in mind that small regional sample sizes may inhibit a reliable representation of geographic differences during certain periods (e.g., during the two Upper Paleolithic periods).

Figure 4.14 shows that South Europeans were taller than North/Central Europeans during the EUP. However, North/Central Europeans have been consistently taller than South Europeans since the LUP. This height difference largely disappeared during the Late Medieval period due to inclusion of the tall-statured North Balkan Medieval sample among South Europeans. The height difference was also quite small during the Early Modern period, but increased again through the Early Modern to Recent periods.

The South European EUP and LUP people were heavier than their North/Central European contemporaries (Fig. 4.15). However, the North/Central Europeans were heavier-bodied than the South Europeans from the Mesolithic period onward, except for the Late Medieval period (again influenced by the inclusion of the large-bodied North Balkan sample), when the South Europeans average slightly heavier, and the Early Modern period, when body masses were about the same.

Table 4.6 Correlations with latitude and longitude in total material. All variables are z-scores.

		Latitude	Longitude
Stature	Correlation	0.157	0.138
	Sig. (2-tailed)	0.000	0.000
	N	2132	2132
Anatomical stature	Correlation	0.172	0.138
	Sig. (2-tailed)	0.000	0.000
	N	1063	1063
Sitting height	Correlation	0.223	0.098
	Sig. (2-tailed)	0.000	0.001
	N	1063	1063
Lower limb length	Correlation	0.095	0.133
	Sig. (2-tailed)	0.002	0.002
	N	1063	1063
Relative sitting height	Correlation	0.107	−0.100
	Sig. (2-tailed)	0.000	0.001
	N	1063	1063
Body mass	Correlation	0.147	0.095
	Sig. (2-tailed)	0.000	0.000
	N	2065	2065
Body mass index	Correlation	0.058	0.012
	Sig. (2-tailed)	0.008	0.591
	N	2044	2044
Biacromial breadth	Correlation	0.084	0.120
	Sig. (2-tailed)	0.003	0.000
	N	1255	1255
Bi-iliac breadth	Correlation	0.030	0.072
	Sig. (2-tailed)	0.298	0.013
	N	1207	1207
Relative (biacromial) shoulder breadth	Correlation	−0.031	0.031
	Sig. (2-tailed)	0.273	0.273
	N	1239	1239
Relative bi-iliac breadth	Correlation	−0.069	−0.016
	Sig. (2-tailed)	0.017	0.590
	N	1202	1202
Crural index	Correlation	−0.196	0.043
	Sig. (2-tailed)	0.000	0.068
	N	1813	1813
Relative lower leg length	Correlation	−0.162	0.066
	Sig. (2-tailed)	0.000	0.032
	N	1063	1063
Brachial index	Correlation	−0.089	0.061
	Sig. (2-tailed)	0.001	0.022
	N	1433	1433

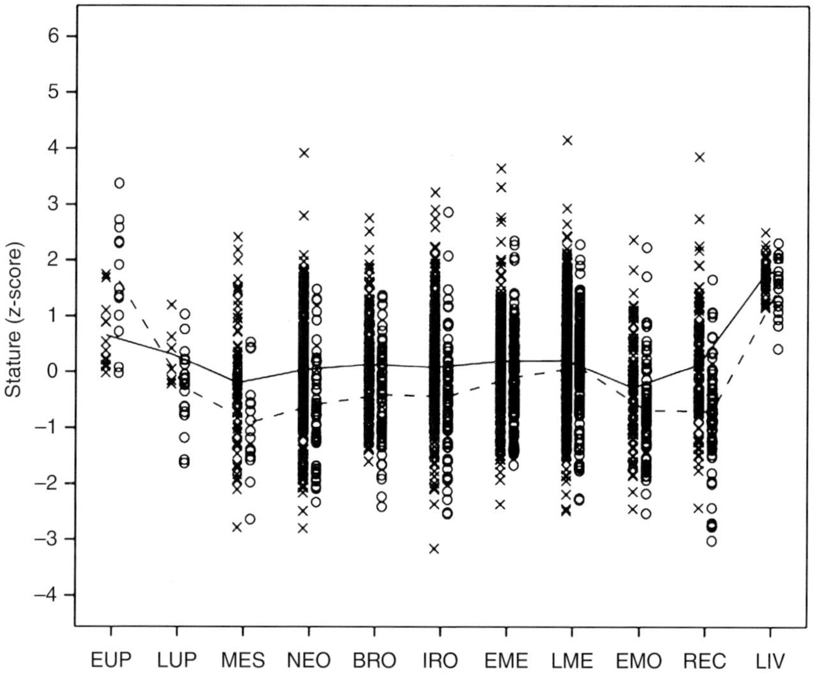

Figure 4.14 Stature (z-scores) plotted against period number. North/Central Europeans: crosses and solid lines. South Europeans: circles and dashed lines. Living Europeans are represented by means of 29 national level population samples from Appendix 3(a).

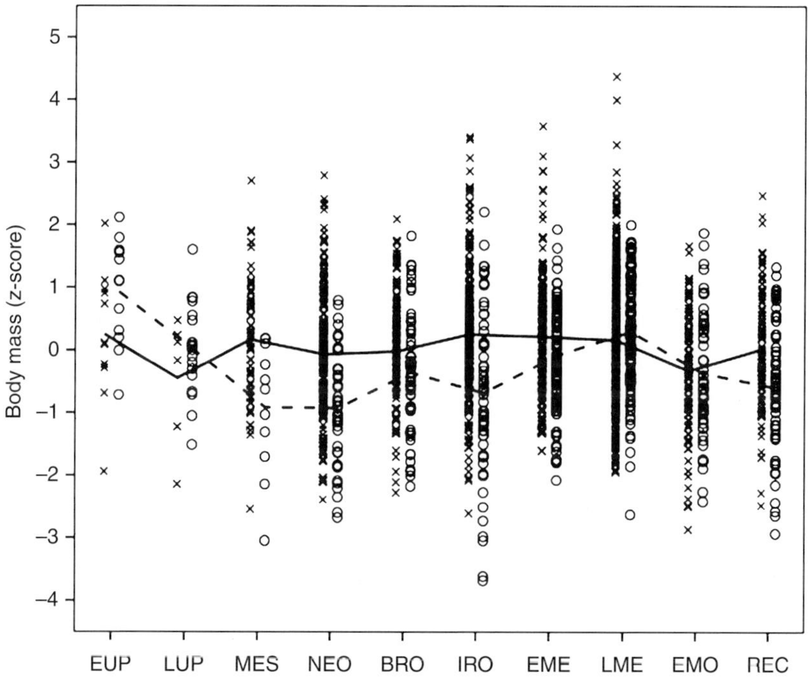

Figure 4.15 Body mass (z-scores) plotted against period number. North/Central Europeans: crosses and solid lines. South Europeans: circles and dashed lines.

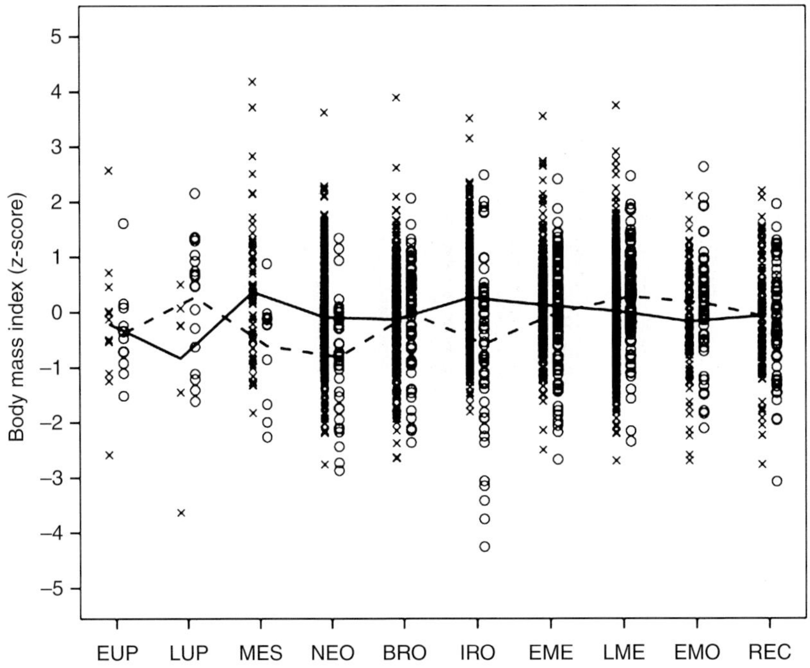

Figure 4.16 Body mass index (z-score) plotted against period number. North/Central Europeans: crosses and solid lines. South Europeans: circles and dashed lines.

Body mass relative to stature does not consistently differentiate the North/Central and South Europeans (Fig. 4.16). The North/Central Europeans averaged relatively heavier during the Mesolithic, Neolithic, and Iron Age, but the South Europeans averaged relatively heavier during the LUP, Late Medieval period, and Early Modern period. Some of the observed differences may reflect sampling because BMI is barely significantly different between higher- and lower-latitude Europeans as a whole.

The North/Central Europeans had equal or greater relative sitting heights and thus relatively shorter lower limbs during all periods except the LUP and Mesolithic (Fig. 4.17). Our Recent sample of North/Central Europeans does not include individuals with estimated absolute and relative sitting heights (i.e., axial skeletal dimensions). The interpolation line is thus a straight line from the Early Modern period to the living samples. Based on anthropometric data for recent and living Europeans (Appendix 3(b)), relative sitting height correlates negatively with latitude in both males ($r = -0.436$, $N = 33$ population samples) and females ($r = -0.320$, $N = 29$ population samples).

Relative lower leg (tibia) length exhibits a very similar (but inverse) geographic and temporal pattern as relative sitting height. The North/Central Europeans had relatively shorter tibiae for stature than their South Europeans contemporaries during all periods except the LUP and Mesolithic (Fig. 4.18). This trend also affected the development of stature estimation equations for Northern and Southern Europeans (see Chapter 2).

Based on selected samples of rural and urban people from the Iron Age to the Early Modern period, the rural people are clearly taller, heavier both absolutely and relatively to stature, absolutely and relatively broader-bodied, and have absolutely and relatively longer lower limbs and longer distal limb segments (Table 4.7). The only non-significant differences between rural and urban people are in sitting height and relative shoulder breadth, which are only slightly greater in rural people.

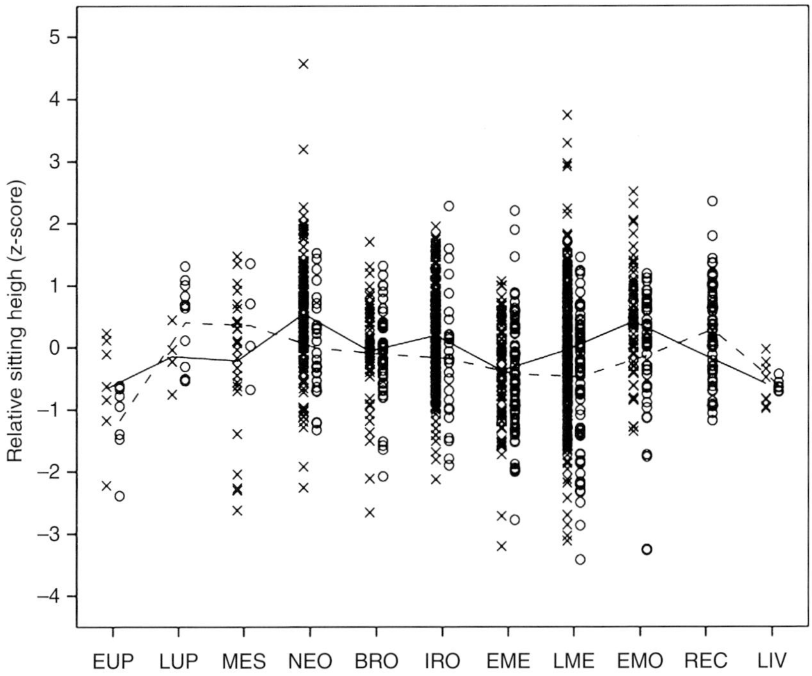

Figure 4.17 Relative sitting height plotted against period number. North/Central Europeans: crosses and solid lines. South Europeans: circles and dashed lines. Recent North/Central Europeans do not have values for this variable. Interpolation line thus runs directly from Early Modern North/Central Europeans to living North/Central Europeans. Living Europeans are represented by sex-specific means of four North/Central European (the Norwegians, Finns, Dutch, and Czechs) and three South European samples (the Basques, Sardinians, and Greeks).

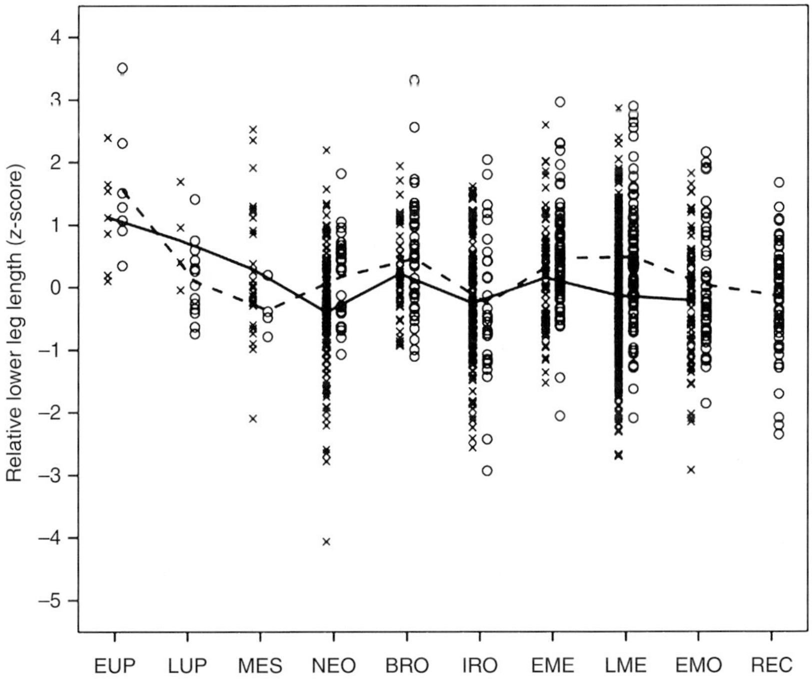

Figure 4.18 Relative lower leg length (z-score) plotted against period number. North/Central Europeans: crosses and solid lines. South Europeans: circles and dashed lines. Recent North/Central Europeans do not have values for this variable.

Table 4.7 ANOVA of rural and urban differences. All variables are z-scores.

	Region	N	Mean	F-ratio	Sig.
Stature	Rural	696	0.1070	7.230	0.007
	Urban	595	−0.0400		
Anatomical stature	Rural	365	0.1855	9.485	0.002
	Urban	360	−0.0377		
Sitting height	Rural	365	0.0334	0.030	0.862
	Urban	360	0.0206		
Lower limb length	Rural	365	0.2470	19.304	0.000
	Urban	360	−0.0728		
Relative sitting height	Rural	365	−0.2376	18.692	0.000
	Urban	360	0.0833		
Body mass	Rural	685	0.1697	11.356	0.001
	Urban	575	−0.0224		
Body mass index	Rural	682	0.1336	6.794	0.008
	Urban	572	−0.0088		
Biacromial breadth	Rural	427	0.0708	6.404	0.012
	Urban	398	−0.1024		
Bi-iliac breadth	Rural	414	0.1718	11.746	0.001
	Urban	387	−0.0738		
Relative biacromial (shoulder) breadth	Rural	414	−0.0238	0.429	0.513
	Urban	392	−0.0700		
Relative bi-iliac breadth	Rural	413	0.0884	5.004	0.026
	Urban	384	−0.0703		
Crural index	Rural	614	−0.0774	14.780	0.000
	Urban	499	−0.2972		
Relative lower leg length	Rural	365	0.1770	26.942	0.000
	Urban	360	−0.2050		
Brachial index	Rural	486	−0.0547	23.815	0.000
	Urban	425	−0.3592		

4.5 Discussion

Mean stature and overall body size have very clearly fluctuated considerably over time in Europe, but body breadths and body proportions have exhibited more stability. Changes in body proportions reflect at least partly stature changes, which in turn are affected by nutrition and health during growth. The very tall mean stature of the EUP Europeans implies good overall nutritional status and health. Their shoulders were very broad, but their bi-iliac breadths were only somewhat larger than the overall European average. Changes in mean stature apparently have little effect on lower trunk breadth (also see Ruff, 1994 for further discussion of this general finding). They had predictably low relative sitting heights because limb lengths change more than trunk lengths with stature changes. They had shorter trunks and longer limbs for stature than all more recent Europeans except the tallest living Europeans, which probably partly reflects their tall stature and partly a genetic predisposition for relatively long limbs inherited from early anatomically modern *Homo sapiens* (see Holliday, 1995, 1997 on this).

Body size very clearly declined and body shape changed across the LGM. While frequent and severe exposure to cold stress may have contributed to relatively short limbs during the LGM, this cold stress cannot explain why foragers living at the end of the Ice Age were shorter and had relatively shorter limbs than those living during the many millennia preceding the LGM. It was simply colder during the Gravettian period than during the Magdalenian and Epigravettian periods (see Clark *et al.*, 2009: their Fig. 4 on climatic reconstructions). It is thus more likely that these body size and shape changes indicate that conditions for somatic growth (including average nutritional status and health) deteriorated and/or there were gene pool changes across the LGM.

Although hunter-gatherers tend to have diverse diets rich in protein, many experience seasonal food shortages and even chronic undernourishment. These in turn result in seasonal fluctuation in weight and nutritional status as well as low BMI (Kelly, 2013:13 and references therein; Jenike, 2001). Chronic undernourishment has a negative effect on children's growth, resulting in short adult stature and relatively short limbs (Bogin *et al.*, 2002; Bogin and Rios, 2003; Smith *et al.*, 2003). Significant decline in stature and some decline in relative limb length between Early and Late Upper Paleolithic periods (as also noted by other researchers e.g., Formicola and Giannecchini, 1999; Meiklejohn and Babb, 2011), as well as increased frequency of enamel hypoplasia and Harris lines across the LGM (Holt and Formicola, 2008), imply that seasonal food shortages and even undernourishment became more common after the LGM.

Reduction in body size across the LGM could have been a microevolutionary adaptive process to reduce energy requirements and hence nutritional demands that took place due to a combination of reduced protein supply (megafauna gradually replaced by smaller prey species) and higher local population density (Formicola and Holt, 2007). Reduced interregional gene flow due to reduced overall mobility may have also resulted in stature decrease (Formicola and Giannecchini, 1999; see Koziel *et al.*, 2011 on effects of marital distance on stature growth). Upper Paleolithic gene pool changes (see Posth *et al.*, 2016; Fu *et al.*, 2016) also very likely partly explain body size and shape changes between the EUP and LUP samples, since there were also considerable changes in cranial configuration (Brewster *et al.*, 2014).

There was a great deal of regional stature variation during the Mesolithic. This variation was distributed from west to east. Mesolithic West Europeans were uniformly short-statured and exhibited very little regional variation from South Scandinavia to the Mediterranean. Males averaged about 160 cm and females about 154 cm. Mesolithic East Europeans (from the North Balkans) were apparently as tall as the EUP Europeans. Mean statures of 12 males and five females from Schela Cladovei of the Danubian Iron Gates region in eastern Romania are 174.9 cm and 163.8 cm, respectively. Stature estimates provided by Formicola and Giannecchini (1999: their Table 4a) derived by using Formicola and Franceschi's (1996) equations (which provide quite similar estimates to our equations, especially for tall-statured individuals) indicate that individuals from Vlasac (another Danubian Iron Gate site) and from the Ukrainian Mesolithic site of Vasilyevka in the lower Dnieper River Drainage and the Northwest Russian Mesolithic site of Olenii Ostrov were equally tall. The Mesolithic inhabitants of Latvia were apparently also tall (males 171.5 cm, N = 17; females 156.0 cm, N = 6) based on estimates provided by Gerhard (2005). Long bone lengths for these Mesolithic Latvians were not provided, and Gerhards' (2005) equations to estimate stature from femoral length produce stature estimates a few centimeters higher in this stature range than those of Formicola and Franceschi (1996) and Ruff *et al.* (2012). However, the Mesolithic Latvian males in particular were still considerably taller than the Mesolithic West Europeans.

Formicola and Giannecchini (1999) suggested that this considerable stature difference between the West and East European Mesolithic people was due to "...differences in terms of nutrition, lifestyle and gene flow." According to Gerhards (2005), this difference may be due to the generally earlier dates of the East European Mesolithic people. That is, their stature had not yet declined as much as that of the West European Mesolithic people. As already mentioned, earlier Mesolithic individuals average taller than later Mesolithic individuals in our data set, but this finding is largely due to the tall-statured Schela Cladovei sample.

Nutrition and life-style are probable reasons why early Mesolithic Europeans from Eastern Europe averaged taller than late Mesolithic Europeans of Western Europe, based on an example provided by the long-limbed and presumably tall-statured representatives of the Mesolithic Lake Cultures of India. These were broad-spectrum foragers, who had adequate diet and a mobile life-style (Lukacs *et al.*, 2014). The East European Mesolithic people had a greater variety of food sources and probably less seasonal food shortage due to their environment. Also, at least some of them (e.g., those from Olenii Ostrov) had a more mobile life-style than the West European Mesolithic people. This mobile life-style could have affected stature growth positively by allowing larger marital networks and thus greater genetic distances between parents (on effects of marital distance, see Koziel *et al.*, 2011).

The Late Mesolithic foragers and their Early Neolithic contemporaries were both very short-statured, but the latter were lighter-bodied. Short average statures of the LBK- and Lengyel Ceramic-associated early Central European farmers have been remarked upon by other researchers (see Piontek and Vančata, 2012). A genome-wide study of ancient DNA indicates that there was actually selection for decreased stature in the Iberian Neolithic and Chalcolithic samples. Predicted 'Genetic heights' based on 180 height-associated SNPs of these South European Neolithic farmers are lower than those of the European hunter-gatherers, Anatolian Neolithic farmers, and Early and Middle Neolithic farmers of Central Europe (Mathieson *et al.*, 2015: their Fig. 4).

A decline in stature and overall body size in the Neolithic is commonly attributed to poorer conditions for growth for children in early agricultural societies (e.g., Larsen, 1997). Prehistoric farmers generally had less diverse diets than many foragers. They also often lived in more crowded conditions than foragers, which resulted in more pathogens affecting overall health and growth (e.g., Cohen and Armegalos, 1984). However, the short statures of Late Mesolithic people show that the 'foundations' of poorer growing conditions were already established during the Mesolithic in at least Western Europe. Furthermore, the Neolithic period foragers of South Scandinavia (the Pitted Ware Culture, the PWC) were considerably shorter-statured than their South Scandinavian farmer contemporaries and neighbors (see Chapter 12). Stature decline associated with agriculture *per se* is thus not necessarily a universal trend (see Mummert *et al.* 2011 on this trend).

We cannot discount the possibility that observed differences in body size and body proportions between the Mesolithic and Neolithic people also had a genetic basis, at least in some locations. Ancient DNA research indicates that there was population discontinuity in all regions of Europe examined so far across the Mesolithic–Neolithic transition (e.g., Bramanti *et al.*, 2009; Malmström *et al.*, 2009; Haak *et al.*, 2010; Fu *et al.*, 2012; Skoglund *et al.*, 2012; Sánchez-Quinto *et al.*, 2012; Brandt *et al.*, 2013).

Considerable stature and overall body size increase from the Early Neolithic to Copper Age was probably due to diet changes associated with the so-called Secondary Products Revolution (see Greenfield, 2010 and references therein). Increased consumption of milk and other dairy products is possibly one of these diet changes. The consumption of dairy products has been found to have positive effects on children's stature growth (Okada, 2004; Berkey *et al.*, 2009). Dairy protein has a much stronger association with height growth, and thus adult stature, than dairy fat, non-dairy animal protein, non-dairy animal fat, vegetable protein, and vegetable fat. In the United States, non-Hispanic white girls who consumed more milk reached nearly 1 inch (2.54 cm) taller adult stature than those who consumed little or no milk (Berkey *et al.*, 2009).

This growth-promoting effect of dairy protein, as well as generally increased protein intake, might explain the stature increase from the Early Neolithic to Copper Age.

Genetic changes from the Middle Neolithic to Late Neolithic/Copper Age also played a role in this stature increase. Increased statures of the Corded Ware people and their more recent descendants in Northern and Central Europe are due to their partial descent from the Yamnaya people (Haak *et al.*, 2015; Allentoft *et al.*, 2015). These steppe people were genetically predisposed to reach taller statures than the West European hunter-gatherers, the Early and Middle Neolithic farmers, current Iberian populations, and Utah residents of North and Central European ancestry (Mathieson *et al.*, 2015: their Fig. 4). Higher frequencies of the so-called 'height-increasing alleles' and taller average statures in the current North/Central Europeans than in the current Mediterranean Europeans (see Turchin *et al.*, 2012) are thus largely due to Neolithic and Early Metal Age population events.

Body size was relatively stable from the Bronze Age until the end of the Medieval period, but there was regional variation. Stature declined in some regions between Early and Late Medieval periods, but increased in other regions. Based on several Scandinavian samples, stature certainly declined in this region during the 14th and 15th centuries (see Chapter 12). It is possible that environmental stress caused by the Little Ice Age (commencing ca. 1275–1300 AD) had more effect on stature growth in Northern Europe than in Southern Europe, because the 14th and 15th century Scandinavians were shorter than the 15th century inhabitants of Bosnia-Herzegovina. Our figures (Fig. 4.1(a,b) and Fig. 4.4(a,b)) do not show this Late Medieval stature decline because all of our Late Medieval samples from Northern Europe predate the Little Ice Age.

Tall average statures and overall body sizes of the Medieval people from Mistihalj (in Bosnia-Herzegovina close to the Montenegro border) are remarkable because this sample is our most recent Medieval sample. All individuals in this sample date to the 15th century AD, when stature had already begun to decline in some regions (e.g., Sweden). More recent inhabitants of this region (the Dinaric Alps) are among the tallest in the world (see Coon, 1939; Pineau *et al.*, 2005; Bjelica *et al.*, 2012). Also, maximum femoral head breadths of current Kosovoans and Bosnians are clearly larger than those of current Euroamericans (Jantz *et al.*, 2008: their Table 2), indicating larger overall body size. We do not have an explanation for the consistently large body size of inhabitants of the Northwest Balkan Peninsula (apparently since the Mesolithic), but its persistent nature implies a strong genetic predisposition to grow large-bodied even in suboptimal conditions (e.g., war, famine).

Stature decline in at least some areas during the Late Medieval period, and especially during the Early Modern period, is not unexpected. Climatic deterioration during the Little Ice Age (Humlum *et al.*, 2011) obviously resulted in widespread nutritional stress by affecting food production. Malnourished bodies were also more prone to infectious diseases (see Kelton, 2007 for an example from Native Americans), which has a further negative influence on children's growth.

The Early Modern period stature decline is also consistent with earlier studies based on historical stature data (e.g., Steckel, 2004; Koepke and Baten, 2005), as well as records on famine (Muroma, 1991). For example, French men had an average height of only 161.7 cm during the 17th century (Komlos *et al.*, 2003). For comparison, males and females in our Early Modern French sample averaged 162.5 cm (N = 24) and 155.4 cm (N = 11), respectively.

Mean heights apparently increased in Europe from very low mean statures of the 17th century to relatively high mean statures of the early 18th century (Komlos *et al.*, 2003; Komlos and Cinnirella, 2007; Komlos and Küchenhoff, 2012). Stature declined again during the later 18th century and during the first half of the 19th century due to overall decline in standards of living associated with the Industrial Revolution (Komlos and Cinnirella, 2007; Cinnirella, 2008a,b;

Komlos and Küchenhoff, 2012; Penttinen *et al.*, 2013). Poor climatic conditions for agriculture (especially during the so-called 'great hunger years' of Finland and northern Sweden in the 1860s) also had an effect at least in Northern Europe (see Chapter 12).

Stature recovery commenced during the second half of the 19th century or early 20th century, depending on the region, as the overall nutritional and health statuses as well as hygiene improved (Tanner, 1992; Komlos, 2007; Komlos and Cinnirella, 2007). This improvement was more noticeable in urban regions that had been characterized by poor health and diet as well poor hygiene during the industrialization process (Tanner, 1992; Martínez-Carrion and Moreno-Lázaro, 2007).

The Industrial Revolution was associated with increased urbanization. Several studies have shown that urban populations were generally shorter than rural ones until the end of the 19th century (see Martínez-Carrion and Moreno-Lázaro, 2007 and references therein). Our findings of larger body size among rural samples are consistent with this. The absolutely and relatively longer lower limbs of rural people also reflect their better growing environment as discussed in the introduction.

Our observation that the stature increase from the Early Modern period to the Recent period was greater in North/Central Europeans than in Mediterranean Europeans is also expected in light of available evidence, which indicates that the stature increase commenced generally earlier in North/Central Europe than in South Europe (Schmidt *et al.*, 1995). This is also reflected in a temporally earlier decrease in postneonatal mortality in North/Central Europe (Schmidt *et al.*, 1995). The same factors that reduced postnatal mortality probably also contributed to an increase in mean stature, because early childhood environment is an important determinant of adult stature (Schmidt *et al.*, 1995; Wadsworth *et al.*, 2002; Li *et al.*, 2007). The North/Central European stature increase has leveled off during the last few decades, but has continued in many South European countries, reducing geographic stature differences (Larnkjær *et al.*, 2006; McEvoy and Visscher, 2009). However, evidence that stature increase has reached a plateau in Italy below the North/Central European stature level (Larnkjær *et al.*, 2006) suggests that at least some South European populations are genetically predisposed to be shorter-statured than most, if not all, North/Central European populations. This is also reflected in longer-term trends. Regardless of regional variation in recent temporal changes, some geographic differences in body size appear to have very deep roots. North/Central Europeans have been consistently taller and heavier than South Europeans at least since the Bronze Age.

Nutritional factors and overall health influence growth of lower limb length more than growth of trunk length (Leitch, 1951; Eveleth and Tanner, 1990; Fredriks *et al.*, 2005; Bogin *et al.*, 2002; Smith *et al.*, 2003; Wadsworth *et al.*, 2002; Bogin and Rios, 2003; Li *et al.*, 2007; Bogin and Varela-Silva, 2010). It is thus not surprising that temporal stature changes are more due to changes in lower limb length than in trunk length in light of both published anthropometric studies (e.g., Udjus, 1964; Ohyama *et al.*, 1987; Manolis *et al.*, 1995) and our osteometric data. Nutritional differences between populations (i.e., between well-nourished recent Europeans and impoverished populations in many developing countries) may diminish ecogeographic variation in relative trunk length and lower limb length (Cowgill *et al.*, 2012). However, relatively longer lower limbs and lower leg (tibial) lengths of the South Europeans regardless of their shorter stature since the Neolithic may indeed reflect ecogeographic differences that have a genetic basis. Past and present South Europeans are apparently genetically predisposed to have relatively longer lower limbs for stature than North/Central Europeans, despite temporal variation in overall health and nutritional level.

Trunk breadth, as indicated by bi-iliac breadth, exhibits relative stability over time regardless of considerable temporal changes in stature, and thus conforms to Ruff's (1994) earlier-mentioned prediction that body breadth should remain constant regardless of stature changes within similar temperature zones. The Early Neolithic Europeans and their absolutely narrow bi-iliac breadths are the most noticeable exception in Europe. Their narrow trunks may reflect Near Eastern genetic affinities, whereas later Neolithic people had mixed European and Near Eastern ancestry. This interpretation is supported by recent ancient DNA studies of prehistoric Europeans' temporal mtDNA lineage changes (Brandt *et al.*, 2013), but is weakened by the observation that Late Mesolithic people had rather narrow bi-iliac breadths, too (although not as narrow as contemporaneous Early Neolithics; see Fig. 4.13b). Anthropometric data on recent and living Europeans (Appendix 3(b)) indicate that this trunk breadth dimension does not differentiate current North/Central and South European populations.

Biacromial breadth variation follows stature variation more closely than does bi-iliac breadth variation over time and space. As mean stature has increased during the last few decades, biacromial breadth has increased roughly proportionately to stature, but there has been little or no change in bi-iliac breadth. As data in Appendix 3(b) demonstrate, the tall-statured North Europeans (e.g., the Norwegians and Finns) have absolutely much broader shoulders than their much shorter-statured South European contemporaries (e.g., the Basque and Sardinians).

4.6 Conclusions

During the past 30,000 years in Europe, there were two major episodes of general body size reduction, which both coincided with – or were at least in some respect associated with – major climatic cooling events: The Last Glacial Maximum and the Little Ice Age. There was a general decline in the health and nutritional status during both of these episodes. The forager-farmer transition resulted in some body size and shape changes, but these changes were not as noticeable as those resulting and/or coinciding with climatic cooling events.

Most of the temporal variation in stature throughout is due to variation in lower limb length. This suggests that: 1) lower limb length is more developmentally labile overall; and 2) that at least some of these changes are due to health and nutritional status changes. Temporal trends in crural and brachial indices are possibly partly attributable to the same causes, but other factors may also be involved here due to the partial independence of intra-limb versus limb/trunk proportions (see Holliday 1995, 1997; Holliday and Ruff, 2001).

There is considerably less change in body breadths than in linear dimensions. Bi-iliac breadth especially exhibits little temporal variation, supporting its relative environmental stability (see Ruff, 1991, 1994). Biacromial breadth exhibits somewhat less stability, tending to follow stature changes. It may be more genetically tied to linear dimensions.

North/Central Europeans have been taller, with relatively shorter tibiae but longer trunks than Southern Europeans since the Neolithic. This may suggest true ecogeographic variation due to climatic selection. This is not the case for the Upper Paleolithic, probably due to long-range seasonal movements of mobile foragers and possibly actual migrations of more southern-derived populations into both Northern and Southern Europe at this time. However, sampling cannot be discounted because Upper Paleolithic sample sizes are small. Local selection probably had some effect on body types of the Mesolithic people, but much larger body sizes of the Mesolithic East Europeans contrast with much smaller body sizes of the Mesolithic West Europeans, probably due to diet and mobility differences. The forager-farmer transition had effects through diet changes, migrations of farmer populations derived from the Near East, various degrees of genetic mixing with the indigenous foragers, and possibly local selection.

There were little or no differences in stature and overall body size between the North/Central and South Europeans during the Late Medieval and Early Modern periods, probably because poor climatic conditions had more impact on food production in more northern latitudes, reducing north–south gradients. The recent secular increase in body size commenced earlier in North/Central Europe than in South Europe, resulting in considerably larger-bodied North/Central Europeans compared to South Europeans. The Northwest Balkan people are an exception among the south Europeans. More recently, stature differences between the northern and southern populations have been reduced due to cessation of secular increases in the north combined with continuing increases in the south.

References

Allen, J.A. (1877) The influence of physical conditions on the genesis of species. *Radical Rev.*, **1**, 108–140.

Allentoft, M.E., Sikora, M., Sjögren, K.-G., Rasmussen, S., Rasmussen, M., Stenderup, J., Damgaard, P.B., Schroeder, H., Ahlström, T., Vinner, L., Malaspinas, A.-S., Margaryan, A., Higham, T., Chivall, D., Lynnerup, N., Harvig, L., Baron, J., Casa, P.D., Dabrowski, P., Duffy, P.R., Ebel, A.V., Epimakhov, A., Frei, K., Furmanek, M., Gralak, T., Gromov, A., Gronkiewicz, S., Grupe, G., Hajdu, T., Jarysz, R., Khartanovich, V., Khokhlov, A., Kiss, V., Kolář, J., Kriiska, A., Lasak, I., Longhi, C., McGlynn, G., Merkevicius, A., Merkyte, I., Metspalu, M., Mkrtchyan, R., Moiseyev, V., Paja, L., Pálfi, G., Pokutta, D., Pospiezny, L., Price, T.D., Saag, L., Sablin, M., Shishlina, N., Smrčka, V., Soenov, V.I., Szeverényi, V., Tóth, G., Trifanova, S.V., Varul, L., Vicze, M., Yepiskoposyan, L., Zhitenev, V., Orlando, L., Sicheritz-Pontén, T., Brunak, S., Nielsen, R., Kristiansen, K., and Willerslev, E. (2015) Population genomics of Bronze Age Eurasia. *Nature*, **522**, 167–172.

Alley, R.B. and Clark, P.U. (1999) The deglaciation of the northern Hemisphere: a global perspective. *Annu. Rev. Earth Planet Sci.*, **27**, 149–182.

Anthony, D.W. and Brown, D.R. (2011) The secondary product revolution, horse-riding, and mounted warfare. *J. World Prehist.*, **24**, 131–160.

Auger, F., Jamison, P.L., Balslev-Jorgensen, J., Lewin, T., de Peña, J.F., and Skrobak-Kaczynski, J. (1980) Anthropometry of circumpolar populations. In: *The Human Biology of Circumpolar Populations* (ed. F.A. Milan), Cambridge University Press, Cambridge. pp. 213–225.

Bang-Andrsen, S. (2012) Colonizing contrasting landscapes. The pioneer coastal settlement and inland utilization in southern Norway 10,000–9500 years before present. *Oxford J. Archaeol.*, **31**, 103–120.

Berkey, C.S., Colditz, G.A., Rockett, A.L., Frazier, A.L., and Willett, W.C. (2009) Dairy consumption and female height growth: perspective cohort study. *Cancer Epidemiol. Biomarkers Prev.*, **18**, 1881–1887.

Bergmann, C. (1847) Uber die verhaltnisse der warmeokononomie der thiere zu ihrer grosse. *Gottingen Stud.*, **1**, 595–708.

Bjelica, D., Popovic, S., Keznovic, M., Petkovic, J., Jurak, G., and Gragruber, P. (2012) Body height and its estimation utilizing arm span measurement in Montenegrin adults. *Anthropological Notebooks*, **18**, 69–83.

Bogin, B. and Rios, L. (2003) Rapid morphological change in living humans: implications for modern human origins. *Comp. Biochem. Physiol. Part A*, **136**, 71–84.

Bogin, B., Smith, P., Orden, A.B., Varela Silva, M.I., and Loucky, J. (2002) Rapid change in height and body proportions of Maya-American children. *Am. J. Hum. Biol.*, **14**, 753–761.

Bogin, B. and Varela-Silva, M.I. (2010) Leg length, body proportion, and health: A review with a note on beauty. *Int. J. Environ. Res. Public Health*, **7**, 1047–1075.

Bowles, G.T. (1932) *New Types of Old Americans at Harvard*. Harvard University Press, Cambridge, Mass.

Bramanti, B., Thomas, M.G., Haak, W., Unterlaender, M., Jores, P., Tambets, K., Antanaitis-Jacobs, I., Haidle, M.N., Jankauskas, R., Kind, C.-J., Lueth, F., Terberger, T., Hiller, J., Matsumura, S., Forster, P., and Burger, J. (2009) Genetic discontinuity between local hunter-gatherers and Central Europe's first farmers. *Science*, **326**, 137–140.

Brandt, G., Haak, W., Adler, C.J., Roth, C., Szécsényi-Nagy, A., Karimnia, S., Möller-Rieker, S., Meller, H., Ganslmeier, R., Friederich, S., Dresely, V., Nicklisch, N., Pickrell, J.K., Sirocko, F., Reich, D., Cooper, A., Alt, K.W., The Geographic Consortium. (2013) Ancient DNA reveals key stages in the formation of Central European mitochondrial genetic diversity. *Science*, **342**, 257–261.

Brewster, C., Meiklejohn, C., von Cramon-Taubadel, N., and Pinhasi, R. (2014) Craniometric analysis of European Upper Paleolithic and Mesolithic samples supports discontinuity at the Last Glacial Maximum. *Nat. Commun.*, **5**, Article Number 4094. doi:10.1038/ncomms5094.

Cinnirella, F. (2008a) On the road to industrialization: nutritional status in Saxony, 1690–1850. *Cliometrica*, **2**, 229–257.

Cinnirella, F. (2008b) Optimism or pessimism? A reconsideration of nutritional status in Britain, 1740–1865. *Eur. Rev. Econ. Hist.*, **12**, 325–354.

Clark, P.U., Dyke, A.S., Shakun, J.D., Carlson, A.E., Clark, J., Wohlfarth, B., Mitrovica, J.X., Hostetler, S.W., and McCabe, A.M. (2009) The Last Glacial Maximum. *Science*, **325**, 710–714.

Cohen, M.N. and Armelagos, G.J. (1984) *Paleopathology at the origins of agriculture*. Academic Press, Orlando, FL.

Coon, C.S. (1939) *The Races of Europe*. The Macmillan Company, New York.

Cowgill, L.W., Eleazer, C.D., Auerbach, B.M., Temple, D.H., and Okazaki, K. (2012) Developmental variation in ecogeographic body proportions. *Am. J. Phys. Anthropol.*, **148**, 557–570.

Csukás, A., Takai, S., and Baran, S. (2006) Adolescent growth in main somatometric traits of Japanese boys: Ogi longitudinal growth study. *HOMO*, **57**, 73–86.

Cullen, K.J. (2010) *Famine in Scotland: The 'Ill Years' of the 1690s*. Edinburgh University Press, Edinburgh.

de Beer, H. (2004) Observations on the history of Dutch physical stature from the late-Middle Ages to the present. *Econ. Hum. Biol.*, **2**, 45–55.

Dolukhanov, P.M. (1996) The Pleistocene–Holocene transition on the East European Plain. In: *Humans at the End of the Ice Age: The Archaeology of the Pleistocene–Holocene Transition* (eds L.G. Straus, B.V. Eriksen, J.M. Erlandson, and D.R. Yesner), Plenum Press, New York, pp. 159–169.

Duyar, I. and Özener, B. (2005) Growth and nutritional status of male adolescent laborers in Ankara, Turkey. *Am. J. Phys. Anthropol.*, **128**, 693–698.

Eveleth, P.B. and Tanner, J.M. (1976) *Worldwide Variation in Human Growth*. Cambridge University Press, Cambridge.

Eveleth, P.B. and Tanner, J.M. (1990) *Worldwide Variation in Human Growth*, 2nd edn. Cambridge University Press, Cambridge.

Formicola, V. and Franceschi, M. (1996) Regression equations for estimating stature from long bones of early Holocene European samples. *Am. J. Phys. Anthropol.*, **100**, 83–88.

Formicola, V. and Giannecchini, M. (1999) Evolutionary trends of stature in Upper Paleolithic and Mesolithic Europe. *J. Hum. Evol.*, **36**, 319–333.

Formicola, V. and Holt, B.M. (2007) Resource availability and stature decrease in Upper Paleolithic Europe. *J. Anthropol. Sci.*, **85**, 147–155.

Fredriks, A.M., van Buuren, S., van Heel, W.J.M., Dijkman-Neerincx, R.H.M., Verloove-Vanhorick, S.P., and Wit, J.M. (2005) Nationwide age references for sitting height, leg length, and sitting height/height ratio, and their diagnostic value for disproportionate growth disorders. *Arch. Dis. Child.*, **90**, 807–812.

Froehlich, J.W. (1970) Migration and the plasticity of physique in the Japanese-Americans in Hawaii. *Am. J. Phys. Anthropol.*, **32**, 429–442.

Fu, Q., Hajdinjak, M., Moldovan, O.T., Constantin, S., Mallick, S., Skoglund, P., Patterson, N., Rohland, N., Lazaridis, I., Nickel, B., Viola, B., Prüfer, K., Meyer, M., Kelso, J., Reich, D., and Pääbo, S. (2015) An early modern human from Romania with a recent Neanderthal ancestor. *Nature*, **524**, 216–219.

Fu, Q., Posth, C., Hajdinjak, M., Petr, M., Mallick, S., Fernandes, D., Furtwängler, A., Haak, W., Meyer, M., Mittnik, A., Nickel, B., Peltzer, A., Rohland, N., Slon, V., Talamo, S., Lazaridis, I., Lipson, M., Mathieson, I., Schiffels, S., Skoglund, P., Derevianko, A.P., Drozdov, N., Slavinsky, V., Tsybankov, A., Cremonesi, R.G., Mallegni, F., Gély, B., Vacca, E., Morales, M.R.G., Straus, L.G., Neugebauer-Maresch, C., Teschler-Nicola, M., Constantin, S., Moldovan, O.T., Benazzi, S., Peresani, M., Coppola, D., Lari, M., Ricci, S., Ronchitelli, A., Valentin, F., Thevenet, C., Wehrberger, K., Grigorescu, D., Rougier, H., Crevecoeur, I., Flas, D., Semal, P., Mannino, M.A., Cupillard, C., Bocherens, H., Conard, N.J., Harvati, K., Moiseyev, V., Drucker, D.G., Svoboda, J., Richards, M.P., Caramelli, D., Pinhasi, R., Kelso, J., Patterson, N., Krause, J., Pääbo, S., and Reich, D. (2016) The genetic history of Ice Age Europe. *Nature*, **534**, 200–205.

Fu, Q., Rudan, P., Pääbo, S., and Krause, J. (2012) Complete mitochondrial genomes reveal Neolithic expansion into Europe. *PLoS One*, **7**(2), e32473–e32473.

Gallagher, A., Gunther, M.M., and Bruchaus, H. (2009) Population continuity, demic diffusion and Neolithic origins in central-southern Germany: The evidence from limb proportions. *HOMO*, **60**, 95–126.

Gerhards, G. (2005) Secular variations in body structure of the inhabitants of Latvia (7th millennium BC – 20th c. AD). *Acta Medica Lituanica*, **12**, 33–39.

Greenfield, H.J. (2010) The secondary products revolution: the past, the present and the future. *World Archaeol.*, **42**, 29–54.

Haak, W., Balanovsky, O., Sanchez, J.J., Koshel, S., Zaporozhchenko, V., *et al.* (2010) Ancient DNA from European Early Neolithic Farmers Reveals Their Near Eastern Affinities. *PLoS Biol.*, **8**(11), e1000536. doi:10.1371/journal.pbio.1000536.

Haak, W., Lozaridis, I., Patterson, N., Rohland, N., Mallick, S., Llamas, B., Brandt, G., Nordenfelt, S., Harney, E., Stewardson, K., Fu, Q., Mittnik, A., Bánffy, E., Economou, C., Francken, M., Friederich, S., Pena, R.G., Hallgren, F., Khartanovich, V., Khokhlov, A., Kunst, M., Kuznetsov, P., Meller, H., Mochalov, O., Moiseyev, V., Nicklisch, N., Pichler, S.L., Risch, R., Guerra, M.A.R., Roth, C., Szécsényi-Nagy, A., Wahl, J., Meyer, M., Krause, J., Brown, D., Anthony, D., Cooper, A., Alt, K.W., and Reich, D. (2015) Massive migration from the steppe was a source for Indo-European languages in Europe. *Nature*, **522**, 207–211.

Heather, P. (2009) *Empires and Barbarians: Migration, Development and the Birth of Europe*. Macmillan, London.

Hendy, E.J., Gagan, M.K., Alibert, C.A., McCulloch, M.T., Lough, J.M., and Isdale, P.J. (2002) Abrupt decrease in tropical Pacific sea surface salinity at the end of Little Ice Age. *Science*, **295**, 1511–1514.

Hiernaux, J. (1985) A comparison of the shoulder-hip-width sexual dimorphism in sub-Saharan Africa and Europe. In: *Human Sexual Dimorphism* (eds J. Ghesquire, R.D. Martin, and F. Newcombe), Taylor and Francis, Philadelphia, pp. 191–206.

Holliday, T.W. (1995) Body Size and Proportions in the Late Pleistocene Western Old World and the Origins of Modern Humans. Ph.D. Diss. The University of New Mexico, Albuquerque, New Mexico.

Holliday, T.W. (1997) Body proportions in Late Pleistocene Europe and modern human origins. *J. Hum. Evol.*, **32**, 423–447.

Holliday, T.W. and Ruff, C.B. (2001) Relative variation in human proximal and distal limb segment lengths. *Am. J. Phys. Anthropol.*, **116**, 26–33.

Holt, B.M. and Formicola, V. (2008) Hunters of the Ice Age: the biology of Upper Paleolithic people. *Yrbk Phys. Anthropol.*, **51**, 70–99.

Hublin, J.-J. (2015) The modern human colonization of western Eurasia: when and where? *Q. Sci. Rev.*, **118**, 194–210.

Hughes, P.D. and Gibbard, P.L. (2015) A stratigraphic basis for the Last Glacial Maximum (LGM). *Quat. Int.*, **383**, 174–185.

Huijzer, B. and Vandenberghe, J. (1998) Climatic reconstruction of the Weichselian Pleniglacial in northwestern and central Europe. *J. Quat. Sci.*, **13**, 391–417.

Humlum, O., Solheim, J.-E., and Stordahl, K. (2011) Identifying natural contributions to late Holocene climate change. *Glob. Planet Change*, **79**, 145–156.

Jacoby, W.G. (2000) Loess: a nonparametric, graphical tool for depicting relationships between variables. *Elect. Stud.*, **19**, 577–613.

Jamison, P.L. and Zegura, S.L. (1970) An anthropometric study of the Eskimos of Wainwright, Alaska. *Arctic Anthropol.*, **7**, 125–143.

Jantz, R.L., Kimmerle, E.H., and Baraybar, J.P. (2008) Sexing and stature estimation criteria for Balkan populations. *J. Forensic Sci.*, **53**, 601–605.

Jenike, M.R. (2001) Nutritional ecology: Diet, physical activity and body size. In: *Hunter-Gatherers: An Interdisciplinary Perspective* (eds C. Panter-Brick, R.H. Layton, and P. Rowley-Conwy), Cambridge University Press, Cambridge, pp. 205–238.

Jordan, W.C. (1996) *The Great Famine: Northern Europe in the Early Fourteenth Century.* Princeton University Press, Princeton.

Katzmarzyk, P.T. and Leonard, W.R. (1998) Climatic influences on human body size and proportions: ecological adaptations and secular trends. *Am. J. Phys. Anthropol.*, **106**, 483–503.

Kelly, R.L. (2013) *The Lifeways of Hunter-Gatherers: The Foraging Spectrum*. Cambridge University Press, Cambridge.

Kelton, P. (2007) *Epidemics and Enslavement: Biological Catastrophe in the Native Southeast 1492–1715*. University of Nebraska Press, Lincoln.

Koepke, N. and Baten, J. (2005) The biological standard of living in Europe during the last two millennia. *Eur. Rev. Econ. Hist.*, **9**, 61–95.

Komlos, J. (2007) Anthropometric evidence on economic growth, biological well-being and regional convergence in the Habsburg Monarchy, c. 1850–1910. *Cliometric*, **1**, 211–237.

Komlos, J., Hau, M., and Bourguinat, N. (2003) An Anthropometric History of Early-Modern France. *Eur. Rev. Econ. Hist.*, **7**, 159–189.

Komlos, J. and Cinnirella, F. (2007) European heights in the early 18th century. *Vierteljahrschrift für Sozial- und Wirtschaftsgeschichte*, **94**. Bd., H.3, pp. 271–284.

Komlos, J. and Küchenhoff, H. (2012) The diminution of the physical stature of the English male population in the eighteenth century. *Cliometrica*, **6**, 45–62.

Koziel, S., Danel, D.P., and Zareba, M. (2011) Isolation by distance between spouses and its effect on children's growth in height. *Am. J. Phys. Anthropol.*, **146**, 14–19.

Kozlowski, J.K. (2015) The origin of the Gravettian. *Quat. Int.*, **359-360**, 3–18.

Larnkjær, A., Schrøder, S.A., Schmidt, I.M., Jørgensen, M.H., and Michaelsen, K.F. (2006) Secular change in adult stature has come to a halt in northern Europe and Italy. *Acta Paediatr.*, **95**, 754–755.

Larsen, C.S. (1997) *Bioarchaeology: Interpreting Behavior from the Human Skeleton*. Cambridge University Press, Cambridge.

Lazaridis, I., Patterson, N., Mittnik, A., Renaud, G., Mallick, S., Kirsanow, K., Sudmant, P.H., Schraiber, J.G., Castellano, S., Lipson, M., Berger, B., Economou, C., Bollongino, R., Fu, Q., Bos, K.I., Nordenfelt, S., Li, H., de Filippo, C., Prüfer, K., Sawyer, S., Posth, C., Haak, W., Hallgren, F., Fornander, E., Rohland, N., Delsate, D., Francken, M., Guinet, J.-M., Wahl, J., Ayodo, G., Babiker, H.A., Bailliet, G., Balanovska, E., Balanovsky, O., Barrantes, R., Bedoya, G., Ben-Ami, H., Bene, J., Berrada, F., Bravi, C.M., Brisighelli, F., Busby, G.B.J., Cali, F., Churnosov, M., Cole, D.E.C., Corach, D., Damba, L., van Driem, G., Dryomov, S., Dugoujon, J.-M., Fedorova, S.A., Romero, I.G., Gubina, M., Hammer, M., Henn, B.M., Hervig, T., Hodoglugil, U., Jha, A.R., Karachanak-Yankova, S., Khusainova, R., Khusnutdinova, E., Kittles, R., Kivisild, T., Klitz, W., Kučinskas, V., Kushniarevich, A., Laredj, L., Litvinov, S., Loukidis, T., Mahley, R.W., Melegh, B., Metspalu, E., Molina, J., Mountain, J., Näkkäläjärvi, K., Nesheva, D., Nyambo, T., Osipova, L., Parik, J., Platonov, F., Posukh, O., Romano, V., Rothhammer, F., Rudan, I., Ruizbakiev, R., Sahakyan, H., Sajantila, A., Salas, A., Starikovskaya, E.B., Tarekegn, A., Toncheva, D., Turdikulova, S., Uktveryte, I., Utevska, O., Vasquez, R., Villena, M., Voevoda, M., Winkler, C.A., Yepiskoposyan, L., Zalloua, P., Zemunik, T., Cooper, A., Capelli, C., Thomas, M.G., Luiz-Linares, A., Tishkoff, S.A., Singh, L., Thangaraj, K., Villems, R., Comas, D., Sukernik, R., Metspalu, M., Meyer, M., Eichler, E.E., Burger, J., Slatkin, M., Pääbo, S., Kelso, J., Reich, D., and Krause, J. (2014) Ancient human genomes suggest three ancestral populations for present-day Europeans. *Nature*, **513**, 409–413.

Leitch, I. (1951) Growth and health. *Br. J. Nutr.*, **5**, 142–151. Reprinted *Int. J. Epidemiol.* (2001), **30**, 212–216.

Li, L., Dangour, A.D., and Power, C. (2007) Early life influences on adult leg and trunk length in the 1958 British birth cohort. *Am. J. Hum. Biol.*, **6**, 836–843.

Lukacs, J.R., Pal, J.N., and Nelson, G.C. (2014) Stature in Holocene foragers of North India. *Am. J. Phys. Anthropol.*, **153**, 408–416.

Malmström, H., Gilbert, M.T.P., Thomas, M.G., Brandström, J., Molnar, P., Andersen, P.K., Bendixen, C., Holmlund, G., Götherström, A., and Willerslev, E. (2009) Ancient DNA reveals lack of continuity between Neolithic hunter-gatherers and contemporary Scandinavians. *Curr. Biol.*, **19**, 1758–1762.

Manolis, S., Neroutsos, A., Zafeiratos, C., and Pentzou-Daponte, A. (1995) Secular changes in body formation of Greek students. *Hum. Evol.*, **10**, 199–204.

Martínez-Carrión, J.-M. and Moreno-Lázaro, J. (2007) Was there an urban height penalty in Spain, 1840–1913? *Econ. Hum. Biol.*, **5**, 144–164.

Mathieseon, I., Lazaridis, I., Rohland, N., Mallick, S., Patterson, N., Roodenberg, S.A.., Harney, E., Stewardson, K., Fernandes, D., Novak, M., Sirak, K., Gamba, C., Jones, E.R., Llamas, B., Dryomov, S., Pickrell, J., Arsuaga, J.L., de Castro, J.M.B., Carbonell, E., Gerritsen, F., Khokhlov, A., Kuznetsov, P., Lozano, M., Meller, H., Mochalov, O., Moiseyev, V., Guerra, M.A.R., Roodenberg, J., Vergès, J.M., Krause, J., Cooper, A., Alt, K.W., Brown, D., Anthony, D., Lalueza-Fox, C., Haak, W., Pinhasi, R., and Reich, D. (2015) Genome-wide patterns of selection in 230 ancient Eurasians. *Nature*, **528**, 499–503.

McEvoy, B.P. and Visscher, P.M. (2009) Genetic of human height. *Econ. Hum. Biol.*, **7**, 294–306.

Meiklejohn, C. and Babb, J. (2011) Long bone length, stature and time in the European Late Pleistocene and Early Holocene. In: *Human Bioarchaeology of the Transition to Agriculture* (eds R. Pinhasi and J.T. Stock), John Wiley & Sons, Ltd, pp. 153–175.

Miller, G.H., Geirsdóttir, Á., Zhong, Y., Larsen, D.J., Otto-Bliesner, B.L., Holland, M.M., Bailey, D.A., Refsnider, K.A., Lehman, S.J., Southon, J.R., Anderson, C., Björnsson, H., and Thordarson, T. (2012) Abrupt onset of the Little Ice Age triggered by volcanism and sustained by sea-ice/ocean feedbacks. *Geophys. Res. Lett.*, **39**, L02708. doi:10.1029/2011GLO50168.

Mummert, A., Esche, E., Robinson, J., and Armelagos, G.J. (2011) Stature and robusticity during the agricultural transition: evidence from the bioarchaeological record. *Econ. Hum. Biol.*, **9**, 284–301.

Muroma, S. (1991) *Suurten kuolovuosien (1696–1697) väestönmenetys Suomessa*. Suomen Historiallinen Seura, Helsinki.

Novembre, J., Johnson, T., Bryc, K., Kutalik, Z., Boyko, A.R., Anton, A., Indap, A., King, K.S., Bergmann, S., Nelson, M.R., Stephens, M., and Bustamante, C.D. (2008) Genes mirror geography within Europe. *Nature*, **456**, 98–101.

Ohyama, S., Hisanaga, A., Inamasu, T., Yamamoto, A., Hirata, M., and Ishinishi, N. (1987) Some secular changes in body height and proportion of Japanese medical students. *Am. J. Phys. Anthropol.*, **73**, 179–183.

Okada, T. (2004) Effects of cow milk consumption on longitudinal height gain in children. *Am. J. Clin. Nutr.*, **80**, 1088–1089.

Otte, M. (2012) Appearance, expansion and dilution of the Magdalenian civilization. *Quat. Int.*, **272-273**, 354–361.

Penttinen, A., Moltchanova, E., and Nummela, I. (2013) Bayesian modeling of the evolution of male height in 18th century Finland from incomplete data. *Econ. Hum. Biol.*, **11**, 405–415.

Paterson, J.D. (1996) Coming to America: acclimation in macaque body structures and Bergmann's rule. *Int. J. Primatol.*, **17**, 585–611.

Paus, A., Velle, G., and Berge, J. (2011) The Late glacial and early Holocene vegetation and environment in the Dovre mountains, central Norway, as signaled in two Late glacial nunatek lakes. *Quat. Sci. Rev.*, **30**, 1780–1796.

Pineau, J.-C., Delamarche, P., and Bozinovic, S. (2005) Les Alpes Dinariques: un people de sujets de grande taille. *C.R. Biol.*, **328**, 841–846.

Piontek, J. and Vančata, V. (2012) Transition to agriculture in central Europe: Body size and body shape amongst the first farmers. *Interdisciplinaria Archaeologica*, **III**, 23–42.

Posth, C., Renauld, G., Mittnik, A., Drucker, D.G., Rougier, H., Cupillard, C., Valentin, F., Thevenet, A., Furtwängler, A., Wißing, C., Francken, M., Malina, M., Bolus, M., Lari, M., Gigli, E., Capecchi, G., Crevecoeur, I., Beauval, C., Flas, D., Germonpré, M., van der Plicht, J., Cottiaux, R., Gély, B., Ronchitelli, A., Wehrberger, K., Grigorescu, D., Svoboda, J., Semal, P., Caramelli, D., Bocherens, H., Harvai, K., Conard, N.J., Haak, W., Powell, A., and Krause, J. (2016) Pleistocene mitochondrial genomes suggest a single major dispersal of non-Africans and a Late Glacial population turnover in Europe. *Curr. Biol.*, **26**, 827–833.

Price, T.D. (2000) *Europe's First Farmers*. Cambridge University Press, Cambridge.

Rankama, T. and Kankaanpää, J. (2011) First evidence of eastern Preboreal pioneers in Arctic Finland and Norway. *Quartät.*, **58**, 183–209.

Raxter, M.H., Auerbach, B.M., and Ruff, C.B. (2006) Revision of the Fully technique for estimating statures. *Am. J. Phys. Anthropol.*, **130**, 374–484.

Roberts, D.F. (1953) Bodyweight, race, and climate. *Am. J. Phys. Anthropol.*, **11**, 85–96.

Roberts, D.F. (1978) *Climate and Human Variability*, 2nd edn. Cummings, Menlo Park, CA.

Ruff, C.B. (1991) Climate, body size and body shape in hominid evolution. *J. Hum. Evol.*, **21**, 81–105.

Ruff, C.B. (1994) Morphological adaptation to climate in modern and fossil hominids. *Yrbk Phys. Anthropol.*, **37**, 65–107.

Ruff, C.B. (2002) Variation in human body size and shape. *Annu. Rev. Anthropol.*, **31**, 211–232.

Ruff, C.B., Holt, B.M., Sladék, V., Berner, M., Garofalo, E., Garvin, H.M., Hora, M., Maijanen, H., Niinimäki, S., Salo, K., Schuplerová, E., and Tompkins, D. (2012) Stature and body mass estimation from skeletal remains in the European Holocene. *Am. J. Phys. Anthropol.*, **148**, 601–617.

Ruff, C.B., Niskanen, M., Junno, J.-A., and Jamison, P. (2005) Body mass prediction from stature and bi-iliac breadth in two high-latitude populations, with application to earlier higher latitude humans. *J. Hum. Evol.*, **48**, 381–392.
Ruff, C.B., Trinkaus, E., and Holliday, T.W. (1997) Body mass and encephalization in Pleistocene Homo. *Nature*, **387**, 173–176.
Sánchez-Quinto, F., Schroeder, H., Ramirez, O., Ávila-Arcos, M.C., Pybus, M., Olalde, I., Velazquez, A.M.V., Marcos, M.E.P., Encinas, J.M.V., Bertranpetit, J., Orlando, L., Gilbert, M.T.P., and Lalueza-Fox, C. (2012) Genomic affinities of two 7,000-year-old Iberian hunter-gatherers. *Curr. Biol.*, **22**, 1494–1499.
Schmidt, I.M., Jørgensen, M.H., and Michaelsen, K.F. (1995) Height of conscripts in Europe: is postneonatal mortality a predictor? *Ann. Hum. Biol.*, **22**, 57–67.
Schultz, A.H. (1923) Fetal growth in man. *Am. J. Phys. Anthropol.*, **6**, 389–399.
Schultz, A.H. (1926a) Fetal growth in man and other primates. *Q. Rev. Biol.*, **1**, 465–521.
Schultz, A.H. (1926b) Variations in man and their evolutionary significance. *Am. Nat.*, **60**, 297–323.
Schönbeck, Y., Talma, H., van Dommelen, P., Bakker, B., Buitendijk, S.E., HiraSing, R.A., and van Buuren, S. (2013) The world's tallest nation has stopped growing taller: the height of Dutch children from 1955 to 2009. *Pediatr. Res.*, **73**, 371–377.
Shakun, J.D. and Carlson, A.E. (2010) A global perspective on Last Glacial Maximum to Holocene climate change. *Quat. Sci. Rev.*, **29**, 1801–1816.
Skoglund, P., Malmström, H., Raghavan, M., Storå, J., Hall, P., Willerslev, E., Thomas, M., Gilbert, P., Götherström, A., and Jakobsson, M. (2012) Origins and genetic legacy of Neolithic farmers and hunter-gatherers in Europe. *Science*, **27**, 466–469.
Smith, P.K., Bogin, B., Varela-Silva, M.I., and Loucky, J. (2003) Economic and anthropological assessment of the health of children in Maya immigrant families in the US. *Econ. Hum. Biol.*, **1**, 145–160.
Steckel, R.H. (2004) New light on "Dark Ages": The remarkably tall stature of North European men during the Medieval Era. *Soc. Sci. Hist.*, **28**, 211–229.
Stini, W.A. (1976) Adaptive strategies of human populations under nutritional stress. In: *Biosocial Interrelations in Population Adaptation* (eds E.S. Watts, F.E. Johnston, and G.W. Lasker), Moutan, The Hague, pp. 19–40.
Stinson, S. (1985) Sex differences in environmental sensitivity during growth and development. *Yrbk Phys. Anthropol.*, **28**, 123–147.
Straus, L.G. (2013) After the deep freeze: confronting "Magdalenian" realities in Cantabrian Spain and beyond. *J. Archaeol. Method Theory*, **20**, 236–255.
Straus, L.G. (2016) Humans confront the Last Glacial Maximum in Western Europe: reflections on the Solutrean weaponry phenomenon in the broader contexts of technological change and cultural adaptation. *Quaternary International*, **425**, 62–68.
Tanner, J.M. (1964) *The physique of the Olympic athlete.* George Allen & Unwin, London.
Tanner, J.M. (1992) Growth as a measure of the nutritional and hygienic status of a population. *Horm. Res.*, **38** (Suppl. 1), 106–115.
Tanner, J.M., Hayashi, T., Preece, M.A., and Cameron, N. (1982) Increase in length of leg relative to trunk in Japanese children and adults from 1957 to 1977: comparison with British and with Japanese Americans. *Ann. Hum. Biol.*, **9**, 411–423.
Turchin, M.C., Chiang, C.W.K., Palmer, C.D., Sankararaman, S., and Reich, D., GIANT consortium, Hirschhorn, J.N. (2012) Evidence of widespread selection on standing variation in Europe at height-associated SNPs. *Nat. Genet.*, **44**, 1015–1019.
Udjus, L.G. (1964) *Anthropometric Changes in Norwegian Men in the Twentieth Century.* Universitetsforlaget.

Verpoorte, A. (2009) Limiting factors on early modern human dispersals: the human biogeography of late Pleniglacial Europe. *Quat. Int.*, **201**, 77–85.

Wadsworth, M.E.J., Hardy, R.J., Paul, A.A., Marshall, S.F., and Cole, T.J. (2002) Leg length and trunk length at 43 years in relation to childhood health, diet and family circumstances; evidence from the 1946 national birth cohort. *Int. J. Epidemiol.*, **31**, 383–390.

Walker, M., Johnsen, S., Rasmusses, S.O., Popp, T., Steffensen, J.-P., Gibbard, P., Hoek, W., Lowe, J., Andrews, J., Björck, S., Cwynar, L.C., Hughen, K., Kershaw, P., Kromer, B., Litt, T., Lowe, D.J., Nakagawa, T., Newnham, R., and Schwander, J. (2009) Formal definition and dating of the GSSP (Global Stratotype Section and Point) for the base of the Holocene using the Greenland NGRIP ice core, and selected auxiliary records. *J. Quat. Sci.*, **24**, 3–17.

Warren, M., Holliday, T.W., and Cole, T.M., III (2002) Ecogeographical patterning in the human fetus. *Am. J. Phys. Anthropol. Suppl.*, **34**, 161–162.

Weedon, M.N. and Frayling, T.M. (2008) Reaching new heights: insights into the genetics of human stature. *Trends Genet.*, **24**, 595–603.

5

Temporal and Geographic Variation in Robusticity

Brigitte Holt[1], Erin Whittey[1], Markku Niskanen[2], Vladimir Sládek[3], Margit Berner[4], and Christopher B. Ruff[5]

[1] Department of Anthropology, University of Massachusetts, Amherst, MA, USA
[2] Department of Archaeology, University of Oulu, Oulu, Finland
[3] Department of Anthropology and Human Genetics, Faculty of Science, Charles University, Prague, Czech Republic
[4] Department of Anthropology, Natural History Museum, Vienna, Austria
[5] Center for Functional Anatomy and Evolution, Johns Hopkins University School of Medicine, Baltimore, MD, USA

5.1 Introduction

Modern human skeletons are less robust than those of our Pleistocene ancestors, probably the result of a temporal decline in physical activity and muscle strength (Ruff *et al.*, 1993; Ruff, 2005; see Figure 5.1). Much of the earlier decline in overall lower limb bone strength relative to body size, prior to the Upper Paleolithic, may be attributable to changes in body shape (Puymerail *et al.*, 2012; Trinkaus and Ruff, 2012). However, more recent declines appear to be linked to changes in subsistence behavior, including in particular the adoption of a more sedentary lifestyle (Frayer, 1981; Jacobs, 1985; Holt, 2003). Over the past 30 years, numerous studies have illustrated changes in skeletal robusticity that have occurred in the context of subsistence or socioeconomic transitions within the Holocene (see Ruff and Larsen, 2014, for a summary; also see below). These studies, however, always cover a relatively limited arc of time. The paucity of well-preserved skeletal samples from most regions of the world prevents extensive chronological studies. Europe, however, represents a notable exception. In Europe, nearly 40,000 years of uninterrupted occupation and a long history of prehistoric research have yielded a uniquely rich skeletal record that spans every economic transition until the present. Our purpose in this chapter is to show how European limb robusticity changed in response to the major socioeconomic transitions of the period spanning approximately 40,000 years to the present.

5.2 Background

5.2.1 Limb Bone Robusticity and Subsistence Changes

Analyses of Late Pleistocene and early Holocene human skeletal remains show that long bone diaphyseal robusticity patterns track changes and variability in subsistence behavior (Ruff, 2000; Stock and Pfeiffer, 2001; Holt *et al.*, 2000; Holt, 2003; Sladek *et al.*, 2006a,b; Stock, 2006; Shackelford, 2007; Marchi, 2008). For instance, differences in postcranial robusticity of recent

Skeletal Variation and Adaptation in Europeans: Upper Paleolithic to the Twentieth Century, First Edition. Edited by Christopher B. Ruff.

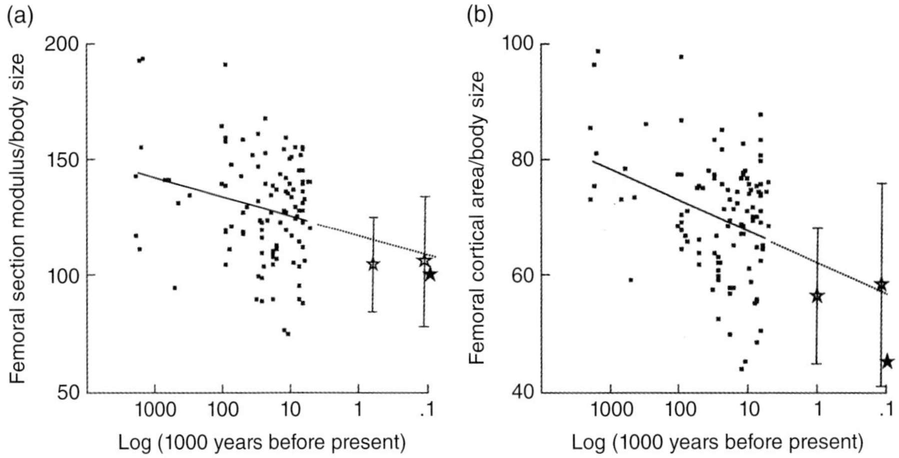

Figure 5.1 Temporal changes in femoral strength (a) and cortical area (b) standardized by body size in early Homo (squares) and means for three modern human samples (open stars: Pecos archeological and modern East African sample means ± 2 SD; filled stars: modern US white sample mean). Reprinted with permission from Ruff, 2005.

hunter-gatherer groups, compared to agriculturalists from the same region, reflect variation in levels of activity and lower limb loading (Larsen, 1995; Ruff *et al.*, 1984; Marchi *et al.*, 2006, 2011). Compared to agricultural groups, foragers tend to have thicker femoral diaphyseal cortices (Ruff *et al.*, 1993) and femoral shafts that have relatively wide antero-posterior (A-P) dimensions (Ruff *et al.*, 1984; Ruff, 1987). Similar results were also reported by Stock and Pfeiffer (2001), who found that predominantly terrestrial foragers had more robust femora and tibiae compared with the stronger humeri and clavicles of marine-adapted island foragers. Similarly, Weiss (2003) reported that native North American male 'ocean-rowers' exhibited significantly more robust humeri than 'river-rowers' and groups that did not paddle at all. These studies demonstrate clearly that diaphyseal cross-sectional rigidity and shape patterns reflect habitual mechanical loads and can, therefore, be used to infer subsistence behavior.

A number of studies have shown that the shift from hunting-gathering to food production is associated with declining upper and lower limb bone strength, probably the result of decreased mobility and workload in agriculturalists (Ruff and Larsen, 2001; Ruff and Hayes, 1983a,b). Other studies, however, found that agriculturalists exhibited decreased relative lower limb strength but more robust humeri than their foraging predecessors (Bridges, 1989; Bridges *et al.*, 2000). In this case, food production reduced mechanical loads on the legs but increased the workload involving upper limb bones. This lack of consistency between subsistence strategies and skeletal robusticity patterns suggests the decline in skeletal robusticity has not been linear, and that simple subsistence categories do not always reflect the types of mechanical loadings applied to bones. While the elevated robusticity of Paleolithic hunter-gatherers clearly reflects their strenuous lifestyles (Shaw and Stock, 2013), cultural factors such as subsistence technology, gender roles and status, as well as environmental characteristics such as terrain may modify observed differences in later groups.

5.2.2 Cultural and Economic Factors

Europe provides an unparalleled opportunity to examine long-term trends in limb bone robusticity in the context of major economic and social changes that have characterized the last 40,000 years (40 kya) in the same geographic region. While such a study cannot take into account individual-level behavior, and must of necessity paint relatively broad strokes, the rich European archeological record allows the reconstruction of important aspects of lifeways and economic and social behavioral characteristics for each time period. Many of these behaviors, such as mobility, craft specialization, food production intensification through mechanized means, and urbanization imply varying physical activity levels that will affect skeletal robusticity. Our goal in this chapter is to outline some of the major temporal changes in skeletal robusticity from the Paleolithic to modern times in light of the unprecedented cultural developments that shaped European populations. We begin by reviewing these events.

The Upper Paleolithic (UP), the period widely associated with the spread of modern humans and of modern patterns of behavior, emerged in Europe during the last Würm-Weichsel Interpleniglacial, around 40 kya, (Straus, 1995; van Andel *et al.*, 2003; Mellars, 2006). During that time, a gradual decline in temperature resulted in the buildup of large ice sheets over most of northern Europe and lower sea levels, a process that cumulated during the Last Glacial Maximum (LGM), around 20 kya, the time of lowest sea levels. Temperature began to rise after 16 kya, as ice sheets retreated. While technology began playing a role in human adaptive strategies around 2.5 million years ago, the proliferation of techno-complexes after 40,000 BP points to the increased prevalence of culture as an environmental buffer (Gamble *et al.*, 2004). Although the UP is generally associated with major technological innovations, none of these innovations figures systematically until after the LGM (Straus, 1995). The spearthrower and harpoon, for instance, do not appear until after the LGM (Schmitt *et al.*, 2003). The appearance of these technologies, and of distinct organizational systems following the LGM, form the basis for the traditional division of the Upper Paleolithic into two distinct phases – Early (EUP), from 40–20 kya, and Late (LUP) Upper Paleolithic, from 19 kya to the end of the Last Glacial, around 10 kya. Numerous indicators suggest that EUP groups were highly mobile, covering large territory for procurement of raw material and maintenance of intergroup relations (Gamble, 1986; Kozlowski, 1991; Svoboda *et al.*, 1996; Flébot-Augustin, 1997; Negrino and Starnini, 2003). Climatic deterioration and shorter growing seasons associated with the LGM, compounded by demographic stress caused by the influx of populations into ice age refugia (Mellars, 1985; Clark and Straus, 1986; Jochim, 1987; Mussi and Zampetti, 1988), led to decreased resource reliability (Straus, 1995). The LUP record suggests that human groups, faced by demographic and environmental pressures, were forced to intensify resource exploitation. The development and refinement of throwing technology, such as points that were probably used as projectiles thrown by the newly introduced spearthrower (atl-atl) indicate diversification of subsistence (Freeman, 1973; Straus, 1990, 1993). Diversification of the food base to include fish, shellfish, and birds suggests a need to cope with demographic stress by adding low-yield/high-cost foods (Straus *et al.*, 1981; Straus, 1986; Stiner, 2001; Stiner *et al.*, 1999). While uneven distribution of resources generally results in higher foraging mobility (Winterhalder, 1981; Keeley, 1988), reduced availability of exploitable territory forced groups to intensify the exploitation of existing resources (Straus and Clark, 1986; Stiner *et al.*, 1999). The presence of relatively sedentary, complex hunter-gatherer groups during the later part of the LUP (Mellars, 1985; Jochim, 1987) attests to further reduction in mobility. Previous work on UP skeletal robusticity showed increased upper limb robusticity (Churchill 1994; Churchill *et al.*, 2000) and increased lower limb diaphyseal circularity (Holt, 2003), possibly linked to resource intensification and reduction in mobility at the LGM.

The major climatic changes initiated around 16 kyr signaled the disappearance of large herds of reindeer, horse, and bison and the onset of the Mesolithic, the period between the end of the Pleistocene and introduction of agriculture, when the last European foragers lived (Milisauskas, 2002a). Systematic incorporation of shellfish allowed for the exploration of new and expanding estuary and riverine environments (Straus, 1995). Although mobility remained high, large residential groups and permanent settlements increased in frequency towards the Late Mesolithic

(Brumfiel, 1987; Price, 1987; Jochim, 2002). This, and the shift away from large game in favor of smaller game and aquatic resources, likely resulted in further decreases in skeletal robusticity.

The European Neolithic ranges from about 7000–4050 cal BP and, as elsewhere in the world, is defined by the appearance of agriculture, ceramics, and markedly increased sedentism (Milisauskas, 2002a,b). Although the use of throwing technologies declined, new implements such as stone-ground tools, hoes, digging sticks, and axes required substantial bimanual physical effort (Milisauskas, 2002a,b; Marchi *et al.*, 2006). In the Southeast United States, the shift from hunting to agricultural technologies in fact correlates with increased upper limb robusticity, probably reflecting a rise in the use of agricultural technology (Bridges, 1989). Starting around the middle part of the Neolithic, the introduction of draft animals and copper allowed for intensification of wheat and barley, and for the development of complex technologies such as the ox-drawn plow and wheeled vehicles (Milisauskas and Kruk, 2002a). While further declines in lower limb robusticity can be expected during the Neolithic, the combined effects of technological improvements on the one hand, and the shift to labor intensive food sources on the other hand, may produce patterns of upper limb change that are more difficult to interpret. Indeed, intensification of crops in Late Neolithic may arguably result in increased upper limb robusticity.

While the use of metal initiates during the later Neolithic, metallurgy became much more systematic in the Bronze Age (4350–2950 cal BP). While the majority of people continued a rural subsistence economy based on Neolithic technologies such as ox-drawn plows, increased metal mining and smithing imply the rise of trade specialization. More efficient metal farming technologies allowed further crop intensification and increased productivity, heralding major socioeconomic and power structure transformations by the middle Bronze Age. Rather than universally improving life conditions, increased food productivity led to increased population size and rising inequality in resource distribution and social stratification. Rich caches of weapons often associated with high-status burials point to increased warfare and the rise of an elite male warrior class (Whitehouse, 2001). Such socioeconomic changes may predict increased variability in skeletal robusticity patterns, especially with respect to the upper limb.

Food staple intensification, increased social complexity and stratification observed during the Bronze Age become much more marked during the Iron Age (2250–1650 cal BP). The adoption of iron resulted in more efficient farming tools, such as hoes, shovels, iron plowshares and axes, making agriculture less physically demanding and more productive and facilitating unprecedented demographic growth (Boserup, 1965, 1975; Peroni, 1979; Collis, 1984; Wells, 1990; Guidi and Piperno, 1992). The Iron Age also sees the appearance of proto-urban centers in the form of large settlements or city-states controlling surrounding farms and villages (Milisauskas, 2002a). The increase in social complexity and stratification that developed in the Iron Age led to craft specialization and unequal distribution of labor (Kristiansen, 1987, 2000; Kristiansen and Rowlands, 1998; Henrich and Boyd, 2008). This is expected to impact the variability in type and intensity of the activities performed by the Iron Age population (Sparacello *et al.*, 2011).

The fall of the Roman Empire in the 5th century signals the beginning of the Medieval period, divided here into early (500–999 AD) and late (1000–1599 AD) periods. Many of the large urban settlements that flourished in late Iron Age disintegrated, and major parts of Europe, such as England and Germany, become almost entirely rural in the early Medieval period (Duby, 1976; Dyer, 1995). The economy of early Medieval Europe remained primarily based on farming and herding. With the ever-growing demand for iron to improve farming technology, craft specialization increased further (Duby, 1976). While the use of oxen-drawn and, later, horse-drawn plows became more prevalent, these required significant manual labor to dig through superficial soil layers into the rich and heavy deeper clays. The eventual addition of wheeled front sections to plows and the spread of water and windmills (Duby, 1976; Campbell, 1995) further reduced the physical demands of farming. Paradoxically, these improvements also raised the cost of farming technology, and more sophisticated plows and large investments such as mills were beyond the means of many peasants, resulting in further social inequalities (Duby, 1976; Campbell, 1995). Even in relatively well-developed areas, mills were not uniformly available and manual milling remained the norm (Duby, 1976). Hence, a combination of high levels of physical labor and unequal access to significant technological improvements may result in little to no decline in skeletal robusticity.

The Early Modern period (1600–1899 AD) encompasses the major social and economic transformations engendered by the Industrial Revolution, a time of unprecedented mechanization and urbanization, and movement of many farming populations into factories. The introduction of mechanized and, eventually, electric devices had profound consequences in all sectors of life. Implements such as steam tractors and washing machines, for instance, reduced the amount of strenuous manual labor in areas such as farming and the household (Cowan, 1999; Musson and Robinson, 1969). The Industrial Revolution, however, also brought about very poor environmental and harsh child labor conditions in some areas of Europe (Humphries, 2010). Elevated levels of physical activity in pre- and post-adolescent periods of active bone formation could result in higher adult skeletal robusticity (Ruff *et al.*, 1994). Nevertheless, we expect that the introduction of mechanized means in all spheres of life should be reflected in significant declines in robusticity in Early Modern Europeans.

This brief survey of the behavioral, social, economic and technological transformations that have shaped the past 40,000 years of European prehistory and history allows the formulation of a number of hypotheses regarding the impacts of these changes on limb bone robusticity:

1) Overall skeletal robusticity and skeletal markers of mobility will decline markedly from Upper Paleolithic to Mesolithic, and again in the Neolithic with the shift from hunting-gathering to food production. Apart from some local geographic variations, there should be little further change in these parameters from Neolithic through Medieval periods.
2) Skeletal robusticity and markers of mobility will decline further in the Early Modern and Recent periods, as urbanization and sedentism increase and physical tasks are increasingly taken over by animal labor and mechanization.
3) Levels of upper limb robusticity should remain relatively high through the Bronze Age, and decrease significantly in Iron Age and again in Early Modern and Recent, reflecting increased use of mechanized means. We also expect that upper limb strength may increase somewhat between Early and Middle/Late Neolithic, reflecting higher physical labor tied to agriculture intensification.
4) The rise in craft specialization, social complexity and inequalities that appeared during the Bronze Age should be reflected in increased within-sample heterogeneity of skeletal robusticity in the Bronze Age, becoming more marked in the Iron Age (Harding, 2002; Milisauskas and Kruk, 2002b).
5) Differences in skeletal robusticity, in particular in the upper limb, between rural and urban samples should increase from the late Iron/Roman period onward as urban centers increase in size, with lower robusticity levels in urban groups (Wells, 2002).
6) Sexual dimorphism in skeletal markers of mobility will decline from the Upper Paleolithic to Neolithic, and then again from Medieval to modern times, reflecting changes in sexual division of labor and gender-specific economic tasks (Ruff, 1987). This hypothesis is addressed in Chapter 6.
7) Asymmetry in upper limb robusticity patterns should decline after the Upper Paleolithic, reflecting the shift from unimanual hunting technology to more bimanual implements. The use of unimanual weapon in warfare in Bronze Age and Iron Age periods, however, may be reflected in increased upper limb asymmetry. These hypotheses are addressed in Chapter 7.

5.2.3 Terrain

Previous studies have shown a correlation between physical terrain and femoral robusticity (Ruff, 1999). Among a set of Native American archeological samples, groups inhabiting mountainous regions exhibited higher relative femur strength than those from flat areas. Terrain, however, had no impact on humeral robusticity, clearly strengthening the causal link between locomotor loads and lower limb bone structure. Higher A-P bending loads produced by hamstring and quadriceps muscles during climbing are expected to result in an A-P-reinforced diaphysis, particularly around the knee region (Ruff, 1987). Given the wide array of European landscapes, from flat coastal and river plains to alpine mountains, it is reasonable to assume that variability in lower limb robusticity should reflect terrain here as well. Based on this assumption, we also tested the following additional hypothesis:

- Populations living in areas with substantially more elevated terrain will exhibit higher lower limb robusticity and markers of mobility.

5.3 Materials

The sample comprises 1834 (1048 males and 786 females) adult skeletons from Europe, divided into seven geographic regions and nine time periods (Table 5.1). The time periods include Upper Paleolithic, Mesolithic, Neolithic, Bronze Age, Iron Age/Roman, Early Medieval, Late Medieval, Early Modern, and Recent (Table 5.2). These divisions provide a somewhat arbitrary but useful chronological framework because they reflect major cultural and/or socioeconomic transitions. While the Upper Paleolithic is traditionally divided into early (EUP) and late (LUP) periods, EUP and LUP samples are combined here because the important changes that took place across the Upper Paleolithic have been published previously (Frayer, 1981, 1984; Jacobs, 1985; Formicola and Giannecchini, 1999; Churchill *et al.*, 2000; Holt, 2003; Holt and Formicola, 2008). In this study, we wish to emphasize the changes that took place in the context of post-Glacial adaptations and beyond. Details about data collection protocol and time periods are given in Chapter 1.

Table 5.1 Sample size by time period and region.

	Region						
Time period	**Britain**	**Scandinavia**	**North-Central Europe**	**France**	**Italy**	**Iberian Peninsula**	**Balkans**
Recent	0	67	26	0	0	52	0
Early Modern	40	50	0	32	22	0	0
Late Medieval	119	198	116	0	8	48	54
Early Medieval	0	0	174	15	34	63	0
Iron Age/Roman	118	63	49	18	17	0	0
Bronze Age	17	0	162	0	33	32	0
Neolithic	19	107	128	30	9	0	0
Mesolithic	0	26	2	12	7	5	15
Upper Paleolithic	2	1	12	13	17	0	0

Table 5.2 Dates for each time period.

Time period	Date range*
Recent	≥1900 AD
Early Modern	1600–1899 AD
Late Medieval	1000–1599 AD
Early Medieval	500–999 AD
Iron Age/Roman	2250–1650 BC
Bronze Age	4350–2950 BC
Neolithic	7000–4050 BC
Mesolithic	10,500–5880 BP
Upper Paleolithic	32,285–10,000 BP

* Dates indicated as BP are calibrated.

5.4 Methods

5.4.1 Aging and Sexing

Age and sex were determined from a number of standard skeletal indicators (Buikstra and Ubelaker, 1994; see Chapter 1).

5.4.2 Reconstruction of Cross-Sectional Dimensions

Long bone diaphyseal cross-sectional dimensions were obtained from either computed tomography (CT) scans or a combination of bi-planar radiographs and molds, following protocols established in previous studies (Trinkaus and Ruff, 1989; O'Neill and Ruff, 2004). Because access to CT technology was possible in only a few cases, the latter method was used for the majority of the sample (see Chapter 3 for details).

Cross-sectional geometric parameters were assessed at three locations: at femoral and tibial midshaft, and at 35% of humeral length from the distal end. These locations provide information about upper and lower limb mechanical loads that reflect some important components of subsistence behavior (Ruff and Hayes, 1983a; Ruff *et al.*, 1993, Trinkaus *et al.*, 1994). Length dimensions used to determine these positions are defined in Ruff (2002). Given low levels of bilateral asymmetry in lower limb diaphyseal dimensions, right or left femur and tibia were used, either from the best-preserved side or at random. Ample evidence shows, however, that asymmetry in upper limb bone cross-sectional properties can be marked, providing important functional information for reconstructing behavior (Trinkaus *et al.*, 1994; Churchill and Formicola, 1997; Rhodes and Knüsel, 2005; Ruff, 2008; Auerbach and Ruff, 2006). Therefore, both right and left humeri were included when available. To ensure consistent orientation of cross-sectional contours, bones were set up in standard positions relative to a set of anatomical axes (Ruff, 2002).

5.4.3 Robusticity Variables

Bone robusticity was evaluated by comparing average bone rigidity or strength, standardized for body size differences (see Chapter 3), and bone shape through comparisons of relative strength in different planes. The polar section modulus, Z_p, was used as a measure of overall bending strength. Section moduli around the M-L and A-P axes were calculated to estimate A-P (Z_x) and M-L (Z_y) bending strength, respectively. Ratios of these variables reflect diaphyseal shape (e.g., Z_x/Z_y). Cortical thickness or percentage cortical area was also examined through ratios of cortical area and total subperiosteal area (CA/TA).

5.4.4 Standardizing Cross-Sectional Dimension for Differences in Body Size

Meaningful comparisons of bone robusticity require accurate body mass reconstruction (Ruff *et al.*, 1993). Body mass was reconstructed using two approaches. When bi-iliac (maximum pelvic) breadth (BIB) could be measured, we used so-called 'cylindrical' equations that combine height (stature) and breadth (BIB) (Ruff, 2000; Auerbach and Ruff, 2004; Ruff *et al.*, 2005; see Chapter 2 for details on body size reconstruction). When this technique could not be used we estimated body mass from femoral head breadth (Ruff *et al.*, 2012; also see Chapter 2).

To control for body size differences prior to analysis, cross-sectional areas (CA, TA) are divided by body mass estimates, and section moduli (Z_p, Z_x, Z_y) by body mass × bone length of the corresponding skeletal element (Ruff, 2000; see also Chapter 3).

5.4.5 Quantification of Terrain

In order to evaluate the impact of topographic relief on lower limb bone robusticity patterns, elevation data were obtained from Digital Elevation Models (DEMs) from the Consortium for Spatial Information (CGIAR; Jarvis *et al.*, 2008) or, for latitudes above 60°N, the Global Land Survey Digital Elevation Model (USGS, 2008). This high-resolution digital elevation model dataset covers most of the globe at a resolution of three arc seconds, or about 90 m. ArcGIS software was used to find the maximum slope between each three arc second pixel and its neighbors. The average, maximum, and standard deviation of these slope values within a 10-km radius of each archeological site's geographic coordinates were calculated using ArcGIS's Focal Statistics tool. Binford (2001) estimates typical daily forager logistical mobility as having a radius of 6–9 km and an average round-trip distance of 15 km. A 10-km radius was used to capture all terrain likely to be covered by a single-day foraging trip. Agricultural and urbanized populations are expected to cover less ground by walking, so this radius should still capture most of the terrain typically traversed. A combination of maximum and average slope values was used to produce a measure of the 'hilliness' of the local area. Resulting elevation values were categorized into a three-code system representing three levels of hilliness. Regions with maximum slope values below 22 degrees were considered 'Flat,' those with maximum slopes between 22 and 44.9 degrees and an average slope below 8 degrees were considered 'Hilly,' and those with a maximum slope of 45 degrees and above or an average slope of 8 degrees or more were considered 'Mountainous.'

5.4.6 Categorization of Urbanization

In order to evaluate the impact of urbanization on robusticity patterns, sites were coded along a rural–urban dichotomy (see Chapter 1). No pre-Iron Age site fell into the 'Urban' category.

5.4.7 Analysis of Robusticity

Overall temporal trends were assessed through linear regressions of upper and lower limb robusticity variables and calibrated dates. As shown previously (Ruff *et al.*, 1993), temporal trends in markers of relative bone strength and bone shape are better fitted by log-linear than linear regressions, hence regressions were carried out against log (dates). In addition, in order to explore these trends both across the entire time period, as well as in the post hunter-gatherer period, regressions were performed both on the entire sample and excluding pre-Neolithic groups. Assessments of temporal changes were also carried out with two-way ANOVA with temporal period and sex as independent variables. If an interaction between sex and period was detected in the two-way ANOVA, differences within sex were examined with one-way ANOVA, with post-hoc pairwise Tukey comparisons between periods ($p < 0.05$). Pairwise differences were assessed through multiple comparisons (Tukey's HSD post-hoc test). A two-way ANOVA (urban status, sex) was also used to examine the impact of urbanization on upper and lower limbs, and of terrain and sex on the lower limb. To examine the impact on robusticity of increased craft specialization, and status and gender roles tied to rising social stratification in Bronze and Iron Ages, the coefficient of variation for upper limb robusticity measures was calculated. The coefficient of variation is expected to be higher in post-Neolithic periods, starting with the Bronze Age, but particularly from the Iron Age on. All analyses were performed using SPSS (Versions 21 and 22).

5.5 Results

5.5.1 Temporal Trends

5.5.1.1 Upper Limb

Summary statistics for humeral robusticity and shape ratios are given in Table 5.3. For the entire sample, there is a significant decrease in right ($p < 0.001$) and left ($p < 0.004$) upper limb robusticity from the oldest groups, dating to around 38 kya, to the most recent, dating to the mid-20th century (Fig. 5.2). Right and left humeral shape (Fig. 5.3) becomes more A-P-oriented over time ($p < 0.001$). These patterns of change remain the same when only post-Mesolithic groups are considered, with significant declines in right ($p < 0.010$) and left ($p < 0.002$) humeral robusticity, and decreased circularity on the right and left sides ($p < 0.001$).

Differences in robusticity and shape were also examined by period with two-way ANOVA (period, sex). Because a significant interaction between the independent variables was found, changes by period were further explored with one-way ANOVA and Tukey HSD post-hoc tests for multiple comparisons, within sex. Following an increase from Upper Paleolithic to Mesolithic, only significant for males and on the left side ($p < 0.01$), and a non-significant decrease between Mesolithic and Neolithic, right and left humeral strength declines sharply for males ($p < 0.001$ and $p < 0.0001$) between the Neolithic and Iron Age (Fig. 5.4). In females, significant declines occur in right and left humeral strength between Neolithic and Early Medieval periods ($p < 0.006$ and $p < 0.0001$). Both sexes then increase in the Medieval period, although this is significant in males only, followed by an insignificant decline in robusticity into the Recent period for both sexes. Percentage cortical area in both arms decreases steadily and significantly for males ($p < 0.001$ and $p < 0.0001$ for right and left sides, respectively) between Neolithic and Iron Age, and between Neolithic and Medieval periods for females ($p < 0.06$ and $p < 0.048$ for right and left sides, respectively) (Fig. 5.5), with a slight increase in the Medieval periods. For both right ($p < 0.009$) and left ($p < 0.0001$) humeri the shape ratio in males decreases significantly between Upper Paleolithic and Mesolithic, and increases steadily in subsequent periods (Fig. 5.6). This may reflect either a shift from a circular shape to a more A-P-elongated cross section, or a decrease in the ridges formed by flexor and extensor muscles (Ruff and Larsen, 2001). Females exhibit similar patterns of decreased circularity in the Mesolithic ($p < 0.0001$) followed by progressive increases in AP elongation in the left humerus, with little change on the right side.

Table 5.3 Upper limb robusticity data, by period and sex.

	Right side									Left side								
Males	Humeral robusticity (Z_p)			Humeral cortical thickness (%CA)			Humeral shape (Z_x/Z_y)			Humeral robusticity (Z_p)			Humeral cortical thickness (%CA)			Humeral shape (Z_x/Z_y)		
Time period	N	Mean	SD	N	Mean	SD	N	Mean	SD	N	Mean	SD	N	Mean	SD	N	Mean	SD
Recent	51	617.8	145.9	53	69.8	9.5	53	1.12	0.11	57	548.8	102.2	58	70.0	9.8	58	1.12	0.10
Early Modern	74	601.8	117.4	75	69.4	11.1	75	1.09	0.16	75	565.4	106.9	77	70.1	10.7	77	1.09	0.13
Late Medieval	147	622.6	127.2	157	73.4	9.2	157	1.09	0.10	147	576.7	112.5	156	73.6	8.9	156	1.12	0.10
Early Medieval	120	632.1	119.1	124	72.3	8.4	124	1.09	0.10	114	578.2	117.0	118	73.0	8.5	118	1.12	0.11
Iron Age/Roman Age	94	572.4	120.2	103	71.6	9.6	103	1.05	0.10	93	512.6	114.5	106	72.0	10.5	106	1.11	0.11
Bronze Age	82	602.1	107.8	90	76.8	7.0	90	1.09	0.08	79	532.3	93.6	87	74.7	8.3	87	1.09	0.09
Neolithic	115	647.4	125.8	126	77.6	8.9	126	1.04	0.11	112	582.2	112.2	119	76.9	8.3	119	1.06	0.10
Mesolithic	16	655.2	166.5	14	71.4	12.5	22	1.01	0.09	13	648.5	116.3	11	73.9	11.2	18	0.90	0.12
Upper Paleolithic	14	627.7	113.3	5	75.3	10.4	15	1.14	0.11	13	493.1	105.5	6	74.7	6.2	14	1.10	0.14

Females	Right side									Left side								
	Humeral robusticity (Z_p)			Humeral cortical thickness (%CA)			Humeral shape (Z_x/Z_y)			Humeral robusticity (Z_p)			Humeral cortical thickness (%CA)			Humeral shape (Z_x/Z_y)		
Time period	N	Mean	SD	N	Mean	SD	N	Mean	SD	N	Mean	SD	N	Mean	SD	N	Mean	SD
Recent	38	458.0	84.0	38	67.0	10.0	38	1.14	0.15	36	423.0	84.3	36	65.1	8.9	36	1.12	0.10
Early Modern	45	445.0	109.3	48	63.1	13.1	48	1.09	0.13	45	422.0	91.6	49	61.8	12.7	49	1.16	0.15
Late Medieval	128	496.2	110.3	137	68.9	10.4	137	1.13	0.10	123	471.0	94.5	131	67.9	10.8	131	1.13	0.10
Early Medieval	88	465.3	79.1	91	65.4	9.8	91	1.10	0.10	82	453.1	84.3	84	66.4	9.2	84	1.17	0.13
Iron Age/ Roman Age	97	479.7	104.7	107	68.1	9.3	107	1.08	0.13	87	455.5	89.3	97	69.1	9.6	97	1.12	0.13
Bronze Age	64	495.0	86.5	71	68.7	10.1	71	1.07	0.11	64	492.7	91.2	68	68.1	11.4	68	1.04	0.08
Neolithic	76	524.3	100.1	79	71.3	9.0	79	1.10	0.12	76	528.5	112.8	81	71.5	9.6	81	1.05	0.11
Mesolithic	6	619.3	213.8	7	70.3	15.0	8	1.11	0.13	10	496.2	161.5	10	70.2	12.8	12	0.98	0.12
Upper Paleolithic	8	550.1	109.6	3	82.3	6.0	8	1.19	0.13	8	486.5	62.5	3	79.8	5.3	9	1.23	0.10

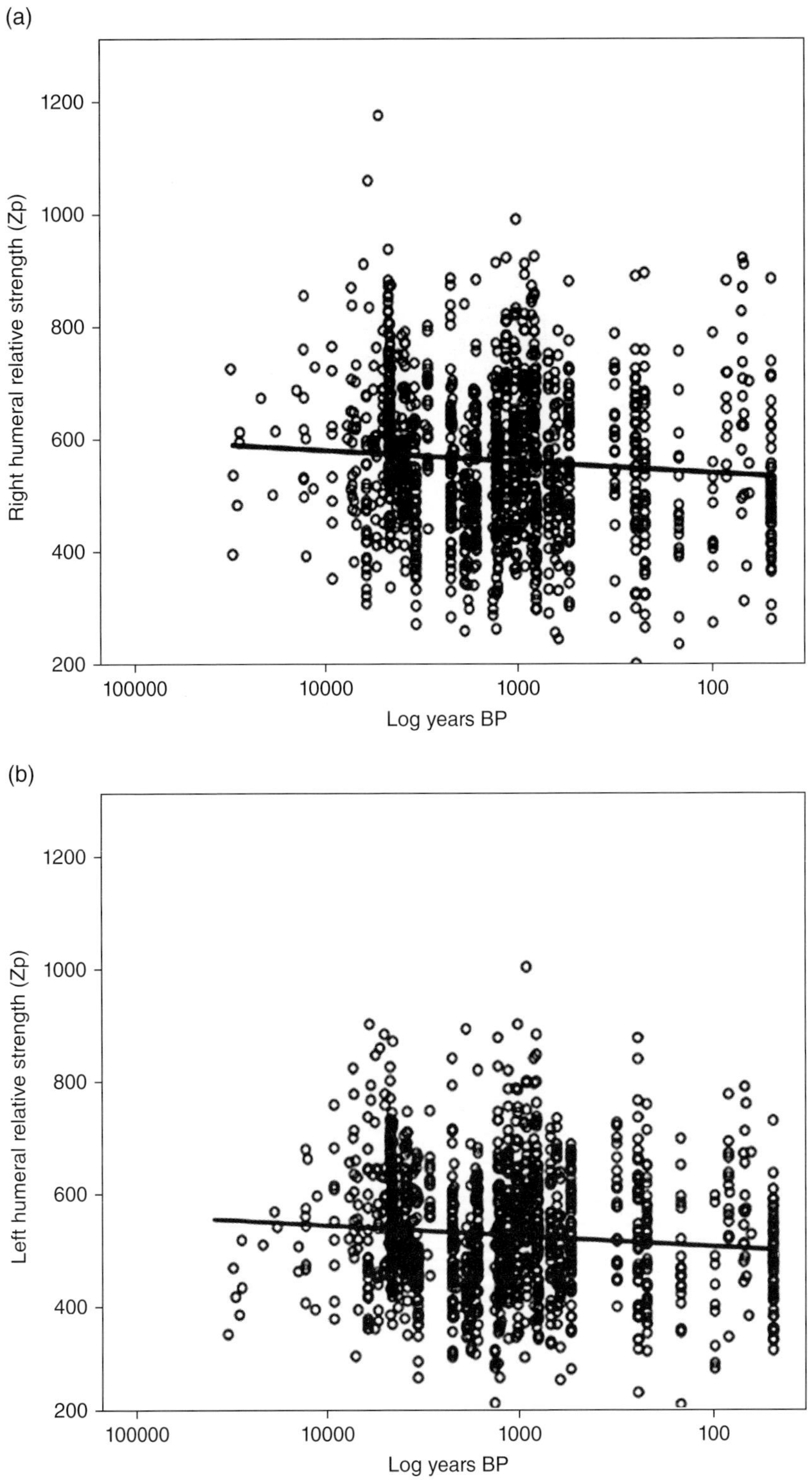

Figure 5.2 Temporal changes in upper limb average bending/torsional strength (Z_p) standardized for body size for right (a) and left (b) humerus, with least square regression lines (Right: $y = 494.300 + 9.362x$, $r = 0.090$, $N = 1263$; Left: $y = 469.553 + 7.341x$, $r = 0.082$, $N = 1233$).

(a)

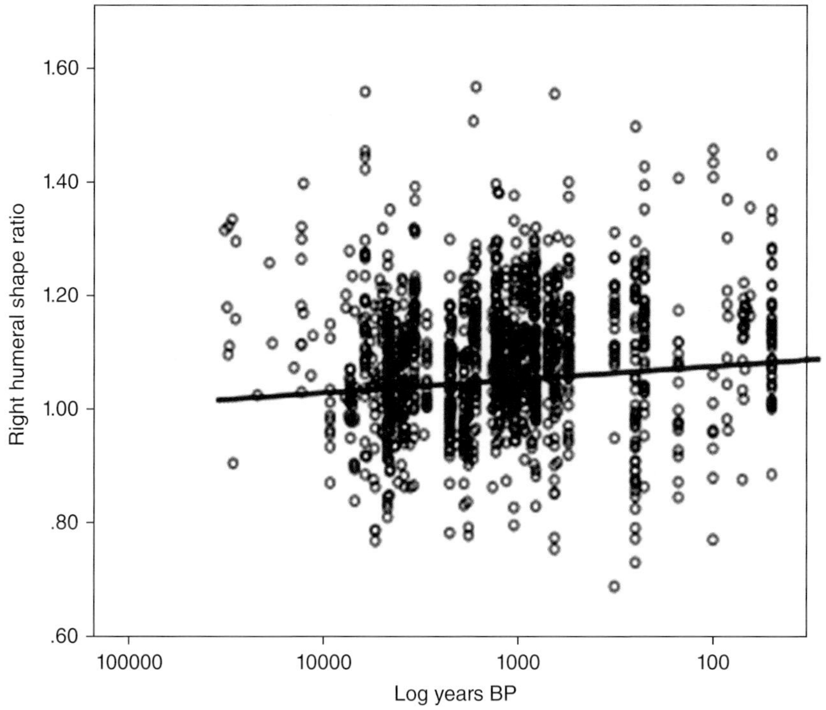

(b)

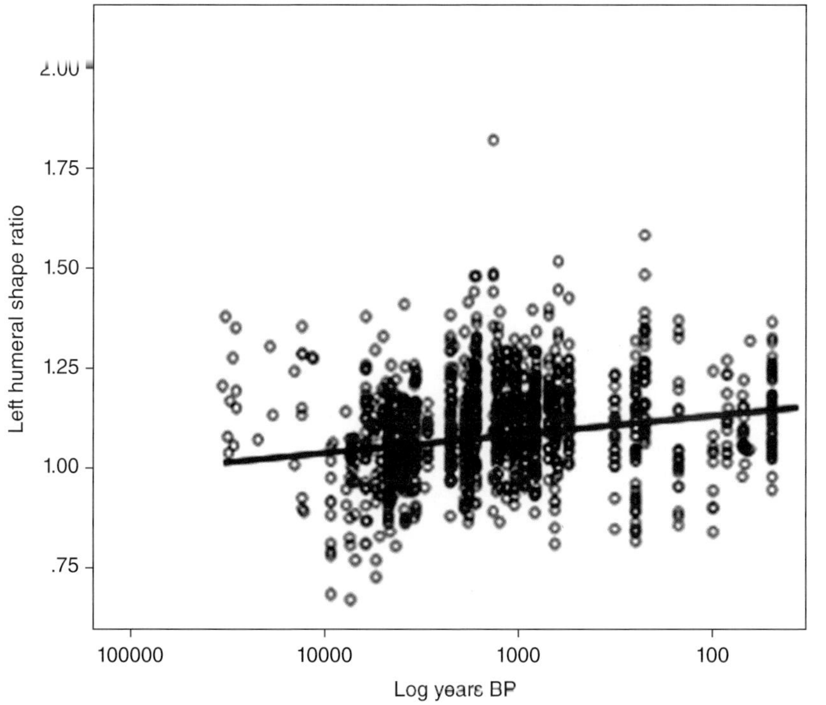

Figure 5.3 Temporal change in A-P to M-L bending strength ratio (Z_x/Z_y) for right (a) and left (b) humerus with least square regression lines (Right: $y = 1.162 - 0.010x$, $r = 0.113$, $N = 1352$; Left: $y = 1.223 - 0.017x$, $r = 0.182$, $N = 1320$).

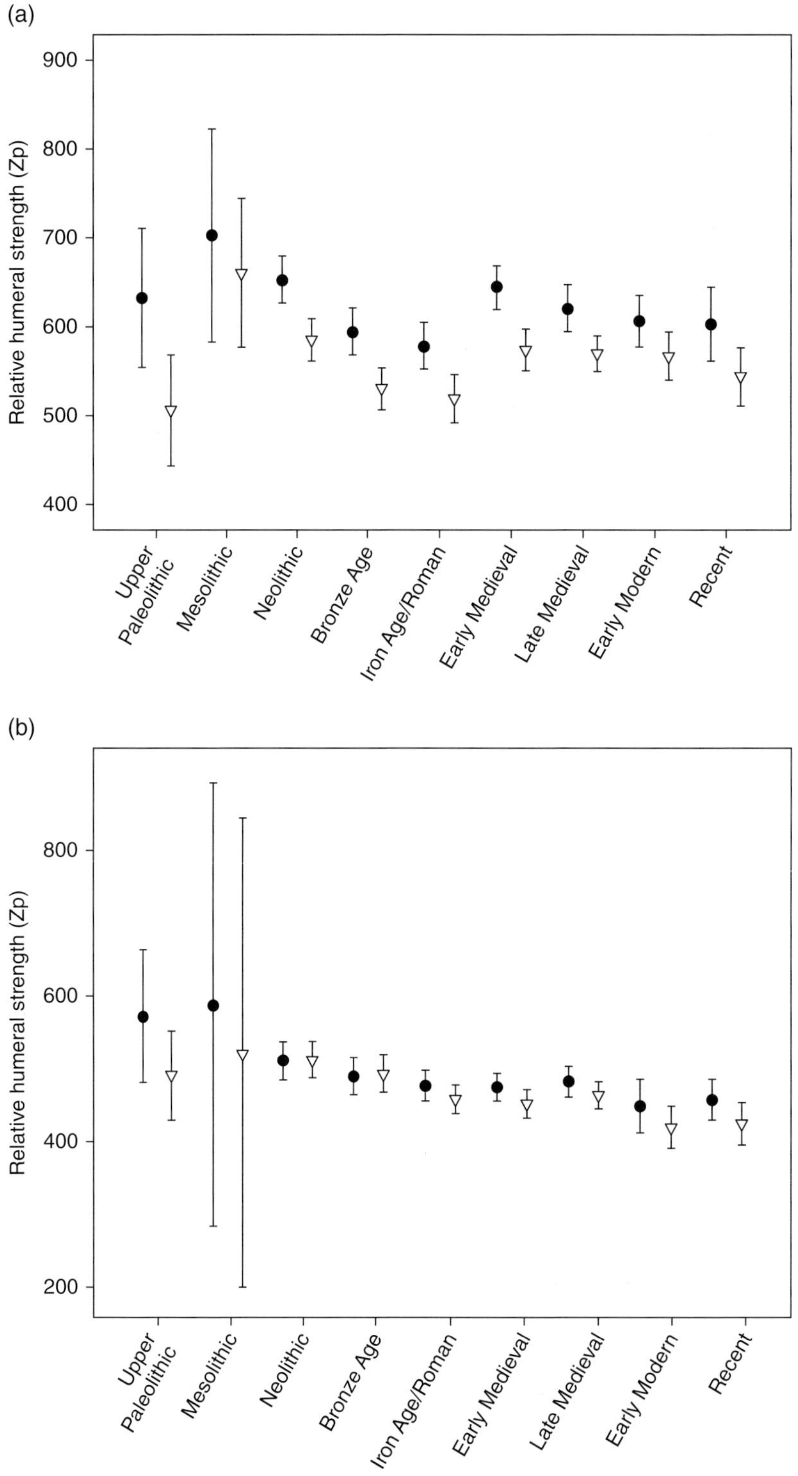

Figure 5.4 Right and left humeral average bending/torsional strength (Z_p) standardized for body size, by period for males (a) and females (b). Circles: right humerus; triangles: left humerus. Central values and bars represent mean and 95% confidence interval, respectively.

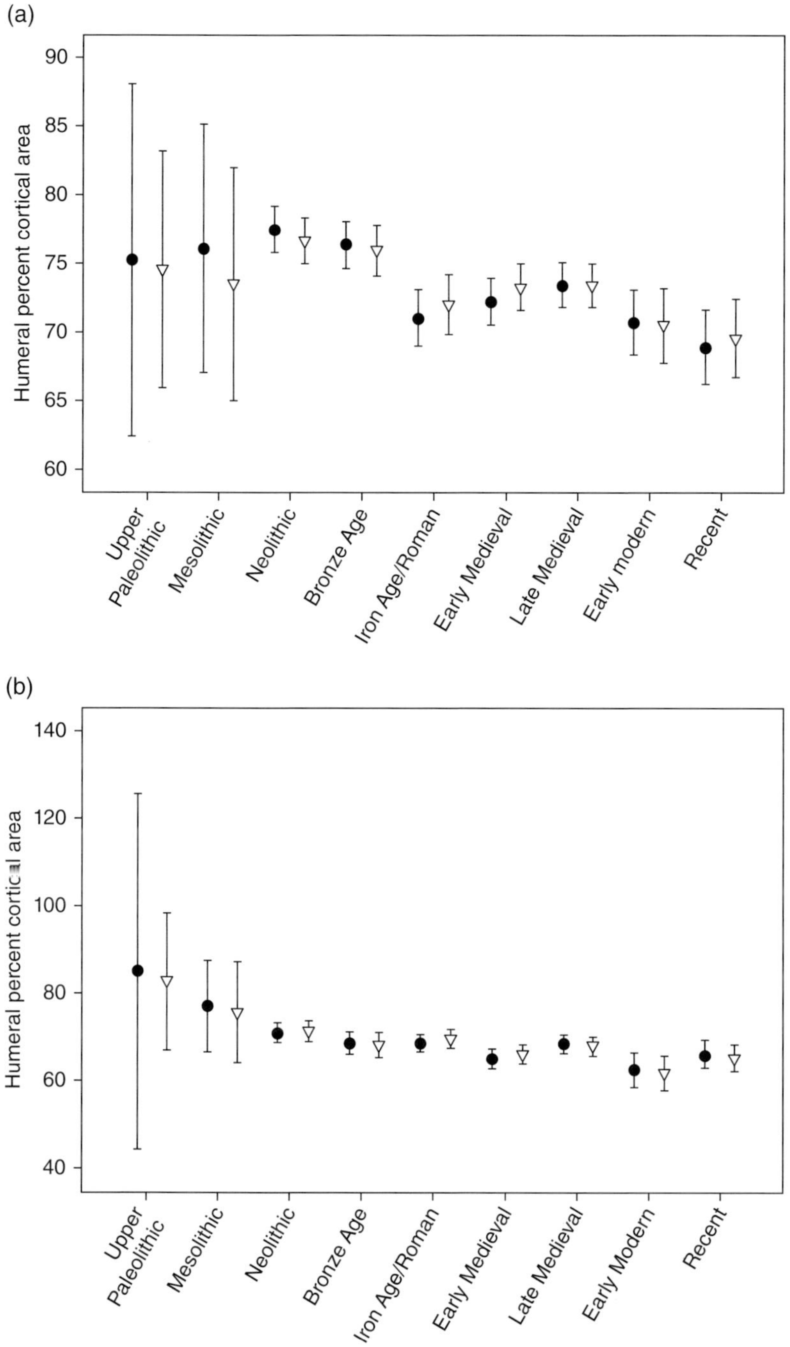

Figure 5.5 Right and left humeral percentage cortical thickness by period for males (a) and females (b). Circles: right humerus; triangles: left humerus. Central values and bars represent mean and 95% confidence interval, respectively.

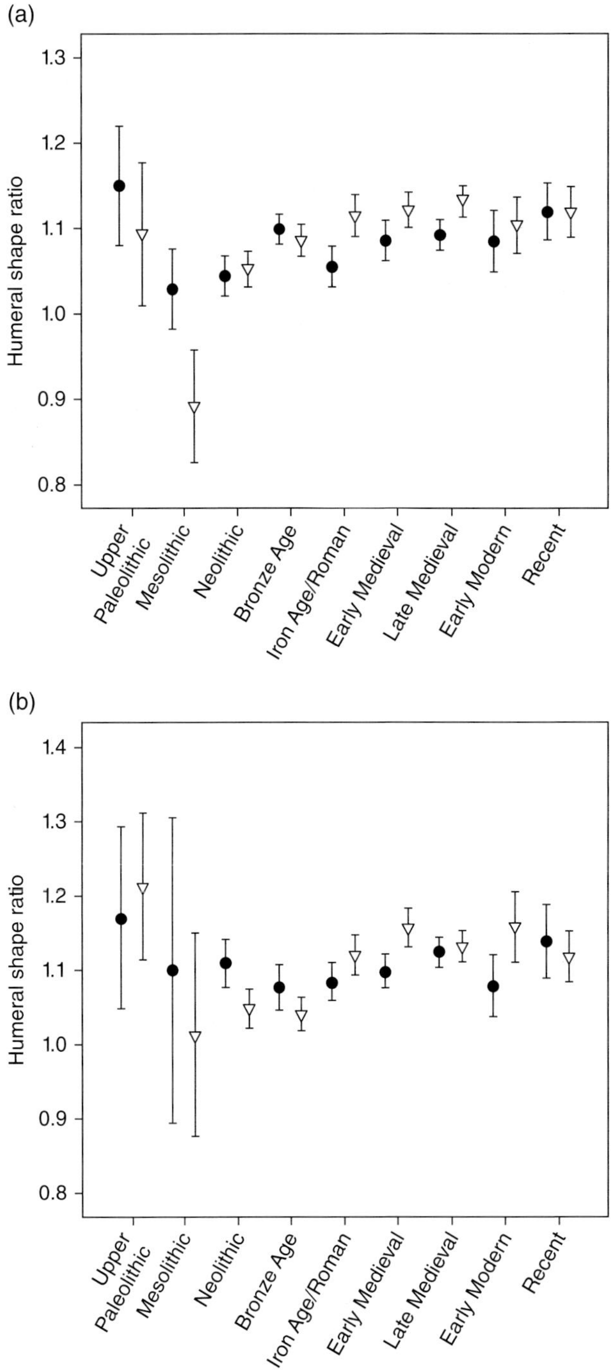

Figure 5.6 Right and left humeral A-P to M-L bending strength ratio (Z_x/Z_y) by period for males (a) and females (b). Circles: right humerus; triangles: left humerus. Central values and bars represent mean and 95% confidence interval, respectively.

Table 5.4 Upper limb robusticity data for Early and Middle/Late Neolithic period.

	Humeral robusticity (Z_p)			Humeral shape (Z_x/Z_y)			Humeral robusticity (Z_p)			Humeral shape (Z_x/Z_y)		
	Right side						Left side					
Period	N	Mean	SD	N	Mean	SD	N	Mean	SD	N	Mean	SD
Early Neolithic	65	559.4	163.3	75	1.08	0.15	63	536.6	138.8	71	1.04	0.13
Middle/Late Neolithic	141	621.6	115.4	146	1.05	0.09	138	572.0	105.9	142	1.05	0.09

To assess the impact of agricultural intensification on upper limb strength within the Neolithic period, we compared humeral strength values for Early Neolithic and Middle/Late Neolithic (in our sample approximately 5051–7300 BP and 4050–5050 BP, respectively). T-tests for pooled sexes show significant increases in right ($p < 0.002$) and left ($p < 0.048$) humeral strength in the later part of the Neolithic (Table 5.4; Fig. 5.7). The least squares regression line (Fig. 5.7) is significant for right humeral strength ($p < 0.001$), but not for the left side. Humeral shape does not change significantly.

5.5.1.2 Lower Limb

Summary statistics for femoral and tibial robusticity and shape variables are given in Table 5.5. Femoral and tibial robusticity (Fig. 5.8) declines significantly across the entire time range of the sample ($p < 0.027$ and $p < 0.001$, respectively), as does diaphyseal shape ($p < 0.001$, both bones) (Fig. 5.9). When pre-Neolithic samples are excluded, these patterns remain the same for tibial robusticity and shape ($p < 0.001$). The changes in femoral robusticity and shape are not significant.

Both males and females follow similar patterns of temporal change in femoral and tibial robusticity, with no significant change between Upper Paleolithic and Mesolithic, but significant declines in femoral bending strength between Mesolithic and Bronze Age (males: $p < 0.0001$, females: $p < 0.03$) (Fig. 5.10). This is followed by significant increases between Bronze Age and Late Medieval ($p < 0.01$, both males and females), and a subsequent slight decline. Declines in tibial bending strength are significant in males between Mesolithic and Bronze Age ($p < 0.009$) and in females between Mesolithic and Early Medieval ($p < 0.01$) (Fig. 5.10). Male tibial strength shows significant declines between Mesolithic and Bronze Age and again between Bronze Age and Early Medieval ($p < 0.0001$ and $p < 0.009$, respectively). Female tibial strength also declines significantly between Mesolithic and Early Medieval ($p < 0.01$).

As shown previously (Holt, 1999, 2003), Upper Paleolithic femora and tibiae exhibit expanded A-P bending rigidity. There is a marked decline in femur bending shape ratios between the Upper Paleolithic and Neolithic ($p < 0.0001$ for both sexes), and again between Neolithic and Iron Age (males: $p < 0.0001$, females: $p < 0.01$) (Fig. 5.11), with little subsequent change for males and a slight but significant increase between Iron Age and Late Medieval for females ($p < 0.02$). Tibiae also become significantly more circular between Upper Paleolithic and Iron Age for males ($p < 0.007$) and between Neolithic and Iron Age for females ($p < 0.009$). No subsequent change is observed in males, but female tibiae continue to increase in circularity, significantly so between Iron Age and Early Modern periods ($p < 0.001$).

As with the upper limb, there are substantial declines in lower limb percentage cortical area throughout the entire period (Fig. 5.12). Upper Paleolithic males and females exhibit much higher femoral and tibial percentage cortical area than their Recent counterparts ($p < 0.0001$ for

(a)

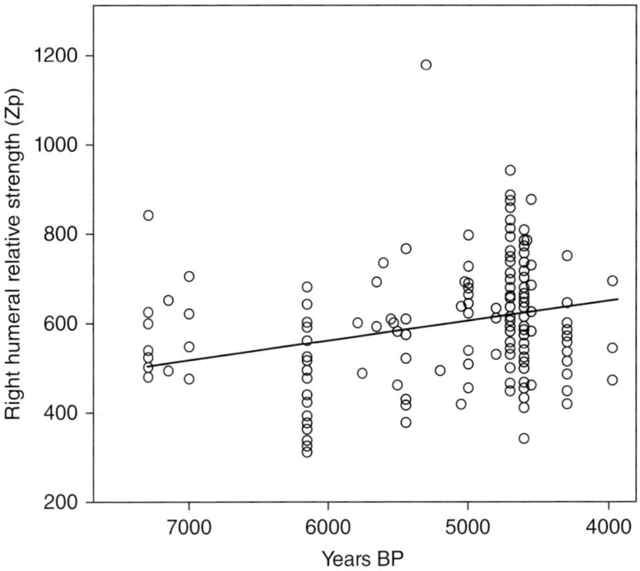

(b)

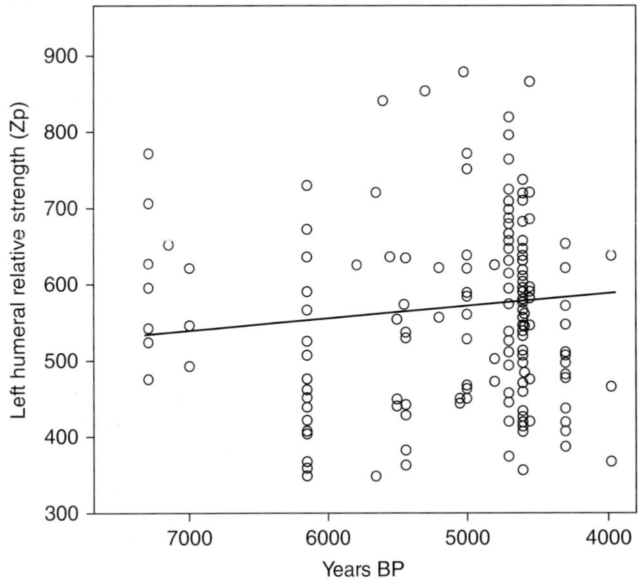

Figure 5.7 Temporal change in right (a) and left (b) upper limb average bending/torsional strength (Z_p) standardized for body size, for the Neolithic period with least square regression lines (Right: $y = 796.793 - 0.039x$, $r = 0.241$, $N = 190$; left: $y = 630.542 - 0.014x$, $r = 0.094$, $N = 187$).

both sexes). Males show the steepest decrease between Upper Paleolithic and Iron Age (femur: $p < 0.004$, tibia: $p < 0.0001$), with a further significant decline for the tibia between Iron Age and Early Modern periods ($p < 0.001$) and only slight declines in the femur. Male tibial percentage cortical area increases slightly but significantly between Early Modern and Recent ($p < 0.04$). Female femoral and tibial percentage cortical area decreases steeply between Upper Paleolithic

Table 5.5 Lower limb robusticity data, by period and sex.

Males	Femoral robusticity (Z_p)			Tibial robusticity (Z_p)			Femoral percentage cortical area (%CA)			Tibial percentage cortical area (%CA)			Femoral shape (Z_x/Z_y)			Tibial shape (Z_x/Z_y)		
Time period	**N**	**Mean**	**SD**	**N**	**Mean**	**SD**	**N**	**Mean**	**SD**	**N**	**Mean**	**SD**	**N**	**Mean**	**SD**	**N**	**Mean**	**SD**
Recent	94	1119.0	159.7	76	909.1	157.[illegible]	96	73.1	6.5	78	72.4	7.8	96	0.93	0.12	78	1.31	0.19
Early Modern	85	1089.8	168.6	78	895.2	208.7	86	72.3	6.0	80	68.8	8.6	86	0.91	0.12	80	1.30	0.27
Late Medieval	204	1152.6	167.9	182	940.2	173.0	209	73.4	6.5	189	73.3	7.2	209	0.93	0.10	189	1.37	0.18
Early Medieval	155	1137.1	168.8	144	851.6	162.8	155	74.7	5.7	147	73.8	6.3	155	0.94	0.11	147	1.33	0.18
Iron Age/ Roman Age	139	1059.4	207.8	124	879.3	197.6	145	74.6	6.4	134	72.9	7.4	145	0.90	0.12	134	1.38	0.18
Bronze Age	114	1079.2	181.5	100	962.0	187.9	120	76.6	5.3	108	76.4	6.4	120	0.96	0.10	108	1.46	0.19
Neolithic	156	1172.2	187.8	149	1007.5	199.8	161	77.4	5.9	153	76.2	6.6	161	0.96	0.12	153	1.45	0.21
Mesolithic	36	1245.2	181.2	27	1107.3	210.1	50	77.8	5.9	41	82.7	4.7	50	1.03	0.12	41	1.52	0.28
Upper Paleolithic	23	1208.0	183.2	19	1070.9	132.9	25	80.0	6.5	19	83.6	6.4	25	1.22	0.21	19	1.57	0.26

(Continued)

Table 5.5 (Continued)

Females	Femoral robusticity (Z_p)			Tibia robusticity (Z_p)			Femoral percentage cortical area (%CA)			Tibial percentage cortical area (%CA)			Femoral shape (Z_x/Z_y)			Tibial shape (Z_x/Z_y)		
Time period	**N**	**Mean**	**SD**	**N**	**Mean**	**SD**	**N**	**Mean**	**SD**	**N**	**Mean**	**SD**	**N**	**Mean**	**SD**	**N**	**Mean**	**SD**
Recent	57	1009.3	156.2	51	805.4	127.9	57	71.4	7.7	52	68.4	9.0	57	0.92	0.12	52	1.25	0.15
Early Modern	50	1004.5	175.2	53	750.1	170.6	52	69.4	9.7	57	65.7	12.0	52	0.87	0.14	57	1.22	0.19
Late Medieval	177	1053.2	171.0	150	788.6	135.4	180	71.3	7.8	156	69.1	9.0	180	0.90	0.10	156	1.34	0.16
Early Medieval	119	1013.2	154.6	111	709.7	118.3	123	73.2	6.5	115	69.3	7.5	123	0.90	0.09	115	1.34	0.16
Iron Age/ Roman Age	131	992.2	173.1	119	741.2	136.6	135	72.5	7.9	122	68.8	7.7	135	0.86	0.11	122	1.33	0.17
Bronze Age	83	972.0	187.5	75	795.4	141.2	87	74.4	7.0	78	71.9	7.1	87	0.93	0.10	78	1.39	0.20
Neolithic	108	1065.4	186.5	94	863.7	162.1	110	74.8	7.8	97	72.5	7.8	110	0.91	0.10	97	1.43	0.20
Mesolithic	14	1131.8	178.0	8	895.6	123.4	23	80.2	6.9	15	79.6	6.1	23	1.02	0.15	15	1.38	0.26
Upper Paleolithic	12	1153.9	211.3	8	1002.6	249.8	14	81.8	6.8	9	83.8	9.7	13	1.09	0.11	9	1.37	0.20

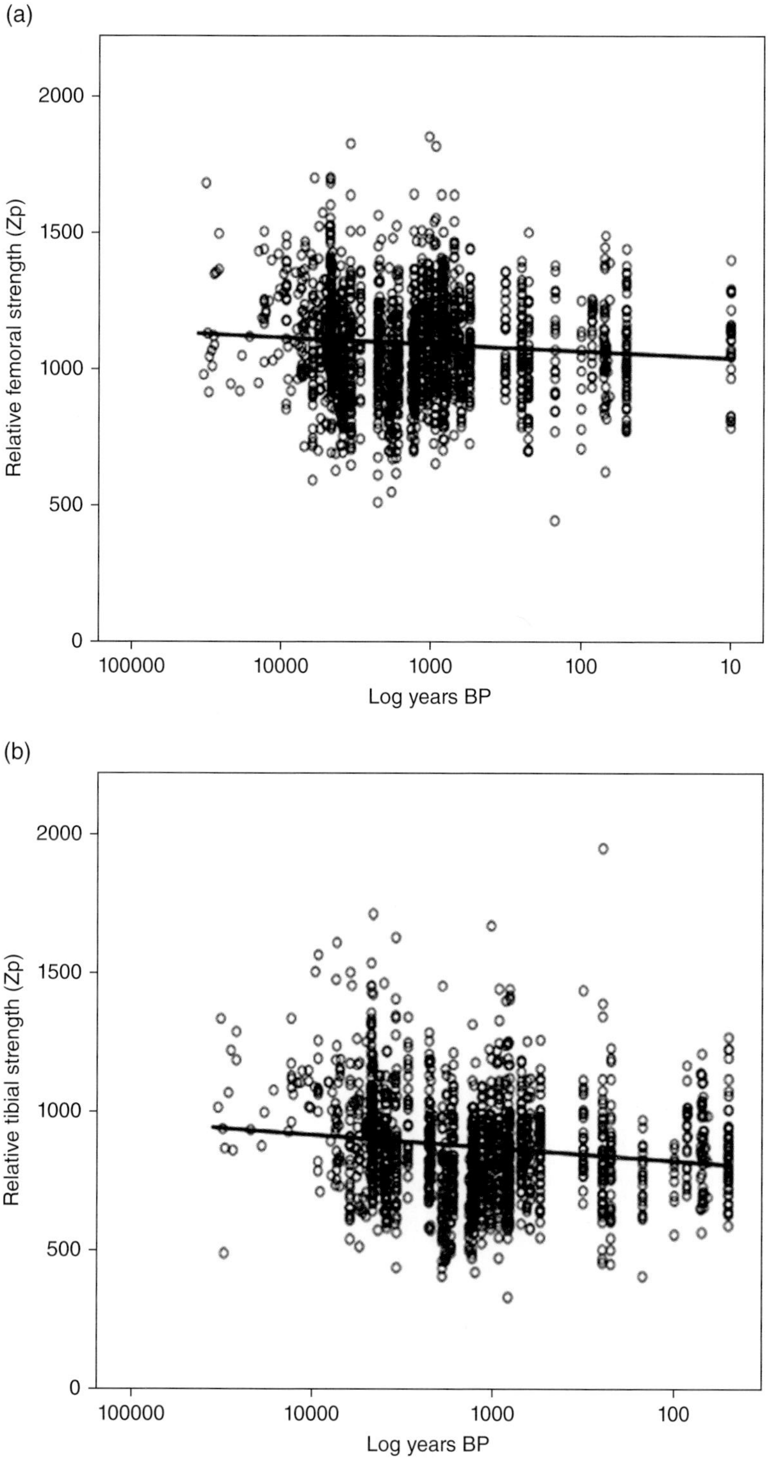

Figure 5.8 Temporal change in femoral (a) and tibial (b) average bending/torsional strength (Z_p) standardized for body size, with least square regression lines (Femur: $y = 1032.321 + 7.290x$, $r = 0.054$, $N = 1757$; Tibia: $y = 707.909 + 22.425x$, $r = 0. 151$, $N = 1567$).

(a)

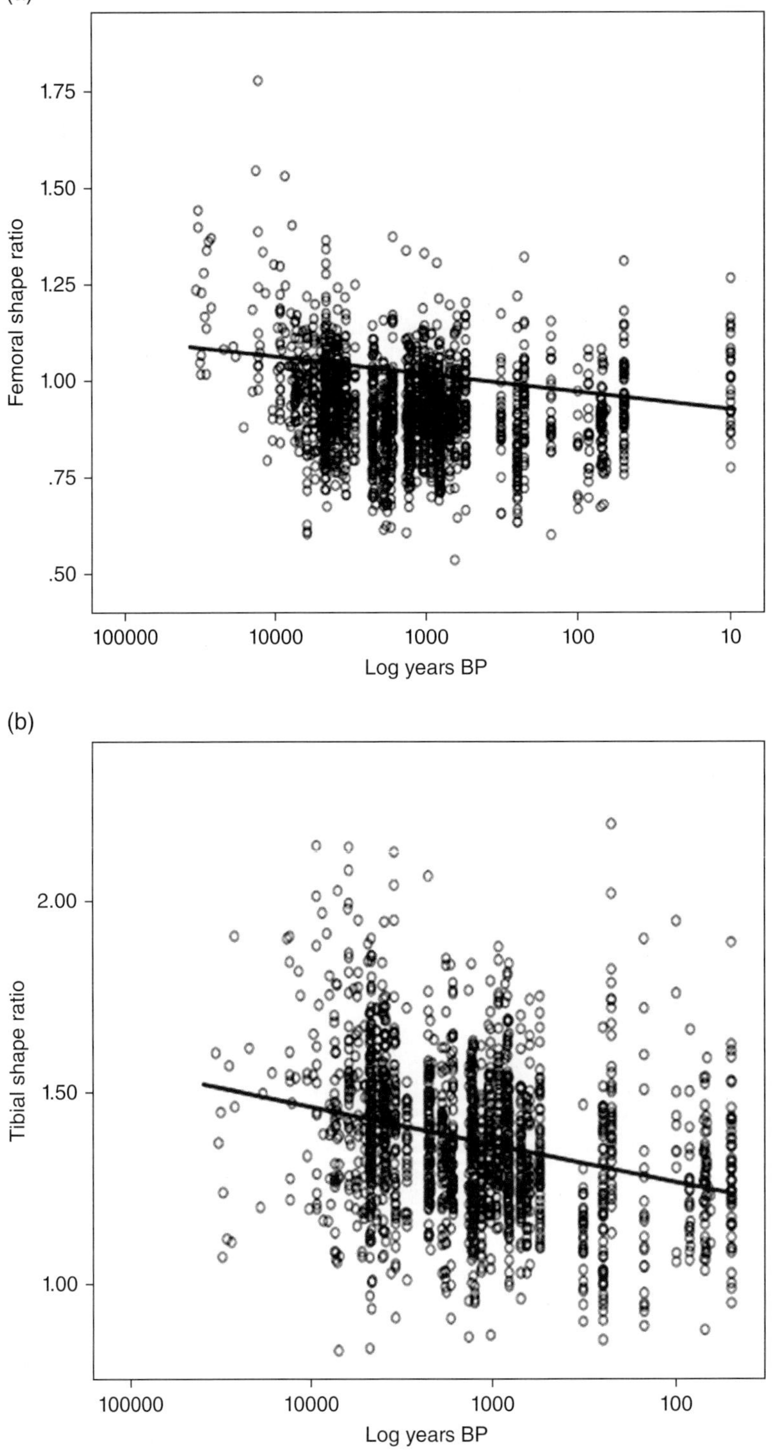

Figure 5.9 Temporal change in femoral (a) and tibial (b) A-P to M-L bending strength ratio (Z_x/Z_y), with least square regression lines (Femur: y = 0.828 + 0.14x, r = 0.159, N = 1826; Tibia: y = 1.065 + 0.042x, r = 0.272, N = 1649).

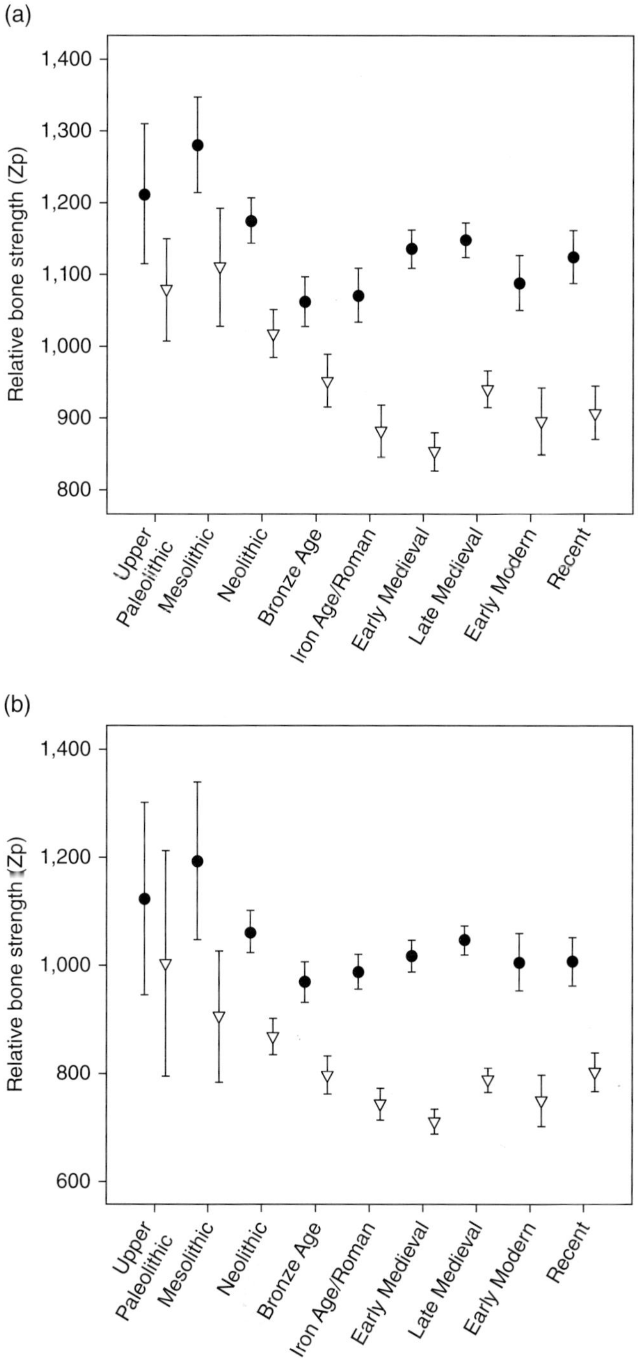

Figure 5.10 Error bar plots showing femoral and tibial average bending/torsional strength standardized for body size, by period, for males (a) and females (b). Circles: femur; triangles: tibia. Central values and bars represent mean and 95% confidence interval, respectively.

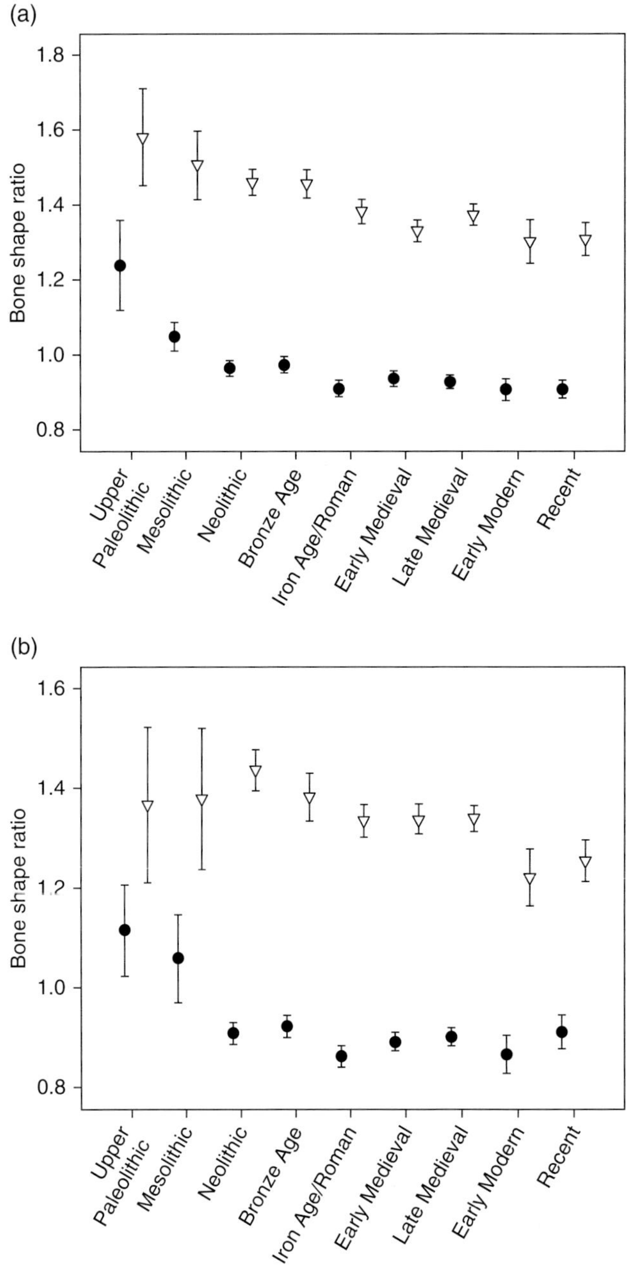

Figure 5.11 Error bar plots showing femoral and tibial A-P to M-L bending strength ratio (Z_x/Z_y) by period, for males (a) and females (b). Circles: femur; triangles: tibia. Central values and bars represent mean and 95% confidence interval, respectively.

and Bronze Age (femur: $p < 0.02$, tibia: $p < 0.002$) and, like males, continues to decline until Early Modern (femur: $p < 0.006$, tibia: $p < 0.001$) and increase between Early Modern and Recent, although the latter change is not significant. In summary, the overall pattern for lower limb robusticity is a net and sharp temporal decline, punctuated by a significant increase in robusticity in Medieval times, and a clear increase in circularity until the Iron Age with little subsequent change.

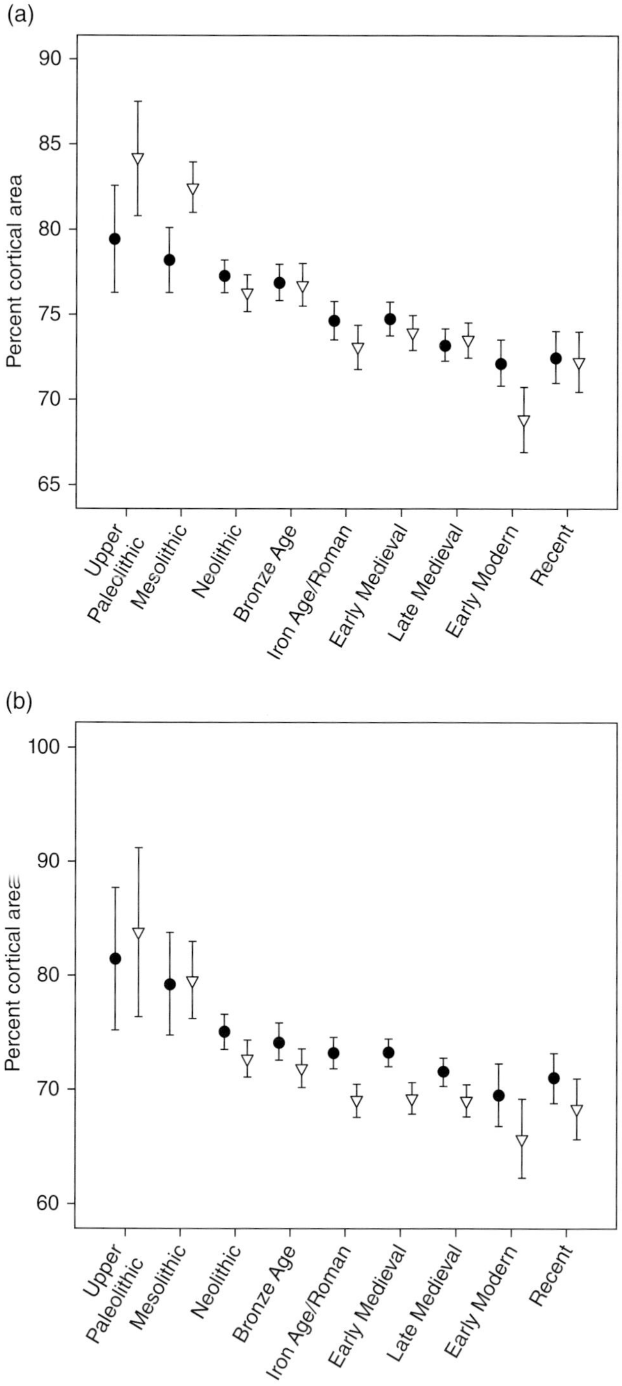

Figure 5.12 Error bar plots showing femoral and tibial percentage cortical area by period, for males (a) and females (b). Circles: femur; triangles: tibia. Central values and bars represent mean and 95% confidence interval, respectively.

5.5.2 Urbanization

Descriptive statistics for upper and lower limb robusticity and shape ratios for rural and urban groups are given in Tables 5.6 and 5.7, respectively. Given that the earliest urban group included in the study dates to the Iron Age, the sample included in this analysis spans the Iron Age to Recent periods. Because an interaction was found between sex and urban status, one-way ANOVA was used to examine differences within sex. There are few significant differences in upper limb properties, apart from significantly higher percentage cortical areas in rural males (Table 5.6; Fig. 5.13), and more circular right diaphyseal cross sections in rural females (Table 5.6).

Few differences in lower limb robusticity are observed, other than significantly higher tibial robusticity in urban females (Table 5.7). Rural males exhibit significantly higher percentage cortical area than their urban counterparts (Fig. 5.14). Rural females also have higher femoral percentage cortical areas (Fig. 5.14), and slightly lower femoral shape ratios than urban females.

5.5.3 Social Complexity

Increased social complexity, craft specialization and unequal distribution of labor in the Iron Age are expected to influence the variability in type and intensity of activities performed by Iron Age populations. The coefficient of variation describes the degree of spread around the mean of a variable, and values above 10 generally characterize heterogeneous populations (Simpson *et al.*, 1960). Coefficients of variation for upper and lower limb robusticity, shape ratios, and cortical thickness for Neolithic, Bronze Age, and Iron Age range from 6.92 to 22.1 (Tables 5.8 and 5.9), indicating relatively high levels of heterogeneity for all periods. Males exhibit a clear pattern of increased variability in the Iron Age for both upper

Table 5.6 Upper limb robusticity data for urban and rural groups by sex (Iron Age to Modern periods).

Property	Rural			Urban			
Males	**N**	**Mean**	**SD**	**N**	**Mean**	**SD**	**Difference***
Right humeral robusticity (Z_p)	236	604.8	119.8	250	617.9	131.2	NS
Left humeral robusticity (Z_p)	219	550.6	120.7	267	567.2	108.3	NS
Right humeral cortical thickness (%CA)	246	72.7	9.7	266	70.9	9.2	<0.030
Left humeral cortical thickness (%CA)	235	73.2	9.8	280	71.4	9.4	<0.039
Right humeral shape (Z_x/Z_y)	246	1.08	0.10	266	1.09	0.12	NS
Right humeral shape (Z_x/Z_y)	235	1.11	0.11	280	1.12	0.11	NS
Females							
Right humeral robusticity (Z_p)	193	471.7	98.4	203	479.7	103.9	NS
Left humeral robusticity (Z_p)	180	449.4	94.1	193	456.2	88.2	NS
Right humeral cortical thickness (%CA)	201	66.9	10.5	220	67.3	10.4	NS
Left humeral cortical thickness (%CA)	187	66.7	10.9	210	67.0	10.1	NS
Right humeral shape (Z_x/Z_y)	201	1.09	0.12	220	1.12	0.11	<0.005
Right humeral shape (Z_x/Z_y)	187	1.13	0.13	210	1.14	0.12	NS

* Results based on one-way Anova.

Table 5.7 Lower limb robusticity data for urban and rural groups by sex (Iron Age to Modern periods).

Property	Rural			Urban			
Males	**N**	**Mean**	**SD**	**N**	**Mean**	**SD**	**Difference***
Femoral robusticity (Z_p)	308	1105.3	187.8	369	1127.4	171.0	NS
Tibial robusticity (Z_p)	277	881.9	192.7	327	909.5	171.3	NS
Femoral cortical thickness (%CA)	314	74.6	6.4	377	73.1	6.1	<0.003
Tibial cortical thickness (%CA)	285	73.6	7.2	343	71.9	7.6	<0.003
Femoral shape (Z_x/Z_y)	314	0.92	0.12	377	0.92	0.11	NS
Tibial shape (Z_x/Z_y)	285	1.34	0.19	343	1.35	0.20	NS
Females							
Femoral robusticity (Z_p)	259	1012.6	160.3	275	1027.1	175.0	NS
Tibial robusticity (Z_p)	237	740.6	144.0	247	771.5	143.1	<0.018
Femoral cortical thickness (%CA)	264	72.6	7.9	283	71.2	7.7	<0.046
Tibial cortical thickness (%CA)	243	69.1	8.7	259	68.2	8.9	NS
Femoral shape (Z_x/Z_y)	264	0.87	0.11	283	0.90	0.11	<0.001
Tibial shape (Z_x/Z_y)	243	1.33	0.17	259	1.30	0.17	NS

* Results based on one-way Anova.

and lower limbs (Figs 5.15(a) and 5.16(a)). The same is true for females, but the patterns are not as marked or consistent (Figs 5.15(b) and 5.16(b)).

5.5.4 Terrain

Given the hunter-gatherer subsistence of pre-Neolithic groups, there should be a substantial difference in mobility between these groups and later ones. To take this factor into account when assessing the effect of terrain, we examined bone properties by terrain across the entire sample as well as in post-Mesolithic groups only. Descriptive statistics for lower limb robusticity and shape ratios for the three terrain groups are given in Table 5.10. Differences among terrain levels were first evaluated through two-way ANOVA, by sex and terrain level. Given the lack of interaction between these two factors, one-way ANOVA was used for pooled sexes, with multiple comparisons through post-hoc Tukey HSD tests (Table 5.10). Although the impact of terrain should be reflected only in the lower limb, upper limb bone robusticity and shape variables were also analyzed, as a kind of 'control.'

Patterns of variability among terrain levels were quite similar for both the total sample and when pre-Neolithic groups were excluded. Groups inhabiting mountainous regions exhibit more robust femora and tibiae (Table 5.10; Fig. 5.17(a)), although the difference is significant only when the whole sample is considered. Femoral and tibial shape differences are highly significant (Table 5.10; Fig. 5.17(b)). Groups inhabiting hilly and mountainous regions have higher femoral A-P/M-L bending strength than those from flat areas. These differences hold when pre-Neolithic groups are excluded, although the difference between hilly and mountainous groups is not significant. Hilly and mountainous groups also have significantly higher tibial A-P bending strength than those from flat areas (Table 5.10; Fig. 5.17(b)) across the whole

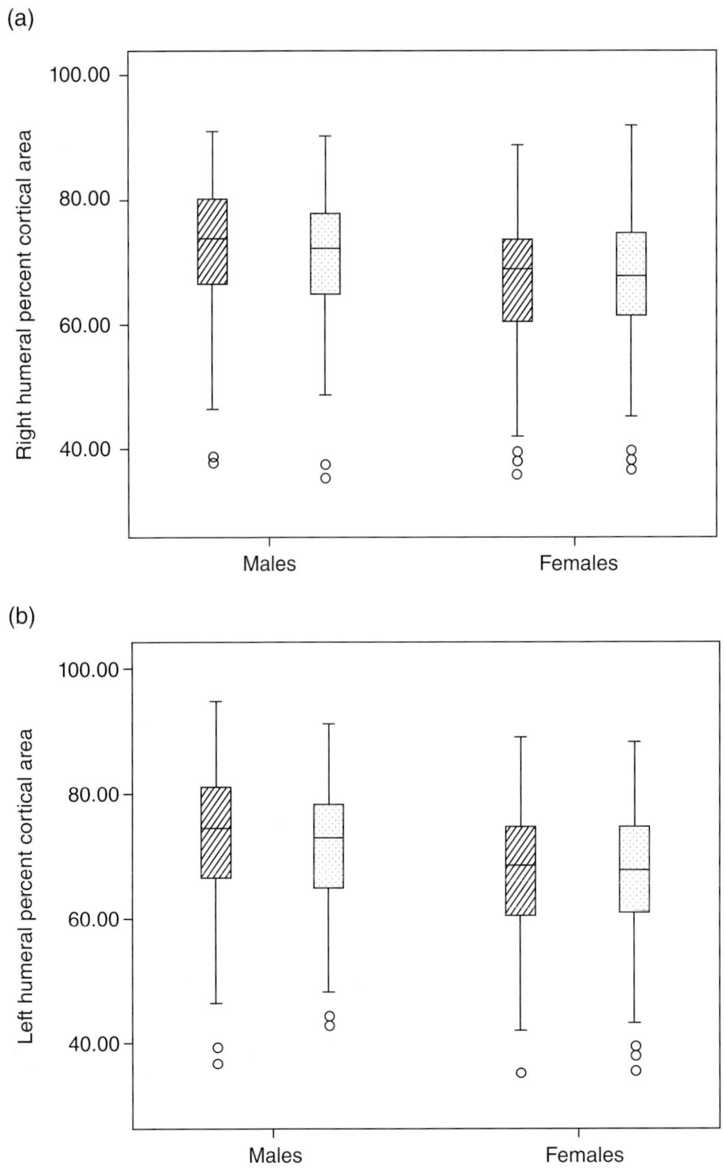

Figure 5.13 Boxplots showing differences in right (a) and left (b) humeral percentage cortical area between rural and urban groups, by sex. Stripes: rural; stipples: urban.

sample. The difference is only significant between flat and hilly groups when pre-Neolithic groups are excluded. As expected, terrain has no effect on upper limb diaphyseal robusticity (Z_p) and shape (data not shown).

5.6 Discussion

This analysis confirms that a decline in lower limb bone strength beginning in the Terminal Pleistocene (Holt, 2003; Ruff *et al.*, 2006) continued throughout at least the early Holocene, most steeply with the shift to food production in the Neolithic and persisting through the Iron

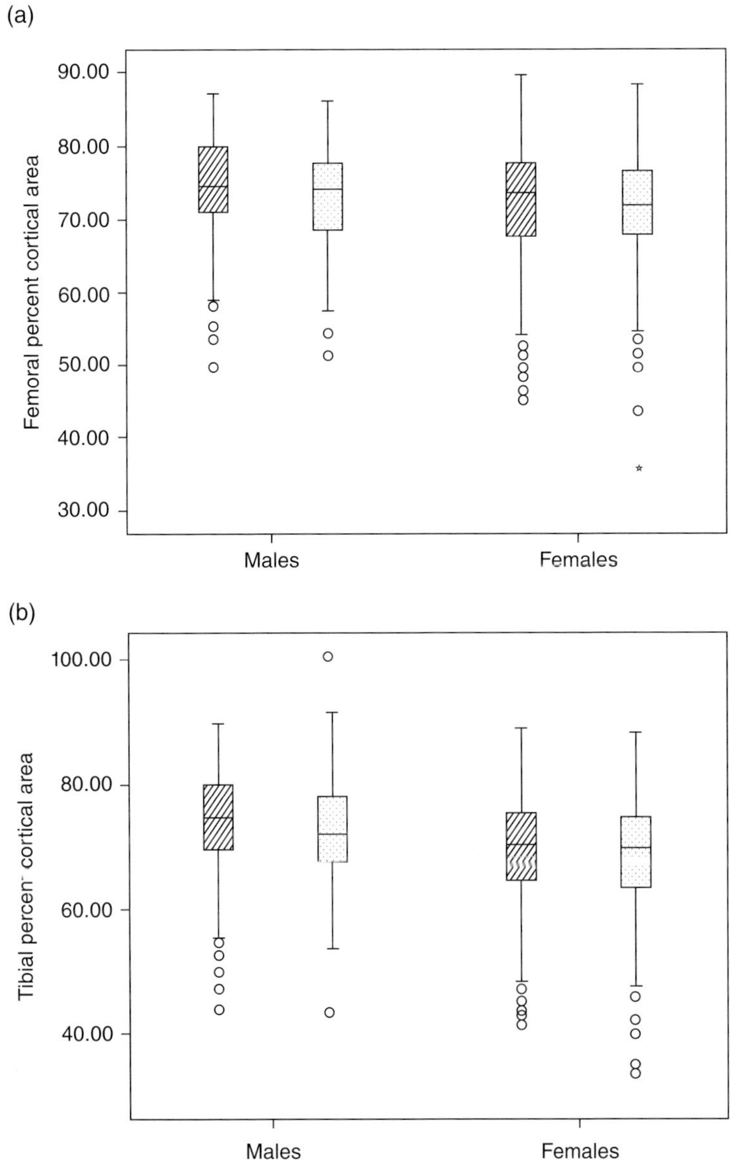

Figure 5.14 Boxplots showing differences in femoral (a) and tibial (b) percentage cortical area between rural and urban groups, by sex. Stripes: rural; stipples: urban.

Age. This decline most probably reflects the more sedentary lives of farmers. Femoral and tibial shape ratios exhibit particularly significant declines, changing from A-P-reinforced cross sections in Paleolithic and Mesolithic hunter-gatherers to more circular diaphyses in Neolithic and later groups. This change confirms those found in previous studies showing a correlation between levels of mobility and relative A-P bending strength in lower limb bones (Ruff *et al.*, 1984; Ruff, 1987; Marchi *et al.*, 2006; Stock, 2006; Sparacello and Marchi, 2008; also see Shaw and Stock, 2009).

While upper limb robusticity decreases over the entire period, steep declines occur earlier than expected, and in parallel with rises in technological complexity and mechanization

Table 5.8 Coefficient of variation* (CV) in upper limb robusticity and shape by sex for Neolithic, Bronze Age and Iron Age.

Males	**Neolithic**	**Bronze Age**	**Iron Age**
Right humerus robusticity (Z_p)	19.4	17.9	21.0
Right humerus cortical thickness	11.4	9.1	13.3
Right humerus shape (Z_x/Z_y)	10.4	7.3	9.5
Left humerus robusticity (Z_p)	19.3	17.6	22.3
Left humerus cortical thickness	10.7	11.2	14.6
Left humerus shape (Z_x/Z_y)	9.6	8.1	10.3
Females	**Neolithic**	**Bronze Age**	**Iron Age**
Right humerus robusticity (Z_p)	19.1	17.5	20.4
Right humerus cortical thickness	12.7	14.6	12.1
Right humerus shape (Z_x/Z_y)	11.3	10.0	11.4
Left humerus robusticity (Z_p)	21.3	18.5	19.4
Left humerus cortical thickness	13.5	16.7	13.7
Left humerus shape (Z_x/Z_y)	10.4	8.1	11.7

* Coefficient of variation = SD/mean × 100.

Table 5.9 Coefficient of variation* (CV) in lower limb robusticity and shape by sex for Neolithic, Bronze Age and Iron Age.

Males	**Neolithic**	**Bronze Age**	**Iron Age**
Femur robusticity (Z_p)	16.0	16.8	18.7
Tibia robusticity (Z_p)	19.8	19.5	22.1
Femur cortical thickness	7.6	6.9	8.3
Tibia cortical thickness	8.7	8.4	10.0
Femurs shape (Z_x/Z_y)	12.8	10.5	13.8
Tibia shape (Z_x/Z_y)	14.7	12.8	12.9
Females	**Neolithic**	**Bronze Age**	**Iron Age**
Femur robusticity (Z_p)	16.0	16.8	18.7
Tibia robusticity (Z_p)	18.8	17.7	20.4
Femur cortical thickness	10.4	9.4	9.8
Tibia cortical thickness	10.7	9.9	10.9
Femurs shape (Z_x/Z_y)	11.0	10.7	12.6
Tibia shape (Z_x/Z_y)	14.0	14.3	12.3

* Coefficient of variation = SD/mean × 100.

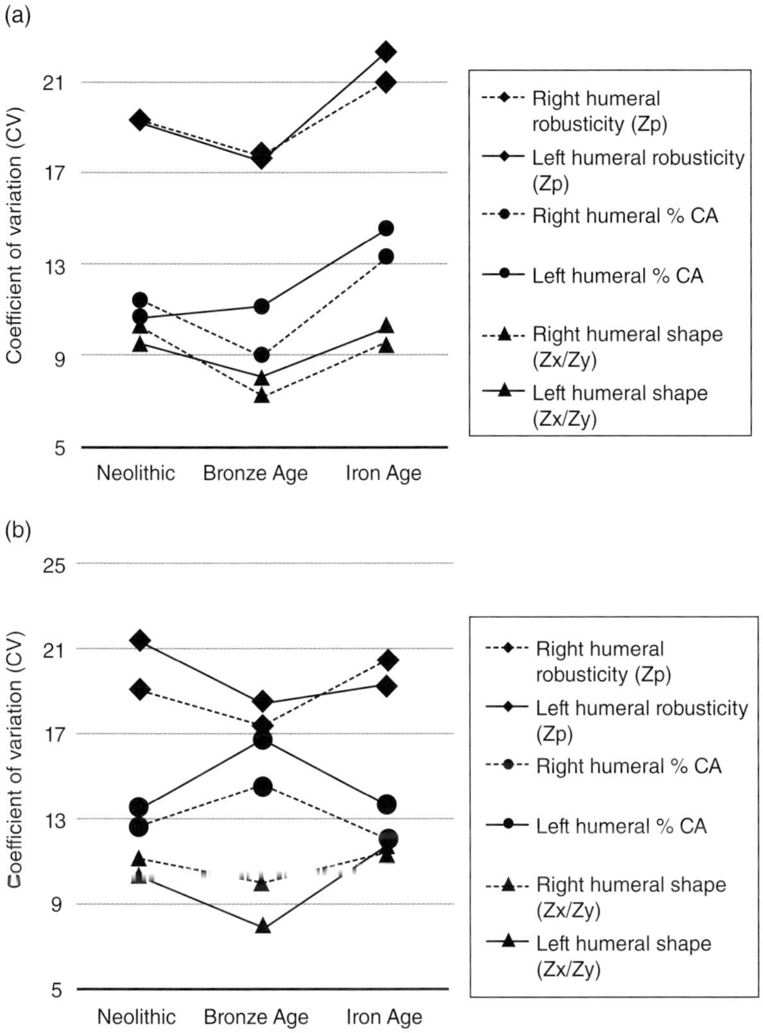

Figure 5.15 Variability (CV) in male (a) and female (b) upper limb average bending/torsional strength (Z_p) standardized for body size (diamonds), percentage cortical area (circles), and A-P to M-L bending strength ratio (Z_x/Z_y, triangles) for Neolithic, Bronze Age, and Iron Age groups. Dashed lines: right humerus; solid lines: left humerus.

observed starting with the Bronze Age. The increase in upper limb robusticity in Middle/Late Neolithic relative to the Early Neolithic is interesting in light of strong evidence for the intensification of grain crops such as wheat and barley. These results parallel those found in other areas such as the New World, where later phases of farming involved much more labor-intensive techniques than initial plant domestication (Bridges, 1989; Bridges *et al.*, 2000). Interestingly, lower limb robusticity does not increase in this period, showing that the changes involve primarily farming technology but that populations remained sedentary.

We found unexpected small but significant increases in upper and lower limb bone robusticity in Medieval Europeans. This trend may reflect de-urbanization following the collapse of the Roman Empire and the infrastructure system of roads and bridges that maintained and connected cities. Many areas became entirely rural and, as mentioned above, while some

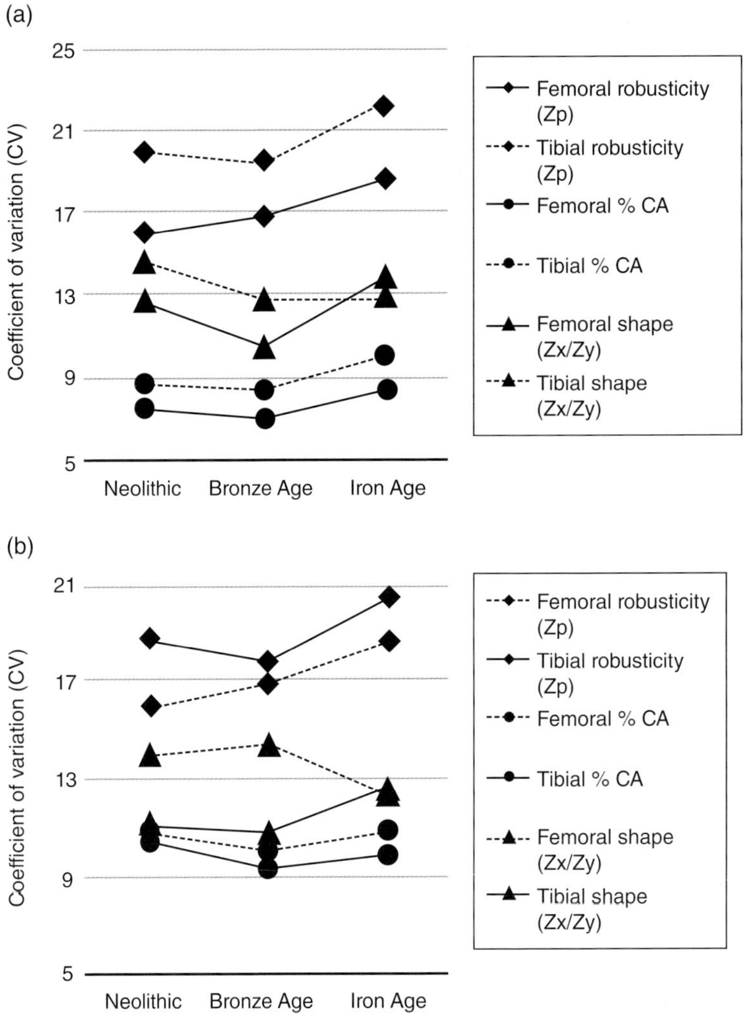

Figure 5.16 Variability (CV) in male (a) and female (b) lower limb average bending/torsional strength (Z_p) standardized for body size (diamonds), percentage cortical area (circles), and A-P to M-L bending strength ratio (Z_x/Z_y, triangles) for Neolithic, Bronze Age, and Iron Age groups. Dashed lines: tibia; solid lines: femur.

technological improvements such as mills and wheeled plows made farming less physically demanding, these were available to few. The periods spanning the 9th–13th centuries saw the expansion of manorial economies, in which most peasants worked long hours in the fields in the service of landed aristocrats. As the price of cereals rose exponentially (Duby, 1976), demand for crop intensification and manual labor increased further (Campbell, 1995). The felling of trees with axes was common, as population growth and wars fueled an ever-growing demand for construction material and arable land. In his classic book *Rural Economy and Country Life in the Medieval West*, Duby (1976) states that:

> "...even in the most favoured sectors of rural life, those of the great farming complexes described by inventories, men used feeble wooden implements. They found themselves ill-equipped to come to grips with nature and worked with their bare hands for a great

Table 5.10 Lower limb robusticity and shape by terrain levels (sex pooled).

Property	Group 1 (≤21° = Flat)			Group 2 (≤44° = Hilly)			Group 3 (≥45° = Mountainous)				Group comparisons		
All periods	N	Mean	SD	N	Mean	SD	N	Mean	SD	F-value and significance*	1 versus 2	1 versus 3	2 versus 3
Femur robusticity (Z_p)	842	1079.4	193.6	708	1079.6	182.0	207	1120.4	180.0	4.35 (<0.013)	NS	**<0.013**	**<0.016**
Tibia robusticity (Z_p)	764	857.29	196.5	619	869.68	183.2	185	903.3	208.2	4.3 (<0.014)	NS	<0.010	NS
Femur shape (Z_x/Z_y)	872	0.91	0.11	736	0.94	0.12	219	0.97	0.16	23.65(<0.0001)	<0.0001	<0.0001	<0.008
Tibia shape (Z_x/Z_y)	797	1.35	0.20	651	1.39	0.20	202	1.39	0.21	6.90 (<0.001)	<0.002	<0.045	NS
Post Mesolithic only													
Femur robusticity (Z_p)	815	1074.7	191.7	692	1077.7	181.4	165	1097.9	174.5	1.07 (NS)	NS	NS	NS
Tibia robusticity (Z_p)	744	852.7	194.1	607	867.0	182.0	155	866.1	193.1	1.05 (NS)	NS	NS	NS
Femur shape (Z_x/Z_y)	826	0.90	0.11	718	0.93	0.11	172	0.93	0.12	15.40 (<.0001)	<0.0001	<0.036	NS
Tibia shape (Z_x/Z_y)	761	1.34	0.20	639	1.39	0.20	166	1.35	0.18	8.48 (<.0001)	<0.0001	NS	NS

* Results based on one-way Anova, with post-hoc Tukey HSD test for group comparisons.

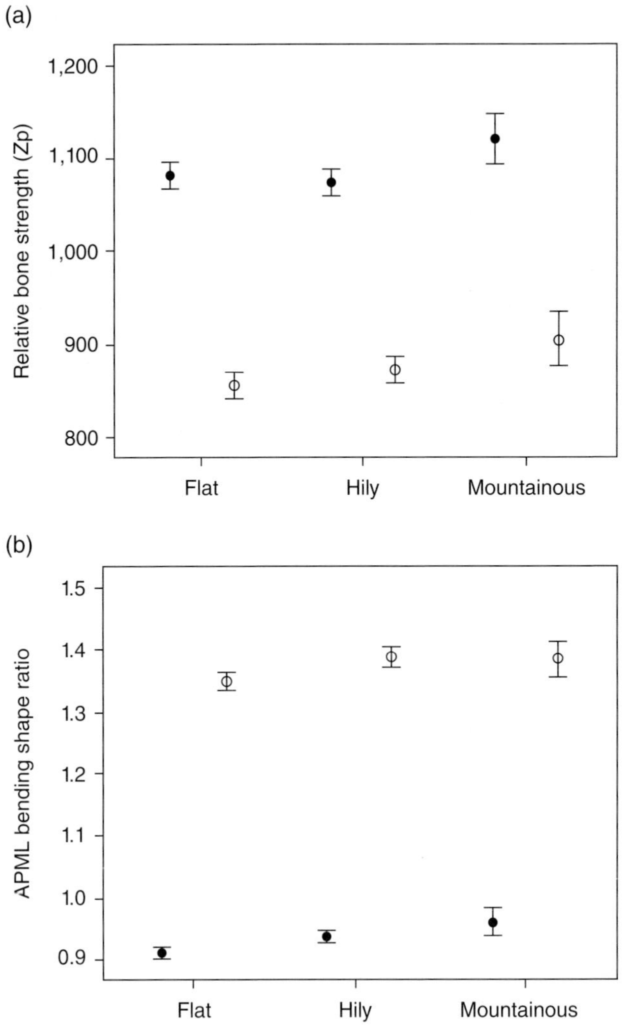

Figure 5.17 Error bar plots showing (a) femoral and tibial average bending/torsional strength standardized for body size and (b) A-P to M-L bending strength ratio (Z_x/Z_y) by terrain category, for pooled sexes (all time periods). Closed circles: femur; open circles: tibia. Central values and bars represent mean and 95% confidence interval, respectively.

> part of the time... Areas of natural vegetation adjoining villages were of course naturally necessary because the cultivation of cereals was so demanding of manpower that each rural community had to supplement its means of livelihood by making the most of the products of the wastelands-animal husbandry, hunting and food gathering." (pp. 21–22).

The wealth of Medieval written records paints a life of tremendous physical labor, with little help from mechanized means. The observed increase in limb strength, more marked in the upper limb in males, may reflect physically demanding agricultural tasks such as tree felling and plowing.

Interestingly, we did not find the predicted higher upper limb robusticity in rural compared to urban groups. One possible explanation for this lack of correlation may be that life in both rural and urban settings involve different but relatively equally strenuous types of activities. During the Industrial Revolution of the Early Modern period, for instance, children routinely worked long hours in factories (Humphries, 2010; Cardoso, 2009), and this labor may have produced mechanical loads similar to those associated with farming tasks. In addition, it could also be that our urban/rural categories do not accurately reflect the realities of urban life in the past. The urban sample comprises groups who lived in a large, agglomerated settlement with evidence for economic specialization, even if some of the people worked in 'non-urban' kinds of occupations, and it is very likely that, until relatively recently, many in smaller urban populations spent the majority of their day in rural settings, working their fields. Rural males and females, however, do exhibit higher upper and lower limb percentage cortical area than their urban counterparts, possibly as a result of better nutrition and life conditions. This is corroborated by the fact that rural males are taller and heavier than urban males (see Chapter 4), paralleling previous studies showing that pre-20th century urban populations were generally shorter and lighter than rural ones (Martínez-Carrion and Moreno-Lázaro, 2007). Access to livestock in rural environments may have afforded better access to calcium- and protein-rich diets. In addition, increased pathogen exposure in unsanitary and crowded urban conditions could also result in greater immune activation and low bone mass.

Another result concerns the temporal decrease in humeral cross-sectional circularity on both right and left sides. Urban groups also exhibit less circular humeral diaphyses than rural ones. While variability in lower limb diaphyseal shape can be reasonably attributed to locomotor patterns (Ruff, 1987; Stock, 2006; Shaw and Stock, 2009), the complexity of upper limb movements in humans makes interpretation of humeral shape more difficult. Attempts to link particular humeral diaphyseal shape to specific activity have yielded few unambiguous results. Analyses of change from elliptical, A-P-elongated, cross-sectional shape in Neandertals and early Upper Paleolithic humeri to a more circular one in later Upper Paleolithic groups have focused on hunting technology developments (Churchill *et al.*, 1996; Churchill, 2002; Schmitt *et al.*, 2003). Specifically, the change was hypothesized to reflect a shift from thrusting spears at close range to the use of long-distance throwing technology after the LGM. Spear thrusting was argued to apply a bending moment on the trailing humerus, causing it to become more robust and elliptical than the forward humerus (Schmitt *et al.*, 2003). Recent experimental work, however, has placed the throwing/thrusting dichotomy in doubt. Electromyography data suggest that the robustness and shape of Neandertal humeri may be better explained by hide scraping than by spear use, and could also explain the bilateral asymmetry observed in Neandertal humeri (Shaw *et al.*, 2012). Scraping is a laborious task, and has been known to take over 8 hours to process one hide (Oakes, 1991). Since each individual would need to process a number of hides each year, this task could leave a significant biomechanical signature (Berthaume, 2014; Berthaume *et al.*, 2014). Finite element analysis of Neandertal and Upper Paleolithic humeri also shows that the circular shape of Neandertal humeri was as efficient at resisting strains associated with spear thrusting as those of spear throwing (Berthaume, 2014; Berthaume *et al.*, 2014). In this analysis, cross-sectional geometry – particularly size and shape – proved ineffective at predicting humeral strains during weapons use (Berthaume, 2014; Berthaume *et al.*, 2014). This, and work by Shaw and colleagues (Shaw *et al.*, 2012), underscores the need for experimental approaches to understand better the complex loading patterns of the humerus, in order to develop more accurate and predictive interpretive models.

Our expectations that rising social complexity, stratification, and craft specialization in the Iron Age would result in increased upper limb robusticity variability were confirmed, although we did not find a change for Bronze Age groups. While archeological evidence of social stratification clearly exists for the Bronze Age, subsistence for most remained primarily based on farming (Harding, 2002), with large urban centers appearing during the Iron Age (Milisauskas, 2002a).

Thus, although the Bronze Age undoubtedly saw a rise in specialized activities, unprecedented changes in metal technology, demographic growth, and unequal access to resources begin primarily with the Iron Age, probably explaining the increased variability in upper limb robusticity.

The analysis by terrain levels demonstrates the strong effect of topography on lower limb robusticity patterns. Similarly to a study of North American prehistoric samples that found elevated relative femur strength in mountainous regions (Ruff 1999), European groups from hilly and mountainous regions exhibit significantly higher femoral and tibial strength than those from flat areas. The most significant difference, however, regards the higher A-P bending strength in groups that inhabited rugged areas, in keeping with biomechanical expectations that travel over rough terrain would engender primarily A-P-oriented bending loads. Similar results were found in a study of Neandertal and Upper Paleolithic tibiae (Higgins, 2014). While the muscles recruited in level walking and climbing are the same, climbing engenders greater moment magnitude and, hence, greater applied muscular force (Levangie and Norkin, 2001), a fact clearly reflected in the A-P adaptive remodeling observed in the limb bones of Europeans inhabiting hilly and mountainous regions. This conclusion is bolstered by the finding that terrain has no impact on upper limb bone robusticity.

5.7 Conclusions

By focusing on a large sample of well-dated human skeletons from a single geographic area, the present study allows more rigorous testing of earlier hypotheses regarding changes in activity level and mechanical loading of the skeleton over the past 30,000 years. Long bone robusticity (strength relative to body size) decreases from the Upper Paleolithic through recent time periods in Europe, although the exact timing and patterning of this reduction varies by skeletal location and property. Femoral and tibial robusticity decline through the Iron and early Medieval periods, respectively, in both sexes. Humeral robusticity shows a more continuous decline among females throughout the entire time range sampled, while male humeri show a pattern more similar to that of the lower limb bones. There is a small gain in relative strength of both limbs during the Medieval periods. Tibiae become more circular in cross-sectional shape throughout the entire temporal span, while femora decline in relative A-P/M-L strength from the Upper Paleolithic through the Iron Age.

Taken together, these results suggest that a combination of factors is responsible for observed changes in skeletal morphology during the past 30,000 years. Increased sedentism with the adoption and elaboration of food production was a major factor, leading to large declines in mobility and A-P bending loads on the lower limb bones that continued from the early Holocene through to pre-Medieval times, with some evidence for more modest declines (in the tibia) over the past 1000 years, possibly due to increased use of alternative modes of transport (i.e., animals and then machines). However, the increase in overall bone strength in the Medieval period relative to immediately preceding periods indicates that despite a decline in mobility, overall workload did not continuously decline. This is consistent with evidence from other studies (e.g., Ruff *et al.*, 1984; Ruff and Larsen, 2001; Sparacello and Marchi, 2008) that not all skeletal indicators of increased or decreased mechanical loading need covary. There is also evidence that lower limb bone shape (A-P/M-L strength) is a better indicator of mobility *per se* than overall bone robusticity (Ruff and Larsen, 2014), although body shape must also be factored into such analyses (Ruff *et al.*, 2006; Shaw and Stock, 2011). Body shape shows relatively minor temporal changes in our sample, at least from the Mesolithic onwards (see Chapter 4), so was probably not a significant factor. Changes in upper limb bone robusticity are also likely multifactorial in nature. Increasingly sophisticated tools and later, mechanization probably played a role in the overall declining trend, but again there are some major departures from this trend, particularly in the Medieval period, and males and females show some contrasting patterns. Temporal trends in sexual dimorphism and upper limb bone bilateral asymmetry are discussed in Chapters 6 and 7, respectively.

Terrain was found to have a significant effect on lower limb, but not upper limb bone robusticity and shape, in the predicted direction of greater robusticity and A-P strength in the lower limb in more hilly terrain. Thus, terrain should be considered in any reconstructions of behavior from skeletal morphology (Ruff, 1999; Marchi *et al.*, 2006; Sparacello and Marchi, 2008; Whittey, 2016). Urbanization was not found to have a significant effect on bone morphology, although samples with more clear urban/rural status differences are needed for a more rigorous test of this factor. We found some evidence for an effect of social complexity on increased variability in morphology, and thus implied mechanical loadings, but again more tightly controlled comparisons would be beneficial.

Acknowledgments

The authors thank Colten Karnedy for help with the tables.

References

Auerbach, B.M. and Ruff, C.B. (2004) Human body mass estimation: a comparison of 'morphometric' and 'mechanical' methods. *Am. J. Phys. Anthropol.*, **125**, 331–342.

Auerbach, B.M. and Ruff, C.B. (2006) Limb bone bilateral asymmetry: variability and commonality among modern humans. *J. Hum. Evol.*, **50**, 203–218.

Berthaume, M.A. (2014) Were Neandertal humeri adapted for spear thrusting or throwing? A finite element study. M.A. Thesis, University of Massachusetts.

Berthaume, M.A., Shaw, C.S., Jewell, C., Hamill, J., Ryan, T.M., and Holt, B. (2014) Were Neandertal humeri adapted for spear thrusting or throwing? A finite element analysis study. *Am. J. Phys. Anthropol.*, **153** (S58), 40.

Binford, L.R. (2001) *Constructing frames of reference – an analytical method for archaeological theory building using hunter-gatherer and environmental data sets.* University of California Press, Berkeley, CA.

Boserup, E. (1965) *The conditions of agricultural growth: the economics of agrarian change underpopulation pressure.* Allen and Unwin, London.

Boserup, E. (1975) The impact of population growth on agricultural output. *Q. J. Agric. Econ.*, **89**, 257–270.

Bridges, P.S. (1989) Changes in activities with the shift to agriculture in the southeastern United States. *Curr. Anthropol.*, **30**, 385–394.

Bridges, P.S., Blitz, J.H., and Solano, M.C. (2000) Changes in long bone diaphyseal strength with horticultural intensification in west-central Illinois. *Am. J. Phys. Anthropol.*, **112**, 217–238.

Brumfiel, E.M. (1987) *Specialization, exchange, and complex societies.* Cambridge University Press, Cambridge.

Buikstra, J.E. and Ubelaker, D.H. (1994) *Standards for Data Collection from Human Skeletal Remains.* Arkansas Archaeological Survey, Fayetteville, Arkansas.

Campbell, B.M.S. (1995) Progressive and backwardness in Thirteenth- and early Fourteenth-Century English agriculture: the verdict of recent research. In: *Peasants and Townsmen in*

Medieval Europe: Studia in Honorem Adriaan Verhulst. Centre Belge d'Histoire Rurale 114, Gent, pp. 541–559.

Cardoso, H.F.V. (2009) The not-so-Dark Ages: Ecology for human growth in Medieval and early Twentieth Century Portugal as inferred from skeletal growth profiles. *Am. J. Phys. Anthropol.*, **138**, 136–147.

Churchill, S.E. (1994) Human upper body evolution in the Eurasian Later Pleistocene. Ph.D. Dissertation, University of New Mexico.

Churchill, S.E. (2002) Of assegais and bayonets: Reconstructing prehistoric spear use. *Evolution. Anthropol. Issues, News, and Reviews*, **11**, 185–186.

Churchill, S.E. and Formicola, V. (1997) A case of marked bilateral asymmetry in the upper limbs of an Upper Palaeolithic male from Barma Grande (Liguria), Italy. *Int. J. Osteoarcheol.*, **7**, 18–38.

Churchill, S.E., Formicola, V., Holliday, T.W., Holt, B.M., and Schumann, B.A. (2000) The Upper Paleolithic population of Europe in an evolutionary perspective. In: *Hunters of the Golden Age: The Mid Upper Paleolithic of Eurasia (30,000–20,000 bp)* (eds W. Roebroeks, M. Mussi, J. Svoboda, and K. Fennema), Leiden University Press, Leiden, pp. 31–57.

Churchill, S.E., Weaver, A.H., and Niewoehner, W.A. (1996) Late Pleistocene human technological and subsistence behavior: functional interpretations of upper limb morphology. *Quat. Nova*, **vi**, 413–447.

Clark, G.A. and Straus, L.G. (1986) Synthesis and conclusions – part I: Upper Paleolithic and Mesolithic hunter-gatherer subsistence in northern Spain. In: *La Riera cave: Stone Age hunter-gatherer adaptations in northern Spain* (eds L.G. Straus and G.A. Clark), Arizona State University Anthropol. Res. Papers, vol. 36, Tempe, pp. 351–365.

Collis, J. (1984) *The European Iron Age*. Routledge, New York.

Cowan, R.S. (1999) The industrial revolution in the home. In: *The Social Shaping of Technology* (eds D. Mackenzie and J. Wajcman), Open University Press, Philadelphia, pp. 281–300.

Duby, G. (1976) *Rural Economy and Country Life in the Medieval West*. University of Carolina Press, Columbia.

Dyer, C. (1995) How urbanized was Medieval England? In: *Peasants and Townsmen in Medieval Europe: Studia in Honorem Adriaan Verhulst*. Centre Belge d'Histoire Rurale 114, Gent, pp. 169–183.

Formicola, V. and Giannecchini, M. (1999) Evolutionary trends of stature in Upper Paleolithic and Mesolithic Europe. *J. Hum. Evol.*, 36, 319–333.

Flébot-Augustin, J. (1997) La circulation des matières premières au Paléolithique. ERAUL 75.

Frayer, D.W. (1981) Body size, weapon use, and natural selection in the European Upper Paleolithic and Mesolithic. *Am. Anthropol.*, **83**, 57–73.

Frayer, D.W. (1984) Biological and cultural changes in the European Late Pleistocene and Early Holocene. In: *The origins of modern humans: A world survey of the fossil evidence* (eds F.H. Smith and F. Spencer), Alan R. Liss, Inc., New York, pp. 211–250.

Freeman, L.G. (1973) The significance of mammalian faunas from Paleolithic occupations in Cantabrian Spain. *Am. Antiq.*, **38**, 3–44.

Gamble, C. (1986) *The Paleolithic settlement of Europe*. Cambridge University Press, Cambridge.

Gamble, C., Davies, W., Pettitt, P., and Richards, M. (2004) Climate change and evolving human diversity in Europe during the Last Glacial. *Philos. Trans. R. Soc. London B*, **359**, 243–254.

Guidi, A. and Piperno, M. (1992) *Italia Preistorica*. Laterza, Rome.

Harding, A.F. (2002) The Bronze Age. In: *European Prehistory: A Survey* (ed. S. Milisauskas), Kluwer Academic/Plenum, New York, pp. 271–334.

Henrich, J. and Boyd, R. (2008) Division of labor, economic specialization, and the evolution of social stratification. *Curr. Anthropol.*, **49**, 715–724.

Higgins, R.W. (2014) The effects of terrain on long bone robusticity and cross-sectional shape in lower limb bones of bovids, Neandertals, and Upper Paleolithic modern humans. In: *Mobility: Interpreting Behavior from Skeletal Adaptations and Environmental Interactions* (eds K. Carlson and D. Marchi), Springer, New York, pp. 227–252.

Holt, B.M. (2003) Mobility in Upper Paleolithic and Mesolithic Europe: evidence from the lower limb. *Am. J. Phys. Anthropol.*, **122**, 200–215.

Holt, B.M., Mussi, M., Churchill, S.E., and Formicola, V. (2000) Biological and cultural trends in Upper Palaeolithic Europe. *Riv. Antropol.*, **78**, 179–182.

Holt, B.M. and Formicola, V. (2008) Hunters of the Ice Age: the biology of Upper Paleolithic people. *Yrbk Phys. Anthropol.*, **51**, 70–99

Jacobs, K. (1985) Evolution in the postcranial skeleton of Late Glacial and Early Postglacial European Hominids. *Z. Morph. Anthropol.*, **75**, 307–326.

Jarvis, A., Reuter, H.I., Nelson, A., and Guevara, E. (2008). Hole-filled SRTM for the globe Version 4, available from the CGIAR-CSI SRTM 90 m Database (http://srtm.csi.cgiar.org).

Harding, A.F. (2002) The Bronze Age. In: *European Prehistory: A Survey* (ed. S. Milisauskas), Kluwer Academic/Plenum, New York, pp. 271–334.

Humphries, J. (2010) *Childhood and Child Labour in the British Industrial Revolution*. Cambridge University Press

Jochim, M. (1987) Late Pleistocene refugia in Europe. In: *The Pleistocene Old World: Regional perspectives* (ed. O. Soffer), Plenum Press, New York, pp. 317–331.

Jochim, M.A. (2002) The Upper Paleolithic. In: *European Prehistory* (ed. S. Milisauskas), Kluwer Academic, New York, pp. 55–114.

Keeley, L.H. (1988) Hunter-gatherer economic complexity and "population pressure": A cross-cultural analysis. *J. Anthropol. Archaeol.*, **7**, 373–411.

Kozlowski, J.K. (1991) Raw material procurement in the Upper Paleolithic of Central Europe. In: *Raw material economies among prehistoric hunter-gatherers* (eds A. Montet-White and S. Holen), Publications in Anthropology 19, University of Kansas, Lawrence, Pp. 187–196.

Kristiansen, K. (1987) From stone to bronze: the evolution of social complexity in Northern Europe, 2300–1200 BC. In: *Specialization, Exchange and Complex Societies* (eds M. Brumfield and T.K. Earle), Cambridge University Press, Cambridge, pp. 30–51.

Kristiansen, K. (2000) *Europe before History*. Cambridge University Press, Cambridge.

Kristiansen, K. and Rowlands, M. (eds) (1998). *Social transformations in archaeology – global and local perspectives*. Routledge, New York.

Larsen, C.S. (1995) Biological changes in human populations with agriculture. *Annu. Rev. Anthropol.*, **24**, 185–213.

Levangie, P.K. and Norkin, C.C. (2011) *Joint Structure and Function: A Comprehensive Analysis*, 5th edn, F.A. Davis Company, Philadelphia.

Marchi, D. (2008). Relationships between lower limb cross-sectional geometry and mobility: The case of a Neolithic sample from Italy. *Am. J. Phys. Anthropol.*, **137**, 188–200.

Marchi, D., Sparacello, V.S., Holt, B.M., and Formicola, V. (2006) Biomechanical approach to the reconstruction of activity patterns in Neolithic Western Liguria, Italy. *Am. J. Phys. Anthropol.*, **131**, 447–455.

Marchi, D., Sparacello, V.S., and Shaw, C.N. (2011) Mobility and lower limb robusticity of a pastoralist Neolithic population from North-Western Italy. In: Human Bioarchaeology of the Transition to Agriculture (eds R. Pinhasi and J. Stock), John Wiley & Sons, Ltd, New York.

Martínez-Carrión, J.-M. and Moreno-Lázaro, J. (2007) Was there an urban height penalty in Spain, 1840–1913. *Econ. Hum. Biol.*, **5**, 144–164.

Mellars, P.A. (1985) The ecological basis of social complexity in the Upper Paleolithic of Southwestern France. In: *Prehistoric Hunter-Gatherers: the Emergence of Cultural Complexity* (eds T.D. Price and J.A. Brown), Academic Press, Orlando,. pp. 271–297.

Mellars, P.A. (2006) Archeology and the dispersal of modern humans in Europe: Deconstructing the "Aurignacian". *Evol. Anthropol.*, **15**, 167–182.
Milisauskas, S. (ed.) (2002a) *European Prehistory: A Survey*. Kluwer Academic/Plenum, New York.
Milisauskas, S. (2002b) Early Neolithic: The first farmers in Europe, 7000–5500/5000 BC. In: *European Prehistory: A Survey* (ed. S. Milisauskas), Kluwer Academic/Plenum, New York, pp. 143–192.
Milisauskas, S. and Kruk, J. (2002a) Late Neolithic, Crises, Collapse, New Ideologies, and Economies, 3500/3000–2200/2000 BC. In: *European Prehistory: A Survey* (ed. S. Milisauskas), Kluwer Academic/Plenum, New York, pp. 247–269.
Milisauskas, S. and Kruk, J. (2002b) Middle Neolithic continuity, diversity, innovations, and greater complexity, 5500/500–3500/3000 BC. In: *European Prehistory: A Survey* (ed. S. Milisauskas), Kluwer Academic/Plenum, New York, pp. 193–246.
Mussi, M. and Zampetti, D. (1988) Frontiera e confini nel Gravettiano e nell'Epigravettiano dell'Italia. Prime considerazioni. *Scienze dell'Antichità*, **2**, 45–78.
Musson, A.E. and Robinson, E. (1969) *Science and Technology in the Industrial Revolution*. University of Toronto Press.
O'Neill, M.C. and Ruff, C.B. (2004) Estimating human long bone cross-sectional geometric properties: a comparison of noninvasive methods. J. Hum. Evol., **47**, 221–235.
Negrino, F. and Starnini, E. (2003) Patterns of lithic raw material exploitation in Liguria from the Palaeolithic to the Copper Age. *Préhistoire du Sud-Ouest Suppl.*, **5**, 235–243.
Oakes, J. (1991) *Skin Preparation Techniques. Copper & Cariboo Inuit skin clothing production.* Canadian Museum of Civilizations, Hull, Quebec, pp. 102–113.
Peroni, R. (1979) From Bronze Age to Iron Age: economic, historical, and social considerations. In: *Italy before the Romans: the Iron Age, Orientalizing, and Etruscan periods* (eds D. Ridgway and R. Francesca), Academic Press, London, pp. 7–30.
Price, T.D. (1987) The Mesolithic of Europe. *J. World Prehistory*, **1**, 225–305.
Puymerail, L., Ruff, C., Bondioli, L., Widianto, H., Trinkaus, E., and Macchiarelli, R. (2012) Structural analysis of the Kresna 11 Homo erectus femoral shaft (Sangiran, Java). *J. Hum. Evol.*, **63**, 741–749.
Rhodes, J.A. and Knüsel, C.J. (2005) Activity-related skeletal change in medieval humeri: Cross-sectional and architectural alterations. *Am. J. Phys. Anthropol.*, **128**, 536–546.
Ruff, C.B. (1987) Sexual dimorphism in human lower limb bone structure: relationship to subsistence strategy and sexual division of labor. *J. Hum. Evol.*, **16**, 391–416.
Ruff, C.B. (1999) Skeletal structure and behavioral patterns of prehistoric Great Basin populations. In: *Prehistoric Lifeways in the Great Basin Wetlands: Bioarchaeological Reconstruction and Interpretation* (eds B.E. Hemphill and C.S. Larsen), University of Utah Press, Salt Lake City, pp. 290–320.
Ruff, C.B. (2000) Body size, body shape, and long bone strength in modern humans. *J. Hum. Evol.*, **38**, 269–290.
Ruff, C.B. (2002) Long bone articular and diaphyseal structure in old world monkeys and apes. I: Locomotor effects. *Am. J. Phys. Anthropol.*, **119**, 305–342.
Ruff, C.B. (2005) Mechanical determinants of bone form: Insights from skeletal remains. *J. Musculoskelet. Neuronal. Interact.*, **5**, 202–212.
Ruff, C.B. (2008) Biomechanical analysis of archeological human skeletons. In: *Biological Anthropology of the Human Skeleton*, 2nd edn (eds M.A. Katzenberg and S.R. Saunders), Wiley-Liss, Inc., New York, pp. 183–206.
Ruff, C.B. and Hayes, W.C. (1983a) Cross-sectional geometry of the Pecos Peublo femora and tibiae–a biomechanical investigation: I. Methods and general pattern of variation. *Am. J. Phys. Anthropol.*, **60**, 359–381.
Ruff, C.B. and Hayes, W.C. (1983b) Cross-sectional geometry of the Pecos Peublo femora and tibiae–a biomechanical investigation. II. Sex, age, and side differences. *Am. J. Phys. Anthropol.*, **60**, 383–400.

Ruff, C.B., Larsen, C.S., and Hayes, W.C. (1984) Structural changes in the femur with the transition to agriculture on the Georgia Coast. *Am. J. Phys. Anthropol.*, **64**, 5–136.
Ruff, C.B. and Larsen, C.S. (2001) Reconstructing behavior in Spanish Florida: The biomechanical evidence. In: *Bioarchaeology of Spanish Florida: The Impact of Colonialism* (ed. C.S. Larsen), University Press of Florida, Gainesville, pp. 113–145.
Ruff, C.B. and Larsen, C.S. (2014) Long bone structural analyses and reconstruction of past mobility: A historical review. In: *Mobility: Interpreting Behavior from Skeletal Adaptations and Environmental Interactions* (eds K. Carlson and D. Marchi), Springer, New York, pp. 13–29.
Ruff, C.B., Trinkaus, E., Walker, A., and Larsen, C.S. (1993) Postcranial robusticity in *Homo*, I: Temporal trends and mechanical interpretation. *Am. J. Phys. Anthropol.*, **91**, 21–53.
Ruff, C.B., Walker, A., and Trinkaus, E. (1994) Postcranial robusticity in *Homo*, III: Ontogeny. *Am. J. Phys. Anthropol.*, **93**, 35–54.
Ruff, C.B., Niskanen, M., Junno, J.A., and Jamison, P. (2005) Body mass prediction from stature and bi-iliac breadth in two high latitude populations, with application to earlier higher latitude humans. *J. Hum. Evol.*, **48**, 381–392.
Ruff, C.B., Holt, B.M., Sládek, V., Berner, M., Murphy, W.A., Nedden, Dz., Seidler, H., and Reicheis, W. (2006) Body size, body shape, and long bone strength of the Tyrolean ice man. *J. Hum. Evol.*, **51**, 91–178.
Ruff, C.B., Holt, B.M., Sladék, V., Berner, M., Garofalo, E., Garvin, H.M., Hora, M., Maijanen, H., Niinimäki, S., Salo, K., Schuplerová, E., and Tompkins, D. (2012) Stature and body mass estimation from skeletal remains in the European Holocene. *Am. J. Phys. Anthropol.*, **148**, 601–617.
Schmitt, D., Churchill, S.E., and Hylander, W.L. (2003) Experimental Evidence Concerning Spear Use in Neandertals and Early Modern Humans. *J. Archaeol. Sci.*, **30**, 103–114.
Shackelford, L.L. (2007) Regional variation in the postcranial robusticity of Late Upper Paleolithic humans. *Am. J. Phys. Anthropol.*, **133**, 655–668.
Shaw, C.N. and Stock, J.T. (2009) Intensity, repetitiveness, and directionality of habitual adolescent mobility patterns influence the tibial diaphysis morphology of athletes. *Am. J. Phys. Anthropol.*, **140**, 149–159.
Shaw, C.N. and Stock, J.T. (2011) The influence of body proportions on femoral and tibial midshaft shape in hunter-gatherers. *Am. J. Phys. Anthropol.*, **144**, 22–29.
Shaw, C.N. and Stock, J.T. (2013) Extreme mobility in the Late Pleistocene? Comparing limb biomechanics among fossil Homo, varsity athletes and Holocene foragers. *J. Hum. Evol.*, **64**, 242–249.
Shaw, C.N., Hofmann, C.L., Petraglia, M.D., Stock, J.T., and Gottschall, J.S. (2012) Neandertal humeri may reflect adaptation to scraping tasks, but not spear thrusting. *PLoS One*, **7**, e40349.
Simpson, G.G., Roe, A., and Lewontin, R.C. (1960) *Quantitative zoology*. Harcourt, Brace & Co, New York.
Sladek, V., Berner, M., and Sailer R. (2006a) Mobility in Central European Late Eneolithic and Early Bronze Age: femoral cross-sectional geometry. *Am. J. Phys. Anthropol.*, **130**, 320–332.
Sladek, V., Berner, M., and Sailer, R. (2006b) Mobility in Central European Late Eneolithic and Early Bronze Age: tibial cross-sectional geometry. *J. Archaeol. Sci.*, **33**, 470–482.
Sparacello, V. and Marchi, D. (2008) Mobility and subsistence economy: a diachronic comparison between two groups settled in the same geographical area (Liguria, Italy). *Am. J. Phys. Anthropol.*, **136**, 485–495.
Sparacello, V.S., Pearson, O.M., Coppa, A., and Marchi, D. (2011) Changes in robusticity in an Iron Age agropastoral group: the Samnites from the Alfedena necropolis (Abruzzo, Central Italy). *Am. J. Phys. Anthropol.*, **144**, 119–130.
Stiner, M.C. (2001) Thirty years on the "Broad spectrum revolution" and Paleolithic demography. *Proc. Natl Acad. Sci. USA*, **98**, 6993–6996.

Stiner, M., Munro, N.D., Surovell, T.A., Tchernov, E., and Bar-Yosef, O. (1999) Paleolithic population growth pulses evidenced by small animal exploitation. *Science*, **283**, 190–194.

Stock, J.T. (2006) Hunter-gatherer postcranial robusticity relative to patterns of mobility, climatic adaptation, and selection for tissue economy. *Am. J. Phys. Anthropol.*, **131**, 194–204.

Stock, J. and Pfeiffer, S. (2001) Linking structural variability in long bone diaphyses to habitual behaviors: Foragers from the southern African Later Stone Age and the Andaman Islands. *Am. J. Phys. Anthropol.*, **115**, 337–348.

Straus, L.G. (1986) Late Würm adaptive systems in Cantabrian Spain: the case of eastern Asturias. *J. Anthropol. Archaeol.*, **5**, 330–368.

Straus, L.G. (1990) The Last Glacial Maximum in Cantabrian Spain: the Solutrean. In: *The World at 18,000 BP: Vol.1 - High latitudes* (eds O. Soffer and C. Gamble), Unwin Hyman, Boston, pp. 89–108.

Straus, L.G. (1993) Upper Paleolithic hunting tactics and weapons in western Europe. In: Hunting and animal exploitation in the Later Palaeolithic and Mesolithic of Eurasia (eds G.L. Peterkin, H.M. Bricker, and P. Mellars), Archeological Papers of the Anthropological Association, vol. **4**, 83–93.

Straus, L.G. (1995) The Upper Paleolithic of Europe: an overview. *Evol. Anthropol.*, **4**, 4–16.

Straus, L. and Clark, G. (1986) La Riera Cave. Anthropological Research Papers 36. Tempe, Arizona.

Straus, L.G., Altuna, G., Clark, G., González, M.M., Laville, H., Leroi-Gourhan, A., Menéndez de la Hoz, M., and Ortea, J. (1981) Paleoecology at La Riera. *Curr. Anthropol.*, **22**, 655–682.

Svoboda, J., Ložek, V., and Vlĭcek, E. (eds) (1996) *Hunters between East and West: the Paleolithic of Moravia*. Plenum, New York.

USGS (2008) GLSDEM, 90 m scene GLSDEM, Global LandCover Facility, University of Maryland, College Park, Maryland.

Trinkaus, E. and Ruff, C.B. (1989) Diaphyseal cross-sectional morphology and biomechanics of the Fond-de-Forêt 1 femur and the Spy 2 femur and tibia. *Bull. Soc. R. Belg. Anthropol. Prehist.*, **100**, 33–42.

Trinkaus, E. and Ruff, C.B. (2012) Femoral and tibial diaphyseal cross-sectional geometry in Pleistocene *Homo*. *PaleoAnthropology*, **2012**, 13–62.

Trinkaus, E., Churchill, S.E., and Ruff, C.B. (1994) Postcranial robusticity in Homo. II: Humeral bilateral asymmetry and bone plasticity. *Am. J. Phys. Anthropol.*, **93**, 1–34.

van Andel, T.H., Davies, W., and Weniger, B. (2003) The human presence in Europe during the Last Glacial Period I: Migrations and the changing climate. In: *Neanderthals and modern humans in the European landscape during the last glaciations* (eds T.H. van Andel and W. Davies), McDonald Institute for Archaeological Research, University of Cambridge, Cambridge, pp. 31–56.

Weiss, E. (2003) The effects of rowing on humeral strength. *Am. J. Phys. Anthropol.*, **121**, 293–302.

Wells PS. 1990. Iron Age temperate Europe: some current research issues. J World Prehist **4**:437–476.

Wells, P.S. (2002) The Iron Age. In: *European Prehistory: A Survey* (ed. S. Milisauskas), Kluwer Academic/Plenum, New York, pp. 335–383.

Whitehouse, R.D. (2001) Exploring gender in prehistoric Italy. *Pap. Br. Sch. Rome*, **69**, 49–96.

Winterhalder, B. (1981) Optimal foraging strategies and hunter-gatherer research in anthropology: Theory and models. In: *Hunter-gatherer foraging strategies* (eds B. Winterhalder and E.A. Smith), University of Chicago Press, Chicago, pp. 13–35.

Whittey, E. (2016) Effects of Terrain on Reconstructions of Mobility in Past Populations. Unpublished M.A. Thesis, University of Massachusetts, Amherst.

6

Sexual Dimorphism

Margit Berner[1], Vladimír Sládek[2], Brigitte Holt[3], Markku Niskanen[4], and Christopher B. Ruff[5]

[1] *Department of Anthropology, Natural History Museum, Vienna, Austria*
[2] *Department of Anthropology and Human Genetics, Faculty of Science, Charles University, Prague, Czech Republic*
[3] *Department of Anthropology, University of Massachusetts, Amherst, MA, USA*
[4] *Department of Archaeology, University of Oulu, Oulu, Finland*
[5] *Center for Functional Anatomy and Evolution, Johns Hopkins University School of Medicine, Baltimore, MD, USA*

6.1 Introduction

Sexual size dimorphism (SD) is a common phenomenon in primates and humans: it is multifactorial in expression and cause, whereby multiple influences may affect its evolution and expression (Plavcan, 2001). Among recent human populations, SD varies and differences have been studied from a number of perspectives including hormonal and nutritional factors, genetic determination, phenotypic plasticity, growth patterns, aging, secular changes, polygyny, and sexual division of labor (Gray and Wolfe, 1980; Wolfe and Gray, 1982; Frayer and Wolpoff, 1985; Stinson, 1985; Holden and Mace, 1999; Plavcan, 2001).

Early hominins were considerably more dimorphic in cranial, skeletal, and tooth dimensions than modern humans, and a decrease in SD has occurred over time (Brace, 1973; Frayer and Wolpoff, 1985). With regard to the postcranium, stature and body mass are among the most studied variables in the past and the present. Stature is a polygenic trait and its expression depends on the interaction of environmental and genetic factors (Lango Allen *et al.*, 2010), although it is not clear whether there is a genetic component for the basic level of sexual size dimorphism (Stinson, 1985). It is assumed that a potential stature is heritable but its expression is dependent on various factors that influence growth and development (Gray and Wolfe, 1980). Males and females are almost of the same size at birth and have similar growth rates until about puberty. At the onset of the growth spurt, size dimorphism increases (Tanner 1990). Male growth may generally be more sensitive to environmental stress than females, although the situation appears to be complex and many studies have produced contradictory results (see for example, Stinson, 1985; Vercellotti *et al.*, 2014).

European Upper Paleolithic groups have been claimed to be more sexually dimorphic in stature than groups of the later Mesolithic and Neolithic periods, attributed to changes in sex-specific activities relating to subsistence strategies (Frayer, 1980, 1981). Despite an overall increase in stature, Frayer found similar levels of SD between the Neolithic and a more modern period (Frayer, 1984). The reduction of SD in the late Pleistocene was

Skeletal Variation and Adaptation in Europeans: Upper Paleolithic to the Twentieth Century, First Edition. Edited by Christopher B. Ruff.

suggested to be the result of a greater decrease in male body size, which was explained in analogy to ethnographic hunter-gatherer societies by higher male involvement in tasks requiring greater body size (Frayer, 1980; Ruff, 1987). However, comparisons between recent ethnographic samples found no association between economic activities and SD in stature (Gray and Wolfe, 1980; Wolfe and Gray, 1982; Holden and Mace, 1999). Differences in stature appear to be related to the overall sexual division of labor, but not the kind of work done (Holden and Mace, 1999). We will re-examine the association between SD in stature and subsistence strategy in the present study sample, in particular, earlier claims that SD declined following the Upper Paleolithic (Frayer, 1980, 1984). If males are more sensitive to environmental influences during growth, we would also expect to see more change in male stature across subsistence transitions (see also Chapter 4).

While linear growth and overall body size seem to be largely controlled by genetics, hormones and nutrition, acquisition of bone mass is influenced by both systemic and localized biomechanical forces (Carter and Beaupré, 2001). With respect to systemic factors, several studies showed that androgen and estrogen receptors differ between the sexes (see references in Auerbach and Ruff, 2006). Sex steroids contribute to sexual dimorphism of the skeleton. It seems that estrogen not only suppresses periosteal apposition, but that it may also reduce the osteogenic response to mechanical strains. There is also evidence that increased exercise during growth affects all of the skeleton, including those parts of the body not directly involved in mechanical loading (Lieberman, 1996). On the other hand, many studies have demonstrated site-specific remodeling of the skeleton in response to localized mechanical forces (see reviews in Trinkaus *et al.*, 1994; Stock and Pfeiffer, 2001; Carlson *et al.*, 2007: Ruff, 2008; see also Chapter 3).

Regarding SD of the postcranium, almost all studies of skeletal remains have revealed SD in bone structural parameters, where male values tend to be greater than in females, but the degree of dimorphism varies between samples/populations. Some sex differences in cross-sectional properties of long bone diaphyses of archaeological skeletons have been explained by differences in activity patterns. Several studies found the highest levels of SD in femoral midshaft shape among hunter-gatherers, somewhat less in agriculturalists, and least in modern industrial societies (Ruff, 1987, 1999, 2008; Ruff and Larsen, 2001). Examinations of North American skeletal remains show that the changes in subsistence and intensification of agriculture are not reflected by a uniform biological response. Rather, they seem to reflect local variation and differences in gender roles (Bridges, 1989; Panter-Brick, 2002; Wescott, 2006). However, both Wescott and Bridges found strong evidence for declining SD in midshaft femur shape in more agricultural populations (Bridges, 1989; Wescott, 2006). Sexual dimorphism in cross-sectional shape of the femur, especially the proximal femur, may also be related to SD in body shape, that is, a greater pelvic breadth in females (Ruff, 1987).

While most of the studies examining temporal trends of SD in cross-sectional properties have been carried out on non-European samples (Ruff, 1987, 2008; Bridges, 1989; Bridges *et al.*, 2000; Ruff and Larsen, 2001; Wescott and Cunningham, 2006; Maggiano *et al.*, 2008), only a few studies analyzed temporal trends in European archaeological remains. Sexual dimorphism of cross-sectional morphology and strength have been examined in Upper Paleolithic and Mesolithic samples (Churchill *et al.*, 2000; Holt, 2003), and for regionally limited Neolithic, Bronze Age, and Medieval samples (Sládek *et al.*, 2006a,b, 2007; Marchi and Sparacello, 2006; Marchi *et al.*, 2011; Macintosh *et al.*, 2014).

Generally, greater sexual dimorphism in femoral or tibial shape has been interpreted as indicating male and female-specific activities that involve different bending loads on the lower limb (Ruff, 1987, 2008). According to Ruff (2008), changes in subsistence strategy might result in differences in activity levels and the mechanical loading of the skeleton, and will be dependent on the particular culture as well as on the physical environment, such as terrain. Differences between the sexes in lower limb shape and strength may be the result of different behavioral characteristics in males and females, where relative size or overall strength is associated more with general activity level, while strength in a particular plane reflects more the type or direction of mechanical loading.

Although the humerus is not weight-bearing and mechanical influences are more complex than in the lower limb (Ruff, 2008), clinical research and studies on archaeological skeletons have shown an impact of mechanical loading on upper extremity dimensions (see references in Trinkaus *et al.*, 1994; Ruff, 2008). Generally, greater male upper body strength is reflected in greater upper limb robusticity, and studies on asymmetry showed greater variation in cross-sectional properties of the humerus in males than in females (Auerbach and Ruff, 2006; Sládek *et al.*, 2007; Weiss, 2009; for asymmetry in the present study see Chapter 7). This observation supports the view that variation in humeral cross-sectional properties reflects greater variability in use of the humerus as a result of habitual and gender-specific behavioral differences. For instance, the increased male humeral robusticity and the greater cross-sectional circularity and higher asymmetry in Early Upper Paleolithic compared to Late Upper Paleolithic were interpreted as shifts in use of hunting weapons (Churchill *et al.*, 2000; Holt and Formicola, 2008; also see Chapter 7).

Human societies are characterized by marked sexual division of labor in subsistence activities. Therefore, it is expected that substantial changes in human societies will have an impact on the expression of SD. This may be particularly the case during the Holocene, with its changes in gender-specific subsistence activities and craft specialization, as well as access to resources. Although we can only rarely link specific activities of males and females in archaeological samples, differences in the degree of SD in cross-sectional properties may indicate differences in general levels or types of physical activities between males and females. While it is acknowledged that pooling samples by broad time periods will group different populations with different cultural and biological characteristics, there may be some general temporal trends that span these time periods and cultural groups and hence reflect broad changes in gender activities. This is explored in the present study.

With regard to propositions of higher male mobility in foraging societies, we predict a reduction of SD at the beginning of the Holocene due to reduced long-distance mobility (Holt, 2003; Ruff, 2008). If correct, this would mainly have resulted in changes in the lower limb, predominantly expressed by a reduction in SD of bone shape. According to studies of more rounded cross-sectional shape and least or no SD femoral shape in 'industrial' societies (Ruff, 1987, 2008), we would expect the lowest SD in femoral and tibial cross-sectional shape in samples that are associated with urban sites. This is based on the observation of a more sedentary life style and improvement in technology replacing physical demand. This should also produce less sexual dimorphism in overall bone strength.

It is more difficult to predict temporal changes in SD of upper limb bone structural properties. Although a transition from foraging to food production might decrease some mechanical loadings of the skeleton, some activities associated with intensification of agriculture (e.g., food grinding) might also increase mechanical loading of the upper limb. This would probably not only be reflected in bilateral asymmetry, that is, changes in physical activities that affect the dominant or both arms (see Chapter 7), but also more generally between temporal groups as a result of changes in agricultural tools and machines, as well as the invention of new technologies and professions which require different stereotypical movements and loading of the arms. Increased specialization and gender-specific tasks could be reflected by greater SD in humeral strength.

6.2 Materials and Methods

A total of 2180 individuals (1229 males, 951 females) from 279 European sites, comprising nine time periods, were analyzed for sexual dimorphism in stature, body mass and cross-sectional properties of femora, tibiae, and humeri (Table 6.1). The time periods are structured along cultural chronology and dating, and encompass the Upper Paleolithic which is the combined Early and Late Upper Paleolithic (33,000–11,000 BP), the Mesolithic (10 500–5900 BP), Neolithic (7300–4000 BP), Bronze Age (4350–2950 BP), the combined Iron Age/Roman sample (2250–1650 BP), further the Early Medieval (600–950 AD), Late Medieval (1000–1450 AD), Early Modern (1500–1850 AD) and very recent (≥1900 AD) time period (for further details see also Chapters 1, 4, 5, and 7). Earlier periods are often represented by few individuals from many sites, while later periods are represented by fewer sites with more individuals. Overall, males are slightly over-represented (56%) compared to females (44%).

The Neolithic sample was further subdivided according to different subsistence strategies: hunter-gatherers, agriculturalists and the Eneolithic or Copper Age period, which is associated with greater social differentiation and social stratification (Peregrine and Ember 2001; also see Chapter 1). For analyses of rural versus urban effects on SD, the Early Medieval through Modern period samples were pooled.

Sex is defined biologically and used as a categorical variable estimated through osteological determination methods (Sofaer, 2006), using pelvic morphology as the primary indicator and cranial morphology as the secondary indicator (see Chapter 1 for details). Stature and body mass were calculated using new formulae derived directly for the European Holocene record (Ruff *et al.*, 2012). For further comparisons the Brachial Index (BI; RBRI, LBRI) and the Crural Index (CI) were calculated. Cross-sectional geometric properties obtained included cortical, total and medullary areas (CA, TA, MA), the section moduli about the x- and y-axis which are antero-posterior (A-P) bending strength (Z_x) and medio-lateral (M-L) bending strength (Z_y), respectively, and their ratio (Z_x/Z_y), and the polar section modulus or average bending/torsional strength (Z_p). Cross-sectional areas were standardized by body mass, and section

Table 6.1 Sample sizes by temporal periods, sex, and sites.

	Total sample			Sites with both sexes			Sites with only females		Sites with only males	
Period	N Sites	nF	nM	N Sites	nF	nM	N Sites	nF	N Sites	nM
Upper Paleolithic (UP)	30	18	29	7	10	11	8	8	15	18
Mesolithic	33	37	56	7	27	36	9	10	17	20
Neolithic	113	122	180	30	88	116	30	34	53	64
Bronze	31	121	146	18	116	128	4	5	9	18
Iron/Roman	38	144	155	22	141	141	3	3	13	14
Early Medieval (EMed)	11	132	187	11	132	187				
Late Medieval (LMed)	13	256	289	13	256	289				
Early modern (EMod)	6	62	89	6	62	89				
Very recent	4	59	98	4	59	98				
Total	**279**	**951**	**1229**	**118**	**891**	**1095**	**54**	**60**	**107**	**134**

F: female; M: male.

moduli Z_p, Z_x and Z_y by the product of body mass and bone length (Ruff, 2008). For further comparison of lower versus upper limb bone proportions, the ratio of femoral to humeral strength was calculated for those individuals who are represented by both bones (Ruff and Larsen, 2001).

For statistical treatment, SPSSX 19 and Excel 2000 were used. Sex differences of absolute values between males and females were tested by applying two-sample *t*-tests. Calculation of sexual dimorphism (SD) is often performed using the mean value of one sex as a percentage of the other. Here, we use $(M - F)/((M + F)/2) \times 100$, which is the same method as for calculating side differences (Auerbach and Ruff, 2006; also see Chapter 7). For comparative purposes we also calculated the median percentage differences.

6.3 Results

6.3.1 Overall Variation in SD

Figure 6.1 shows the overall variation of SD in the total combined sample across different properties. Means and medians are given in Tables 6.2–6.4. Except for CI, all variables were statistically significant in *t*-tests. SD in body mass averages about 17%, and in stature about 6%, while SD in body proportions is lower. The brachial index (RBRI, LBRI) is higher in males (about 2%), and bi-iliac breadth/stature (BIB/Stat) is higher in females (about 3%), while the crural index (CI) shows no SD. Males have consistently greater size-standardized cross-sectional properties. The largest SD was found in Z_p, Z_x, Z_y and CA in the right humerus (about 20%), followed by the left humerus and tibia (about 15–20%), with the lowest SD in the femur (about 10%). Males have higher Z_x/Z_y ratios in the femur and tibia (3–4%), and females in the right humerus (2%).

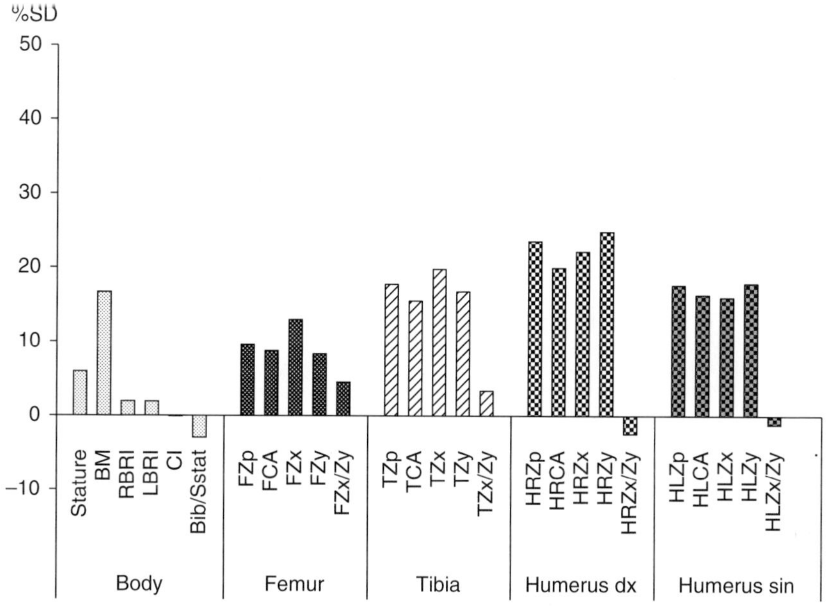

Figure 6.1 Comparison of % SD of stature, body mass (BM), brachial index (RBRI, LBRI), crural index (CI), bi-iliac/stature index (BIB/stat), and standardized femoral, tibial and humeral cross-sectional properties of the total sample. All male–female differences are statistically significant, except for crural index.

Table 6.2 Summary statistics and % SD of mean and median and *t*-test for stature, body mass, and body indices.

	Stature		Body mass		BIB/Stat		CI		RBRI		LBRI	
	% SD Mean (Med.)	*t*-Test	% SD Mean (Med.)	*t*-Test	% SD Mean (Med.)	*t*-Test	% SD Mean (Med.)	*t*-Test	% SD Mean (Med.)	*t*-Test	% SD Mean (Med.)	*t*-Test
UP	5.53 (5.46)	***	14.02 (16.10)	***	−1.09 (−1.97)		−1.15 (−0.73)		−0.25 (−0.16)		−0.89 (−0.44)	
Mesolithic	5.26 (5.72)	***	11.02 (10.84)	**	−0.16 (3.25)		1.39 (1.19)		3.23 (3.54)	*	4.74 (4.76)	***
Neolithic	6.44 (5.85)	***	18.13 (16.96)	***	−0.58 (−1.12)	***	0.24 (−0.20)		1.34 (1.73)	*	1.58 (1.74)	**
Bronze	6.65 (6.73)	***	17.64 (17.29)	***	−3.83 (−4.87)	***	0.50 (0.67)		2.32 (2.09)	***	2.75 (3.09)	***
IA/Roman	5.26 (4.67)	***	17.00 (17.25)	***	−1.58 (−1.73)	***	−0.06 (0.02)		2.28 (2.66)	***	2.31 (2.30)	***
EMed	5.78 (6.00)	***	16.40 (15.14)	***	−3.15 (−2.46)	***	−0.77 (−1.00)	*	0.93 (0.60)		1.12 (0.91)	*
LMed	6.11 (6.03)	***	18.14 (18.64)	***	−2.66 (−4.04)	***	−0.34 (−0.39)		1.69 (1.48)	***	1.65 (0.99)	***
EMod	4.96 (4.10)	***	12.58 (12.56)	***	−5.13 (−6.47)	*	−0.36 (−0.97)		2.96 (3.53)	***	2.23 (2.34)	**
Recent	7.52 (6.97)	***	16.02 (16.32)	***	−7.36 (−6.51)		0.17 (−0.49)		2.07 (1.99)	**	2.86 (2.83)	***
Total	5.98 (5.84)	***	16.67 (17.19)	***	−2.94 (−3.43)	***	−0.01 (0.03)		1.96 (1.82)	***	1.89 (1.86)	***

* $p < 0.05$; ** $p < 0.01$; *** $p < 0.001$.

Table 6.3 Summary statistics and % SD of mean and median and *t*-test for femoral and tibial cross-sectional properties of temporal periods.

	FZ_{xstd}		FZ_{ystd}		FZ_{pstd}		FZ_x/Z_y		FCA_{std}		FTA_{std}		FMA_{std}	
	% SD Mean (Med.)	*t*-Test	% SD Mean (Med.)	*t*-Test	% SD Mean (Med.)	*t*-Test	% SD Mean (Med.)	*t*-Test	% SD Mean (Med.)	*t*-Test	% SD Mean (Med.)	*t*-Test	% SD Mean (Med.)	*t*-Test
UP	8.11 (11.88)		−0.15 (5.54)		4.16 (9.57)		10.29 (14.31)	*	0.93 (5.64)		2.27 (4.07)		−4.11 (−8.74)	
Mesolithic	13.15 (14.57)	*	9.78 (7.72)	*	9.65 (10.86)	*	1.68 (2.60)		8.32 (7.12)		9.48 (14.93)	**	4.48 (6.69)	
Neolithic	12.48 (11.27)	***	6.93 (4.93)	**	9.55 (7.4)	***	5.78 (6.50)	***	8.41 (6.24)	***	4.78 (5.87)	***	7.26 (7.34)	
Bronze	12.31 (11.47)	***	8.62 (11.8)	**	10.45 (11.29)	***	3.45 (3.87)	*	9.65 (8.73)	***	6.33 (5.43)	**	9.19 (6.46)	
IA/Roman	10.51 (7.67)	***	5.57 (7.23)	*	6.55 (6.06)	**	4.78 (3.44)	**	6.25 (6.31)	**	3.73 (4.41)	*	3.94 (7.87)	
EMed	15.87 (16.75)	***	11.6 (12.51)	***	11.52 (13.62)	***	4.62 (4.40)	**	9.94 (9.79)	***	8.28 (10.18)	***	8.97 (8.22)	
LMed	11.55 (12.1)	***	8.56 (8.23)	***	9.01 (8.88)	***	2.82 (3.10)	**	8.75 (8.54)	***	5.75 (4.90)	***	8.45 (8.78)	
EMod	12.83 (15.18)	***	8.07 (6.83)	*	8.15 (6.72)	**	3.64 (3.08)		9.76 (5.12)	***	5.82 (7.28)	***	12.39 (12.46)	
Recent	11.72 (12.7)	***	9.52 (9.18)	**	10.3 (9.77)	***	1.22 (−1.09)		8.8 (7.44)	***	6.22 (5.10)	***	7.33 (6.13)	
Total	12.99 (13.66)	***	8.32 (9.32)	***	9.59 (10.56)	***	4.53 (3.75)	***	8.65 (8.73)	***	5.77 (6.43)	***	−2.28 (−0.64)	

(*Continued*)

Table 6.3 (Continued)

	TZ_{xstd}		TZ_{ystd}		TZ_{pstd}		TZ_x/Z_y		TCA_{std}		TTA_{std}		TMA_{std}	
	% SD Mean (Med.)	*t*-Test	% SD Mean (Med.)	*t*-Test	% SD Mean (Med.)	*t*-Test	% SD Mean (Med.)	*t*-Test	% SD Mean (Med.)	*t*-Test	% SD Mean (Med.)	*t*-Test	% SD Mean (Med.)	*t*-Test
UP	7.44 (4.10)		−4.02 (−11.97)		7.36 (6.59)		12.81 (7.05)	*	1.55 (4.73)		1.1 (−5.51)		−4.11 (−8.74)	
Mesolithic	23.07 (19.59)	*	16.22 (10.97)	*	20.86 (20.04)	*	9.07 (12.25)		19.03 (17.68)	***	14.43 (16.26)	**	4.48 (6.69)	
Neolithic	17.39 (16.19)	***	16.92 (14.9)	***	15.37 (13.72)	***	1.16 (1.34)		13.84 (12.56)	***	8.81 (9.67)	***	7.26 (7.34)	
Bronze	22.19 (18.65)	***	16.94 (18.01)	***	18.96 (14.25)	***	4.61 (2.14)	*	16.19 (13.28)	***	9.97 (7.90)	***	9.19 (6.46)	
IA/Roman	19.67 (18.53)	***	16.5 (17.43)	***	17.04 (17.01)	***	3.84 (5.28)	**	14.39 (15.84)	***	8.77 (7.57)	***	3.94 (7.87)	
EMed	18.47 (18.67)	***	18.45 (18.1)	***	18.18 (18.53)	***	−0.72 (1.2)		15.53 (17.97)	***	8.87 (10.34)	***	8.97 (8.22)	*
LMed	19.3 (17.74)	***	16.43 (15.08)	***	17.54 (15.75)	***	2.52 (2.13)	*	15.29 (13.94)	***	9.22 (9.61)	***	8.45 (8.78)	
EMod	18.07 (17.73)	***	14.23 (12.59)	**	17.53 (12.51)	***	5.99 (1.33)		15.2 (13.69)	***	11.67 (11.17)	***	12.39 (12.46)	
Recent	15.15 (13.64)	***	11.37 (12.67)	***	12.1 (9.32)	***	4.43 (4.26)	*	12.19 (8.83)	***	6.64 (5.34)	**	7.33 (6.13)	
Total	19.79 (18.87)	***	16.74 (16.26)	***	17.76 (17.01)	***	3.37 (3.80)	***	15.08 (14.84)	***	9.3 (9.10)	***	−5.35 (−4.00)	**

F: femoral; T: tibial.

Table 6.4 Summary statistics and % SD of mean and median and *t*-test for humeral cross-sectional properties of temporal periods.

	RHZ_{xstd}		RHZ_{ystd}		RHZ_{pstd}		RHZ_x/Z_y		$HRCA_{std}$		$HRTA_{std}$		$HRMA_{std}$	
	% SD Mean (Med.)	*t*-Test	% SD Mean (Med.)	*t*-Test	% SD Mean (Med.)	*t*- Test	% SD Mean (Med.)	*t*-Test	% SD Mean (Med.)	*t*-Test	% SD Mean (Med.)	*t*-Test	% SD Mean (Med.)	*t*-Test
UP	12.05 (18.80)		16.62 (13.15)		11.59 (17.16)		−3.56 (−9.15)		11.17 (18.36)		18.68 (19.79)		40.73 (26.68)	
Mesolithic	2.7 (14.66)		12.63 (17.63)		5.64 (11.38)		−9.06 (−11.72)	*	5.92 (10.71)		7.12 (5.85)		26.70 (44.21)	
Neolithic	18.53 (17.83)	***	24.61 (22.32)	***	21.02 (18.59)	***	−5.85 (−4.66)	***	18.89 (19.92)	***	10.06 (10.55)	***	−15.97 (−16.33)	**
Bronze	19.42 (20.43)	***	17.78 (20.1)	***	19.53 (19.78)	***	1.17 (0.50)		18.28 (15.98)	***	7.41 (9.12)	***	−21.8 (−22.95)	***
IA/Roman	17.67 (19.17)	***	19.53 (24.38)	***	17.64 (19.29)	***	−2.67 (−2.57)		14.24 (13.28)	***	9.59 (10.06)	***	−1.30 (−2.64)	
EMed	29.78 (28.03)	***	31.22 (29.66)	***	30.4 (28.06)	***	−1.25 (−1.40)		25.63 (26.29)	***	15.81 (15.41)	***	−6.26 (−7.55)	
LMed	19.57 (20.96)	***	23.85 (25.94)	***	22.6 (27.78)	***	−3.6 (−3.03)	**	18.18 (21.87)	***	11.80 (13.01)	***	−4.10 (0.18)	
EMod	28.17 (26.81)	***	29.04 (31.29)	***	29.97 (29.73)	***	0.26 (1.69)		27.08 (28.19)	***	17.97 (17.34)	***	0.01 (4.86)	
Recent	29.74 (27.89)	***	31.29 (30.73)	***	29.71 (28.09)	***	−1.08 (−1.22)		22.98 (23.08)	***	18.58 (17.57)	***	9.18 (4.40)	
Total	22.18 (23.73)	***	24.84 (25.96)	***	23.53 (24.45)	***	−2.41 (−1.95)	***	20.1 (19.79)	***	12.46 (12.84)	***	−6.12 (−6.68)	**

(*Continued*)

Table 6.4 (Continued)

	LHZ_{xstd}		LHZ_{ystd}		LHZ_{pstd}		LHZ_x/Z_y		$HLCA_{std}$		$HLTA_{std}$		$HLMA_{std}$	
	% SD Mean (Med.)	*t*-Test	% SD Mean (Med.)	*t*-Test	% SD Mean (Med.)	*t*-Test	% SD Mean (Med.)	*t*-Test	% SD Mean (Med.)	*t*-Test	% SD Mean (Med.)	*t*-Test	% SD Mean (Med.)	*t*-Test
UP	−3.05 (−2.89)		9.08 (4.74)		3.40 (0.45)		−10.4 (−8.67)		4.53 (−3.05)		15.56 (14.18)		16.7 (10.76)	
Mesolithic	18.58 (21.59)		33.84 (33.17)	**	26.6 (24.12)	*	−8.84 (−13.16)		20.31 (20.94)	*	16.21 (15.64)	**	8.28 (−1.88)	
Neolithic	10.57 (10.71)	***	9.95 (9.08)	**	9.69 (9.31)	**	1.21 (0.47)		11.16 (10.60)	***	9.96 (11.10)	*	−17.94 (−21.07)	**
Bronze	9.73 (8.03)	**	4.67 (3.5)		7.72 (8.11)	*	5.04 (5.51)	***	9.95 (4.99)	***	13.12 (13.63)		−21.95 (−27.50)	**
IA/Roman	11.05 (12.24)	**	11.99 (13.87)	***	11.79 (11.66)	***	−1.04 (−0.99)		8.14 (5.56)	**	4.44 (2.04)	*	−4.31 (−7.53)	
EMed	22.14 (20.16)	***	26.62 (23.39)	***	24.25 (22.87)	***	−4.00 (−4.44)	**	22.39 (22.19)	***	0.85 (−0.56)	***	−8.62 (−5.31)	
LMed	17.8 (18.99)	***	18.39 (20.42)	***	20.18 (19.99)	***	−0.41 (−0.28)		18.53 (19.49)	***	3.67 (2.16)	***	−10.78 (−11.62)	*
EMod	24.73 (25.17)	***	30.05 (29.07)	***	29.07 (28.38)	***	−5.95 (−5.44)	**	28.19 (29.66)	***	−1.14 (22.33)	***	−7.35 (−12.76)	
Recent	27.32 (23.99)	***	27.8 (28.53)	***	25.89 (24.58)	***	−0.05 (−2.54)		22.84 (22.53)	***	−3.71 (−1.44)	***	0.40 (1.59)	
Total	16.00 (17.48)	***	17.89 (18.24)	***	17.65 (18.51)	***	−1.15 (−1.09)	***	16.45 (15.41)	***	8.68 (9.00)	***	−9.95 (−12.22)	***

HR: right humerus; HL: left humerus.

6.3.2 Temporal Trends in Sexual Dimorphism

Summary statistics, the results of t-tests, and mean and median percentage sex differences for the nine temporal subsamples and the total combined sample are also presented in Tables 6.2–6.4. Temporal trends in SD, along with male and female values within periods, are plotted in Figures 6.2–6.5.

6.3.3 Body Size and Body Proportions

There is little change in SD of stature between the Upper Paleolithic (UP) and Mesolithic, while SD in body mass declines slightly (Fig. 6.2). There is no indication that UP or Mesolithic samples have higher SD in body size than succeeding time periods; in fact, the opposite is generally

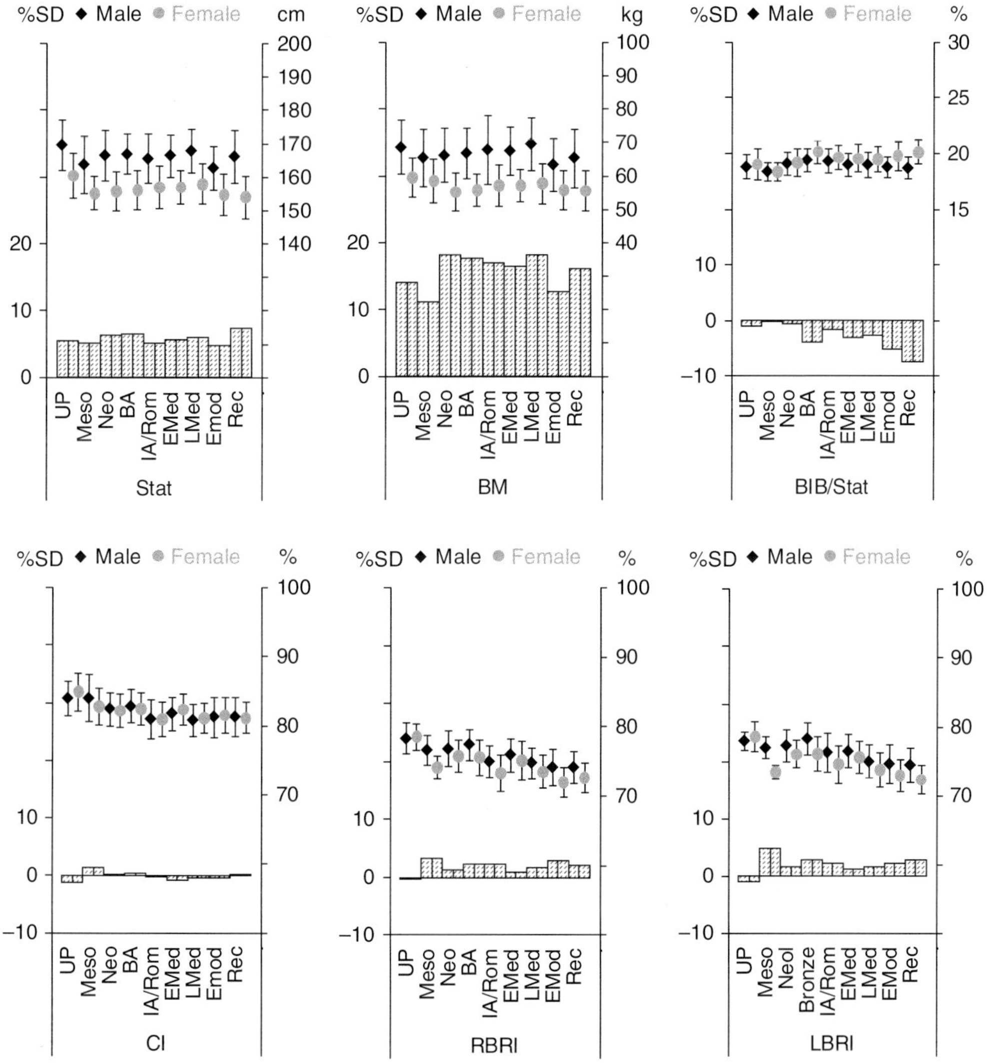

Figure 6.2 Temporal variation of % SD and mean ± SD of males and females in stature, body mass, bi-iliac breadth/stature index, crural index and right and left brachial indices.

true, especially for body mass. The increase in SD in stature in the Neolithic is the result of higher average male values, while the increase in SD in body mass is based on decreased average female values. SD in stature and body mass varies little in the following temporal periods through the Late Medieval. In the Early Modern period SD in body size decreases due to lower average male values. This is followed by an increase in male values and SD in the recent period. Very little SD was found in the BIB/Stature Index in the three earliest periods. SD in BIB/Stature (females larger) is greatest in the Early Modern and Recent periods due to an increase in average female values, especially in the most recent period. SD in the crural index is always quite small, while brachial indices are consistently larger in males (except in the UP), with the largest SD in the Mesolithic and smallest SD in the Early Medieval.

6.3.4 Average Strength

Figure 6.3 illustrates temporal trends in SD for femoral, tibial, and humeral midshaft polar section modulus (Z_p), a measure of torsional and average bending strength. SD in femoral and tibial average strength is relatively constant through time, except for reduced levels in the UP. The increase in SD between the UP and Mesolithic is mainly the result of higher average male values for the femur, and decreased average female values for the tibia. Thereafter, male and female values fluctuate in parallel, with slight declines in femoral SD in the Iron Age/Roman and in the recent period in the tibia.

Sexual dimorphism in average strength shows more temporal variability in humeri than in the lower limb bones, as well as distinctive differences between right and left sides (Fig. 6.3). SD in the UP is very small in the left humerus and moderately small in the right humerus. During the Mesolithic, SD increases greatly on the left side due to an increase in strength in males, and decreases on the right side due to an increase in females. SD then increases on the right side in

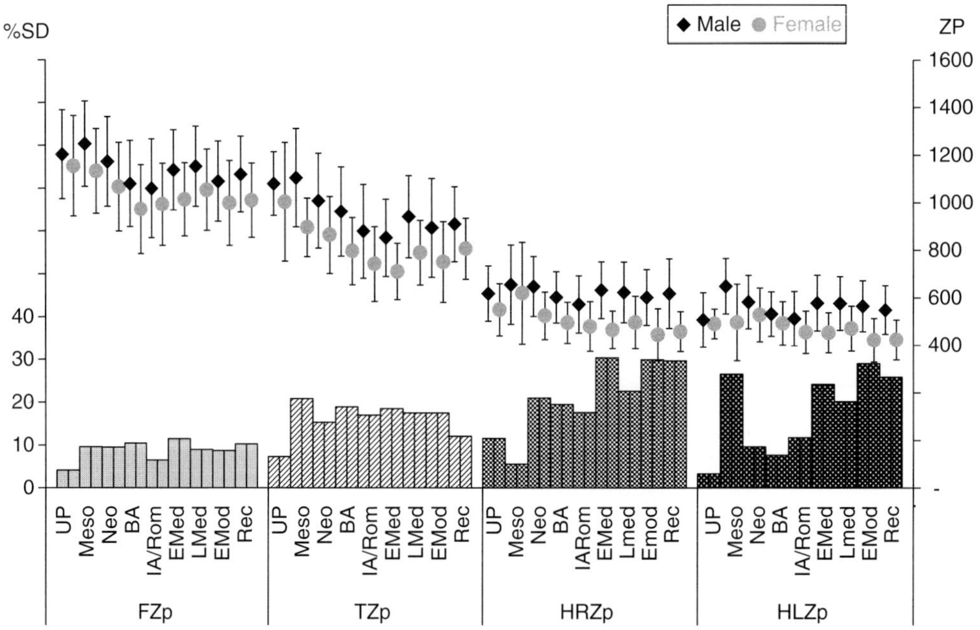

Figure 6.3 Temporal variation of % SD and mean ± SD of males and females in femoral, tibial and humeral average strength (Z_p).

the Neolithic (and remains high thereafter) due to female declines, and decreases on the left side due to female increases, remaining low through the Iron Age/Roman. These contrasting patterns can be interpreted in light of changes in humeral bilateral asymmetry among males and females during the Terminal Pleistocene and early Holocene, as described in Chapter 7 (also see Discussion below). Strength asymmetry declines among males between the UP and Mesolithic because of increases in strength on the left side (and little change on the right side), leading to greatly increased SD in strength on the left side. During the early agricultural periods (Neolithic through Iron Age/Roman) strength asymmetry is very small among females, again because of increases on the left side, which also decreases SD in left humeri. From the Early Medieval through recent periods, SD in average humeral strength is relatively constant and high compared to that in the lower limb. Right humeri generally show slightly higher SD than left humeri, reflecting slightly higher bilateral asymmetry among males (see Chapter 7).

6.3.5 Cross-Sectional Shape

Changes in femoral, tibial, and humeral Zx/Zy ratios are shown in Figure 6.4. Sexual dimorphism in femoral and tibial shape (males larger ratios) is very high in the UP, and also high in the tibia in the Mesolithic. Both sexes decline in femoral Z_x/Z_y between the UP and Mesolithic, but males decline much more, leading to very little SD in the Mesolithic. SD in femoral shape then slightly increases and remains relatively constant through the rest of the Holocene, with a decline in the recent period. Males show an almost continuous temporal decline in tibial Z_x/Z_y from the UP on, while females show no decline (and even a slight increase) from the UP through the Neolithic, resulting in almost no SD in tibial shape in the Neolithic. SD in tibial shape increases in the Bronze Age to moderately low values and shows only minor fluctuations thereafter, with a dip in the Early Medieval.

The most obvious pattern for humeral Z_x/Z_y shape ratio is a negative SD in most periods, indicating absolute higher female values (Fig. 6.4). Males are only larger in both sides in the

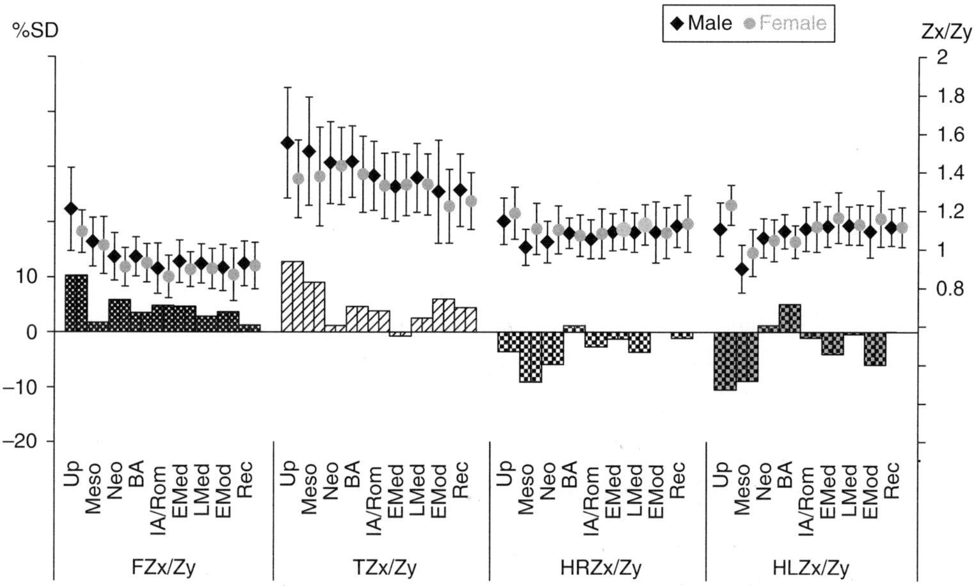

Figure 6.4 Temporal variation of % SD and mean ± SD of males and females in femoral, tibial, and humeral shape (Z_x/Z_y).

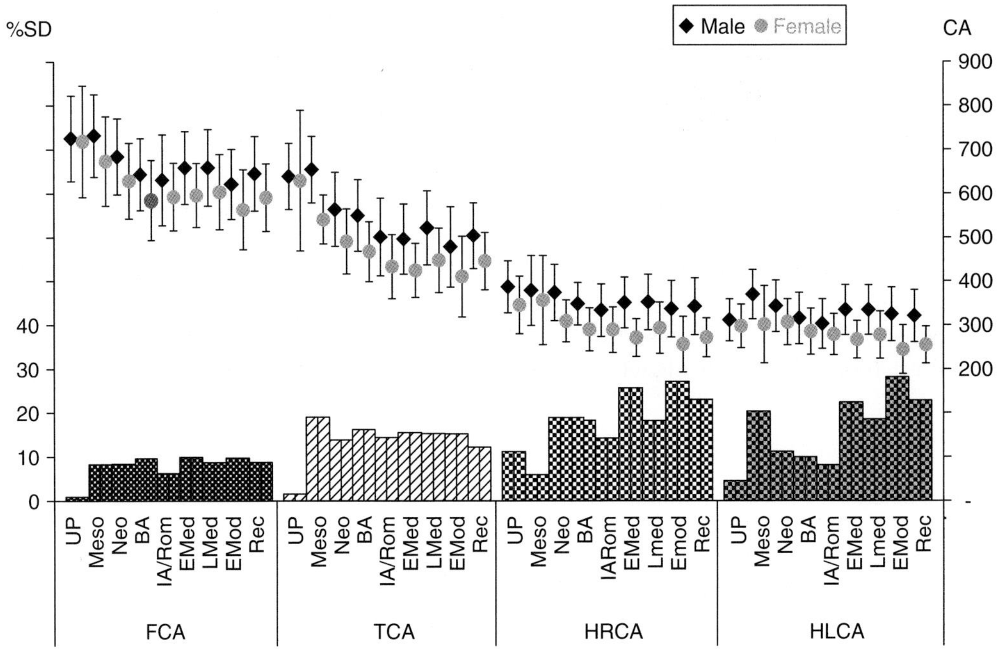

Figure 6.5 Temporal variation of % SD and mean ± SD of males and females in femoral, tibial, and humeral cortical area (CA).

Bronze Age. Only for a few temporal periods were statistically significant differences found using *t*-tests (Table 6.4). The highest SD in humeral shape occurs on the right side in the Mesolithic and Neolithic periods, and on the left side in the UP and Mesolithic periods. Both sexes decline in humeral Z_x/Z_y between the UP and Mesolithic (more so on the left) and then increase (on the right side, only among females). The smaller SD in most later time periods is mainly due to greater increases in the ratio among males.

6.3.6 Cross-Sectional Area

The temporal patterns of SD and average values of male and female femoral, tibial, and humeral cortical areas (CA) presented in Figure 6.5 are similar to those found for the polar section modulus, Z_p. In the lower limb, SD in the UP is low, increases in the Mesolithic, and remains very similar throughout the Holocene. For the humerus, SD of CA is lowest in the UP as well as on the right side in the Mesolithic. SD is low on the left side during the early agricultural periods and increases on both sides during the later Holocene.

6.3.7 Comparison of Neolithic Subsistence Groups

Further analyses of lower limb cross-sectional properties of the three different Neolithic subsistence groups (Neolithic hunter-gatherers, agriculturalists, and late Neolithic or Copper Age) were carried out, with results presented in Table 6.5 and Figure 6.6. The figures compare median SD in femoral and tibial cross-sectional shape and average strength of each Neolithic subgroup to the Upper Paleolithic, Mesolithic, and Bronze Age samples. In terms of SD in cross-sectional shape, the Neolithic hunter-gatherers are most similar to Mesolithics, with

Table 6.5 Summary statistics and results of *t*-test of the Neolithic subsamples.

	Neolithic Hunter Gatherer				Neolithic Agriculturalists				Copper Age			
	N F/M	Mean F/M	% SD Mean (Med.)	*t*-Test	N F/M	Mean F/M	% SD Mean (Med.)	*t*-Test	N F/M	Mean F/M	% SD Mean (Med.)	*t*-Test
FZ_p	26/44	1224.0/1298.9	5.94 (1.24)		44/56	997.8/1153.7	14.5 (13.32)	***	39/56	1033.1/1091.2	5.47 (−0.17)	
FZ_x/FZ_y	27/44	0.906/0.913	0.80 (0.05)		45/59	0.912/0.969	6.03 (6.81)	**	39/58	0.911/0.994	8.75 (7.60)	**
FZ_x	26/44	628.3/662.2	5.26 (4.58)		44/56	514.9/614.1	17.56 (16.9)	***	39/56	522.8/584.7	11.18 (11.21)	**
F_{zy}	26/44	694.4/729.4	4.92 (0.15)		44/56	569.7/633.4	10.58 (11.86)	**	39/56	577.5/593.8	2.79 (−2.50)	
FCA	26/44	689.1/726.1	5.24 (4.64)		44/56	607.5/683.7	11.8 (15.03)	***	39/57	611.9/652	6.34 (4.52)	*
TZ_p	22/43	967.0/1150.4	17.31 (9.55)	***	40/50	816.7/952.8	15.39 (14.39)	**	33/57	848.1/943.1	10.61 (11.00)	**
TZ_x/TZ_y	23/43	1.309/1.414	7.75 (8.88)	*	41/53	1.471/1.469	−0.19 (3.88)		34/58	1.47/1.457	−0.88 (−1.07)	
TZ_x	22/43	636.1/790.0	21.59 (17.04)	***	40/50	569.3/675.8	17.11 (16.12)	**	33/57	599.1/675.5	12.00 (14.59)	**
T_{zy}	22/43	488.4/567.7	15.02 (11.17)	**	40/50	391.9/472.3	18.6 (20.71)	***	33/57	408.4/467.2	13.44 (11.72)	***
TCA	22/43	507.6/597.8	16.31 (15.86)	***	40/50	453.5/555.2	13.81 (16.37)	***	33/57	486.1/544	11.23 (8.96)	***

F: female; M: male.

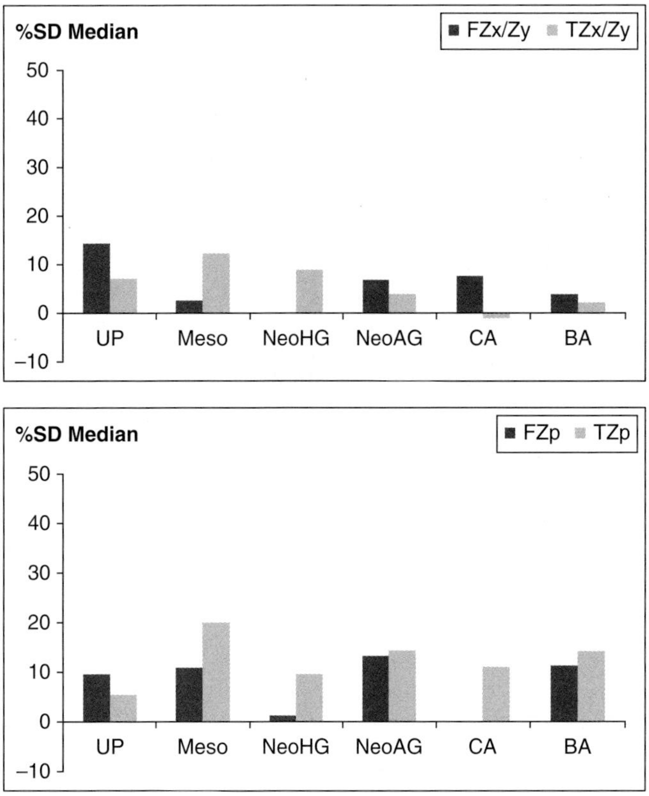

Figure 6.6 Variation in % SD of femoral and tibial shape ratios (Z_x/Z_y) and average strength (Z_p) among Neolithic subgroups (hunter-gatherers, agriculturalists, and Copper Age) in comparison to the Upper Paleolithic, Mesolithic, and Bronze Age samples.

relatively high SD in tibial shape and little SD in femoral shape. Neolithic agriculturalists appear more similar to Copper and Bronze Age samples, with smaller SD in tibial shape and moderately low SD in femoral shape. Copper and Bronze Age samples are similar to each other in this respect, and unlike UP and Mesolithic samples. SD in overall average strength (Z_p) does not as clearly distinguish different subsistence groups., with overlapping patterns in foragers and agriculturalists (Fig. 6.6).

6.3.8 Urban Versus Rural

Figure 6.7 gives the comparison of SD of cross-sectional properties between urban and rural samples, with summary statistics presented in Table 6.6. With the exception of tibial and right humeral Z_x/Z_y shape, the rural sample exhibits generally higher SD. The patterning of SD is similar in the two groups, and, both are significantly sexually dimorphic for almost all traits.

6.3.9 Femoro-Humeral Strength Ratio (FZ_p/HZ_p)

Because males and females show different patterns of SD in the upper and lower limb bones, lower to upper limb bone strength variation was also explored, using an index of femoral to humeral average strength (Z_p) in the nine temporal periods, presented in Figure 6.8 and Table 6.7. The generally negative SD values are the result of higher female femoral/humeral ratios. This pattern is more consistent for the right humerus, with more fluctuation in the left

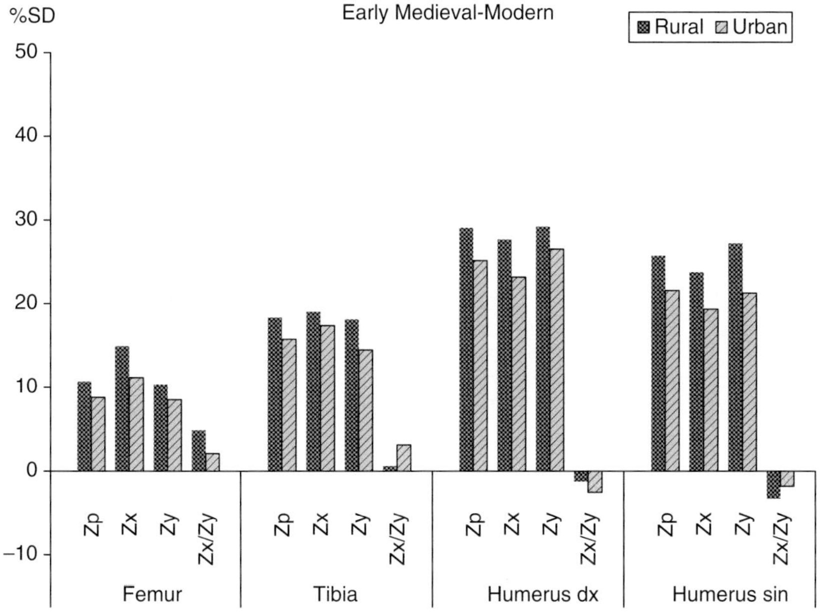

Figure 6.7 Comparison of % SD of cross-sectional properties for rural and urban groups of the combined Early Medieval to Modern sample.

because of the marked temporal changes in left humeral strength noted above. For both humeri, SD in femoral to humeral Z_p increases from the Bronze Age to the two most recent periods, due to increases in the ratio among females.

6.4 Discussion

During the past 30,000 years in Europe, substantial changes in subsistence strategies have occurred, which can be used to roughly define three broader periods. These are foraging societies of the Upper Paleolithic and the Mesolithic, predominantly farming societies starting with the appearance of agriculture in the Neolithic and its intensification in the Bronze and Iron Ages, and finally the Medieval to modern period, which represents a time span with increasing urbanization and sedentary lifestyle, as well as mechanization of work. The intensification of agriculture beginning in the Late Neolithic and extending into the Bronze and Iron Ages included innovations such as plows, wagons, and horse riding as well as dairy and wool production (Secondary Products Revolution), pottery, the beginning of metallurgy and increased craft specialization, long-distance trading, and social differentiation and stratification (Bogucki and Crabtree, 2004). By the Iron/Roman period, increasing specialization in production and crafts, as well as increasing non-food-producing populations had emerged (Hoffmann, 2014). Until the Early Medieval the majority of the population lived in rural societies and small villages as subsistence farmers with some surplus production. With increasing population size, this settlement pattern shifted in the Middle Ages to larger villages; large parts of European woodland became converted into agricultural landscape. This development was accompanied by intensive woodworking and new farming techniques such as heavy plowing and water drainage (Hoffmann, 2014). Despite a heavy workload in both sexes and overlap of activities – particularly during harvesting time – there is evidence that in medieval peasants, household and childcare were seen as women's tasks while males were working more in fields and forests, for example while carting (Beattie, 2006). Although roles of men and women varied in time and place according to

Table 6.6 Summary statistics and *t*-tests between rural and urban in the combined Early Medieval to Modern sample.

	Rural				Urban			
	N F/M	Mean F/M	% SD Mean (Med.)	*t*-Test	N F/M	Mean F/M	% SD Mean (Med.)	*t*-Test
Stature	231/265	157.2/166.6	5.86 (5.82)	***	268/386	156.6/166.4	6.08 (6.07)	***
Body mass	226/264	57.1/68.1	17.7 (16.94)	***	260/377	57.2/66.9	15.6 (16.96)	***
FZ_p	164/207	1027.4/1143.6	10.71 (14)	***	239/331	1030.3/1125.1	8.8 (7.46)	***
FZ_x	164/207	516.2/599.6	14.95 (16.56)	***	239/331	517.1/578.3	11.17 (10.73)	***
FZ_y	164/207	581/644.3	10.33 (11.04)	***	239/331	579/630.6	8.54 (8.73)	***
FZ_xZ_y	168/208	0.891/0.936	4.88 (4.94)	***	244/338	0.902/0.922	2.11 (1.08)	**
FCA	164/207	594.9/658.7	10.17 (9.83)	***	239/332	594.5/645.3	8.21 (7.69)	***
TZ_p	151/188	743.1/893.2	18.35 (19.13)	***	214/292	774.1/906.6	15.76 (15.6)	***
TZ_x	151/188	493.4/597.3	19.04 (18.88)	***	214/292	505.2/601.4	17.39 (16.91)	***
TZ_y	151/188	378.9/454.4	18.14 (17.44)	***	214/292	390.9/451.9	14.48 (13.93)	***
TZ_xZ_y	156/192	1.32/1.328	0.61 (−0.61)		224/302	1.3/1.342	3.13 (3.43)	**
TCA	151/189	432/505	15.58 (16.32)	***	214/293	437.2/503.7	14.13 (12.42)	***
HRZ_p	119/166	465.8/624.4	29.1 (29.55)	***	180/226	480.2/618.3	25.14 (27.05)	***
HRZ_x	119/166	263.7/348.3	27.65 (27.24)	***	180/226	272.4/343.8	23.16 (26.2)	***
HRZ_y	119/166	238.5/320.2	29.24 (29.46)	***	180/226	243.2/317.5	26.51 (27.92)	***
HRZ_xZ_y	123/171	1.11/1.097	−1.22 (−1.22)		191/238	1.12/1.093	−2.51 (−2.46)	**
HRCA	119/167	273.7/351.4	24.85 (24.84)	***	184/233	282/345.2	20.15 (21.67)	***
HLZ_p	117/152	445.3/576.8	25.72 (24.48)	***	169/241	456.7/567.2	21.57 (21.23)	***
HLZ_x	117/152	258.5/328.1	23.74 (23.62)	***	169/241	265.9/322.8	19.34 (19.97)	***
HLZ_y	117/152	225.2/296.1	27.19 (24.79)	***	169/241	234.8/290.7	21.26 (21.7)	***
HLZ_xZ_y	120/157	1.151/1.114	−3.27 (−3.45)	**	180/252	1.138/1.118	−1.78 (−1.94)	
HLCA	117/153	262.6/334.6	24.09 (23.33)	***	173/245	269.6/328.5	19.69 (20.56)	***

F: female; M: male.

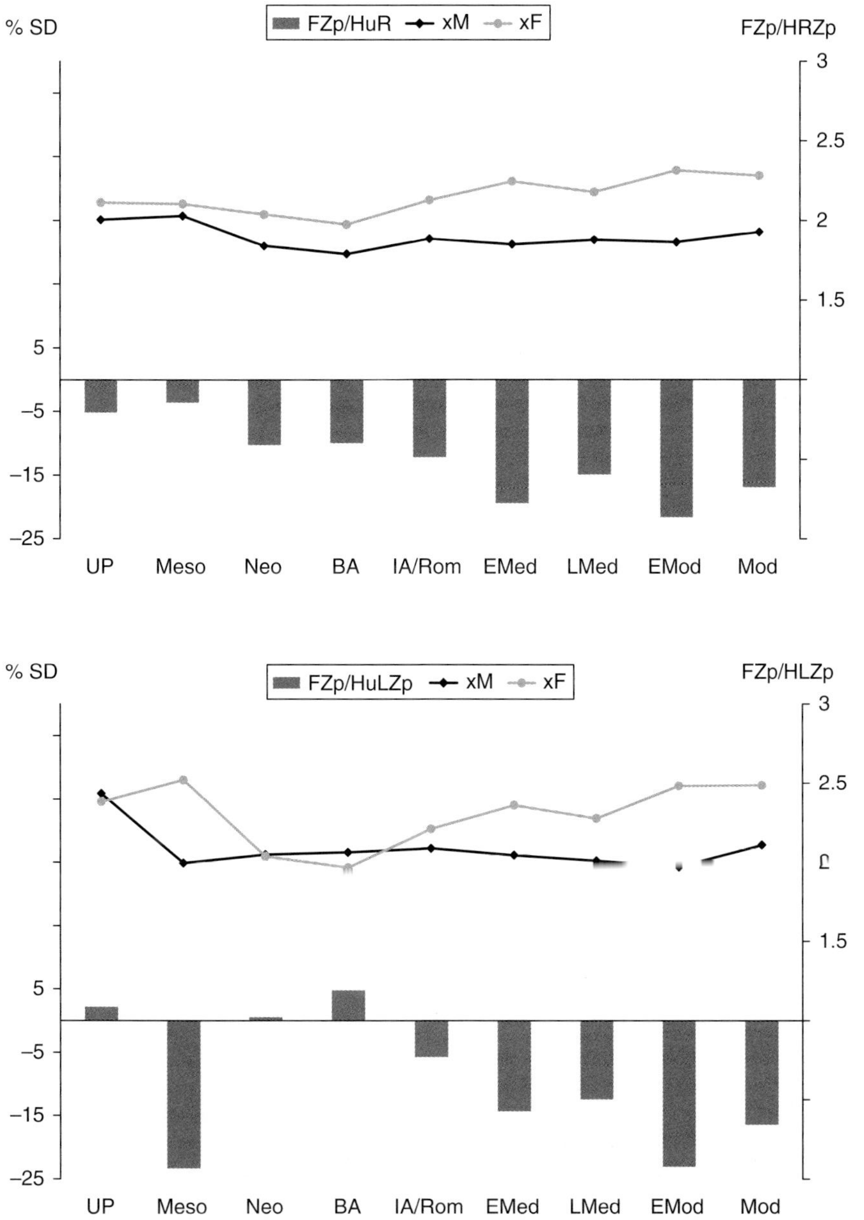

Figure 6.8 Temporal variation of % SD of the ratio of femoral to humeral robusticity (FeZ_p/HuZ_p).

the context, and despite enormous economic changes, much of this pattern of rural societies continued into the early modern period, as the majority of the population continued to live in the country, producing agricultural products for their own or their landlord's use (Hoffmann, 2014). Delegated work was often performed by persons of the same sex. In the early modern period many occupations were professionalized with specialized trainings and early onset in life, although this was restricted much more to males. Women's work was essential for rural and urban economies, but their economic activities were increasingly restricted and the vast majority of women's work was regarded as low-status and badly paid (Wiesner, 1993).

Table 6.7 Summary statistics and *t*-tests between the sexes for the ratio of femoral to humeral robusticity (FeZ_p/HZ_p) in individuals who are represented by both.

	$FeZ_p/HuRZ_p$				$FeZ_p/HuLZ_p$			
	N F/M	Mean F/M	% SD Mean (Med.)	*t*-Test	N F/M	Mean F/M	% SD Mean (Med.)	*t*-Test
UP	8/15	2.112/2.005	−5.19 (−6.31)		8/14	2.383/2.435	2.18 (−1.08)	
Meso	4/14	2.104/2.029	−3.63 (−10.21)		7/13	2.522/1.995	−23.31 (−25.77)	*
Neol	73/106	2.038/1.840	−10.26 (−8.57)	***	74/105	2.037/2.048	0.56 (3.09)	
Bronze	57/78	1.976/1.788	−9.96 (−8.42)	**	60/76	1.966/2.063	4.79 (4.08)	
IA/Roman	93/90	2.129/1.885	−12.16 (−9.45)	***	83/91	2.211/2.088	−5.72 (−4.84)	*
EMed	87/116	2.246/1.850	−19.33 (−18.04)	***	81/110	2.360/2.045	−14.33 (−15.64)	***
LMed	124/144	2.179/1.878	−14.84 (−19.26)	***	119/144	2.276/2.009	−12.45 (−13.4)	***
EMod	43/72	2.322/1.864	−21.88 (−21.37)	***	42/73	2.483/1.970	−23.03 (−19.88)	***
Recent	37/50	2.282/1.927	−16.86 (−15.37)	***	35/55	2.487/2.109	−16.44 (−14.61)	***
Total	527/685	2.157/1.866	−14.48 (−14.18)	***	510/681	2.244/2.050	−9.05 (−9.73)	***

F: female; M: male.

These changes in subsistence strategies are associated with changes in locomotor patterns, such as the reduction in regular long-distance travel of hunter-gatherer societies to more sedentary lifestyles in modern societies, as well as changes in different patterns of activities and manipulative behavior in males and females. The purpose of the present comparisons was to see how these various environmental influences might be reflected in changing patterns of SD in both body size and shape, and mechanical loading of the limb bones.

6.4.1 Body Size and Shape

As expected, male body size was larger overall and in almost all temporal subgroups. Average SD in body mass (17%) and stature (6%) are similar to means reported for modern humans generally (Stini, 1976; Gray and Wolfe, 1980; Ruff, 2002; Garvin, 2012).

Our results showed essentially no change in SD in stature, and only a slight decrease in SD of body mass between the Upper Paleolithic and Mesolithic periods, followed by a more marked increase in the Neolithic period. A decrease in SD in stature and body mass between the Upper Paleolithic and the Mesolithic, has been reported in some previous studies (Frayer 1980, 1981, 1984). In these studies a greater and more pronounced decrease was observed in male body size than in female body size following the Upper Paleolithic, which was explained by better hunting technology, leading to a reduction of risk and strength requirements of males (Frayer, 1980, 1981, 1984; also see Holt and Formicola, 2008). In our sample, both males and females declined in body size following the UP, but the decline was similar, casting some doubt on this hypothesis (or suggesting the concurrent operation of other factors, including dietary and demographic factors – see Formicola and Giannecchini, 1999; Formicola and Holt, 2007). It should be noted that other studies have also reported only small changes in SD of stature or long bone lengths between the UP and Mesolithic in Europe (Meiklejohn *et al.*, 1984; Meiklejohn and Babb, 2011).

We found a small increase in SD of stature and a larger increase in SD of body mass during the transition from the Mesolithic to Neolithic, resulting from an increase in male average stature and a decrease in female average body mass. Frayer (1980, 1984) found little change in SD of stature between the Mesolithic and Neolithic, while Meiklejohn *et al.* (1984) reported a small increase, similar in magnitude to ours. As noted earlier, evidence for consistent effects of agricultural versus foraging subsistence strategies on stature SD in living human populations is mixed. Except for body mass in females, our results do not indicate a decline in body size with the introduction of agriculture in Europe, nor a decline in SD of body size. Neither observation supports a general scenario of decreased health (at least as reflected in growth in body size) at the introduction of food production in Europe, or the 'female buffering' hypothesis in which males show greater environmental sensitivity to such changes (Stinson, 1985; Cohen and Crane-Kramer, 2007). Nor is there any evidence in our data for significant changes in body size SD during agricultural intensification in the Bronze Age, although there is a slight decline in the Iron Age/Roman period.

There is some evidence for more male variability in body size in our three most recent samples, from the Late Medieval to early modern to recent periods. Males decline in body size more than females in the early modern period, then increase more in the recent period, leading to a reduction followed by an increase in SD. The lower SD of body size in the early modern period might be a result of increased nutritional stress and generally lower living standards in this period of increasing industrialization (see Chapter 4), if the female buffering hypothesis applies to these populations. Greater recovery of male body size, resulting in an increase in SD in the 20th century, may reflect the same phenomena. This trend is apparent in some analyses that include anthropometric data, for example, in Britain (Chapter 8).

The lack of temporal change in SD of linear body proportions, despite some overall temporal trends (reduction in brachial and crural indices through the Holocene; see Chapter 4), implies that males and females responded similarly to whatever environmental effects (e.g., nutritional, climatic) produced these trends. The increase in SD of bi-iliac breadth/stature during the Holocene is the result of a temporal decrease in this index among females after the Bronze Age (also see Chapter 4).

6.4.2 Lower Limb Strength

Differences in locomotion and general activity level are reflected in lower limb diaphyseal morphology. Running and climbing over rough terrain should preferentially increase A-P bending loads on the lower limb bones, and thus mobile populations (or subpopulations) should show higher A-P relative to M-L bending strength in femoral and tibial midshaft sections (see discussion in Ruff, 1987, 2008).

The most significant change in SD of femoral and tibial Z_x/Z_y in our sample occurs between the UP and Neolithic. SD in bone shape declines – in the femur between the UP and Mesolithic, and in the tibia from the UP through the Neolithic – due to greater declines among males (females show no decline in the tibia). This trend is consistent with earlier findings that SD in bone shape, and implied mobility, decreases between foragers and food producers (Ruff, 1987, 2008). While occupational tasks were still gender-defined throughout much of the Holocene (see above), sex differences in mobility *per se* likely declined following the UP, as long-distance hunting (by males) was replaced by the exploitation of more local resources, and eventually agriculturally related tasks (see Holt, 2003; Holt and Formicola, 2008; also see Chapters 5 and 7). Our conclusions here are somewhat different from those of an earlier paper (Holt, 2003), in which less evidence for temporal changes in SD between the Early UP, Late UP, and Mesolithic were found. However, sample sizes in that study were smaller, particularly when the Early and Late UP were divided. Later periods following the Mesolithic were also not available for comparison.

The lack of any consistent change in SD of bone shape after the Neolithic suggests that sex differences in mobility also did not vary significantly, despite some continuing overall decline in relative A-P bending strength (Z_x/Z_y), especially in the tibia. Thus, while occupational specialization and gender roles certainly varied during this time, these changes apparently did not affect mobility as much as other behavioral characteristics such as asymmetry in upper limb use (see below, and Chapter 7). The further decline in femoral Z_x/Z_y in the recent period may reflect even smaller sex-related differences in mobility associated with industrialization (Ruff, 1987), although the tibia does not show the same trend as strongly.

SD in overall femoral and tibial strength (Z_p) does not show the same temporal trend as SD in bone shape, with an increase in SD after the UP and then no further significant change. This highlights the specificity of bone mechanical loading effects, that is, A-P versus average strength. Differential changes in mobility would be expected to preferentially affect A-P bending, but not necessarily M-L bending or torsional loading of the lower limb bones; thus, SD in Z_p would not be predicted to follow the same temporal trends as in Z_x/Z_y. This interpretation is supported by studies of living humans. Shaw and Stock (2009a) compared tibial cross-sectional properties associated with different athletic activities, and found a more A-P strengthened tibial shape in runners, corresponding to a more unidirectional locomotor pattern, and a more equal triangular-shaped tibia in hockey players, corresponding to a more multidirectional pattern of locomotion. Similarly, clinical studies on professional basketball players found higher M-L force during turning and side movements (McClay *et al.*, 1994), and higher tibial compressive strength in a strain gauge study during vigorous zig-zag movements (Burr *et al.*, 1996).

Our results for femoral and tibial Z_p also suggest that SD in overall activity level, muscularity, and so forth did not vary significantly throughout the Holocene (and may even have increased somewhat between the UP and Mesolithic).

Comparisons between Neolithic foragers and agriculturalists also indicate that bone shape rather than overall average strength better distinguish between different subsistence groups. Neolithic hunter-gatherers exhibit relatively high SD in tibial Z_x/Z_y, more similar to that of UP and Mesolithic samples than to later agricultural samples. SD in tibial Z_p is again not diagnostic of subsistence differences. SD in femoral shape, while declining after the UP in all groups, does not distinguish between Mesolithic and Neolithic samples (hunter-gatherer or agricultural). It has been proposed that the tibia is a better indicator of mobility than the femur, in part because of potentially confounding effects of body shape on the femur (Ruff *et al.*, 2006; Stock, 2006). If this is true, then it indicates that SD in mobility among Neolithic foragers was more similar to those of Mesolithic than to agricultural Neolithic populations. Because our Neolithic foragers are restricted to Scandinavia, it is not clear whether this represents a general pattern or a more regional one. However, Marchi *et al.* (2011) reported similar findings for a Neolithic sample from Liguria, a region in northern Italy where the local population practiced a mainly pastoral rather than agricultural subsistence strategy. In their sample, SD in femoral and tibial bone shape was quite high – most similar to that of European UP samples (they compared maximum to minimum rather than A-P to M-L properties, and rigidity rather than strength, so it is difficult to make precise comparisons to those of the present study). Their findings are consistent with a greater male role in herding activities in this population, which involved more long-distance travel. The ruggedness of the local terrain also likely increased SD in lower limb loadings (also see Chapter 5). Thus, the use of broad subsistence categories may mask more subtle differences in gender-specific activity patterns, which are reflected in bone morphology.

As expected, the comparison of the urban and rural groups in the merged Early Medieval to Modern sample exhibited, with the exception of tibial shape (where SD was low in both groups), a higher SD in the rural group. This may be related to an increased homogeneity of activity patterns among urban populations – that is, less sexual division of labor (at least with regard to large differences in mechanical loadings of the limbs), which is consistent with findings for the lower limb in broader samplings (Ruff, 1987, 2008).

6.4.3 Humerus

From ethnographic observations and archaeological studies it is assumed that foraging societies are characterized by an active lifestyle and gender-based division of labor, with little task specialization within sex (Stock and Pfeiffer, 2001). Early agricultural societies are associated with greater task overlap in males and females in fieldwork and food processing, but still little occupational specialization (Bogucki and Crabtree, 2004). Later agricultural and industrial societies are characterized by increasing task variation, occupational training and craft specialization (Wiesner, 1993; Hoffmann, 2014). It is predicted that higher levels of gender-based task specialization involving differences in manipulative behavior between males and females will result in higher SD in upper limb bone strength.

In general, humeri reflect multifunctional adaptations and are more difficult to interpret in terms of mechanical loadings than bones of the lower limb, but studies of athletes have shown that habitual manual activities and repetitive stereotypical use of the upper limbs are reflected in humeral cross-sectional geometry (e.g., Trinkaus *et al.*, 1994; Haapasalo *et al.*, 1996; Shaw and Stock, 2009b). Based on these results, a number of studies have interpreted variation of SD in humeral strength and cross-sectional geometric properties in skeletal remains as a possible result of differences in habitual manual activities in males and females (Ruff and Larsen, 1990, 2001; Trinkaus *et al.*, 1994; Ruff, 1999; Bridges *et al.*, 2000; Stock and Pfeiffer, 2001; Weiss, 2003, 2009; Sládek *et al.*, 2007). Many biomechanical studies of humeri concentrated on humeral bilateral asymmetry, including sex differences (Trinkaus *et al.*, 1994; Ruff, 1999; Stock and Pfeiffer, 2004; Auerbach and Ruff, 2006; Sládek *et al.*, 2007; Carlson *et al.*, 2007; Ruff, 2008; Weiss, 2009; see also Chapter 7). For instance, higher male bilateral asymmetry in humeral strength in Upper Paleolithic populations was interpreted to indicate regular use of hunting weapons, and greater cross-sectional circularity in EUP males compared to LUP males was explained by a shift in habitual hunting technology from hand-held thrusting spears and higher A-P stress to spear throwing associated with higher torsional strength (Churchill *et al.*, 2000). But even in females, higher asymmetry was noted when compared to recent foragers, suggesting different behavior possibly as a result of regular weapon usage to some extent (Churchill *et al.*, 2000). Studies with findings of less asymmetry in cross-sectional properties in females were linked to greater symmetrical use of the upper limbs, possibly as a result of bimanual activities such as grain processing and digging activities (Bridges, 1989; Fresia *et al.*, 1990; Weiss, 2009). Decreasing SD in asymmetry of upper limb bone dimensions was found in European industrial groups in comparison to preindustrial groups, possibly reflecting less sex differentiation in manipulative behavior (Auerbach and Ruff, 2006). Other studies addressed SD of humeral strength in light of general patterns of habitual behavior and SD in relation to subsistence (Bridges *et al.*, 2000; Stock and Pfeiffer, 2001; Weiss, 2003).

Males have generally stronger humeri relative to body size than females, but there is marked temporal variation in SD of humeral strength, which is strongly dependent on the degree of bilateral asymmetry in each sex. As noted above, in the UP, males have very marked humeral strength asymmetry ('right dominant') and females less asymmetry; this leads to more SD in right humeral strength. Males then decline greatly in asymmetry in the Mesolithic due to increases in strength on the left side (related to changes in hunting strategy), while females do not (see Chapter 7), leading to greater SD in left humeri. This is followed in the early agricultural periods (Neolithic through Iron/Roman) by a decrease in right humeral strength in females associated with symmetric use of the upper limbs in two-handed food grinding (Chapter 7), leading to more SD on the right side again, and less on the left. With the transition to one-handed food grinding, females become more asymmetric, and more similar in that respect to males; SD in humeral strength then becomes equivalent on the two sides, and relatively high, through the rest of the Holocene.

The overall high levels of SD in humeral average strength in the later Holocene are interesting and suggest that, despite increasing urbanization and eventually mechanization, males were still loading their upper limbs significantly more than females, who decline in relative humeral strength throughout the later Holocene (Chapter 7). Despite a heavy workload for both sexes, males may have been characterized by tasks requiring more upper limb strength along with early training. Females were responsible for a variety of tasks with greater overlap in urban and rural societies (Howell, 1986; Wiesner, 1993). Keeping in mind that most 'manufacturing' work took place in Medieval and early modern urban and rural households (Beattie, 2006; Wiesner, 1993) and women took part to a greater extent in the unskilled labor market, the temporal reduction in humeral strength among females might be a result of a change in manipulative activities to ones involving more prolonged but lower peak maximum strains.

Variation in upper and lower limb bone strength in males and females is further demonstrated by comparing SD in femoral/humeral strength across temporal periods. SD in this inter-limb index progressively increases after the Mesolithic for right humeri, and for left humeri undergoes large fluctuations from the UP through the Bronze Age, then also progressively increases. This illustrates the changing patterns of humeral bilateral asymmetry in the early Holocene in both sexes, as well as the general decrease in upper limb strength among females in the later Holocene.

Changes in humeral cross-sectional shape (i.e., Z_x/Z_y) are less straightforward to interpret, since these reflect in part both variation in A-P/M-L bending strength as well as localized changes dependent on muscle insertion morphology, for example the distal humeral flexor and extensor ridges (Ruff and Larsen, 2001; their Fig. 5.13). The variable position of the upper limb during different activities also makes functional anatomical planes more difficult to define. However, the generally high SD in humeral shape in the UP through Neolithic periods suggests more differentiation in the direction of loading of the upper limb in males and females during this time range, possibly the result of changing subsistence-related activities. In general, males have rounder humeri during this period, which could reflect more torsional loading of the humerus associated with throwing activities (see above), at least during the UP and Mesolithic. Greater development of the medial and lateral flexor and extensor ridges in males, associated with forearm and hand movements, may also increase M-L relative to A-P section moduli, producing smaller Z_x/Z_y ratios. In any event, the decrease in SD in humeral shape following the Neolithic suggests more homogeneous loading of the upper limb in terms of directionality, but not overall magnitude (i.e., Z_p).

6.5 Conclusions

The interpretation of variation in SD in past populations remains challenging. Change in mechanical loading is only one among various factors influencing SD, including genetics, hormones, growth patterns, and nutrition. Nevertheless, our results show that sex has a significant effect on almost all of the structural properties examined here, and that many observed trends can be interpreted in mechanical/behavioral terms.

The largest changes in SD of lower limb bone strength properties occurs between the Upper Paleolithic and Neolithic, corresponding to major changes in sex-related subsistence roles, particularly among males. Declining mobility, more marked in males, leads to more similarity in lower limb bone cross-sectional shape. Changes in sexual dimorphism in the lower limb are relatively small after the Neolithic, reflecting relatively minor changes in mobility patterns. In contrast, SD in right and left humeral strength varies throughout the Holocene as a function of changing manipulative tasks in males and females. In the early Holocene these are largely driven by contrasting patterns of bilateral asymmetry in upper limb use related to food procurement and/or processing, and in the later Holocene by a progressive decline in female humeral strength that is not paralleled in males. SD in body size shows relatively minor fluctuations during the Holocene, although there is a decline during the early modern period that may be related to poorer health conditions.

Acknowledgments

In addition to those people listed in Chapter 1, we would like to thank Patrik Galeta, Kristýna Farkašová, Martin Hora, Daniel Sosna and Eliška Schuplerová for their help with technical support and data processing.

References

Auerbach, B.M. and Ruff, C.B. (2006) Limb bone bilateral asymmetry: variability and commonality among modern humans. *J. Hum. Evol.*, **50**, 203–218.

Beattie, C. (2006) Division of Labor. In: *Women and Gender in Medieval Europe: An Encyclopedia* (ed. M. Schaus), Routledge, New York, pp. 213–214.

Bogucki, P.I. and Crabtree, P.C. (eds) (2004) *Ancient Europe 8000 BC–1000AD: An Encyclopedia of The Barbarian World*. Charles Scribner/Thomson Gale, New York.

Brace, C.L. (1973) Sexual dimorphism in human evolution. *Yrbk Phys. Anthropol.*, **16**, 31–79.

Bridges, P.S. (1989) Changes in activities with the shift to agriculture in the southeastern United States. *Curr. Anthropol.*, **30**, 385–394.

Bridges, P.S., Blitz, J.H., and Solano, M.C. (2000) Changes in long bone diaphyseal strength with horticultural intensification in West-Central Illinois. *Am. J. Phys. Anthropol.*, **112**, 217–238.

Burr, D.B., Milgrom, C., Fyhrie, D., Forwood, M., Nyska, M., Finestone, A., Hoshaw, S., Saiag, E.,and Simkin, A. (1996) In vivo measurement of human tibial strains during vigorous activity. *Bone*, **18**, 405–410.

Carlson, K.J., Grine, F.E., and Pearson, O.M. (2007) Robusticity and sexual dimorphism in the postcranium of modern hunter-gatherers from Australia. *Am. J. Phys. Anthropol.*, **134**, 9–23.

Carter, D.R. and Beaupré, G.S. (2001) *Skeletal function and form. Mechanobiology of Skeletal Development, Aging, and Regeneration*. Cambridge University Press, Cambridge.

Churchill, S.E., Formicola, V., Holliday, T.W., Holt, B.M., and Schumann, B.A. (2000) The Early Upper Paleolithic population of Europe in an evolutionary perspective. In: *Hunters of the Golden Age: The Mid Upper Paleolithic of Eurasia 30,000–20,000 BP* (eds W. Roebroeks, M. Mussi, J. Svoboda, and K. Fennema), Leiden University Press, Leiden, pp. 31–57.

Cohen, M.N. and Crane-Kramer, G.M.M. (2007) Editors' summation. In: *Ancient Health: Skeletal Indicators of Agricultural and Economic Intensification* (eds M.N. Cohen and G.M.M. Crane-Kramer), University Press of Florida, Gainesville, pp. 320–343.

Formicola, V. and Giannecchini, M. (1999) Evolutionary trends of stature in Upper Paleolithic and Mesolithic Europe. *J. Hum. Evol.*, **36**, 319–333.

Formicola, V. and Holt, B.M. (2007) Resource availability and stature decrease in Upper Paleolithic Europe. *J. Anthropol. Sci.*, **85**, 147–155.

Frayer, D.W. (1980) Sexual dimorphism and cultural evolution in the late Pleistocene and Holocene of Europe. *J. Hum. Evol.*, **9**, 399–415.

Frayer, D.W. (1981) Body size, weapon use, and natural selection in the European Upper Paleolithic and Mesolithic. *Am. Anthropol.*, **83**, 57–73.

Frayer, D.W. (1984) Biological and cultural change in the European late Pleistocene and early Holocene. In: *The Origin of Modern Humans: A World Survey of the Fossil Evidence* (eds F.H. Smith and F. Spencer), Alan R. Liss, New York, pp. 211–250.

Frayer, D.W. and Wolpoff, M.H. (1985) Sexual Dimorphism. *Annu. Rev. Anthropol.*, **14**, 429–473.

Fresia, A.E., Ruff, C.B., and Larsen, C.S. (1990) Temporal decline in bilateral asymmetry of the upper limb on the Georgia coast. In: *The Archaeology of Mission Santa Catalina de Guale: 2, Biocultural interpretations of a population in transition* (ed. C.S. Larsen), Anthropological Papers of the American Museum of Natural History, 68, New York, pp. 121–132.

Formicola, V. and Holt, B.M. (2007) Resource availability and stature decrease in Upper Paleolithic Europe. *J. Anthropol. Sci.*, **85**, 147–155.

Garvin, H.M. (2012) The Effects of Living Conditions on Human Cranial and Postcranial Sexual Dimorphism. Ph.D. Thesis, Johns Hopkins University, Baltimore.

Gray, J.P. and Wolfe, L.D. (1980) Height and sexual dimorphism of stature among human societies. *Am. J. Phys. Anthropol.*, **53**, 441–456.

Haapasalo, H., Sievanen, H., Kannus, P., Heinonen, A., Oja, P., and Vuori, I. (1996) Dimensions and estimated mechanical characteristics of the humerus after long-term tennis loading. *J. Bone Miner. Res.*, **11**, 864–72.

Hoffmann, R. (2014) *An Environmental History of Medieval Europe*. Cambridge University Press, Cambridge.

Holden, C. and Mace, R. (1999) Sexual dimorphism in stature and women's work: a phylogenetic cross-cultural analysis. *Am. J. Phys. Anthropol.*, **110**, 27–45.

Holt, B.M. (2003) Mobility in the Upper Paleolithic and Mesolithic Europe: Evidence from the lower limb. *Am. J. Phys. Anthropol.*, **122**, 200–215.

Holt, B.M. and Formicola, V. (2008) Hunters of the Ice Age: the biology of Upper Paleolithic people. *Yrbk Phys. Anthropol.*, **51**, 70–99.

Howell, M.C. (1986) *Women, Production and Patriarchy in Late Medieval Cities*. University of Chicago Press, Chicago.

Lango Allen, H., Estrada, K., Lettre, G., Berndt, S.I., Weedon, M.N., Rivadeneira, F., Willer, C.J., Jackson, A.U., Vedantam, S., Raychaudhuri, S., *et al.* (2010) Hundreds of variants clustered in genomic loci and biological pathways affect human height. *Nature*, **467**, 832–838.

Liebermann, D.E. (1996). How and why humans grow thin skulls: Experimental evidence for systemic cortical robusticity. *Am. J. Phys. Anthropol.*, **101**, 217–236.

Macintosh, A.A., Pinhasi, R., and Stock, J.T. (2014) Lower limb skeletal biomechanics track long-term decline in mobility across 6150 years of agriculture in Central Europe. *J. Archaeol. Sci.*, **52**, 376–390.

Maggiano, I.S., Schultz, M., Kierdorf, H., Sosa, T.S., Maggiano, C.M., and Blos, V.T. (2008) Cross-sectional analysis of long bones, occupational activities and long-distance trade of the classic Maya from Xcambó – Archaeological and osteological evidence. *Am. J. Phys. Anthropol.*, **136**, 470–477.

Marchi, D. and Sparacello, V. (2006) Cross-sectional geometry of the humerus of a Western Liguria Neolithic sample. In: *Il processo di umanizzazione* (eds A. Guerci, St. Consigliere and S. Castagno), Atti del XVI Congresso degli Antropologi Italiani, pp. 631–640.

Marchi, D., Sparacello, V., and Shaw, C. (2011) Mobility and lower limb robusticity of a pastoralist Neolithic population from North-Western Italy. In: *Human Bioarchaeology of the Transition to Agriculture* (eds R. Pinhasi and J.T. Stock), Wiley-Blackwell, Chichester, pp. 317–346.

McClay, I.S., Robinson, J.R., Andriacchi, T.P., Frederick, E.C., Gross, T., Martin, P., Valiant, G., Williams, K.R., and Cavanagh, P.R. (1994) A profile of ground reaction forces in professional basketball. *J. Appl. Biomech.*, **10**, 222–236.

Meiklejohn, C., Schentag, C., Venema, A., and Key, P. (1984) Socioeconomic change and patterns of pathology and variation in the Mesolithic and Neolithic of Western Europe: Some suggestions. In: *Paleopathology at the Origins of Agriculture* (eds M.N. Cohen and G.J. Armelagos), Academic Press, Orlando, pp. 75–100.

Meiklejohn, C. and Babb, J. (2011) Long bone length, stature and time in the European Late Pleistocene and Early Holocene. In: *Human Bioarchaeology of the Transition to Agriculture* (eds R. Pinhasi and J.T. Stock), Wiley-Blackwell, Chichester, pp. 153–175.

Panter-Brick, C. (2002) Sexual dimorphism of labor: Energetic and evolutionary scenarios. *Am. J. Hum. Biol.*, **14**, 627–640.

Peregrine, P.N. and Ember, M. (eds) (2001) *Encyclopedia of Prehistory. Vol. 4, Europe*. Plenum, New York.

Plavcan, J.M. (2001) Sexual dimorphism in primate evolution. *Yrbk Phys. Anthropol.*, **33**, 25–53.

Ruff, C.B. (1987) Sexual Dimorphism in human lower limb bone structure: relationships to subsistence strategy and sexual division of labor. *J. Hum. Evol.*, **16**, 396–416.

Ruff, C.B. (1999) Skeletal structure and behavioral patterns of prehistoric Great Basin populations. In: *Prehistoric Lifeways in the Great Basin Wetlands: Bioarchaeological Reconstruction and Interpretation* (eds B.E. Hemphill and C.S. Larsen), University of Utah Press, Salt Lake City, pp. 290–320.

Ruff, C.B. (2002) Variation in human body size and shape. *Annu. Rev. Anthropol.*, **31**, 211–232.

Ruff, C.B. (2008) Biomechanical analyses of archaeological human skeletons. In: *Biological Anthropology of the Human Skeleton*, 2nd edn (eds M.A. Katzenberg and S.R. Saunders), Wiley-Liss, New York, pp. 183–206.

Ruff, C.B. and Larsen, C.S. (1990) Postcranial biomechanical adaptations to subsistence strategy changes in the Georgia coast. In: *The Archaeology of Mission Santa Catalina de Guale 2, Biocultural interpretations of a population in transition* (ed. C.L. Larsen), Anthropological Papers of the American Museum of Natural History, 68, New York, pp. 94–118.

Ruff, C.B. and Larsen, C.S. (2001) Reconstructing Behavior in Spanish Florida. The Biomechanical Evidence. In: *Bioarchaeology of Spanish Florida. The Impact of Colonialism* (ed. C.S. Larsen), University Press Florida, Gainesville, pp. 113–145.

Ruff, C.B., Holt, B.M., Sládek, V., Berner, M., Murphy, W.A., zur Nedden, D., Seidler, H., and Reicheis, W. (2006) Body size, body shape, and long bone strength of the Tyrolean 'Iceman'. *J. Hum. Evol.*, **51**, 91–101.

Ruff, C.B., Holt, B.M., Niskanen, M., Sládek, V., Berner, M., Garofalo, E., Garvin, H.M., Hora, M., Maijanen, H., Niinimäki, S., Salo, K., and Schuplerová, E. (2012) Stature and body mass estimation from skeletal remains in the European Holocene. *Am. J. Phys. Anthropol.*, **148**, 601–617.

Shaw, C.N. and Stock, J.T. (2009a) Intensity, repetitiveness, and directionality of habitual adolescent mobility patterns influence the tibial diaphysis morphology of athletes. *Am. J. Phys. Anthropol.*, **140**, 149–159.

Shaw, C.N. and Stock, J.T. (2009b) Habitual throwing and swimming correspond with upper limb diaphsyseal strength and shape in modern human athletes. *Am. J. Phys. Anthropol.*, **140**, 160–172.

Sládek, V., Berner, M., and Sailer, R. (2006a) Mobility in Central European Late Eneolithic and Early Bronze Age: Femoral cross-sectional geometry. *Am. J. Phys. Anthropol.*, **130**,320–332.

Sládek, V., Berner, M., and Sailer, R. (2006b) Mobility in Central European Late Eneolithic and Early Bronze Age: Tibial cross-sectional geometry. *J. Archaeol. Sci.*, **33**, 470–482.

Sládek, V., Berner, M., Sosna, D., and Sailer, R. (2007) Human manipulative behavior in the Central European Late Eneolithic and Early Bronze Age: humeral bilateral asymmetry. *Am. J. Phys. Anthropol.*, **133**, 669–681.

Sofaer, J.R. (2006) *The Body as Material Culture. A Theoretical Osteoarchaeology.* Cambridge University Press, Cambridge.

Stini, W.A. (1976) Adaptive strategies of human populations under nutritional stress. In: *Biosocial Interrelations in Population Adaptation* (eds E.S. Watts, F.E. Johnston, and G.W. Lasker), Moutan, The Hague, pp. 19–40.

Stinson, S. (1985) Sex differences in environmental sensitivity during growth and development *Yrbk Phys. Anthropol.*, **28**, 123–147.

Stock, J.T. (2006) Hunter-gatherer postcranial robusticity relative to patterns of mobility, climatic adaptation, and selection for tissue economy. *Am. J. Phys. Anthropol.*, **131**, 194–204.

Stock, J. and Pfeiffer, S. (2001) Linking structural variability in long bone diaphyses to habitual behaviors: foragers from the Southern African Later Stone Age and the Andaman Islands. *Am. J. Phys. Anthropol.*, **115**, 337–348.

Stock, J.T. and Pfeiffer, S.K. (2004) Long bone robusticity and subsistence behavior among Later Stone Age foragers of the forest and fynbos biomes of South Africa. *J. Archaeol. Sci.*, **31**, 999–1013.

Tanner, J.M. (1990) *Fetus into man: Physical growth from conception to maturity.* Harvard University Press, Cambridge.

Trinkaus, E., Churchill, S.E., and Ruff, C.B. (1994) Postcranial robusticity in Homo. II. Humeral bilateral asymmetry and bone plasticity. *Am. J. Phys. Anthropol.*, **93**, 1–34.

Vercellotti, G., Piperata, B.A., Agnew, A.M., Wilson, W.M., Dufour, D.L., Reina, J.C., Boano, R., Justus, H.M., Larsen, C.S., Stout, S.D., and Sciulli, P.W. (2014) Exploring the multidimensionality of stature variation in the past through comparisons of archaeological and living populations. *Am. J. Phys. Anthropol.*, **155**, 229–242.

Weiss, E. (2003) Effects of rowing on humeral strength. *Am. J. Phys. Anthropol.*, **121**, 293–302.

Weiss, E. (2009) Sex differences in humeral bilateral asymmetry in two hunter-gatherer populations: California Amerinds and British Columbia Amerinds. *Am. J. Phys. Anthropol.*, **140**, 19–24.

Wescott, D.J. (2006) Effect of mobility on femur midshaft external shape and robusticity. *Am. J. Phys. Anthropol.*, **130**, 201–213.

Wescott, D.J. and Cunningham, D.L. (2006) Temporal changes in Arikara humeral and femoral cross-sectional geometry associated with horticultural intensification. *J. Archaeol. Sci.*, **33**, 1022–1036.

Wiesner, M.E. (1993) *Women and Gender in Early Modern Europe*. Cambridge University Press, Cambridge.

Wolfe, L.D. and Gray, J.P. (1982) Subsistence practice and human sexual dimorphism and stature. *J. Hum. Evol.*, **11**, 575–580.

7

Past Human Manipulative Behavior in the European Holocene as Assessed Through Upper Limb Asymmetry

Vladimír Sládek[1], Margit Berner[2], Brigitte Holt[3], Markku Niskanen[4], and Christopher B. Ruff[5]

[1] *Department of Anthropology and Human Genetics, Faculty of Science, Charles University, Prague, Czech Republic*
[2] *Department of Anthropology, Natural History Museum, Vienna, Austria*
[3] *Department of Anthropology, University of Massachusetts, Amherst, MA, USA*
[4] *Department of Archaeology, University of Oulu, Oulu, Finland*
[5] *Center for Functional Anatomy and Evolution, Johns Hopkins University School of Medicine, Baltimore, MD, USA*

7.1 Introduction

Upper limb bone bilateral asymmetry can supply important information that can be used to assess past manipulative differences within or between temporal groups (Bridges, 1985, 1989, 1991; Fresia *et al.*, 1990; Trinkaus *et al.*, 1994; Albert and Greene, 1999; Trinkaus and Ruff, 1999; Ledger *et al.*, 2000; Stock and Pfeiffer, 2001, 2004; Rhodes and Knüsel, 2005; Auerbach and Ruff, 2006; Marchi *et al.*, 2006; Sládek *et al.*, 2007; Kujanová *et al.*, 2008), as well as gender differences (Fresia *et al.*, 1990; Bridges, 1991; Weiss, 2003), ontogenetic changes (Stirland, 1993; Steele and Mays, 1995; Blackburn, 2011), and the effects of socioeconomic status (Constandse-Westermann and Newell, 1989). Skeletal bilateral asymmetry among humans is most pronounced in the upper limb skeleton (see details about other skeletal/dental asymmetries in Auerbach and Ruff, 2006; Steele, 2000), since the upper limb is directly influenced by the lateralized effects of manipulation and is free from the relatively symmetrical loading of the lower limbs caused by bipedal locomotion (but see details about 'crossed symmetry' in Auerbach and Ruff, 2006; Plochocki, 2004). Across different eco-geographic human populations, skeletal asymmetry is largest and most consistent in the humerus, and is somewhat less marked in the ulna and radius (see Appendix in Auerbach and Ruff, 2006). This probably indicates that asymmetrical patterns of manipulation have different effects on the arm and forearm, and that the humerus is generally the most affected bone in asymmetrical loading (Watson, 1974; Bridges *et al.*, 2000; Auerbach and Ruff, 2006). Upper limb bone asymmetry is lower and more variable between individuals and populations for articular breadths or bone lengths, but is significantly larger in magnitude and variability for diaphyseal breadths and cross-sectional geometric (CSG) parameters of the bone shaft (Trinkaus *et al.*, 1994; Auerbach and Ruff, 2006). Other evidence indicates that long bone lengths and breadths are less environmentally plastic and more subject to genetic and developmental constraints compared to CSG properties (Lieberman *et al.*, 2001).

Skeletal Variation and Adaptation in Europeans: Upper Paleolithic to the Twentieth Century, First Edition. Edited by Christopher B. Ruff.

The relatively large bilateral asymmetry found in CSG parameters is consistent with the high plasticity of long bone shaft structural properties in response to environmental/behavioral factors related to mechanical stimuli (see review and discussion about plasticity of CSG in Ruff *et al.*, 2006; and Skedros, 2011). Thus, it is expected that CSG asymmetry of long bone shafts provides direct evidence of functional adaptation to past manipulative behavior (e.g., Trinkaus *et al.*, 1994; Haapasalo *et al.*, 2000; Steele, 2000; Ruff, 2008; Shaw and Stock, 2009; Skedros, 2011). Assessment of asymmetry in CSG parameters has other advantages for the reconstruction of past human behavior. Comparing right and left limbs of the same individuals minimizes the influence of other possible non-mechanical factors that may have an effect on cortical tissue, such as diet (Watson, 1974; Trinkaus *et al.*, 1994). Body mass is also constant, so that possible biases induced by estimation of body mass are avoided since non-size-standardized data can be used for computation of asymmetry (Ruff, 2002).

Asymmetry in mechanical loading of the human upper limb is shaped, at least in part, by handedness (Perelle and Ehrman, 1994; Trinkaus *et al.*, 1994; Shaw, 2011). Handedness is usually defined as the systematic preference of one hand for a particular function, where the dominant hand is associated with higher dexterity and skill acquisition (Steele, 2000; Raymond and Pontier, 2004). Handedness is expected to be established early in ontogeny (see review in Fagard, 2013), and has also been associated with genetic background (e.g., Levy and Nagylaki, 1972; Hicks and Kinsbourne, 1976; Annett, 1981). However, several other studies showed that genetic assessments of handedness are biased by environmental factors such as mimicking of manipulative behavior in twin studies (Perelle and Ehrman, 1994) and strong cultural influences (Bakan, 1978; Corballis and Morgan, 1978; Perelle and Ehrman, 1982).

Human handedness is strongly right-biased, with left-handers constituting a significantly lower proportion among modern humans. The proportion of left-handers varies among recent populations according to the criteria used for determining handedness, ranging from 1–11% established by certain behavioral measures (Hardyck and Petrinovich, 1977) to 4–28% by a more generalized motor skill test (Raymond and Pontier, 2004). Left-handedness exhibits significant variation in geographic distribution and varies in writing hand preference between 2.5% in Mexico to 13% in Canada (Perelle and Ehrman, 1994: their Table I). Furthermore, familial similarities (left-handed writers have a significantly higher frequency of left-handed mothers and/or left-handed fathers, and a significantly higher probability of left-handed siblings), and variation in gender distribution (left-handers are more frequent among males than females) have been observed (see review in Perelle and Ehrman, 1994). The geographic variation in left-handedness was also supported for manipulative activities other than writing, but the pattern was less marked, and was also influenced by sex and age (Raymond and Pontier, 2004). Right-handed writers perform other activities at a higher frequency with their dominant right hand (between 66.5% and 93.4%) than left-handed writers (57.3–75%) (Perelle and Ehrman, 1994: their Table V), and a higher frequency of ambidexterity among left-handers was also suggested by other studies (e.g., Kimura, 1973).

Handedness has been associated with several other behavioral parameters, but the results are to some degree controversial. Raymond and Pontier (2004) suggest that education and socio-economic status (SES) must be considered as possible confounding variables explaining the apparent discrepancy between studies within countries. The impact of SES on hand preference has been supported by Annett and Kilshaw (1983), Leiber and Axelrod (1981), and Perelle and Ehrman (1994), though in another study SES was found to be non-significant when explaining variation in hand preference (Brito *et al.*, 1989). Similarly, contradictory results have been found when education level was assessed relative to hand preference. In hand writing preference there was no significant relationship between education level and hand preference, but at the same time there was a significant relationship between hand preference and type of primary school (higher frequency of left-handed writers in public than in private) (Perelle and Ehrman, 1994). This contradiction highlights another important feature of handedness – the cultural/social negative perception of left-handers. There is reported evidence of negative perceptions in the past (e.g., negative presentation of left-handers in the Old and New Testament), suggestions that left-handedness is an evolutionary retrogression, and a tendency to switch hand preference in writing during elementary school education (see review in Hardyck and Petrinovich, 1977).

Reconstruction of past human manipulative behavior has to take into account that hand dominance is not the only factor involved in asymmetric use of the upper limbs (Bridges, 1991). Human manipulative tasks create different demands for lateralization; involvement of the dominant hand in manipulation can vary based on the skill demands of the activity and the magnitude of load produced during employment of the hand. Thus, the usage of limbs in manipulative behavior is much more complex and the maximum load is not always directly imposed on the dominant hand. There are highly unilateral activities that directly subject the dominant hand to the highest mechanical loading, such as professional tennis playing (Trinkaus *et al.*, 1994), woodworking and warfare-related weapon training (Rhodes and Knüsel, 2005; Sparacello *et al.*, 2011), but there are also activities that employ both hands equally without respect to the dominant hand, such as maize pounding using a saddle quern (Bridges, 1991; Sládek *et al.*, 2016a) or digging activities (Marshall, 1976; Lee, 1979; Ledger *et al.*, 2000). There are also activities that involve lower loading of the dominant hand, since the dominant hand is used for better skill control with low loading compared to the non-dominant limb. For example, Shaw *et al.* (2012: their Table 1) showed that maximum muscle activity was higher on the left (non-dominant) upper limb compared to the right upper limb, when the bilateral activity pattern was assessed for underhand spear thrusting.

Patterns of manipulation should be associated with subsistence economy and gender roles, although in a complex way. We can predict that right-biased asymmetry dominates over Holocene Europe, but the degree of asymmetry does not need to be constant and can decrease to near 0 based on specific subsistence demands and gender roles. The transition among Holocene foragers to agriculture has been more often characterized by a decrease of asymmetry among females than changes in patterns of asymmetry among males (Bridges, 1985; Fresia *et al.*, 1990). Females are expected to be more involved with agricultural activities and food preparation (i.e., mainly pounding), whereas males continue during the early agricultural period with previous manipulative tasks associated with hunting (Bridges, 1985; Fresia *et al.*, 1990; Bridges *et al.*, 2000). However, the expected pattern has to reflect the complexity of changes associated with agriculture and new manipulative tasks. Archeologists have demonstrated that agriculture undergoes a significant intensification through the Holocene, well demonstrated for example in changes in food preparation (Mauldin, 1993; Wright, 1994; Adams, 1999; Holodňák, 2001). Food grinding starts in the early Neolithic with two-handed grinding using saddle quern (i.e., using ground and hand stone), but developed in the early Iron Age into specific stone mills which were possible to control only by one hand or were scaled to incorporate animal or water power into the process (Fig. 7.1) (Lynch and Rowland, 2005). In the Early Neolithic such food preparation is expected to be limited to females, as also suggested from archaeological and ethnographic data (Bridges, 1989, 1991; Haaland, 1995; Ogilvie, 2004; Hamon and Le Gall, 2013). However, with intensification of agriculture later in the Holocene, food grinding responsibilities tended to pass to specialized groups dominated by males (at the same time, there is evidence that bi-manual grinding can be preserved as an alternative small-scale system limited to the household during later periods of the Holocene (e.g., Venclová *et al.*, 2008). Moreover, similarities in the degree of upper limb directional asymmetry among males between hunter-gather and early agricultural periods need not be evidence of continuity of

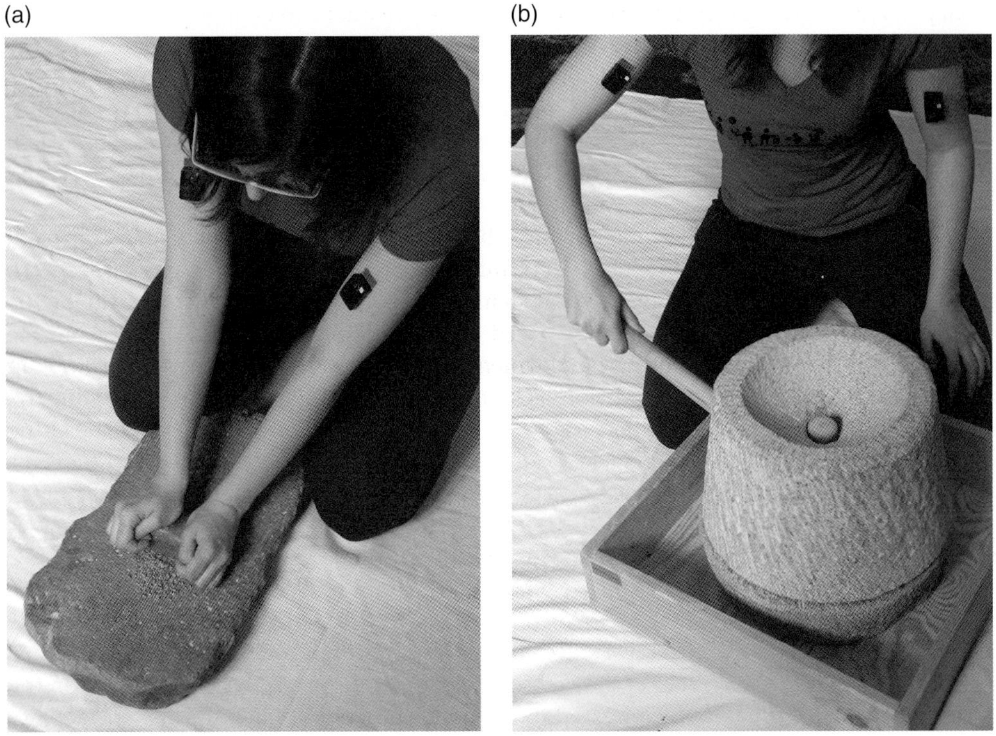

Figure 7.1 Experimental grinding with two-handed saddle quern (a) and one-handed rotary quern (b). Photograph by Vladimír Sládek, reprinted with permission from Sládek *et al.*, 2016b.

subsistence roles between periods, but could be due to the appearance of new unilateral activities for males, for example associated with stone hatchets and woodworking (Sparacello *et al.*, 2011). Thus, new subsistence demands associated with early agriculture may be more easily discerned by a decrease of directional asymmetry in females rather than any changes among males. Increased reliance on agricultural subsistence is also accompanied during the Holocene by other processes such as economic specialization and increased social complexity. Thus, manipulative tasks can reflect opposing processes associated with higher unilateral loading related to gender-specific and non-agricultural activities (e.g., emergence of elites and armed cohort and their uni-manual each day training activity with weapons; Sparacello *et al.*, 2011). Industrialization may also affect the degree of bilateral asymmetry, by reducing sexual dimorphism in activities (Auerbach and Ruff, 2006).

The main goal of the chapter is to investigate the impact of European Holocene changes in subsistence economy on upper limb bone bilateral asymmetry. We study right and left upper limb variation in bone lengths, articulations, breadths, and CSG parameters across both temporal and geographic divisions. We also consider patterns of sexual dimorphism in asymmetry among European Holocene groups. In general, we expect a decrease in asymmetry between Paleolithic/Mesolithic hunter-gatherers and Neolithic agriculturists, since agriculture is a less unilaterally dominated activity. However, later in the Holocene the pattern of asymmetry can be more complicated and can reflect improvements in technology (e.g., grain milling intensification and decreased symmetric use of the arms among females), decreased gender-specific roles (e.g., with industrialization), and increased economic specialization, formation of hierarchical societies, and urbanization, which may lead to an increase of variation in asymmetry within populations. Although the primary focus of this chapter is on bilateral asymmetry, we also include some analyses of temporal changes in each limb separately at the end, in order to help interpret the asymmetry results. Some of these analyses overlap with those in Chapter 5, but involve a different set of structural parameters and somewhat different temporal subdivisions, as explained below.

7.2 Materials and Methods

7.2.1 Samples

The composition of the sample used in this chapter is presented in Table 7.1. The maximum number of bones from the right side is 1523, and from the left side 1505. For asymmetry analyses *per se* only the paired side sample is used, with a reduced total number of matched pairs of 1250. Due to differences in preservation, the number of individuals varies somewhat according to the particular analysis (see Tables for sample sizes in each analysis).

The skeletal sample was taken from across most of Europe and subdivided into seven geographic regions: Britain; Scandinavia; North-Central Europe; France; Italy; Iberian Peninsula; and Balkans (see details about geographic composition of the sample in Chapter 1 and Appendix 1). The temporal periods and their abbreviations are shown in Table 7.1. The pooled pan-European sample includes three pre-agricultural groups (EUP, LUP, Meso) and eight agricultural groups, with calibrated C^{14} BP range for pre-agricultural groups between 33 kyr to 6 kyr BP and for agricultural groups between 7 kyr BP to 20th century. As indicated by the range of maximum and minimum C^{14} BP estimates, the periods slightly overlap in the earlier part of the Holocene. Therefore, the main criterion for individual/site classification was based on the archaeological culture, which should limit bias due to differences in time of appearance of each culture across the European continent. The only exception is to include the early Copper Age sample from Fontenoce Recatini (5438 cal. BP) in the Neolithic and to reserve the Copper Age period for the Bell Beaker and Corded Ware Cultures from Central Europe and Scandinavia, which experienced one of the major subsistence changes in the early agricultural period (i.e., Secondary Products Revolution; Sherratt, 1981; Greenfield, 2010). Further, there are some regionally limited groups which cannot be compared over all European regions, including Central European Avars and Scandinavian Neolithic foragers. We incorporate such regional-specific groups into the most similar temporal group; for example, Avars are assimilated into the Early Medieval period. The most important regionally limited group is the Scandinavian Neolithic foragers, who are not found in other European regions. Since the foraging to farming transition is one of the key subsistence changes in the beginning of the Holocene, we provide additional analyses for Scandinavian Neolithic foragers (S Neol HG) and for European Neolithic farmers (Neol Ag; no data available for Scandinavian Neolithic farmers; see details in Table 7.1) and also provide detailed data for both groups in each summary table separately from the pooled general Neolithic sample. Finally, the Recent group includes individuals from pre-industrial 17th century to industrial of the 19th and 20th centuries. Given the fact that recent pre-industrial and recent industrial samples are non-significantly different in asymmetry variables we combine all the recent groups into one period (see Discussion).

In the following analyzes we use the terms pre-Holocene for the EUP and LUP samples, and Holocene for Mesolithic to Recent samples. Further, we also distinguish two major subsistence groups: pre-agricultural (EUP, LUP, Meso) and agricultural (Neol-Recent). In addition, since the results indicate that agricultural groups can be distinguished by several upper limb skeletal features and implied manipulative behaviors into two temporally limited periods, we separate

Table 7.1 Structure of European sample used for upper limb comparison.[a]

	Abbrev.	Mid C^{14} BP[b]	Max C^{14} BP[c]	Min C^{14} BP[d]	Right[e]			Left[f]			R+L[g]		
					Total	Male	Female	Total	Male	Female	Total	Male	Female
Early Upper Paleolithic	EUP	30 439	33 388	26 406	13	8	5	18	10	8	13	8	5
Late Upper Paleolithic	LUP	13 789	21 922	11 367	17	12	5	18	12	6	14	9	5
Mesolithic	Meso	8134	10 500	6025	38	24	14	43	25	18	31	20	11
Neolithic	Neol	5354	7300	4581	137	82	55	133	77	56	113	66	47
Scan. Neol. (foragers)[h]	S Neol HG	4686	4700	4581	53	33	20	51	30	21	43	25	18
Neol. (farmers)[h]	Neol Ag	5835	7300	4800	84	49	35	82	47	35	72	41	31
Copper Age	CopA	4491	4600	3975	74	44	30	76	47	29	54	32	22
Bronze Age	BA	3660	4350	2950	184	102	82	185	98	87	150	79	71
Iron Age	IA	2051	2250	1700	133	67	66	126	69	57	108	54	54
Roman	Roman	1388	1900	950	156	85	71	152	77	75	67	33	34
Early Medieval	EMed	1179	1300	950	161	96	65	147	94	53	184	111	73
Late Medieval	LMed	758	925	550	391	204	187	386	211	175	320	172	148
Recent	Recent	143	320	10	219	129	90	221	135	86	196	115	81
Total					**1523**	853	670	**1505**	855	650	**1250**	699	551

[a] The number of individuals indicates the maximum individuals preserved in each period. Number of individuals varies according to preservation of respective parameter; final number of individuals used in each analysis is provided in the results and summary tables.
[b] Mid C^{14}: mean calibrated C^{14} BP date for period.
[c] Max C^{14}: maximum calibrated C^{14} BP date for period.
[d] Min C^{14}: minimum calibrated C^{14} BP date for period.
[e] Maximum number of individuals for right upper limb.
[f] Maximum number of individuals for left upper limb.
[g] Maximum number of individuals with both right and left side preserved.
[h] Scandinavian Neolithic forager and European farmer subsamples from general pooled Neolithic group (see Materials for further details); not separately included in Total Ns.

the total agricultural sample into early agricultural (Neol-BA) and late agricultural periods (IA-Recent). To simplify descriptions, the Recent sample is treated as falling among the late agricultural period, but specific details concerning the non-agricultural origin for the majority of 20th century individuals are included in the discussion.

7.2.2 Linear Measurements and CSG Parameters

A list of the selected linear measurements taken on upper limb bones is provided together with abbreviations of Martin (1928) in Table 7.2. All the linear measurements are taken in millimeters and the osteometric protocol follows the standardized approach for measuring bones presented in Bräuer (1988). Dimensions include maximum lengths of the clavicle, humerus, radius, and ulna, articular breadths of the humeral head and distal humerus, and diaphyseal breadths at 35% of biomechanical length (see Chapter 3) from the distal end of the humerus and at 50% of biomechanical length of the radius and ulna.

CSG parameters were studied using humeral cross-sectional areas and section moduli (see Chapter 3 for a list and details). Parameters for CSG were taken at 35% of humeral biomechanical length either by X-ray or computed tomography (CT) scan image protocols. Details for the X-ray and CT protocols are described in Chapter 3. Derived CSG parameters were used to estimate manipulative differences in axial compressive/tensile strength (CA), bending strength in A-P and M-L planes (Z_x, Z_y), and average bending strength (Z_P).

For calculation of asymmetry we take advantage of the constant effect of body size on right and left sides, so that asymmetry is computed using raw non-standardized values. Leaving out size standardization also helps to increase sample size. When right and left sides for humeral CSG are analyzed separately, then CSG areas are standardized to body mass (BM) and section moduli to BM and biomechanical length (BML) (see Table 7.2 for details). Methods for estimation of BM are provided in Ruff *et al.* (2012) and in Chapter 2.

7.2.3 Computation of Asymmetry and Sexual Dimorphism

Both maximum and directional asymmetry were calculated as standardized percentage values. Maximum asymmetry (%MaxA) is used to express the overall amount of asymmetry without respect to side and was computed as:

$$\%\text{MaxA} = \left[(\text{max} - \text{min})/((\text{max} + \text{min})/2)\right] \times 100,$$

where 'max' = side with larger value and 'min' = side with smaller value. A value of zero indicates bilateral symmetry; deviation from zero describes the magnitude of asymmetry.

Directional asymmetry (%DirA) describes the magnitude of laterality and was computed as:

$$\%\text{DirA} = \left[(\text{R} - \text{L})/((\text{R} + \text{L})/2)\right] \times 100,$$

where R = right side and L = left side. Asymmetry deviated to the right side is expressed as a positive value and asymmetry deviated to the left side as a negative value, with values near to zero indicating absence of directional asymmetry.

Finally, we also provide comparisons of significant laterality – that is, percentages of individuals with a right or left side bias – for CSG measurements. To avoid biologically non-significant side biases, for these comparisons we include only individuals with %DirA greater than ±0.5% and consider individuals within differences less this as non-lateralized (see also Auerbach and Ruff, 2006). Further details about statistical characteristics of our approach and

Table 7.2 Osteometric and CSG measurement used for European Holocene upper limb comparison.

Measurements	Abbrev.	Description
Clavicular maximum length	CML	Maximum distance between the most extreme ends of the clavicle (M[Cl]#1)
Humeral maximum length	HML	Direct distance from the most superior point on the head of the humerus to the most inferior point on the trochlea. Humerus shaft should be positioned parallel to the long axis of the osteometric board (M[Hu]#1)
Radial maximum length	RML	Distance from the most proximally positioned point on the head of radius to the tip of the styloid process without regard for the long axis of the bone (M[Ra]#1)
Humeral head S-I breadth	HHDSI	Direct distance between the most superior and inferior points on the border of the articular surface on proximal humeral head (M[Hu]#10)
Humeral distal articular breadth	HDML	Distance between the midpoint of the outer edge of trochlea and the midpoint of the outer edge of the capitulum (M[Hu]#12a)
A-P humeral 35% diaphyseal breadth	H35AP	Sagittal diameter of humeral diaphysis at 35% of biomechanical length
M-L humeral 35% diaphyseal breadth	H35ML	Transverse diameter of humeral diaphysis at 35% of biomechanical length
A-P radial 50% diaphyseal breadth	R50AP	Sagittal mid-shaft diameter of radial diaphysis (M[Ra]#5a)
M-L radial 50% diaphyseal breadth	R50ML	Transverse mid-shaft diameter of radial diaphysis (M[Ra]#4a)
Total area	H35TA	Humeral total subperiosteal area at 35% of biomechanical length (adjusted by BM, multiplied by 100)[a]
Cortical area	H35CA	Humeral cortical area between periosteal and endoosteal margin at 35% of biomechanical length; compressive/ tensile strength (adjusted by BM, multiplied by 100)[a]
A-P section modulus	$H35Z_x$	Humeral A-P bending strength at 35% of biomechanical length (adjusted by BM*BML, multiplied by 10^4)[a]
M-L section modulus	$H35Z_y$	Humeral M-L bending strength at 35% of biomechanical length (adjusted by BM*BML, multiplied by 10^4)[a]
Index of circularity	$H35Z_x/Z_y$	Ratio of A-P section modulus vs. M-L section modulus (only not adjusted data used)
Polar section modulus	$H35Z_p$	Humeral torsional and average bending strength at 35% of biomechanical length (adjusted by BM*BML, multiplied by 10^4)[a]

[a] Body mass standardization used only when individual right and left CSG properties are compared (see details in Materials and Methods).

discussion of the biological significance of %MaxA and %DirA are provided elsewhere (Van Valen, 1962; Mays, 2002; Auerbach and Ruff, 2006; Sládek *et al.*, 2007;).

Sexual dimorphism was computed using the formula:

$$\text{SexD} = \%\text{Asymmetry}_{\text{Male}} - \%\text{Asymmetry}_{\text{Female}}.$$

SexD is sensitive to stochastic unequally distributed males and females in the compared groups, which limits some regional analyses with relatively small available sample sizes.

7.2.4 Statistical Techniques

For CSG standardized properties within sides we expect normal distributions, and therefore summary statistics for these measurements are reported using the mean and SE or ±95% CI for the mean

Distributions of %MaxA and %DirA generally deviate from normality; moreover, as indicated also by other studies (e.g., Auerbach and Ruff, 2006), transformation of raw % asymmetry data does not improve the normal distribution. Thus, we employ robust statistical treatments for all % asymmetry comparisons. Summary statistics are provided using the median and ±95% CI of median. Since medians do not have a simple sampling distribution and computation of CIs can be difficult by standard statistical procedures (Efron and Tibshirani, 1994; Haukoos and Lewis, 2005), we employed resampling with 10,000 replicates and provide CI of median as ±95% bootstrap confidence interval. The bootstrap ±95% CI was computed in SYSTAT 13 using the BCa method (i.e., Bias corrected and accelerated method; see Efron and Tibshirani, 1994). Similarly, bootstrap resampling was also employed for ±95% CI of mean for the Z_x/Z_y ratio, since in general this may not be normally distributed. Non-parametric statistical tests are used to test for statistical significance, including the Mann–Whitney *U*-test for two independent samples to test for differences between males and females, and the Sign median test to test for significant directional asymmetry, that is, that directional asymmetry is significantly greater than 0. Because the great majority of comparisons indicated significant directional asymmetry, non-significant asymmetry is highlighted in the tables through underlining. The sign median test and bootstrap ±95% CI can be biased by small sample size; therefore, we do not provide a test of significance for regional variation in Table 7.14, where a high number of unequal and small sample sizes are assessed. Chi-square tests were used to test the significance for proportions of individuals with a right or left bias. A significance level of 0.05 was employed in all tests.

Tables and raw data were created in Excel Microsoft Office Professional 2010 (2010, Microsoft Corporation), statistical tests were computed mainly in Statistica 64 bits 12 for Windows (StatSoft, Inc., 1984–2013), and bootstrap estimates for median ±95% CI in SYSTAT 32 bits 13 for Windows (SYSTAT Software, Inc. 2009).

7.3 Results

7.3.1 Asymmetry Variation

7.3.1.1 General Differences Between Structural Properties

Figure 7.2 shows a summary of general differences in maximum asymmetry between several upper limb bone properties – humeral length, head breadth, diaphyseal breadth, CA, and Z_p – by sex and temporal period. Overall, differences in the relative magnitude of asymmetry between different properties are similar across periods and between the sexes. In general, maximum asymmetry in humeral length is the lowest (0.9–2.8%), followed by somewhat larger asymmetry in humeral articular breadth (1.5–2.9%), humeral diaphyseal breadth (1.9–4.7%), and HCA (4.2–10%). Finally, humeral Z_p has the largest asymmetry across all the periods (5.3–16%). The difference in asymmetry between length and section modulus is even more pronounced when pre-Holocene males and females are compared (the HZ_p increases up to 40% in the pre-Holocene male sample and 15% in the female sample, whereas length asymmetry is similar to that of Holocene periods). Moreover, humeral length and humeral articular breadth asymmetry are less variable between all periods compared to relatively large between-period variation in CSG properties, that is, HCA and HZ_p. The increase in asymmetry from humeral length to HZ_p is more pronounced in males than in females. Thus, the results indicate

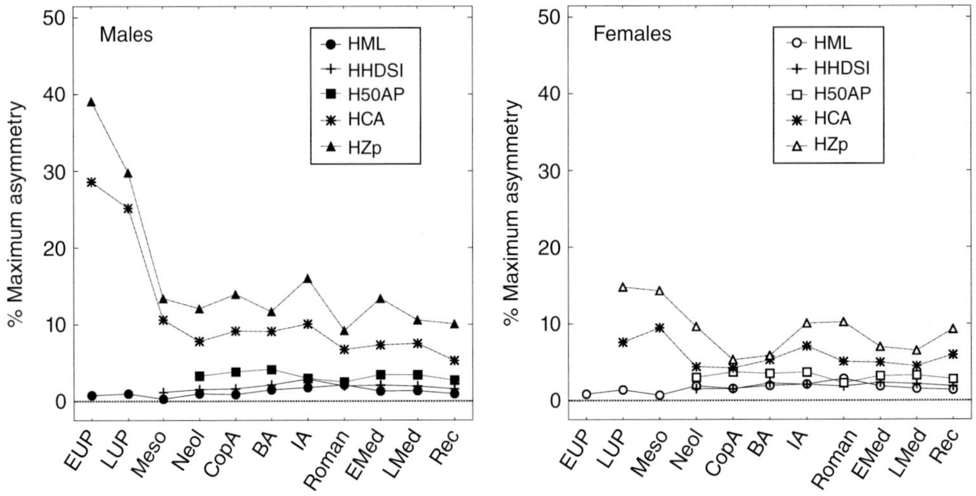

Figure 7.2 A summary of maximum asymmetry (medians).

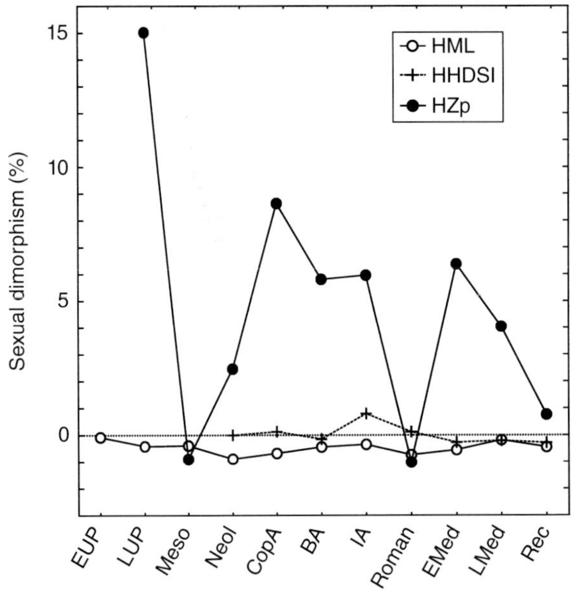

Figure 7.3 A summary of sexual dimorphism in maximum asymmetry.

that expected low asymmetry in lengths and breadths and high asymmetry in CSG parameters (Trinkaus *et al.*, 1994) is consistent across all the studied Holocene subsistence and cultural groups, and can be seen as a general feature of human postcranial variation. The low variation observed in lengths and articular dimensions contra high variation observed for CSG parameters also supports the view that asymmetry in lengths are more related to fluctuating asymmetry, probably due to systemic stress (Van Valen, 1962; but see Smith *et al.*, 1982), but CSG dimensions are more influenced by directional asymmetry induced by behavioral differences in mechanical loading (see also Tables 7.10 and 7.12 and text below).

A summary of sexual dimorphism in maximum asymmetry for humeral length, head breadth, and Z_p is plotted in Figure 7.3. In general, maximum asymmetry in HZ_p is greater in males, except in the Mesolithic and Roman periods, while asymmetry in length is greater in females. Sexual dimorphism in maximum asymmetry of humeral head breadth is very small (<0.3%) in most periods. Sexual dimorphism in HZ_p asymmetry is relatively variable, with values ranging between −1% and 15%. Sexual dimorphism in length generally falls between −0.9% and −0.1%.

7.3.1.2 Temporal Trends in Asymmetry

7.3.1.2.1 Length Asymmetry

Temporal trends in asymmetry in lengths of the upper limb bones are shown in Tables 7.3 and 7.4 and Figure 7.4. Maximum and directional asymmetry in lengths shows generally similar temporal patterns. The clavicle shows slightly greater maximum asymmetry than the humerus and radius in both male and female samples, and clavicular length is also the only property with

Table 7.3 Maximum asymmetry in diaphyseal lengths for European pre-Holocene and Holocene upper limb bones [Median (N; −95%CI – +95%CI)].

	Males	Females	SexD
Clavicle			
EUP	1.92 (4; 0.00–2.50)	–	–
LUP	3.69 (5; 0.72–5.45)	–	–
Mesolithic	2.00 (11; 0.00–2.50)	1.90 (5; 0.68–2.12)	0.10
Neolithic	1.62 (33; 0.69–2.26)	2.21 (23; 0.40–2.88)	−0.59
S Neol HG	2.25 (15; 0.00–3.39)	1.90 (11; 0.00–2.85)	0.35
Neol Ag	1.38 (18; 0.64–1.44)	2.40 (12; 0.00–3.20)	−1.02
CopA	2.52 (14; 0.65–3.71)	4.02 (9; 1.57–5.10)	−1.50
BA	2.36 (39; 1.29–3.01)	1.74 (33; 1.09–2.22)	0.62
IA	1.44 (14; 0.68–1.79)	2.07 (18; 0.70–2.80)	−0.63
Roman	2.04 (14; 0.96–3.18)	3.61 (15; 1.07–4.40)	−1.57
EMed	2.05 (84; 1.35–2.26)	2.89 (53; 1.43–3.08)	−0.84
LMed	1.90 (121; 1.33–2.13)	2.08 (101; 1.37–2.21)	−0.18
Recent	2.31 (82; 1.34–2.69)	1.54 (46; 0.81–1.78)	0.77
Humerus			
EUP	0.76 (8; 0.00–1.40)	0.84 (5; 0.00–1.56)	−0.08
LUP	0.96 (9; 0.00–1.58)	1.38 (5; 0.00–2.71)	−0.42
Mesolithic	0.29 (20; 0.00–0.48)	0.68 (11; 0.33–1.37)	−0.39*
Neolithic	0.99 (61; 0.65–1.26)	1.88 (47; 1.05–2.00)	−0.89*
S Neol HG	0.85 (24; 0.34–1.19)	2.36 (18; 1.38–2.81)	−1.51*
Neol Ag	1.10 (37; 0.34–1.29)	1.53 (29; 0.56–1.94)	−0.43*
CopA	0.87 (31; 0.30–1.21)	1.55 (22; 0.32–1.69)	−0.68
BA	1.47 (79; 0.79–1.66)	1.91 (71; 1.35–2.27)	−0.44*
IA	1.76 (45; 1.24–2.06)	2.11 (50; 1.32–2.38)	−0.35
Roman	2.09 (32; 1.25–2.33)	2.83 (32; 1.80–3.28)	−0.74
EMed	1.30 (111; 0.95–1.49)	1.85 (73; 1.56–2.06)	−0.55*
LMed	1.34 (172; 1.13–1.49)	1.52 (146; 1.25–1.69)	−0.18
Recent	0.94 (115; 0.51–1.08)	1.38 (81; 0.99–1.55)	−0.44*

(Continued)

Table 7.3 (Continued)

	Males	Females	SexD
Radius			
EUP	0.43 (4; 0.00–0.74)	–	–
LUP	–	–	–
Mesolithic	0.55 (11; 0.00–0.85)	1.33 (4; 0.48–1.93)	−0.78
Neolithic	0.69 (51; 0.43–0.87)	1.31 (31; 0.45–1.44)	−0.62*
S Neol HG	0.76 (17; 0.21–1.15)	1.52 (13; 0.00–1.74)	−0.76
Neol Ag	0.69 (34; 0.40–0.86)	0.96 (18; 0.43–1.38)	−0.27
CopA	0.69 (26; 0.00–1.14)	0.90 (20; 0.46–1.09)	−0.21
BA	0.85 (75; 0.43–1.12)	1.39 (62; 0.49–1.74)	−0.54*
IA	0.89 (41; 0.43–1.17)	0.78 (40; 0.43–0.92)	0.11
Roman	0.92 (30; 0.43–1.24)	1.77 (32; 0.50–2.09)	−0.85*
EMed	0.71 (102; 0.40–0.83)	1.14 (72; 0.47–1.41)	−0.43*
LMed	1.12 (157; 0.81–1.21)	1.29 (142; 0.48–1.33)	−0.17
Recent	0.90 (117; 0.46–1.10)	1.05 (74; 0.55–1.26)	−0.15

* Mann–Whitney *U*-test for sexual dimorphism (SexD) significant.

directional asymmetry shifted toward the left side in all of the compared periods. Humeral and radial lengths are consistently right-biased. There is no overall temporal change in length asymmetry between pre-agricultural (i.e., from EUP to Mesolithic) and agricultural (i.e., from Neolithic to Recent) groups, and only a limited temporal trend among the Holocene groups. A moderate trend in the Holocene is observed only for humeral length asymmetry, which gradually increases from the Mesolithic to Roman period and decreases afterwards to the Recent period. However, the magnitude of maximum asymmetry for humeral length does not exceed 3% and the range of between-period variation is usually on average only 0.5–2%.

Sexual dimorphism for maximum and directional asymmetry in upper limb lengths is also given in Tables 7.3 and 7.4. With few exceptions (mainly in the clavicle), females have more length asymmetry than males, often reaching statistical significance. Larger between-period variation in sexual dimorphism is observed for the clavicle than for the humerus, with no consistent temporal trends.

7.3.1.2.2 Articular Breadth Asymmetry

Temporal differences and sexual dimorphism in maximum and directional asymmetry in articular breadths are given in Tables 7.5 and 7.6 and Figure 7.5. Both the proximal and distal humerus shows a general right-bias, which is more consistent in the distal humerus. There are no available data for the pre-Holocene periods, but temporal trends within the Holocene are relatively minor and inconsistent. Sexual dimorphism is non-significant in all cases and there is no apparent temporal trend in sexual dimorphism.

7.3.1.2.3 Diaphyseal Breadth Asymmetry

Temporal differences and sexual dimorphism in maximum and directional asymmetry of diaphyseal breadths are given in Tables 7.7 and 7.8 and Figure 7.6. With few exceptions, breadths are right-biased in both sexes throughout the Holocene (no breadth data are available for the pre-Holocene samples, or for the radius in the Mesolithic). The exceptions are humeral M-L

Table 7.4 Directional asymmetry in diaphyseal lengths for European pre-Holocene and Holocene upper limb bones [Median (N; –95%CI – +95%CI)].

	Males	Females	SexD
Clavicle			
EUP	−0.64 (4; −2.40–2.60)	–	–
LUP	−3.40 (5; −8.34–1.39)	–	–
Mesolithic	−1.21 (11; −4.14–1.38)	−1.92 (5; −7.94– −0.68)	0.71
Neolithic	−1.36 (33; −2.69– −0.87)	−1.34 (23; −2.82–0.00)	−0.02
S Neol HG	−2.03 (15; −3.56– −0.40)	−1.67 (11; −3.01–0.40)	−0.36
Neol Ag	−1.03 (18; −1.54–0.68)	−1.02 (12; −2.34–2.92)	−0.01
CopA	−1.83 (14; −3.28–1.28)	−4.00 (9; −5.66– −2.22)	2.17*
BA	−0.85 (39; −1.76–1.86)	−1.18 (33; −1.54–0.74)	0.33
IA	−0.08 (14; −1.29–1.38)	−1.88 (18; −3.01–0.71)	1.80
Roman	−1.08 (14; −2.86–1.31)	−2.55 (15; −4.38–1.52)	1.47
EMed	−1.57 (84; −2.05– −0.63)	−1.77 (53; −2.94–0.00)	0.20
LMed	−1.15 (121; −1.56– −0.67)	−0.91 (101; −2.10–0.00)	−0.24
Recent	−1.70 (82; −2.55– −0.55)	−1.20 (46; −1.52–0.00)	−0.50
Humerus			
EUP	0.64 (8; −0.61–1.40)	0.59 (5; 0.00–1.87)	0.05
LUP	0.62 (9; −1.65–1.25)	1.38 (5; 0.90–2.71)	−0.76
Mesolithic	0.02 (20; −4.14–0.01)	0.64 (11; −0.36–1.37)	−0.62*
Neolithic	0.85 (61; 0.34–1.08)	1.87 (47; 0.73–2.00)	−1.02*
S Neol HG	0.74 (24; 0.33–0.97)	2.36 (18; 1.38–2.97)	−1.62*
Neol Ag	0.98 (37; 0.00–1.28)	1.50 (29; 0.35–1.94)	−0.52
CopA	0.62 (31; −0.29–0.94)	1.55 (22; 0.32–1.69)	−0.93*
BA	1.41 (79; 0.65–1.65)	1.91 (71; 1.35–2.27)	−0.50*
IA	1.73 (45; 0.96–2.06)	2.11 (50; 1.32–2.35)	−0.38
Roman	2.10 (32; 1.21–2.31)	2.84 (32; 1.82–3.22)	−0.74
EMed	1.17 (111; 0.61–1.31)	1.85 (73; 1.41–2.06)	−0.68*
LMed	1.25 (172; 0.94–1.40)	1.49 (146; 1.20–1.63)	−0.24
Recent	0.86 (115; 0.33–0.99)	1.34 (81; 0.89–1.54)	−0.48*
Radius			
EUP	0.03 (4; −1.08–0.39)	–	–
LUP	–	–	–
Mesolithic	−0.12 (11; −0.43–0.85)	0.45 (4; −1.93–1.86)	−0.57
Neolithic	0.33 (51; 0.00–0.77)	1.31 (31; 0.44–1.43)	−0.98*
S Neol HG	0.73 (17; −0.44–1.15)	1.52 (13; 0.00–1.74)	−0.79*
Neol Ag	0.16 (34; −0.38–1.13)	0.97 (18; 0.42–1.38)	−0.81*
CopA	0.24 (26; −0.40–0.75)	0.29 (20; −0.91–0.86)	−0.05
BA	0.40 (75; −0.20–0.43)	1.16 (62; 0.46–1.55)	−0.76*
IA	−0.12 (41; −0.55–0.43)	0.27 (40; −0.46–0.44)	−0.39
Roman	0.83 (30; 0.00–1.23)	1.77 (32; 0.48–2.07)	−0.94*
EMed	0.40 (102; 0.00–0.66)	0.95 (72; 0.43–1.38)	−0.55*
LMed	0.87 (157; 0.40–1.11)	1.18 (142; 0.47–1.32)	−0.31
Recent	0.47 (117; 0.00–0.64)	0.91 (74; 0.24–1.09)	−0.44*

* Mann–Whitney *U*-test for sexual dimorphism (SexD) significant.
Underlined values indicate one sample Wilcoxon Signed-Rank Test for asymmetry *not significant.*

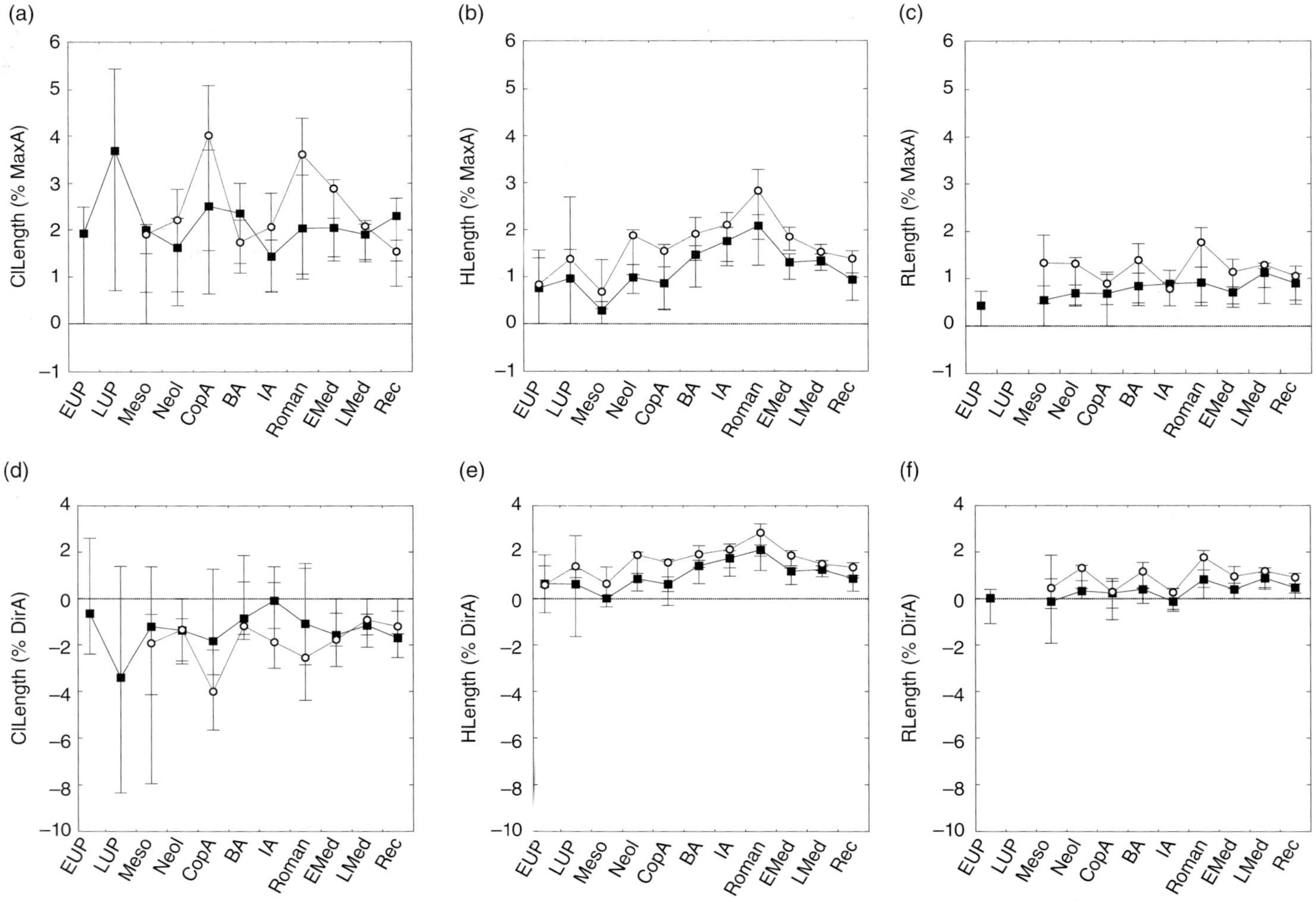

Figure 7.4 Maximum (a–c) and directional (d–f) asymmetry in upper limb lengths by sex (median ± 95% CI; males: solid symbols; females: open symbols).

Table 7.5 Maximum asymmetry in articular breadths for European pre-Holocene and Holocene humeri [Median (N; −95%CI – +95%CI)].

	Males	Females	SexD
HHDSI			
EUP	–	–	–
LUP	–	–	–
Mesolithic	1.16 (6; 0.41–1.52)	–	–
Neolithic	1.52 (46; 0.49–1.85)	1.52 (34; 0.76–2.02)	0.00
S Neol HG	1.45 (19; 0.23–1.86)	1.48 (10; 0.61–1.98)	−0.03
Neol Ag	1.51 (27; 0.46–1.99)	1.62 (24; 0.51–2.12)	−0.11
CopA	1.61 (15; 0.00–2.14)	1.48 (11; 0.00–2.20)	0.13
BA	2.11 (61; 1.07–2.20)	2.26 (49; 0.00–2.41)	−0.15
IA	2.87 (39; 1.29–3.66)	2.07 (34; 0.94–2.57)	0.80
Roman	1.96 (26; 0.62–2.41)	1.84 (20; 0.24–3.31)	0.12
EMed	2.06 (87; 1.55–2.18)	2.33 (54; 0.71–2.48)	−0.27
LMed	1.94 (163; 1.10–2.05)	2.15 (148; 1.25–2.35)	−0.21
Recent	1.57 (91; 0.99–1.89)	1.85 (73; 1.00–2.43)	−0.28
HDML			
EUP	–	–	–
LUP	–	–	–
Mesolithic	2.63 (9; 0.21–3.54)	2.38 (5; 0.00–3.69)	0.25
Neolithic	1.75 (48; 0.87–2.33)	2.56 (37; 0.51–3.39)	−0.81
S Neol HG	1.23 (20; 0.69–1.83)	1.56 (14; 0.26–2.29)	−0.33
Neol Ag	2.28 (28; 0.65–2.79)	3.79 (23; 0.51–5.16)	−1.51
CopA	2.50 (17; 0.64–2.95)	1.73 (13; 0.27–2.80)	0.77
BA	3.25 (60; 1.50–4.38)	2.15 (52; 1.28–3.19)	1.10
IA	2.63 (25; 0.73–3.28)	2.38 (32; 1.14–3.04)	0.25
Roman	3.60 (26; 1.71–4.22)	2.95 (18; 1.32–3.97)	0.65
EMed	2.52 (99; 1.73–3.08)	2.45 (63; 1.41–2.87)	0.07
LMed	2.30 (141; 1.81–2.57)	1.83 (114; 1.40–2.18)	0.47
Recent	3.05 (101; 1.69–3.38)	2.49 (60; 1.40–3.41)	0.56

* Mann–Whitney *U*-test for sexual dimorphism (SexD) significant.

and radial A-P breadths for Neolithic through Bronze Age females, which show almost no asymmetry or are slightly left-biased, and humeral M-L breadth for Mesolithic males, which is left-biased. Mediolateral breadths of both bones show slightly more maximum and directional asymmetry than A-P breadths in most temporal periods.

The magnitude of sexual dimorphism in diaphyseal breadths varies between periods, but almost every significant difference shows males to have more asymmetry (right-biased). The exception is humeral M-L breadth in the Mesolithic period, where females have more directional asymmetry. Sexual dimorphism in directional asymmetry tends to be greater in the early Holocene, especially for the humerus, and declines in more recent periods.

Table 7.6 Directional asymmetry in articular breadths for European pre-Holocene and Holocene humeri [Median (N; −95%CI – +95%CI)].

	Males	Females	SexD
HHDSI			
EUP	–	–	–
LUP	–	–	–
Mesolithic	0.82 (6; −1.32–1.52)	–	–
Neolithic	1.13 (46; 0.00–1.74)	0.48 (34; −0.83–1.38)	0.65
S Neol HG	1.21 (19; −0.23–1.82)	0.34 (10; −1.64–1.76)	0.87
Neol Ag	1.04 (27; −0.43–1.85)	0.58 (24; −0.87–1.56)	0.46
CopA	−0.30 (15; −1.04–2.14)	0.03 (11; −1.78–1.48)	−0.33
BA	1.65 (61; 0.00–2.16)	0.11 (49; −1.26–2.41)	1.54
IA	2.38 (39; 0.80–3.39)	1.75 (34; 0.36–2.38)	0.63
Roman	0.88 (26; −1.82–1.76)	0.67 (20; −0.77–2.26)	0.21
EMed	0.52 (87; 0.00–1.55)	1.57 (54; 0.00–2.37)	−1.05
LMed	0.69 (163; 0.00–1.03)	0.80 (148; 0.00–1.22)	−0.11
Recent	0.55 (91; −0.22–0.99)	0.16 (73; −1.86–1.86)	0.39
HDML			
EUP	–	–	–
LUP	–	–	–
Mesolithic	1.27 (9; −2.87–3.54)	2.34 (5; 0.00–3.69)	−1.07
Neolithic	0.92 (48; −0.45–1.31)	1.14 (37; −0.54–2.29)	−0.22
S Neol HG	0.49 (20; −1.48–0.98)	0.15 (14; −0.54–2.29)	0.34
Neol Ag	1.57 (28; −0.46–2.41)	2.60 (23; −1.29–4.84)	−1.03
CopA	1.90 (17; −1.98–2.83)	0.95 (13; −2.01–1.37)	0.95
BA	1.96 (60; 0.00–2.53)	1.59 (52; 0.00–2.57)	0.37
IA	1.88 (25; −0.89–2.49)	1.34 (32; −1.05–2.83)	0.54
Roman	1.72 (26; −2.14–3.88)	2.90 (18; 1.06–4.12)	−1.18
EMed	1.21 (99; −0.43–1.67)	1.81 (63; 0.26–2.52)	−0.60
LMed	1.11 (141; 0.22–1.53)	1.29 (114; 0.51–1.58)	−0.18
Recent	1.71 (101; 0.53–2.21)	1.60 (60; −0.52–2.35)	0.11

* Mann–Whitney *U*-test for sexual dimorphism (SexD) significant.
Underlined values indicate one sample Wilcoxon Signed-Rank Test for asymmetry *not significant.*

7.3.1.2.4 Asymmetry in Cross-Sectional Properties

Results for asymmetry comparisons of humeral CSG areas are shown in Tables 7.9 and 7.10 and Figure 7.7, and for section moduli in Tables 7.11 and 7.12 and Figure 7.8. In general, there is a consistent pattern of temporal variation among all of the compared humeral CSG properties. For males, there is a major decline in maximum and directional CSG asymmetry from the EUP to the Mesolithic, that is, prior to the introduction of agriculture. The decrease in asymmetry is about 20% for areas and 35% for section moduli, and clearly exceeds between-period variation among all of the agricultural groups. Thus, the major change in CSG for males appears

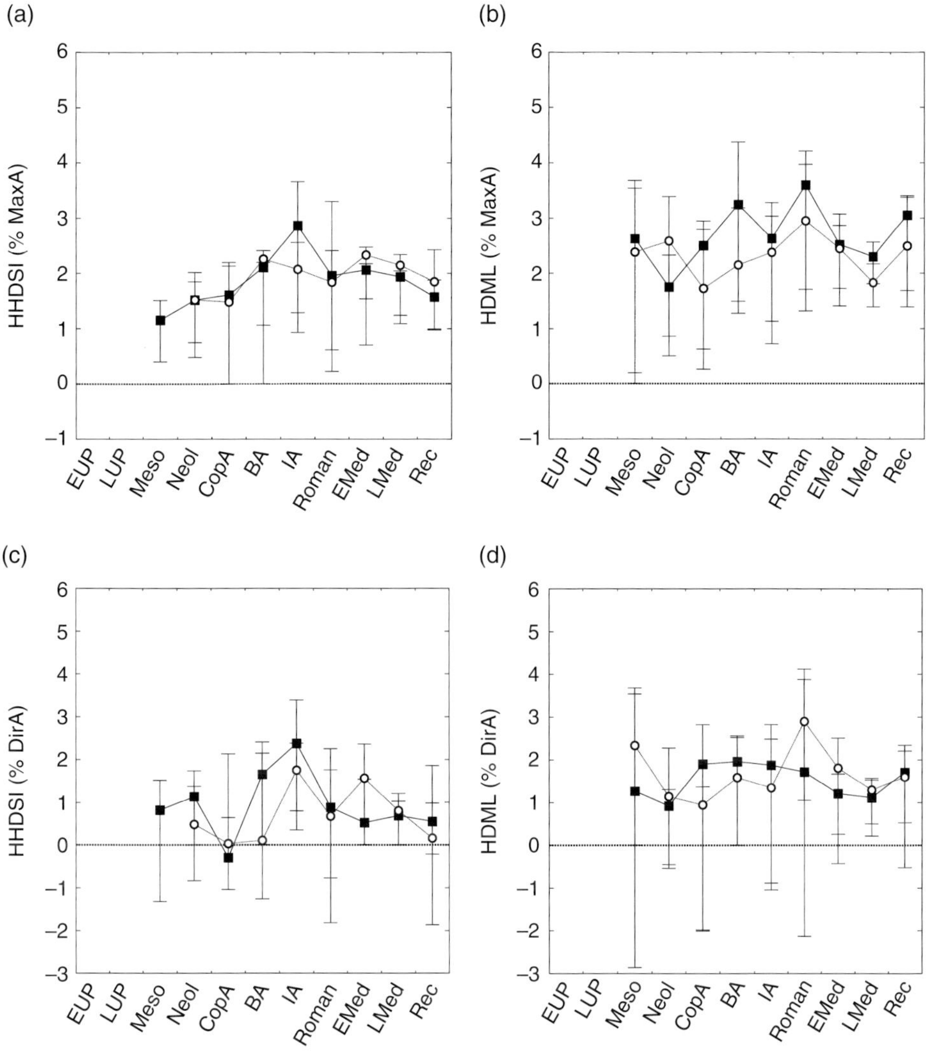

Figure 7.5 Maximum (a,b) and directional (c,d) asymmetry in humeral articular breadths by sex (median ± 95% CI; males: solid symbols; females: open symbols).

prior to the adoption of agriculture and only to a lesser degree, if at all, in the transition between Mesolithic foragers and Neolithic groups. Interestingly, though, among contemporaneous Scandinavian Neolithic foragers and European Neolithic farmers, there is a large difference in humeral Z_p asymmetry among males, with farmers having less asymmetry (Fig. 7.9), demonstrating that there may be an effect of agricultural subsistence on male upper limb asymmetry, at least in certain regions. This result is discussed further in the Discussion section. Changes among males after the Mesolithic are relatively smaller, although there is an increase in maximum and directional asymmetry in the Iron Age, and a decrease from the Early Medieval to Recent periods.

For females, there is no preserved sample for the EUP and only limited sample sizes for the LUP and Mesolithic (for both periods n = 4–5), but it is still apparent that maximum asymmetry in CSG parameters continually decreases between the Mesolithic, Neolithic, and Copper

Table 7.7 Maximum asymmetry in diaphyseal breadths for European pre-Holocene and Holocene upper limb bones [Median (N; −95%CI – +95%CI)].

	Males	Females	SexD
H35AP			
EUP	–	–	–
LUP	–	–	–
Mesolithic	4.75 (18; 1.37–6.26)	2.62 (8; 1.00–3.90)	2.13
Neolithic	1.90 (66; 1.03–2.65)	3.12 (50; 1.69–3.50)	−1.22
S Neol HG	2.43 (26; 0.48–3.43)	3.28 (19; 0.94–4.03)	−0.85
Neol Ag	1.69 (40; 0.99–2.65)	2.96 (31; 1.14–3.55)	−1.27
CopA	3.43 (31; 0.95–3.80)	2.61 (21; 0.57–3.11)	0.82
BA	4.71 (74; 2.99–5.22)	2.21 (64; 0.57–2.93)	2.50*
IA	2.94 (56; 1.89–3.59)	2.07 (60; 1.07–2.50)	0.87
Roman	2.37 (35; 1.40–2.97)	2.28 (33; 0.56–3.51)	0.09
EMed	3.01 (109; 1.89–3.36)	2.01 (72; 1.12–2.38)	1.00*
LMed	2.21 (133; 1.42–2.60)	1.96 (108; 1.34–2.20)	0.25
Recent	2.08 (116; 1.43–2.37)	2.33 (80; 1.68–2.66)	−0.25
H35ML			
EUP	–	–	–
LUP	–	–	–
Mesolithic	3.96 (18; 1.15–6.42)	3.83 (8; 0.63–4.83)	0.13
Neolithic	4.84 (66; 3.47–5.60)	2.76 (50; 1.54–3.60)	2.08*
S Neol HG	5.99 (26; 3.73–6.69)	3.05 (19; 1.12–3.66)	2.94*
Neol Ag	4.03 (40; 2.24–4.80)	2.53 (31; 0.60–3.95)	1.50
CopA	5.24 (31; 1.25–6.64)	3.49 (21; 0.91–4.55)	1.75
BA	4.35 (74; 2.81–5.41)	2.85 (64; 1.06–3.62)	1.50
IA	7.24 (56; 5.34–9.27)	4.66 (60; 2.21–5.82)	2.58*
Roman	4.78 (35; 1.54–5.57)	2.91 (33; 1.23–3.56)	1.87
EMed	4.00 (109; 2.86–4.36)	4.58 (72; 2.95–4.99)	−0.58
LMed	4.98 (133; 2.97–5.55)	3.24 (108; 2.33–3.69)	1.74*
Recent	4.34 (116; 3.07–4.79)	2.95 (80; 1.95–3.55)	1.39
R50AP			
EUP	–	–	–
LUP	–	–	–
Mesolithic	–	–	–
Neolithic	2.89 (10; 0.78–3.45)	2.21 (6; 0.00–4.17)	0.68
S Neol HG	–	–	–
Neol Ag	2.89 (10; 0.78–3.45)	2.28 (6; 0.00–4.17)	0.61
CopA	2.40 (21; 0.81–3.72)	2.63 (20; 0.84–4.54)	−0.23
BA	4.86 (55; 3.13–6.40)	3.70 (46; 1.13–4.17)	1.16*
IA	5.85 (22; 3.34–7.54)	4.18 (28; 1.00–4.84)	1.67*

Table 7.7 (Continued)

	Males	Females	SexD
Roman	4.92 (25; 0.89–5.72)	3.08 (27; 0.90–3.96)	1.84*
EMed	3.22 (85; 1.62–3.96)	3.10 (58; 1.82–3.81)	0.12
LMed	2.59 (94; 1.52–2.78)	3.14 (82; 1.82–3.78)	−0.55
Recent	3.95 (39; 2.56–4.57)	2.72 (39; 1.06–3.93)	1.23
R50ML			
EUP	–	–	–
LUP	–	–	–
Mesolithic	–	–	–
Neolithic	6.67 (10; 2.86–9.00)	6.99 (6; 3.54–9.03)	−0.32
S Neol HG	–	–	–
Neol Ag	6.68 (10; 2.44–9.00)	6.94 (6; 3.54–9.03)	−0.26
CopA	5.35 (21; 1.81–7.27)	5.27 (20; 2.95–6.19)	0.08
BA	4.92 (54; 2.62–5.72)	2.79 (46; 1.42–3.76)	2.13*
IA	8.77 (22; 6.41–9.47)	7.28 (28; 4.62–7.95)	1.49
Roman	5.13 (25; 1.22–7.06)	5.22 (27; 2.12–7.47)	−0.09
EMed	5.57 (85; 3.64–6.31)	5.66 (58; 3.55–7.30)	−0.09
LMed	4.47 (94; 3.03–4.99)	4.97 (82; 3.72–5.97)	−0.50
Recent	5.01 (39; 2.84–6.77)	3.95 (39; 2.45–4.68)	1.06

* Mann-Whitney U test for sexual dimorphism (SexD) significant.

Table 7.8 Directional asymmetry in diaphyseal breadths for European pre-Holocene and Holocene upper limb bones [Median (N; −95%CI – +95%CI)].

	Males	Females	SexD
H35AP			
EUP	–	–	–
LUP	–	–	–
Mesolithic	3.45 (18; −1.20–5.71)	2.14 (8; −2.64–3.90)	1.31
Neolithic	1.50 (66; 0.00–2.32)	0.71 (50; −1.80–1.69)	0.79*
S Neol HG	1.79 (26; −0.45–2.94)	−0.27 (19; −3.83–1.05)	2.06
Neol Ag	1.42 (40; −0.43–2.55)	1.21 (31; −2.40–2.47)	0.21
CopA	1.71 (31; −2.00–3.36)	1.82 (21; −1.46–2.64)	−0.11
BA	4.20 (74; 2.24–5.17)	0.16 (64; −0.81–3.07)	4.04*
IA	2.20 (56; −0.98–2.60)	0.46 (60; −1.05–1.07)	1.74*
Roman	0.51 (35; −1.64–1.46)	0.94 (33; −1.10–1.70)	−0.43
EMed	2.12 (109; 0.46–2.81)	1.11 (72; 0.00–1.65)	1.01
LMed	0.94 (133; 0.00–1.38)	0.85 (108; −0.47–1.55)	0.09
Recent	0.68 (116; −0.45–1.27)	1.19 (80; −0.54–1.74)	−0.51

(Continued)

Table 7.8 (Continued)

	Males	Females	SexD
H35ML			
EUP	–	–	–
LUP	–	–	–
Mesolithic	−2.44 (18; −4.98–1.15)	3.65 (8; −1.67–4.83)	−6.09*
Neolithic	4.16 (66; 1.91–4.96)	0.40 (50; −3.06–3.33)	3.76*
S Neol HG	5.99 (26; 3.69–6.69)	2.52 (19; −1.12–3.25)	3.47*
Neol Ag	2.44 (40; −0.57–3.80)	−0.72 (31; −2.15–0.71)	3.16*
CopA	3.88 (31; 0.49–5.65)	−0.13 (21; −5.38–1.25)	4.01*
BA	3.90 (74; 2.60–5.13)	−0.18 (64; −3.29–1.69)	4.08*
IA	7.24 (56; 5.25–9.27)	3.97 (60; 0.56–5.75)	3.27*
Roman	2.46 (35; −1.69–4.24)	2.70 (33; 0.00–3.46)	−0.24
EMed	3.32 (109; 1.55–4.02)	3.51 (72; 0.63–4.58)	−0.19
LMed	4.19 (133; 1.90–5.11)	1.84 (108; 0.00–2.84)	2.35*
Recent	3.61 (116; 1.89–4.22)	2.00 (80; 0.59–2.71)	1.61
Ra50AP			
EUP	–	–	–
LUP	–	–	–
Mesolithic	–	–	–
Neolithic	2.05 (10; −3.08–3.45)	0.17 (6; −10.63–1.98)	1.88
S Neol HG	–	–	–
Neol Ag	2.10 (10; −3.08–3.45)	0.11 (6; −10.63–1.98)	1.99
CopA	0.50 (21; −2.34–1.74)	0.80 (20; −1.81–2.05)	−0.30
BA	4.15 (55; 0.43–5.21)	−1.34 (46; −3.58–0.96)	5.49*
IA	1.84 (22; −4.45–6.62)	3.04 (28; 0.00–4.79)	−1.20
Roman	2.71 (25; −3.16–4.42)	1.10 (27; −2.77–1.91)	1.61
EMed	1.66 (85; 0.00–2.30)	1.47 (58; −0.94–1.98)	0.19
LMed	0.91 (94; −0.38–1.60)	1.01 (82; 0.00–1.73)	−0.10
Recent	3.55 (39; 0.00–4.22)	2.08 (39; −0.96–2.45)	1.47
Ra50ML			
EUP	–	–	–
LUP	–	–	–
Mesolithic	–	–	–
Neolithic	3.42 (10; −9.71–6.86)	6.99 (6; 3.54–9.03)	−3.57
S Neol HG	–	–	–
Neol Ag	3.36 (10; −9.71–6.86)	7.02 (6; 3.54–9.03)	−3.66
CopA	5.00 (21; 0.88–7.27)	5.26 (20; 2.95–6.19)	−0.26
BA	3.27 (54; 1.16–4.99)	1.19 (46; −0.95–1.85)	2.08
IA	8.77 (22; 6.99–9.47)	7.29 (28; 4.62–7.95)	1.48
Roman	4.85 (25; −0.56–7.06)	3.57 (27; −1.93–6.58)	1.28
EMed	4.71 (85; 2.53–6.16)	5.28 (58; 2.50–7.29)	−0.57
LMed	3.12 (94; 0.62–4.20)	4.64 (82; 2.20–5.56)	−1.52
Recent	4.16 (39; −0.64–4.94)	2.92 (39; −1.54–4.32)	1.24

* Mann–Whitney *U*-test for sexual dimorphism (SexD) significant.
Underlined values indicate one sample Wilcoxon Signed-Rank Test for asymmetry *not significant.*

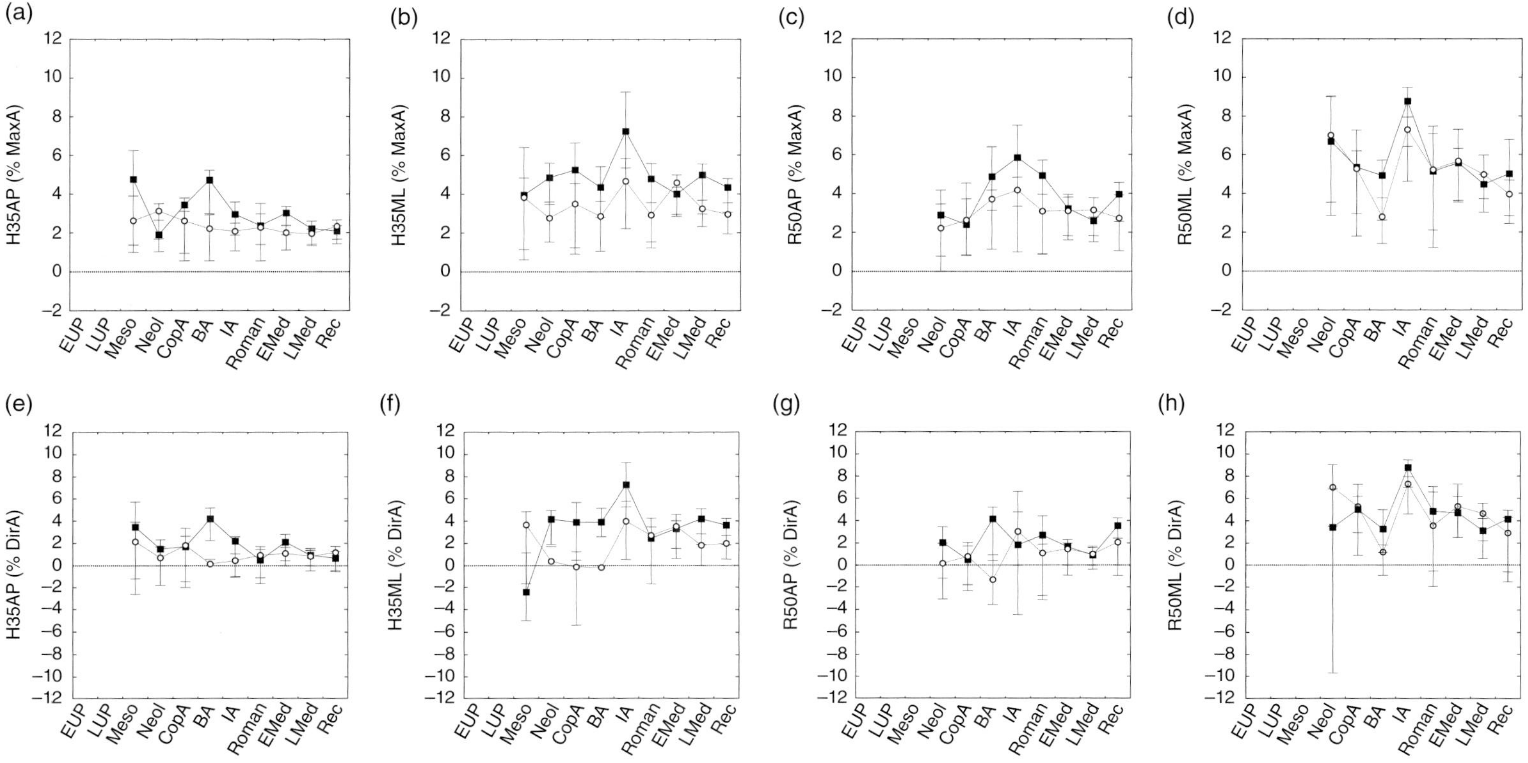

Figure 7.6 Maximum (a–d) and directional (e–h) asymmetry in upper limb diaphyseal breadths (median ± 95% CI; males: solid symbols; females: open symbols).

Table 7.9 Maximum asymmetry in cross-sectional areas for European pre-Holocene and Holocene humeri [Median (N; −95%CI – +95%CI)].

	Males	Females	SexD
HTA			
EUP	26.95 (5; 15.66–30.86)	–	–
LUP	–	–	–
Mesolithic	5.09 (9; 2.28–7.47)	8.58 (4; 3.55–17.84)	−3.49
Neolithic	8.00 (66; 4.18–9.74)	3.70 (44; 1.12–5.54)	4.30*
S Neol HG	10.52 (25; 5.74–11.04)	3.67 (13; 0.30–5.87)	6.85*
Neol Ag	5.42 (41; 2.41–6.57)	3.74 (31; 0.93–5.75)	1.68
CopA	8.12 (32; 4.83–9.78)	2.60 (21; 0.79–3.47)	5.52*
BA	8.09 (70; 5.11–9.89)	3.32 (57; 1.62–4.23)	4.77*
IA	10.42 (54; 7.27–12.08)	5.28 (54; 2.87–6.17)	5.14*
Roman	5.61 (33; 1.66–6.59)	4.40 (34; 1.65–6.55)	1.21
EMed	8.44 (104; 7.15–8.96)	4.85 (73; 3.72–5.38)	3.59*
LMed	6.35 (122; 4.83–7.13)	3.22 (103; 2.32–3.62)	3.13*
Recent	5.16 (114; 3.74–6.29)	4.14 (77; 2.64–5.19)	1.02
HCA			
EUP	28.62 (5; 6.74–41.94)	–	–
LUP	25.18 (9; 3.43–27.33)	7.62 (5; 0.80–25.13)	17.56*
Mesolithic	10.62 (16; 4.39–12.71)	9.48 (5; 3.84–13.26)	1.14
Neolithic	7.83 (66; 4.86–10.01)	4.39 (44; 1.79–6.04)	3.44*
S Neol HG	9.95 (25; 6.05–12.01)	3.59 (13; 0.19–5.47)	6.36*
Neol Ag	5.89 (41; 2.77–7.20)	4.99 (31; 1.33–6.41)	0.90
CopA	9.17 (32; 2.10–10.81)	4.21 (21; 1.82–5.68)	4.96*
BA	9.10 (70; 6.13–10.89)	5.32 (57; 2.93–6.42)	3.78*
IA	10.06 (54; 6.20–10.95)	7.10 (54; 3.95–8.71)	2.96*
Roman	6.76 (33; 3.19–8.65)	5.09 (34; 2.17–6.94)	1.67
EMed	7.35 (104; 4.37–8.77)	5.00 (73; 2.27–5.95)	2.35*
LMed	7.52 (122; 5.18–8.94)	4.50 (103; 3.13–5.01)	3.02*
Recent	5.32 (114; 3.88–6.14)	5.99 (77; 4.12–6.79)	−0.67

* Mann–Whitney *U*-test for sexual dimorphism (SexD) significant.

Age, whereas directional asymmetry abruptly declines at the beginning of the Holocene between the Mesolithic and Neolithic. Directional asymmetry in the female sample is close to 0 in the early agricultural periods (i.e., Neol, CopA, BA) and increases in the later agricultural periods. In fact, the three early agricultural female samples are the only ones without right-dominated asymmetry in CSG properties among all the temporal/sex groups in the study. Maximum asymmetry in Neolithic females is not particularly low compared to later agricultural groups, suggesting a two-step process during the early agricultural period, with first a reduction of directional asymmetry in the early Neolithic, followed by a reduction in all asymmetry in the late Neolithic, that is, Copper Age. These trends are also supported by comparisons

Table 7.10 Directional asymmetry in cross-sectional areas for European pre-Holocene and Holocene humeri [Median (N; −95%CI – +95%CI)].

	Males	Females	SexD
HTA			
EUP	27.15 (5; 15.66–35.39)	–	–
LUP	–	–	–
Mesolithic	<u>4.36 (9; −4.03–7.86)</u>	<u>8.56 (4; 3.55–17.84)</u>	−4.20
Neolithic	7.66 (66; 3.13–9.72)	<u>0.31 (44; −2.46–0.93)</u>	7.35*
S Neol HG	10.52 (25; 4.87–11.04)	<u>1.51 (13; −7.95–3.28)</u>	9.01*
Neol Ag	4.96 (41; 1.08–6.48)	<u>−0.19 (31; −4.94–0.80)</u>	5.15*
CopA	7.48 (32; 1.87–9.78)	<u>0.43 (21; −2.23–1.87)</u>	7.05*
BA	7.90 (70; 5.05–9.25)	<u>0.37 (57; −2.37–1.27)</u>	7.53*
IA	10.41 (54; 7.27–11.86)	4.11 (54; 0.41–5.29)	6.30*
Roman	5.61 (33; 1.58–6.59)	3.08 (34; 0.89–4.74)	2.53
EMed	8.18 (104; 6.36–8.94)	4.64 (73; 2.98–5.34)	3.54*
LMed	5.87 (122; 4.07–6.53)	2.68 (103; 1.75–3.21)	3.19*
Recent	4.39 (114; 3.27–5.39)	3.46 (77; 1.89–4.32)	0.93
HCA			
EUP	28.76 (5; 0.00–41.94)	–	–
LUP	<u>22.38 (9; 18.26–25.75)</u>	<u>6.83 (5; −0.80–10.05)</u>	15.55
Mesolithic	10.51 (16; 0.14–12.71)	9.42 (5; 3.84–13.26)	1.09
Neolithic	7.07 (66; 3.54–8.83)	<u>0.02 (44; −2.62–1.33)</u>	7.05*
S Neol HG	9.32 (25; 6.03–10.83)	<u>−0.99 (13; −3.57–6.04)</u>	10.31*
Neol Ag	4.51 (41; 0.25–6.97)	<u>0.56 (31; −2.72–2.22)</u>	3.95*
CopA	8.30 (32; 1.52–10.64)	<u>0.89 (21; −7.64–2.42)</u>	7.41*
BA	9.01 (70; 4.95–10.82)	<u>0.09 (57; −2.15–4.55)</u>	8.92*
IA	9.77 (54; 3.74–10.63)	<u>2.65 (54; −2.72–4.68)</u>	7.12*
Roman	5.53 (33; 1.24–7.51)	2.87 (34; −1.30–4.45)	2.66
EMed	5.56 (104; 3.04–7.53)	1.78 (73; −1.02–2.92)	3.78*
LMed	6.20 (122; 3.32–7.74)	3.86 (103; 1.76–4.63)	2.34*
Recent	4.14 (114; 1.51–5.02)	4.90 (77; 2.54–6.15)	−0.76

* Mann–Whitney *U*-test for sexual dimorphism (SexD) significant.
Underlined values indicate one sample Wilcoxon Signed-Rank Test for asymmetry *not significant.*

between Scandinavian Neolithic foragers and European Neolithic farmers (Fig. 7.9), where female farmers show more of a reduction in directional asymmetry of humeral Z_p than in maximum asymmetry. Interestingly, unlike males, female Scandinavian Neolithic foragers are closer in asymmetry to Neolithic farmers than to Mesolithics, a result that is addressed further in the Discussion.

Temporal changes in Z_y (M-L bending strength) asymmetry among agricultural groups tend to be greater than those in Z_x (A-P bending strength), in both sexes (see Fig. 7.8). Thus, changes in overall strength asymmetry (Z_p) of the humerus are due more to M-L rather than A-P dimensional changes. This is not generally true for the pre-agricultural periods, however.

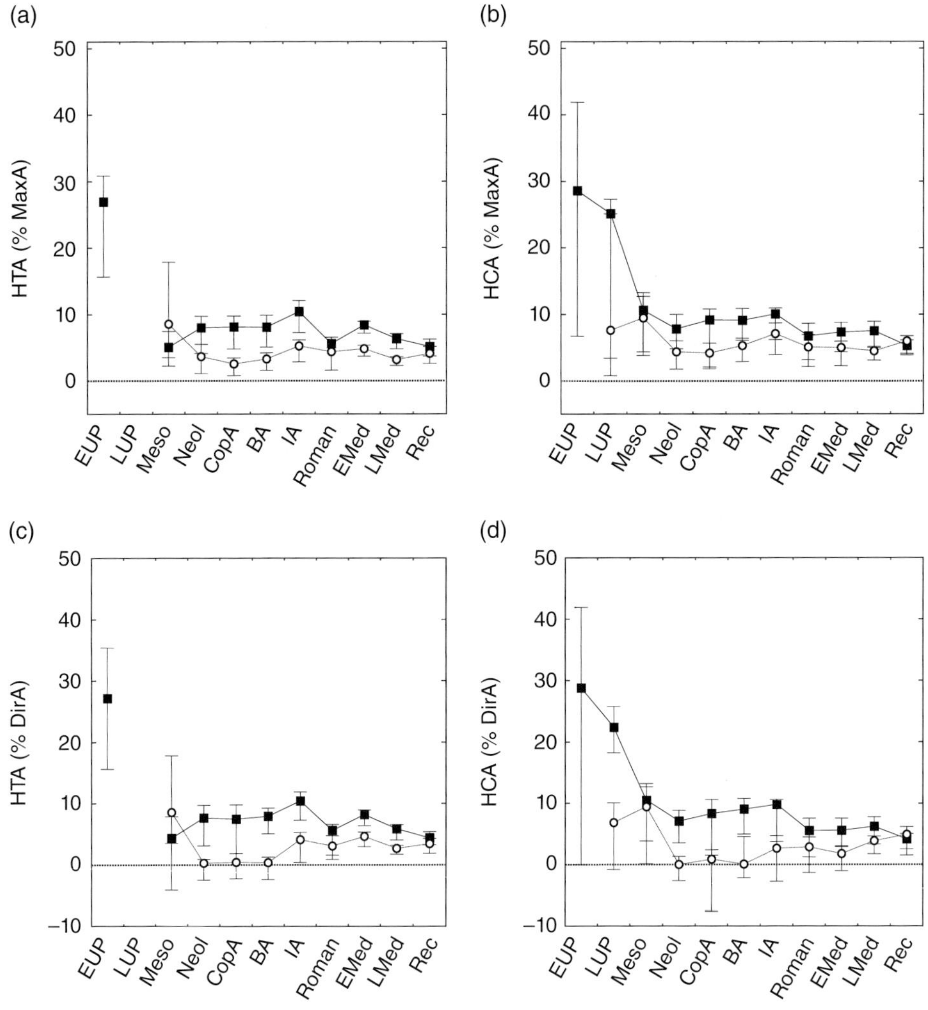

Figure 7.7 Maximum (a,b) and directional (c,d) asymmetry in humeral CSG areas (median ± 95% CI; males: solid symbols; females: open symbols).

Changes in sexual dimorphism in CSG asymmetry are shown in Tables 7.9–7.12 and summarized for humeral Z_p in Figure 7.10 (the maximum asymmetry results repeat some data shown earlier in Fig. 7.3). The very large sex-related difference in directional asymmetry of humeral strength in the early agricultural periods is highlighted in Figure 7.10. Sexual dimorphism in maximum asymmetry of strength shows smaller temporal fluctuations except in the LUP sample. The Mesolithic is the only time period with more directional asymmetry of Z_p in females, although the small female sample size here should be noted. The Roman period shows particularly low sexual dimorphism in asymmetry. There is a general reduction of sexual dimorphism in both maximum and directional asymmetry from the Early Medieval through the Recent period, due to slightly greater decreases in asymmetry among males.

In summary, in males the shift in CSG asymmetry after the Pleistocene is well established earlier in the Mesolithic for both maximum and directional asymmetry, and remains relatively constant during the rest of the Holocene. For females, the shift in asymmetry in CSG is still in process through the Neolithic in directional asymmetry and through the Copper Age in

Table 7.11 Maximum asymmetry in diaphyseal strength (section moduli) for European pre-Holocene and Holocene humeri [Median (N; −95%CI – +95%CI)].

	Males	Females	SexD
HZ_x			
EUP	40.97 (5; 25.48–50.02)	–	–
LUP	25.05 (9; 9.83–29.16)	11.79 (5; 2.30–34.66)	13.26
Mesolithic	14.43 (16; 7.32–16.22)	14.22 (5; 3.13–19.21)	0.21
Neolithic	11.18 (66; 8.53–12.45)	7.02 (44; 3.72–9.45)	4.16
S Neol HG	12.97 (25; 7.03–15.41)	7.04 (13; 3.50–9.32)	5.93
Neol Ag	10.43 (41; 4.65–11.15)	7.50 (31; 1.22–12.67)	2.93
CopA	11.53 (32; 5.93–15.22)	3.80 (21; 0.46–5.21)	7.73*
BA	12.61 (70; 7.90–14.68)	5.55 (57; 3.44–7.01)	7.06*
IA	14.46 (54; 10.74–15.84)	8.09 (54; 5.87–9.37)	6.37*
Roman	9.45 (33; 3.81–10.82)	8.59 (34; 2.60–10.83)	0.86
EMed	10.32 (104; 7.43–11.27)	7.35 (73; 4.68–9.51)	2.97*
LMed	8.78 (122; 5.88–10.20)	5.61 (103; 3.62–6.34)	3.17*
Recent	8.16 (114; 5.72–9.94)	8.40 (77; 4.50–9.82)	−0.24
HZ_y			
EUP	35.30 (5; 21.29–41.61)	–	–
LUP	32.49 (9; 25.44–34.84)	11.91 (5; 0.73–31.99)	20.58*
Mesolithic	11.42 (16; 4.82–17.15)	9.14 (5; 4.47–13.84)	2.28
Neolithic	13.50 (66; 9.11–15.75)	9.74 (44; 4.34–12.45)	3.76*
S Neol HG	16.59 (25; 10.09–19.82)	13.25 (13; 1.85–16.2)	3.34
Neol Ag	11.74 (41; 4.70–15.37)	7.96 (31; 2.17–11.28)	3.78
BA	13.60 (32; 9.69–16.74)	5.01 (21; 0.85–7.73)	8.59*
CopA	11.89 (70; 8.20–14.26)	7.15 (57; 3.86–8.82)	4.74*
IA	19.70 (54; 11.17–21.86)	12.73 (54; 6.24–15.39)	6.97*
Roman	10.64 (33; 6.37–13.23)	8.96 (34; 4.81–12.99)	1.68
EMed	14.95 (104; 12.58–16.43)	8.25 (73; 5.26–10.99)	6.70*
LMed	11.87 (122; 7.22–13.82)	6.80 (103; 4.79–7.64)	5.07*
Recent	9.15 (114; 6.24–11.06)	7.76 (77; 3.70–9.10)	1.39
HZ_p			
EUP	39.08 (5; 24.47–47.01)	–	–
LUP	29.81 (9; 6.11–31.90)	14.79 (5; 7.59–34.07)	15.02
Mesolithic	13.41 (16; 8.02–14.51)	14.31 (5; 5.54–18.98)	−0.90
Neolithic	12.09 (66; 6.67–17.02)	9.63 (44; 5.64–11.21)	2.46*
S Neol HG	16.99 (25; 8.28–18.36)	7.92 (13; 0.84–11.21)	9.07*
Neol Ag	7.62 (41; 2.68–10.99)	10.21 (31; 3.07–12.10)	−2.59*
CopA	13.95 (32; 6.98–16.46)	5.32 (21; 1.55–6.70)	8.63*
BA	11.68 (70; 7.13–14.64)	5.87 (57; 3.07–6.81)	5.81*
IA	16.02 (54; 10.55–19.58)	10.06 (54; 5.81–11.96)	5.96*
Roman	9.23 (33; 3.78–11.29)	10.24 (34; 2.50–15.86)	−1.01
EMed	13.40 (104; 11.08–15.26)	7.02 (73; 4.06–8.75)	6.38*
LMed	10.57 (122; 7.75–12.18)	6.52 (103; 4.54–7.92)	4.05*
Recent	10.08 (114; 8.03–11.14)	9.31 (77; 6.06–11.75)	0.77

* Mann–Whitney *U*-test for sexual dimorphism (SexD) significant.

Table 7.12 Directional asymmetry in diaphyseal strength (section moduli) for European pre-Holocene and Holocene humeri [Median (N; –95%CI – +95%CI)].

	Males	Females	SexD
HZ_x			
EUP	41.30 (5; 25.48–60.60)	–	–
LUP	21.88 (9; 16.31–27.51)	8.04 (5; –2.30–12.54)	13.84
Mesolithic	13.85 (16; 6.93–16.22)	14.21 (5; 3.13–19.21)	–0.36
Neolithic	10.68 (66; 6.36–11.56)	1.44 (44; –4.03–3.87)	9.24*
S Neol HG	12.13 (25; 4.90–15.00)	5.14 (13; –7.10–7.74)	6.99
Neol Ag	9.53 (41; 1.59–11.03)	0.22 (31; –1.92–6.15)	9.31*
CopA	10.76 (32; 3.81–14.64)	1.03 (21; –4.18–2.53)	9.73*
BA	12.29 (70; 7.67–14.37)	0.70 (57; –3.70–2.36)	11.59*
IA	13.62 (54; 5.52–15.49)	2.20 (54; –5.87–5.48)	11.42*
Roman	6.68 (33; –3.04–9.64)	4.30 (34; –2.60–8.50)	2.38
EMed	9.61 (104; 6.49–10.87)	3.03 (73; –1.77–5.56)	6.58*
LMed	6.77 (122; 3.47–8.39)	3.38 (103; 1.24–4.53)	3.39*
*Recent	5.86 (114; 2.97–7.22)	5.37 (77; –0.17–7.06)	0.49
HZ_y			
EUP	35.62 (5; 0.00–41.61)	–	–
LUP	28.62 (9; 28.28–32.12)	9.83 (5; –0.73–19.42)	18.79
Mesolithic	0.83 (16; –17.15–7.06)	8.38 (5; –4.47–13.84)	–7.55
Neolithic	11.29 (66; 4.12–14.81)	–3.11 (44; –8.17–2.10)	14.40*
S Neol HG	15.34 (25; 6.03–19.37)	–9.14 (13; –17.61–8.01)	24.48*
Neol Ag	7.31 (41; –2.15–11.05)	–1.55 (31; –6.41–2.64)	8.86*
CopA	12.84 (32; 4.56–15.41)	–0.31 (21; –1.87–5.38)	13.15*
BA	11.19 (70; 7.14–13.88)	–1.61 (57; –4.55–2.24)	12.80*
IA	19.14 (54; 10.61–21.63)	7.45 (54; 0.28–13.10)	11.69*
Roman	10.55 (33; 5.28–13.23)	6.15 (34; 0.39–8.56)	4.40
EMed	14.23 (104; 11.07–16.20)	7.29 (73; 4.22–8.87)	6.94*
LMed	10.18 (122; 5.09–12.68)	4.18 (103; 0.41–5.88)	6.00*
Recent	6.18 (114; 2.72–8.53)	4.98 (77; –0.75–8.87)	1.20
HZ_p			
EUP	39.41 (5; 24.47–54.44)	–	–
LUP	26.07 (9; 21.85–30.36)	13.52 (5; 10.05–15.73)	12.55
Mesolithic	9.83 (16; –11.95–14.51)	14.21 (5; 5.54–19.83)	–4.38
Neolithic	10.84 (66; 4.90–15.14)	0.24 (44; –3.07–8.17)	10.60*
S Neol HG	16.22 (25; 8.28–18.06)	2.88 (13; –13.01–7.43)	13.34*
Neol Ag	6.69 (41; 2.04–9.09)	–0.96 (31; –7.28–8.17)	7.65*
CopA	12.38 (32; 3.00–16.46)	–0.32 (21; –3.50–5.74)	12.70*
BA	11.25 (70; 6.49–14.48)	–0.23 (57; –1.64–3.92)	11.48*
IA	15.59 (54; 8.68–19.58)	5.86 (54; –2.06–9.61)	9.73*
Roman	7.64 (33; –1.80–9.52)	6.22 (34; –1.33–10.40)	1.42
EMed	13.07 (104; 10.56–15.14)	5.90 (73; 1.45–7.73)	7.17*
LMed	8.97 (122; 6.15–10.77)	5.07 (103; 2.84–5.98)	3.90*
Recent	8.73 (114; 4.52–10.09)	7.61 (77; 2.28–9.32)	1.12

* Mann–Whitney *U*-test for sexual dimorphism (SexD) significant.
Underlined values indicate one sample Wilcoxon Signed-Rank Test for asymmetry *not significant*.

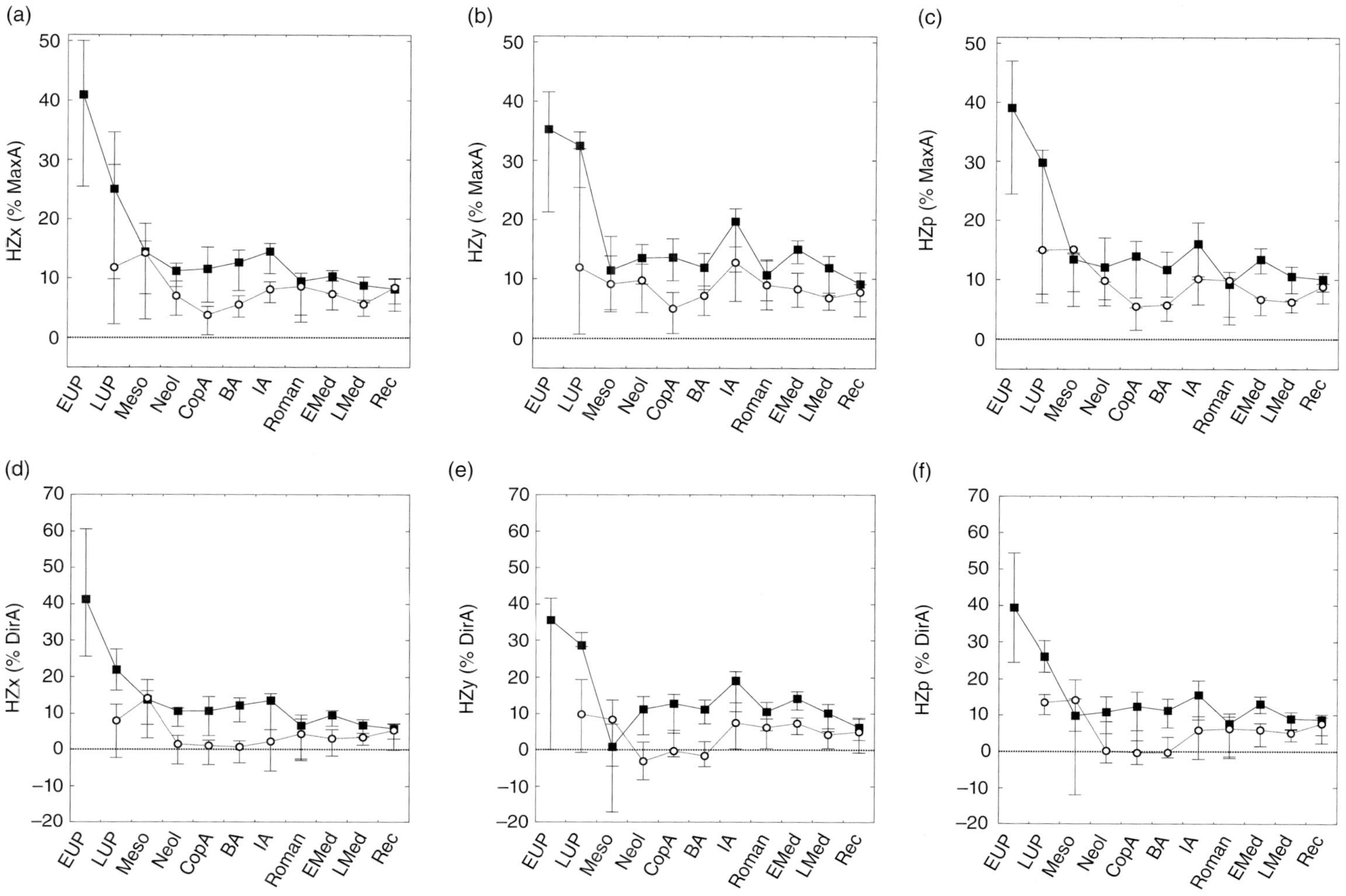

Figure 7.8 Maximum (a–c) and directional (e–f) asymmetry in humeral section moduli (median ± 95% CI; males: solid symbols; females: open symbols).

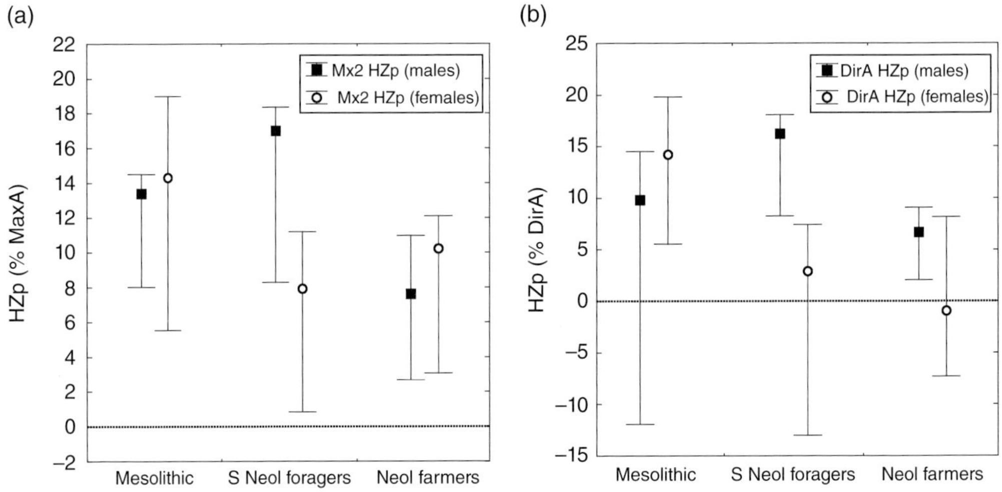

Figure 7.9 Range plots of maximum (a) and directional (b) asymmetry in section moduli in Mesolithic foragers, Scandinavian Neolithic foragers and European Neolithic farmers (median ± 95% CI; males: solid symbols; females: open symbols).

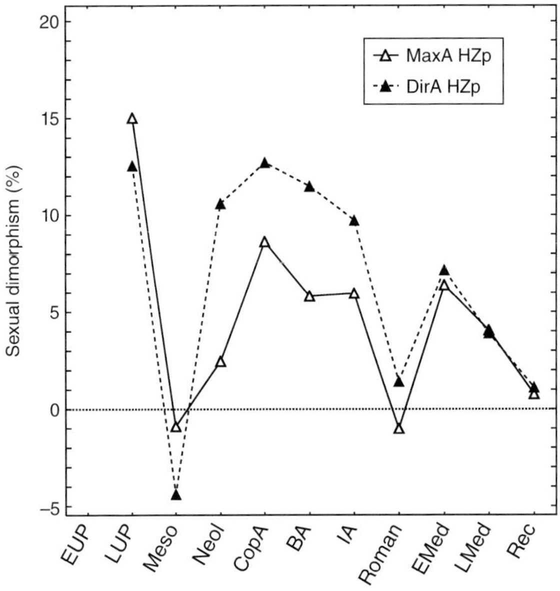

Figure 7.10 Sexual dimorphism for maximum and directional asymmetry in polar section modulus.

maximum asymmetry. Females, but not males, have a distinct early agricultural period characterized by an absence of CSG directional asymmetry, followed by later agricultural groups with relatively higher CSG directional asymmetry, although still below the values observed before the Holocene. The difference in timing of these changes in males and females leads to high sexual dimorphism in the early agricultural periods, with less dimorphism afterwards.

7.3.1.3 Regional Differences in Asymmetry

Comparisons between regions for maximum and directional asymmetry of humeral Z_p are shown in Tables 7.13 and 7.14, with a summary for directional asymmetry in Figure 7.11.

Table 7.13 Percentage maximum asymmetry in humeral diaphyseal strength (HZ_p) for regional variation [Median (N; −95%CI – +95%CI)].

	Neolithic	BA	IA	Roman	EMed	LMed	Recent
Males							
Britain	23.87 (6; 4.80–35.25)	11.15 (9; 8.35–15.11)	16.21 (29; 6.31–21.26)	8.58 (22; 6.27–15.42)	–	9.32 (25; 5.58–14.10)	8.80 (15; 5.26–19.26)
Scandinavian	17.30 (34; 8.17–21.42)	–	19.57 (15; 8.68–20.06)	–	–	10.55 (20; 8.20–16.13)	8.27 (38; 3.86–11.64)
NC Europe	6.67 (7; 2.38–16.86)	14.55 (39; 5.92–19.26)	20.38 (4; 12.30–22.45)	8.63 (6; 3.78–10.89)	12.38 (66; 6.50–17.68)	7.43 (22; 2.16–17.04)	–
France	4.90 (13; 2.44–7.13)	–	12.93 (6; 6.46–15.11)	–	17.02 (7; 10.99–43.33)	11.72 (18; 8.23–18.11)	–
Italy	16.11 (6; 12.71–21.95)	4.93 (16; 2.72–12.74)	–	13.95 (5; 6.14–21.34)	18.34 (7; 15.26–35.24)	18.83 (4; 13.34–32.57)	11.21 (17; 7.59–17.73)
Iberia	–	22.91 (6; 17.04–28.04)	–	–	13.21 (24; 8.03–18.50)	14.09 (24; 5.54–22.90)	13.98 (25; 4.52–19.35)
Balkans	–	–	–	–	–	10.52 (27; 6.49–15.97)	–
Females							
Britain	–	6.81 (5; 6.62–21.92)	11.57 (27; 5.28–14.77)	9.24 (24; 2.14–19.21)	–	7.91 (26; 2.88–11.38)	8.08 (17; 6.46–15.64)
Scandinavian	7.42 (13; 1.99–11.60)	–	10.26 (20; 4.63–15.15)	–	–	8.40 (12; 2.29–16.30)	8.59 (19; 3.11–14.67)
NC Europe	3.04 (9; 0.65–9.74)	3.92 (27; 0.98–8.33)	–	5.67 (4; 2.25–11.40)	5.37 (41; 2.66–11.17)	5.18 (19; 3.00–8.29)	–
France	11.69 (17; 8.17–21.27)	–	6.60 (5; 4.16–11.76)	–	11.53 (4; 3.38–24.87)	–	15.35 (8; 4.18–30.64)
Italy	–	–	5.73 (15; 3.78–11.61)	20.66 (6; 11.47–27.18)	17.42 (6; 5.37–39.29)	9.94 (8; 3.36–18.99)	13.26 (9; 11.58–21.94)
Iberia	–	9.66 (10; 2.93–18.13)	–	–	10.71 (22; 5.85–13.63)	8.30 (13; 4.66–9.99)	7.23 (24; 2.58–13.03)
Balkans	–	–	–	–	–	5.37 (25; 3.06–10.54)	–

Table 7.14 Percentage directional asymmetry in humeral diaphyseal strength (HZ_p) for regional variation [Median (N; –95%CI – +95%CI)].

	Neolithic	BA	IA	Roman	EMed	LMed	Recent
Males							
Britain	23.87 (6; 4.80–35.25)	11.15 (9; 8.35–15.11)	16.21 (29; 4.92–21.26)	7.76 (22; 1.69–14.70)	–	9.32 (25; 4.39–14.10)	8.80 (15; 0.26–19.26)
Scandinavian	17.14 (34; 7.64–21.42)	–	19.57 (15; 6.52–20.06)	–	–	9.17 (20; –2.59–13.37)	8.27 (38; 3.60–11.64)
NC Europe	6.67 (7; –0.52–16.86)	12.69 (39; 5.85–19.26)	20.38 (4; 12.30–22.45)	8.63 (6; –1.33–10.89)	11.99 (66; 5.02–17.68)	5.20 (22; –0.64–16.54)	–
France	3.70 (13; 1.68–7.13)	–	8.84 (6; –5.71–15.11)	–	17.02 (7; 10.99–43.33)	–	10.23 (18; 6.14–15.77)
Italy	13.06 (6; 10.99–18.81)	4.93 (16; 2.72–12.74)	–	5.16 (5; –6.14–13.95)	18.34 (7; 15.26–35.24)	18.83 (4; 1.75–32.57)	3.35 (17; –9.40–17.73)
Iberia	–	22.91 (6; 17.04–28.04)	–	–	13.21 (24; 6.47–18.50)	8.12 (24; 1.07–22.08)	11.87 (25; 3.04–17.80)
Balkans	–	–	–	–	–	10.52 (27; 6.49–15.97)	–
Females							
Britain	–	6.81 (5; 6.62–21.92)	6.28 (27; –5.28–14.19)	7.14 (24; –0.36–17.41)	–	7.65 (26; 2.25–11.38)	6.83 (17; –0.84–15.64)
Scandinavian	1.99 (13; –1.96–11.21)	–	5.45 (20; –2.20–13.30)	–	–	2.29 (12; –4.34–12.56)	5.74 (19; 1.35–14.21)
NC Europe	–0.63 (9; –3.07–2.86)	–0.61 (27; –4.97–2.06)	–	5.67 (4; 2.25–11.40)	4.06 (41; –0.08–9.42)	5.05 (19; 2.10–8.29)	–
France	–6.04 (17; –11.69–8.18)	–	4.15 (5; –3.60–6.61)	–	8.15 (4; –3.38–24.87)	–	15.35 (8; 4.18–30.64)
Italy	–	–2.52 (15; –5.29–7.09)	–	4.48 (6; –27.18–20.31)	17.42 (6; –3.67–39.29)	3.04 (8; –6.84–12.30)	8.32 (9; –12.50–13.27)
Iberia	–	0.38 (10; –8.19–11.15)	–	–	9.25 (22; 0.95–13.32)	5.41 (13; 0.10–9.99)	7.23 (24; 2.03–13.03)
Balkans	–	–	–	–	–	4.54 (25; 2.52–7.94)	–

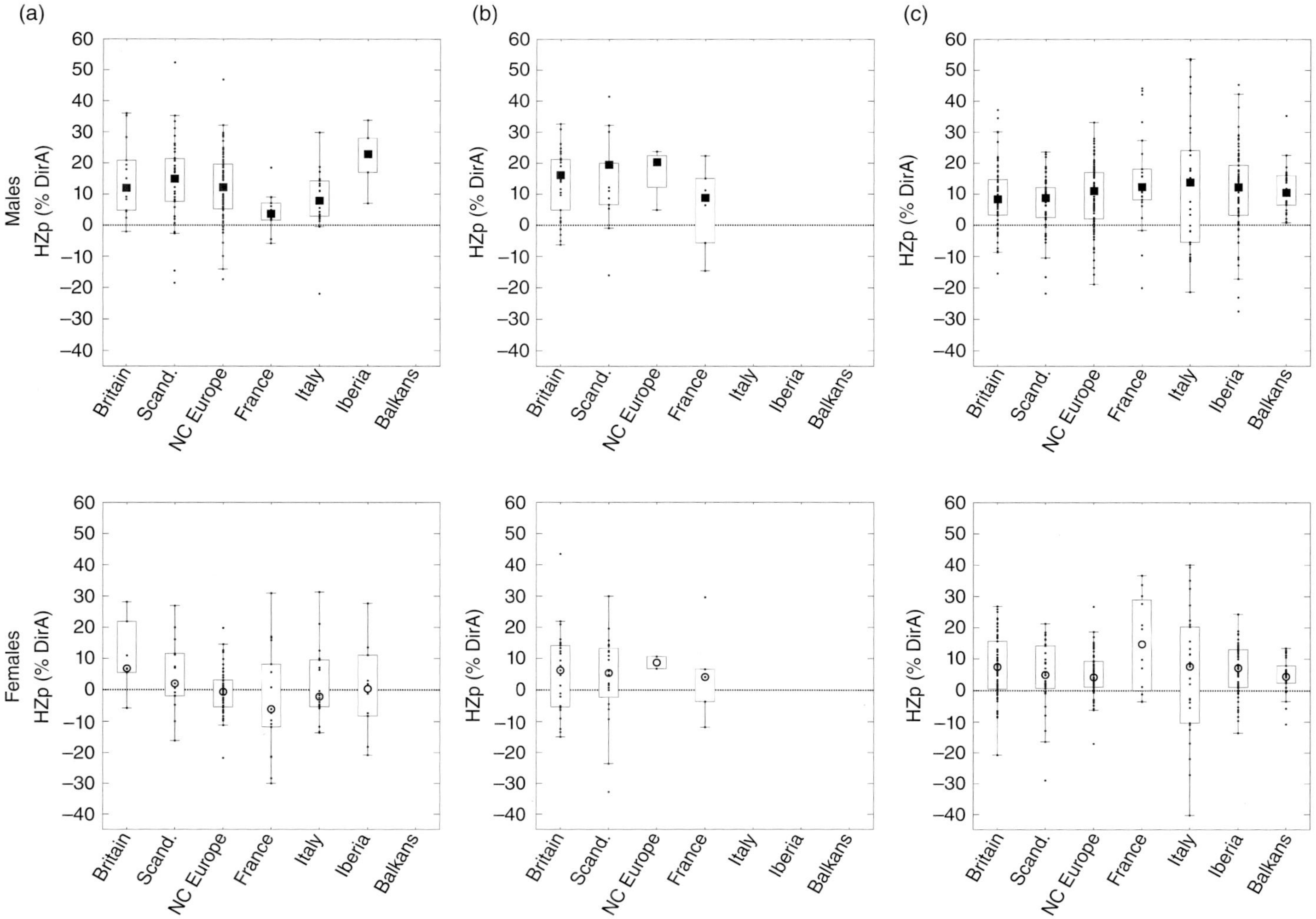

Figure 7.11 Range plots of regional variation in % directional asymmetry in HZ_p by sex and temporal period [median; (a) early agriculture (Neol–BA), (b) Iron Age, and (c) late agriculture (Roman–Rec)].

The description of results is based on those samples that have at least four individuals and for which at least three regions are represented within a time period. Thus, Tables 7.13 and 7.14 do not summarize values for pre-Holocene periods or for the Copper Age, since only a limited sampling from the Copper Age is available outside of North-Central Europe. Figure 7.11 further consolidates some time periods to increase sampling across regions within broader periods: a) early agricultural (Neolithic–Bronze Age); b) Iron Age; and c) later agricultural (Roman–Recent).

Regional patterns of variation are similar for maximum and directional asymmetry. Both asymmetries show no consistent geographic trends associated with either a North-South or West-East clinal distribution. In contrast, comparisons within regions support the previously described general pan-European temporal trends. Inter-regional comparisons provide strong evidence that changes in directional asymmetry associated with the adoption of agriculture and with technological progress in the Iron Age and later Holocene are not restricted to one specific region, but can be seen as the global reflection of subsistence effects on CSG properties across all European regions. Comparing early agricultural female groups in Figure 7.11(a), directional asymmetry is very low across all regions except Britain (which has a sample size of only five individuals; see Table 7.14). All regions show positive directional asymmetry for females in the Iron Age and more recent periods. Thus, reduced directional asymmetry in females can be seen as a specific feature associated with early agricultural subsistence technology throughout most of Europe. Similarly, the inter-regional comparisons support the view that the Iron Age represents a change in manipulative behavior among females from the earlier agricultural periods. Males maintain more similar levels of asymmetry across all regions during this time range. Where they can be compared within regions, males also show a slight decline in directional asymmetry in most regions from the Iron through more recent periods (Fig. 7.11(c)).

7.3.2 Right Versus Left Dominance

Proportions of individuals exhibiting right and left bias or no side bias in asymmetry of humeral CSG properties are shown by sex and temporal group in Table 7.15. In both male and female samples the comparison is affected by small sample sizes for pre-agricultural groups (see earlier tables for sample sizes), and inconsistency in the hunter-gatherers may be due to stochastic factors rather than biological and behavioral effects. However, even given that sample sizes for pre-agricultural periods are limited, the data support previous findings of a high percentage of right handedness in pre-agricultural populations.

Due to the small sample sizes for pre-agriculturalists, detailed analyses of results are limited to agricultural periods. In males, all the compared agricultural periods exhibit significantly right-biased directional asymmetry. This observation is consistent among all the studied CSG properties. However, the earlier agricultural groups have a slightly higher proportion of right-biased individuals than later agricultural groups, with a maximum in the Bronze Age (84–94%) and a minimum in the Recent period (68–77%), which supports the view that directional asymmetry slightly decreases among males during the Holocene (see also Figures 7.7 and 7.8).

In females, all the early agricultural groups (i.e., Neol, CopA, and BA) have a non-significant side bias, whereas later agricultural groups are significantly dominated by right-biased individuals. This temporal trend is more marked for Z_y than for Z_x, supporting previous observations that Z_x shows less temporal variation across agricultural samples.

Periods and properties with significant sexual dimorphism in side bias are also indicated in Table 7.15. Most of these occur during the early agricultural periods, consistent with earlier observations of much reduced asymmetry among females only during these periods. Males

Table 7.15 Percentage directional asymmetry bias in CSG properties for pre-Holocene and Holocene humeri.[a]

	Males					Females				
	HTA	HCA	HZ_x	HZ_y	HZ_p	HTA	HCA	HZ_x	HZ_y	HZ_p
EUP	0-0-100	0-0-100	0-0-100	0-0-100	0-0-100	–	–	–	–	–
LUP	–	22-0-78	11-0-89*	22-0-78	22-0-78	–	40-0-60	40-0-60	40-0-60	20-0-80
Mesolithic	22-0-78	12-13-75*	13-0-87*	50-0-50	38-0-62	0-0-100	0-0-100	0-0-100	20-0-80	0-0-100
Neolithic	**14-6-80***	**15-3-82***	**15-5-80***	**24-0-76***	**12-0-88***	**39-16-45**	**48-7-45**	**37-11-52**	**54-7-39**	**52-0-48**
CopA	**16-3-81***	**12-10-78***	19-0-81*	**16-3-81***	**16-0-84***	**43-9-48**	**38-10-52**	29-19-52	**48-14-38**	**52-0-47**
BA	**7-0-93***	**11-0-89***	**7-2-92***	**13-3-84***	**4-2-94***	**42-11-47**	**47-5-48**	**47-0-53**	**54-4-42**	**49-7-44**
IA	**5-2-93***	18-6-76*	**20-4-76***	**7-0-93***	**15-0-85***	**24-6-70***	32-9-59*	**42-4-54**	**24-7-69***	**33-2-65***
Roman	12-3-85*	15-3-82*	30-0-70*	12-0-88*	27-0-73*	18-6-76*	35-0-67	38-3-59	21-6-73*	26-0-74*
EMed	11-2-87*	21-7-72*	12-2-86*	11-0-89*	11-2-87*	14-3-84*	36-4-60*	40-4-56	19-0-81*	25-4-71*
LMed	11-5-84*	23-4-73*	22-7-71*	18-4-78*	16-3-81*	17-10-73*	17-8-75*	28-7-65*	32-3-65*	17-6-77*
Recent	18-5-77*	25-7-68*	26-3-71*	26-4-70*	21-3-76*	21-5-74*	21-2-77*	31-5-64*	34-4-62*	19-3-78*

[a] First value = % of individuals with left side asymmetry; middle value = % of individuals with directional asymmetry < ±0.5; last value = % of individuals with right side asymmetry.
* Indicates Chi-Square test significant for right versus left side bias (non-asymmetric individuals not included).
Bold text indicates significant sexual dimorphism (Chi-Square test) for side bias (non-asymmetric individuals not included).

also show significantly more right bias in the Iron Age, and generally show slightly (although non-significantly) more right bias in later time periods, which may reflect slightly more right dominated manipulative tasks.

7.3.3 Variation in Right and Left Humeral CSG

Temporal trends in right and left side humeral CSG properties, standardized for body size, are shown for areas in Table 7.16 and Figure 7.12, and for section moduli in Table 7.17 and Figure 7.13. General temporal trends are similar in areas and section moduli. For males, the

Table 7.16 Right and left total and cortical area for European pre-Holocene and Holocene humeri [Mean (N; SE)].

	Right		Left	
	Males	Females	Males	Females
HTA				
EUP	513.8 (4; 27.14)	424.6 (3; 32.97)	360.7 (5; 14.45)	367.4 (3; 37.44)
LUP	–	–	–	–
Mesolithic	515.4 (14; 15.64)	479.9 (5; 41.78)	499.5 (11; 16.40)	424.8 (9; 23.28)
Neolithic	492.0 (79; 7.08)	423.8 (50; 6.58)	458.1 (71; 6.94)	423.5 (51; 7.37)
S Neol HG	514.0 (33; 10.24)	439.4 (16; 10.86)	469.7 (27; 11.53)	428.1 (17; 10.95)
Neol Ag	476.1 (46; 9.08)	416.4 (34; 8.01)	451.1 (44; 8.61)	421.3 (34; 9.69)
CopA	460.8 (44; 7.28)	454.4 (28; 9.17)	426.9 (45; 8.11)	441.6 (28; 9.72)
BA	454.6 (85; 6.34)	422.1 (66; 6.01)	422.3 (79; 6.28)	418.8 (65; 6.49)
IA	445.4 (65; 7.74)	426.7 (65; 7.40)	397.3 (67; 7.23)	406.4 (57; 7.06)
Roman	504.6 (34; 9.26)	417.1 (39; 10.78)	471.6 (36; 8.14)	402.5 (38; 7.71)
EMed	484.4 (131; 5.19)	414.6 (91; 5.13)	457.0 (119; 5.21)	401.8 (87; 5.38)
LMed	479.6 (143; 5.09)	425.7 (128; 4.93)	454.8 (146; 5.02)	411.4 (121; 4.73)
Recent	489.5 (126; 6.45)	407.8 (84; 5.78)	462.1 (133; 5.49)	393.9 (82; 5.15)
HCA				
EUP	387.8 (4; 41.73)	350.0 (3; 33.80)	269.4 (5; 9.70)	294.9 (3; 38.27)
LUP	386.7 (11; 15.95)	343.7 (5; 34.05)	334.9 (9; 14.04)	299.2 (6; 18.99)
Mesolithic	378.8 (18; 18.72)	357.0 (6; 41.35)	370.0 (14; 14.94)	301.8 (10; 27.79)
Neolithic	380.9 (79; 8.00)	301.4 (50; 7.14)	353.0 (71; 7.24)	300.1 (51; 7.32)
S Neol HG	385.1 (33; 12.48)	306.8 (16; 10.09)	352.1 (27; 12.29)	305.4 (17; 10.33)
Neol Ag	377.8 (46; 10.52)	298.8 (34; 9.42)	353.6 (44; 9.03)	297.5 (34; 9.76)
CopA	362.0 (44; 7.18)	324.1 (28; 7.38)	328.2 (45; 7.72)	319.8 (28; 9.63)
BA	348.6 (85; 5.23)	290.2 (66; 6.04)	315.6 (79; 5.12)	285.7 (65; 6.51)
IA	327.1 (65; 7.82)	291.2 (65; 6.41)	293.5 (67; 6.68)	279.3 (57; 6.86)
Roman	346.5 (34; 8.82)	286.5 (39; 8.40)	320.4 (36; 9.44)	279.0 (38; 6.32)
EMed	350.4 (131; 5.12)	273.2 (91; 4.61)	335.0 (119; 5.23)	267.5 (87; 4.83)
LMed	353.1 (143; 5.26)	292.8 (128; 5.18)	334.5 (146; 4.65)	278.3 (121; 4.78)
Recent	339.2 (126; 5.72)	263.5 (84; 6.08)	323.2 (133; 5.21)	249.3 (82; 5.53)

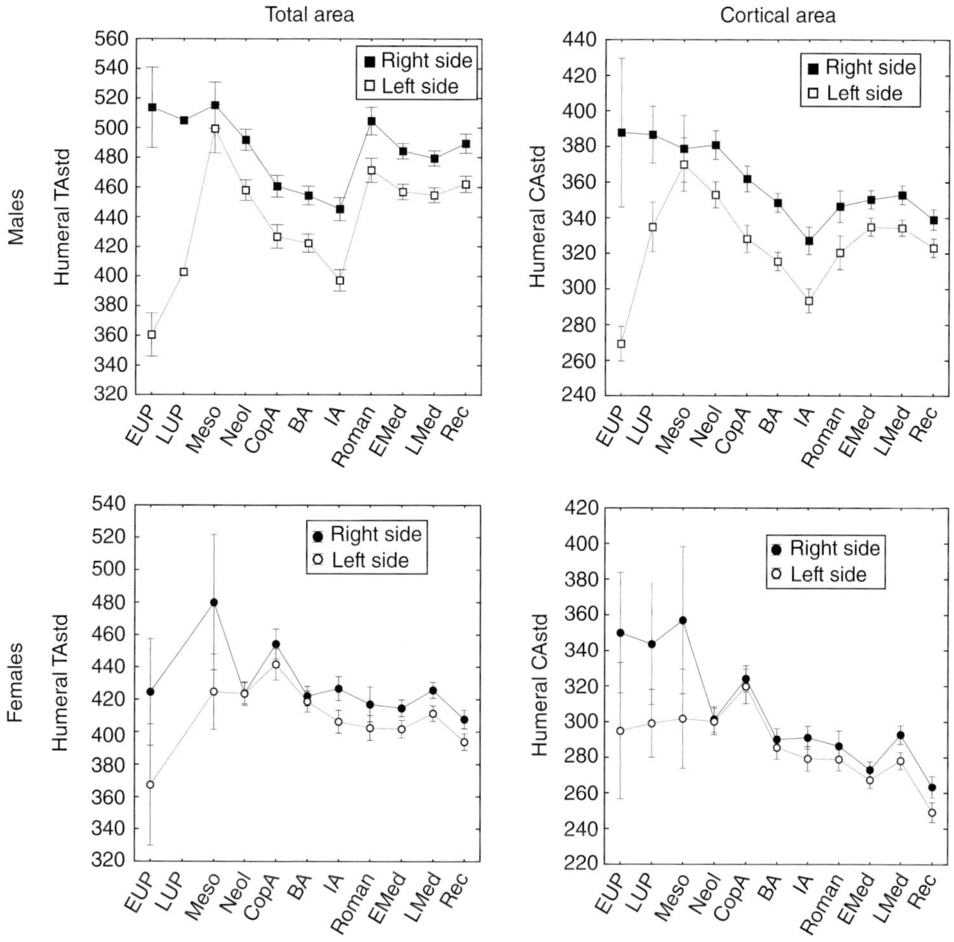

Figure 7.12 Range plots of variation in right and left standardized humeral cross-sectional areas by sex (mean ± SE; right humerus: solid symbols; left humerus: open symbols).

decrease in asymmetry between the EUP and Mesolithic is due to a large increase in values on the left side, whereas the right side remains almost unchanged. After the Mesolithic, changes within right and left sides are mostly parallel. Females do not show this trend, and instead appear to increase in robusticity more on the right side from the LUP to Mesolithic; however, details are difficult to evaluate since only a small number of female UP individuals is available. Females then decline in asymmetry in the Neolithic through a large decrease on the right side, while the left side remains relatively constant. Thus, during the transition from foraging to farming, changes in robusticity are more marked on the left side (increase) in males, and on the right side (decrease) in females.

Right and left humeri are very similar in robusticity among early agricultural females, consistent with their very low asymmetry. However, despite this low level of asymmetry, both sides are relatively robust (especially during the Copper Age) in females compared to later agricultural time periods. Both males and females show declines in humeral robusticity from the Copper to the Iron Age, but the decline is greater on the left side among females, leading to increased asymmetry. Males show a large increase in robusticity on both sides from the Iron

Table 7.17 Right and left diaphyseal strength (section moduli) for European pre-Holocene and Holocene humeri [Mean (N; SE)].

	Right		Left	
	Males	Females	Males	Females
HZ_x				
EUP	360.1 (3; 33.17)	278.9 (3; 30.95)	230.4 (4; 11.91)	262.8 (2;31.19)
LUP	354.4 (11; 19.46)	330.1 (5; 27.32)	294.3 (9; 15.30)	296.2 (6; 10.48)
Mesolithic	354.9 (16; 21.09)	345.4 (6; 49.08)	323.9 (13; 11.92)	268.8 (10; 32.05)
Neolithic	357.0 (73; 8.95)	280.2 (48; 8.03)	330.0 (68; 7.74)	278.5 (49; 8.85)
S Neol HG	382.1 (31; 14.14)	307.2 (15; 14.36)	342.5 (27; 13.06)	292.6 (17; 13.90)
Neol Ag	338.5 (42; 10.81)	267.9 (33; 9.04)	321.7 (41; 9.44)	270.9 (32; 11.29)
CopA	336.8 (42; 7.58)	307.7 (28; 7.49)	305.1 (44; 8.59)	305.4 (27; 9.57)
BA	329.1 (82; 6.36)	270.9 (64; 5.71)	292.8 (79; 5.84)	265.7 (64; 5.90)
IA	294.5 (62; 8.42)	257.1 (62; 7.93)	263.4 (58; 7.89)	252.3 (53; 7.77)
Roman	355.2 (32; 11.01)	276.2 (35; 7.98)	331.4 (35; 9.55)	268.9 (34; 7.13)
EMed	348.4 (130; 5.43)	260.5 (91; 4.47)	325.7 (119; 5.75)	260.9 (87; 5.22)
LMed	347.0 (137; 5.97)	283.9 (125; 6.03)	328.7 (142; 5.74)	275.5 (118; 5.47)
Recent	341.7 (125; 6.57)	255.9 (83; 6.57)	320.1 (132; 5.79)	246.7 (81; 6.26)
HZ_y				
EUP	301.2 (3; 25.53)	252.5 (3; 34.76)	201.0 (4; 14.69)	262.4 (2; 16.55)
LUP	318.4 (11; 22.09)	271.0 (5; 25.47)	278.0 (9; 22.46)	239.0 (6; 11.94)
Mesolithic	356.7 (16; 24.19)	314.4 (6; 44.73)	377.1 (13; 22.81)	268.0 (10; 23.95)
Neolithic	351.1 (73; 9.36)	253.3 (48; 8.09)	318.3 (68; 8.41)	269.1 (49; 9.82)
S Neol HG	386.6 (31; 13.46)	277.4 (15; 14.12)	336.7 (27; 12.57)	295.7 (17; 13.53)
Neol Ag	324.8 (42; 11.39)	242.3 (33; 9.38)	306.2 (41; 10.94)	255.0 (32; 12.65)
CopA	323.2 (42; 8.23)	288.4 (28; 7.34)	285.1 (44; 8.11)	289.3 (27; 9.74)
BA	303.9 (82; 6.24)	254.3 (64; 6.08)	269.5 (79; 5.68)	257.2 (64; 6.32)
IA	287.6 (62; 8.03)	248.3 (62; 7.49)	241.3 (58; 7.61)	231.3 (53; 6.67)
Roman	324.9 (32; 9.61)	244.3 (35; 9.91)	292.5 (35; 8.32)	230.7 (34; 6.92)
EMed	322.7 (130; 5.57)	237.2 (91; 4.47)	293.4 (119; 5.54)	224.8 (87; 4.74)
LMed	320.1 (137; 5.97)	251.1 (125; 4.97)	293.5 (142; 5.36)	244.7 (118; 4.94)
Recent	312.8 (125; 6.26)	231.2 (83; 5.48)	291.6 (132; 5.50)	217.5 (81; 5.34)
HZ_p				
EUP	627.8 (3; 54.69)	494.2 (3; 57.72)	406.1 (4; 25.10)	440.7 (2; 67.12)
LUP	627.8 (11; 36.81)	583.7 (5; 49.56)	531.9 (9; 33.79)	501.8 (6; 20.60)
Mesolithic	655.3 (16; 41.63)	619.3 (6; 87.28)	648.5 (13; 32.27)	496.3 (10; 51.09)
Neolithic	660.6 (73; 16.21)	503.3 (48; 15.56)	596.3 (68; 14.17)	510.1 (49; 16.92)
S Neol HG	717.9 (31; 23.55)	553.9 (15; 26.54)	624.9 (27; 22.63)	539.1 (17; 21.44)
Neol Ag	618.3 (42; 19.99)	480.4 (33; 18.01)	577.5 (41; 17.78)	494.6 (32; 23.03)
CopA	624.7 (42; 15.02)	560.4 (28; 14.00)	560.6 (44; 15.40)	562.0 (27; 18.29)
BA	602.2 (82; 11.91)	495.0 (64; 10.82)	532.3 (79; 10.53)	492.8 (64; 11.41)
IA	547.9 (62; 15.29)	475.3 (62; 13.74)	473.8 (58; 14.06)	451.3 (53; 13.23)
Roman	620.1 (32; 18.78)	487.5 (35; 16.81)	576.9 (35; 16.45)	462.2 (34; 13.42)
EMed	630.6 (130; 10.39)	467.4 (91; 8.41)	579.0 (119; 10.67)	453.9 (87; 9.41)
LMed	623.4 (137; 10.98)	495.5 (125; 9.91)	576.0 (142; 9.48)	471.2 (118; 8.53)
Recent	608.2 (125; 11.58)	450.9 (83; 10.79)	558.2 (132; 9.13)	422.3 (81; 9.78)

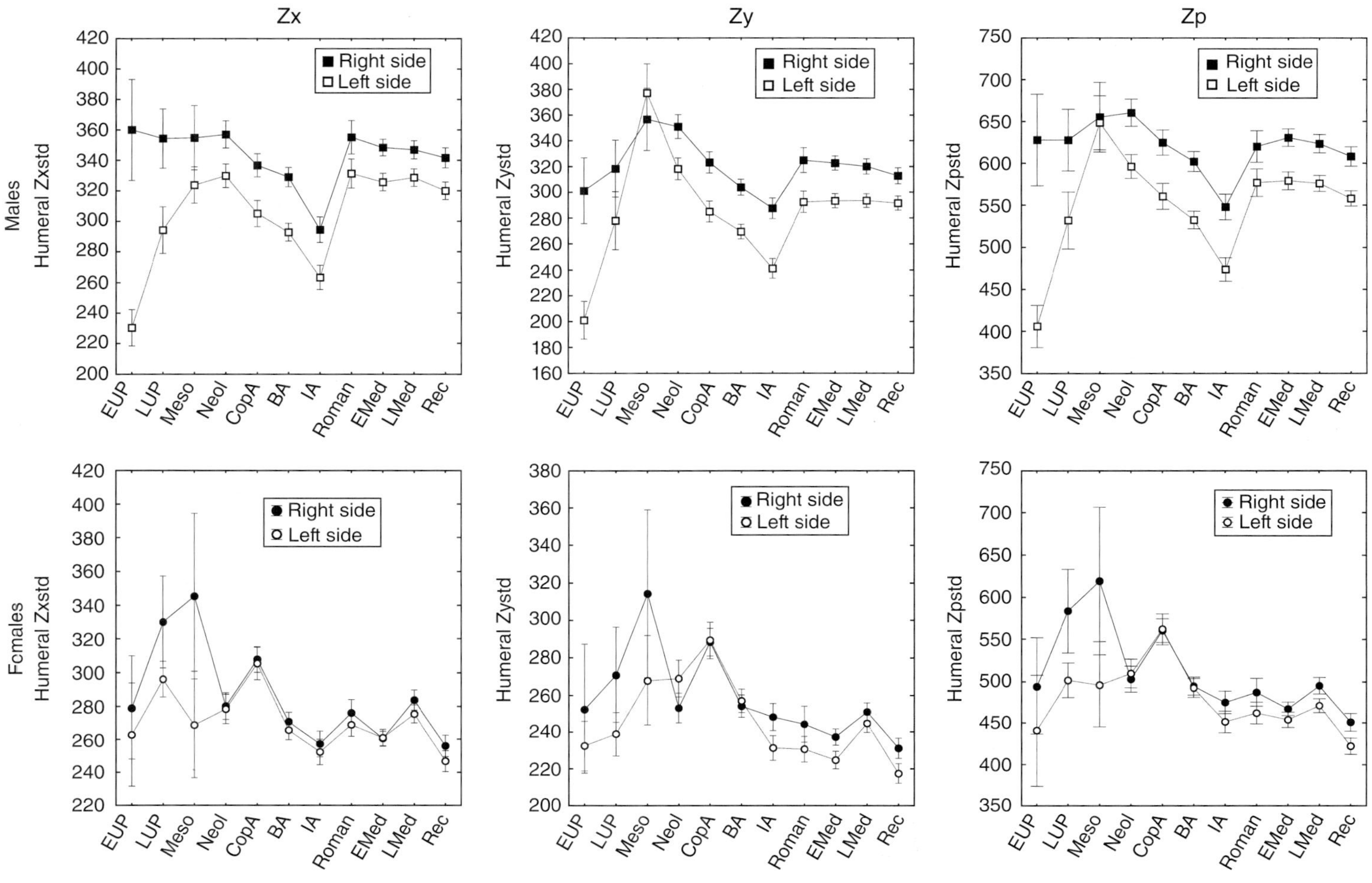

Figure 7.13 Range plots of variation in right and left standardized humeral section moduli by sex (mean ± SE; right humerus: solid symbols; left humerus: open symbols).

Age to Roman period. In contrast, females show an almost continuous decline in robusticity after the Copper Age. Males decline slightly after the Roman period.

The bootstrap means and mean ±95% CIs for humeral diaphyseal shape (Z_x/Z_y) are shown in Table 7.18 and Figure 7.14. The index is high in both sexes and on both sides during the pre-Holocene periods, indicating a relatively A-P-reinforced humeral shaft. Both sexes subsequently show a large decrease in the index on both sides in the Mesolithic, which is due to relatively

Table 7.18 Right and left diaphyseal shape (Z_x/Z_y) for European pre-Holocene and Holocene humeri [Mean (N; ±95%CI].

	Right		Left	
	Males	**Females**	**Males**	**Females**
EUP	1.204 (5; 1.10–1.27)	1.159 (3; 0.9–1.32)	1.155 (6; 1.07–1.21)	1.195 (3; 1.06–1.35)
LUP	1.131 (11; 1.06–1.20)	1.225 (5; 1.12–1.27)	1.087 (9; 0.95–1.17)	1.247 (6; 1.15–1.30)
Mesolithic	1.012 (22; 0.96–1.05)	1.108 (8; 0.96–1.17)	0.901 (18; 0.82–0.95)	0.985 (12; 0.88–1.04)
Neolithic	1.037 (82; 1.00–1.06)	1.121 (51; 1.07–1.16)	1.051 (74; 1.02–1.07)	1.045 (53; 1.00–1.07)
S Neol HG	0.990 (33; 0.95–1.02)	1.116 (16; 1.04–1.16)	1.020 (27; 0.98–1.05)	1.000 (18; 0.94–1.04)
Neol Ag	1.069 (49; 1.02–1.10)	1.123 (35; 1.06–1.17)	1.069 (47; 1.02–1.10)	1.069 (35; 1.01–1.10)
CopA	1.047 (44; 1.02–1.07)	1.070 (28; 1.04–1.09)	1.077 (45; 1.05–1.10)	1.053 (28; 1.00–1.08)
BA	1.086 (90; 1.07–1.10)	1.073 (71; 1.04–1.10)	1.095 (87; 1.07–1.11)	1.041 (68; 1.02–1.06)
IA	1.032 (67; 1.00–1.05)	1.041 (66; 1.01–1.06)	1.095 (69; 1.06–1.12)	1.092 (57; 1.05–1.12)
Roman	1.097 (36; 1.05–1.13)	1.151 (41; 1.10–1.19)	1.130 (37; 1.08–1.16)	1.157 (40; 1.10–1.19)
EMed	1.089 (134; 1.07–1.11)	1.106 (94; 1.08–1.13)	1.120 (123; 1.10–1.14)	1.165 (89; 1.13–1.19)
LMed	1.091 (147; 1.07–1.11)	1.130 (134; 1.11–1.15)	1.125 (151; 1.11–1.14)	1.129 (126; 1.11–1.15)
Recent	1.104 (128; 1.07–1.13)	1.109 (86; 1.06–1.13)	1.104 (135; 1.08–1.12)	1.143 (85; 1.11–1.17)

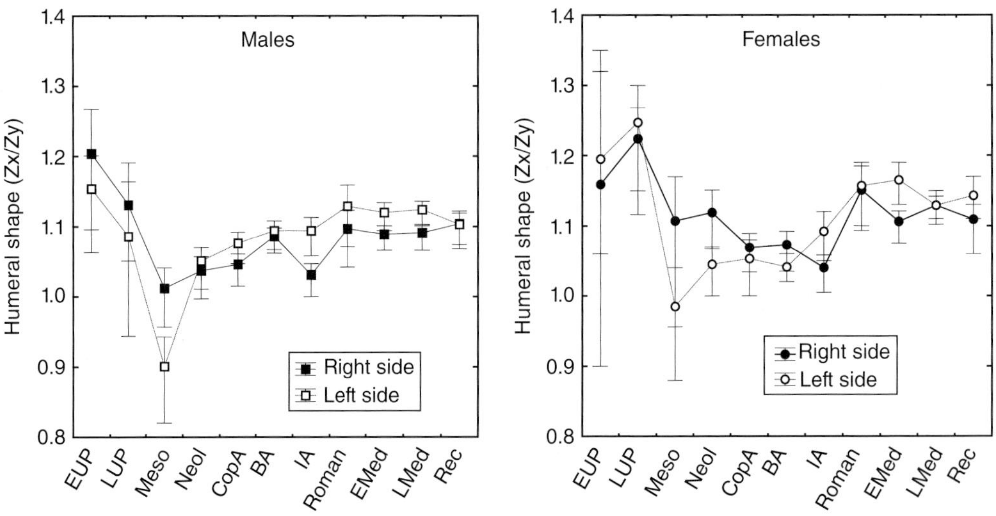

Figure 7.14 Range plots of variation in right and left humeral diaphyseal shape Z_x/Z_y by sex (mean ± 95% CI; right humerus: solid symbols; left humerus: open symbols).

more positive changes in Z_y than Z_x, especially in males (see Fig. 7.13). In the later Holocene both sexes increase in the index, due to greater declines (or smaller increases) in Z_y. This is more marked on the right side, leading to more circular right humeri, especially among males.

7.4 Discussion

The overall comparisons of maximum asymmetry in different upper limb bone dimensions indicate that asymmetry in CSG properties is larger and more variable between periods than asymmetries in lengths and articular breadths. This result is consistent with previous observations of various Upper Paleolithic and Holocene samples (Trinkaus *et al.*, 1994; Auerbach and Ruff, 2006; Sládek *et al.*, 2007). This general pattern is apparent in both sexes and across all pre-Holocene and Holocene samples. Thus, relatively high asymmetry in CSG properties versus low asymmetry in lengths, and articular breadths can be seen as a general feature of human upper limb postcranial variation. This also supports the view that CSG properties show high environmental plasticity associated with variation in mechanical loading (Trinkaus *et al.*, 1994), whereas lengths and articular breadths are less environmentally plastic, probably due to greater genetic control during growth and a reduced response to loading differences (see below). The similarity of articular and length asymmetry across the entire time span sampled here, despite large changes in CSG asymmetry, also implies that genetic effects on asymmetry remained relatively constant from the Upper Paleolithic through the Holocene.

This does not imply, however, that asymmetries in bone length and articular breadth are caused by the same biological processes, as indicated by differences in sexual dimorphism of asymmetries in the two types of properties. Results presented here suggest that maximum asymmetry in length exhibits sex-linked variation whereas asymmetry in articular breadth is not sexually dimorphic. Thus, the genetic control of articular breadth appears to be similar in males and females whereas asymmetry in length, while also genetically controlled, has an additional sex-specific component. Moreover, patterns of maximum asymmetry also show that asymmetry in CSG properties and length have opposite patterns of sexual dimorphism: CSG asymmetries are significantly larger among males, whereas length asymmetries are larger among females. This contrasting pattern between CSG and length has been observed in other comparisons (Schultz, 1937; Auerbach and Ruff, 2006; Sládek *et al.*, 2007) but the biological basis of this observation is still under debate. It can be hypothesized that higher maximum asymmetry in lengths in females than in males is related to endogenous factors such as androgens, estrogens, and their receptors, influencing either the expression or/and timing of longitudinal bone growth of the upper limb (see discussion in Auerbach and Ruff, 2006). However, if sexual dimorphism in bone length asymmetry is purely under endogenous control or if it can also be associated with environmentally induced gender differences or metabolic stress during prenatal and/or early postnatal development is unclear.

Our results for CSG properties indicate that changes in upper limb use during the European transition to agricultural were gender-specific. Asymmetry in CSG properties strongly declines from the EUP to the Mesolithic in males, whereas in females this decline is absent prior to the Mesolithic (however, assessment of values for females in the Upper Paleolithic is limited by a small sample). In contrast, females undergo a sharp decrease in CSG asymmetry after the Mesolithic and the decrease in maximum asymmetry continues through the Copper Age. Since hunting is an important activity for males in preagricultural groups (Lee and DeVore, 1969; Lee, 1979; Binford, 2001), our results suggest that prior to the Neolithic the change in males was mainly related to alterations in hunting strategy, likely resulting from a shift in focus from big game hunting, characterized by strongly unilateral activities (Schmitt *et al.*, 2003; but see discussion about involvement of dominant and non-dominant hand in Shaw *et al.*, 2012), to exploitation of small-scale local resources involving less asymmetric use of the upper limbs (i.e., the 'Broad Spectrum Hypothesis'; Flannery, 1969; Stiner, 2001; Weiss *et al.*, 2004) and more symmetric use of both arms (e.g., bow-and-arrow) (Brues, 1959, 1977; Frayer, 1981; Price, 1987; Holt and Formicola, 2008). In contrast, manipulative behavior related to gathering as the primary subsistence activity for preagricultural females would be expected to remain relatively constant during the UP and Mesolithic, at least with respect to asymmetric use of the upper limbs. Reduction of asymmetry in females came later with the adoption of agriculture in the Neolithic, probably due to new bimanual tasks related to food processing, that is, pounding or grinding; no similar change is observed among Neolithic males. This gender-specific response to subsistence changes is also supported by variation within side in CSG properties. Our results indicate that males experienced a strong increase of robusticity on the left side during the LUP and Mesolithic whereas the right side remained relatively constant. In contrast, the reduction in asymmetry among Neolithic females was caused by a decrease in robusticity on the right side, with no change on the left side.

Regional comparisons show that the reduction of right-biased directional asymmetry among Neolithic females is not regionally specific and exhibits a pan-European occurrence. Moreover, the decrease of asymmetry among females has been observed also for other regions, for example among Native Americans, and is always associated with early agricultural subsistence (Bridges *et al.*, 2000). Thus, we expect that a decrease of directional asymmetry in CSG is one of the key features of upper limb morphology for early agricultural females. It is very unlikely that the increase of left-biased asymmetry in these females is due to an increase of individuals with neurologically determined left-handedness in early agricultural societies. Our results indicate that the left-handed bias in early agricultural females of close to 50% is more than the maximum expected left-handed frequency of 30% based on recent studies (Hardyck and Petrinovich, 1977; Raymond and Pontier, 2004). Moreover, recent studies also support a relatively constant proportion of left-handers across a broad geographic spectrum of human populations. Thus, the reduction of directional asymmetry among early agricultural females must be explained as the effect of equal involvement of right and left hand in new manipulative tasks probably associated with food processing. Archeological evidence shows that food processing of grain in early agricultural societies is based mainly on grinding using saddle querns (e.g., Pavlů and Zápotocká, 2007). We have shown experimentally that grinding through use of early agricultural pounding stones is a strong bimanual activity that produces equal loads on right and left upper limbs (Sládek *et al.*, 2016a).

The impact of food processing techniques is also supported by the subsequent increase in directional asymmetry of CSG properties during the Iron Age in females. The Iron Age is associated with new technological improvements for food processing and the appearance of rotary querns (Wefers, 2011) As indicated by other (currently unpublished) experiments, grinding with rotary querns causes asymmetrical loading of the upper limbs, and this impact may account for the increase of directional asymmetry observed in females subsequent to the early agricultural periods. However, whether changes in food pounding/grinding techniques alone are responsible for the reduction and then increase of directional asymmetry in agricultural females must be further investigated.

There is larger sexual dimorphism in asymmetry among Scandinavian Neolithic foragers than in either Mesolithics or European Neolithic farmers, largely attributable to low asymmetry among Scandinavian forager females (similar to farmer Neolithics) but continued high asymmetry among forager males (similar to Mesolithics). It is unlikely that similarities between Scandinavian forager and European farmer females are due to identical patterns of subsistence-related behavior in these populations, especially since Scandinavian Neolithic male foragers exhibit relatively high asymmetry. It is difficult to explain these observations without more detailed knowledge of the specific lifestyle and cultural and biological interactions of these forager and farmer groups. For example, the similarities between forager and farmer females may indicate a specific pattern of assimilation of farmer females by foragers, as is discussed in Bollongino *et al.* (2013: p. 481) using mtDNA, although such a direction of assimilation is not supported by a recent genomic diversity study (Skoglund *et al.*, 2014). We also need asymmetry data for Scandinavian Neolithic farmers, as well as more data for Scandinavian Mesolithic samples (at present only two males and one female have paired humeral strength data).

Our results show that after the adoption of agriculture the changes in humeral CSG asymmetry are less marked than in prior periods. However, a detailed consideration of asymmetry and variation within each side indicates that upper limb CSG properties are not constant in the later Holocene. The Iron Age marks an interesting transition, where females are no longer as symmetric as they are in the early agricultural periods, and after which both males and females generally decline slightly in asymmetry through the Recent period. The decline is greater in males, perhaps reflecting a greater decline in unilateral tasks in later times. The Iron Age also represents a low point during the Holocene for humeral robusticity on both sides among males, after which there is some increase. Humeral robusticity also declines during the Iron Age in females, but then shows no subsequent increase. It is possible that work involving the upper limb increased during the Medieval (and Roman) period in males (also see Chapter 5), even as asymmetry in use of the limbs declined. The slight decline in robusticity in the most recent periods in both sexes may reflect a general reduction in mechanical loading of the upper limbs with increasing mechanization and urbanization.

It has been shown here that despite the decrease in magnitude of asymmetry between the pre-Holocene and Holocene periods, the relative frequency of right-side-dominated individuals remains similar, except among early agricultural females. For most CSG properties and temporal periods, the frequency of right side bias varies between about 60% and 90%. This is similar to the overall frequency of right bias in humeral shaft breadth of 76% for a worldwide Holocene sample (Auerbach and Ruff, 2006) and 78% in an Upper Paleolithic sample (Churchill and Formicola, 1997), and is also broadly consistent with results of generalized motor skill tests in living humans (see above). However, our results for early agricultural females in particular also argue for caution in using such percentages as indicators of 'handedness,' at least in a neurological sense. Frequency and intensity of mechanical loading are likely to be much more important than motor control *per se* in producing bilateral asymmetry in upper limb bone strength. Thus, the neurologically 'dominant' hand may not be the mechanically dominant limb in all cases. A similar explanation was advanced to explain the finding of slightly left-biased shaft breadths in the lower limb, despite the fact that most people are 'right-footed' – that is, they have more motor control over their right lower limb. In this case the left side acts as more of a stabilizer, supporting body weight while the right side is used in 'manipulative' activities (Auerbach and Ruff, 2006). Specific behaviors can produce more symmetrical use of the upper limbs regardless of normal handedness. Thus, deviations from the expected frequency of side bias, in addition to magnitudes of asymmetry, can provide important information that can be used to help reconstruct past behavioral patterns.

7.5 Summary and Conclusions

Our analyses of bilateral asymmetry in the upper limb skeleton support previous findings that CSG properties are more variable between populations than lengths and breadths. This trend is supported both among pre-Holocene and Holocene periods, and can be therefore seen as a general feature of human upper limb postcranial variation. Our results suggest that maximum asymmetry in lengths is sex-linked, with longer left sides in females, whereas asymmetry in articular breadths is not sex-specific. Asymmetry in CSG properties also varies by sex, but in contrast to length generally exhibits larger asymmetry for males than for females.

Asymmetry in CSG properties shows a gender-specific response to subsistence changes at the end of the Pleistocene in Europe. Both sexes decline in asymmetry between the Upper Paleolithic and Neolithic, but at different time points. For males, asymmetry declines from the EUP to the Mesolithic and then remains stable, whereas females decline only in the Neolithic. We suspect that changes in asymmetry in males prior to the Neolithic were associated with changes in hunting strategy between the UP and Mesolithic, with decreasing emphasis on big game hunting. For females, we hypothesize that the decline in asymmetry associated with the agricultural transition was induced by new manipulative tasks related to food processing, in particular bimanual food pounding or grinding. Directional asymmetry in humeral CSG is virtually absent in early agricultural females, the only observed sex/temporal grouping to show this pattern. Asymmetry in females then increases in the Iron Age, possibly due to the introduction of more sophisticated food grinding techniques that involved less bimanual use of the upper limbs. These observations are not localized to particular regions but are found across Europe. Asymmetry declines slightly after the Iron Age, possibly associated with increasing mechanization and urbanization.

Acknowledgments

In addition to those people listed in Chapter 1, we would like to thank Patrik Galeta, Kristýna Farkašová, Martin Hora, Daniel Sosna and Eliška Makajevová for their help with technical support and data processing.

References

Adams, J.L. (1999) Refocusing the role of food-grinding tools as correlates for subsistence strategies in the US Southwest. *Am. Antiq.*, **64** (3), 475–498.

Albert, A.M. and Greene, D.L. (1999) Bilateral asymmetry in skeletal growth and maturation as an indicator of environmental stress. *Am. J. Phys. Anthropol.*, **110** (3), 341–349.

Annett, M. (1981) The genetics of handedness. *Trends Neurosci.*, **4**, 256–258.

Annett, M. and Kilshaw, D. (1983) Right- and left-hand skill II: Estimating the parameters of the distribution of L-R differences in males and females. *Br. J. Psychol.*, **74** (2), 269–283.

Auerbach, B.M. and Ruff, C.B. (2006) Limb bone bilateral asymmetry: variability and commonality among modern humans. *J. Hum. Evol.*, **50** (2), 203–218.

Bakan, P. (1978) Why left-handedness? *Behav. Brain Sci.*, **1** (2), 279–280.

Binford, L.R. (2001) *Constructing frames of reference: an analytical method for archaeological theory building using ethnographic and environmental data sets.* University of California Press, Berkeley.

Blackburn, A. (2011) Bilateral asymmetry of the humerus during growth and development. *Am. J. Phys. Anthropol.*, **145** (4), 639–646.

Bollongino, R., Nehlich, O., Richards, M.P., Orschiedt, J., Thomas, M.G., Sell, C., Fajkošová, Z., Powell, A., and Burger, J. (2013) 2000 years of parallel societies in Stone Age Central Europe. Science, **342** (6157), 479–481.

Bräuer, G. (1988) Osteometrie. In: *Anthropologie: Handbuch der Vergleichenden Biologie des Menschen, Band 1; Wesen und Methoden der Anthropologie, Teil 1. Wissenschaftstheorie, Geschichte, Morphologische Methoden* (ed. R. Knussman), Gustav Fischer Verlag, Stuttgart, pp. 160–232.

Bridges, P.S. (1985) Changes in long bone structure with the transition to agriculture: Implications for prehistoric activities. Ph. D. Dissertation, University of Michigan, Ann Arbor.

Bridges, P.S. (1989) Changes in activities with the shift to agriculture in the southeastern United States. *Curr. Anthropol.*, **30** (3), 385–394.

Bridges, P.S. (1991) Skeletal evidence of changes in subsistence activities between the Archaic and Mississippian time periods in northwestern Alabama. In: *What mean these bones? Studies in Southeastern Bioarchaeology* (eds M.L. Powell, P.S. Bridges, and A.M.W. Mires), University of Alabama Press, Tuscaloosa, pp. 89–101.

Bridges, P.S., Blitz, J.H., and Solano, M.C. (2000) Changes in long bone diaphyseal strength with horticultural intensification in West-Central Illinois. *Am. J. Phys. Anthropol.*, **112**, 217–238.

Brito, G.N., Brito, L.S., Paumgartten, F.J., and Lins, M.F. (1989) Lateral preferences in Brazilian adults: an analysis with the Edinburgh Inventory. *Cortex*, **25** (3), 403–415.

Brues, A. (1959) The Spearman and the Archer – An Essay on Selection in Body Build. *Am. Anthropol.*, **61** (3), 457–469.

Brues, A.M. (1977) *People and races*. Macmillan, New York.

Churchill, S.E. and Formicola, V. (1997) A case of marked bilateral asymmetry in the upper limb of an Upper Palaeolithic male from Barma Grande (Liguria), Italy. *Int. J. Osteoarch.* **7** (1), 18–38.

Constandse-Westermann, T.S. and Newell, R.R. (1989) Limb lateralization and social stratification in western European Mesolithic societies. In: *People and Culture in Change* (ed. I. Hershkovitz), Archaeopress, Oxford, pp. 405–433.

Corballis, M.C. and Morgan, M.J. (1978) On the biological basis of human laterality: I. Evidence for a maturational left–right gradient. *Behav. Brain Sci.*, **1** (2), 261–269.

Efron, B. and Tibshirani, R.J. (1994) *An introduction to the bootstrap*. CRC Press, New York.

Fagard, J. (2013) Early development of hand preference and language lateralization: Are they linked, and if so, how? *Dev. Psychobiol.*, **55**,596–607.

Flannery, K.V. (1969) Origins and Ecological Effects of Early Domestication in Iran and the Near East. In: *The Domestication and Exploitation of Plants and Animals* (eds P.J. Ucko and G.W. Dimbleby), Aldine Publishing Co., Chicago, pp. 73–100.

Frayer, D.W. (1981) Body Size, Weapon Use, and Natural Selection in the European Upper Paleolithic and Mesolithic. *Am. Anthropol.*, **83** (1), 57–73.

Fresia, A.E., Ruff, C.B., and Larsen, C.S. (1990) Temporal decline in bilateral asymmetry of the upper limb on the Georgia coast. In: *The Archaeology of Mission Santa Catalina de Guale: 2. Biocultural interpretations of a population in transition* (ed. C.S. Larsen), Anthropological Papers of the American Museum of Natural History, 68, New York, pp. 121–132.

Greenfield, H.J. (2010) The Secondary Products Revolution: the past, the present and the future. *World Archaeol.*, **42** (1), 29–54.

Haaland, R. (1995) Sedentism, cultivation, and plant domestication in the Holocene Middle Nile region. *J. Field Archaeol.*, **22** (2), 157–174.

Haapasalo, H., Kontulainen, S., Sievanen, H., Kannus, P., Jarvinen, M., and Vuori, I. (2000) Exercise-induced bone gain is due to enlargement in bone size without a change in volumetric bone density: A peripheral quantitative computed tomography study of the upper arms of male tennis players. *Bone*, **27** (3), 351–357.

Hamon, C. and Le Gall, V. (2013) Millet and sauce: The uses and functions of querns among the Minyanka (Mali). *J. Anthropol. Archaeol.*, **32** (1), 109–121.

Hardyck, C. and Petrinovich, L.F. (1977) Left-handedness. *Psychol. Bull.*, **84** (3), 385–404.

Haukoos, J.S. and Lewis, R.J. (2005) Advanced statistics: bootstrapping confidence intervals for statistics with 'difficult' distributions. *Acad. Emerg. Med.*, **12** (4), 360–365.
Hicks, R.E. and Kinsbourne, M. (1976) On the genesis of human handedness: A review. *J. Mot. Behav.*, **8** (4), 257–266.
Holodňák, P. (2001) Experiment s mletím obilnin na žernovech tzv. řeckého typu. *Archelogické rozhledy*, **53** (1), 31–44.
Holt, B.M. and Formicola, V. (2008) Hunters of the Ice Age: the biology of Upper Paleolithic people. *Am. J. Phys. Anthropol.*, **137** (S47), 70–99.
Kimura, D. (1973) Manual activity during speaking – II. Left-handers. *Neuropsychologia*, **11** (1), 51–55.
Kujanová, M., Bigoni, L., Velemínská, J., and Velemínský, P. (2008) Limb bones asymmetry and stress in medieval and recent populations of Central Europe. *Int. J. Osteoarchaeol.*, **18** (5), 476–491.
Ledger, M., Holtzhausen, L.-M., Constant, D., and Morris, A.G. (2000) Biomechanical beam analysis of long bones from a late 18th century slave cemetery in Cape Town, South Africa. *Am. J. Phys. Anthropol.*, **112**, 207–216.
Lee, R.B. (1979) *The !Kung San. Men, women, and work in a foraging society.* Cambridge University Press, Cambridge.
Lee, R.B. and DeVore, I. (1969) *Man the hunter.* Transaction Publishers, New Jersey.
Leiber, L. and Axelrod, S. (1981) Intra-familial learning is only a minor factor in manifest handedness. *Neuropsychologia*, **19** (2), 273–288.
Levy, J. and Nagylaki, T. (1972) A model for the genetics of handedness. *Genetics*, **72** (1), 117–128.
Lieberman, D.E., Devlin, M.J., and Pearson, O.M. (2001) Articular area responses to mechanical loading: Effects of exercise, age, and skeletal location. *Am. J. Phys. Anthropol.*, **116** (4), 266–277.
Lynch, A.J. and Rowland, C.A. (2005) *The history of grinding.* Society for Mining Metallurgy & Exploration, Littleton, Colorado.
Marchi, D., Sparacello, V.S., Holt, B.M., and Formicola, V. (2006) Biomechanical approach to the reconstruction of activity patterns in Neolithic Western Liguria, Italy. *Am. J. Phys. Anthropol.*, **131** (4), 447–455.
Marshall, L.J. (1976) *The !Kung of Nyae Nyae.* Harvard University Press, Cambridge.
Martin, R. (1928) *Lehrbuch der Anthropologie.* Fischer, Jena.
Mauldin, R. (1993) The relationship between ground stone and agricultural intensification in western New Mexico. *Kiva*, **58** (3), 317–330.
Mays, S.A. (2002) Asymmetry in metacarpal cortical bone in a collection of British post-mediaeval human skeletons. *J. Archaeol. Sci.*, **29** (4), 435–441.
Ogilvie, M.D. (2004) *Mobility and the locomotor skeleton at the foraging to farming transition. From Biped to Strider.* Springer, New York, pp. 183–201.
Pavlů, I. and Zápotocká, M. (2007) Archeologie pravěkých Čech 3: Neolit. Archeologický ústav AV ČR, Praha.
Perelle, I.B. and Ehrman, L. (1982) What is a lefthander? *Cell Mol. Life Sci.*, **38** (10), 1256–1258.
Perelle, I.B. and Ehrman, L. (1994) An International Study of Human Handedness – the Data. *Behav. Genet.*, **24** (3), 217–227.
Plochocki, J.H. (2004) Bilateral variation in limb articular surface dimensions. *Am. J. Hum. Biol.*, **16** (3), 328–333.
Price, T.D. (1987) The Mesolithic of Western Europe. *J. World Prehistory*, **1** (3), 225–305.
Raymond, M. and Pontier, D. (2004) Is there geographical variation in human handedness? *Laterality*, **9** (1), 35–51.
Rhodes, J.A. and Knüsel, C.J. (2005) Activity-related skeletal change in medieval humeri: Cross-sectional and architectural alterations. *Am. J. Phys. Anthropol.*, **128** (3), 536–546.

Ruff, C.B. (2002) Variation in human body size and shape. *Annu. Rev. Anthropol.*, **31**, 211–232.
Ruff, C.B. (2008) Biomechanical analysis of archaeological human skeletons. In: *Biological anthropology of the human skeleton*, 2nd ed (eds M.A. Katzenberg and S.R. Saunders), Wiley-Liss, Inc., New York, pp. 183–206.
Ruff, C.B., Holt, B., and Trinkaus, E. (2006) Who's Afraid of the Big Bad Wolff?: 'Wolff's Law' and Bone Functional Adaptation. *Am. J. Phys. Anthropol.*, **129** (4), 484–498.
Ruff, C.B., Holt, B.M., Niskanen, M., Sladek, V., Berner, M., Garofalo, E., Garvin, H.M., Hora, M., Maijanen, H., Niinimaki, S., *et al.* (2012) Stature and body mass estimation from skeletal remains in the European Holocene. *Am. J. Phys. Anthropol.*, **148** (4), 601–617.
Shaw, C.N. (2011) Is 'hand preference' coded in the hominin skeleton? An in-vivo study of bilateral morphological variation. *J. Hum. Evol.*, **61** (4), 480–487.
Shaw, C.N. and Stock, J.T. (2009) Intensity, repetitiveness, and directionality of habitual adolescent mobility patterns influence the tibial diaphysis morphology of athletes. *Am. J. Phys. Anthropol.*, **140** (1), 149–159.
Shaw, C.N., Hofmann, C.L., Petraglia, M.D., Stock, J.T., and Gottschall, J.S. (2012) Neandertal humeri may reflect adaptation to scraping tasks, but not spear thrusting. *PLoS One*, **7** (7), e40349.
Sherratt, A. (1981) *Plough and pastoralism: aspects of the Secondary Products Revolution.* Cambridge University Press, London.
Schmitt, D., Churchill, S.E., and Hylander, W.L. (2003) Experimental evidence concerning spear use in Neandertals and early modern humans. *J. Archaeol. Sci.*, **30** (1), 103–114.
Schultz, A.H. (1937) Proportions, variability and asymmetries of the long bones of the limbs and the clavicles in man and apes. *Hum. Biol.*, **9** (3), 281–328.
Skedros, J. (2011) Interpreting load history in limb-bone diaphyses: important considerations and their biomechanical foundations. In: *Bone Histology: an Anthropological Perspective* (eds C.M. Crowder and S.D. Stout), CRC Press, New York, pp. 153–220.
Skoglund, P., Malmström, H., Omrak, A., Raghavan, M., Valdiosera, C., Günther, T., Hall, P., Tambets, K., Parik, J., and Sjögren, K.-G. (2014) Genomic Diversity and Admixture Differs for Stone-Age Scandinavian Foragers and Farmers. *Science*, **344** (6185), 747–750.
Sládek, V., Berner, M., Sosna, D., and Sailer, R. (2007) Human manipulative behavior in the Central European Late Eneolithic and Early Bronze Age: Humeral bilateral asymmetry. *Am. J. Phys. Anthropol.*, **133** (1), 669–681.
Sládek V, Hora M, Farkasova K, Rocek T. (2016a) Impact of grinding technology on bilateral asymmetry in muscle activity of the upper limb. *J. Archaeol. Sci.*, **72**, 142–156.
Sládek, V., Ruff, C.B., Berner, M., Holt, B., Niskanen, M., Schuplerová, E., and Hora, M. (2016b) The impact of subsistence changes on humeral bilateral asymmetry in Terminal Pleistocene and Holocene Europe. *J. Hum. Evol.*, **92**, 37–49.
Sparacello, V.S., Pearson, O.M., Coppa, A., and Marchi, D. (2011) Changes in skeletal robusticity in an Iron Age Agropastoral Group: The Samnites from the Alfedena Necropolis (Abruzzo, Central Italy). *Am. J. Phys. Anthropol.*, **144** (1), 119–130.
Steele, J. (2000) Handedness in past human populations: skeletal markers. *Laterality*, **5** (3), 193–220.
Steele, J. and Mays, S. (1995) Handedness and directional asymmetry in the long bones of the human upper-limb. *Int. J. Osteoarchaeol.*, **5** (1), 39–49.
Stiner, M.C. (2001) Thirty years on the "Broad Spectrum Revolution" and paleolithic demography. *Proc. Natl Acad. Sci. USA*, **98** (13), 6993–6996.
Stirland, A.J. (1993) Asymmetry and activity related change in the male humerus. *Int. J. Osteoarchaeol.*, **3**, 105–113.
Stock, J.T. and Pfeiffer, S. (2001) Linking structural variability in long bone diaphyses to habitual behaviors: Foragers from the Southern African Later Stone Age and the Andaman Islands. *Am. J. Phys. Anthropol.*, **115** (4), 337–348.

Stock, J.T. and Pfeiffer, S. (2004) Long bone robusticity and subsistence behaviour among Later Stone Age foragers of the forest and fynbos biomes of South Africa. *J. Archaeol. Sci.*, **31** (7), 999–1013.

Trinkaus, E. and Ruff, C.B. (1999) Diaphyseal cross-sectional geometry of Near Eastern Middle Palaeolithic humans: the tibia. *J. Archaeol. Sci.*, **26** (10), 1289–1300.

Trinkaus, E., Churchill, S.E., and Ruff, C.B. (1994) Postcranial robusticity in Homo II: Humeral bilateral asymmetry and bone plasticity. *Am. J. Phys. Anthropol.*, **93**, 1–34.

Van Valen, L. (1962) A study of fluctuating asymmetry. *Evolution*, **16** (2), 125–142.

Venclová, N., Drda, P., Michálek, J., Militký, J., Salač, V., Sankot, P., and Vokolek, V. (2008) Archeologie pravěkých Čech 7: Doba laténská. Archeologický ústav AV ČR, Praha.

Watson, R.C. (1974) Bone growth and physical activity in young males. In: *International Conference on Bone Mineral Measurement* (eds R.B. Mazess and J.R. Cameron), Washington, DC, US Department of HEW Publications (NIH) 75-683, pp. 380–386.

Wefers, S. (2011) Still using your saddle quern? A compilation of the oldest known rotary quern in western Europe. In: *Bread for the people: the archaeology of mills and milling* (eds D. Williams and D. Peacock), Archaeopress, Oxford, pp. 67–76.

Weiss, E. (2003) Effects of rowing on humeral strength. *Am. J. Phys. Anthropol.*, **121**, 293–302.

Weiss, E., Wetterstrom, W., Nadel, D., and Bar-Yosef, O. (2004) The broad spectrum revisited: evidence from plant remains. *Proc. Natl Acad. Sci. USA*, **101** (26), 9551–9555.

Wright, K.I. (1994) Ground-stone tools and hunter-gatherer subsistence in southwest Asia: Implications for the transition to farming. *Am. Antiq.*, **59** (2), 238–263.

8

Britain

Christopher B. Ruff[1], Evan Garofalo[1], and Sirpa Niinimäki[2]

[1] *Center for Functional Anatomy and Evolution, Johns Hopkins University School of Medicine, Baltimore, MD, USA*
[2] *Department of Archaeology, University of Oulu, Oulu, Finland*

8.1 Introduction

This regional chapter includes analyses of skeletal samples from Great Britain, although virtually all of the specimens originate from within present-day England only (the exception being the Early Upper Paleolithic Paviland 1 skeleton, from Wales). The sites are shown in Figure 8.1, and a list is given in Table 8.1. A total of 317 individuals were analyzed, almost equally divided between males and females. Assignment of cultural periods follows that outlined in Chapter 1. No samples from the Early Medieval or Very Recent periods were measured for this region.

As would be expected, all of the sites prior to the Iron/Roman period are quite small in terms of number of burials (seven or less), although in total the sample sizes are reasonable for the Neolithic (n = 19) and Bronze Age (n = 17). The Upper Paleolithic and Mesolithic periods are represented by only one individual each. Data for these individuals are given in the tables, but they are not included in statistical comparisons. (For comparisons including these early periods across Europe, see Chapters 4 and 5.) The three latest periods represented in our sample are primarily composed of large cemetery samples of 40–61 individuals (Wetwang Slack, Poundbury, York, Blackgate, Spitalfields), although with a few smaller sites in the Iron/Roman period. Some comparisons within temporal period are carried out between the larger site samples. The only 'urban' sites are those from the Late Medieval and Early Modern periods, plus Poundbury; thus, possible rural–urban comparisons are largely confounded by temporal/cultural differences and so are not included here, except for a few comparisons between Poundbury and other Iron/Roman samples. Terrain is relatively uniform across the sites, with a few classified as 'hilly' but not 'mountainous' (see Chapters 1 and 5), so is also not included as a factor in analyses. More details on the sample composition, sources of samples, and so forth are given in Appendix 1.

It is fully realized that some of the samples or individuals included as 'British' here were actually relatively recent migrants to the British Isles. The Spitalfields sample, for example, is largely composed of the descendants of Huguenots from western France who had arrived in England within a few generations, although there was later admixture with the non-French population (individuals in the present study sample were born between 1709 and 1816 AD) (Molleson and Cox, 1993). There is evidence that the well-known Bronze Age 'Amesbury Archer' (Fitzpatrick, 2011), the subject of special analyses later in this chapter, grew up in central Europe

Skeletal Variation and Adaptation in Europeans: Upper Paleolithic to the Twentieth Century,
First Edition. Edited by Christopher B. Ruff.

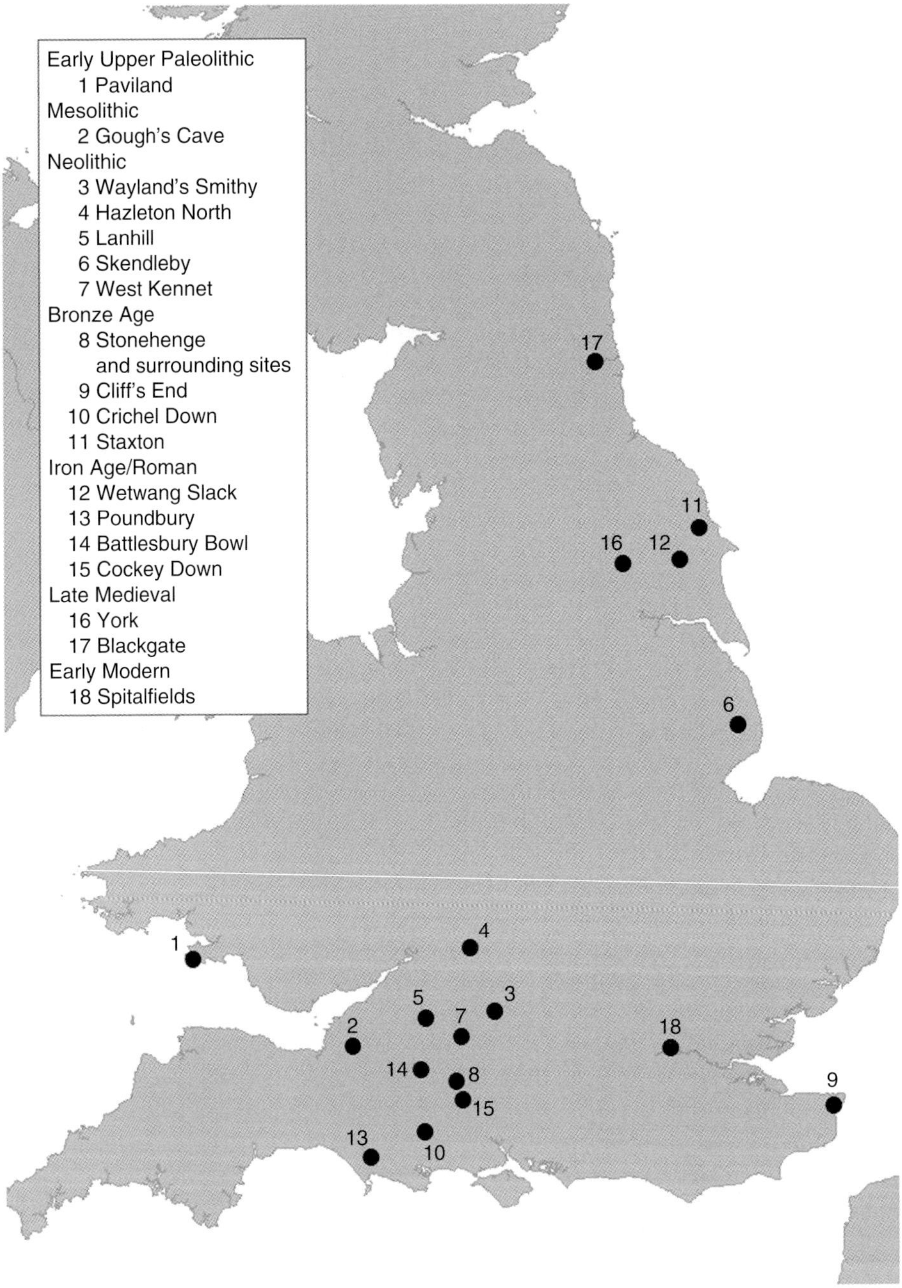

Figure 8.1 Map showing locations of sites included in the study. Numbers refer to the list in Table 8.1.

(Chenery and Evans, 2011). However, given that there have been numerous population migrations into the British Isles from continental Europe since the Neolithic (Leslie *et al.*, 2015), it seems artificial to attempt to distinguish between relatively recent migrants and more long-standing inhabitants. Whether the Spitalfields sample is broadly representative of its time period within England is further considered below.

Table 8.1 British samples.

Map no.[a]	Site	Period	Date Range (cal.)	Male N	Female N	Total N
1	Paviland	Early UP	33,400–34,000 BP	1	0	1
2	Gough's Cave	Mesolithic	8000–8450 BC	1	0	1
3	Wayland's Smithy I	Neolithic	3500–3600 BC	5	2	7
4	Hazelton North	Neolithic	3300–3700 BC	1	1	2
5	Lanhill	Neolithic	2500–3500 BC	2	2	4
6	Skendleby	Neolithic	2800–3300 BC	1	2	3
7	West Kennet	Neolithic	3600–3700 BC	2	1	3
8	A303	Bronze	2000–2500 BC	1	0	1
8	Amesbury	Bronze	2300–2500 BC	1	1	2
8	Boscombe	Bronze	2200–2500 BC	1	0	1
8	Shrewton	Bronze	1900–2250 BC	3	1	4
8	Stonehenge	Bronze	2300–2400 BC	1	0	1
8	Wilsford S. Lake	Bronze	1750–2000 BC	0	2	2
9	Cliff's End	Bronze	920–1190 BC	0	2	1
10	Crichel Down	Bronze	1500–2400 BC	2	0	2
11	Staxton	Bronze	1800–2600 BC	2	1	3
12	Wetwang Slack	Iron/Roman	100–400 BC	30	30	60
13	Poundbury	Iron/Roman	300–400 AD	20	20	40
13	Poundbury Farm	Iron/Roman	0–400 AD	3	8	11
14	Battlesbury Bowl	Iron/Roman	200 BC–0 AD	2	1	3
15	Cockey Down	Iron/Roman	400 BC–50 AD	2	1	4
16	York	Late Medieval	1000–1350 AD	30	30	60
17	Blackgate	Late Medieval	1000–1150 AD	30	31	61
18	Spitalfields	Early Modern	1700–1850 AD	20	20	40
Total				161	156	317

[a] See Figure 8.1.

In addition to the skeletal samples, stature and body mass data for a number of living British anthropometric samples were gleaned from the literature, to serve as very recent comparisons. The literature sources, specific samples, and sex-specific means are given in Table 8.2. Data for other anthropometric dimensions in living British adult samples are much more scarce and are not included here. Only national or other large ($n > 2000$) surveys were included. When several subsamples had been measured (e.g., in Kemsley, 1950) their means were averaged. An attempt was made to sample both the most recent (post-2000 AD) and earlier 20th century periods, for use in recent secular change analyses. All of the statures listed in Table 8.2 were obtained without shoes (except for Kemsley, 1950, where a downward adjustment of 1 inch (2.5 cm) for males and 1½ inches (4 cm) for females was made by the authors) and should be broadly comparable, between themselves and to the archaeological samples. However, several of the studies measured body mass (weight) with subjects wearing varying amounts of clothing, and reported data with or without different correction factors. In two studies (Montegriffo, 1968; Khosla and

Table 8.2 Living British anthropometric samples.

Reference	Sample	Meas. date	Sex	Height (cm)	Weight (kg)	Birth year	Years BP
Kemsley, 1950	Industrial workers, housewives (25–34 yr)	1943	F	157.6	54.4	1914	86
Joint Clothing Council, 1957[a]	National (?) (used 27 yr)	1957	F	161.1	57.6	1930	70
Montegriffo, 1968[d]	BP staff (24–39 yr)	1965	F	162.8	58.5	1934	66
Thompson *et al.*, n.d.[a]	National (20 yr)	1970	F	163.3	59.5	1950	50
Rosenbaum *et al.*, 1985	National (25–34 yr)	1980	F	161.8	60.6	1950	50
National survey[b,d]	National (25–34 yr)	2012	F	163.3	70.2	1982	18
Kemsley, 1950[c]	Industrial workers (25–34 yr)	1943	M	169.3	64.0	1914	86
Montegriffo, 1968[d]	BP staff (24–39 yr)	1965	M	175.7	71.8	1934	66
Khosla and Lowe, 1968[d]	Port Talbot steel workers (25–34 yr)	1965	M	171.8	67.0	1936	64
Rosenbaum *et al.*, 1985	National (25–34 yr)	1980	M	175.0	74.0	1950	50
National survey[b,d]	National (25–34 yr)	2012	M	177.8	82.2	1982	18

[a] Cited in Eveleth and Tanner, 1976 (see text on age assigned).
[b] http://www.hscic.gov.uk/catalogue/PUB13219.
[c] Weight adjusted upwards by 1 kg females, 2 kg males (see text).
[d] Weight adjusted downwards by 1 kg females, 1.5 kg males (see text).

Lowe, 1968), no adjustment was made by the authors for light indoor clothing without shoes; in these cases weight was adjusted downwards by 1 kg in females and 1.5 kg in males. Kemsley (1950) measured subjects in 'working clothing,' including shoes or boots, and subtracted 10 pounds (4.5 kg) for males and 6 pounds (2.7 kg) for females. This seemed excessive, so weight was adjusted upwards by 1 kg in females and by 2 kg in males. Because both height and weight are affected by age among adults (as demonstrated in many of these studies), data reported for young adults, aged 25 to 34 or 39 years (as available), were used whenever possible (averaging across smaller age groups subdivisions); in two cases the reported age was younger (20 years, Thompson *et al.*, n.d.) or not given (Joint Clothing Council, 1957). In the latter case an average age of 27 years was assigned since the sample was listed in Eveleth and Tanner's 1976 compendium (their Appendix Tables 8.5a,b) as 'young' adults. The average age of the samples and the study date was used to calculate average birth year (Table 8.2 here).

In addition to these sources, Steckel's (1995: their Table 6) summary of male UK statures at 50-year intervals from 1750 to 1950 (birth years) was also used in recent secular trend analyses. Comparisons of earlier stature data were carried out with those provided by Roberts and Cox (2003: their Table 8.1) for a very large sample of British archaeological remains. In addition, statures for two of the present archaeological sites (Poundbury and Spitalfields) could be compared to those derived previously for larger samples from the same sites (Farwell and Molleson, 1993; Molleson and Cox, 1993; respectively).

Body size and shape parameters used in comparisons include stature, body mass, (living) bi-iliac breadth, crural and brachial indices, relative sitting height, relative bi-iliac breadth (bi-iliac breadth/stature), and relative clavicular length (summed clavicular lengths/stature) (see Chapter 2). Statures were derived from anatomical reconstruction (i.e., modified Fully method) in about three-quarters of the sample, and body masses from the stature/bi-iliac method in almost 80% of the sample. Specimens from earlier periods were less complete on average, but in every period from the Neolithic onward, more than 60% of individuals could be estimated using these more 'anatomical' methods (except for stature in the Neolithic, with 42% reconstructed anatomically).

Bone cross-sectional parameters include body size-standardized A-P bending strength (Z_x) of the femur and tibia, and torsional/average bending strength (Z_P) of the femur, tibia, and right and left humerus, as well as percentage cortical area of all four bones (see Chapter 3). Variation in humeral bilateral asymmetry in Z_P is also examined.

Comparisons of these properties between temporal periods are carried out using pairwise Tukey tests, using a 0.05 significance level. Comparisons between British and other European samples, within temporal period, are carried out using *t*-tests with Bonferroni correction for multiple comparisons (with five temporal periods, 0.05/5 = 0.01 significance level). All comparisons are performed within sex. Comparisons between sex are performed using *t*-tests, again with Bonferroni correction. Average sexual dimorphism is calculated using male (M) and female (F) means as:

$$\left[\left(M-F\right)/\left(\left(M+F\right)/2\right)\right]\times 100$$

8.2 Body Size and Shape

8.2.1 Body Size

Summary statistics for stature, body mass, and bi-iliac breadth by temporal period in the British samples are given in Table 8.3. Box plots of temporal changes in stature and body mass from the Neolithic to Early Modern periods are shown in Figure 8.2, along with averages for recent living British samples (from Table 8.2). The latter are represented as distributions of sample means rather than individuals. Roberts and Cox's (2003) average statures for the Neolithic, Bronze, Iron/Roman, and Late Medieval periods are also plotted for comparison. (The Iron/Roman values are averages of their Iron and Roman period means. Their post-Medieval sample, ranging from 1550 to 1850 AD, is not comparable in time to our one post-Medieval sample (Spitalfields; see Table 8.1), so was not included here.)

Among the archaeological samples, Bronze Age males stand out as taller than the other periods (all Tukey tests significant) (Fig. 8.2(a)). Thus, there is a large increase in stature among males from the Neolithic to the Bronze Age, followed by a decline in the Iron/Roman period. As shown in the figure, this is also the pattern observed by Roberts and Cox (2003). Bronze Age females do not show as marked a trend. Iron/Roman and Late Medieval samples are similar in height, and the Early Modern Spitalfields sample tends to be shorter, although not significantly so. Living samples average larger than archaeological samples, except for Bronze Age males.

Body mass trends are similar to those in stature (Fig. 8.2(b)), although there is more overlap of Bronze Age males with Iron/Roman and Late Medieval males. This analysis highlights the small size of Neolithic and Early Modern males (significant compared to the other three periods). Females are non-significantly different between periods, although Early Modern females are near-significantly smaller than Bronze and Late Medieval females. Living samples overlap broadly with archaeological samples in body mass.

Table 8.3 Temporal differences in body size.

Property	Sex		Up. Pal.	Mesol.	Neol.	Bronze	Iron/ Roman	Late Med.	Early Mod.
Stature	Males	N	1	1	11	11	57	60	20
(cm)		Mean	172.7	164.8	163.31	172.32	164.14	166.33	163.51
		SE			1.80	1.50	0.79	0.76	1.65
	Females	N			8	6	61	61	20
		Mean			158.03	159.70	156.51	158.22	153.58
		SE			2.44	2.31	0.70	0.80	1.20
Body	Males	N	1	1	11	11	57	59	20
mass		Mean	75.2	67.1	61.86	72.74	67.57	68.55	61.60
(kg)		SE			1.84	1.72	1.10	0.98	1.91
	Females	N			8	6	61	60	20
		Mean			55.28	60.99	56.62	57.69	54.17
		SE			2.58	1.35	0.64	0.77	1.10
Bi-iliac	Males	N	1	1	8	8	34	49	20
bd.		Mean	30.5	28.9	27.45	29.58	28.98	28.95	27.35
(cm)		SE			0.40	0.57	0.38	0.27	0.42
	Females	N			4	4	44	55	20
		Mean			26.57	29.53	27.90	27.98	27.36
		SE			0.68	0.73	0.28	0.30	0.39

Bi-iliac breadth is wide in Bronze Age through Late Medieval males (Table 8.3), which accounts for the smaller difference in body mass than in stature between these samples (since body mass is determined by both stature and body breadth). Bi-iliac breadth is narrower in both Neolithic and Early Modern males (significantly so in Early Moderns), which contributes to the small body masses of these periods. Females show similar trends, although no inter-period differences reach significance.

The one Early Upper Paleolithic specimen, Paviland 1, is typical for his time period, being both tall and heavy compared to later periods (see Chapter 4), although not taller (or much wider) than British Bronze Age males (Fig. 8.2(a)). The one Mesolithic male, Gough's Cave 1, is much smaller and lighter than Paviland 1, although with a nearly equivalent bi-iliac breadth, again both typical for European Mesolithic males (see Chapter 4). It fits almost directly on the means for Iron/Roman period males (Table 8.3).

Roberts and Cox's (2003) mean stature values (as plotted in Figure 8.2(a)) are similar to those of our samples for the Neolithic and Bronze periods, but tend to be larger in Iron/Roman males and females and Late Medieval males (also see Table 8.3). However, it should be noted that statures in the samples that they included in their collation were generally calculated using Trotter and Gleser's (1952, 1958) US white estimation equations (C. Roberts, personal communication). As shown elsewhere (Ruff *et al.*, 2012), Trotter and Gleser's femoral formulae systematically overestimate stature in European Holocene samples, by an average of

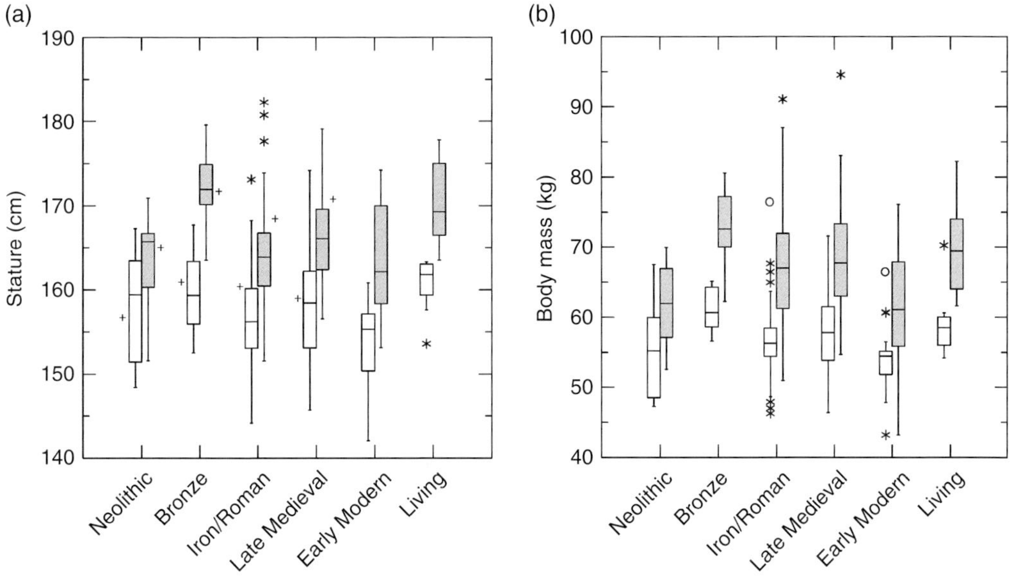

Figure 8.2 Boxplots showing temporal variation in (a) stature and (b) body mass. Open boxes: females; gray boxes: males. 'Living' refers to the samples listed in Table 8.2, with box plots representing the distribution of sample means rather than individuals. Summary statistics for archaeological samples are given in Table 8.3. Small crosses in (a) are means from Roberts and Cox (2003) (see text).

almost 2 cm in males and 1 cm in females, compared to anatomically reconstructed statures. (Note that there is a typographical error in Table 4 of Ruff *et al.*, 2012 – results there are compared with Trotter and Gleser, 1952, not 1958.) Trotter and Gleser's tibial formulae (for the 1952 equations) are also subject to apparent mis-measurement of the tibia, the average effect of which is again an overestimation of stature of about 3–4 cm (Jantz *et al.*, 1994). Therefore, depending on which bone(s) were actually used in each of the samples included in Roberts and Cox's summary, stature was overestimated by between 1 cm and 4 cm (at least – the compounded effects of the tibial mis-measurement and general mis-matching of Trotter and Gleser's modern US sample with European Holocene samples were not investigated in Ruff *et al.*, 2012), with likely greater effects in males. If this is taken into consideration, Roberts and Cox's mean values are much more similar to those of the present study. As noted earlier, the general temporal patterning of variation in stature that they found is also similar to that demonstrated here.

It is also of interest to compare the present study estimates of stature for the Poundbury and Spitalfields samples to those determined previously for much larger samples from the same sites, in part to check on whether our subsamples are representative of the entire site samples, and also to investigate the effects of using different stature reconstruction methods. Farwell and Molleson (1993) tried several formulae on the Poundbury sample, but ended by using Trotter and Gleser's 1952 formulae, averaging results for the femur and tibia. They noted that the tibia gave larger estimates than the femur when using the Trotter and Gleser formulae, but not other researchers' formulae, which can now be explained as above. Their mean statures for 341 males and 360 females from the site were 166.2 cm and 160.9 cm, respectively (Farwell and Molleson, 1993: their Table 28). Applying downward corrections of 3 cm for females and 4 cm for males, based on the figures given above, these are 163.2 cm and 157.9 cm. The mean statures for our sample of 20 males and 20 females from Poundbury are 164.3 cm and 154.7 cm, respectively, with almost all statures (90%) calculated using anatomical reconstruction. Thus, the males are very close on average, while our females are somewhat shorter than Farwell and Molleson's average, although the various approximations incorporated into the adjustments above make any precise comparisons tenuous. The present study sample was also derived from only the late Roman main cemetery, while Farwell and Molleson incorporated individuals from other cemeteries within the site. Thus, on the whole our sample appears to be fairly representative of the total Poundbury sample.

Molleson and Cox (1993) used Trotter and Gleser's 1952 formulae for all long bones to estimate stature in the Spitalfields sample. They again noted that tibial estimates were the highest. Mean stature estimates of 167.9–170.3 cm for 114 males and 154.1–158.5 cm for 116 females were obtained, depending on the particular bones used. Applying the same general correction factors as above, mean statures would be reduced to about 164–166 cm for males and 151–155 cm for females. The mean statures for our Spitalfields sample are 163.5 cm in males and 153.6 cm in females. (All but three individuals, or 92%, were calculated using anatomical reconstruction.) These are within the range of mean estimates for females, and just below the range of mean estimates for males, indicating that, like Poundbury, they are fairly representative of the overall site sample. One other factor that should be noted with regard to these comparisons is that we purposely eliminated very old (>50 years) individuals in the Spitalfields and Poundbury samples (known age in Spitalfields and estimated in Poundbury), to avoid biasing bone robusticity comparisons with other large sites and time periods, where average ages fell between about 34 and 40 years (also see below). Since previous investigators included individuals of all ages for Spitalfields and Poundbury, this may have slightly biased statures in our subsamples, perhaps towards smaller individuals (e.g., Gunnell *et al.*, 2001), although if so the effects were inconsistent and apparently minor.

No differences in body mass or stature were found between the two temporally paired larger samples – that is, Wetwang Slack and Poundbury, and York and Blackgate (see Table 8.1). Comparisons were also carried out between the more urban (large settlement) Poundbury sample and the broadly contemporaneous but more rural (smaller settlement) Poundbury Farm, Battlesbury Bowl, and Cockey Down samples (Table 8.1). No differences were found among males, but among females, those from the more urban Poundbury sample were significantly shorter (154.7 ± 1.0 cm (SE) versus 159.3 ± 1.6 cm).

Changes in average stature and body mass in British population samples since 1750, including the Spitalfields sample, are shown in Figure 8.3. The clear increase in both parameters and in both sexes during the 20th century is evident, consistent with recent secular trends in northern European populations (see Chapter 4). Changes in male stature prior to this period are much more modest (Fig. 8.3(a); similar data for females are not available), although the 50-year interval groupings from Steckel (1995) may obscure more fine-grained trends (Komlos and Kuchenhoff, 2012). Males from the Spitalfields sample are quite close in stature to Steckel's (1995) mean for 1750 birth year males, indicating that our sample is not unusually small for its time period. There are no comparable data for females, but given that sexual dimorphism in stature within our Spitalfields sample is also quite normal by average modern human standards (see below), this suggests that both sexes were not abnormal in stature for their time. This is also supported by comparisons of body mass (Fig. 8.3(b)), in which Spitalfields males and females are just slightly smaller than early 20th century British samples, similar to the stature comparisons. The most recent living samples are larger than the Bronze Age sample (compare with Fig. 8.2), similar to general European comparisons

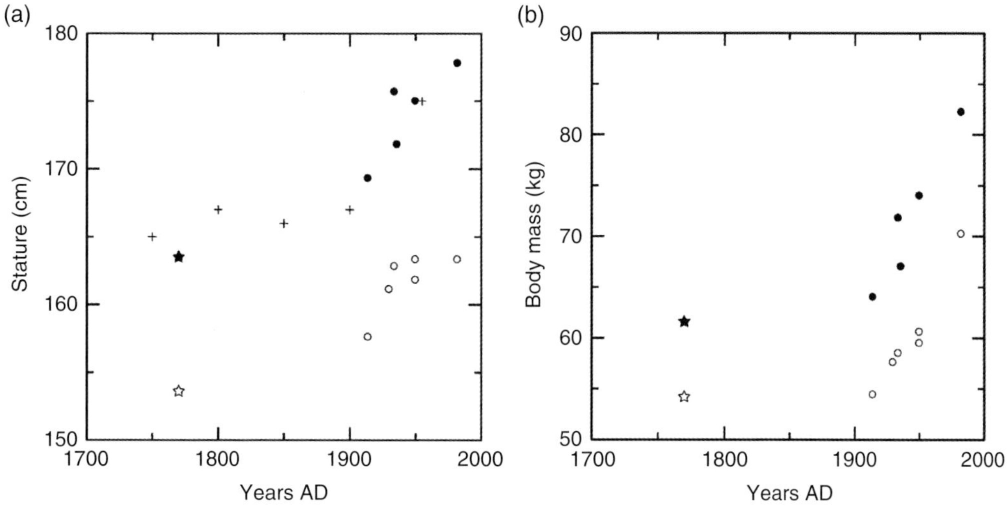

Figure 8.3 Variation by birth year in (a) stature and (b) body mass in British samples from 1750 to 1982. Open symbols: females; filled or solid symbols: males. Stars: present Spitalfields sample; crosses: Steckel (males only); circles: samples in Table 8.2.

(Chapter 4). However, early 20th century populations are not, illustrating the relatively recent increase in body size relative to some earlier populations. Data for other body dimensions among modern British samples are unfortunately very limited, although Coon (1939: 392) cites a range of bi-iliac breadth means of 28–30 cm for recent British males, which is quite similar to the means for most of our samples (Table 8.3). This is consistent with evidence for relatively little change in bi-iliac breadth through time in Europe generally (Chapter 4).

Sexual dimorphism (SD) in average stature, body mass, and bi-iliac breadth for each period is shown in Table 8.4. Sexual dimorphism in stature ranges between about 3% and 8%, with the highest value during the Bronze Age. Average SD in stature among modern human populations is about 7%, with a range between about 4% and 10% (e.g., Stini, 1976; Gray and Wolfe, 1980), so several of the values in Table 8.4, especially that for the Neolithic (3.3%), seem low (although sample sizes for this period are small; see Table 8.3). Roberts and Cox's (2003: their Table 8.1) male and female means for larger samples give average SD in stature values between 5.0% and 7.3% for the same periods, with the Neolithic period again having the lowest value, but closer to the modern worldwide mean. Variation in SD between periods is largely due to greater variation among males than among females – that is, both sexes show similar patterns of increase or decrease between periods, but the changes are more marked among males (see Fig. 8.2(a)). This could reflect so-called 'female buffering' during growth to environmental influences such as nutritional level (Stini, 1976; Gray and Wolfe, 1980), although evidence for this among modern populations is somewhat ambiguous (Stinson, 1985). Within the four matched samples of living British data in Table 8.2 (Kemsley, 1950; Montegriffo, 1968; Rosenbaum *et al.*, 1985; Health Survey for England, 2013), SD in stature averages 7.8%, with a slight but progressive increase from the earliest birth year (1914) to the latest (1982) – namely 7.2%, 7.6%, 7.8%, and 8.5% – which may also support the female buffering hypothesis (assuming that living conditions in Britain progressively improved during the 20th century). The similarity in both mean male and female statures, and SD in stature, between the most recent groups and the Bronze Age sample is again intriguing.

Table 8.4 Sexual dimorphism.

	Neolithic	Bronze	Iron/Roman	Late Med.	Early Mod.
Stature	3.28	**7.60**	**4.76**	**5.00**	**6.26**
Body mass	11.25	**17.57**	**17.63**	**17.22**	**12.83**
Bi-iliac bd.	3.25	0.20	**3.80**	**3.41**	−0.04
Crural	−0.89	1.02	−0.25	0.47	0.21
Rt. brachial		0.54	**2.40**	1.28	0.31
Lt. brachial		2.15	**2.17**	1.38	0.08
Rel. sit. ht.			−1.64	−2.98	−0.66
Bi-iliac/Stat	−0.20	−6.60	−1.52	−0.89	**−6.38**
Rel. clav. ln.		0.69	**3.75**		2.54

Sexual dimorphism = [(M − F)/(M + F)/2] × 100.
Bold text = p <0.05, with Bonferroni correction.

Sexual dimorphism in body mass varies between 11% and 18% (Table 8.4), which is normal for modern human populations (mean about 15%; Ruff, 2002). Neolithic and Early Modern samples have the least SD, with a non-significant difference (with Bonferroni correction) between the sexes in the Neolithic (although again, samples sizes are small). Bi-iliac breadth tends to be slightly larger in males, but shows no significant difference between the sexes in any period (Table 8.4).

8.2.2 Body Shape

Period-specific means for linear proportions (crural, brachial, and relative sitting height indices) are given in Table 8.5, and for body breadth proportions (bi-iliac/stature and relative clavicular length indices) in Table 8.6. Because there are no significant sex differences within period in crural and relative sitting height indices (see below), the two sexes were combined for statistical comparisons between periods. Crural indices fall within a fairly circumscribed range of about 80–82 (except for the one Mesolithic specimen), although Neolithics tend to be higher (significant relative to Iron/Roman and Late Medieval) and Iron/Romans lower (significant relative to Early Modern as well as Iron/Roman, near-significant relative to Bronze Age). These patterns are to some extent reflected in relative sitting height, where Neolithics have the lowest values, or relatively longest lower limbs (Table 8.5; significant relative to Iron/Roman and near-significant relative to Early Modern). As noted above, the one Mesolithic specimen has a very high crural index (86.1), although only a moderate relative sitting height. The Early Upper Paleolithic specimen has a lower crural index (82.9), although still high compared to most of the post-Mesolithic specimens. Both fall well within the ranges of pan-European samples for these time periods (see Chapter 4).

Brachial indices show SD (see below), and so are analyzed separately by sex. Similar to pan-European trends (Chapter 4), there is a definite temporal decline in brachial indices following the Bronze Age, in both sexes and on both sides, with most inter-period differences reaching significance or near-significance. Unlike the crural index, however, brachial indices are not particularly high during the Neolithic. No data were available for our pre-Neolithic specimens.

Table 8.5 Temporal differences in linear body proportions.

Property	Sex		Up. Pal.	Mesol.	Neol.	Bronze	Iron/ Roman	Late Med.	Early Mod.
Crural	Males	N	1	1	8	11	54	50	18
		Mean	82.9	86.1	82.34	81.89	79.88	80.88	81.37
		SE			0.61	0.70	0.36	0.34	0.64
	Females	N			5	5	57	48	20
		Mean			83.07	81.05	80.08	80.50	81.20
		SE			1.08	0.79	0.26	0.28	0.59
Rt. brachial	Males	N			3	11	41	29	15
		Mean			75.72	77.31	75.02	74.61	71.74
		SE			0.82	0.79	0.40	0.46	0.54
	Females	N				4	48	39	19
		Mean				76.89	73.24	73.66	71.52
		SE				1.24	0.34	0.33	0.48
Lt. brachial	Males	N			6	8	37	23	15
		Mean			75.62	78.41	76.39	75.18	72.27
		SE			0.51	0.79	0.42	0.50	0.63
	Females	N				4	40	29	16
		Mean				76.75	74.75	74.14	72.21
		SE				1.45	0.45	0.40	0.60
Rel. sit. ht.	Males	N		1	3	5	20	5	9
		Mean		38.8	38.64	39.43	40.02	39.49	40.18
		SE			0.77	0.43	0.20	0.40	0.38
	Females	N				1	28	9	15
		Mean				39.52	40.69	40.68	40.44
		SE					0.16	0.64	0.20

Table 8.6 Temporal differences in body width proportions.

Property	Sex		Up. Pal.	Mesol.	Neol.	Bronze	Iron/ Roman	Late Med.	Early Mod.
Bi-iliac/ Stat.	Males	N	1	1	8	8	34	49	20
		Mean	17.6	17.6	16.66	17.34	17.62	17.51	16.72
		SE			0.16	0.28	0.16	0.17	0.16
	Females	N			4	4	44	55	20
		Mean			16.69	18.53	17.89	17.67	17.82
		SE			0.34	0.59	0.15	0.17	0.23
Rel. clav. len.	Males	N		1	5	8	34	40	11
		Mean		16.4	17.09	17.93	17.98	17.89	17.62
		SE			0.83	0.14	0.14	0.16	0.26
	Females	N				1	23	45	18
		Mean				18.08	17.86	17.23	17.18
		SE					0.20	0.15	0.18

Relative body breadths (Table 8.6) are also sexually dimorphic, so were compared separately by sex. Bi-iliac breadth/stature varies within a small range between periods, but is lowest in the Neolithic and Early Modern periods in males (significantly smaller in Early Modern relative to Iron/Roman and Late Medieval) and the Neolithic in females (non-significantly). Relative clavicular length shows no apparent temporal patterning. Sample sizes are quite small, however, for both indices in the earlier periods.

8.2.3 Comparisons to other Europeans

Comparisons of stature and body mass between British and other pooled European samples, by temporal period and sex, are shown in Figure 8.4. The larger size of British Bronze Age males – in both stature and body mass – is clearly apparent ($p = 0.003$, significant with Bonferroni correction). Bronze Age British males are 6 cm taller and 7 kg heavier on average than their continental counterparts. Bronze Age females are also significantly larger in body mass (+5 kg on average) than non-British Europeans. In other time periods, British samples fall fairly close to continental samples, with no significant or near-significant differences. British Neolithic males tend to be somewhat smaller in body mass (–4 kg), but the difference does not reach significance ($p = 0.039$ without Bonferroni correction). The single Early Upper Paleolithic and Mesolithic males are close to their respective pan-European means.

Because northern Europeans tend to be taller and heavier than southern Europeans (Chapter 4), comparisons were also carried out with just continental northern Europeans (from Scandinavia/Finland and Central Europe). Bronze Age British males are still significantly taller (+5 cm) and heavier (+6 kg) in these comparisons; females are near-significantly heavier (+5 kg). These differences can be explained in part, but not completely, by temporal variation within the broader Bronze Age throughout Europe; this is considered further in the Discussion section. Neolithic British males are near-significantly lighter than other northern Europeans

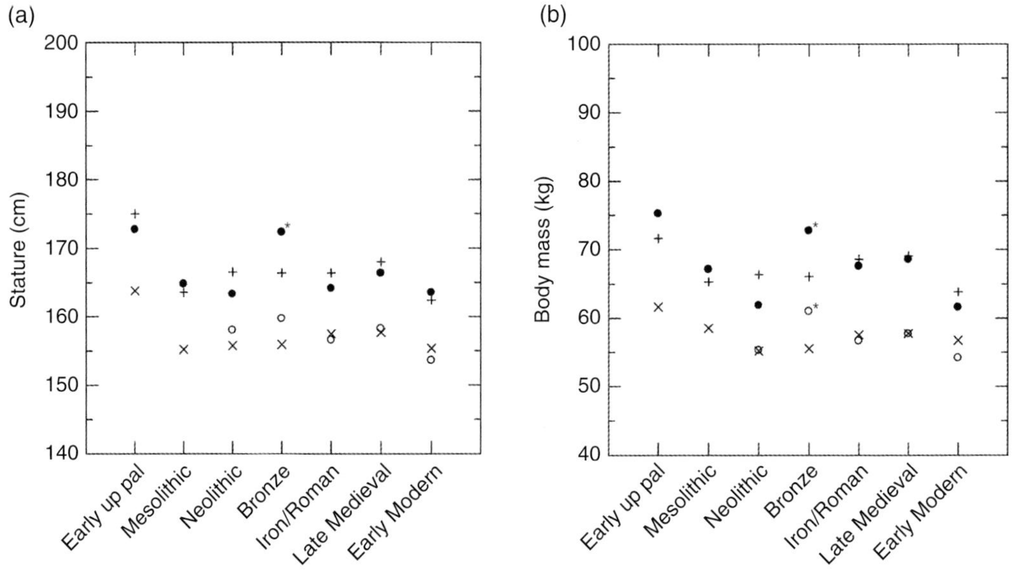

Figure 8.4 Mean values for (a) stature and (b) body mass in British and non-British samples. Filled circles: British males; open circles: British females; +: other European males; x: other European females. Asterisks indicate that British values are significantly different from non-British values (with Bonferroni correction).

($p = 0.019$ without Bonferroni correction). Iron/Roman British males and females are significantly or near-significantly shorter and lighter than other northern Europeans.

In terms of other body dimensions and indices, compared to their continental counterparts, Neolithic British males are significantly narrower-bodied (smaller bi-iliac breadth and bi-iliac breadth/stature), and Iron/Roman British males and females have smaller crural indices and larger relative sitting heights, indicating relatively shorter limbs. Early Modern British males have smaller brachial indices. When considered as a whole and compared only to northern continental Europeans (because of ecogeographic effects on body shape; see Chapter 4), the British samples have significantly smaller crural indices and brachial indices (except near-significant for females, left side), although average differences are small – about 1%. No other body shape indices are significantly different in such comparisons.

8.3 Cross-Sectional Properties

8.3.1 Bone Strength

Bone strength parameters of the lower limb, standardized for body size, are shown in Table 8.7 and Figures 8.5 and 8.6 by sex and temporal period. The anteroposterior (A-P) bending strength (Z_x) of the tibia declines after the Neolithic in both sexes (Fig. 8.5(a)). The decline continues through the Early Modern period in females, but levels off after the Bronze Age in males. Because of wide intra-period variability, inter-period statistical comparisons are all non-significant. However, comparisons of the Neolithic with all pooled post-Neolithic periods are significant in both sexes ($p < 0.05$ in males, $p < 0.005$ in females), illustrating the post-Neolithic decline. Femoral A-P bending strength shows no consistent temporal pattern in either sex (Fig. 8.5(b)).

Tibial and femoral average bending/torsional strengths (Z_p) show temporal patterns similar to those of A-P bending strength (Fig. 8.6(a)), but with a less marked decline in the tibia. Comparisons of Neolithic with pooled post-Neolithic periods are significant for females ($p < 0.05$), but not for males. Temporal changes in the femur are again inconsistent (Fig. 8.6(b)).

Average bending/torsional strength of the right and left humeri, by sex and temporal period, are shown in Table 8.8 and Figure 8.7. The most marked temporal difference among males is the high strength value for both sides in the Late Medieval period (significantly greater than all other periods for the right side, and all periods except the Early Modern period for the left side). Otherwise, males show no temporal trends in humeral strength. Females appear to show a decline in right (but not left) humeral strength after the Neolithic, although the only significant difference is with the Early Modern sample. There is again an increase in the Late Medieval period, significant compared to Iron/Roman (right) and Early Modern (both sides) samples, with the Early Modern period relatively low compared to other periods. Overall, in both sexes, right humeri show more temporal variability than left humeri.

Bilateral asymmetry in humeral strength is plotted in Figure 8.8 with summary statistics given in Table 8.8. While Neolithic males appear to show somewhat higher asymmetry than those in other periods, no inter-temporal differences are significant in either sex, likely in part because of small sample sizes in the earlier groups (Table 8.8). Males are overall more asymmetric than females ($p < 0.005$, *t*-test on pooled sample), but differences within periods, with Bonferroni correction, are not significant.

In comparisons between the two temporally paired larger samples – that is, Wetwang Slack and Poundbury, and York and Blackgate (see Table 8.1) – very few significant differences in bone strength are apparent. Poundbury males (but not females) tend to be somewhat stronger

Table 8.7 Temporal differences in lower limb bone strengths.

Bone	Prop.	Sex		Up. Pal.	Meso.	Neol.	Bronze	Iron/ Roman	Late Med.	Early Mod.
Tibia	Z_{xstd}	Males	N	1	1	8	11	54	50	18
			Mean	721.5	663.0	751.3	609.7	651.8	641.2	642.1
			SE			44.4	26.5	16.2	21.0	21.9
		Females	N			5	4	57	47	20
			Mean			617.1	568.4	547.4	515.0	479.7
			SE			18.6	36.1	11.4	17.6	25.6
Femur	Z_{xstd}	Males	N	1	1	11	11	57	57	20
			Mean	578.9	771.7	611.1	543.4	571.3	598.7	558.1
			SE			17.6	32.2	14.1	13.7	21.7
		Females	N			8	6	61	60	20
			Mean			527.1	495.9	497.4	528.8	496.1
			SE			26.1	9.7	9.5	11.5	21.9
Tibia	Z_{pstd}	Males	N	1	1	8	11	54	50	18
			Mean	1017.2	1015.8	1027.0	878.9	948.0	928.3	994.3
			SE			55.5	44.6	24.5	27.7	72.0
		Females	N			5	4	57	47	20
			Mean			893.0	857.1	811.1	762.0	724.3
			SE			34.3	35.5	14.6	23.5	36.6
Femur	Z_{pstd}	Males	N	1	1	11	11	57	57	20
			Mean	981.1	1293.4	1136.9	1027.2	1095.0	1164.6	1044.3
			SE			36.7	47.9	23.1	23.1	38.7
		Females	N			8	6	61	60	20
			Mean			1016.3	934.1	995.5	1042.4	962.6
			SE			52.6	42.0	20.4	24.2	42.9

than Wetwang Slack males, reaching significance in A-P bending strength of the femur and bending/average torsional strength of the left humerus, but in general differences between the groups are relatively minor.

Comparisons between the more urban Poundbury sample and other more rural broadly contemporaneous samples are shown in Table 8.9, and illustrated graphically for tibial Z_x and right humeral Z_p in Figure 8.9. No significant differences were found between rural and urban males. However, among females, the Poundbury urban sample is significantly weaker than rural samples for several properties, including tibial strengths and right humeral strength, with differences averaging about 20%. Because of these contrasting patterns in males and females, SD in bone strength increases in the urban sample (Fig. 8.9). The decline in humeral strength in urban females is much larger on the right side; thus, asymmetry in humeral strength among females also declines in the urban sample (Table 8.9). A smaller decline in asymmetry also characterizes urban males (Table 8.9), but does not reach statistical significance.

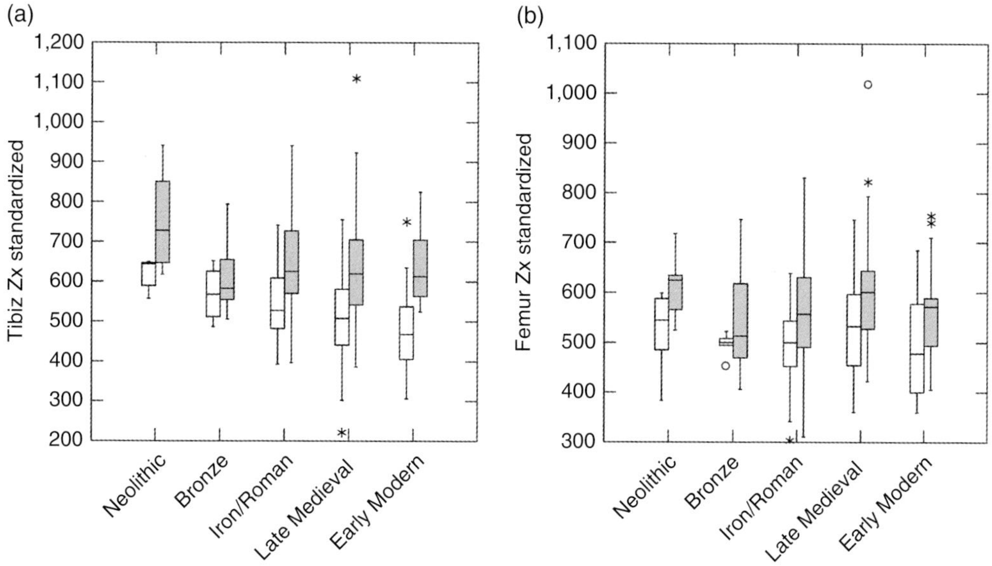

Figure 8.5 Boxplots showing temporal variation in lower limb bone A-P bending strength, standardized for body size: (a) tibia; (b) femur. Open boxes: females; gray boxes: males.

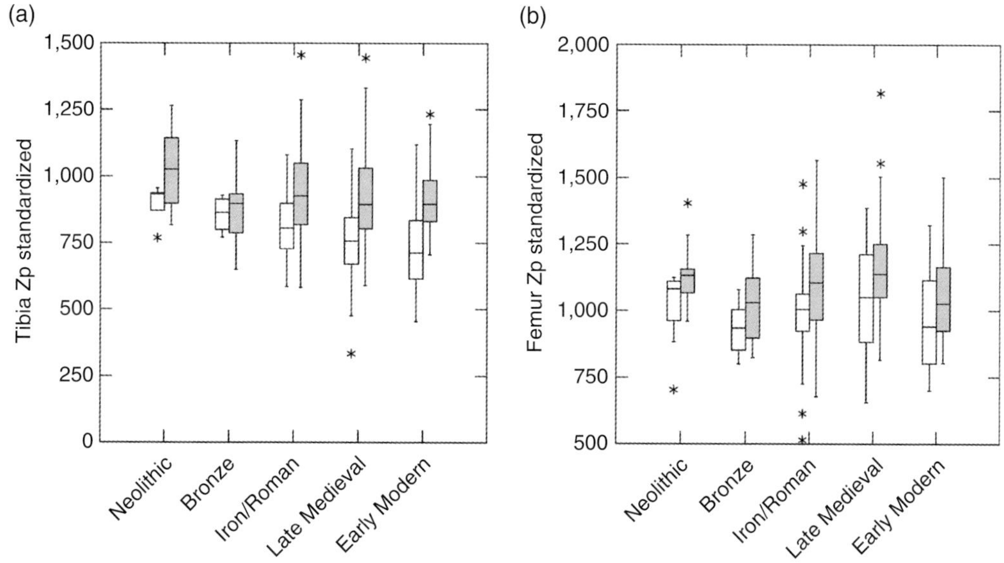

Figure 8.6 Boxplots showing temporal variation in lower limb bone average bending/torsional strength, standardized for body size: (a) tibia; (b) femur. Open boxes: females; gray boxes: males.

In summary, among British samples, tibial A-P bending strength shows a temporal decline after the Neolithic, with declines and then a leveling off in males but continuing through later periods in females. The femur shows no consistent temporal change, while humeri increase in strength in the Late Medieval period compared to earlier and later periods. More urban Iron/Roman period females show a decline in tibial and right humeral strength relative to rural females.

Table 8.8 Temporal differences in upper limb bone strengths and bilateral asymmetry.

Bone	Prop.	Sex		Up. Pal.	Neol.	Bronze	Iron/ Roman	Late Med.	Early Mod.
Rt. humerus	Z_{pstd}	Males	N		7	11	54	36	17
			Mean		579.7	586.0	596.9	698.5	605.2
			SE		25.2	28.6	17.1	17.5	30.7
		Females	N		4	5	58	43	19
			Mean		610.7	545.4	502.7	559.2	441.8
			SE		65.0	29.5	13.0	17.6	24.6
Lt. humerus	Z_{pstd}	Males	N	1	8	9	53	36	18
			Mean	338.1	512.4	525.4	536.7	646.7	566.4
			SE		35.0	26.7	16.8	19.7	20.2
		Females	N		3	5	52	36	18
			Mean		460.5	473.9	478.2	516.5	427.0
			SE		69.7	10.3	13.0	15.5	17.2
Asymmetry	Z_p	Males	N		6	9	51	25	15
			Mean		21.0	10.8	12.0	9.9	10.6
			SE		6.0	2.3	1.9	1.7	3.0
		Females	N		2	5	51	26	17
			Mean		2.6	13.8	6.4	7.2	6.8
			SE		8.4	4.7	1.8	1.4	2.6

Asymmetry = $[(R - L)/((R + L)/2)] \times 100$.

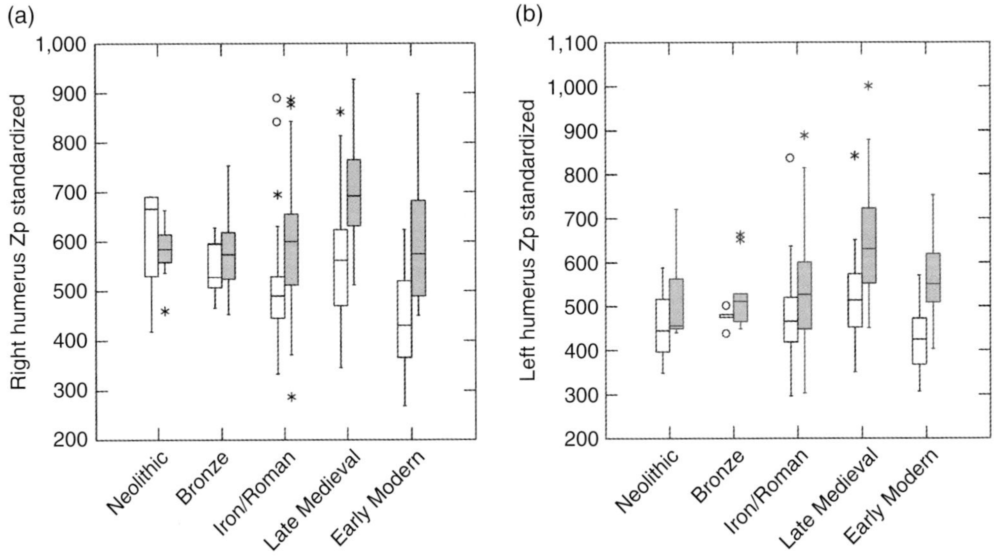

Figure 8.7 Boxplots showing temporal variation in humeral average bending/torsional strength, standardized for body size: (a) right humerus; (b) left humerus. Open boxes: females; gray boxes: males.

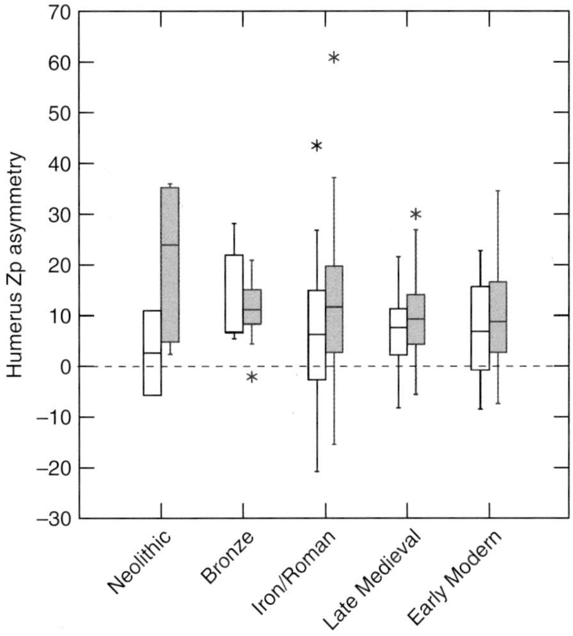

Figure 8.8 Humeral asymmetry in average bending/torsional strength. Open boxes: females; gray boxes: males.

8.3.2 Percentage Cortical Area

Percentage cortical area (%CA), representing the relative thickness of the cortices, is plotted by temporal period and sex in Figure 8.10, with summary statistics given in Table 8.10. Temporal declines in %CA are apparent in all sections and both sexes, and in most cases continue through the entire temporal sequence. Declines are more consistent in the lower limb (Fig. 8.10(a,b)), with all pairwise comparisons significant among males, except between some adjacent periods, and about half of comparisons significant among females. Regressions of %CA on Log_{10}(time) are all highly significant ($p < 0.001$). (Logarithmic transformation of time is used because of unequal sample distributions across the Holocene.) The temporal decline in humeri is less marked (Fig. 8.10(c,d)), with only a few significant inter-period differences, and no apparent decline among female left humeri until the Early Modern period. Regressions against Log_{10}(time) are still highly significant, however, for each bone and sex ($p < 0.005$).

Thus, temporal changes in bone strength and %CA in these samples are generally not parallel, with the exception of tibial strength in females. All other strength comparisons show little change (or an actual increase) among post-Neolithic periods, while %CA generally declines throughout. The significance of this observation is further discussed below. In comparisons between more urban and rural Iron/Roman period samples, the urban samples show a significant decline in %CA in the tibia (both sexes) and right humerus (males only) (data not shown).

8.3.3 Comparisons to other Europeans

In most temporal periods and for most cross-sectional properties, British samples do not depart significantly from non-British European samples. However, British Iron/Roman period

Table 8.9 Bone strengths in Iron/Roman period rural and urban (Poundbury) samples.

	Rural					
	Males			Females		
Property	**N**	**Mean**	**SE**	**N**	**Mean**	**SE**
Tibia Z_{xstd}	6	665.0	62.9	9	627.6	29.6
Femur Z_{xstd}	7	557.0	38.8	11	512.7	18.1
Tibia Z_{pstd}	6	983.2	108.1	9	910.0	23.4
Femur Z_{pstd}	7	1023.4	73.2	11	1047.7	55.7
Rt. hum. Z_{pstd}	5	601.4	65.8	10	562.9	34.9
Lt. hum. Z_{pstd}	6	540.4	75.9	8	478.5	25.4
Hum. Z_p asym.(%)	5	16.3	5.4	8	14.5	2.9

	Urban					
	Males			Females		
Property	**N**	**Mean**	**SE**	**N**	**Mean**	**SE**
Tibia Z_{xstd}	18	653.6	25.9	20	516.9*	11.6
Femur Z_{xstd}	20	617.1	22.0	20	515.5	14.8
Tibia Z_{pstd}	18	947.8	39.0	20	773.5*	19.8
Femur Z_{pstd}	20	1117.6	32.4	20	978.9	29.2
Rt. hum. Z_{pstd}	20	623.8	24.7	19	470.0*	13.5
Lt. hum. Z_{pstd}	20	573.9	20.8	19	465.7	15.5
Hum. Z_p asym.(%)	20	9.8	3.7	18	5.2*	2.9

* Significantly different from rural.

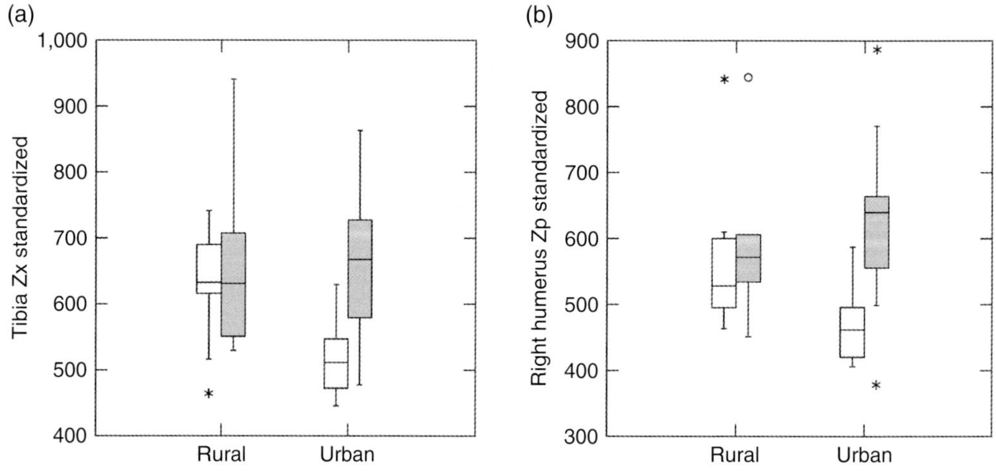

Figure 8.9 Rural versus urban samples in late Iron/Roman period: (a) tibial A-P bending strength; (b) Right humeral average bending/torsional strength. Open boxes: females; gray boxes: males.

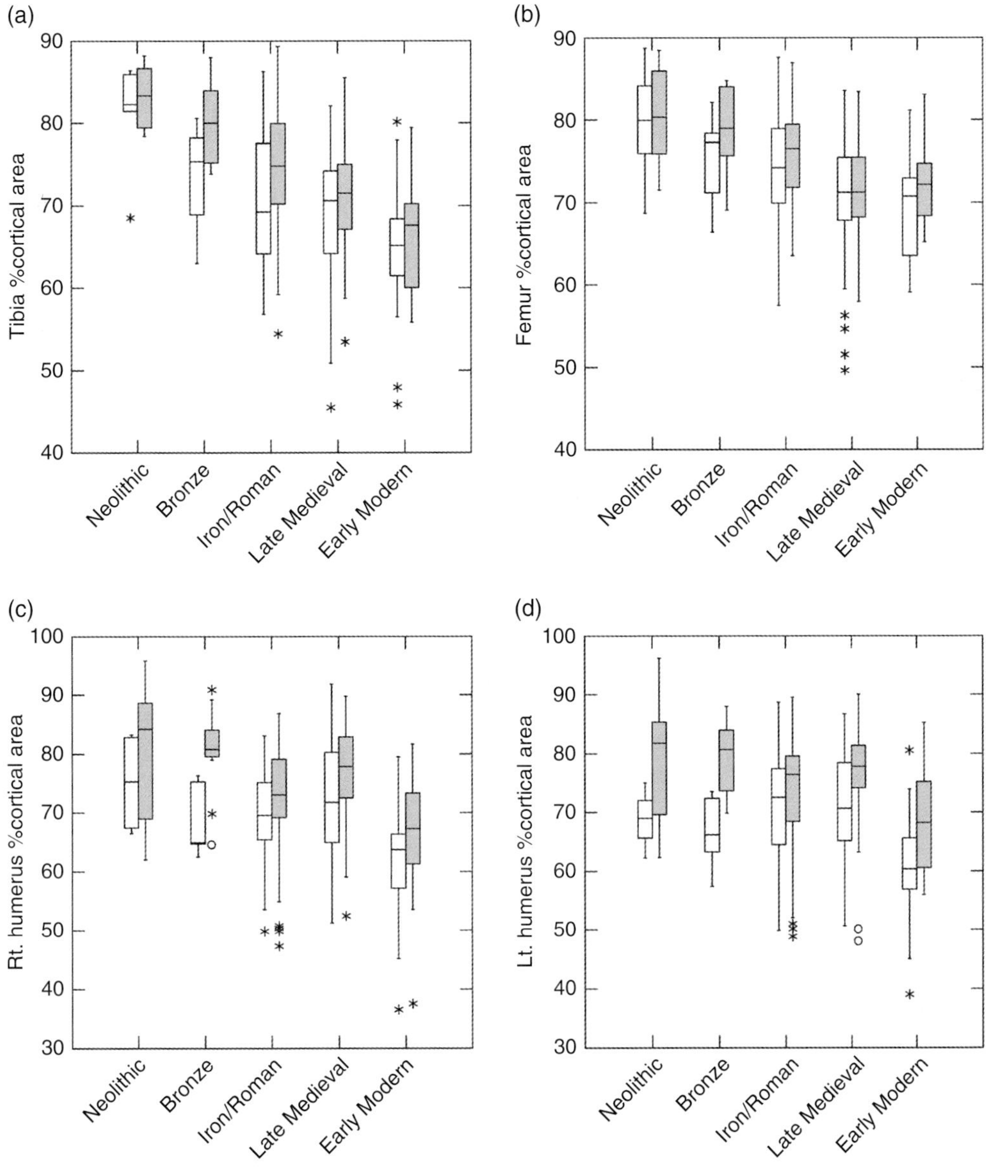

Figure 8.10 Temporal changes in percentage cortical area: (a) tibia; (b) femur; (c) right humerus; (d) left humerus. Open boxes: females; gray boxes: males.

tibiae and humeri (both sides) are significantly stronger than continental samples in both sexes (near-significant with Bonferroni correction in male humeri). Comparative data for tibial Z_x and humeral Z_p are shown in Figure 8.11. Unlike other European samples (Fig. 8.11; also see Chapter 5), there is no decline in tibial or humeral strength between the Bronze and Iron/Roman periods in Britain, thus resulting in higher values in the Iron/Roman period. British humeri (both sides and sexes) are also stronger in the Late Medieval period. Femora show no significant differences between British and non-British samples, although they are near-significantly stronger in British Iron/Roman males. Early Modern males from Spitalfields have significantly stronger tibiae (Z_x only) than non-British males. The %CA shows no consistent

Table 8.10 Temporal differences in percentage cortical area.

Bone	Sex		Up. Pal.	Mesol.	Neol.	Bronze	Iron/ Roman	Late Med.	Early Mod.
Femur	Males	N	1	1	11	11	57	58	20
		Mean	76.6	85.2	80.4	79.1	75.2	71.3	72.0
		SE			1.9	1.6	0.8	0.7	1.0
	Females	N			8.0	6.0	61.0	61.0	19.0
		Mean			79.7	75.4	74.2	70.7	69.5
		SE			2.2	2.3	0.8	0.9	1.5
Tibia	Males	N	1	1	8	11	56	51	17
		Mean	77.7	76.5	83.2	80.0	74.3	70.8	66.6
		SE			1.4	1.5	0.9	1.0	1.6
	Females	N			5.0	4.0	57.0	48.0	20.0
		Mean			80.9	73.6	70.7	69.0	64.5
		SE			3.2	3.7	1.0	1.1	1.9
Rt.	Males	N			7	11	54	38	17
humerus		Mean			79.6	80.6	71.9	76.8	65.8
		SE			5.1	2.3	1.3	1.3	2.5
	Females	N			4.0	5.0	58.0	45.0	19.0
		Mean			75.1	68.8	69.6	71.9	61.2
		SE			4.5	2.9	1.0	1.5	2.4
Lt.	Males	N	1		8	9	53	38	18
humerus		Mean	75.4		79.0	79.5	73.3	76.4	67.6
		SE			4.1	2.2	1.4	1.4	1.9
	Females	N			3.0	5.0	52.0	36.0	18.0
		Mean			68.8	66.5	70.9	71.3	60.1
		SE			3.7	3.0	1.2	1.6	2.3

differences within temporal period between British and non-British samples. Although included in plots and tables, the sample size of one individual for British male Early Upper Paleolithic and Mesolithic periods is too small for any meaningful conclusions to be drawn.

8.3.4 The 'Amesbury Archer'

The 'Amesbury Archer' is a famous early Bronze Age (Bell Beaker) burial excavated from Boscombe Down, Amesbury, close to and contemporaneous with the building of Stonehenge (see Table 8.1 and Fig. 8.1) (Fitzpatrick, 2011). It was discovered in 2002, and contains the richest collection of associated artifacts of any Bronze Age burial in Britain including, of course, many arrowheads (Fitzpatrick, 2002). As noted earlier, dental isotopic evidence indicates that the individual, a middle-aged or older male, spent his childhood in central Europe rather than in Britain (Chenery and Evans, 2011).

The Archer's skeleton has been the subject of a detailed osteological analysis (McKinley, 2011a,b), which has identified a number of pathologies. The most striking of these is the

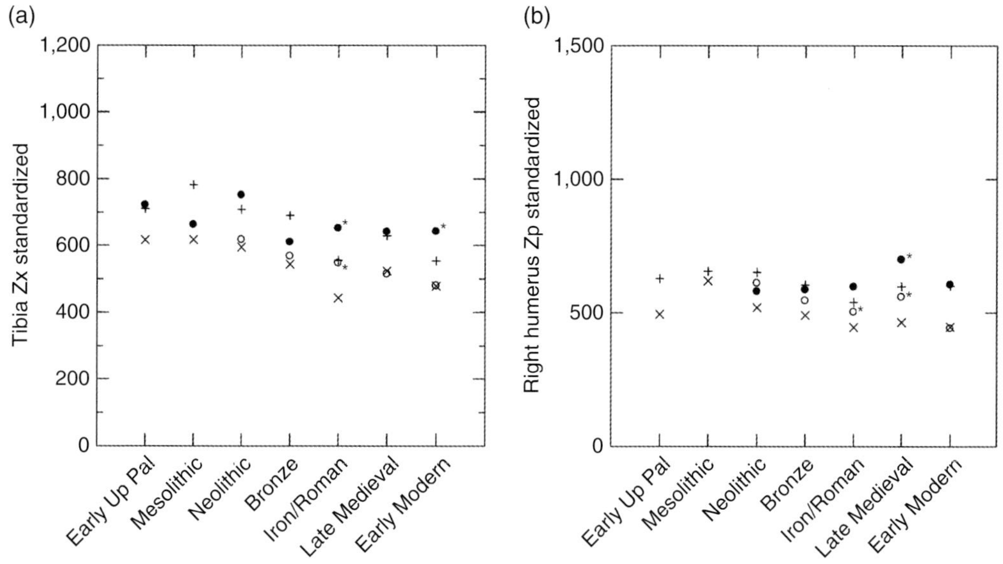

Figure 8.11 Mean values for (a) tibial A-P bending strength and (b) right humeral average bending/torsional strength in British and non-British samples. Filled circles: British males; open circles: British females; +: other European males; x: other European females. Asterisks indicate that British values are significantly different from non-British values (with Bonferroni correction).

absence of a left patella, accompanied by major changes in morphology of the distal left femur and large asymmetry in shaft breadths of the right and left femora and tibiae (left smaller). These observations, and others, are consistent with long-standing absence of, or injury to, the left patella, with subsequent mechanical unloading of the left lower limb. McKinley (2011b) reviews several known syndromes involving congenital patellar agenesis, but notes that none of them completely characterizes the Archer. She notes that evidence for a severe infection (osteomyelitis) in the distal femur and proximal tibia suggests the possibility of a traumatic injury to this region, which might also explain or have contributed to the other morphological changes observed in the lower limbs. The upper limb bones were thought to be 'not markedly robust' compared to other individuals from nearby sites of a similar time period.

The structural data collected for the present study provide the opportunity to revisit these issues within a more mechanical context. Because of his obvious lower limb asymmetry, both lower limbs were measured in the Archer, although only the 'normal' right side was used in general analyses. Here the two sides are compared. In addition, the availability of similar data for 10 other Bronze Age British males (Table 8.1) allows more complete assessments relative to broadly contemporaneous specimens, both in terms of relative bone strength and body size/shape. Thus, for example, we can address questions such as whether the increased lower limb asymmetry in the Archer is due to atrophy of the left side, hypertrophy of the right side, or both, relative to his body size, and what this might signify regarding the origin of his anomalies. Only a few of the more important results are given here; a fuller treatment of this individual will be given in a separate publication. The present analyses also provide the opportunity to reconsider some of the basic biological issues underlying the overall design of this study, that is, the relationships between long bone morphology, mechanical loading of the limbs, and behavior (see Chapters 1 and 3).

In terms of body size and shape, the Amesbury Archer is generally quite typical of other British Bronze Age (BA) males. His estimated stature using the anatomical method is 172.2 cm, almost precisely on the mean for the other 10 BA males (172.3 cm). His estimated body mass (from stature and bi-iliac breadth) is 71.3 kg, again quite close to the mean for the comparative group (72.9 kg). Living bi-iliac breadth (29.3 cm), relative bi-iliac breadth (17.4 cm), relative sitting height (39.9 cm), and crural index (81.8) are similarly close to those for other British BA males (see Tables 8.3, 8.5, and 8.6). He does have relatively small brachial indices (short forearms) – 72.6 cm (right) and 74.6 cm (left), which are below the minimum observed for other British BA males – and relatively long clavicles/humeri – 18.4 cm – above the maximum observed for the comparative sample. This might suggest that he has relatively short humeri, but that is not the case since he has normal humeral/femoral length proportions. In any event, despite his origin in Central Europe, in most respects he is very similar to typical British males of his time period. In particular, he is relatively large-bodied, a characteristic of British BA males in general (see above and also Discussion). These results also suggest that his various skeletal abnormalities did not affect overall growth.

Comparisons of size-standardized strength of the Archer's femora, tibiae, and humeri are shown in Figure 8.12, along with values from the 10 other British BA males. Within this group, the three male skeletons recovered from the Boscombe Lower Camp, Stonehenge (the 'Stonehenge Archer'), and A303 (Normanton Down) sites, which are both geographically and temporally very close to the Amesbury Archer (Chenery and Evans, 2011; also see Table 1), are also discriminated. Bilateral asymmetry in A-P bending and average bending/torsional strengths of the Archer's femora and tibiae range between 40% and 70%. These are calculated using the mean of right and left sides as the denominator, as in analyses of bilateral asymmetry of the humeri throughout this volume. Considered as simple ratios, A-P bending strength of the lower limb bones is close to twice as great on the right side, and average bending/torsional strength about 60–70% greater on the right side. Comparisons in Figure 8.12 demonstrate that this very large asymmetry is due to both increased relative strength on the right (normal) side, and decreased relative strength on the left (abnormal) side – the Archer's left lower limb bones fall below those of all other British BA males, and his right lower limb bones fall above or

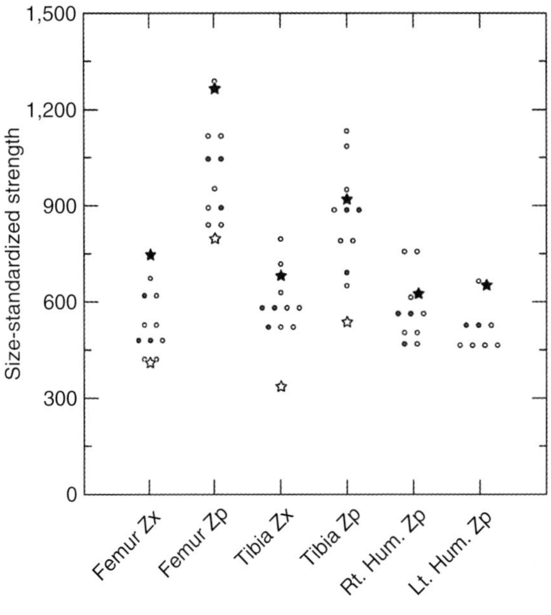

Figure 8.12 Femoral, tibial, and humeral strengths, standardized for body size, in the Amesbury Archer (stars) and other British Bronze Age males (small circles). In the lower limb, filled stars indicate the Archer's 'normal' (right) side, and open stars the abnormal (left) side. Gray circles are the three males closest in time and space to the Archer (see text).

within the highest few values. This indicates both mechanical unloading of the abnormal limb, as well as compensatory overloading of the normal limb.

In terms of humeral strength values, contrary to McKinley's (2011b) impressions, the Archer has relatively strong humeri compared to other British BA males (Fig. 8.12). His right humerus is the third strongest of the combined group, while his left humerus is almost equal to the strongest (note that only eight males had preserved left humeri). The slight left 'dominance' of the Archer's humeri (2.1% larger) is at variance with every other British BA male, all of whom have stronger right humeri (mean 12.4%, range 4.4–20.9%). Left 'dominance' in humeral strength is relatively rare among males in our sample (see Chapter 7); only three out of 69 (4%) other BA males across Europe are characterized in this way. The three other most closely associated males from the Amesbury region fall within the general strength distributions of the British BA males, or even nearer the lower end of some distributions (Fig. 8.12), so these results are not a consequence of any general regional/temporal variation within the British BA.

The degree of bilateral asymmetry in lower limb bone strength evidenced in the Amesbury Archer is a vivid demonstration of the plasticity of diaphyseal cortical bone. Asymmetries this large have been noted in right and left upper limb strength comparisons of highly selected but 'normal' individuals, such as professional tennis players and baseball pitchers (Trinkaus *et al.*, 1994; Warden *et al.*, 2014), as well as a few non-pathological Neandertal and Upper Paleolithic individuals (Trinkaus *et al.*, 1994; Churchill and Formicola, 1997; also see Chapter 7). However, non-pathological asymmetry in lower limb bone strength is always much lower (Auerbach and Ruff, 2006), due to the requirements of normal gait. It is obvious, then, that the gait and overall mechanical loading of the limbs in the Archer were severely affected by his condition. It is also obvious that this altered loading occurred for many years prior to his death, in order to produce such a large change. The pattern of change is most consistent with this condition existing from childhood. In other data not shown here, it can be seen that the asymmetry in Z_x and Z_p is due mainly to a larger total subperiosteal area on the right side (30–35% asymmetry) in combination with a smaller medullary cavity on that side (10–36% asymmetry). Mechanical unloading of diaphyses in adults has never been reported to lead to resorption of bone on the periosteal surface; thus, the small periosteal dimensions of the Archer's left lower limb bones must have occurred during growth. Also, following general growth models based on observations of human subjects (Ruff *et al.*, 1994; Kontulainen *et al.*, 2002), the periosteal surface of long bones is most responsive to altered mechanical loading during growth, while the endosteal surface is most responsive during adulthood. Thus, the periosteal expansion and medullary contraction of the Archer's right lower limb bones relative to his left side are consistent with increased loading of the right side during both childhood and adulthood – that is, a congenital condition existing throughout life. Therefore, despite the lack of other associated effects elsewhere in the skeleton corresponding to modern known syndromes, it is apparent that the Archer suffered from the effects of congenital absence of the left patella, a diagnosis that is also consistent with other morphological changes in his distal femur (McKinley, 2011b). The infection in that region may have increased the mechanical imbalance between his lower limbs close to his death (or even contributed to his death), but it cannot possibly explain the changes observed in his diaphyses.

Interestingly, despite the large diaphyseal strength asymmetry, there is no evidence for large asymmetries in either bone length or articular (femoral head) breadth in the Archer's lower limbs. Bilateral asymmetry in femoral maximum length is 2.4%, in tibial maximum length 0%, and in femoral head breadth 3.4%. The femoral asymmetries (larger on the normal right side) are on the high end, but within the range of normal modern human variation (Auerbach and

Ruff, 2006). As in many past studies, this demonstrates again the much greater developmental plasticity of long bone cortical dimensions than either bone length or articular dimensions (see Trinkaus *et al.*, 1994 for a review). It also re-emphasizes that the underlying cause of the Archer's lower limb asymmetry was mechanical in nature, and not a general growth disruption.

The strong humeri and slight left 'dominance' of the Amesbury Archer are also interesting. As noted by McKinley (2011b), use of the bow and arrow places large mechanical loads on both the dominant and non-dominant limbs. This may in fact be partly responsible for the decline in upper limb bone strength asymmetry observed among males in our study in the transition from the Upper Paleolithic to the Mesolithic, when bows became more commonly used (see Chapter 7). However, bilateral asymmetry in humeral strength is still generally present among males from later temporal periods, greatly favoring the right side (84% of post-Upper Paleolithic males). This is true of the two skeletons most closely associated with the Amesbury Archer – the 'Stonehenge Archer' (Z_p asymmetry of 8.3%) and the Boscombe Lower Camp individual (Z_p asymmetry of 15.1%), both stronger on the right side (the left humerus of the third individual, from A303, was not preserved well enough to measure cross-sectional properties). The particularly strong left humerus in the Amesbury Archer may be related to the need for additional bracing of the upper body due to reduced support of this kind in his lower limbs, that is, their asymmetric weight-bearing capacity. The very high anteversion angle of the left femoral neck (McKinley, 2011b) also suggests abnormal positioning of this limb during weight support, which again may have decreased stability. The Amesbury Archer also shows the condition bilaterally of *os acromialie* (non-fusion of the tip of the acromion process in the scapula) (McKinley, 2011b), which may be linked to activity-related use, including archery (Stirland, 1987). This also suggests very high mechanical loads on both of his shoulders, and is consistent with the humeral strength results.

As with the lower limb, asymmetries in length and articular breadths of the Archer's humeri do not parallel the slight left side dominance in strength: these linear measurements are slightly larger (<2%) on the right side (McKinley, 2011b, and our measurements). This is normal in humans (Auerbach and Ruff, 2006), but does not necessarily indicate right handedness, since these dimensions are not particularly developmentally plastic (see above and Chapter 7). However, it is possible that the Archer was right-handed, but that the increased loading of his left humerus during use of the bow and arrow (exacerbated by his condition) overwhelmed any additional loading of the right side during other activities (also see Chapter 7). This would be consistent with many experimental studies demonstrating that bone diaphyseal structure preferentially adapts to vigorous, that is, high-magnitude loadings (Ruff *et al.*, 2006). It also provides indirect support for the general approach of this study, whereby changes in bone strength are assumed to be linked to vigorous use of the limbs in specific ways, that is, major behavioral differences. It may also indicate that less vigorous behaviors will not be clearly 'recorded' in bone structure, which also has implications for interpreting temporal and other trends, or the lack thereof (see Chapter 5 and other chapters in this volume).

Considering his physical disability, it is certainly remarkable that the Amesbury Archer traveled as far as he did – from Central Europe to Great Britain at some point in his life – and that he achieved the apparently high social status that he did. He also lived to a relatively advanced age – his age is given as 35–45 years in Fitzpatrick (2011) (without details regarding the aging methods applied), while our estimate, based on pubic symphyseal and auricular surface morphology, is 50+ years. His very strong upper limbs and right lower limb indicate that he was still active until close to death, perhaps indicating a link between physical prowess and social status in this society.

8.4 Discussion

8.4.1 Body Size and Shape

Body sizes for the samples included in this study appear to be broadly representative of those determined for much larger British samples, when account is taken of methodological differences (Farwell and Molleson, 1993; Molleson and Cox, 1993; Roberts and Cox, 2003). Temporal trends in body size generally parallel those observed in Europe as a whole, with the exception of the large increase in stature and body mass in the Bronze Age. However, part of this apparent difference is likely attributable to the composition of the British Bronze Age sample. With the exception of the Cliff's End sample, all of the British BA samples were associated with Beakers and can be considered, both culturally and temporally, as early Bronze Age or Eneolithic ('Copper Age'). As noted in Chapter 4, Eneolithic populations tend to be tall compared to earlier Neolithics (see Chapter 4: Fig. 4.12). If only Eneolithic north-central continental European populations are compared to the early British BA samples, differences in stature, in both sexes, become non-significant (continental sample males, 171.3 ± 0.8 cm, females, 158.0 ± 0.9 cm, both slightly lower but within 2% of the British means). However, differences in body mass remain significant, with British samples averaging 8-9% larger in both sexes (continental sample males, 67.5 ± 1.0 kg, females, 56.4 ± 0.8 kg). This is also concordant with data presented in Chapter 4, in which Eneolithic samples did not differ as much in body mass as in stature compared to earlier Neolithics, especially middle Neolithics (Chapter 4; Fig. 4.13(a)). This implies a relatively narrower body among continental Eneolithics, but this was not the case among British early BA samples, who had the largest bi-iliac breadths and among the largest bi-iliac/stature ratios of all of the British samples (see Tables 8.5 and 8.6). Thus, they were both tall and heavy, and in the latter characteristic differed from their continental counterparts.

One possible explanation for the increase in stature among Eneolithic or early Bronze Age populations may be nutritional – that is, the increased consumption of dairy products associated with the Secondary Products Revolution during the late Neolithic (Greenfield, 2010; also see Chapter 4). Health indicators do not improve in the transition from the Neolithic to Bronze Age among British samples (Roberts and Cox, 2003, 2007), so a general decline in disease or metabolic 'stress' is not a likely contributor. It is also possible that population movements, namely immigration, contributed to the increase in body size in British early BA samples. The Bell Beaker culture was associated with major changes in the genetic composition of some Western Europe populations (Brandt *et al.*, 2013), although the extent to which population movement versus cultural diffusion was responsible for spread of the culture to all regions is debated (Milisauskas and Kruk, 2002). It is apparent that some early BA individuals in the Stonehenge area, including the Amesbury Archer and several other specimens not in our sample, were immigrants from continental Europe, although other individuals in our sample (Stonehenge Archer, Boscombe Down, Normanton Down (A303)) were local in origin (Evans *et al.*, 2006; Chenery and Evans, 2011). Of course, growing up locally does not guarantee that some recent ancestors did not migrate to Britain. There is evidence for a substantial genetic contribution of larger-bodied Asian steppe populations (i.e., Yamnaya) in Western continental Europe during the Late Neolithic (Haak *et al.*, 2015; Mathieson *et al.*, 2015). Whether the increase in body size in Britain was the result of *in situ* environmental (e.g., nutritional) changes and/or immigration cannot be determined from our data, but in any event the British early BA sample appears to fall comfortably within the range of other European samples of the same time period in terms of stature.

However, the wide body and greater body mass of the British early BA sample suggest that additional factors may also have been operative. One possibility is that the present sample is biased towards larger individuals through a focus on burials with richer grave goods, if larger individuals are indeed associated with higher status. All sufficiently well-preserved early BA individuals recovered in the Wiltshire area around Stonehenge and Amesbury were included in the study (as well as three from Staxton, Yorkshire East Riding, who were of similar body size to the Wiltshire specimens); all were associated with some cultural remains (e.g., a beaker) but most were not particularly elaborately furnished. One would also have to explain why British early BA females were also large-bodied, since ostensibly body size and status differences would not be as strongly correlated. Better preservation of larger individuals may also be a factor, although among adults this seems less likely as an explanation. Future inclusion of additional individuals not sufficiently preserved for the present analyses of mechanical properties, but with other measurable features (such as the femoral head), could help to better establish possible temporal and geographic variation in body size among British BA populations and their underlying causes.

The large body size of the British BA samples stands in contrast to both immediately earlier (Neolithic) and later (Iron/Roman Age) British samples, which tend to be somewhat smaller than other contemporary northern Europeans. This suggests, possibly, that nutrition and living conditions were less optimal in Britain during these time periods. The specific samples included in the study may have influenced these results to some degree, although it appears to be a fairly consistent pattern through the several sites/samples included for Britain. The effects of a more 'urban' lifestyle on the Poundbury sample appear to have depressed body size in females, but not males (also see below). Because only one continental European Iron/Roman age sample (Quadrella, Italy) could be considered 'urban,' while the Poundbury sample made up a substantial portion of the British Iron/Roman total sample (see Table 8.1), this may have influenced results for Iron/Roman females, though not males, who showed the same pattern.

The decline among British samples in stature during the Early Modern period, followed by a recovery during the 20th century, is typical for European populations generally (see Chapter 4), and with historic data for the UK (Kuh *et al.*, 1991; Komlos, 1993; Komlos and Kuchenhoff, 2012). Although anthropometric data before the 20th century are not available for body mass, the results shown here indicate that the same general temporal patterns pertained as for stature. The decline has been attributed to deleterious effects of the Industrial Revolution and demographic trends on health in the majority of the population, particularly during the late 18th and early 19th centuries (Komlos, 1993; Komlos and Kuchenhoff, 2012). The progressive increase beginning in the late 19th or early 20th century is characteristic of northern European populations (see Chapter 4), and probably reflects more rapid improvements in health status, including childhood diet and disease prevalence (Tanner, 1992). Even so, average body size did not surpass that of the British BA until the late 20th century. Interestingly, SD in body size is also most similar (i.e., largest) between the 20th century samples and the BA sample. If SD in body size does reflect in part greater environmental sensitivity of males during the growth period – that is, 'female buffering' (Stini, 1976; Gray and Wolfe, 1980) – then this could be another indication of relatively good environmental conditions for growth in both the BA and very recent periods in Britain.

Except for the decline in brachial index since the Neolithic, as part of a pan-European trend (Chapter 4), no consistent or particularly marked temporal trends in body shape are apparent among the British samples since the Neolithic. The crural index is somewhat high in the Neolithic sample as well as in our two pre-Neolithic specimens, also similar to Europeans as a whole. The temporal changes in limb proportions may reflect a combination of environmental and genetic (population history) effects (Chapter 4). The significance of the generally lower crural and brachial indices in British versus other northern continental samples is not clear, unless it again reflects slightly less advantageous living conditions in Britain during most periods (see above and Chapter 4).

8.4.2 Bone Structure

The most apparent temporal trend in bone strength among the British samples is a decline in size-standardized bending strength of the tibia after the Neolithic. Among males, bending strength then stabilizes, while in females it continues to decline until the Early Modern period. The femur also shows some evidence of a decline in bending strength after the Neolithic, but strength values then vacillate, particularly among males. In part these temporal trends may reflect a reduction in mobility after the Neolithic. A number of authors have argued that, despite the adoption of food production, significant residential and logistical mobility continued to characterize early Neolithic populations within Europe as a whole (Whittle, 1996: 160, 365; Milisauskas, 2002: 159), and Britain specifically (Roberts and Cox, 2003: 55; Smith and Brickley, 2009: 114). The intensification of agriculture, the development of larger settlements, and innovations such as the introduction of wheeled vehicles and the domesticated horse, led to reduced (human-powered) mobility in later periods. Why British females continued to decline through the Early Modern period while males did not is unclear. This implies increasing SD in use of the lower limbs through time, possibly due to more sedentary lifestyles among females in later periods, and/or other activities in males that increased lower limb loadings in the later periods. Interestingly in this regard, in the one possible comparison of more urban and rural samples within the same temporal period, town-living (i.e., in Poundbury) had more of an effect on females in terms of both body size and relative bone strength (upper as well as lower), suggesting both greater nutritional/health as well as mechanical effects on females. It is quite possible, of course, that many of the individuals buried at Poundbury actually worked on surrounding farms (Farwell and Molleson, 1993: 239). If so, this may have involved males more than females, if the work included plowing of fields, and so forth. A similar pattern might explain sex-related differences in the Late Medieval samples, although it is difficult to invoke such an explanation for the completely urban sample from Spitalfields.

Right (but not left) humeral strength also tends to decline among females through time, also possibly reflecting a progressive decrease in workload on the dominant arm. Because of sampling issues (many fewer individuals with preserved bones from both arms in earlier time periods), this is not reflected in temporal reductions in bilateral asymmetry, which remains relatively constant in both sexes. The moderately high asymmetry in British BA females is at variance with general pan-European trends (Chapter 7), in which Copper and Bronze Age females have very low asymmetry, probably due to the use of two-handed saddle querns for food grinding. However, the higher distribution of British BA females is due mainly to very high values recorded for two (of the five) specimens (22% and 28%), with the other three specimens recording low values (5–7%). One of the high values is from a late Bronze Age site (Cliff's End, 920–1190 BC), close to the beginning of the Iron Age in Britain (700 BC; Wells, 2002); however, this is still prior to the first appearance of one-handed rotary querns in Britain (Wefers, 2011). The variation in asymmetry among British BA females may simply reflect variation in food-processing techniques at this time, or even differential responsibility for such activities within the general population.

Increased humeral strength in the Late Medieval period, in both sexes but particularly in males, parallels trends observed throughout Europe (Chapter 5), and may reflect a mechanically demanding lifestyle during this time period. Neither of the two available British samples has detailed contextual information at present (see Appendix 1), but they are both from church cemeteries on the outskirts of towns, and it is quite possible that many of the individuals worked outside the towns, that is, in fields, or in otherwise physically demanding jobs (little evidence for any status differences was noted in the available archaeological reports). In fact, high rates of degenerative joint disease were described for the York sample, suggesting that "... both men and women ... were very physically active from a young age" (Anon., 2010: 34). The subsequent decline in strength in the Early Modern Spitalfields sample from London is consistent with their generally less rigorous occupations (Molleson and Cox, 1993). Many of the men (including most in the present sample) were weavers, dyers, or merchants, and their wives would also have been relatively shielded from hard physical labor (occupations were recorded for most of the males in the sample). There were several individuals with more physically demanding recorded occupations, but unfortunately too few to carry out any statistical comparisons.

In contrast to the generally modest and variable direction of temporal change in bone strengths after the Bronze Age (except tibial strength in females), %CA shows consistent declines in all bones and both sexes (except the left humerus in females) through the entire temporal sequence. The declines between the Neolithic and Bronze Age, and in the tibia and right humerus in females, could be explained in part by declines in mechanical loading, since higher mechanical loadings lead to endosteal deposition of bone, that is, relatively thicker cortices, among adults (Ruff *et al.*, 1994). However, this does not apply to post-BA males or to the femur in either sex. There is evidence that relative cortical thickness declines with poor nutrition due to insufficient endosteal bone apposition during growth (Garn *et al.*, 1969). The effects may also be observed in archaeological samples (Ruff, 1999). Age at death among adults also affects %CA of long bones, since endosteal resorption of bone is a normal feature of adult aging (Garn, 1970; Ruff and Hayes, 1983). However, there are no significant differences in age among our samples ($p > 0.05$, ANOVA and all pairwise comparisons between temporal periods).

Thus, the continuing temporal decline in %CA in most bones may reflect gradual deterioration of living conditions following the Bronze Age. This would be concordant with the observed temporal declines in stature and body mass over the same time period. It is also consistent with physiological 'stress indicators' in British archaeological samples, which show a general increase through time since the Neolithic (Roberts and Cox, 2003, 2007). These observations also serve to emphasize the multifactorial nature of bone tissue, which serves a number of physiological functions, both mechanical and non-mechanical (Ruff *et al.*, 2006). Distinguishing between such factors is an important component of bone structural analysis and past behavioral reconstruction.

8.5 Conclusions

Among British samples, body size increases from the early Neolithic to the Eneolithic/early Bronze Age, and then declines through the Early Modern period. These trends may track changes in nutrition and living conditions, which may have improved with the Secondary Products Revolution in the late Neolithic and then gradually deteriorated. This interpretation receives support from trends in %CA, which generally declines throughout the entire temporal sequence. Relative bone strength of the tibia declines from the Neolithic to the Bronze Age in both sexes, probably reflecting substantial degrees of mobility in early Neolithic populations and a subsequent decline with agricultural intensification. Bone strength then stabilizes in males but continues to decline in females, possibly related to sex-related differences in occupational tasks. Intensification of physical labor may explain an increase in humeral strength during the Late Medieval period, particularly in males. An analysis of the well-known Eneolithic/early Bronze age 'Amesbury Archer' demonstrates the extent of plasticity possible in diaphyseal cortical bone under conditions of altered mechanical loading, in his case very likely due to a congenital absence of the left patella.

Acknowledgments

We thank Trang Diem Vu and Heather Garvin for help in data processing, Andrew Chamberlain, Rob Kruszynski, Jay Stock, Mercedes Okumura, Jane Ellis-Schön, Jacqueline McKinley, Lisa Webb, Jillian Greenaway, Alison Brookes, Jo Buckberry, and Chris Knüsel for providing access to collections and facilitating data acquisition, and Erik Trinkaus and Steven Churchill for sharing data for the Paviland and Gough's Cave specimens.

References

Anon. (2010) Excavating All Saints – a Medieval church rediscovered. *Curr. Archaeol.*, **245**, 30–37.

Auerbach, B.M. and Ruff, C.B. (2006) Limb bone bilateral asymmetry: variability and commonality among modern humans. *J. Hum. Evol.*, **50**, 203–218.

Brandt, G., Haak, W., Adler, C.J., Roth, C., Szecsenyi-Nagy, A., Karimnia, S., Moller-Rieker, S., Meller, H., Ganslmeier, R., Friederich, S., Dresely, V., Nicklisch, N., Pickrell, J.K., Sirocko, F., Reich, D., Cooper, A., and Alt, K.W. (2013) Ancient DNA reveals key stages in the formation of central European mitochondrial genetic diversity. *Science*, **342**, 257–261.

Chenery, C.A. and Evans, J.A. (2011) A summary of the strontium and oxygen isotope evidence for the origins of Bell Beaker individuals found near Stonehenge. In: *The Amesbury Archer and the Boscombe Bowmen* (ed. A.P. Fitzpatrick), Wessex Archaeology Ltd (Wessex Archaeology Report 27), Salisbury, UK, pp. 185–190.

Churchill, S.E. and Formicola, V. (1997) A case of marked bilateral asymmetry in the upper limbs of an Upper Palaeolithic male from Barma Grande (Liguria), Italy. *Int. J. Osteoarchaeol.*, 7, 18–38.

Coon, C.S. (1939) *The Races of Europe*. MacMillan, New York.

Evans, J.A., Chenery, C.A., and Fitzpatrick, A.P. (2006) Bronze age childhood migration of individuals near Stonehenge, revealed by strontium and oxygen isotope tooth enamel analysis. *Archaeometry*, **48**, 309–321.

Farwell, D.E. and Molleson, T.I. (1993) *Excavations at Poundbury, Dorset 1966–1982. Volume II: The Cemeteries*. Dorset Natural History and Archaeological Society Monograph Series No. 7.

Fitzpatrick, A.P. (2002) 'The Amesbury Archer': A well-furnished Early Bronze Age burial in southern England. *Antiquity*, **76**, 629–630.

Fitzpatrick, A.P. (2011) *The Amesbury Archer and Boscombe Bowmen*. Bell Beaker Burials on Boscombe Down, Amesbury, Wiltshire. Wessex Archaeology Ltd (Wessex Archaeology Report 27), Salisbury, UK.

Garn, S.M. (1970) *The Earlier Gain and the Later Loss of Cortical Bone*. Charles C. Thomas, Springfield.

Garn, S.M., Guzman, M.A., and Wagner, B. (1969) Subperiosteal gain and endosteal loss in protein-calorie malnutrition. *Am. J. Phys. Anthropol.*, **30**, 153–155.

Gray, J.P. and Wolfe, L.D. (1980) Height and sexual dimorphism of stature among human societies. *Am. J. Phys. Anthropol.*, **53**, 441–456.

Greenfield, H.J. (2010) The Secondary Products Revolution: the past, the present and the future. *World Archaeol.*, **42**, 29–54.

Gunnell, D., Rogers, J., and Dieppe, P. (2001) Height and health: predicting longevity from bone length in archaeological remains. *J. Epidemiol. Community Health*, **55**, 505–507.

Haak, W., Lazaridis, I., Patterson, N., Rohland, N., Mallick, S., Llamas, B., Brandt, G., Nordenfelt, S., Harney, E., Stewardson, K., Fu, Q., Mittnik, A., Banffy, E., Economou, C., Francken, M., Friederich, S., Pena, R.G., Hallgren, F., Khartanovich, V., Khokhlov, A., Kunst, M., Kuznetsov, P., Meller, H., Mochalov, O., Moiseyev, V., Nicklisch, N., Pichler, S.L., Risch, R., Rojo Guerra, M.A., Roth, C., Szecsenyi-Nagy, A., Wahl, J., Meyer, M., Krause, J., Brown, D., Anthony, D., Cooper, A., Alt, K.W. and Reich, D. (2015) Massive migration from the steppe was a source for Indo-European languages in Europe. *Nature* **522**, 207–211.

Health Survey for England (2013) Available at: http://www.hscic.gov.uk/catalogue/PUB13219.

Jantz, R.L., Hunt, D.R., and Meadows, L. (1994) Maximum length of the tibia: how did Trotter measure it? *Am. J. Phys. Anthropol.*, **93**, 525–528.

Joint Clothing Council (1957) *Women's Measurements and Sizes*. Her Majesty's Stationery Office, London (cited in Eveleth and Tanner, 1976, Appendix Table 5b).

Kemsley, W.F.F. (1950) Weight and height of a population in 1943. *Ann. Eugenics*, **15**, 161–183.

Khosla, T. and Lowe, C.R. (1968) Height and weight of British men. *Lancet* **1**, 742.

Komlos, J. (1993) The secular trend in the biological standard of living in the United Kingdom, 1730–1860. *Econ. Hist. Rev.*, **46**, 115–144.

Komlos, J. and Kuchenhoff, H. (2012) The diminution of the physical stature of the English male population in the eighteenth century. *Cliometrica*, **6**, 45–62.

Kontulainen, S., Sievanen, H., Kannus, P., Pasanen, M., and Vuori, I. (2002) Effect of long-term impact-loading on mass, size, and estimated strength of humerus and radius of female racquet-sports players: a peripheral quantitative computed tomography study between young and old starters and controls. *J. Bone Miner. Res.*, **17**, 2281–2289.

Kuh, D.L., Power, C., and Rodgers, B. (1991) Secular trends in social class and sex differences in adult height. *Int. J. Epidemiol.*, **20**, 1001–1009.

Leslie, S., Winney, B., Hellenthal, G., Davison, D., Boumertit, A., Day, T., Hutnik, K., Royrvik, E.C., Cunliffe, B., Wellcome Trust Case Control C, International Multiple Sclerosis Genetics C, Lawson, D.J., Falush, D., Freeman, C., Pirinen, M., Myers, S., Robinson, M., Donnelly, P., and Bodmer, W. (2015) The fine-scale genetic structure of the British population. *Nature*, **519**, 309–314.

Mathieson, I., Lazaridis, I., Rohland, N., Mallick, S., Patterson, N., Roodenberg, S.A., Harney, E., Stewardson, K., Fernandes, D., Novak, M., Sirak, K., Gamba, C., Jones, E.R., Llamas, B., Dryomov, S., Pickrell, J., Arsuaga, J.L., de Castro, J.M., Carbonell, E., Gerritsen, F., Khokhlov, A., Kuznetsov, P., Lozano, M., Meller, H., Mochalov, O., Moiseyev, V., Guerra, M.A., Roodenberg, J., Verges, J.M., Krause, J., Cooper, A., Alt, K.W., Brown, D., Anthony, D., Lalueza-Fox, C., Haak, W., Pinhasi, R., and Reich, D. (2015) Genome-wide patterns of selection in 230 ancient Eurasians. *Nature*, **528**, 499–503.

McKinley, J.I. (2011a) Grave 25000: The Boscombe Bowmen. Human Remains. In: *The Amesbury Archer and the Boscombe Bowmen* (ed. A.P. Fitzpatrick), Wessex Archaeology Ltd (Wessex Archaeology Report 27), Salisbury, UK, pp. 18–32.

McKinley, J.I. (2011b) Graves 1236 and 1289: The Amesbury Archer and 'companion'. Human Remains. In: *The Amesbury Archer and the Boscombe Bowmen* (ed. A.P. Fitzpatrick), Wessex Archaeology Ltd (Wessex Archaeology Report 27), Salisbury, UK, pp. 77–87.

Milisauskas, S. (2002) Early Neolithic, The first farmers in Europe. In: *European Prehistory: A Survey* (ed. S. Milisauskas), Kluwer Academic/Plenum, New York, pp. 143–192.

Milisauskas, S. and Kruk, J. (2002) Late Neolithic, Crises, Collapse, New Ideologies, and Economies, 3500/3000–2200/2000 BC. In: *European Prehistory: A Survey* (ed. S. Milisauskas), Kluwer Academic/Plenum, New York, pp. 247–269.

Molleson, T. and Cox, M. (1993) *The Spitalfields Project, Vol. 2*. CBA Research report 86. Council for British Archaeology, York.

Montegriffo, V.M.E. (1968) Height and weight of a United Kingdom adult population with a review of anthropometric literature. *Ann. Hum. Genet.*, **31**, 389.

Roberts, C. and Cox, M. (2003) *Health and Disease in Britain: From Prehistory to the Present Day*. Sutton Publishing, Stroud, UK.

Roberts, C. and Cox, M. (2007) The impact of economic intensification and social complexity on human health in Britain from 6000 BP (Neolithic) and the introduction of farming to the mid-nineteenth century AD. In: *Ancient Health: Skeletal Indicators of Agricultural and Economic Intensification* (eds M.N. Cohen and G.M.M. Crane-Kramer), University Press of Florida, Gainesville, pp. 149–163.

Rosenbaum, S., Skinner, R.K., Knight, I.B., and Garrow, J.S. (1985) A survey of heights and weights of adults in Great Britain, 1980. *Ann. Hum. Biol.*, **12**, 115–127.

Ruff, C.B. (1999) Skeletal structure and behavioral patterns of prehistoric Great Basin populations. In: *Prehistoric Lifeways in the Great Basin Wetlands: Bioarchaeological Reconstruction and Interpretation* (eds B.E. Hemphill and C.S. Larsen), University of Utah Press, Salt Lake City, pp. 290–320.

Ruff, C.B. (2002) Variation in human body size and shape. *Annu. Rev. Anthropol.*, **31**, 211–232.

Ruff, C.B. and Hayes, W.C. (1983) Cross-sectional geometry of Pecos Pueblo femora and tibiae – a biomechanical investigation. II. Sex, age, and side differences. *Am. J. Phys. Anthropol.*, **60**, 383–400.

Ruff, C.B., Holt, B.H., and Trinkaus, E. (2006) Who's afraid of the big bad Wolff? Wolff's Law and bone functional adaptation. *Am. J. Phys. Anthropol.*, **129**, 484–498.

Ruff, C.B., Holt, B.M., Niskanen, M., Sladek, V., Berner, M., Garofalo, E., Garvin, H.M., Hora, M., Maijanen, H., Niinimaki, S., Salo, K., Schuplerova, E., and Tompkins, D. (2012) Stature and body mass estimation from skeletal remains in the European Holocene. *Am. J. Phys. Anthropol.*, **148**, 601–617.

Ruff, C.B., Walker, A., and Trinkaus, E. (1994) Postcranial robusticity in *Homo*, III: Ontogeny. *Am. J. Phys. Anthropol.*, **93**, 35–54.

Smith, M. and Brickley, M. (2009) *People of the Long Barrows. Life, Death and Burial in the Earlier Neolithic*. The History Press, Stroud, UK.

Steckel, R.H. (1995) Stature and the standard of living. *J. Econ. Lit.*, **33**, 1903–1940.

Stini, W.A. (1976) Adaptive strategies of human populations under nutritional stress. In: *Biosocial Interrelations in Population Adaptation* (eds E.S. Watts, F.E. Johnston, and G.W. Lasker), Moutan, The Hague, pp. 19–40.

Stinson, S. (1985) Sex differences in environmental sensitivity during growth and development. *Yrbk Phys. Anthropol.*, **28**, 123–147.

Stirland, A. (1987) A possible correlation between Os acromialie and occupation in the burials from the Mary Rose. In: *Proceedings of the Fifth European meeting of the Palaeopathology Association, 1984* (eds V. Capecchi and E.R. Massa), Tipografia Senese, Siena, pp. 327–334.

Tanner, J.M. (1992) Growth as a measure of the nutritional and hygienic status of a population. *Horm. Res.*, **38** (Suppl. 1), 106–115.

Thompson, D., Barden, J.D., Kirk, N.S., Mitchelson, D.L. and Ward, J.S. (n.d.) *Anthropometry of British Women*. Institute for Consumer Ergonomics Ltd, Leicestershire (cited in Eveleth and Tanner, 1976, Appendix Table 5b).

Trinkaus, E., Churchill, S.E., and Ruff, C.B. (1994) Postcranial robusticity in *Homo*, II: Humeral bilateral asymmetry and bone plasticity. *Am. J. Phys. Anthropol.*, **93**, 1–34.

Trotter, M. and Gleser, G.C. (1952) Estimation of stature from long bones of American whites and Negroes. *Am. J. Phys. Anthropol.*, **10**, 463–514.
Trotter, M. and Gleser, G.C. (1958) A re-evaluation of estimation of stature based on measurements of stature taken during life and of long bones after death. *Am. J. Phys. Anthropol.*, **16**, 79–123.
Warden, S.J., Mantila Roosa, S.M., Kersh, M.E., Hurd, A.L., Fleisig, G.S., Pandy, M.G., and Fuchs, R.K. (2014) Physical activity when young provides lifelong benefits to cortical bone size and strength in men. *Proc. Natl Acad. Sci. USA*, **111**, 5337–5342.
Wefers, S. (2011) Still using your saddle quern? A compilation of the oldest known rotary quern in western Europe. In: *Bread for the People: the Archaeology of Mills and Milling* (eds D. Williams and D. Peacock), Archaeopress, Oxford, pp. 67–76.
Wells, P. (2002) The Iron Age. In: *European Prehistory: A Survey* (ed. S. Milisauskas), Kluwer Academic/Plenum, New York, pp. 335–383.
Whittle, A. (1996) *Europe in the Neolithic; The Creation of New Worlds.* Cambridge University Press, Cambridge.

9

France and Italy

Brigitte Holt, Erin Whittey, and Dannielle Tompkins

Department of Anthropology, University of Massachusetts, Amherst, MA, USA

9.1 Introduction

This chapter presents analyses of skeletal samples from France and Italy. The majority of French sites are distributed across an east–west central band and southern France, with the exception of Loschbour 1 (Luxembourg). Italian sites range relatively evenly from northernmost regions (alpine regions and Ligurian coast) to Sicily, Sardinia, and Puglia (Fig. 9.1 and Table 9.1). For the purpose of this chapter, Modern and Post-Medieval periods were pooled into Modern, Early and Late Medieval into Medieval, and Early and Late Upper Paleolithic into Upper Paleolithic (UP), resulting in seven cultural periods (Table 9.1). A total sample of 341 individuals is relatively well represented across periods (Table 9.1). The relatively even distribution of rural and urban sites in Modern, Medieval and Iron Age periods (Table 9.2) allowed evaluation of urbanization during those periods. The impact of terrain was also assessed, given the substantial variability in topography across the entire region. A 'pan-European' comparative sample comprises data from other regions of Europe (see Chapter 1, Fig. 1.1), exclusive of France and Italy.

9.2 Samples

All skeletal material comprises adults only, with fused epiphyses. The Upper Paleolithic and Mesolithic samples include material from France and Italy, as well as the skeleton from Loschbour (Luxemburg), and range from about cal. 33,000 to 6000 Before Present (BP).

The Neolithic group comprises a Middle Neolithic sample from the Auvergne region in France (Pontcharaud, 6150 cal. BP) and a sample of Eneolithic agro-pastoralists from the Marche region in Italy (Fontenoce Recatini, 5438 cal. BP). The Bronze Age sample comprises only the late Bronze age site from Olmo di Nogara (Veneto, Italy, 3400 cal. BP). The inclusion of swords with male burials and ornaments with female burials suggests the Olmo population belonged to a relatively stratified society (Salzani, 2005).

The Iron Age sample comprises material from a rural Gallo-Roman necropolis from Southwest France (Rue Jacques Brel, 1825 cal. BP), and two urban Roman Imperial sites of Lucus Feroniae (Latium, Italy, 1850 cal. BP) and Quadrella (Molise, Italy, 1800 cal. BP).

Skeletal Variation and Adaptation in Europeans: Upper Paleolithic to the Twentieth Century, First Edition. Edited by Christopher B. Ruff.

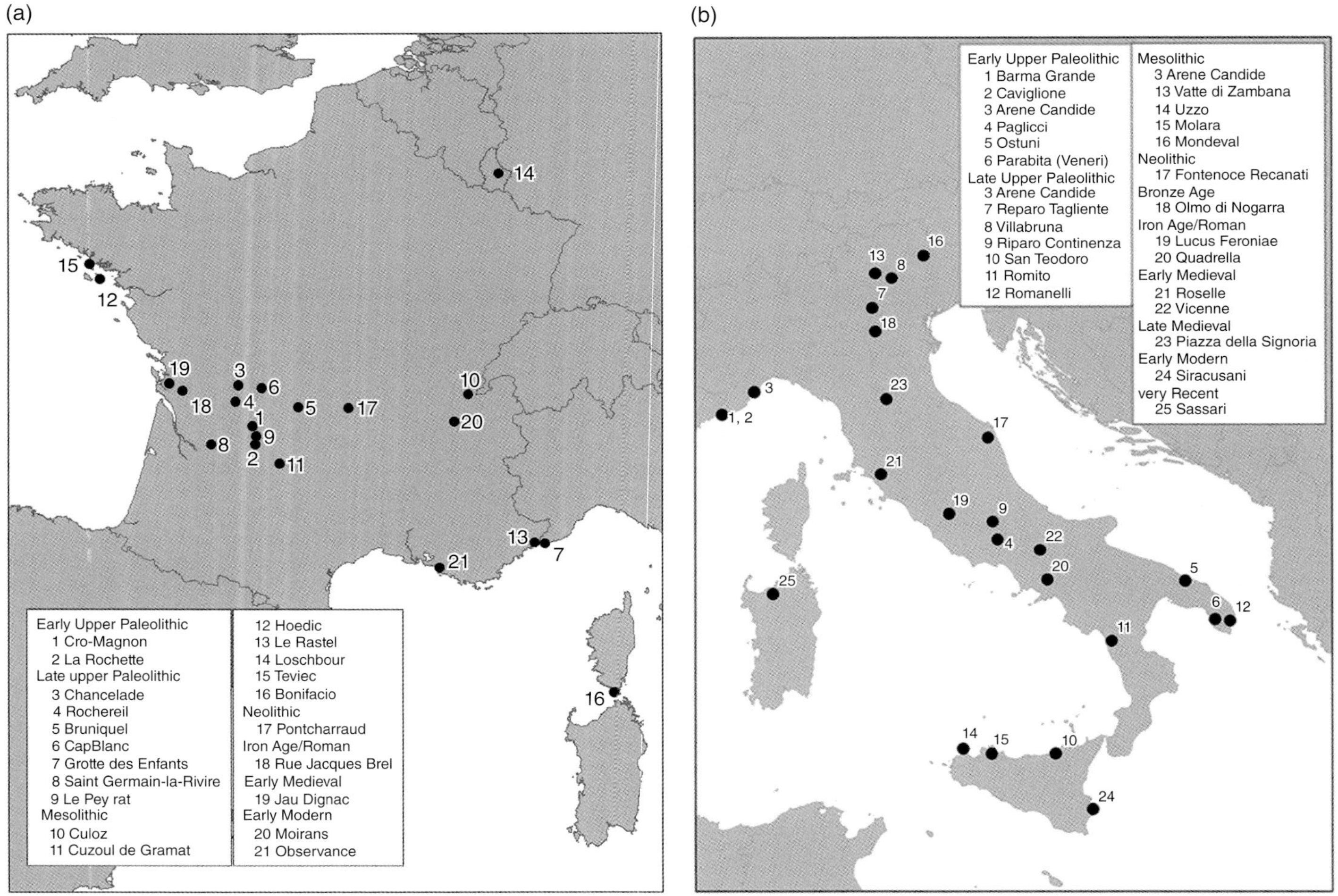

Figure 9.1 Maps of France (a) and Italy (b) with location of sites used in analysis.

Table 9.1 Dates and sample sizes for each time period.

Time period	Date range*
Modern	1600 AD to ≥1900 AD
Medieval	500–1599 AD
Iron Age/Roman	2250–1650 BC
Bronze Age	4350–2950 BC
Neolithic	7000–4050 BC
Mesolithic	10,500–5880 BC
Upper Paleolithic	33,388–11,367 BP

*Dates indicated as BP are calibrated.

Table 9.2 Sample by time period and urban status.

Time period	Modern	Medieval	Iron Age/ Roman
Rural	25	57	19
Urban	43	14	45
Total	68	71	64

The lack of grave goods suggests that Lucus Feroniae people may have been freed slaves and retired soldiers (Manzi *et al.*, 1997). Burial context and tombstone inscriptions at Quadrella indicate these may have been freed slaves as well (Belcastro *et al.*, 2007). The Medieval group comprises two rural Early Medieval samples from coastal sites in southwest France (Jau Dignac, 1350 cal. BP) and central Italy (Roselle, 1250 cal. BP), a rural Early Medieval site from northern Italy (Vicenne Campochiaro, 1300 cal. BP), and an urban Late Medieval Italian sample from Florence (Piazza della Signoria, 650 cal. BP). The association of rich grave goods with most of the burials suggests the individuals buried at Jau Dignac belonged to the aristocracy (Cartron and Castex, 2006). Males at the Longobardo site of Vicenne are buried with horses, harnesses and weapons, and biological distance and archeological evidence suggests the graves may comprise individuals of multi-ethnic origins (Ceglia and Genito, 1991; Belcastro and Facchini, 2001).

The Modern sample comprises one rural (Moirans, Isère, 250 cal. BP) and one urban (L'Observance, Marseille, 250 cal. BP) site from France, and one rural (Sassari, 100 cal. BP) and one urban (Siracusani, 150 cal. BP) site from Italy. L'Observance is a plague cemetery dating to the 1722 Marseille plague (Signoli *et al.*, 1997), while the individuals from Moirans were buried in the cemetery associated with the rural church of Saint-Pierre de Moirans (Isère) (Diverrez *et al.*, 2012). The Siracuni sample comprises peasants, laborers, carpenters, and housewives of known sex, age, and occupation from a dismantled graveyard from the Siracusa district in Sicily (Monica Zavataro, personal communication), while the Sassari sample includes individuals of known sex, age, and occupation from rural Sardinia (Belcastro *et al.*, 2008). As with Siracusani, the Sassari sample is composed primarily of farm laborers and housewives from low socioeconomic environments (Belcastro *et al.*, 2008).

9.3 Methods

9.3.1 Body Shape

Measures of body size and shape include stature, body mass, body mass index (BMI), relative shoulder (summed clavicular length/stature) and hip breadth (living bi-iliac breadth/stature), and brachial and crural indices (see Chapter 2 for details). Stature for about 60% of individuals was reconstructed via the anatomical method (modified Fully method; see Chapter 2), and body mass using the cylindrical approach (stature/bi-iliac breadth; see Chapter 2) in around 75% of cases. Otherwise, stature and body mass were estimated using previously described methods utilizing long bone lengths and femoral head breadth, respectively (Ruff *et al.*, 2012).

9.3.2 Robusticity

Bone robusticity parameters for femur, tibia, and right and left humeri include torsional or average bending strength (Z_p), cross-sectional shape ratio of AP (Z_x) and ML (Z_y) bending strength, and percentage cortical area (cortical area/total area). All measures were standardized for body size (see Chapter 2).

9.3.3 Statistical Analysis

Temporal changes within the French-Italian (FI) sample were evaluated using one-way ANOVA with post-hoc pair-wise Tukey comparisons between periods ($p < 0.05$). *T*-Tests with Bonferroni correction for multiple comparisons (alpha = 0.007) were used to evaluate differences between sexes within periods. Comparisons of the FI group against the pan-Europeans within period were carried out through *t*-tests, again with Bonferroni correction. Upper limb asymmetry and the impact of urbanization (for post-Bronze Age samples) were assessed using paired *t*-tests within sex ($p < 0.05$). Finally, lower limb robusticity patterns were compared by terrain level (Table 9.3; see Chapter 5) through one-way ANOVA within sex with post-hoc pair-wise Tukey comparisons ($p < 0.05$).

Table 9.3 Sample by region and terrain.

Time period	Flat	Hilly	Mountainous
France	54	68	18
Italy	59	69	75
Total	113	137	93

9.4 Body Size and Shape

9.4.1 Temporal Changes Within the FI group

Male and female stature (Table 9.4; Fig. 9.2(a)) decreases between UP and Mesolithic for males, and again into the Neolithic for females. Bronze Age females are shorter than their UP predecessors. Following a slight increase into Neolithic and Bronze Age, male stature fluctuates slightly until the Medieval period, and decreases in Modern. Female stature changes little after

Table 9.4 Body size and shape data, by period and sex.

		Males						
		Upper Paleolithic	Mesolithic	Neolithic	Bronze Age	Iron Age/ Roman Age	Medieval	Modern
Stature	N	18	20	20	17	28	37	46
	Mean	168.5	161.6	162.2	165.3	162.1	165.1	160.6
	SD	9.1	6.8	7.1	5.8	5.3	7.1	6.0
Body mass	N	18	14	17	17	24	33	45
	Mean	69.8	63.3	60.0	67.8	61.2	66.3	62.7
	SD	6.8	6.3	8.1	7.8	9.6	7.1	7.4
Body mass index	N	18	14	17	17	24	33	45
	Mean	24.6	23.8	22.6	24.8	23.3	24.0	24.2
	SD	1.8	1.1	2.5	2.1	3.2	2.3	2.2
Relative shoulder breadth	N	11	5	11	12	16	26	39
	Mean	178.0	173.2	181.7	177.4	184.9	174.3	183.8
	SD	8.5	6.5	7.8	7.8	13.5	9.3	8.7
Relative hip breadth	N	10	11	12	13	5	14	28
	Mean	192.7	188.4	186.8	194.8	195.4	191.3	188.8
	SD	10.0	6.2	9.5	8.6	8.9	12.1	9.2
Brachial index	N	–	–	12	16	5	10	28
	Mean	–	–	78.8	77.1	73.9	74.9	75.4
	SD	–	–	2.1	1.9	3.0	2.3	2.4
Crural index	N	16	20	19	17	22	29	46
	Mean	83.7	84.9	82.9	81.6	80.5	81.0	80.3
	SD	2.6	3.5	1.8	2.7	3.2	2.5	3.3

(*Continued*)

Table 9.4 (Continued)

		Females						
		Upper Paleolithic	Mesolithic	Neolithic	Bronze Age	Iron Age/ Roman Age	Medieval	Modern
Stature	N	14	9	23	16	32	29	22
	Mean	161.1	153.9	153.2	152.7	155.3	155.4	152.1
	SD	7.5	3.0	6.1	7.3	7.8	5.9	7.7
Body mass	N	12	4	22	16	29	24	19
	Mean	59.8	54.0	51.1	55.7	54.5	56.5	54.8
	SD	5.8	5.7	5.3	5.3	7.4	4.7	7.7
Body mass index	N	12	4	22	16	29	24	19
	Mean	22.8	23.0	21.7	23.9	22.5	23.3	23.7
	SD	1.9	3.0	1.4	1.2	1.8	1.4	1.7
Relative shoulder breadth	N	7	–	15	12	15	17	16
	Mean	172.2	–	175.9	179.9	168.2	176.7	177.8
	SD	10.0	–	8.1	7.7	11.7	7.7	5.2
Relative hip breadth	N	5	1	15	5	2	12	11
	Mean	189.3	178.7	187.8	194.9	198.1	195.1	203.0
	SD	15.0	–	12.4	11.0	2.8	13.1	11.3
Brachial index	N	–	–	9	13	6	15	14
	Mean	–	–	77.3	75.8	73.8	74.6	72.5
	SD	–	–	2.1	2.8	2.7	3.0	2.4
Crural index	N	11	7	18	15	24	28	20
	Mean	85.0	82.3	81.2	81.0	80.4	82.0	81.1
	SD	3.1	2.5	1.6	1.1	2.9	2.6	2.7

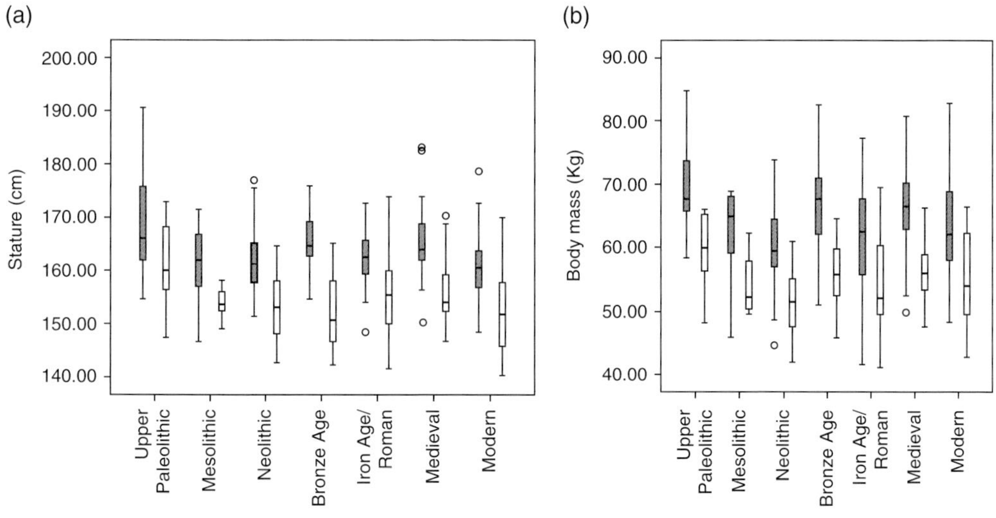

Figure 9.2 Boxplots showing temporal variation in (a) stature and (b) body mass. Gray boxes: males; white boxes: females.

the Bronze Age and, as with males, declines into Modern. None of these changes reaches statistical significance, apart from the marked difference between UP and Modern for both males and females (Bonferroni adjusted $p < 0.001$ and $p < 0.003$, respectively). This decrease in height is clearly caused by declining lower limb length, as sitting height does not change much through time, while patterns in lower limb length change mirror stature (data not shown; also see Chapter 4). Body mass (Table 9.4, Fig. 9.2(b)) for males and females follows identical patterns, although few of these changes are significant for females. Body mass decreases steadily between UP and Neolithic ($p < 0.004$ for males and $p < 0.003$ for females), followed by increases in Bronze Age (males, $p < 0.05$). As with stature, body mass fluctuates into the Medieval period and declines into Modern, although none of these changes is significant. Because the post-Neolithic gains in body mass exceed those in stature, the BMI (Table 9.4) increases substantially from Neolithic to Modern, particularly for females ($p < 0.004$). BMI also increases between Neolithic and Bronze Age ($p < 0.002$, females). None of the changes is significant for males.

Due to very small Mesolithic sample sizes, statistical significance for body breadth variables (relative shoulder and hip breadth; Table 9.4 and Fig. 9.3(a,b)) was evaluated for post-Mesolithic groups only. Female relative shoulder breadth decreases between Bronze and Iron Age ($p < 0.005$) and increases in the Medieval period ($p < 0.04$). Medieval males, on the other hand, have relatively narrower shoulders than both Iron Age and Modern males ($p < 0.01$ and $p < 0.002$, respectively). Female relative hip breadth increases between Neolithic and Modern ($p < 0.02$), while male hip breadth fluctuates with no significant change.

Intra-limb proportions (Table 9.4; Fig. 9.4(a,b)) also change through time (brachial index values could not be evaluated for UP and Mesolithic). The brachial index declines steadily between Neolithic and Iron Age for both sexes, although the difference is significant only for males ($p < 0.001$). Males change little subsequently, but females decrease further into Modern ($p < 0.001$ for Neolithic-Modern decline). The change in brachial index results primarily from increases in humeral length, with little change in radial length (not shown). After a slight increase between UP and Mesolithic for males, the crural index also declines steadily until the Iron Age, followed by slight fluctuation. The female crural index decreases continuously between UP and Iron Age. Differences in crural index between UP and Iron Age are significant

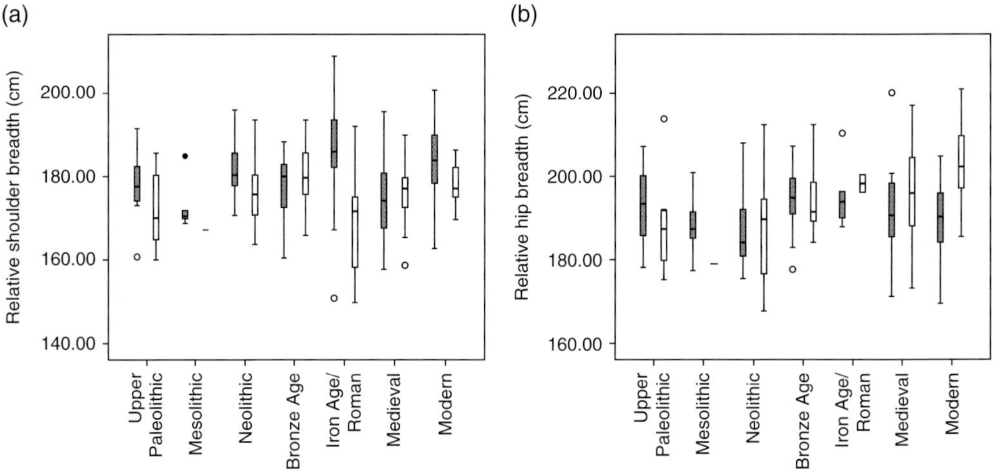

Figure 9.3 Boxplots showing temporal variation in (a) relative shoulder breadth and (b) relative hip breadth. Gray boxes: males; white boxes: females.

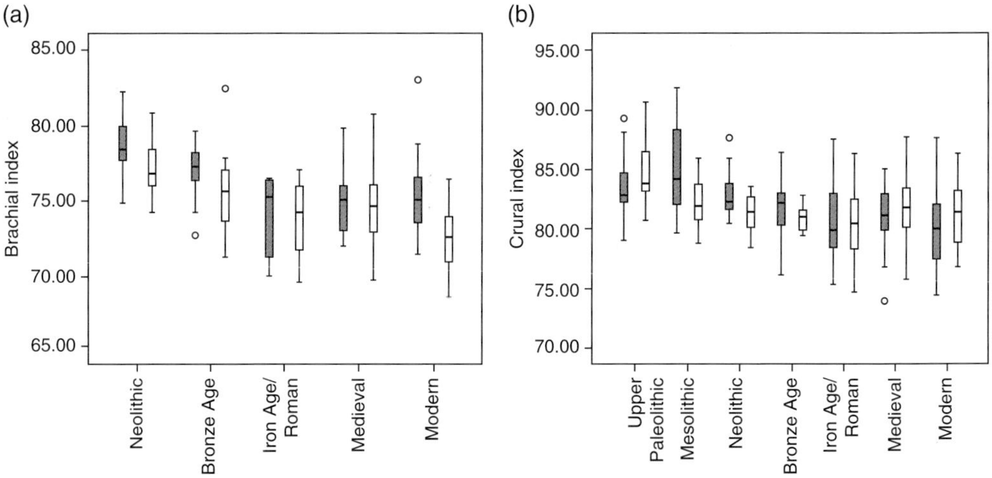

Figure 9.4 Boxplots showing temporal variation in (a) brachial index and (b) crural index. Gray boxes: males; white boxes: females.

(males: $p < 0.02$; females: $p < 0.0001$), with most change (especially in males) occurring after the Mesolithic. As with the brachial index, the decline in crural index after the Mesolithic stems from temporal increases in the length of the proximal element, that is, the femur, between Neolithic and Iron Age (not shown).

9.4.2 Body Size and Shape Sexual Dimorphism

Males exceed females in height and weight in all periods, with highly significant differences in most periods (Fig. 9.2(a,b); Table 9.5). Sexual dimorphism in relative hip and shoulder breadth remains low throughout, except in the Modern period when female hip breadth exceeds male values significantly (Fig. 9.3(b); Table 9.4), and the Iron Age, when male shoulder breadth

Table 9.5 Body size and shape sexual dimorphism.

	Upper Paleolithic	Mesolithic	Neolithic	Bronze Age	Iron Age/ Roman Age	Medieval	Modern
Stature	4.50	**4.89**	**5.70**	**7.89**	**4.28**	**6.10**	**5.42**
Body mass	15.51	15.83	**16.03**	**19.61**	**11.61**	**15.95**	**13.48**
Body mass index	**7.85**	3.26	3.89	3.85	3.77	2.92	2.36
Relative shoulder breadth	3.32	_	3.26	−1.42	**9.43**	−1.40	3.36
Relative hip breadth	1.75	5.27	−0.53	−0.07	−1.39	−1.96	**−7.25**
Brachial index	_	_	1.86	1.62	0.05	0.44	**3.84**
Crural index	−1.55	3.09	**2.04**	0.71	0.08	−1.25	−1.04

Bold text = p <0.05, with Bonferroni correction.

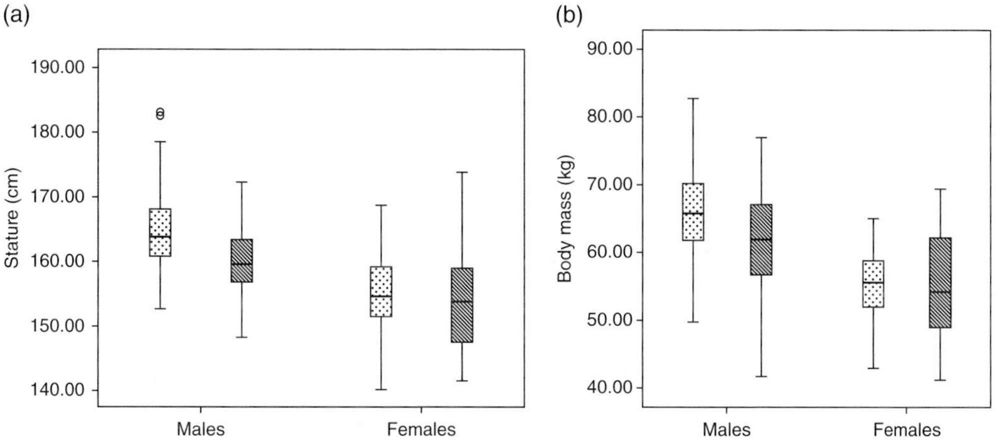

Figure 9.5 Boxplots showing differences in (a) stature and (b) body mass between rural and urban males and females (post-Bronze Age groups only). Stippled boxes: rural; stripped boxes: urban.

exceeds female values (Fig. 9.3(a); Table 9.4). Males exhibit higher brachial index values than females in all periods, but the difference is only significant in Moderns (Fig. 9.4(a); Table 9.4). Sex differences in crural index values are less patterned, with males generally exceeding females slightly in early periods, but falling below starting with the Iron Age (Fig. 9.4(b); Table 9.4). The difference is only significant in the Iron Age.

9.4.3 Impact of Urbanization

We also explored the impact urbanization may have had on body size and shape. As mentioned in the Methods section, pre-Iron Age samples were excluded in this analysis since there are no urban samples prior to that time. Rural males are taller and heavier (Fig. 9.5(a,b); Table 9.6) than urban males (stature: $p < 0.0001$; body mass: $p < 0.003$), with slightly lower brachial indices ($p < 0.05$) and relatively narrower shoulders ($p < 0.002$). Rural females have slightly higher

Table 9.6 Body size and shape by urban/rural status.

Males		Rural	Urban	Rural vs. Urban
Stature	N	55	56	
	Mean	165.0	159.9	**<0.0001**
	SD	6.3	5.7	
Body mass	N	51	51	
	Mean	65.9	61.2	**<0.0030**
	SD	7.1	8.4	
Body mass index	N	51	51	
	Mean	24.1	23.8	NS
	SD	2.1	2.8	
Relative shoulder breadth	N	41	40	
	Mean	177.4	184.7	**<0.0020**
	SD	9.0	11.6	
Relative hip breadth	N	26	21	
	Mean	190.9	189.5	NS
	SD	10.7	9.5	
Brachial index	N	20	23	
	Mean	74.3	75.8	**<0.0500**
	SD	1.9	2.6	
Crural index	N	47	50	
	Mean	80.9	80.2	NS
	SD	3.1	3.0	
Females		**Rural**	**Urban**	**Rural vs. Urban**
Stature	N	44	39	
	Mean	154.6	154.3	NS
	SD	6.1	8.3	
Body mass	N	39	33	
	Mean	55.3	55.2	NS
	SD	5.7	7.8	
Body mass index	N	39	33	
	Mean	23.2	22.9	NS
	SD	1.3	2.1	
Relative shoulder breadth	N	22	26	
	Mean	176.3	172.8	NS
	SD	6.2	11.3	
Relative hip breadth	N	17	8	
	Mean	198.6	199.2	NS
	SD	11.5	14.2	
Brachial index	N	18	17	
	Mean	74.0	73.2	NS
	SD	3.1	2.5	
Crural index	N	41	31	
	Mean	81.8	80.5	NS
	SD	2.9	2.5	

longer tibiae relative to femora than their urban counterparts, although the difference is not statistically significant ($p < 0.06$) (Table 9.6). Most of these differences parallel those found in the broader European sample (see Chapter 4).

9.4.4 Comparisons with Europe

Relative to the pan-European sample, the FI group is shorter across all periods, with the exception of FI UP females that exceed their European counterparts (Fig. 9.6(a)). Modern FI males, in particular, are distinctly shorter than European males ($p < 0.0001$). Body mass of Europeans also either exceed or closely approximate that of FI groups in most periods (Fig. 9.6(b)), except in the UP when FI males and female body mass falls slightly above European means. Neolithic and Iron Age FI males are particularly lighter than their European counterparts ($p < 0.001$ and $p < 0.0001$, respectively).

Analysis of relative hip breadth shows that, until the Iron Age, European females have wider pelves than FI females and that this is reversed in Moderns, but none of the comparisons are statistically significant, likely a result of small sample sizes. Relative shoulder breadth comparisons reveal few clear patterns, with the exceptions of the UP, Mesolithic and Iron Age, when European females distinctly exceed FI females, but the difference is only significant in the Iron Age ($p < 0.003$).

Few clear patterns in brachial index differences emerge, except for the Neolithic FI males and females who have relatively longer radii than other Europeans (Fig. 9.7(a)), but the difference is only significant in Neolithic males ($p < 0.004$). UP and Mesolithic differences could not be evaluated because of low sample sizes. The post-Mesolithic decline in crural index values observed in FI males and females mirrors the overall European pattern (Fig. 9.7(b)). Within

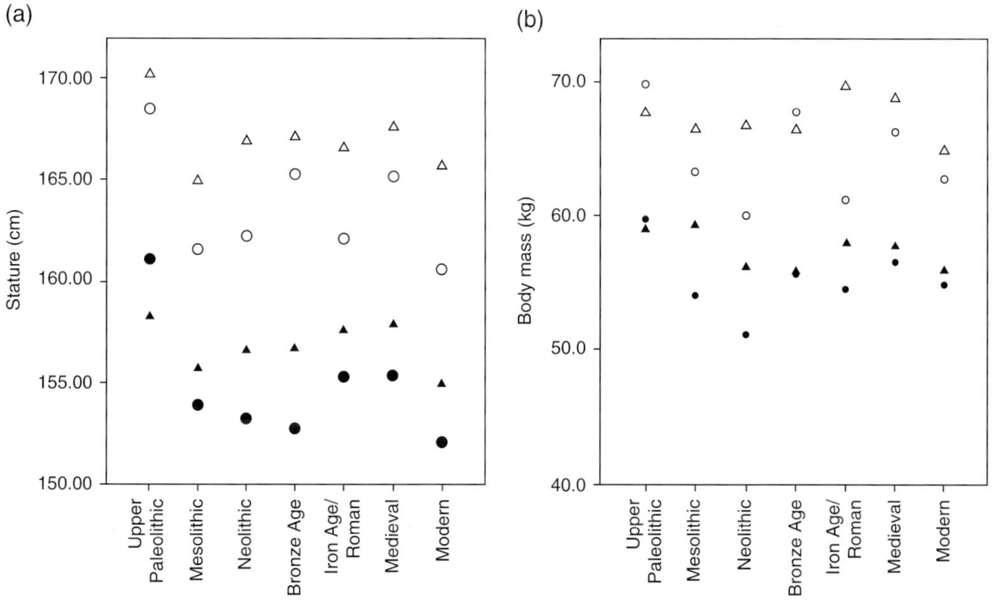

Figure 9.6 Scatterplots showing differences in temporal variation between FI and other European males and females in (a) stature and (b) body mass. Empty circles: FI males; filled circles: FI females; empty triangles: Pan-European males; filled triangles: Pan-European females.

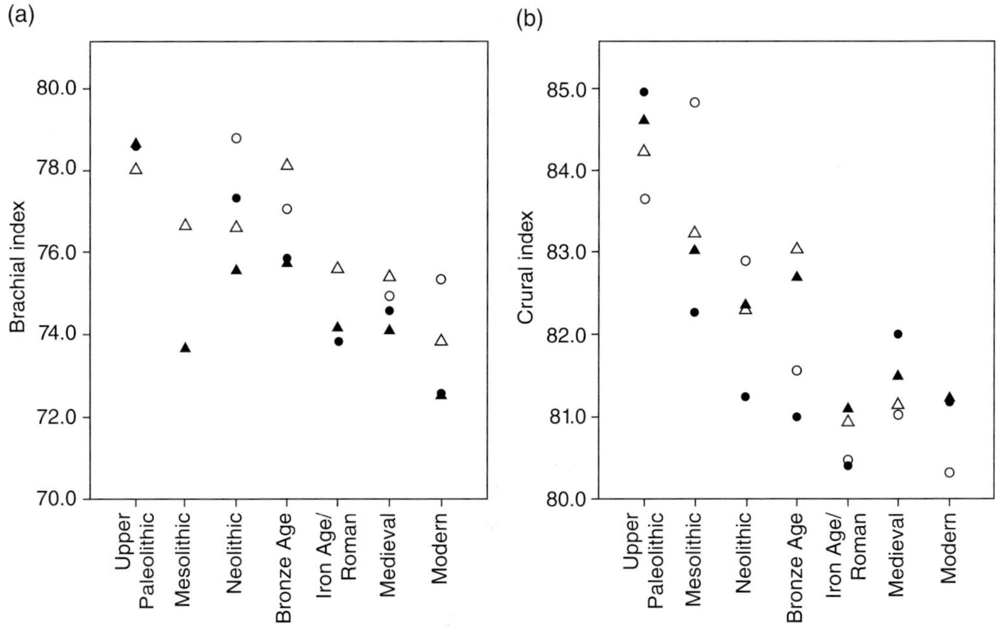

Figure 9.7 Scatterplots showing differences in temporal variation between FI and other European males and females in (a) brachial index and (b) crural index. Empty circles: FI males; filled circles: FI females; empty triangles: Pan-European males; filled triangles: Pan-European females.

periods, few patterns emerge, with the exception of the Bronze Age, when FI males and females exhibit shorter relative tibial length than their Europe counterparts (Fig. 9.7(b)), significantly in females ($p < 0.007$).

9.5 Long Bone Robusticity

9.5.1 Temporal Trends

9.5.1.1 Upper Limb

After non-significant increases from UP to Meso, male right and left humeral robusticity (Table 9.7; Fig. 9.8(a,b)) declines markedly until the Bronze Age ($p < 0.002$ and $p < 0.001$, respectively), followed by steady increases on the right side throughout each subsequent period, although the difference is only significant between Bronze Age and Modern groups ($p < 0.02$). Left humeral strength also increases slightly, but not significantly, between Bronze Age and Modern. Although the lack of female Mesolithic data constrained statistical comparisons to post-Mesolithic groups, UP females clearly exhibit stronger right and left humeri than subsequent periods (Table 9.7; Fig. 9.8(a,b)), with sharp drops in Neolithic on both sides, and significant further declines between Neolithic and Modern ($p < 0.02$) on the left.

Male humeral shape (Fig. 9.9(a,b)) becomes more circular between UP and Mesolithic, although this is significant only on the left ($p < 0.003$). This increased circularity continues on the right side between Bronze Age and Modern ($p < 0.005$). Following an increase between Mesolithic and Bronze Age ($p < 0.008$), the left humerus shape ratio also declines slightly into

Table 9.7 Upper limb robusticity data, by period and sex.

Males			Upper Paleolithic	Mesolithic	Neolithic	Bronze Age	Iron Age/ Roman Age	Medieval	Modern
Right side	Humeral robusticity (Z_p)	N	10	4	15	17	17	15	37
		Mean	624.4	715.6	564.1	477.7	477.7	556.5	582.7
		SD	127.3	110.1	104.1	81.6	81.6	106.5	116.6
	Humeral cortical thickness (%CA)	N	2	0	20	17	17	21	38
		Mean	81.2	–	73.8	78.4	78.4	68.2	65.9
		SD	9.8	–	7.8	5.8	5.8	9.8	11.2
	Humeral shape (Z_x/Z_y)	N	10	8	20	17	15	21	38
		Mean	1.17	1.04	1.10	1.15	1.03	1.02	1.00
		SD	0.12	0.06	0.17	0.10	0.13	0.13	0.16
Left side	Humeral robusticity (Z_p)	N	9	3	15	16	10	16	39
		Mean	523.3	719.8	545.4	442.4	488.5	459.3	532.3
		SD	107.5	87.9	105.2	75.3	108.2	100.1	117.7
	Humeral cortical thickness (%CA)	N	2	0	19	16	14	22	40
		Mean	79.8	–	73.9	77.6	67.5	65.2	67.1
		SD	8.2	–	7.6	7.3	9.8	8.7	11.3
	Humeral shape (Z_x/Z_y)	N	9	7	19	16	14	22	40
		Mean	1.10	0.88	1.02	1.10	1.04	1.07	1.01
		SD	0.17	0.12	0.16	0.11	0.10	0.16	0.12

(Continued)

Table 9.7 (Continued)

Females			Upper Paleolithic	Mesolithic	Neolithic	Bronze Age	Iron Age/ Roman Age	Medieval	Modern
Right side	Humeral robusticity (Z_p)	N	7	1	18	15	8	18	16
		Mean	571.7	514.3	424.8	423.2	451.3	433.8	386.1
		SD	98.2	–	73.1	95.0	170.8	98.3	86.8
	Humeral cortical thickness (%CA)	N	2	0	20	16	14	24	18
		Mean	85.2	–	67.8	65.4	60.0	66.3	57.2
		SD	4.5	–	9.6	10.9	11.8	10.4	12.7
	Humeral shape (Z_x/Z_y)	N	7	1	20	16	14	24	18
		Mean	1.17	1.28	1.14	1.18	1.10	1.05	0.99
		SD	0.13	–	0.19	0.08	0.17	0.17	0.16
Left side	Humeral robusticity (Z_p)	N	7	1	17	14	10	15	18
		Mean	491.5	488.3	417.9	421.3	394.4	388.7	341.8
		SD	65.7	–	52.0	93.4	75.9	73.7	66.2
	Humeral cortical thickness (%CA)	N	3	0	20	15	15	20	19
		Mean	79.8	–	67.3	61.9	63.4	64.5	56.8
		SD	5.3	–	11.9	11.4	10.2	11.2	13.3
	Humeral shape (Z_x/Z_y)	N	8	2	20	15	15	20	19
		Mean	1.23	1.06	1.08	1.07	1.10	1.14	1.10
		SD	0.11	0.02	0.13	0.09	0.15	0.24	0.16

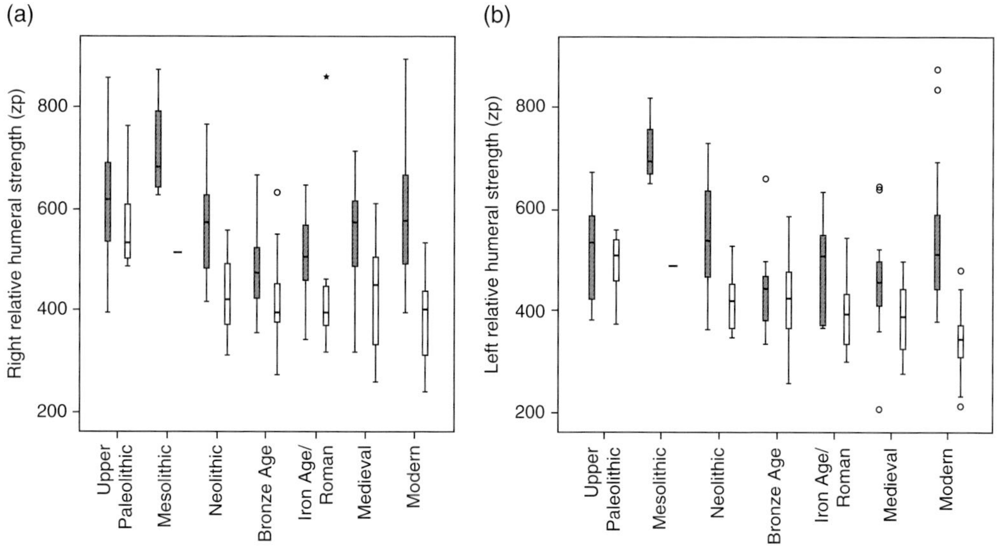

Figure 9.8 Boxplots showing temporal variation in upper limb strength standardized for body size for (a) right humerus and (b) left humerus. Gray boxes: males; white boxes: females.

Modern. As with robusticity, female shape ratios were compared for post-Mesolithic groups only. Female right humeri become significantly more circular between Neolithic and Modern ($p < 0.004$), with little change on the left side. UP females exhibit clearly less circular humeri than any subsequent groups.

Due to very low Mesolithic and UP sample sizes, humeral percentage cortical thickness statistical analysis comprises post-Mesolithic groups only, although UP males and females clearly exhibit thicker cortices than all subsequent periods (Fig. 9.10(a,b)). Males see significant declines on the right side between Bronze Age and Iron Age ($p < 0.02$) and Bronze Age and Medieval ($p < 0.008$), with a net decrease between Neolithic and Modern ($p < 0.02$). A similar pattern obtains on the left side, with declines in cortical thickness between Bronze Age and Iron Age ($p < 0.04$) and Bronze Age and Modern ($p < 0.003$). Females also decrease, but the change is only significant on the right side between Neolithic and Modern ($p < 0.03$).

9.5.1.2 Lower Limb

Male femoral and tibial robusticity (Table 9.8; Fig. 9.11(a,b) exhibits substantial declines between UP and Modern ($p < 0.002$ and $p < 0.0001$, respectively, with the biggest changes either between Mesolithic and Bronze Age for the femur ($p < 0.0001$), or between UP and Neolithic for the tibia ($p < 0.01$). Despite a small rise in femur robusticity between Bronze Age and Medieval, significant only in males ($p < 0.03$), femur and tibia robusticity remains lower in all post-Mesolithic periods for both sexes.

Male and female femoral shape (Table 9.8; Fig. 9.12(a)) follows similar patterns, becoming markedly more circular following the UP, reflecting significant declines in A-P bending strength between UP and Mesolithic ($p < 0.0001$). Females see another slight decrease in the Neolithic ($p < 0.05$), with little subsequent change in males. Examination of changes in A-P and M-L bending strength separately shows clearly that, for both sexes, the increases in femur cross-sectional circularity are driven primarily by declining A-P bending, with M-L bending strength fluctuating through time (Fig. 9.13(a,b)). There is also a net decline in tibial A-P/M-L bending strength

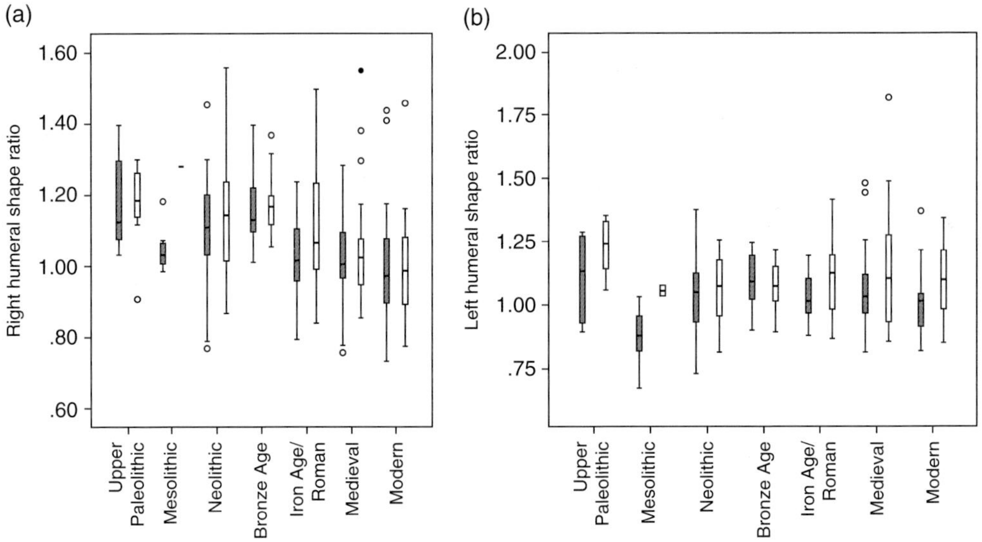

Figure 9.9 Boxplots showing temporal variation in right (a) and left (b) humeral A-P relative to M-L bending strength ratio. Gray boxes: males; white boxes: females.

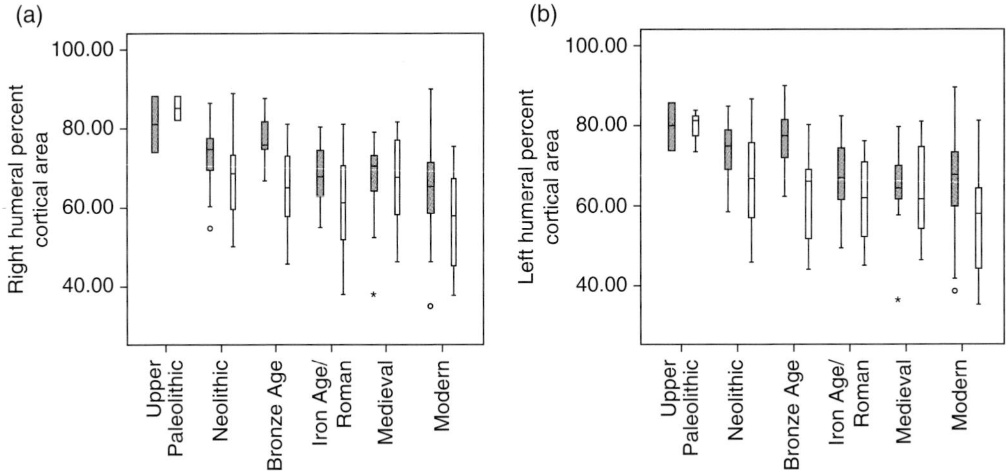

Figure 9.10 Boxplots showing temporal variation in right (a) and left (b) humeral percentage cortical area. Gray boxes: males; white boxes: females.

(Table 9.8; Fig. 9.12(b)) for both sexes between UP and Modern periods (males: p <0.001; females: p <004), with significant decreases also between Neolithic and Iron Age (p <0.003 and p <0.0001, respectively). The slight increase between Mesolithic and Neolithic is not significant, and little change is observed after the Bronze Age.

Femoral percentage cortical thickness (Table 9.8; Fig. 9.14(a)) declines precipitously between UP and Neolithic for both males and females (p <0.0001), increases slightly in the Bronze Age (p <0.04 and p <0.06, respectively), and declines again in the Iron Age (p <0.025 and p <0.008,

Table 9.8 Lower limb robusticity data, by period and sex.

Males		Upper Paleolithic	Mesolithic	Neolithic	Bronze Age	Iron Age/ Roman Age	Medieval	Modern
Femoral robusticity (Z_p)	N	15	13	17	17	21	33	44
	Mean	1261.2	1245.3	1092.3	913.0	1055.1	1077.8	1057.3
	SD	176.0	151.1	190.4	129.5	216.9	155.9	174.3
Tibial robusticity (Z_p)	N	11	13	17	15	22	29	42
	Mean	1161.9	1108.5	896.3	877.0	809.3	800.4	829.7
	SD	87.5	206.1	214.9	121.2	183.5	188.4	233.3
Femoral percentage cortical area (%CA)	N	16	19	20	17	24	34	45
	Mean	82.5	78.9	72.6	78.7	72.6	70.7	70.5
	SD	4.6	5.5	6.2	4.8	6.7	6.5	6.1
Tibial percentage cortical area (%CA)	N	11	19	20	15	27	33	43
	Mean	87.1	83.0	71.7	83.6	69.6	69.0	66.2
	SD	5.8	4.1	6.2	4.6	7.6	6.7	8.2
Femoral shape (Z_x/Z_y)	N	16	19	20	17	24	34	45
	Mean	1.22	1.00	0.94	1.00	0.88	0.95	0.88
	SD	0.24	0.10	0.16	0.08	0.10	0.14	0.13
Tibial shape (Z_x/Z_y)	N	11	19	20	15	27	33	43
	Mean	1.63	1.46	1.59	1.39	1.34	1.27	1.31
	SD	0.23	0.22	0.23	0.16	0.22	0.20	0.26

(*Continued*)

Table 9.8 (Continued)

Females		Upper Paleolithic	Mesolithic	Neolithic	Bronze Age	Iron Age/ Roman Age	Medieval	Modern
Femoral robusticity (Z_p)	N	10	3	21	14	26	23	19
	Mean	1198.1	1073.9	936.9	912.9	985.2	950.7	980.0
	SD	203.6	197.7	161.3	197.6	185.6	149.0	197.8
Tibia robusticity (Z_p)	N	7	2	19	15	21	23	18
	Mean	1075.6	987.2	764.1	699.7	714.4	696.7	705.6
	SD	151.9	63.4	174.7	114.9	207.3	132.7	196.5
Femoral percentage cortical area (%CA)	N	12	6	22	14	29	28	21
	Mean	82.4	80.5	67.9	76.6	66.6	69.4	67.8
	SD	7.1	11.1	7.6	7.4	8.6	6.6	11.8
Tibial percentage cortical area (%CA)	N	8	5	20	15	23	28	21
	Mean	85.6	80.6	68.4	75.1	64.9	66.8	59.6
	SD	8.6	6.0	7.4	5.9	7.5	7.4	14.2
Femoral shape (Z_x/Z_y)	N	11	6	22	14	29	28	21
	Mean	1.11	1.06	0.88	0.89	0.88	0.87	0.80
	SD	0.11	0.24	0.14	0.13	0.09	0.12	0.13
Tibial shape (Z_x/Z_y)	N	8	5	20	15	23	28	21
	Mean	1.40	1.42	1.57	1.24	1.22	1.27	1.15
	SD	0.18	0.31	0.23	0.18	0.19	0.19	0.17

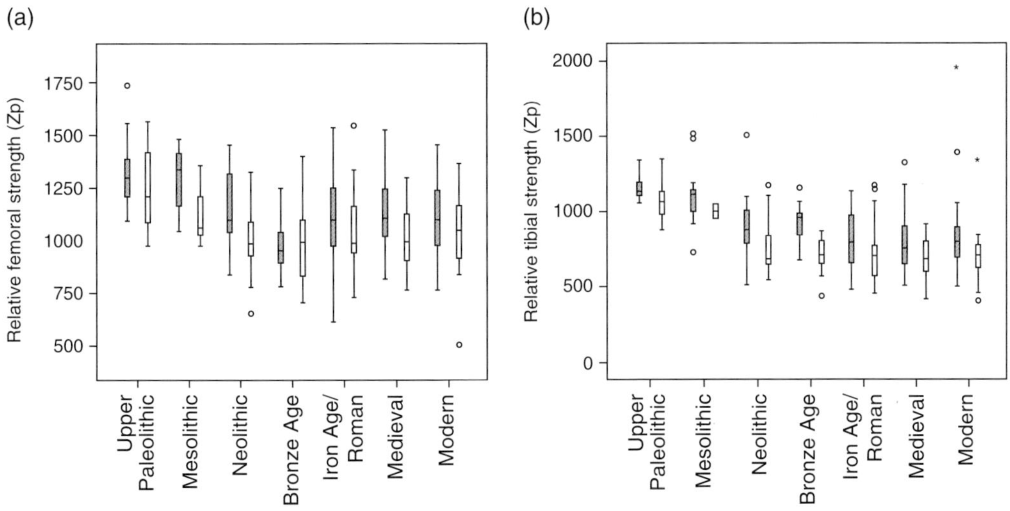

Figure 9.11 Boxplots showing temporal variation in femoral (a) and tibial (b) strength standardized for body size. Gray boxes: males; white boxes: females.

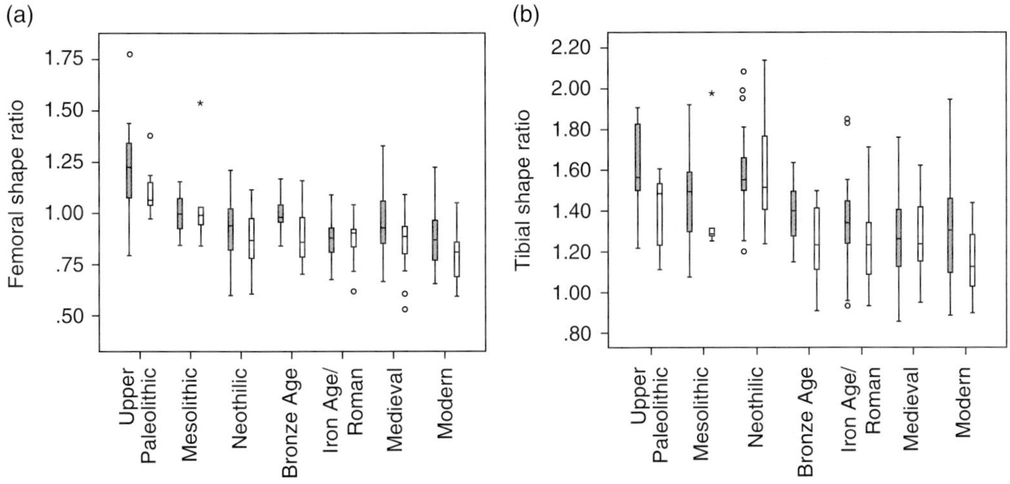

Figure 9.12 Boxplots showing temporal variation in femoral (a) and tibial (b) A-P relative to M-L bending strength ratio. Gray boxes: males; white boxes: females.

respectively), with little subsequent change. Tibial cortical thickness (Table 9.8; Fig. 9.14(b)) follows similar patterns in both sexes, with sharp declines between UP and Neolithic ($p < 0.0001$), followed by increases in the Bronze Age (males, $p < 0.0001$) and further declines in the Iron Age ($p < 0.0001$ and $p < 0.01$, respectively) and little change thereafter.

9.5.2 Sexual Dimorphism

Due to very small female sample sizes for UP and Mesolithic periods, analysis of sexual dimorphism in upper limb robusticity focuses on post-Mesolithic periods. Males exhibit higher upper limb strength and percentage cortical thickness than females in all periods, although the

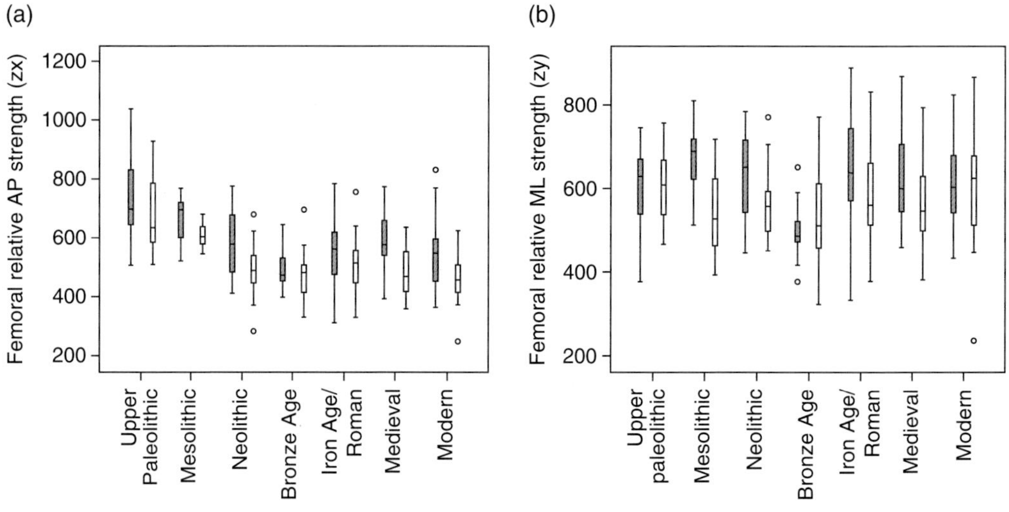

Figure 9.13 Boxplots showing temporal variation in femoral (a) A-P strength standardized for body size and (b) M-L strength standardized for body size. Gray boxes: males; white boxes: females.

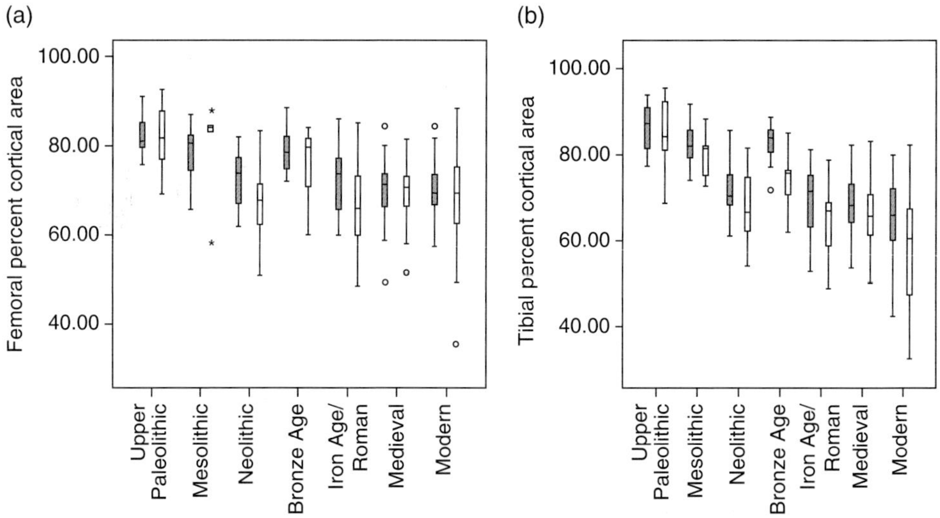

Figure 9.14 Boxplots showing temporal variation in femoral (a) and tibial (b) percentage cortical area. Gray boxes: males; white boxes: females.

differences are significant only in a few instances (Table 9.9; Figs 9.8 and 9.10). The upper limb strength dimorphism observed in the Neolithic sample, for instance, supports a previous analysis showing elevated postcranial male robusticity and marked sexual division of labor for the site of Pontcharaud (Civetta *et al.*, 2009). Males have higher lower limb strength, shape ratios, and percentage cortical thickness than females in almost all periods (Table 9.10) but, as with the upper limb, these differences reach statistical significance in few periods (Table 9.10; Figs 9.11, 9.12, and 9.14).

Table 9.9 Upper limb robusticity sexual dimorphism.

		Neolithic	Bronze Age	Iron Age/ Roman Age	Medieval	Modern
Right side	Z_p	**28.17**	12.11	5.70	**24.78**	**40.59**
	%CA	8.52	**18.12**	26.65	2.91	**14.16**
	Z_x/Z_y	−4.23	−2.36	−6.30	−2.98	0.72
Left side	Z_p	**26.47**	4.87	21.31	16.64	**43.59**
	%CA	9.27	**22.43**	6.26	1.06	**16.64**
	Z_x/Z_y	−5.17	2.25	−5.79	−7.00	−8.77

Bold text = p <0.05, with Bonferroni correction.

Table 9.10 Lower limb robusticity sexual dimorphism.

	Neolithic	Bronze Age	Iron Age/ Roman Age	Medieval	Modern
Femoral robusticity (Z_p)	15.32	0.00	6.86	**12.53**	7.58
Tibial robusticity (Z_p)	15.92	**22.49**	12.46	13.85	16.16
Femoral percentage cortical area (%CA)	6.72	2.73	**8.65**	1.82	3.86
Tibial percentage cortical area (%CA)	4.62	**10.74**	6.90	3.20	10.50
Femoral shape (Z_x/Z_y)	5.78	**11.92**	−0.06	8.89	10.05
Tibial shape (Z_x/Z_y)	1.21	10.93	8.84	0.39	13.11

Bold text = p <0.05, with Bonferroni correction.

9.5.3 Bilateral Asymmetry

Male and female bilateral asymmetry in humeral robusticity decreases between UP and Neolithic (Table 9.11; Fig. 9.15). Following little change in Bronze Age and Iron Age, asymmetry increases markedly in Medievals, with a slight subsequent decline. The Neolithic-Medieval increase is statistically significant (p <0.05) for males. Female asymmetry also decreases between UP and Neolithic, with a slight increase in Medieval and Modern, but the overall change is not statistically significant. Increased asymmetry in Medievals is due primarily to high values in two individuals, both from the Vicenne Campochiaro site. Asymmetry values for the FI males and females fall below or very close to the pan- European ones in all periods, except for the Medieval period for which asymmetry is significantly higher for the FI group, again mostly due to the two individuals from Vicenne. None of the other differences is significant.

9.5.4 Rural/Urban Status

Post Iron-Age rural and urban groups differ little in upper limb robusticity and shape, although rural males have significantly lower left humeral robusticity and higher shape ratios in the right side (Table 9.12). Both, rural males and females exhibit higher robusticity asymmetry, although

Table 9.11 Upper limb robusticity asymmetry, by period and sex.

Males		**Upper Paleolithic**	**Mesolithic**	**Neolithic**	**Bronze Age**	**Iron Age/ Roman Age**	**Medieval**	**Modern**
Humeral robusticity (Z_p)	N	9	7	19	16	11	18	35
	Mean	18.9	7.8	6.3	7.0	6.5	22.5	9.8
	SD	29.6	12.3	10.9	5.7	18.4	17.1	17.1

Females		**Upper Paleolithic**	**Mesolithic**	**Neolithic**	**Bronze Age**	**Iron Age/ Roman Age**	**Medieval**	**Modern**
Humeral robusticity (Z_p)	N	7	1	20	15	10	18	17
	Mean	15.9	5.5	−0.8	0.2	−3.7	8.2	10.3
	SD	13.1	–	18.1	9.4	26.7	19.8	18.9

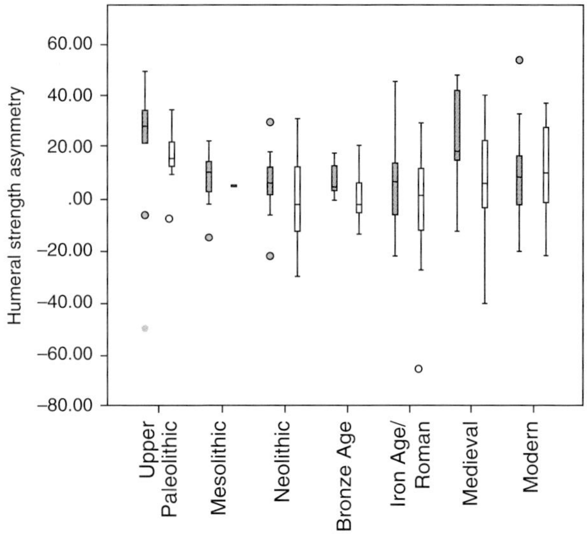

Figure 9.15 Humeral asymmetry in average bending strength (Z_p). Gray boxes: males; white boxes: females.

the difference is only significant in females (Table 9.12). Rural and urban groups do not exhibit any significant differences in lower limb robusticity and shape patterns (Table 9.13).

9.5.5 Terrain

To assess the impact of topographic variability on lower limb robusticity and shape, the sample was divided into flat, hilly, and mountainous terrain categories (see Chapter 5 for details). Males inhabiting hilly and mountainous regions have significantly higher femoral and tibial robusticity than those from flat areas (Table 9.14; Fig. 9.16(a,b)), but shape does not vary (Table 9.14). No differences were found for females.

9.5.6 Comparisons with Europe

9.5.6.1 Upper Limb

Small sample sizes for pre-Neolithic samples prevent meaningful analysis, so results focus on post-Mesolithic periods. In all periods following the Mesolithic, FI male and female humeral strength and percentage cortical area fall below their pan-European counterparts (Fig. 9.17(a,b) and Fig. 9.18(a,b)). In males, the difference is statistically significant for right humeral strength in Neolithic and Bronze Age ($p < 0.005$ and $p < 0.0001$, respectively), left humeral strength in Bronze Age and Medieval ($p < 0.0001$) and left humeral percentage cortical area in Medieval ($p < 0.0001$). FI females exhibit significantly lower right and left humeral strength than Europeans in Neolithic and Bronze Age ($p < 0.0001$), on the left side in Medieval ($p < 0.001$), and right and left sides in Modern ($p < 0.003$ and $p < 0.0001$, respectively). Modern FI females also have significantly lower right and left cortical thickness ($p < 0.002$ and $p < 0.004$, respectively), and on the left side in Iron Age ($p < 0.0001$).

FI male right and left humeral strengths follow similar trends as pan-Europeans, decreasing between Neolithic and Iron Age, followed by increases into Modern (Fig. 9.17(a,b)). Both, FI and European females, on the other hand, decrease between Neolithic and Modern, although the change is more pronounced in the FI group. Male and female humeral percentage cortical

Table 9.12 Upper limb robusticity by urban/rural status.

Males		**Rural**	**Urban**	**Rural vs. Urban**
Right humeral robusticity (Z_p)	N	32	30	
	Mean	551.6	574.8	NS
	SD	111.3	115.4	
Left humeral robusticity (Z_p)	N	30	35	
	Mean	477.5	533.3	**p <0.05**
	SD	115.0	110.3	
Right humeral shape (Z_x/Z_y)	N	37	37	
	Mean	1.06	0.97	**p <0.006**
	SD	0.13	0.14	
Left humeral shape (Z_x/Z_y)	N	36	40	
	Mean	1.05	1.02	NS
	SD	0.14	0.12	
Humeral asymmetry	N	31	33	
	Mean	16.6	9.3	NS
	SD	17.4	18.4	
Females		**Rural**	**Urban**	**Rural vs. Urban**
Right humeral robusticity (Z_p)	N	27	15	
	Mean	412.0	431.6	NS
	SD	85.2	150.8	
Left humeral robusticity (Z_p)	N	24	19	
	Mean	361.3	381.9	NS
	SD	73.1	74.8	
Right humeral shape (Z_x/Z_y)	N	30	26	
	Mean	1.05	1.04	NS
	SD	0.15	0.18	
Left humeral shape (Z_x/Z_y)	N	26	28	
	Mean	1.16	1.08	NS
	SD	0.23	0.14	
Humeral asymmetry	N	24	21	
	Mean	13.5	-1.8	**p <0.01**
	SD	17.3	23.1	

thickness also decreases through time for both FI and Europe (Fig. 9.18(a,b)), although the trends in the FI group are not as linear, as seen in the large increase in males between Neolithic and Bronze Age, and in females between Iron Age and Medieval. Also noteworthy is the sharp decrease in FI Modern females.

Table 9.13 Lower limb robusticity by urban/rural status.

Males		Rural	Urban	Rural vs. Urban
Femoral robusticity (Z_p)	N	49	49	
	Mean	1053.6	1073.8	NS
	SD	178.8	176.2	
Tibial robusticity (Z_p)	N	46	47	
	Mean	815.5	815.9	NS
	SD	243.9	166.3	
Femoral shape (Z_x/Z_y)	N	52	51	
	Mean	0.92	0.89	NS
	SD	0.15	0.11	
Tibial shape (Z_x/Z_y)	N	50	53	
	Mean	1.31	1.30	NS
	SD	0.23	0.24	
Females		**Rural**	**Urban**	**Rural vs. Urban**
Femoral robusticity (Z_p)	N	38	30	
	Mean	950.7	999.1	NS
	SD	155.4	198.1	
Tibial robusticity (Z_p)	N	37	25	
	Mean	676.8	747.5	NS
	SD	150.8	205.4	
Femoral shape (Z_x/Z_y)	N	42	36	
	Mean	0.85	0.86	NS
	SD	0.12	0.12	
Tibial shape (Z_x/Z_y)	N	42	30	
	Mean	1.24	1.20	NS
	SD	0.19	0.18	

9.5.6.2 Lower Limb

Low sample sizes for UP groups limit the analysis to post-UP periods. In general, FI temporal trends in femoral and tibial robusticity mirror those for Europeans, but pan-Europeans exhibit either higher or equal femur and tibial robusticity than FI groups in most periods (Fig. 9.19(a,b)), significantly for male femur robusticity in the Bronze Age ($p < 0.0001$), and tibial robusticity in Medieval and Moderns ($p < 0.001$ and $p < 0.003$, respectively). The decreases in femoral and tibial robusticity between Mesolithic and Neolithic are particularly steep for FI males and females. Interestingly, all groups increase slightly in femoral robusticity either between Bronze Age and Iron Age or Medieval periods. Note that the mean value for FI Bronze Age male femora is hidden behind the female mean. Differences in female femoral and tibial robusticity are

Table 9.14 Lower limb robusticity and shape, by terrain levels and sex (pooled time periods).

Males		Flat	Hilly	Mountainous	Flat versus hilly	Flat versus Mountainous	Hilly versus Mountainous
	N	53	58	49			
Femur robusticity (Z_p)	Mean	1019.5	1111.6	1121.2	**<0.03**	**<0.02**	NS
	SD	201.0	197.8	163.0			
	N	50	55	44			
Tibia robusticity (Z_p)	Mean	817.4	877.4	961.8	NS	**<0.005**	NS
	SD	190.7	203.4	260.3			
	N	60	63	52			
Femoral shape (Z_x/Z_y)	Mean	0.97	0.92	0.97	NS	NS	NS
	SD	0.13	0.17	0.20			
	N	58	60	50			
Tibial shape (Z_x/Z_y)	Mean	1.34	1.43	1.39	NS	NS	NS
	SD	0.25	0.26	0.25			

Females		Flat	Hilly	Mountainous	Flat versus hilly	Flat versus Mountainous	Hilly versus Mountainous
	N	41	51	24			
Femur robusticity (Z_p)	Mean	952.8	991.4	1005.5	NS	NS	NS
	SD	209.1	165.3	207.6			
	N	40	45	20			
Tibia robusticity (Z_p)	Mean	701.5	779.7	754.9	NS	NS	NS
	SD	177.9	191.8	207.3			
	N	45	57	29			
Femoral shape (Z_x/Z_y)	Mean	0.88	0.88	0.93	NS	NS	NS
	SD	0.14	0.13	0.20			
	N	43	50	27			
Tibial shape (Z_x/Z_y)	Mean	1.24	1.35	1.32	NS	NS	NS
	SD	0.18	0.28	0.22			

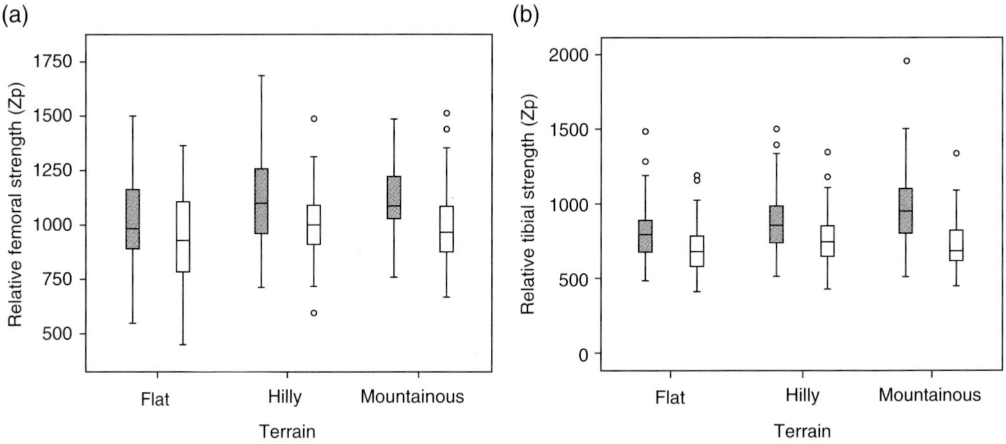

Figure 9.16 Boxplots showing variation among terrain categories in femoral (a) and tibial (b) strength standardized for body size (for pooled temporal periods). Gray boxes: males; white boxes: females.

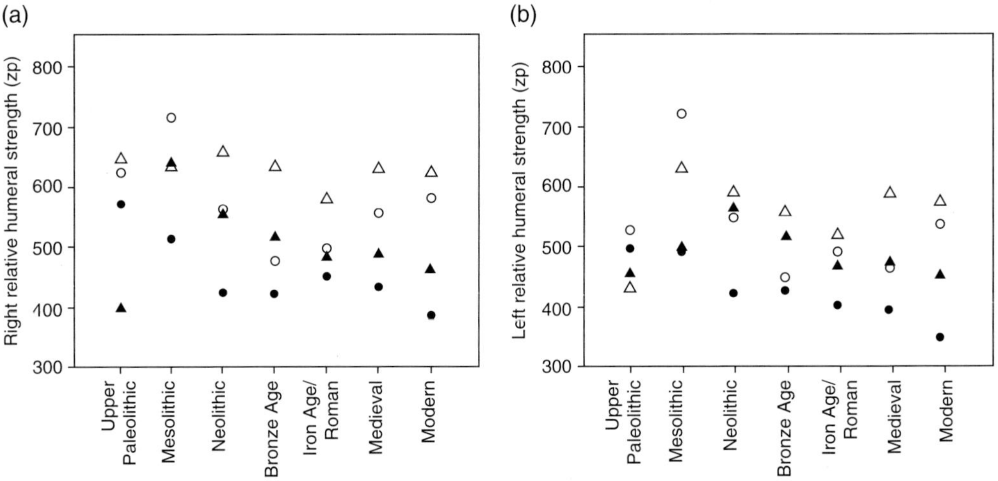

Figure 9.17 Scatterplots showing differences in temporal variation between FI and other European males and females in (a) right and (b) left humeral strength relative to body size. Empty circles: FI males; filled circles: FI females; empty triangles: Pan-European males; filled triangles: Pan-European females.

significant for Neolithic groups ($p < 0.0001$ and $p < 0.002$, respectively), and for tibial robusticity in the Bronze Age ($p < 0.003$). Femoral and tibial percentage cortical thickness is also significantly higher in pan-European males in most periods (Fig. 9.20(a,b)). A notable exception concerns the Bronze Age, when both male and female FI groups exceed Europeans in femoral and tibial cortical thickness, significantly for male tibial cortical thickness ($p < 0.0001$).

With the exception of the Neolithic, femur A-P/M-L bending strength for FI males falls either slightly above or very close to the pan-European sample, whereas the tibial shape ratio lies either slightly below or very close to pan-European males in all post- Neolithic periods (Fig. 9.21(a,b)). Neolithic FI males have significantly higher tibial shape ratios than their European counterparts

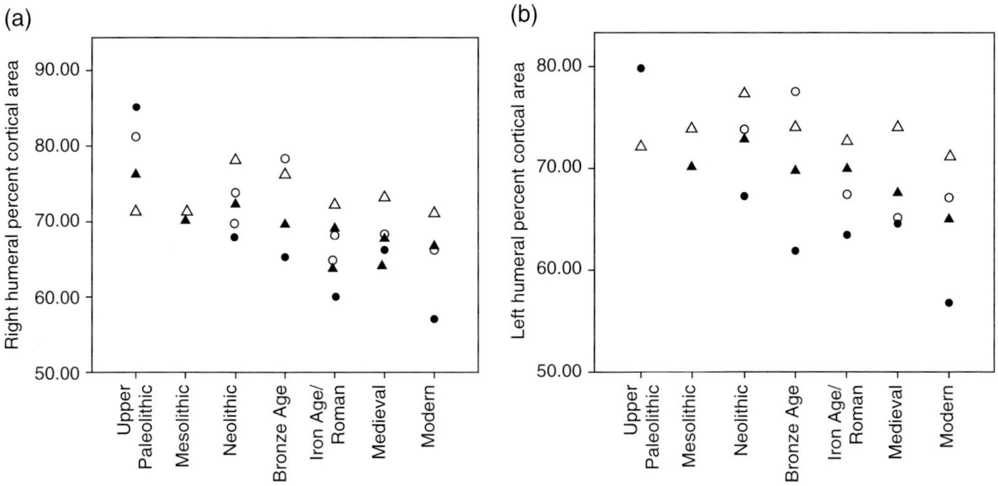

Figure 9.18 Scatterplots showing differences in temporal variation between FI and other European males and females in (a) right and (b) left humeral percentage cortical area. Empty circles: FI males; filled circles: FI females; empty triangles: Pan-European males; filled triangles: Pan-European females.

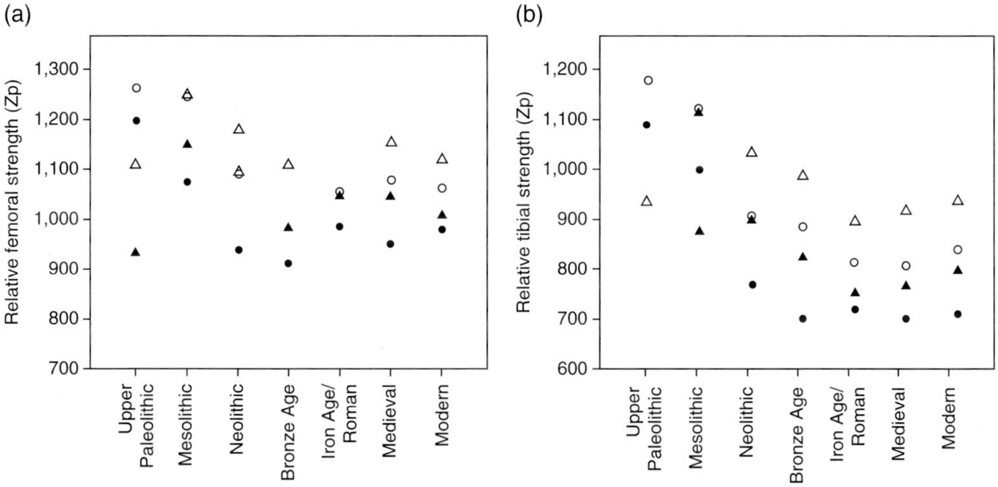

Figure 9.19 Scatterplots showing differences in temporal variation between FI and other European males and females in (a) femoral and (b) tibial strength relative to body size. Empty circles: FI males; filled circles: FI females; empty triangles: Pan-European males; filled triangles: Pan-European females.

($p < 0.002$). All post-Mesolithic females have lower femur shape ratios than pan-European females, although the difference is significant only for Moderns ($p < 0.0001$). As in males, Neolithic FI females exhibit much higher tibial shape ratios than their pan-European counterparts ($p < 0.0001$) (Fig. 9.21(b)). In the Bronze Age and Iron Age, the tibial shape ratio for FI females falls significantly below other females ($p < 0.001$), as well as in Moderns, although the latter difference is not quite significant with Bonferroni correction ($p < 0.009$).

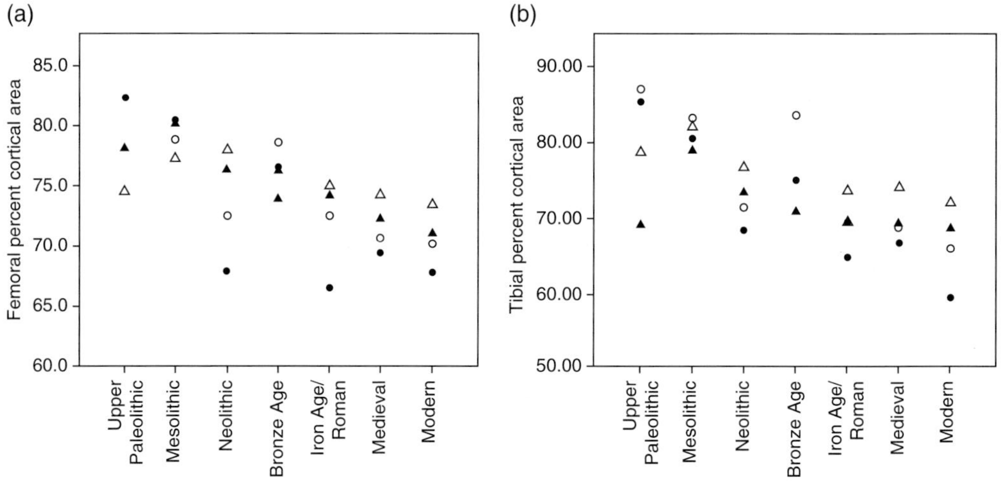

Figure 9.20 Scatterplots showing differences in temporal variation between FI and other European males and females in (a) femoral and (b) tibial percentage cortical area. Empty circles: FI males; filled circles: FI females; empty triangles: Pan-European males; filled triangles: Pan-European females.

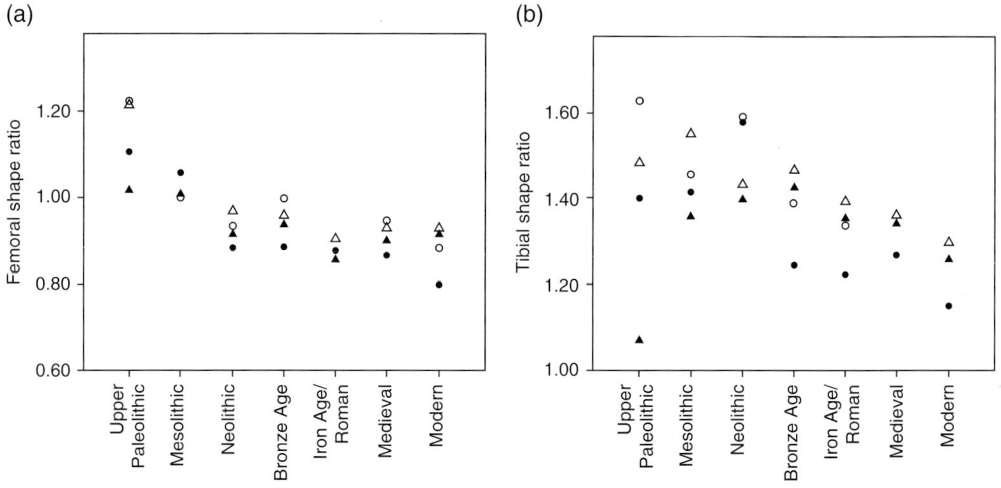

Figure 9.21 Scatterplots showing differences in temporal variation between FI and other European males and females in (a) femoral and (b) tibial A-P relative to M-L strength. Empty circles: FI males; filled circles: FI females; empty triangles: Pan-European males; filled triangles: Pan-European females.

9.6 Discussion

The FI sample lies below the European sample in stature and limb length in all periods, and either close to or slightly below in body mass values. Trends in body size for the FI sample follow those of the broader European sample, with a steep decline in stature from Upper Paleolithic to Mesolithic, and a fair amount of fluctuation in subsequent periods. The notable decrease in stature between Upper Paleolithic and Mesolithic parallels that found across Europe, and probably reflects a combination of reduced protein availability, resource stress related to higher population density, and decreased gene flow. Indicators of health from skeletal remains as well as faunal data suggest that, relative to Holocene populations – particularly those relying on food-producing economies – Upper Paleolithic groups enjoyed good health and nutritional conditions (Holt and Formicola, 2008). The fact that temporal changes in height result primarily from declines in lower limb length rather than trunk height suggests that nutritional factors played an important role, as relatively longer limbs generally reflect better environmental conditions (Eveleth and Tanner, 1990; Bogin and Rios, 2003). The concomitant decreases in crural index between Mesolithic and Iron Age and in brachial index between Neolithic and Iron Age parallel some of the changes observed in the pan-European sample (see Chapter 4). The fact that increased length in proximal elements (humerus and femur) account for the changes in these indices does not point to climatic adaptation during this time period. Instead, the decline in percentage cortical thickness over the same time span points to the role of nutrition, in particular protein availability, as the probably main cause for the change in body proportions during that time.

Unlike in other European groups, stature for Neolithic FI males stays relatively low, rebounding only in the Bronze Age, and FI female stature continues declining until the Bronze Age. FI males and females also see sharp declines in body mass in both Mesolithic and Neolithic, unlike other Europeans, who change relatively little. The pattern of steadily decreasing height and body mass in Neolithic French/Italians suggests a decline in nutritional and living conditions, in keeping with numerous studies documenting stature decrease with the adoption of agriculture or agricultural intensification (e.g., Angel, 1984; Kennedy, 1984; Larsen, 1984; Meiklejohn *et al.*, 1984). The mean stature values for our Neolithic sample match those from Neolithic sites located on the Italian Ligurian coast (Sparacello, 2014). A number of studies document a rise in caries between Neolithic and Iron Age in Italian samples, pointing to a decline in dietary protein levels relative to carbohydrates (Borgognini-Tarli and Canci, 1992; Manzi *et al.*, 1997). This is corroborated by archeological evidence of progressive socioeconomic stratification, intensified urbanization and an overall decline in the quality of life (Peroni, 1989; Manzi *et al.*, 1989). Interestingly, the rise in stature, body mass and percentage cortical thickness in Bronze Age males are seen in the British sample as well (see Chapter 8), possibly reflecting the increased consumption of milk and other dairy products in Late Neolithic (see Greenfield, 2010 and references therein). In Italy, for instance, there is evidence of intensification of pastoralism during the transition between Neolithic and Copper Age (Skeates, 2013). It should be noted, however, that our Bronze Age sample, Olmo di Nogara, comprises individuals buried with rich ornaments and weapons, suggesting relatively high status (Salzani, 2005), and a possibly high quality lifestyle.

Our Neolithic sample stands out in a number of ways, in particular low BMI and relative bi-iliac breadth. While it comprises some Eneolithic individuals from Italy, the bulk of the sample comes from the large Middle Neolithic mortuary assemblage of Pontcharaud in central France (Loison, 1998). A recent analysis of dietary isotope data (Goude *et al.*, 2013) shows that people at Pontcharaud enjoyed a diet high in meat products, confirming archeological evidence of mixed agriculture and pastoralism, and suggesting that the low BMI (as well as stature) values are not tied to diet, but may reflect genetic factors. Recent analyses of ancient DNA from European Early Neolithic Farmers point to substantial contributions from Near Eastern farmers to the post-Mesolithic European gene pool (Haak *et al.*, 2010; Bollongino *et al.*, 2013). The body size and shape characteristics of our Neolithic sample may thus reflect this admixture.

Our Iron Age sample also falls significantly below the pan-European sample in stature and body mass. Our stature means for males and females (162.08 cm and 155.28 cm, respectively), however, lie well below those from another Iron Age site from Italy, Alfedena (Sparacello, 2014). The Iron Age is generally seen as a period of strong socioeconomic and status differentiation for both men and women (Guidi and Piperno, 1992; Cunliffe, 1994). At Alfedena, burial context for many of the graves clearly reflect differentiated status. High status individuals are taller than low status ones (170.93 cm versus 167.37 cm for males and 162.10 cm versus 158.33 cm for females) (Sparacello, 2014), but all values exceed our Iron Age means. It is possible that our sample, comprising skeletal material from Southwest France (Rue Jacques Brel) and Italian Roman sites (Quadrella and Lucus Feroniae), represent low-status individuals. It should be noted that all three sites are late Iron Age or Romanized sites (100–400 AD). A previous study (Giannechini and Moggi-Cecchi, 2008) showed a marked decline in stature between Italian Iron Age and Roman age samples. Alternatively, the difference may reflect genetic variability associated with widespread migration characteristic of the Iron Age (Cunliffe, 1994). Finally, Quadrella and Lucus Feroniae were relatively urbanized populations, and our analysis suggests that urban males were shorter than rural ones (see below), which may also contribute to the stature values of our Iron Age sample.

In most Western European regions, stature declined during the 17th–19th centuries as a result of famines and poor life conditions associated with urbanization and the Industrial Revolution (see Chapter 4). As health conditions and nutrition improved during the later half of the 19th and mid-20th centuries, height levels recovered in many regions. In the FI sample, however, stature values continue declining in the Modern period. This could be due in part to the composition of the FI Modern sample that includes several collections from Sardinia and Sicily, areas that were characterized with extreme poverty for the greater part of the 19th to mid-20th century. When the Modern sample is broken into early Moderns (17th—19th centuries) and Recent (late 19th–20th centuries) subgroups, the decline remains apparent for FI females, whereas FI male stature remains stable after the late Medieval period (Fig. 9.22). It should be noted, however, that sample sizes for FI Modern samples are quite small, and the two Italian samples, Sassari (Sardinia) and Siracusani (Sicily), come from two regions where stature was relatively low, in particular Sardinia (Sanna, 2002). Figure 9.23 shows stature trends for French and Italian males from 1750 to 1950, including values for the FI Modern sample broken into French and Italian sub-samples, as well as mean stature for Italian and French male army

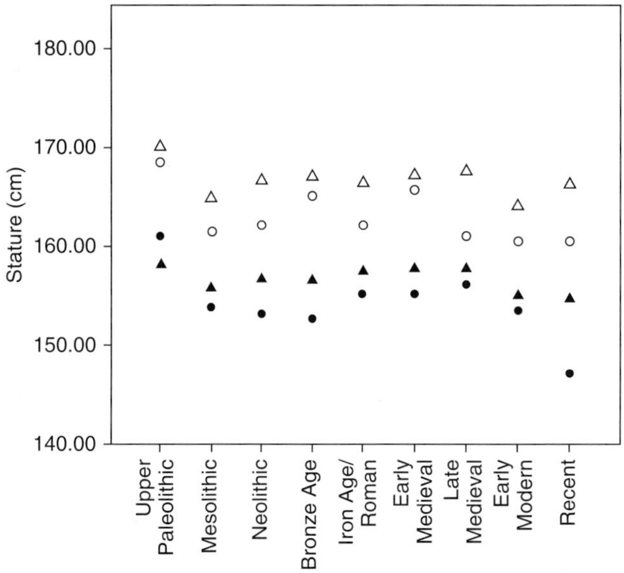

Figure 9.22 Scatterplots showing differences in temporal variation in stature between FI and other European males and females. Empty circles: FI males; filled circles: FI females; empty triangles: Pan-European males; filled triangles: Pan-European females.

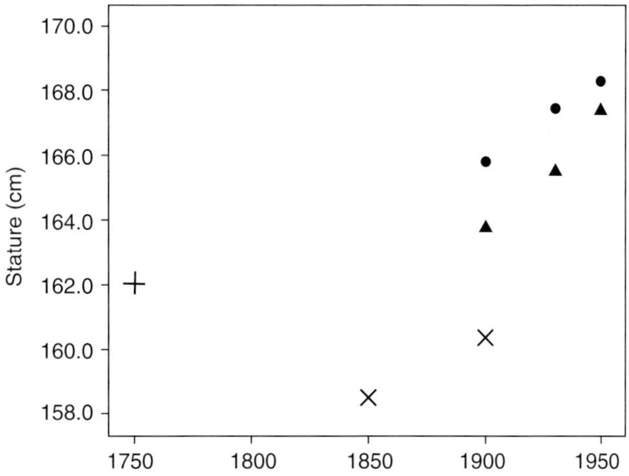

Figure 9.23 Mean stature for French and Italian males from 1750 to 1950. Data from this study: French early Modern (+), Italian early Modern and Recent (x); and French (circles) and Italian (triangles) military recruits (data from Chamla, 1964).

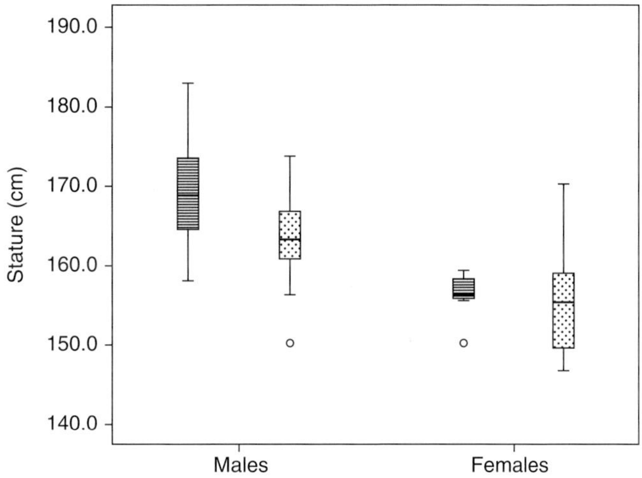

Figure 9.24 Boxplots showing differences in stature between Vicenne Campo Chiaro Medievals and Other European males and females. Stripped boxes: Vicenne; stippled boxes: other Europeans.

recruits for 1900, 1930, and 1950 (Chamla, 1964). While Italians remain below the French, both groups exhibit clearly the recent secular trends observed in most of Europe (see Chapter 4).

The increases in stature and body mass in the Medieval sample parallel trends observed for the pan-European sample. One sample stands out, that from the Early Medieval necropolis (7th–8th C.) of Vicenne-Campochiaro (Molise, Italy). Male and female values for stature for the Vicenne sample fall above those for the rest of the FI sample, for all periods combined or Medieval only (Fig. 9.24). The site is noteworthy for the presence of graves including human and horse skeletons, armors and horse harnesses (Ceglia, 2000). These cultural elements recall Iron Age rituals from Asian regions of Siberia and Mongolia, and are thought to originate in Central Asia (Kurylëv *et al.*, 1989). Studies of both archeological context and skeletal remains point to the presence of multi-ethnic groups at the site (Belcastro and Facchini, 2001). In addition, the presence of horses and riding implements in association with some of the graves indicates that some males were high-status horsemen (Belcastro and Facchini, 2001), which may explain the high stature for Vicenne males. Interestingly, high status at the Italian Iron Age site of Alfadena is also associated with increased stature (Sparacello, 2014). Belcastro *et al.* (2001) have argued that the well-developed

m. *pectoralis major* and mm. *latissimus dorsi/teres major* insertions on the right side observed in some Vicenne males may reflect the use of armors. Low sample sizes did not allow us to corroborate these findings with humeral strength data. The fact that Vicenne female stature also exceeds that of other FI temporal groups, however, suggests that migration may also help explain the high stature of the Vicenne people.

As with the broader European sample, FI rural males are taller and heavier than urban males, perhaps reflecting more optimal nutritional conditions, in keeping with previous studies showing that, prior to the 20th century, urban populations were generally shorter and lighter than rural ones (Martínez-Carrion and Moreno-Lázaro, 2007). This may be due to the poorer life conditions associated with both Medieval and Industrial Era urban areas. Steckel (1995), for instance, shows a clear association between declining health and industrialization, and Lewis' (2002) study of child health in Medieval and Industrialized England finds few health differences between urban and rural groups prior to industrialization. Significant increases in stature parallel improvements in health care, hygiene, and diet during the second half of the 19th century or early 20th in many urban areas of Europe (see Chapter 4).

In the FI sample, however, the difference in stature and weight between rural and urban groups affects only males (see Table 9.6). Examination of samples comprised in rural and urban groups (Table 9.15) shows that, with two exceptions, the rural male stature means range is 163.5–165.1 cm, compared with a range of 158.5 to 161.4 cm for urban males. Two rural samples stand out: The Modern group from Sassari (Sardinia, Italy, mean = 160.4 cm) and the Early Medieval sample (7th–8th C.) from Vicenne-Campochiaro (Molise, Italy, mean = 169.8 cm). As mentioned above, Sardinians are known for their particularly short stature (Sanna, 2002). The particularly high stature of Vicenne males and females may reflect migration from northern groups into Northern Italy (see above).

Table 9.15 Stature for rural and urban males, by sample (Iron Age to Modern).

Sample (N)	Time period	Stature (cm)[a]
Rural		
Sassari (5)	Late 19th–early 20th Century	160.4, 3.2
Moirans (10)	17th–19th Century	164.9, 7.4
Jau Dignac (10)	Medieval	163.5, 5.1
Roselle (12)	Medieval	163.8, 3.9
Vicenne Campochiaro (11)	Medieval	169.8, 7.9
Rue Jacques Brel (8)	Iron Age	165.1, 4.9
Urban		
Siracusani (17)	19th Century	158.5, 5.1
L' Observance (14)	18th Century	159.9, 5.5
Piazza della Signoria (6)	Medieval	161.0, 10.4
Quadrella (12)	Iron Age	161.4, 4.9
Lucus Feroniae (11)	Iron Age	159.2, 5.2

[a] Values are Mean, SD.

Both cultural and biological factors could account for the lack of stature differences between rural and urban females. Analysis of stature and social status (SES) in the Medieval sample from Trino Vercellese (Italy), for instance, shows a positive effect of high SES on male stature, but none for females (Vercellotti *et al.*, 2014). Dietary isotope data underscore these differences (Reitsema and Vercellotti, 2012). High SES male diets included more animal protein, but no such difference was observed for females, leading the authors to state that "...Perhaps, due to their role in reproduction and child rearing, females were afforded consistent access to foods without regard to their socioeconomic status and without risky fluctuations year-to-year or season-to-season". (Reitsema and Vercellotti, 2012: 598). While such cultural factors undoubtedly played a role in some cases, it is difficult to imagine they could explain the consistent pattern observed in our samples, given that they span the Iron Age to early 20th century. If negative life conditions associated with urban settings did contribute to stature variability, sex differences may also result from higher male susceptibility to environmental insults.

Numerous studies have shown sex differences in morbidity and mortality, with males often exhibiting much higher rates in mortality (Graunt, 1975; De Witte, 2010), and higher frequencies of stress markers such as periostitis (Larsen, 1998, Šlaus, 2008) and linear enamel hypolasias (Guatelli-Steinberg and Lukacs, 1999). Under the so-called 'female buffering' model, girls' bodies are evolutionarily designed to harness nutrients such as fat and protein during growth, buffering them during times of stress (Stini, 1985; Stinson, 1985; Guatelli-Steinberg and Lukacs, 1999). Biological factors underlying enhanced male sensitivity to poor life conditions include increased susceptibility to some viruses, bacteria and parasites (Hoff *et al.*, 1979; Brabin and Brabin, 1992; Jansen *et al.*, 2007; Taylor *et al.*, 2009), the disproportionate impact of X-linked immune diseases on males (Waldron, 1984), and the link between sex hormones and immune system depression and degenerative diseases (Grossman, 1985; Klein, 2000; Roberts *et al.*, 2001).

Upper and lower limb robusticity and percentage cortical area decline sharply between Mesolithic and Bronze Age, and with a continued downward trend for upper limb values in the Iron Age, and little subsequent change. Following the Mesolithic, FI groups fall below the European sample in upper and lower limb bone strength. As with the pan-European sample (see Chapter 5), the decline in limb robusticity appears to parallel development in technological complexity and mechanization observed starting with the Bronze Age. The introduction of draft animals in the Neolithic (Milisauskas and Kruk, 2002) undoubtedly helped reduce the intensity of field labor. The decline in robusticity probably reflects lower levels of mechanical loads in parallel with improvements in farming technology (see Chapter 5), but the temporal change in percentage cortical area may also reflect the impact of poor nutrition on relative cortical thickness (Garn *et al.*, 1969). Cortical thickness also declines with age (Garn, 1970; Ruff and Hayes, 1983), but this should not be a factor in our sample. Mean age at death for the Modern groups is higher than all other periods, but all other pairwise comparisons between periods are non-significant, so clearly not a factor in the post-Mesolithic decline in cortical bone. This suggests that the temporal decline in percentage cortical area and limb robusticity between Mesolithic and Bronze Age may reflect a combination of protein-poor diets and decreased mechanical loads. As expected, urbanization appears to have impacted upper limb asymmetry, with rural males and females exhibiting greater bilateral asymmetry, although significantly so only in females. This may reflect the prevalence of lateralized farming tasks in rural regions. The absence of a greater impact of urbanization on limb robusticity is surprising, however, considering the development after the Industrial Revolution of mechanized means in all sectors of life (Cowan, 1999; Musson and Robinson, 1969). Our classification of certain Iron Age and Medieval sites as 'urban,' however, may not take into account the fact that many people lived in small towns, but farmed during the day. In addition, some activities performed in more urban settings may have involved equally strenuous tasks, such as metal smithing and factory work (see Chapter 5).

The general decrease in lower limb bending strength in the FI sample between Upper Paleolithic and Mesolithic, and in the Neolithic in the pan-European group, clearly reflects declining mobility (see Chapter 5). Finally, the role of terrain in shaping lower limb diaphyseal structure is underscored by the fact that femoral and tibial robusticity values are significantly higher in males from hilly and mountainous areas.

9.7 Conclusions

Trends in body size for the French/Italian sample follow those of the broader European sample, with a steep decline in stature from Upper Paleolithic to Mesolithic. Unlike in some (but not all) other European groups, stature for Neolithic FI males stays relatively low, rebounding only in the Bronze Age, and FI female stature continues declining until the Bronze Age. FI males and females also see sharp declines in body mass in both Mesolithic and Neolithic, unlike some other Europeans, who change relatively little. The pattern of steadily decreasing height and body mass in Neolithic French/Italians suggests a decline in nutritional and living conditions, possibly reflecting the transition to agriculture. Concomitant decline in percentage cortical area supports this interpretation. As elsewhere in Europe, stature declined during the 19th century, reflecting poor life conditions and urbanization associated with the Industrial Revolution. We did not observe the late 19th—20th century secular trend observed in other parts of Europe, however, possibly reflecting the inclusion of several samples from very poor areas of Italy. Several studies also suggest a later start of the recent positive secular trend in body size in other parts of southern Europe (Cardoso and Gomes, 2009; María-Dolores and Martínez-Carrión, 2011).

Male relative upper limb bone strength declines markedly between the Upper Paleolithic and Bronze Age, with some fluctuation until Modern, while females decline continuously between the Upper Paleolithic and Modern periods. For both males and females, lower limb bones decline sharply in relative strength and become more circular either in the Mesolithic or Neolithic, clearly reflecting declining mobility. Analysis of lower limb robusticity patterns by terrain also shows a significant impact of local topography on femora and tibial strength, stressing the complexity of loading environments affecting long bone structure.

Acknowledgments

We thank Sarah Reedy for help in data processing, and Colten Karnedy for help with the tables. We are grateful to the following persons for access to skeletal collections and for facilitating data acquisition: Maria Giovanna Belcastro, Alessandro Riga, Nico Radi, Giorgio Manzi, Maryanne Tafuri, Pascal Murail, Patrice Courtaud, Dominique Castex, Fredérik Léterlé, Emilie Thomas, Aurore Schmitt, Aurore Lambert, Sandy Parmentier, Alessandro Canci, Gino Fornaciari, Davide Caramella, Monica Zavattaro, Elsa Pacciani, Fulvia Lo Schiavo, Eligio Vacca, and Donato Coppola. We thank Erik Trinkaus, Steven Churchill, and Trent Holliday for generously making available data for some of the Upper Paleolithic and Mesolithic samples.

References

Angel, J.L. (1984) Health as a crucial factor in the changes from hunting to developed farming in the eastern Mediterranean. In: *Paleopathology at the Origins of Agriculture* (eds M.N. Cohen and G.J. Armelagos), Academic Press, Orlando, FL, pp. 51–73.

Belcastro, M.G. and Facchini, F. (2001) Anthropological and cultural features of a skeletal sample of horseman from the medieval necropolis of Vicenne-Campochiaro (Molise, Italy). *Collegium Antropologicum*, **25** (2), 387–401.

Belcastro, M.G., Facchini, F., Neri, R., and Mariotti, V. (2001) Skeletal markers of activity in the early Middle Ages Necropolis of Vicenne-Campochiaro (Molise, Italy). *J. Paleopathol.*, **13**, 9–20.

Belcastro, G., Rastelli, E., Mariotti, V., Consiglio, C., Facchini, F., and Bonfiglioli, B. (2007) Continuity or discontinuity of the life-style in Central Italy during the Roman Imperial Age-Early Middle Ages transition: diet, health, and behavior. *Am. J. Phys. Anthropol.*, **132**, 381–394.

Belcastro, M., Rastelli, E., and Mariotti, V. (2008) Variation of the degree of sacral vertebral body fusion in adulthood in two European modern skeletal collections. *Am. J. Phys. Anthropol.*, **135**, 149–160.

Bogin, B. and Rios, L. (2003) Rapid morphological change in living humans: implications for modern human origins. *Comp. Biochem. Physiol. Part A*, **136**, 71–84.

Bollongino, R., Nehlich, O., Richards, M., Orschiedt, J., Thomas, M., Sell, C., Fajkosová, Z., Powell, A., and Burger, J. (2013) 2000 years of parallel societies in Stone Age Central Europe. *Sci. New York NY*, **342**, 479–81.

Borgognini-Tarli, S.M. and Canci, A. (1992) Aspetti antropologici e paleodemografici dal Paleolitico superiore alla prima eta' del Ferro. In: *Italia preistorica* (eds A. Guidi and M. Piperno), Laterza, Bari, pp. 239–273.

Brabin, L. and Brabin, B.J. (1992) Parasitic infections in women and their consequences. *Adv. Parasitol.*, **31**, 1–60.

Cardoso, H.F.V. and Gomes, J.E.A. (2009) Trends in adult stature of peoples who inhabited the modern Portuguese territory from the Mesolithic to the late 20th century. *Int. J. Osteoarchaeol.*, **19** (6), 711–725.

Cartron, I. and Castex, D. (2006) L'occupation d'un ancien îlot de l'estuaire de la Gironde: Du temple antique à la chapelle Saint-Siméon (Jau-Dignac et Loirac). *Aquitania*, **22**, 253–282.

Ceglia, V. (2000) Campochiaro (CB). La Necropoli di Vicenne. In: *L'oro degli Avari. Popolo delle Steppe in Europa* (eds E.A. Arslan and M. Buora). InForm Edizioni, Milan, pp. 212–221.

Ceglia, V. and Genito, B. (1991) La Necropoli Altomedievale di Vicenne Campochiaro. In: *Samnium Archeologia Archeologia del Molise* (eds S. Capini and A. Di Niro), Quasar ed., Rome, pp. 329–334.

Chamla, M.C. (1964) L'accroissement de la stature en France de 1880 à 1960; comparaison avec les pays d'Europe occidentale. In: *Bulletins et Mémoires de la Société d'anthropologie de Paris, XI° Série*. Tome 6, fascicule 2, 1964. pp. 201–278.

Civetta, A., Schmitt, A., Saliba-Serre, B., Gisclon, J.-L., and Loison, G. (2009) Comparaison de deux 'populations' de la deuxième moitié du Ve millénaire avant notre ère: approche anthropométrique. *Bull. Mém. Soc. Anthropol. Paris*, **21** (3-4), 141–158.

Cowan, R.S. (1999) The industrial revolution in the home. In: *The Social Shaping of Technology* (eds D. Mackenzie and J. Wajcman), Open University Press, Philadelphia, pp. 281–300.

Cunliffe, B. (1994) *The Oxford Illustrated Prehistory of Europe*. Oxford University Press, New York.

DeWitte, S. (2010) Sex differentials in frailty in medieval England. *Am. J. Phys. Anthropol.*, **143**, 285–297.

Diverrez, F., Poulmarc'h, M., and Schmitt, A. (2012) Nouvelles données sur les inhumations *ad sanctos* à l'époque modern en milieu rural: Le cas de l'église Saint-Pierre de Moirans (Isère). *Bull. Mém. Soc. Anthropol. Paris*, 1–12.

Eveleth, P.B. and Tanner, J.M. (1990) *Worldwide Variation in Human Growth*. Cambridge University Press, Cambridge.

Garn, S.M. (1970) *The Earlier Gain and the Later Loss of Cortical Bone*. Charles C. Thomas, Springfield.

Garn, S.M., Guzman, M.A., and Wagner, B. (1969) Subperiosteal gain and endosteal loss in protein-calorie malnutrition. *Am. J. Phys. Anthropol.*, **30** (1), 153–155.

Giannechini, M. and Moggi-Cecchi, J. (2008) Stature in archeological samples from Central Italy: Methodological issues and diachronic changes. *Am. J. Phys. Anthropol.*, **135**, 284–292.

Goude, G., Schmitt, A., Herrscher, E., Loison, G., Cabut, S., and Andre, G. (2013) Pratiques alimentaires au Neolithique moyen: nouvelles données sur le site de Pontcharaud 2 (Puy-de-Dome, Auvergne, France). *Bull. de la Societé Préhistorique Française*, **110**, 299–317

Graunt, J. (1975) *Natural and political observations mentioned in a following index and made upon the bills of mortality*. Arno Press, New York.

Greenfield, H.J. (2010) The Secondary Products Revolution: the past, the present and the future. *World Archaeol.*, **42** (1), 29–54.

Grossman, C.J. (1985) Interactions between the gonadal steroids and the immune system. *Science*, **227**, 257–261.

Guatelli-Steinberg, D. and Lukacs, J.R. (1999) Interpreting sex differences in enamel hypoplasia in human and non-human primates: developmental, environmental, and cultural considerations. *Am. J. Phys. Anthropol.*, **110**, 73–126.

Guidi, A. and Piperno, M. (1992) *Italia Preistorica*. Laterza, Roma.

Haak, W., Balanovsky, O., Sanchez, J., Koshel, S., Zaporozhchenko, V., Adler, C., Sarkissian, C., Brandt, G., Schwarz, C., Nicklisch, N., *et al.* (2010) Ancient DNA from European early Neolithic farmers reveals their near eastern affinities. *PLoS Biol.*, **8**, e1000536.

Hoff, R., Mott, K.E., Silva, J.F., Menezes, V., Hoff, J.N., Barrett, T.V., and Sherlock, I. (1979) Prevalence of parasitaemia and seroreactivity to *Trypanosoma cruzi* in a rural population of northeast Brazil. *Am. J. Trop. Med. Hyg.*, **28**, 461–466.

Holt, B.M. and Formicola, V. (2008) Hunters of the Ice Age: the biology of Upper Paleolithic people. *Yrbk Phys. Anthropol.*, **51**, 70–99.

Jansen, A., Stark, K., Schneider, T., and Schoneberg, I. (2007) Sex differences in clinical leptospirosis in Germany: 1997–2005. *Clin. Infect. Dis.*, **44**, e69–e72.

Kennedy, K.A.R. (1984) Growth, nutrition, and pathology in changing paleodemographic settings in South Asia. In: *Paleopathology at the Origins of Agriculture* (eds M.N. Cohen and G.J. Armelagos), Academic Press, Orlando, FL, pp. 169–192.

Klein, S.L. (2000) The effects of hormones on sex differences in infection: from genes to behavior. *Neurosci. Biobehav. Rev.*, **24**, 627–638.

Kurylëv, V.P., Pavlinskaya, L.R., and Simakov, G.N. (1989) Harness and weaponry. In: *Nomads of Eurasia* (ed. V.N. Basilov), Natural History Museum of Los Angeles County, Academic of Sciences of the USSR, Los Angeles, pp. 137–152.

Larsen, C.S. (1984) Health and disease in prehistoric Georgia: the transition to agriculture. In: *Paleopathology at the Origins of Agriculture* (eds M.N. Cohen and G.J. Armelagos), Academic Press, Orlando, FL, pp. 367–392.

Larsen, C.S. (1998) Gender, health, and activity in foragers and farmers in the American southeast: implications for social organization in the Georgia Bight. In: *Sex and Gender in Paleopathological Perspective* (eds A.L. Grauer and P. Stuart-Macadam), Cambridge University Press, Cambridge, pp. 165–187.

Lewis, M. (2002) Impact of industrialization: Comparative study of child health in four sites from medieval and postmedieval England (A.D. 850–1859). *Am. J. Phys. Anthropol.*, **119**, 211–223.

Loison, G. (1998) La nécropole de Pontcharaud en Basse Auvergne. In: *Sépultures d'Occident et genèse des mégalithismes (9000–3500 av. notre ère)* (ed. J. Guilaine), Errance, Paris, pp. 187–206.

Manzi, G., Santandrea, E., and Passarello P. (1997) Dental size and shape in the Roman Imperial Age: two examples from the area of Roma. *Am. J. Phys. Anthropol.*, **102**, 469–479

Manzi, G., Censi, L., Sperduti, A., and Passarello, P. (1989) Linee di Harris e ipoplasia dello smalto nei rest scheletrici delle popolazioni umane di Isola Sacra e Lucus Feroniae (Roma, I-III sec. d.C.). *Riv. di Antro.*, **67**, 129–148.

María-Dolores, R. and Martínez-Carrión, J.M. (2011) The relationship between height and economic development in Spain, 1850–1958. *Econ. Hum. Biol.*, **9** (1), 30–44.

Martínez-Carrión, J.-M. and Moreno-Lázaro, J. (2007) Was there an urban height penalty in Spain, 1840–1913. *Econ. Hum. Biol.*, **5**, 144–164.

Meiklejohn, C., Schentag, C., Venema, A., and Key, P. (1984) Socioeconomic change and patterns of pathology and variation in the Mesolithic and Neolithic of western Europe: some suggestions. In: *Paleopathology at the Origins of Agriculture* (eds M.N. Cohen and G.J. Armelagos), Academic Press, Orlando, FL, pp. 75–100.

Milisauskas, S. and Kruk, J. (2002) Late Neolithic, Crises, Collapse, New Ideologies, and Economies, 3500/3000–2200/2000 BC. In: *European Prehistory: A Survey* (ed. S. Milisauskas), Kluwer Academic/Plenum, New York, pp. 247–269.

Musson, A.E. and Robinson, E. (1969) *Science and Technology in the Industrial Revolution.* University of Toronto Press.

Peroni, R. (1989) *Protostoria dell'Italia continentale. La penisola italiana nelle eta' del Bronzo e del Ferro.* Biblioteca di Storia Patria, Roma.

Reitsema, L. and Vercellotti, G. (2012) Stable isotope evidence for sex- and status-based variations in diet and life history at medieval Trino Vercellese, Italy. *Am. J. Phys. Anthropol.*, **148**, 589–600.

Roberts, C.W., Walker, W., and Alexander, J. (2001) Sex-associated hormones and immunity to protozoan parasites. *Clin. Microbiol. Rev.*, **14**, 476–488.

Ruff, C.B. and Hayes, W.C. (1983) Cross-sectional geometry of the Pecos Pueblo femora and tibiaea biomechanical investigation: I. Methods and general pattern of variation. *Am. J. Phys. Anthropol.*, **60**, 359–381.

Ruff, C.B., Holt, B.M., Sladék, V., Berner, M., Garofalo, E., Garvin, H.M., Hora, M., Maijanen, H., Niinimäki, S., Salo, K., Schuplerová, E., and Tompkins, D. (2012) Stature and body mass estimation from skeletal remains in the European Holocene. *Am. J. Phys. Anthropol.*, **148**, 601–617.

Salzani, L. (a cura di) (2005) *La necropoli dell'Et. del Bronzo all'Olmo di Nogara'.* Museo Civico di Storia Naturale di Verona, Comune di Verona.

Sanna, E. (2002) Il secular trend in Italia. *Antropo*, **3**, 23–49.

Signoli, M., Leonetti, G., and Dutour, O. (1997) The great plague of Marseille (1720–1722): new anthropological data. *Acta Biol.*, **42**, 123–133.

Skeates, R. (2013) Neolithic Italy at 4004 BC: People and places. In: *Rethinking the Italian Neolithic* (eds M. Pearce and R.D. Whitehouse), Accordia Research Papers, vol. 13, Special Issue, pp. 1–29.

Šlaus, M. (2008) Osteological and dental markers of health in the transition from the late antique to the early medieval period in Croatia. *Am. J. Phys. Anthropol.*, **136**, 455–469.

Sparacello, V. (2014) The bioarchaeology of changes in social stratification, warfare, and habitual activities among Iron Age Samnites of central Italy. Doctoral Dissertation, University of New Mexico.

Steckel, R.H (1995) Stature and the Standard of Living. J. Econ. Lit., **33** (4), 1903–1940.

Stini, W.A. (1985) Growth rates and sexual dimorphism in evolutionary perspective. In: *The analysis of prehistoric diets* (eds R.I. Gilbert and J.H. Mielke), Academic Press, Orlando, FL, pp. 191–226.

Stinson, S. (1985) Sex differences in environmental sensitivity during growth and development. *Yrbk Phys. Anthropol.*, **28**, 123–147.

Taylor, B.C., Yuan, J.M., Shamliyan, T.A., Shaukat, A., Kane, R.L., and Wilt, T.J. (2009) Clinical outcomes in adults with chronic hepatitis B in association with patient and viral characteristics: a systematic review of evidence. *Hepatology*, **49** (5 Suppl.), S85–S95.

Vercellotti, G., Piperata, B., Agnew, A., Wilson, W., Dufour, D., Reina, J., Boano, R., Justus, H., Larsen, C., Stout, S., *et al.* (2014) Exploring the multidimensionality of stature variation in the past through comparisons of archaeological and living populations. *Am. J. Phys. Anthropol.*, **155**, 229–242.

Waldron, I. (1984) The role of genetic and biological factors in sex differences in mortality. In: *Sex differentials in mortality: trends, determinants, and consequences* (eds A.D. Lopez and L.T. Ruzicka), Department of Demography, Australian National University, Canberra, pp. 141–164.

10

Iberia

Christopher B. Ruff and Heather Garvin

Center for Functional Anatomy and Evolution, Johns Hopkins University School of Medicine, Baltimore, MD, USA

10.1 Introduction

The sites included in this regional analysis are listed in Table 10.1 and shown in Figure 10.1. They include four samples from present-day Portugal and six samples from Spain. Although there is cultural and some biological variation among populations from these two present-day countries (see below), for the purposes of the present analyses they are considered together, in part to increase temporal period coverage. It is also likely that broader temporal trends were similar in the two regions, or at least not more dissimilar than between different parts of the same present-day countries, such as coastal versus inland locations. Fortunately, there is some geographic variation within broader temporal periods, with Mesolithic and Medieval samples from more than one sub-region (Fig. 10.1). As described further below, the unique very recent sample from the Luis Lopez Collection in Lisbon also allows comparisons of modern high and low socio-economic status individuals, with each other and with earlier population samples.

The three Mesolithic samples of Moita de Sebastiao, Muge Arruda, and Los Canes are all from coastal or near-coastal locations. In fact, almost all known Iberian Mesolithic sites are coastal, which may reflect ecological factors following the end of the Ice Age in this region, that is, a lack of consistent water resources further inland (Zilhao, 2000). The three sites included here are all relatively close in age and date from the late Mesolithic, not long before food production was introduced to the Iberian peninsula. Moita de Sebastiao and Muge Arruda are large shellmound sites from the lower Tagus River Valley on the west coast of Portugal (Jackes and Lubell, 1999a), while Los Canes is from a small cave located about 12 km from the Atlantic coast of northern Spain (although probably several times this distance in terms of traveling distance; Arias and Garralda, 1996; Straus, 2008). Archaeological evidence for at least Moita de Sebastiao and Muge Arruda indicates "a low degree of residential mobility bordering on sedentism" (Straus, 2008), with semi-permanent structures, large cemeteries, and probable occupation of the sites throughout the year. The subsistence economy was very diversified and concentrated on resources near the coast, although periodic trips farther inland for the purposes of hunting and flint acquisition were also undertaken. Los Canes Cave is located on the side of hill overlooking a valley connected to the Atlantic, with surrounding mountains (Arias and Garralda, 1996). Despite its position, the cave is characterized by an "astonishing frequency of marine shells" (ibid.: 133), suggesting a close association with the coast, although whether

Skeletal Variation and Adaptation in Europeans: Upper Paleolithic to the Twentieth Century,
First Edition. Edited by Christopher B. Ruff.

Table 10.1 Iberian samples.

Map no.[a]	Site	Period	Date range (cal.)	Male N	Female N	Total N
1	Moita de Sebastiao	Mesolithic	5710–5910 BC	5	2	7
2	Muge Arruda	Mesolithic	5370–5840 BC	1		1
3	Los Canes	Mesolithic	5210–5830 BC	2	1	3
4	Terrera del Reloj	Bronze	1710–1200 BC	2	2	4
5	Castellon Alto	Bronze	1710–1200 BC	15	15	30
6	Santa Maria de Hito	Early Med.	500–1200 AD	19	21	40
7	Villanueva de Soportiva	Early Med.	850–1100 AD	13	10	23
8	San Baudelio de Berlanga	Late Med.	1100–1200 AD	12	10	22
9	Leiria	Late Med.	1200–1550 AD	16	10	26
10	Luis Lopez Collection	Very Recent	1933–1973 AD	26	25	51
	(Luis Lopez Low-SES)			(13)	(7)	(20)
	(Luis Lopez High-SES)			(13)	(18)	(31)
Total				111	96	207

[a] See Figure 10.1.

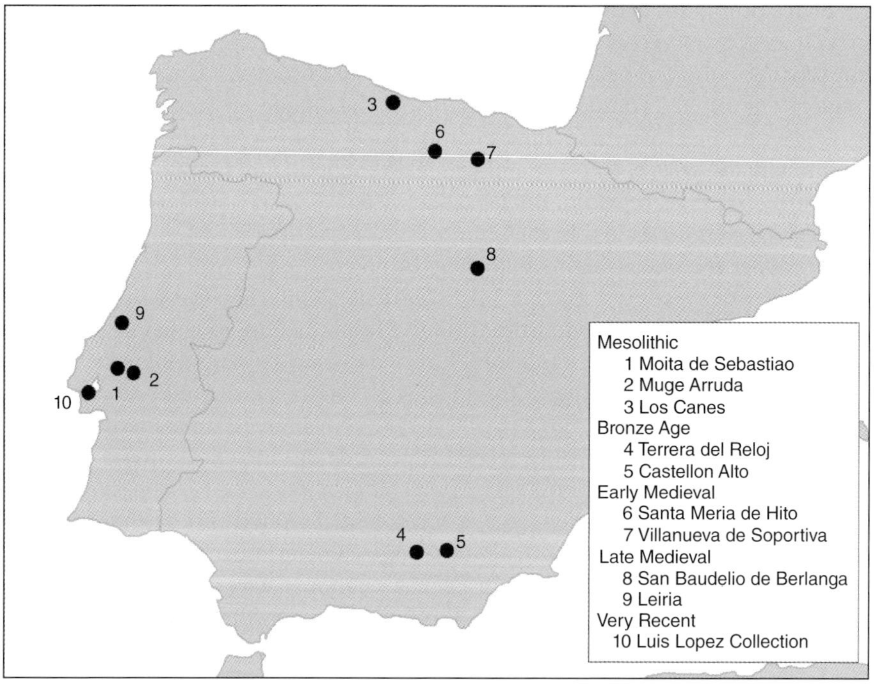

Figure 10.1 Map of site locations included in the study. Numbers refer to the list in Table 10.1.

this involved temporary visits to the coast or actual physical and cultural continuity with coastal 'Asturian' populations is unknown.

The two Bronze Age sites included here, from southeastern Spain – Terrera del Reloj and Castellón Alto – both represent the Argar Culture, which is well known for its extensive metallurgical production (al-Oumaoui *et al.*, 2004; Lull *et al.*, 2011). Subsistence was based primarily on animal husbandry and agriculture. Settlements were situated on the tops of steep hills in rugged terrain. There is evidence for class structure and sexual division of labor, with males more likely to be involved in strenuous activities (such as herding) (al-Oumaoui *et al.*, 2004; Jiménez-Brobeil *et al.*, 2010; Lull *et al.*, 2011).

Two Early Medieval sites from northern Spain are sampled – Santa Maria de Hito and Villanueva de Soportiva. Both represent small rural villages with an agrarian economy based on cattle-raising and agriculture (Galera and Garralda, 1993; al-Oumaoui *et al.*, 2004). Both sites are located in river valleys, surrounded by mountains. Males at Villanueva de Soportiva also participated as warriors in regional conflicts, and have been described as "...peasant-soldiers, who alternated the plough and the sword" (Galera and Garralda, 1993: 345).

San Baudelio de Berlanga was a small rural livestock farm associated with a Late Medieval monastery in a hilly region of northern Spain. Given the terrain and emphasis on herding, it can be expected that the lifestyle was rigorous, particularly for males (al-Oumaoui *et al.*, 2004). In contrast, our other Late Medieval sample, from a church cemetery in the city of Leiria, central Portugal, was urban and inhabited by a heterogeneous population including multiple social classes (Cardoso and Garcia, 2009). The city was relatively prosperous during this time period (about 1200–1550 AD), with the area around the church occupied by individuals involved in the service sector, such as merchants, as well as members of the local aristocracy.

A very recent osteological sample was obtained from the Luís Lopes Collection in Lisbon. These specimens were obtained from cemeteries in Lisbon in the late 20th and early 21st centuries following customary exhumation of material from graves after natural skeletonization (Cardoso, 2006). Individuals are associated with medical and other legal records giving sex, age, and other demographic information, including area of residence and occupation. Using this information, it is possible to reconstruct, to varying degrees of accuracy, socioeconomic status (SES). Details are given in Cardoso (2007). A modified version of this procedure was used here (H. Cardoso, personal communication) to categorize individuals into either 'High' or 'Low' SES, corresponding roughly to non-manual and manual workers. The categorization of most of the males is more certain that that of the females, the great majority of whose occupations were listed as simply 'doméstica' (i.e., housewife). In four females it was impossible to even estimate SES. For the purposes of the present analyses they were grouped with High status females, first, because most (14/21) of the other females in the sample were categorized that way, making it more likely on probability grounds that these were as well, and second, because the average reconstructed stature of the uncategorized females was closer to that of the High-SES group. As will be shown later, however, there is reason to believe that SES analyses are probably only valid for males. Because of the possibility of different lifestyles leading to different anthropometric as well as bone strength characteristics, the High- and Low-SES individuals are treated as separate groups in temporal comparisons within Iberia. Average and range of birth and death years were very similar in the two groups: for High-SES individuals, year of birth averaged 1912 (range 1890–1928), and year of death 1953 (range 1933–1974); corresponding figures for Low-SES individuals were 1911 (1900–1938) and 1952 (1934–1973), respectively. Thus, the two groups can be considered to be essentially exactly contemporaneous, and living within the early-mid-20th century in Lisbon. The sample was also selected to avoid very old individuals, to minimize possible effects of osteoporosis: the maximum age was 67 years, with a mean age of 41 years in both SES groups (High-SES range 21–67 years; Low-SES range 20–65 years).

In addition to these skeletal samples, comparative anthropometric data for recent Spanish and Portuguese living populations were also obtained from the literature. Only stature could be included, with other anthropometric data too sparse for comparisons. Also, primarily because of the availability of military records, data for males were much more numerous than for females. Table 10.2 lists the living samples with their sources, measurement dates, and average birth years. The military conscript data (Lacerda, 1904; Sobral, 1990; Padez, 2003; María-Dolores and Martínez-Carrión, 2011) extend back to the 19th century (birth date). In some cases (Sobral, 1990; María-Dolores and Martínez-Carrión, 2011) they were derived for only selected regions of the country, but they include both rural and urban areas and are argued to be broadly representative of national distributions by the investigators. The data from Sobral (1990: their Table 5) were re-analyzed to combine urban and rural subsamples and all four regions, with the average difference between urban and rural used to calculate missing values for the most recent rural subsamples. National Health Survey data for Spain were derived from graphical representations in Cavelaars *et al.* (2000: their Fig. 1), and are based on a total of almost 25,000 individuals. Unlike the military personnel studies, which were limited to individuals about 20 years of age, this survey included adults ranging from 20 to 74 years, with the data then subdivided by birth year. Thus, there is a definite possibility that average stature in some of the earlier birth years for this data set is slightly reduced due to normal loss of stature with aging (e.g., see Sorkin *et al.*, 1999). The data for Portuguese females were obtained from a maternity hospital in Lisbon, and thus represent a primarily urban population (Conceição *et al.*, 2012). More than 30,000 individuals participated, and these were relatively evenly distributed between five educational levels, ranging from very little schooling to university educated. Therefore, the sample should be broadly representative of multiple SES groups, and thus comparable to the (pooled) Luís Lopes Collection sample.

Body size (stature, body mass, living bi-iliac breadth) and shape parameters (limb length proportions, relative bi-iliac breadth and clavicular length) were determined as described in Chapter 2. Almost two-thirds of the total sample had statures calculated using the anatomical method, and body mass calculated using the stature/bi-iliac technique. However, these percentages are somewhat inflated by the very recent Lisbon sample, where almost all individuals were calculated in this way. Only one Mesolithic individual (Los Canes III) was complete enough to apply these methods, and in fact due to incomplete preservation of bone ends, stature could only be calculated for nine of 11 individuals, and body mass (using the femoral head, except Los Canes III) in five individuals. Because of the relative lack of body mass data for this group, and their importance in the overall analyses, we include some additional temporal analyses of cross-sectional shape (Z_x/Z_y in the femur and tibia) that do not depend on body mass estimation; this more than doubled the available Mesolithic sample. Stature and body mass of the Bronze Age sample were calculated using anatomical techniques in 42% and 28% of that sample, respectively; all other samples had about half or more individuals estimated anatomically.

Cross-sectional robusticity parameters include A-P bending strength (Z_x) of the femur and tibia, and average bending/torsional strength (Z_p) of the femur, tibia, and right and left humeri, all standardized for body size, as well as A-P/M-L bending strength (Z_x/Z_y) of the lower limb bones, as noted above. Percentage cortical area of all four bones was also compared.

Because of possible variation between samples within temporal periods (e.g., the two Late Medieval samples and the two Very Recent samples), the different sites and SES groups in the Very Recent sample were analyzed separately in temporal analyses, except that the Mesolithic and Bronze Age sites were each combined due to sample size limitations (although possible differences between the Los Canes and Portuguese Mesolithic sites are examined separately). Tukey tests are used to compare across these samples, using a 0.05 significance level. All analyses are

Table 10.2 Living Iberian anthropometric samples.

Country	Sample	Meas. date	Sex	Height (cm)	Birth year	Years BP
Portugal	Maternity hospital[a]	1991–2004	F	160.2	1966	34
Portugal	Maternity hospital[a]	1991–2004	F	160.9	1974	26
Portugal	Maternity hospital[a]	1991–2004	F	160.9	1982	18
Portugal	Military conscripts[b]	1904	M	163.2	1884	116
Portugal	Military conscripts[c]	1930	M	163.8	1910	90
Portugal	Military conscripts[c]	1940	M	164.7	1920	80
Portugal	Military conscripts[c]	1950	M	165.4	1930	70
Portugal	Military conscripts[c]	1960	M	166.0	1940	60
Portugal	Military conscripts[c]	1970	M	167.4	1952	48
Portugal	Military conscripts[c]	1980	M	168.1	1962	38
Portugal	Military conscripts[d]	1985	M	169.2	1967	35
Portugal	Military conscripts[d]	1990	M	171.4	1972	28
Portugal	Military conscripts[d]	2000	M	172.1	1982	18
Spain	National Health Survey[e]	1987	F	158.0	1917	83
Spain	National Health Survey[e]	1987	F	159.5	1927	73
Spain	National Health Survey[e]	1987	F	160.0	1937	63
Spain	National Health Survey[e]	1987	F	160.5	1947	53
Spain	National Health Survey[e]	1987	F	161.0	1957	43
Spain	National Health Survey[e]	1987	F	162.0	1965	35
Spain	Military conscripts[f]	1875	M	161.9	1855	145
Spain	Military conscripts[f]	1885	M	162.1	1865	135
Spain	Military conscripts[f]	1895	M	162.2	1875	125
Spain	Military conscripts[f]	1905	M	163.1	1885	115
Spain	Military conscripts[f]	1915	M	163.4	1895	105
Spain	Military conscripts[f]	1925	M	164.2	1905	95
Spain	Military conscripts[f]	1935	M	165.0	1915	85
Spain	Military conscripts[f]	1945	M	164.4	1925	75
Spain	Military conscripts[f]	1955	M	165.5	1935	65
Spain	Military conscripts[f]	1965	M	166.7	1945	55
Spain	Military conscripts[f]	1975	M	168.8	1955	45
Spain	Military conscripts[f]	1985	M	171.9	1965	35
Spain	Military conscripts[f]	1995	M	174.3	1975	25
Spain	National Health Survey[e]	1987	M	167.0	1917	83
Spain	National Health Survey[e]	1987	M	167.5	1927	73
Spain	National Health Survey[e]	1987	M	169.0	1937	63
Spain	National Health Survey[e]	1987	M	170.5	1947	53
Spain	National Health Survey[e]	1987	M	172.0	1957	43
Spain	National Health Survey[e]	1987	M	173.0	1965	35

[a] Conceicao *et al.*, 2012; [b] Lacerda, 1904; [c] Sobral, 1990; [d] Padex, 2003; [e] Cavelaars *et al.*, 2000; [f] Maria-Dolores and Martinez-Carrion, 2011.

carried out within sex. Sexual dimorphism within samples is tested using *t*-tests with Bonferroni correction, as are comparisons with non-Iberian Europeans. The latter are carried out within broader temporal periods (i.e., Mesolithic, Bronze, Early Medieval, Late Medieval, Very Recent). Average sexual dimorphism is calculated using male (M) and female (F) means as:

$$\left[\left(M-F\right)/\left(M+F\right)/2)\right]\times 100.$$

Humeral bilateral asymmetry is assessed using paired *t*-tests. An urban–rural comparison between the Late Medieval Leiria (urban) and San Baudelio de Berlanga samples is also carried out, again using (paired) *t*-tests. Because almost all of the samples derive from hilly or mountainous regions (the exceptions being some of the Mesolithic sites), no comparisons across terrain were performed.

10.2 Body Size and Shape

10.2.1 Body Size

Temporal period comparisons of stature, body mass, and bi-iliac breadth are given in Table 10.3, with boxplots of stature and body mass shown in Figure 10.2. The boxplots of stature also include mean statures for living Iberian populations (Spain and Portugal), derived from Table 10.2. For this comparison, only living samples with post-World War II birth years (i.e., 1945 or later) were included. Secular changes in stature among all the living samples are considered below.

Stature among the archaeological/osteological samples is relatively constant across temporal periods, except for Mesolithic males, who are very short, and Very Recent (VR) Low-SES males, who are also short (Fig. 10.2(a)). In fact, Mesolithic males are statistically significantly shorter than all other temporal samples, except for VR Low-SES males. VR Low-SES males are significantly or near-significantly shorter than some of the Medieval samples (Santa Maria de Hito, San Baudelio de Berlanga, and Leiria; $p = 0.064$) and the VR High-SES sample. VR High-SES males are equivalent in stature to Bronze Age and Medieval samples. Living (post-1945) males are the tallest, averaging 170.3 cm, which is still only 2–4% higher than earlier pre-20th century samples (except Mesolithic males).

Females show less consistent patterns of temporal variation in stature, although this may in part be due to the lack of data for Mesolithic females ($n = 1$), and problems in assigning SES to the VR females, as discussed above. Very Recent 'High-SES' females are actually (non-significantly) shorter here than 'Low-SES' females, and are significantly shorter than two of the Medieval samples (Santa Maria de Hito and Leira), while 'Low-SES' females are not. The only other significant temporal difference in stature among females is that between the Bronze and Santa Maria de Hito (taller) samples. Living females are 1–5% taller than those of earlier periods.

Temporal trends in body mass are similar to those in stature (Fig. 10.2(b)), with small-bodied Mesolithic and VR Low-SES males, and otherwise no significant differences between periods (among other male samples, and any female samples). The differences in body mass between the two smaller male samples and other samples is perhaps even more pronounced overall than the differences in stature, with less overlap in distributions, particularly for the VR Low-SES sample. The two Late Medieval male samples also tend to be heavier (although non-significantly so) than Early Medieval and VR High-SES males, while statures are very similar in these five groups, which suggests that they were also somewhat wider-bodied (also see below). Very Recent female samples are again reversed in expected direction between 'High-' and 'Low-' SES groups.

Table 10.3 Temporal differences in body size.

Prop.	Sex		Mesol.	Bronze	EMed. S.M.H.	EMed. Vill.	LMed. S.B.	LMed. Leiria	VR Low	VR High
Stature (cm)	Males	N	8	17	20	13	12	16	13	13
		Mean	155.84	163.87	165.89	163.61	166.47	165.74	159.48	167.08
		SE	1.64	1.23	1.23	1.56	0.95	1.61	2.02	1.48
	Females	N	1	16	19	10	10	10	7	18
		Mean	153.38	153.67	158.72	155.00	153.69	157.15	154.58	151.00
		SE		0.88	0.95	1.41	1.94	1.31	1.31	1.04
Body mass (kg)	Males	N	4	16	21	13	12	16	13	13
		Mean	53.06	62.80	66.68	66.22	70.20	68.46	58.01	65.77
		SE	5.01	1.50	1.48	1.61	1.31	2.33	2.46	2.57
	Females	N	1	16	19	10	10	10	7	18
		Mean	47.03	52.00	57.08	57.00	54.52	56.15	57.12	53.24
		SE		1.11	0.94	1.95	1.95	1.44	1.71	1.16
Bi-iliac bd. (cm)	Males	N	1	3	8	10	7	13	12	12
		Mean	26.25	28.32	28.71	28.89	29.85	29.36	26.84	28.69
		SE		0.27	0.66	0.47	0.46	0.71	0.63	0.55
	Females	N		6	11	8	5	5	7	17
		Mean		27.92	27.18	28.88	29.25	28.64	28.72	27.65
		SE		0.47	0.54	0.84	0.45	0.68	0.65	0.49

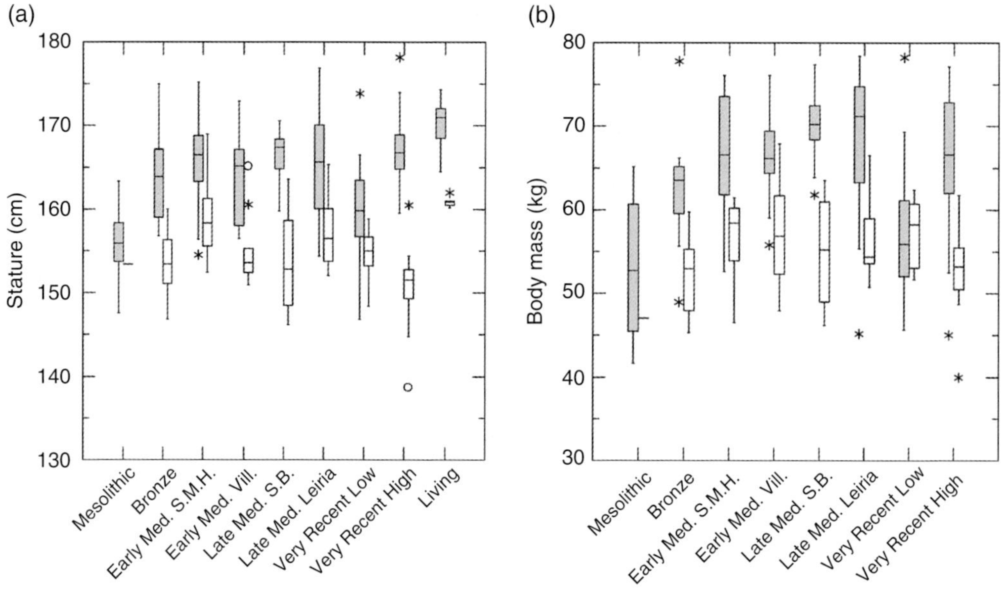

Figure 10.2 Boxplots showing temporal variation in (a) stature and (b) body mass. Open boxes: females; gray boxes: males. 'Living' refers to the samples listed in Table 10.2 with birth dates ≥1945, with distributions based on sample means. Summary statistics for archaeological samples are given in Table 10.3.

Bi-iliac breadth shows no significant temporal variation in females, and relatively little variation in males, although the one available Mesolithic male is narrow-bodied, as are the VR Low-SES males (Table 10.3). Thus, the reduction in body mass of the Low-SES males is due to both a reduction in body breadth as well as stature.

The two Los Canes males with stature and body mass estimates fall close (within about 1%) to the mean of the available Portuguese shellmound sites (Moita de Sebatiao and Muge Arruda), indicating no regional differences, although sample sizes for body mass in particular are very small (Table 10.3). There is no evidence for any significant urban–rural difference in body size between the Leiria and San Baudelio de Berlanga samples, although urban (Leiria) females are slightly larger and urban males slightly smaller (Table 10.3).

Secular trends in stature since the late 19th century are shown in Figure 10.3. A positive temporal trend is apparent in all of the data sets, in general concordance with Europe as a whole (e.g., Steckel, 1995; Cavelaars *et al.*, 2000; Danubio and Sanna, 2008). The longer-term military data also demonstrate the slower increases characteristic of the late 19th and early 20th centuries, and the more rapid pace of increase in the later 20th century, which is consistent with previous studies of Iberian samples (Padez, 2003; María-Dolores and Martínez-Carrión, 2011), and socioeconomic data showing the most marked improvements in standards of living in this region occurring since the middle of the last century (also see Cardoso and Garcia, 2009). These results are also consistent with pan-European comparisons, in which most northern European countries have leveled off in stature since the late 20th century, while Portugal and Spain have continued to increase (Larnkjaer *et al.*, 2006). The two military samples are similar in height until the late 20th century, when the Spanish sample moves ahead of the Portuguese sample, matching other recent data (Larnkjaer *et al.*, 2006). The Spanish female sample is also somewhat taller than the Portuguese female sample at the one time point in common between them (1965–1966 birth years). Interestingly, the Spanish National Health Survey data for males indicate a higher average stature than Spanish military conscripts until the 1960s, when the gap

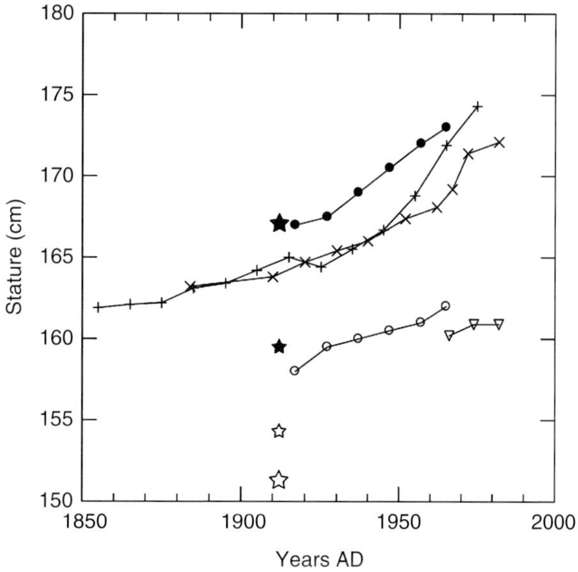

Figure 10.3 Temporal changes in stature among living Iberian samples (see Table 10.2) and the Lisbon osteological sample, by birth year. Filled circles: males; open circles: females; +: Spanish military conscripts; x: Portuguese military conscripts; circles: Spanish NHS; triangles: Portuguese maternity hospital; stars: Lisbon osteological sample – larger = High-SES, smaller = Low-SES.

closes, suggesting some possible selection in the military sample, although it is also possible that some recruits were still growing (Padez, 2003). There is an apparent increase in sexual dimorphism of stature in both Spain and Portugal in the most recent samples, a trend that has also been noted elsewhere in Europe (Danubio and Sanna, 2008; also see Chapters 4 and 8).

The High-SES Lisbon sample males are about as tall as the Spanish NHS males of the same time period, and slightly taller than the two military samples. In contrast, the Low-SES Lisbon sample males are only slightly taller than contemporaneous Spanish NHS *females*, and well below the other male samples. Thus, the outlier position of the Lisbon Low-SES males relative to both earlier and later samples (Fig. 10.2(a)) is confirmed. Both Lisbon female samples are shorter than contemporaneous Spanish females. Without more data it is difficult to tell if they would also be significantly shorter than other contemporaneous Portuguese females, although given the average difference between Spanish and Portuguese samples at this and later time periods, it seems likely, at least for the 'High'-SES group.

Sexual dimorphism in stature, body mass, and bi-iliac breadth by temporal period is shown in Table 10.4. Sexual dimorphism (SD) in stature in the Bronze Age and Medieval samples varies between about 4% and 8%, which is normal for modern human populations (Stini, 1976; Gray and Wolfe, 1980) (there are insufficient data to calculate SD in the Mesolithic sample). The level of SD in stature in the Very Recent Lisbon samples depends on how the male and female samples are paired. As shown earlier, there is evidence from both the quality of the demographic records as well as the patterning of biological differences, that SES classifications of the females are likely not accurate. Therefore, SD in stature for the Very Recent groups were calculated in two ways: 1) by pairing ostensibly similar SES groups; and 2) by pairing Low- and High-SES males with an average of all Lisbon females (i.e., pooled SES). The latter approach seems to produce more realistic SD estimates, of 5% and 9% for Low- and High-SES groups, respectively. (The corresponding figures for 'matched SES' comparisons are 3% and 10%, with a non-significant difference in stature in the 'Low-SES' group.) Using either method, however, the High-SES group has much more SD in stature than the Low-SES group, which is consistent with expectations of the 'female buffering' hypothesis (Stini, 1976; Gray and Wolfe, 1980; although also see Stinson, 1985).

Table 10.4 Sexual dimorphism.

Property	Bronze	EMed. S.M.H.	EMed. Vill.	LMed. S.B.	LMed. Leiria	VR Low*	VR High*	Pooled
Stature	**6.42**	**4.42**	**5.40**	**7.98**	**5.32**	3.13 (**4.81**)	**10.11 (9.45)**	**5.92**
Body Mass	**18.82**	**15.52**	**14.96**	**25.16**	**19.76**	1.54 (6.56)	**21.05 (19.05)**	**16.87**
Bi–iliac Bd.	1.42	5.48	0.04	2.05	2.51	−6.76 (−4.07)	3.69 (2.56)	1.72
Crural	0.06	0.12	−0.41	−0.14	−0.41	−0.11	1.62	−0.02
Rt. Brachial	0.74	2.92	1.52	3.30	2.64	1.46	2.41	**2.10**
Lt. Brachial	1.55	2.83	3.36	1.31	2.06	2.14	2.67	**2.38**
Rel. Sit. Ht.	−0.59	−1.66	−6.93		0.11	1.09	0.73	−0.54
Bi–iliac/Stat.	−4.62	0.55	−4.75	−3.29	−1.63	**−9.66**	**−6.77**	**−4.26**
Rel. Clav. Ln.	4.67	3.03	0.92	7.00	5.65	5.12	1.99	**3.81**

* Values in parentheses calculated using pooled SES Very Recent female sample (see text).
Sexual dimorphism = [(M − F)/(M + F)/2)] × 100
Bold text = p <0.05, with Bonferroni correction.

Sexual dimorphism in body mass prior to the Very Recent period varies between 15% and 25%, which is moderate to fairly high for modern humans (Ruff, 2002). No particular patterning is discernible, although the Late Medieval San Baudelio sample has the highest level. Sexual dimorphism in the Lisbon High-SES group (calculated either way) is again much greater (19–21%) than in the Low-SES group (2–7%). Bi-iliac breadth shows relatively low levels of SD, as expected (Ruff, 1994). Lisbon Low-SES males show particularly low values for bi-iliac breadth (thus, negative values for SD). Average SD values for stature, body mass, and bi-iliac breadth over all Iberian samples (Table 10.4) are similar to those reported worldwide (see references above).

10.2.2 Body Shape

Linear body proportions of the Iberian samples are shown in Table 10.5. There is no statistically significant variation in crural index between temporal groups within sex. In pooled-sex comparisons, the (male) Mesolithic sample has a significantly lower crural index than the pooled Late Medieval San Baudelio sample, and is absolutely lower than any other period. Thus, unlike Europe generally (see Chapter 4), Mesolithic males in Iberia do not have relatively high crural indices compared to later Holocene samples. Other than this, however, there is no discernible general temporal trend.

Brachial indices show the typical temporally declining pattern of the European Holocene in general (see Chapter 4), although differences between pairs of samples only reach significance in a few cases – Very Recent High-SES is lower than Early Medieval Villanueva and Bronze Age (left side), and the two Late Medieval samples are lower than the Bronze Age sample on the right side in females. Mesolithic samples are very small and not consistently high or low.

Relative sitting height is high (i.e., lower limbs are relatively short) in both High- and Low-SES Very Recent male samples, compared to all earlier groups (statistically significant except for comparisons to Leiria). Very Recent females of either SES group do not show this pattern, however. No other temporal patterning is apparent.

Body breadth proportions are shown in Table 10.6. Except for relatively smaller bi-iliac breadths in Early Medieval Santa Maria females, there is no temporal patterning in these proportions in either sex. There are no significant urban–rural differences in body shape between

Table 10.5 Temporal differences in linear body proportions.

Prop.	Sex		Mesol.	Bronze	EMed. S.M.H.	EMed. Vill.	LMed. S.B.	LMed. Leiria	VR Low	VR High
Crural	Males	N	5	10	18	13	10	13	13	13
		Mean	80.59	84.15	83.00	83.11	84.28	82.09	82.92	83.65
		SE	0.45	0.89	0.46	0.55	0.76	0.67	1.06	0.62
	Females	N		12	12	10	7	9	7	18
		Mean		84.09	82.90	83.45	84.40	82.43	83.02	82.31
		SE		0.88	0.88	0.42	0.67	0.48	0.89	0.50
Rt. brachial	Males	N		8	11	11	10	13	13	12
		Mean		77.46	75.95	77.33	74.25	75.44	74.65	74.04
		SE		0.45	0.90	0.60	0.82	0.65	0.83	0.75
	Females	N		10	10	7	4	8	7	18
		Mean		76.89	73.76	76.16	71.84	73.48	73.57	72.27
		SE		0.75	0.80	0.51	1.03	1.27	0.70	0.46
Lt. brachial	Males	N	2	7	10	10	10	15	13	13
		Mean	78.08	77.80	77.34	78.07	74.66	75.34	74.80	74.06
		SE	2.46	0.66	0.46	0.87	0.65	0.65	0.80	0.81
	Females	N	1	7	13	8	3	9	5	18
		Mean	72.44	76.60	75.18	75.49	73.69	73.80	73.22	72.11
		SE		1.12	0.74	0.57	0.42	1.17	1.19	0.40
Rel. sit. ht.	Males	N		4	1	2	6	6	13	13
		Mean		38.56	38.64	38.24	39.02	39.40	40.32	40.62
		SE		0.47		1.31	0.32	0.24	0.25	0.23
	Females	N		2	1	2		3	7	16
		Mean		38.79	39.29	40.99		39.36	39.88	40.33
		SE		0.85		1.46		0.21	0.32	0.21

Table 10.6 Temporal differences in body width proportions.

Property	Sex		Mesol.	Bronze	EMed. S.M.H.	EMed. Vill.	LMed. S.B.	LMed. Leiria	VR Low	VR High
Bi-iliac/	Males	N	1	3	8	10	7	13	12	12
Stat.		Mean	16.53	17.38	17.20	17.79	17.98	17.71	16.86	17.12
		SE		0.34	0.45	0.30	0.29	0.40	0.30	0.30
	Females	N		6	11	8	5	5	7	17
		Mean		18.20	17.11	18.65	18.58	18.00	18.57	18.32
		SE		0.22	0.32	0.47	0.46	0.31	0.33	0.29
Rel. clav.	Males	N		8	10	9	9	10	11	12
len.		Mean		18.23	17.88	17.68	18.77	18.11	17.99	18.10
		SE		0.29	0.26	0.24	0.17	0.34	0.24	0.21
	Females	N	1	9	11	8	4	8	5	16
		Mean	16.43	17.40	17.34	17.51	17.50	17.11	17.09	17.74
		SE		0.29	0.21	0.25	0.14	0.27	0.28	0.23

the Leiria and San Baudelio de Berlanga samples, except that the Leiria (urban) sample has a higher crural index (pooled sex).

Similar to other Europeans (Chapter 6), there is no SD in crural index, while males have higher brachial indices and relative clavicular breadths, and females relatively wider bi-iliac breadths (Table 10.4). There is no marked temporal patterning to these differences, except that SD in relative bi-iliac breadth is particularly pronounced in the Very Recent samples, especially the Low-SES comparison, again highlighting the narrow bodies of the Low-SES males.

10.2.3 Comparisons to other Europeans

Stature and body mass of the Iberian samples, pooled within general temporal period, are compared to non-Iberian Europeans from our study sample in Figure 10.4. In general, as with living populations (Cavelaars *et al.*, 2000; Larnkjaer *et al.*, 2006), Iberians are shorter on average than other Europeans, although the differences only reach significance (with Bonferroni correction) for Mesolithic males (stature and body mass) and Bronze Age females (stature). There is some temporal variation in this general pattern, however. In particular, Iberian Early Medieval females and Late Medieval males are very similar in average height and weight to other Europeans of these time periods. The greater reduction in body size, especially body mass, between Late Medieval and Very Recent Iberian males compared to other Europeans is striking, although part of this apparent reduction may be due to inclusion of only Portuguese in our Very Recent group, who tend to be shorter than recent Spanish samples (Fig. 10.3). On the other hand, the Portuguese Late Medieval sample from Leiria is as large-bodied as the Late Medieval Spanish sample from San Baudelio (Fig. 10.2). This recent trend towards smaller body size is not as marked for females (Fig. 10.4). The small body size of Mesolithic Iberian males, noted earlier, is further emphasized by comparisons with other Mesolithic males, who are much taller and heavier on average. This is true even if non-Iberian Mesolithics are limited

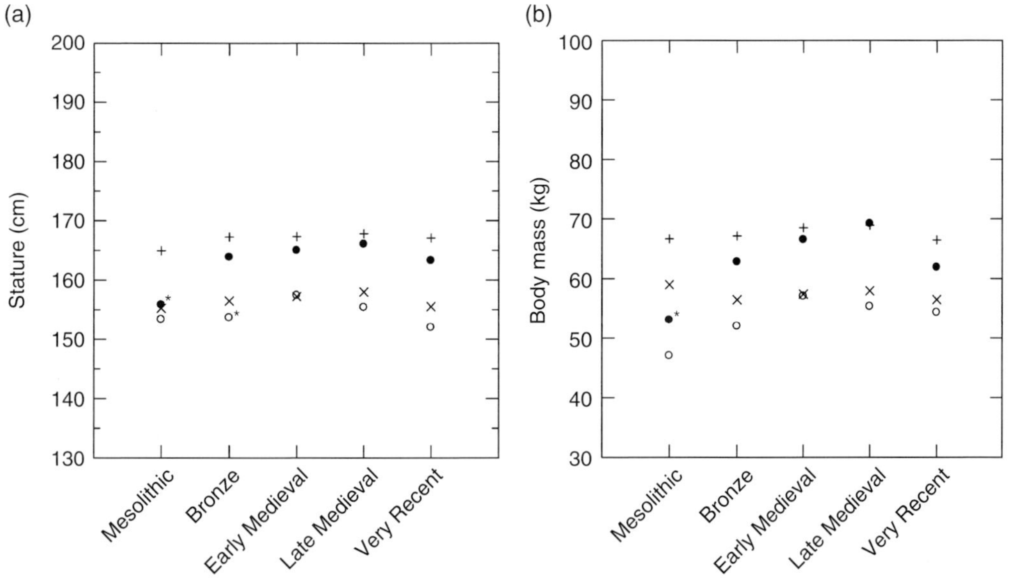

Figure 10.4 Comparisons of (a) stature and (b) body mass in Iberians and other Europeans by temporal period. Filled circles: Iberian males; open circles: Iberian females; +: other European males; x: other European females. Asterisks indicate significant difference of Iberian from other European samples.

to Southern Europe. Other geographic and temporal factors bearing on this finding are discussed further below.

Bi-iliac breadth (data not shown) is non-significantly different between Iberian and other European samples within temporal periods. Thus, despite being generally shorter than average Europeans, Iberians were not particularly narrow-bodied.

Comparisons of linear body proportions with other Europeans (not shown) demonstrate a few significant differences, mainly higher crural indices in Iberian Late Medieval and Very Recent samples, which is consistent with general Southern/Northern European differences (Ruff *et al.*, 2012; also see Chapter 4). This is not true for Iberian Mesolithic males, however, who have small crural indices compared to other European Mesolithics (Southern or Northern). Very Recent Iberian males have near-significantly higher relative sitting height (with Bonferroni correction, $p < 0.041$) than other Europeans. There are no significant differences in body breadth proportions with other Europeans, except for the relatively low bi-iliac/stature index in the Santa Maria female sample noted earlier.

10.3 Cross-Sectional Properties

10.3.1 Bone Strength

Temporal period comparisons of lower limb bone strength parameters, standardized for body size, are shown in Table 10.7 and Figures 10.5 and 10.6. For the tibia (Figs 10.5(a) and 10.6(a)), the main temporal distinction is the greater strength of Bronze Age males compared to all other temporal periods. This reaches significance or near-significance ($p < 0.10$) for A-P bending strength (Z_x) in all comparisons, and for average bending/torsional strength (Z_p) in comparisons with the Early Medieval Santa Maria, Late Medieval Leiria, and two Very Recent samples. Bronze Age females also have somewhat elevated tibial strengths, but comparisons with other periods do not reach statistical significance or near-significance. Thus, SD in tibial strength is greater in the Bronze Age (34–35%) than in any other period (all <30%). The pattern for the femur is not nearly as marked (Figs. 10.5(b) and 10.6(b)), with Bronze Age males only significantly or near-significantly greater than the Late Medieval Leiria and VR High-SES samples (Z_x only). In fact, a more prominent pattern in the femur is for Leiria males to be particularly low in strength compared to other groups (significant or near-significant for Z_p in all comparisons except Mesolithic and VR High-SES). Leiria females are also low in femoral strength compared to most time periods, reaching significance or near-significance for Z_x and Z_p with the other Late Medieval sample (San Baudelio) and for Z_p with the Early Medieval Villaneuva sample. Mesolithic femora and tibiae are not particularly strong, falling near the middle or even lower end of variation of most more recent groups.

As noted earlier, because there are relatively few Mesolithic femora or tibiae associated with sufficient material to estimate body mass, we also calculated Z_x/Z_y values, which do not require body size standardization. The results are shown in Table 10.7 and Figure 10.7. The tibia shows a general downward progression in values from the Bronze Age onward, with linear regressions on date that are significant in both sexes ($p < 0.01$). Pairwise tests between periods reach significance for several female comparisons (Bronze Age > Early Medieval Santa Maria and VR High- and Low-SES, Early Medieval Villaneuva and Late Medieval San Baudelio > VR High-SES). There are no clear temporal trends in femoral shape, however (Fig. 10.7(b)), with non-significant differences between all periods or regressions on date. Mesolithic specimens are not consistently higher or lower than those of other periods, with somewhat high values for the femur but low values for female tibiae. The two Los Canes males with data do not show any systematic differences in tibial or femoral shape from the shellmound sites (Moita de Sebatiao and

Table 10.7 Temporal differences in lower limb bone strengths.

Bone	Prop.	Sex		Mesol.	Bronze	EMed. S.M.H.	EMed. Vill.	LMed. S.B.	LMed. Leiria	VR Low	VR High
Tibia	Z_{xstd}	Males	N	3	10	16	13	10	13	13	13
			Mean	608.93	864.07	595.19	701.42	687.95	578.39	665.89	582.03
			SE	73.31	61.87	27.23	37.83	49.10	30.68	30.44	37.51
		Females	N	1	9	12	10	7	9	7	18
			Mean	607.20	606.11	514.44	547.62	518.61	517.63	508.79	513.66
			SE		29.13	25.79	32.74	32.73	29.22	26.72	16.31
Femur	Z_{xstd}	Males	N	4	14	17	13	12	16	13	13
			Mean	612.73	678.91	642.95	682.32	665.06	545.92	632.77	573.58
			SE	47.67	29.34	17.69	35.86	27.64	20.16	28.51	18.51
		Females	N		15	18	10	10	10	7	18
			Mean		550.41	540.29	568.48	620.73	489.58	507.84	543.68
			SE		32.98	22.22	27.67	38.04	24.91	28.40	13.29
Tibia	Z_{pstd}	Males	N	3	10	16	13	10	13	13	13
			Mean	913.43	1217.26	890.54	1031.72	1015.23	875.28	977.14	840.59
			SE	86.95	71.54	33.67	60.05	64.73	42.94	50.54	50.04
		Females	N	1	9	12	10	7	9	7	18
			Mean	826.60	864.86	780.32	818.30	766.51	767.54	807.01	811.39
			SE		43.77	30.82	46.36	49.36	41.07	42.34	23.03
Femur	Z_{pstd}	Males	N	4	14	17	13	12	16	13	13
			Mean	1054.90	1303.56	1178.95	1215.63	1270.16	1009.58	1185.70	1072.45
			SE	40.27	39.96	35.69	68.14	52.26	30.01	51.52	37.77
		Females	N		15	18	10	10	10	7	18
			Mean		1088.23	1061.63	1153.65	1201.41	939.05	952.09	1025.03
			SE		64.88	34.35	58.81	47.23	40.86	52.27	37.28

(*Continued*)

Table 10.7 (Continued)

Bone	Prop.	Sex		Mesol.	Bronze	EMed. S.M.H.	EMed. Vill.	LMed. S.B.	LMed. Leiria	VR Low	VR High
Tibia	Z_x/Z_y	Males	N	8	11	17	13	10	13	13	13
			Mean	1.43	1.55	1.39	1.47	1.44	1.34	1.36	1.35
			SE	0.09	0.08	0.03	0.05	0.05	0.04	0.04	0.07
		Females	N	2	10	12	10	7	9	7	18
			Mean	1.15	1.58	1.36	1.42	1.47	1.41	1.26	1.22
			SE	0.32	0.07	0.04	0.05	0.06	0.07	0.06	0.03
Femur	Z_x/Z_y	Males	N	7	15	17	13	12	16	13	13
			Mean	1.07	0.96	0.99	1.02	0.92	0.97	0.95	0.96
			SE	0.07	0.03	0.02	0.04	0.03	0.02	0.03	0.04
		Females	N	3	16	18	10	10	10	7	18
			Mean	1.06	0.92	0.92	0.90	0.92	0.94	0.96	0.97
			SE	0.06	0.03	0.02	0.02	0.05	0.05	0.02	0.03

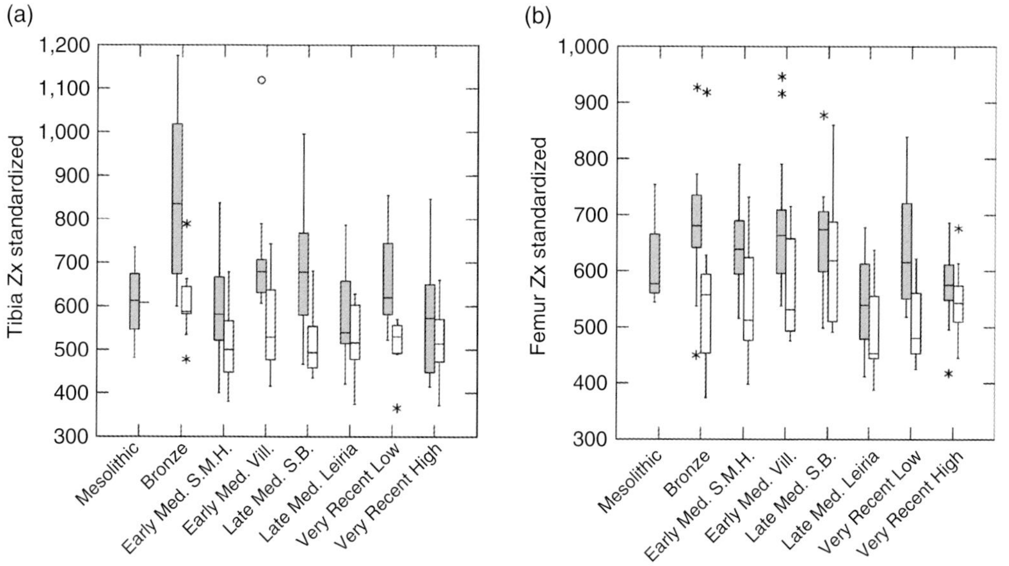

Figure 10.5 Boxplots of temporal variation in (a) tibial and (b) femoral anteroposterior (A-P) bending strength, standardized for body size. Open boxes: females; gray boxes: males.

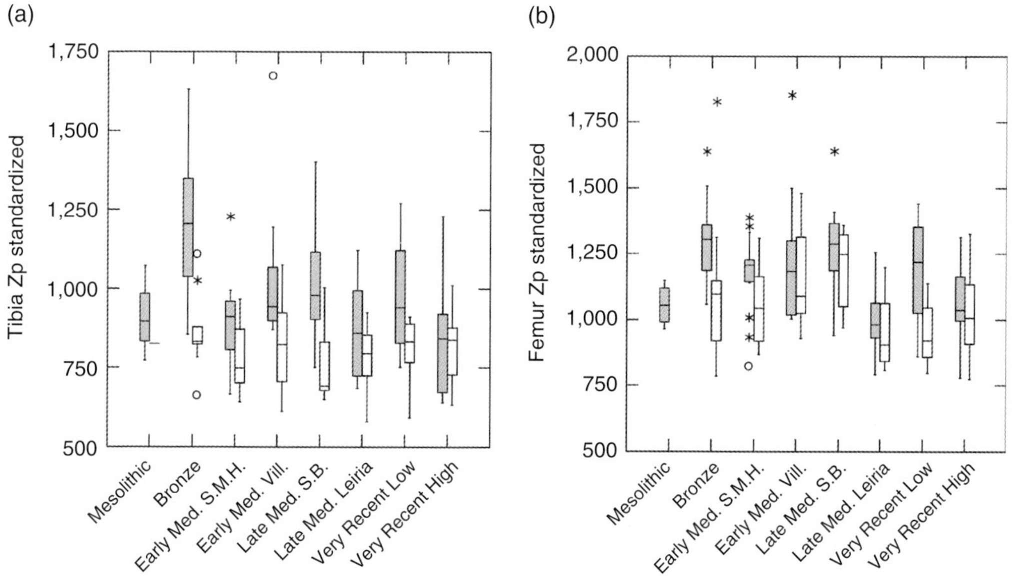

Figure 10.6 Boxplots of temporal variation in (a) tibial and (b) femoral average bending/torsional strength, standardized for body size. Open boxes: females; gray boxes: males.

Muge Arruda, n = 6 and 5 for the tibia and femur, respectively), being somewhat lower in Z_x/Z_y values for the tibia (means of 1.332 versus 1.462) and higher for the femur (1.246 versus 1.003). Female Mesolithic samples are too small for meaningful comparisons between sites.

Humeral average bending/torsional strength results are shown in Table 10.8 and Figure 10.8. As with the lower limb, temporal variation in relative strength among males is greater than that among females. The Late Medieval Leiria and Very Recent male samples are lower in strength

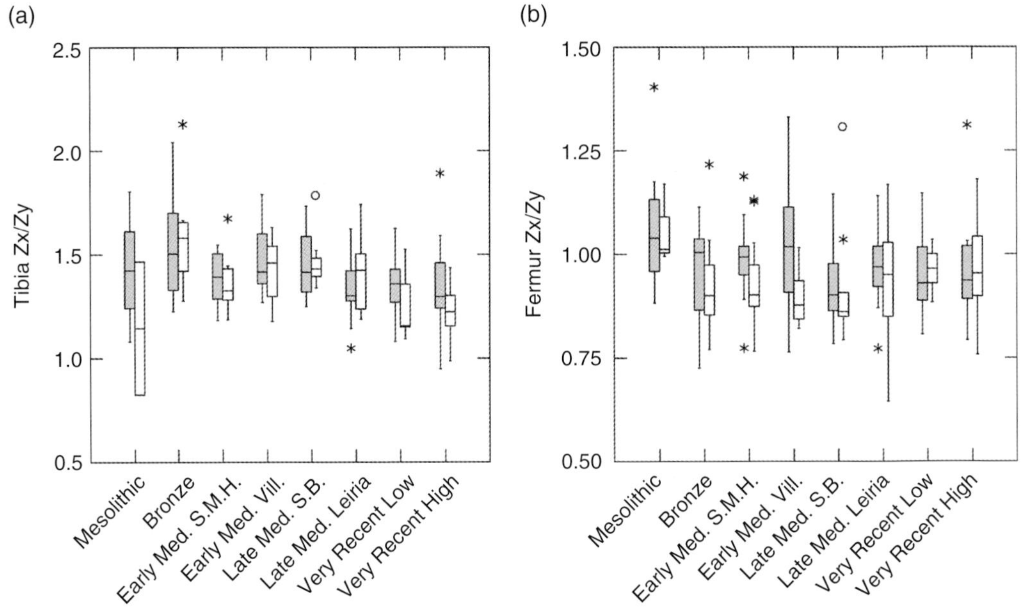

Figure 10.7 Boxplots of temporal variation in (a) tibial and (b) femoral A-P/M-L bending strength. Open boxes: females; gray boxes: males.

than most earlier periods, reaching significance for both humeri in comparisons of Leiria and VR High-SES samples with the other three Medieval samples (except on the right side with Santa Maria). The one exception is Bronze Age left humeri, which are also relatively low (significant compared to Villaneuva and San Baudelio). Females show less variation, with no significant pairwise differences between periods, although they do show a general temporal decline in relative humeral strength, significant for the left side ($p < 0.05$, regression on date). Thus, both sexes exhibit some reduction in humeral strength over time, although the change is gradual in females and abrupt in males, that is, only in the Leiria and Very Recent samples.

Bilateral asymmetry in humeral average bending/torsional strength is shown in Figure 10.9. No inter-period comparisons are significant, although there is a tendency for males to reduce in asymmetry over time after the Bronze Age ($p < 0.05$, regression on date). Females show generally lower and more constant levels of asymmetry, with a particularly low level in the Bronze Age (and consequently high SD in asymmetry in that period), similar to Europe as a whole (see Discussion and Chapter 7).

Urban–rural comparisons of bone strength are shown in Table 10.9. As would be expected from the previously presented results, the Late Medieval urban Leiria sample is reduced in relative strength compared to the rural San Baudelio de Berlanga sample in many properties, especially in males. Differences reach statistical significance in both sexes for femoral properties and in humeri for males, with average differences of 22–28%. Male tibiae average 16–19% larger in San Baudelio ($p < 0.10$). However, female tibiae and humeri show no differences between samples. Results for femoral Z_x and right humeral Z_p are illustrated in Figure 10.10. There are no marked effects on humeral bilateral asymmetry (Table 10.9).

Because of their special relationship, it is also of interest to compare relative bone strengths directly between the VR High- and Low-SES groups alone (using *t*-tests). When this is done, in males all lower limb relative strengths are near-significantly ($p < 0.10$) higher, and right humeri are significantly higher ($p < 0.05$) in the Low-SES group. Females show no differences between SES groups.

Table 10.8 Temporal differences in humeral strengths and bilateral asymmetry.

Bone(s)	Prop.	Sex		Mesol.	Bronze	EMed. S.M.H.	EMed. Vill.	LMed. S.B.	LMed. Leiria	VR Low	VR High
Rt.	Z_{pstd}	Males	N	1	8	15	12	10	14	13	12
humerus			Mean	604.10	609.21	642.20	699.63	725.15	525.09	600.67	499.73
			SE		26.17	25.96	37.86	37.89	37.30	34.53	32.14
		Females	N	1	11	14	9	5	9	7	18
			Mean	385.10	484.29	517.81	492.57	477.46	452.47	464.14	456.50
			SE		16.51	20.43	25.77	48.97	30.12	36.30	15.07
Lt.	Z_{pstd}	Males	N	2	10	15	11	10	16	13	13
humerus			Mean	547.65	489.62	586.61	648.04	652.31	477.21	12.68	477.92
			SE	37.55	23.41	27.82	36.76	35.60	20.87	2.59	22.63
		Females	N	1	11	18	9	4	9	6	18
			Mean	298.70	488.74	503.64	474.70	423.58	431.27	6.84	427.11
			SE		22.63	18.42	34.14	28.14	26.47	2.25	16.31
Asymmetry	Z_p	Males	N	2	6	13	11	10	14	13	12
			Mean	8.84	21.95	16.43	10.24	11.21	9.77	12.68	5.77
			SE	5.66	3.77	3.45	3.11	5.93	3.73	2.59	4.82
		Females	N	1	10	14	8	4	9	6	18
			Mean	24.55	0.17	5.33	9.64	2.73	5.65	6.84	7.82
			SE		4.68	2.30	3.45	4.27	2.63	2.25	1.65

Asymmetry = [(R − L)/(R + L)/2)] × 100.

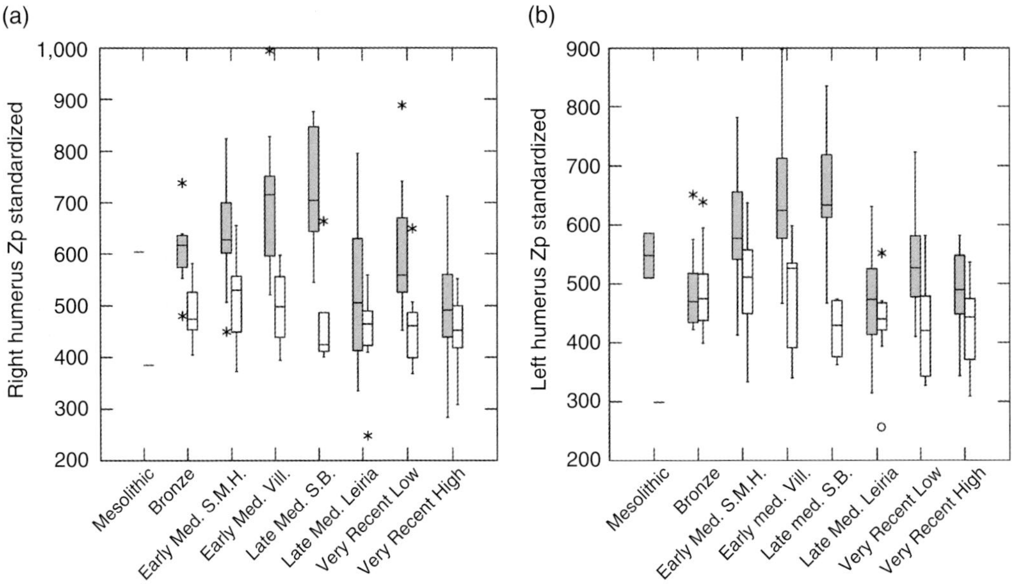

Figure 10.8 Boxplots of temporal variation in (a) right and (b) left humeral average bending/torsional strength, standardized for body size. Open boxes: females; gray boxes: males.

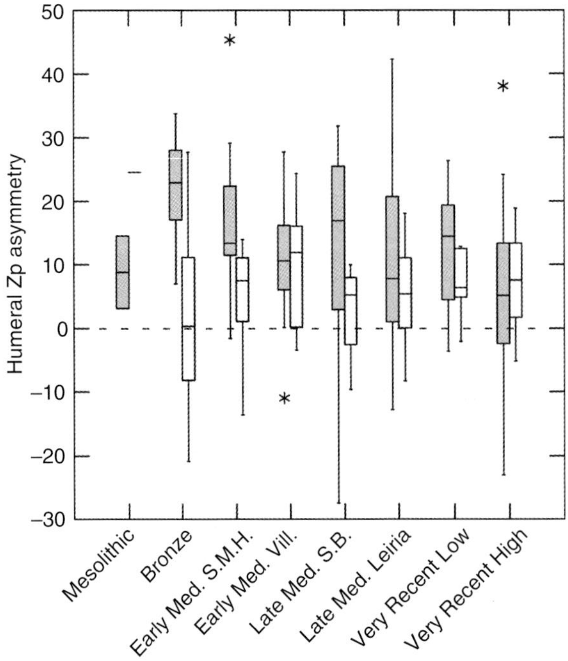

Figure 10.9 Humeral asymmetry in average bending/torsional strength. Open boxes: females; gray boxes: males.

10.3.2 Percentage Cortical Area

Temporal changes in percentage cortical area (%CA) of the femoral, tibial, and two humeral sections are shown in Figure 10.11, with data listed in Table 10.10. The only bone showing evidence of significant temporal variation in this parameter is the tibia, in males (Fig. 10.11(a)). Mesolithic males are significantly greater in tibial %CA than all other periods (near-significant

Table 10.9 Bone strengths in Late Medieval rural (San Baudelio de Berlanga) and urban (Leiria) samples.

	Rural					
	Males			Females		
Property	**N**	**Mean**	**SE**	**N**	**Mean**	**SE**
Tibia Z_{xstd}	10	688.0	49.1	7	518.6	32.7
Femur Z_{xstd}	12	665.1	27.6	10	620.7	38.0
Tibia Z_{pstd}	10	1015.2	64.7	7	766.5	49.4
Femur $_{Zpstd}$	12	1270.2	52.3	10	1201.4	47.2
Rt. Hum. Z_{pstd}	10	725.2	37.9	5	477.5	49.0
Lt. Hum. $_{Zpstd}$	10	652.3	35.6	4	423.6	28.1
Hum. Z_p asym.	10	11.2	5.9	4	2.7	4.3
	Urban					
	Males			**Females**		
Property	**N**	**Mean**	**SE**	**N**	**Mean**	**SE**
Tibia Z_{xstd}	13	578.4	30.7	9	517.6	29.2
Femur Z_{xstd}	16	545.9*	20.2	10	489.6*	24.9
Tibia Z_{pstd}	13	875.3	42.9	9	767.5	41.1
Femur Z_{pstd}	16	1009.6*	30.0	10	939.0*	40.9
Rt. Hum. Z_{pstd}	14	525.1*	37.3	9	452.5	30.1
Lt. Hum. Z_{pstd}	16	477.2*	20.9	9	431.3	26.5
Hum. Z_p asym.	14	9.8	3.7	9	5.7	2.6

* Significantly different from rural.

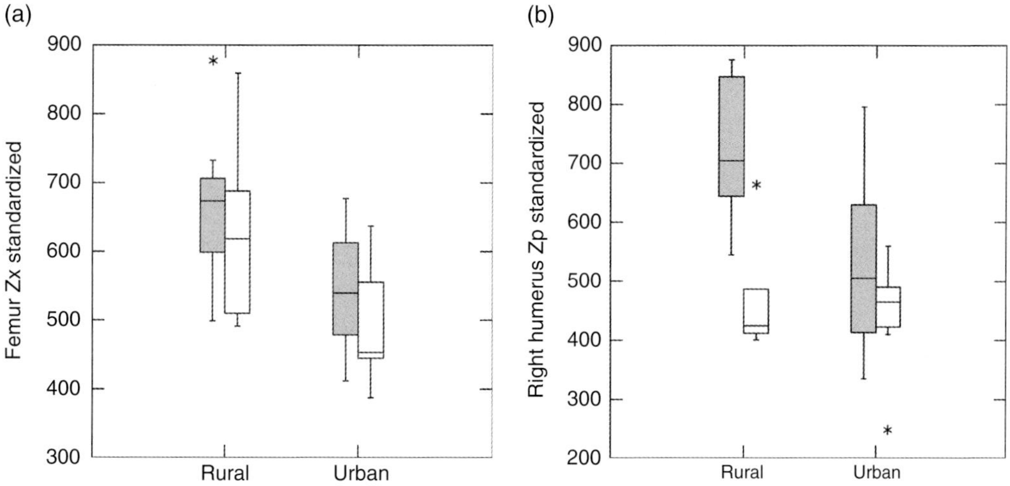

Figure 10.10 Rural versus urban Late Medieval samples: (a) tibial A-P bending strength; (b) Right humeral average bending/torsional strength. Open boxes: females; gray boxes: males.

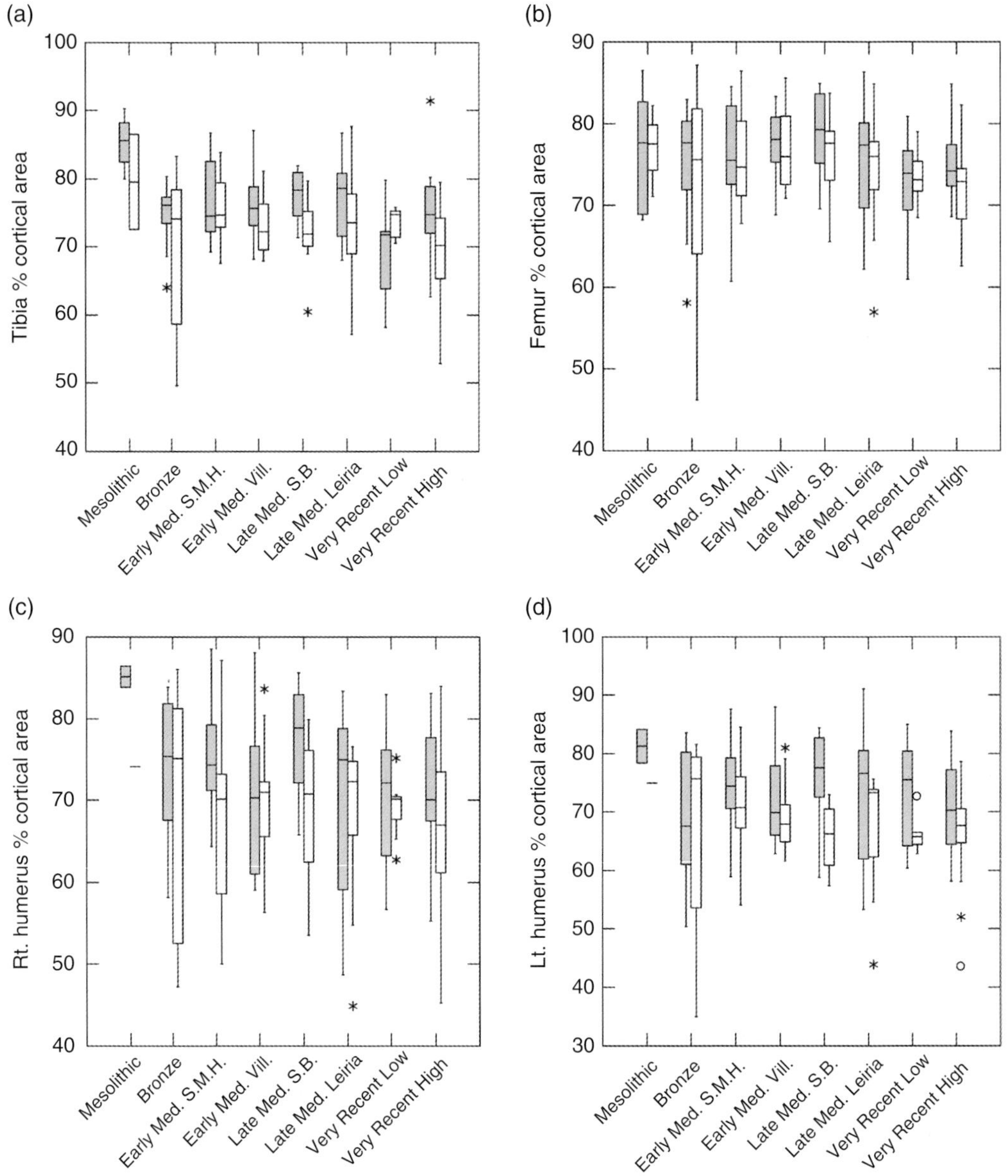

Figure 10.11 Temporal changes in percentage cortical area: (a) tibia; (b) femur; (c) right humerus; (d) left humerus. Open boxes: females; gray boxes: males.

for San Baudelio), and VR Low-SES males are significantly lower than all other periods (near-significant for VR High-SES males). Mesolithic females are also high in tibial %CA, but inter-period differences do not reach statistical significance. These patterns are not as apparent in the femur or humeri, although the small samples of Mesolithic male humeri have consistently high values (Fig. 10.11(c,d)). In *t*-tests between just the High- and Low-SES samples, Low-SES males have significantly lower tibial %CA than High-SES males.

Table 10.10 Temporal differences in percentage cortical area.

Bone	Sex		Mesol.	Bronze	EMed. S.M.H.	EMed. Vill.	LMed. S.B.	LMed. Leiria	VR Low	VR High
Femur	Males	N	7	15	17	13	12	16	13	13
		Mean	76.49	75.12	76.04	77.32	78.85	75.67	72.28	74.86
		SE	3.04	1.88	1.46	1.25	1.42	1.75	1.62	1.13
	Females	N	3	16	18	10	10	10	7	18
		Mean	76.93	72.78	75.23	76.94	76.34	73.94	73.55	72.28
		SE	3.22	2.96	1.30	1.55	1.76	2.43	1.35	1.37
Tibia	Males	N	8	11	17	13	10	13	13	13
		Mean	85.36	74.78	76.34	76.16	77.43	77.16	68.75	75.15
		SE	1.25	1.46	1.35	1.51	1.16	1.60	1.92	1.94
	Females	N	2	10	12	10	7	9	7	18
		Mean	79.55	69.33	75.94	73.57	71.80	73.22	73.48	69.51
		SE	6.97	3.70	1.44	1.50	2.30	2.94	0.88	1.68
Rt. humerus	Males	N	2	9	15	13	10	14	13	12
		Mean	85.12	74.00	75.36	70.79	77.22	70.76	70.19	71.44
		SE	1.29	2.97	1.76	2.64	2.18	3.06	2.41	2.24
	Females	N	1	12	14	9	5	9	7	18
		Mean	74.15	68.10	68.33	70.00	68.56	67.06	69.19	66.43
		SE		4.34	2.83	2.82	4.77	3.59	1.53	2.09
Lt. humerus	Males	N	2	11	15	11	10	16	13	13
		Mean	81.29	69.26	74.48	71.99	76.38	73.45	72.08	70.70
		SE	2.87	3.68	1.94	2.41	2.44	2.83	2.44	2.19
	Females	N	1	12	18	9	4	9	6	18
		Mean	74.93	66.62	70.21	69.06	65.69	67.02	66.29	65.98
		SE		4.66	1.90	2.32	3.27	3.71	1.37	1.90

10.3.3 Comparisons to other Europeans

Two representative properties – tibial A-P bending strength and right humeral average bending/torsional strength, both size-standardized – are plotted in Figure 10.12 for Iberian and non-Iberian European samples. The Iberian Mesolithic female sample and male humeral sample are too small to include here. Iberian Mesolithic males have relatively reduced lower limb bone strength compared to other European males – significant (with Bonferroni correction) for femoral Z_p, but consistent across all properties. In contrast, Iberian Bronze Age males have much stronger lower limb bones than other Bronze Age Europeans ($p < 0.001$, all femoral and tibial properties). Iberian Bronze Age females also show this tendency, but to a lesser extent, significant only for femoral $Z_{p.}$ Early Medieval Iberian males and females are also stronger in the lower limb sections (significant in all properties except for femoral Z_p in males). Tibial Z_x/Z_y is significantly higher in Iberian Bronze Age females and Early Medieval males (along with femoral Z_x/Z_y in Early Medieval males). Humeri do not show as strong a contrast in these temporal periods, with only Iberian Early Medieval females being significantly stronger than non-Iberian Europeans.

The two more recent temporal periods do not show any differences between Iberian and non-Iberian samples in lower limb bone strength, although this is in part a result of combining the different individual Iberian samples. For example, if the Late Medieval sample is divided between Leiria and San Baudelio, the more rural and more robust San Baudelio sample is significantly or near-significantly stronger than non-Iberian Late Medieval samples for femoral properties, in both sexes. However, these other European Late Medieval samples include a mix of urban and rural sites (see other chapters in this volume), so it is appropriate to include both types of Iberian sites in comparisons. Similarly, like the Lisbon pooled Very Recent sample, other VR European samples likely include a mix of socioeconomic statuses as well, although most of them appear to be primarily of lower SES (see other chapters). If only the Lisbon

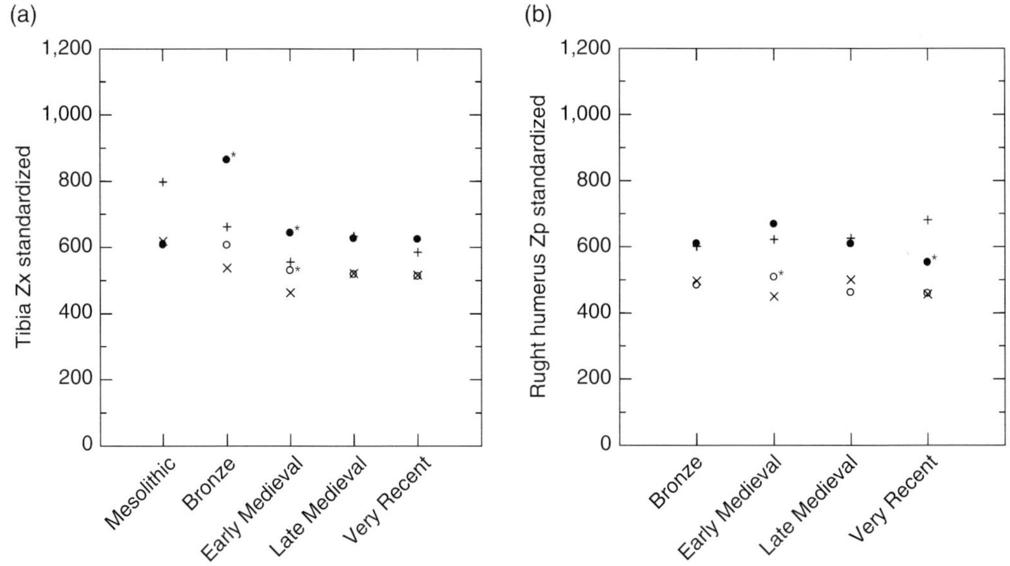

Figure 10.12 Mean values for (a) tibial A-P bending strength and (b) right humeral average bending/torsional strength in Iberian and non-Iberian samples. Filled circles: Iberian males; open circles: Iberian females; +: other European males; x: other European females. Asterisks indicate that Iberian values are significantly different from non-Iberian values (with Bonferroni correction).

Low-SES males are compared to other VR Europeans, the Lisbon sample has significantly higher tibial Z_x and near-significantly higher femoral Z_x than non-Iberian males (again with Bonferroni correction). Humeri (both sides) show no differences between Iberian and non-Iberian samples for the Late Medieval period and for VR females, and are actually significantly weaker in VR Iberian males (although this becomes non-significant when only Low-SES Lisbon males are included). Thus, the most consistent differences between Iberian and other European samples are in lower limb bone strength in the earlier periods, with Iberian Mesolithic samples relatively weaker, and Iberian Bronze Age and Early Medieval samples relatively stronger.

Percentage cortical area tends to be higher in Iberian Medieval samples compared to other Europeans, significant in Early Medieval female tibiae, femora, and left humeri, and Late Medieval male tibiae and femora, and female femora.

10.4 Discussion

10.4.1 Body Size and Shape

The smaller average body size of modern Iberian populations relative to the rest of Europe has been noted by a number of researchers (Schmidt *et al.*, 1995; Cavelaars *et al.*, 2000; Larnkjaer *et al.*, 2006). These comparisons have been based on military conscript data, and are thus limited to males and may be affected by some slight selection bias (see above). However, comparison of both military and non-military data for very recent (post-1960 birth date) samples also indicates that Iberians are among the shortest of living European populations: for this time period Iberian males average about 173 cm and females about 162 cm (see Fig. 10.3), which is at the low end of variation for Europeans generally (see Chapter 4: Fig. 4.1). The general explanation for body size differences among recent Europeans has centered around variation in living conditions (Schmidt *et al.*, 1995; Cavelaars *et al.*, 2000; Larnkjaer *et al.*, 2006). In particular, the relatively poor living conditions of many Iberian populations during much of the 19th and early 20th centuries have been discussed by a number of authors (Padez, 2003; Cardoso and Garcia, 2009; Cardoso and Gomes, 2009; María-Dolores and Martínez-Carrión, 2009). The other Very Recent samples in our study from Finland, Germany, and Italy were also of generally low SES (see Chapters 9, 11, and 12), yet the Low-SES Lisbon sample is still smaller. The later improvement in conditions on the Iberian Peninsula since the mid-20th century would also explain the continuing secular increase in stature in Spanish and Portuguese populations over the past several decades, while most northern European populations have leveled off (Larnkjaer *et al.*, 2006). This also implies that differences in body size between Iberian and other European populations will continue to diminish in the next few decades.

However, there is also evidence from both living (Turchin *et al.*, 2012) and paleogenomic studies (Mathieson *et al.*, 2015) for long-standing differences in genes controlling stature between northern and southern European populations, dating back to the Neolithic, that could also partially explain past and present patterns of variation. In that context, it is interesting to examine temporal changes in body size in Iberia versus other parts of Europe in our sample. One of the largest differences occurs in the Mesolithic, where Iberian males are much smaller than other Mesolithic males. This could argue in favor of long-term genetic effects. However, it has also been noted that there is a marked East–West difference in stature in the European Mesolithic, with Western Europeans being significantly shorter (Formicola and Giannecchini, 1999, and see Chapters 4 and 11) – that is, this phenomenon is not limited to Iberian samples alone. In addition, there is a decrease in stature between the Early and Late Mesolithic in our European samples (Chapter 4: Fig. 4.12). All of the Iberian Mesolithic samples are late (5900–5200 cal BC), so this may also have affected our results. However, even when limiting the non-Iberian samples to Late Mesolithic (post-6200 cal BC), Iberian males are still shorter and lighter than other Europeans ($p = 0.067$ for stature, $p = 0.013$ for body mass) (female samples are insufficient for comparisons). Further complicating interpretations, there is a spatial/temporal relationship in that eastern European Mesolithic samples tend to be earlier than western European Mesolithics. The explanation for the general East–West difference is not clear, possibly involving both genetic and environmental effects (Formicola and Giannecchini, 1999). Interestingly, recent evidence indicates that there is East–West genetic differentiation in the European Mesolithic – that is, a distinct Holocene 'Western European hunter-gatherer' group, which includes samples from Iberia (Lazaridis *et al.*, 2014; Haak *et al.*, 2015). This may further support long-term genetic effects on body size, although if so, the greater stature of recent northwestern European populations relative to Iberians would need to be explained.

Other studies that have included Iberian Neolithic samples have shown little change in stature between Mesolithic and Neolithic samples; that is, Iberian Neolithics are also short (Fox, 1996; Cardoso and Gomes, 2009). Body size increases dramatically in our Bronze Age sample, and continues to increase slightly in the Medieval period (except for stature in females), paralleling post-Neolithic changes reported by others (Fox, 1996; Cardoso and Gomes, 2009). Medieval Iberians are virtually the same size as other Medieval Europeans, which argues against a consistent genetic explanation for the regional differences observed in other periods, unless one postulates better living conditions for Iberian Medieval populations (thus counteracting genetic tendencies to be smaller). In fact, Iberian Medieval samples do exhibit elevated %CA in many bone sections compared to other Europeans, which could indicate a better nutritional environment (see below).

The large decline in male body size from the Late Medieval Leiria sample to the early 20th century Lisbon Low-SES sample is also not unexpected given previous findings (Cardoso and Garcia, 2009; Cardoso and Gomes, 2009). The High-SES male sample is equivalent in body size to the Leiria (and any other Medieval) sample, however. This would seem to argue for very strong environmental effects on growth in the two Lisbon samples, as detailed in the above references (also see Padez, 2003; Conceição *et al.*, 2012), and these doubtless contributed to the differences observed. Interestingly, however, *both* VR SES groups show relatively short lower limbs (i.e., high relative sitting height) compared to earlier samples, with no difference between SES groups. Reduced relative lower limb length under conditions of poor health and an increase in relative lower limb length with improvements in living conditions have been observed in a number of studies (Tanner *et al.*, 1982; Bogin *et al.*, 2002; Bogin and Varela-Silva, 2010). This would seem to indicate that both Lisbon SES groups had poorer growth environments than earlier Iberian populations. The general secular increase in stature over the past 100 years in Portugal (see Fig. 10.3) would support this to some degree, although this has produced a population that is even taller than the previous Medieval peak. It would be interesting in this regard to compare relative lower limb lengths in living populations (or even more recent skeletal samples) to those for the early 20th century sample included here.

The male Low-SES group was also narrower-bodied than the High-SES group and earlier groups (except Mesolithics), which contributed to their low body masses. Bi-iliac breadth does not appear to be environmentally sensitive, at least short of outright pathology (Ruff, 1994), so this seems to be better explained as a genetic effect. Thus, while body size differences between the two SES groups are almost certainly affected by their different growth environments, other factors, including possible genetic differences, also appear to be involved. The results for VR Low- and High-SES females do not parallel those for males, or expectations based on other studies. As noted earlier, the SES designations for females are much less secure than those for males, so this is not surprising.

In general, except for VR Low-SES males (and the one Mesolithic male with data), all Iberian samples have bi-iliac breadths that are wide as those of other Europeans, despite differences in stature. Thus, despite being shorter than other Europeans in some sex/temporal groups, they are not narrower-bodied. This is consistent with general patterns of variation within broadly similar ecological zones, in which stature varies more than body breadth between populations or ancestor-descendant groups (Ruff, 1994; also see Chapter 4).

The low crural index of Iberian Mesolithics is consistent with their shorter stature, and differs from that found among Mesolithic Europeans in general, including those from western Europe (Holliday, 1999; see Chapter 4). This reinforces their status as morphological outliers to some extent. Samples sizes are too small to really evaluate brachial indices in the Mesolithic. The general temporal decline in brachial index throughout the Holocene, found here and elsewhere in Europe (see Chapter 4), is unexplained.

As in some other regions of Europe (e.g., Britain; see Chapter 8), temporal variation in body size is greater in males than in females, and largely accounts for variation in sexual dimorphism in body size. Sexual dimorphism is greatest in the Bronze and Medieval periods, when stature and body mass also peak (at least in males). This in itself provides indirect support for the 'female buffering' hypothesis, whereby males are more affected by changes in the growth environment (Stini, 1976; Wolfe and Gray, 1982; although see Stinson, 1985). The very low SD in body size in the Low-SES Lisbon sample – whether 'Low-SES' females or all Lisbon females are included – and much higher SD in the High-SES sample also supports this interpretation. Sexual dimorphism in stature and body mass in the Low-SES sample is far below the averages for Very Recent Europe as a whole, while SD in the High-SES sample is comparable to VR non-Iberian samples (see Chapter 6). Trends among living Iberian populations in the later 20th century indicate more of a secular increase in body size in males, and thus an increase in SD, again similar to trends in other parts of Europe (e.g., Britain; Chapter 8). Interestingly, there is no difference in body size or SD in body size in the Late Medieval urban Leiria compared to rural San Baudelio sample, suggesting that growth conditions in the two environments – at least at this point in history – were not significantly different, although other factors such as genetic variation between the populations, must also be considered. In other respects, however (see below), these two samples are distinct.

10.4.2 Bone Structure

The very strong lower limb bones of males from the Bronze Age Argar sample are consistent with their lifestyle as reconstructed from archaeological remains. As noted earlier, settlements were located on the tops of hills in a very rugged environment, and males engaged in livestock raising and herding, which would have involved high mechanical loading of their lower limbs (al-Oumaoui *et al.*, 2004; Jiménez-Brobeil *et al.*, 2010; Lull *et al.*, 2011). Interestingly in this regard, the Argar sample males do *not* possess strong humeri; in fact, their left humeri are significantly weaker than those of some Medieval samples. Increased mechanical loading appears to be limited to the lower limb, and perhaps weight-bearing elements in general, including vertebrae (Jiménez-Brobeil *et al.*, 2010). Bilateral asymmetry of the humeri is somewhat increased among Argar males compared to later groups, but Argar females are virtually symmetric. The latter is consistent with general trends in Europe as a whole, and is likely due to food-processing techniques among early agricultural societies involving two-handed grinding (Sládek *et al.*, 2016; see Chapter 7). The somewhat elevated bilateral asymmetry in Argar males may be due in part to metallurgically related activities (Macintosh *et al.*, 2014), although the extent to which most males participated in such activities in this region has been debated (Lull *et al.*, 2011). Argar females are more similar to subsequent populations, although there is evidence for somewhat elevated tibial strength. Overall, a marked sexual differentiation of economic roles inferred on the basis of archaeological evidence (Jiménez-Brobeil *et al.*, 2010; Lull *et al.*, 2011), is supported by the skeletal analyses.

In contrast to the Bronze Age samples, Iberian Mesolithic males do not have either strong lower or upper limb bones compared to later time periods. This is contrary to general trends throughout Europe, where Mesolithics tend to be intermediate between (stronger) Upper Paleolithic and later Holocene samples (Ruff *et al.*, 2015; see Chapter 5). The lack of increased tibial and femoral A-P bending strength (Z_x) or tibial Z_x/Z_y in Iberian Mesolithic males is particularly striking, since these show some of the strongest temporal trends in pan-European comparisons (ibid.). The deviation can very likely be attributed to the quite sedentary lifestyle and relatively flat terrain (at least locally) characteristic of these populations (Lubell *et al.*, 1994; Straus, 2008). While some logistical forays into the surrounding hills may have been undertaken occasionally, for example to obtain flint or for hunting, the main food resources were marine, and settlements were probably occupied year-round (ibid.). Even the slightly more inland Los Canes cave site contained a predominance of marine shell remains (Arias and Garralda, 1996). As noted earlier, no differences in body size or long bone structure were detected between this site and the coastal Portuguese sites, although samples sizes are very limited. Our results for the Iberian Mesolithic in general provide additional evidence for the importance of localized behavioral adaptations to environmental conditions, modifying expectations based on broader subsistence classifications (e.g., Marchi, 2008).

Jackes and Lubell (1999a,b) and Zilhão (2000) presented data on femoral and tibial cross-sectional size and shape in Portuguese Mesolithic and Neolithic samples. Both assessed external breadths and breadth ratios; Jackes and Lubell (1999a) also measured density and cortical thickness of direct sections from the anterior femur in the Moita de Sebastiao and Muge Arruda Mesolithic samples. They found the Moita femora to have thicker and materially denser cortices and higher pilasteric (midshaft A-P/M-L breadth) indices than the Arruda sample, which they interpreted as indicating lower mechanical stresses in the latter, possibly as a result of greater sedentariness (the sample is also somewhat later in time than Moita; see Table 10.1). However, with regard to cortical thickness, or any absolute dimension, it is important to control for body size differences when comparing between samples. We were able estimate body mass (from the femoral head) for only two individuals from these two samples in our analyses, but the results are informative. Both are males; femoral Z_p in the Arruda male is 2666 mm^3, much larger than that of the Moita male, 1736 mm^3. However, the Arruda male was also much larger in estimated body size: 65.2 kg versus 41.7 kg for the Moita male. When this, and their difference in femoral length are taken into account, the standardized FZ_p values are actually slightly greater for Moita (1148) than for Arruda (1017). Absolute or relative (%) cortical thickness alone is also of limited mechanical relevance; for example, as shown above, Mesolithic tibiae have high %CA, but low relative bending/torsional strength. Cortical bone density in archaeological remains is heavily influenced by diagensis, so cannot be used to infer living bone density; also, bone tissue density does not change in response to activity differences (Ruff, 2008). The difference in femoral pilasteric index between the two sites is interesting; however, their results for a similar external breadth index in the tibia gave opposite results (the Moita sample had a higher cnemic index, indicating a more circular, i.e., less A-P-strengthened shaft). In a subsequent analysis that also included humeri, Jackes and Lubell (1999b) found a complicated pattern of variation among Mesolithic and Neolithic samples, with no evidence for clustering of either time period, or for distinct changes across the transition. Based on paleodemographic data, the Mesolithic samples were considered to represent "...small relatively sedentary ... bands" (Jackes and Lubell, 1999b), which would be consistent with the present results.

Zilhão (2000) examined Mesolithic–Neolithic differences in external size and shape of femora and tibiae, and illustrated a difference in tibial shape between the two time periods, with Mesolithics having more A-P-buttressed tibiae (lower cnemic indices). Changes in femoral size and shape were more gradual or non-existent. No specific behavioral interpretation was offered, although "...replacement of the hunter-fisher-gatherer way of life of the local Mesolithic by farming and ovicaprid herding" (Zilhao, 2000: 180) was suggested as a possibility. We saw no evidence for a difference in tibial shape between our Mesolithic and later samples, although Mesolithics did have a somewhat (but non-significantly) high femoral Z_x/Z_y ratio. More true cross-sectional data, appropriately size-standardized, for both Mesolithic and Neolithic Iberian samples would be useful for further testing proposed ideas regarding population continuity, replacement, and cultural change in this region (Fox, 1996; Jackes *et al.*, 1997; Zilhao, 2000).

There is a general decline in tibial Z_x/Z_y after the Bronze Age in both sexes, which parallels that found throughout Europe (see Chapter 5) and is probably indicative of a general (although not uniform) decrease in mobility through the Holocene. In terms of comparisons within the Medieval samples, perhaps the most interesting results are for those between the Late Medieval Leiria and other samples, particularly the Late Medieval San Baudelio sample. As noted earlier, the urban Leiria sample is not small in body size, which matches its relatively prosperous economic situation (Cardoso and Garcia, 2009). However, males of the sample have significantly weaker femora and humeri relative to body size (tibiae overlap more, but are weaker on average as well). Females show the same trend for the femur, but not the humeri. These observed bone structural differences are consistent with behavioral differences between the samples. As noted earlier, the Leiria sample is probably representative of multiple social classes, including merchants and members of the local aristocracy. Although manual labor was likely still performed by at least a portion of the sample, the overall average workload would be expected to be considerably reduced from that of the rural San Baudelio sample, where everyone participated in the running of the livestock farm in rugged terrain. In fact, both Leiria males and females are most similar in relative bone strength to the modern High-SES Lisbon sample, which may be the best behavioral match for them, rather than to any of the other Medieval samples. The reduction of humeral strength in only Leiria males appears to reflect a broader pattern whereby humeral strength varies between temporal periods much more in males than in females, even though both sexes generally decline through time in relative strength. This may indicate more variation in use of the upper limb among males in general. The very large reduction between the San Baudelio and Leiria samples appears reasonable considering the replacement of manual farming tasks with commercially related work in much of the latter. The resulting reduction in SD of humeral strength in the Leiria sample is also consistent with these behavioral changes.

The greater relative bone strengths of the Lisbon Low-SES versus High-SES males are also explicable in terms of work requirements in the two groups – that is, males in the Low-SES group were much more likely to be manual laborers (either from direct knowledge of occupation or indirectly through place of residence) (Cardoso, 2007). The effect is most marked (statistically significant) for the right humerus, which might also be expected to show some of the largest differences in mechanical loading between the groups (given that most people are right-handed). In contrast to the Leiria/San Baudelio comparison, in this case there is a significant difference in body size, in the opposite direction to relative bone strength – that is, Low-SES males are smaller, yet relatively stronger. This nicely illustrates the relative independence of two physiological influences on skeletal morphology – a general systemic influence (nutrition, health) on overall body size, and localized mechanical loading of particular skeletal elements. Females in the two SES groups do not show either of these effects, further reinforcing the poor SES classification of females.

In comparisons with other Europeans, Bronze Age and Early Medieval Iberians have relatively stronger limb bones, perhaps reflecting the relatively strenuous lifestyle of these populations, including animal husbandry in rugged environments. The same is actually also true of the Late Medieval San Baudelio sample (which had similar living conditions) – if this sample alone is compared to other European Late Medieval samples, without combining it with the Leiria sample, it is also significantly stronger in several sections (femur and humeri in males, femur in females). The exception to this general pattern is the Iberian Mesolithic sample, which inhabited a relatively more benign terrain and was likely more sedentary than many of our other European Mesolithic samples. The mechanically demanding lifestyle of our later, more interior populations does not imply that they were generally impoverished or nutritionally deprived, though – stature and body mass in the Medieval samples are similar to those in other parts of Europe, and %CA is actually higher for several sections in Medieval Iberians. As noted in Chapter 8, a lower relative cortical thickness can reflect a poorer nutritional environment during growth (Garn *et al.*, 1969). Consistent with this interpretation, Low-SES males in the Lisbon sample have a reduced tibial %CA compared to High-SES males, despite having generally stronger bones relative to body size. The opposite is true for Mesolithic Iberian males, who have high tibial %CA but low relative bone strength. The relative independence, and different functional interpretation of variation in relative cortical thickness and variation in bone strength is well illustrated by these two cases (for other examples, see Ruff *et al.*, 1984; Ruff, 1999).

Despite their description as 'peasant-soldiers' (Galera and Garralda, 1993), males of the Early Medieval Villaneuva de Soportiva sample do not show evidence of increased upper limb strength or humeral bilateral asymmetry, nor is there evidence for increased SD in bone strength compared to other Medieval samples (except Leiria). Thus, they do not exhibit the more extreme characteristics reported for warrior classes in other regions (e.g., Sparacello *et al.*, 2011). It is possible that their agrarian responsibilities outweighed periodic military obligations, and/or that military training was not as standardized or intensive as that in other populations.

10.5 Conclusions

Population samples from the Iberian peninsula exhibit changes in skeletal morphology reflecting some of the same broad temporal trends, for example, a general decrease in mobility during the Holocene, as other European regions, but superimpose on this specific variations in response to local environmental conditions and cultural adaptations. Mesolithic samples are less skeletally robust than most other European Mesolithics, which is probably attributable to their coastal environment and semi-sedentary subsistence strategy. In contrast, Bronze Age Argars are extremely robust, in keeping with their very vigorous lifestyle and rugged environment, with high sexual dimorphism corresponding to archaeological evidence for culturally well-defined sex roles. Medieval Iberians in general are skeletally robust, except for the urban Leiria sample, which likely had a much less physically demanding lifestyle. Socioeconomic status played an important role in early 20th century samples from Lisbon, with Low-SES manual laborers showing increased bone strength relative to High-SES males. Temporal trends in body size and shape do not necessarily parallel trends in skeletal robusticity, reflecting the different physiological factors operative in each case (nutrition and health versus mechanical loading). Body size reaches a peak in the Medieval period, including the Leiria sample, and declines precipitously in early 20th century Lisbon Low-SES males, reflecting a decline in living conditions. High-SES males are similar in body size to (although less skeletally robust than) Medieval samples. Living Iberians show evidence for continuing secular increases in body size over the past several decades. There are also possible long-term genetic influences on body size that may partly account for differences between Iberian and other European samples, although Medieval Iberians were not much – if at all – smaller than Medieval Europeans generally, and show other evidence for a relatively good nutritional environment.

Acknowledgments

For assistance in data collection and processing, we thank Trang Diem Vu. For allowing access to collections and facilitating data collection, we thank Hugo Cardoso, Sylvia Jiménez-Brobeil, and Maria Dolores Garralda.

References

al-Oumaoui, I., Jimenez-Brobeil, S., and du Souich, P. (2004) Markers of activity patterns in some populations of the Iberian Peninsula. *Int. J. Osteoarchaeol.*, **14**, 343–359.

Arias, P. and Garralda, M.D. (1996) Mesolithic burials in Los Canes cave (Asturias, Spain). *Hum. Evol.*, **11**, 129–138.

Bogin, B., Smith, P., Orden, A.B., Varela Silva, M.I., and Loucky, J. (2002) Rapid change in height and body proportions of Maya American children. *Am. J. Hum. Biol.*, **14**, 753–761.

Bogin, B. and Varela-Silva, M.I. (2010) Leg length, body proportion, and health: a review with a note on beauty. *Int. J. Environ. Res. Public Health*, **7**, 1047–1075.

Cardoso, H.F. (2006) Brief communication: the collection of identified human skeletons housed at the Bocage Museum (National Museum of Natural History), Lisbon, Portugal. *Am. J. Phys. Anthropol.*, **129**, 173–176.

Cardoso, H.F. (2007) Environmental effects on skeletal versus dental development: Using a documented subadult skeletal sample to test a basic assumption in human osteological research. *Am. J. Phys. Anthropol.*, **132**, 223–233.

Cardoso, H.F. and Garcia, S. (2009) The Not-so-Dark Ages: ecology for human growth in medieval and early twentieth century Portugal as inferred from skeletal growth profiles. *Am. J. Phys. Anthropol.*, **138**, 136–147.

Cardoso, H.F.V. and Gomes, J.E.A. (2009) Trends in adult stature of peoples who inhabited the modern Portuguese territory from the Mesolithic to the late 20th century. *Int. J. Osteoarchaeol.*, **19**, 711–725.

Cavelaars, A.E., Kunst, A.E., Geurts, J.J., Crialesi, R., Grotvedt, L., Helmert, U., Lahelma, E., Lundberg, O., Mielck, A., Rasmussen, N.K., Regidor, E., Spuhler, T., and Mackenbach, J.P. (2000) Persistent variations in average height between countries and between socio-economic groups: an overview of 10 European countries. *Ann. Hum. Biol.*, **27**, 407–421.

Conceição, E.L.N., Garcia, S., Padez, C., and Cardoso, H.F.V. (2012) Changes in stature of Portuguese women born between 1966 and 1982, according to educational level. *Antropologia Portuguesa*, **29**, 81–96.

Danubio, M.E. and Sanna, E. (2008) Secular changes in human biological variables in Western Countries: an updated review and synthesis. *J. Anthropol. Sci.*, **86**, 91–112.

Formicola, V. and Giannecchini, M. (1999) Evolutionary trends of stature in upper Paleolithic and Mesolithic Europe. *J. Hum. Evol.*, **36**, 319–333.

Fox, C.L. (1996) Physical anthropological aspects of the Mesolithic–Neolithic transition in the Iberian peninsula. *Curr. Anthropol.*, **37**, 689–695.

Galera, V. and Garralda, M.D. (1993) Enthesopathies in a Spanish Medieval population: anthropological, epidemiological, and ethnohistorical aspects. *Int. J. Anthropol.*, **8**, 247–258.
Garn, S.M., Guzman, M.A., and Wagner, B. (1969) Subperiosteal gain and endosteal loss in protein-calorie malnutrition. *Am. J. Phys. Anthropol.*, **30**, 153–155.
Gray, J.P. and Wolfe, L.D. (1980) Height and sexual dimorphism of stature among human societies. *Am. J. Phys. Anthropol.*, **53**, 441–456.
Haak, W., Lazaridis, I., Patterson, N., Rohland, N., Mallick, S., Llamas, B., Brandt, G., Nordenfelt, S., Harney, E., Stewardson, K., Fu, Q., Mittnik, A., Banffy, E., Economou, C., Francken, M., Friederich, S., Pena, R.G., Hallgren, F., Khartanovich, V., Khokhlov, A., Kunst, M., Kuznetsov, P., Meller, H., Mochalov, O., Moiseyev, V., Nicklisch, N., Pichler, S.L., Risch, R., Rojo Guerra, M.A., Roth, C., Szecsenyi-Nagy, A., Wahl, J., Meyer, M., Krause, J., Brown, D., Anthony, D., Cooper, A., Alt, K.W., and Reich, D. (2015) Massive migration from the steppe was a source for Indo-European languages in Europe. *Nature*, **522**, 207–211.
Holliday, T.W. (1999) Brachial and crural indices of European Late Upper Paleolithic and Mesolithic humans. *J. Hum. Evol.*, **36**, 549–566.
Jackes, M. and Lubell, D. (1999a) Human biological variability in the Portuguese Mesolithic. *Arqueologia*, **24**, 25–42.
Jackes, M. and Lubell, D. (1999b) Human skeletal biology and the Mesolithic–Neolithic transition in Portugal. In: *Europe des Derniers Chasseurs Épipaléolithique et Mésolithique* (ed. A. Thévenin), Actes du 5e Colloque International UISPP, Commission XII, Grenoble, 18–23 Septembre 1995. Éditions du CTHS, Paris, pp. 59–64.
Jackes, M., Lubell, D., and Meiklejohn, C. (1997) On physical anthropological aspects of the Mesolithic–Neolithic transition in the Iberian Peninsula. *Curr. Anthropol.*, **38** (5), 839–846.
Jiménez-Brobeil, S.A., Al Oumaoui, I., and Du Souich, P. (2010) Some types of vertebral pathologies in the Argar Culture (Bronze Age, SE Spain). *Int. J. Osteoarchaeol.*, **20**, 36–46.
Lacerda, J. (1904) Estatura do Português Adulto. Ph.D. Thesis, University of Coimbra, Coimbra.
Larnkjaer, A., Schroder, S.A., Schmidt, I.M., Jorgensen, M.H., and Michaelsen, K.F. (2006) Secular change in adult stature has come to a halt in northern Europe and Italy. *Acta Paediatr.*, **95**, 754–755.
Lazaridis, I., Patterson, N., Mittnik, A., Renaud, G., Mallick, S., Kirsanow, K., Sudmant, P.H., Schraiber, J.G., Castellano, S., Lipson, M., Berger, B., Economou, C., Bollongino, R., Fu, Q.M., Bos, K.I., Nordenfelt, S., Li, H., de Filippo, C., Prufer, K., Sawyer, S., Posth, C., Haak, W., Hallgren, F., Fornander, E., Rohland, N., Delsate, D., Francken, M., Guinet, J.M., Wahl, J., Ayodo, G., Babiker, H.A., Bailliet, G., Balanovska, E., Balanovsky, O., Barrantes, R., Bedoya, G., Ben-Ami, H., Bene, J., Berrada, F., Bravi, C.M., Brisighelli, F., Busby, G.B.J., Cali, F., Churnosov, M., Cole, D.E.C., Corach, D., Damba, L., van Driem, G., Dryomov, S., Dugoujon, J.M., Fedorova, S.A., Romero, I.G., Gubina, M., Hammer, M., Henn, B.M., Hervig, T., Hodoglugil, U., Jha, A.R., Karachanak-Yankova, S., Khusainova, R., Khusnutdinova, E., Kittles, R., Kivisild, T., Klitz, W., Kucinskas, V., Kushniarevich, A., Laredj, L., Litvinov, S., Loukidis, T., Mahley, R.W., Melegh, B., Metspalu, E., Molina, J., Mountain, J., Nakkalajarvi, K., Nesheva, D., Nyambo, T., Osipova, L., Parik, J., Platonov, F., Posukh, O., Romano, V., Rothhammer, F., Rudan, I., Ruizbakiev, R., Sahakyan, H., Sajantila, A., Salas, A., Starikovskaya, E.B., Tarekegn, A., Toncheva, D., Turdikulova, S., Uktveryte, I., Utevska, O., Vasquez, R., Villena, M., Voevoda, M., Winkler, C.A., Yepiskoposyan, L., Zalloua, P., Zemunik, T., Cooper, A., Capelli, C., Thomas, M.G., Ruiz-Linares, A., Tishkoff, S.A., Singh, L., Thangaraj, K., Villems, R., Comas, D., Sukernik, R., Metspalu, M., Meyer, M., Eichler, E.E., Burger, J., Slatkin, M., Paabo, S., Kelso, J., Reich, D., and Krause, J. (2014) Ancient human genomes suggest three ancestral populations for present-day Europeans. *Nature*, **513**, 409.
Lubell, D., Jackes, M., Schwartz, H., Knfy, M., and Meiklejohn, C. (1994) The Mesolithic–Neolithic transition in Portugal: Isotopic and dental evidence of diet. *J. Archaeol. Sci.*, **21**, 201–216.

Lull, V., Micó, R., Herrada, C.R., and Risch, R. (2011) El Argar and the beginning of class society in the western Mediterranean. *Archäologie in Eurasien*, **24**, 381–414.

Macintosh, A.A., Pinhasi, R., and Stock, J.T. (2014) Divergence in male and female manipulative behaviors with the intensification of metallurgy in Central Europe. *PLoS ONE*, **9**, e112116.

Marchi, D. (2008) Relationships between lower limb cross-sectional geometry and mobility: the case of a Neolithic sample from Italy. *Am. J. Phys. Anthropol.*, **137**, 188–200.

María-Dolores, R. and Martínez-Carrión, J.M. (2009) The relationship between height and economic development in Spain: A historical perspective. Documentos de Trabajo (DT-AEHE) 0912:1–33.

María-Dolores, R. and Martínez-Carrión, J.M. (2011) The relationship between height and economic development in Spain, 1850–1958. *Econ. Hum. Biol.*, **9**, 30–44.

Mathieson, I., Lazaridis, I., Rohland, N., Mallick, S., Patterson, N., Roodenberg, S.A., Harney, E., Stewardson, K., Fernandes, D., Novak, M., Sirak, K., Gamba, C., Jones, E.R., Llamas, B., Dryomov, S., Pickrell, J., Arsuaga, J.L., de Castro, J.M., Carbonell, E., Gerritsen, F., Khokhlov, A., Kuznetsov, P., Lozano, M., Meller, H., Mochalov, O., Moiseyev, V., Guerra, M.A., Roodenberg, J., Verges, J.M., Krause, J., Cooper, A., Alt, K.W., Brown, D., Anthony, D., Lalueza-Fox, C., Haak, W., Pinhasi, R., and Reich, D. (2015) Genome-wide patterns of selection in 230 ancient Eurasians. *Nature*, **528**, 499–503.

Padez, C. (2003) Secular trend in stature in the Portuguese population (1904–2000). *Ann. Hum. Biol.*, **30**, 262–278.

Ruff, C.B. (1994) Morphological adaptation to climate in modern and fossil hominids. *Yrbk Phys. Anthropol.*, **37**, 65–107.

Ruff, C.B. (1999) Skeletal structure and behavioral patterns of prehistoric Great Basin populations. In: *Prehistoric Lifeways in the Great Basin Wetlands: Bioarchaeological Reconstruction and Interpretation* (eds B.E. Hemphill and C.S. Larsen), University of Utah Press, Salt Lake City, pp. 290–320.

Ruff, C.B. (2002) Variation in human body size and shape. *Annu. Rev. Anthropol.*, **31**, 211–232.

Ruff, C.B. (2008) Biomechanical analyses of archaeological human skeletal samples. In: *Biological Anthropology of the Human Skeleton*, 2nd edn (eds M.A. Katzenburg and S.R. Saunders), John Wiley & Sons, Inc., New York, pp. 183–206.

Ruff, C.B., Holt, B.M., Niskanen, M., Sladek, V., Berner, M., Garofalo, E., Garvin, H.M., Hora, M., Junno, J.-A., Schuplerova, E., Vilkama, R., and Whittey, E. (2015) Gradual decline in mobility with the adoption of food production in Europe. *Proc. Natl Acad. Sci. USA*, **112**, 7147–7152.

Ruff, C.B., Holt, B.M., Niskanen, M., Sladek, V., Berner, M., Garofalo, E., Garvin, H.M., Hora, M., Maijanen, H., Niinimaki, S., Salo, K., Schuplerova, E., and Tompkins, D. (2012) Stature and body mass estimation from skeletal remains in the European Holocene. *Am. J. Phys. Anthropol.*, **148**, 601–617.

Ruff, C.B., Larsen, C.S., and Hayes, W.C. (1984) Structural changes in the femur with the transition to agriculture on the Georgia coast. *Am. J. Phys. Anthropol.*, **64**, 125–136.

Schmidt, I.M., Jorgensen, M.H., and Michaelsen, K.F. (1995) Height of conscripts in Europe – Is postneonatal mortality a predictor? *Ann. Hum. Biol.*, **22**, 57–67.

Sládek, V., Ruff, C.B., Berner, M., Holt, B., Niskanen, M., Schuplerová, E., and Hora, M. (2016) The impact of subsistence changes on humeral bilateral asymmetry in Terminal Pleistocene and Holocene Europe. *J. Hum. Evol.*, **92**, 37–49.

Sobral, F. (1990) Secular changes in stature in southern Portugal between 1930 and 1980 according to conscript data. *Hum. Biol.*, **62**, 491–504.

Sorkin, J.D., Muller, D.C., and Andres, R. (1999) Longitudinal change in height of men and women: implications for interpretation of the body mass index: the Baltimore Longitudinal Study of Aging. *Am. J. Epidemiol.*, **150**, 969–977.

Sparacello, V.S., Pearson, O.M., Coppa, A., and Marchi, D. (2011) Changes in skeletal robusticity in an iron age agropastoral group: the Samnites from the Alfedena necropolis (Abruzzo, Central Italy). *Am. J. Phys. Anthropol.*, **144**, 119–130.

Steckel, R.H. (1995) Stature and the standard of living. *J. Econ. Lit.*, **33** (4), 1903–1940.

Stini, W.A. (1976) Adaptive strategies of human populations under nutritional stress. In: *Biosocial Interrelations in Population Adaptation* (eds E.S. Watts, F.E. Johnston, and G.W. Lasker), Moutan, The Hague, pp. 19–40.

Stinson, S. (1985) Sex differences in environmental sensitivity during growth and development. *Yrbk Phys. Anthropol.*, **28**, 123–147.

Straus, L.G. (2008) The Mesolithic of Atlantic Iberia. In: *Mesolithic Europe* (eds G. Bailey and P. Spikins), Cambridge University Press, Cambridge, pp. 302–327.

Tanner, J.M., Hayashi, T., Preece, M.A., and Cameron, N. (1982) Increase in length of leg relative to trunk in Japanese children and adults from 1957 to 1977: Comparison with British and with Japanese Americans. *Ann. Hum. Biol.*, **9**, 411–423.

Turchin, M.C., Chiang, C.W., Palmer, C.D., Sankararaman, S., Reich, D., Genetic Investigation of ATC, and Hirschhorn, J.N. (2012) Evidence of widespread selection on standing variation in Europe at height-associated SNPs. *Nat. Genet.*, **44**, 1015–1019.

Wolfe, L.D. and Gray, J.P. (1982) A cross-cultural investigation into the sexual dimorphism of stature. In: *Sexual dimorphism in* Homo sapiens (ed. R.L. Hall), Praeger, New York, pp. 197–230.

Zilhao, J. (2000) From the Mesolithic to the Neolithic in the Iberian peninsula. In: *Europe's First Farmers* (ed. T.D. Price), Cambridge University Press, Cambridge, pp. 144–182.

11

Central Europe

Vladimír Sládek[1], Margit Berner[2], Eliška Makajevová[1], Petr Velemínský[3], Martin Hora[1], and Christopher B. Ruff[4]

[1] *Department of Anthropology and Human Genetics, Faculty of Science, Charles University, Prague, Czech Republic*
[2] *Department of Anthropology, Natural History Museum, Vienna, Austria*
[3] *Department of Anthropology, National Museum, Prague, Czech Republic*
[4] *Center for Functional Anatomy and Evolution, Johns Hopkins University School of Medicine, Baltimore, MD, USA*

11.1 Introduction

In this chapter we summarize temporal variation in postcranial properties in the Central European region from the Terminal Pleistocene through the Holocene. We briefly review some background for Central European bioarchaeological and archaeological research. The results deal with regional variation of postcranial features such as stature, body mass, body mass index, relative bi-iliac breadth, and crural and brachial indices. We also compare femoral, tibial, and humeral cortical areas and section moduli and bilateral directional asymmetry in humeral cross-sectional properties. Methodological details regarding data collection are given in Chapters 2 and 3, and more general European trends are covered in Chapters 4–7.

11.1.1 Central Europe: Geography and Paleoenvironment

Central Europe is divided in paleoanthropological research into south-central and north-central Europe (see review in Sládek, 2000). Since our Late Pleistocene and Holocene samples here consist of sites excavated in Bohemia, Moravia, north-eastern Austria, and (mainly) eastern Germany (see Fig. 11.1), we will limit our description of paleoenvironmental and archaeological background to only this narrow Central European region.

The studied temporal span covers the end of the Late Pleistocene (MIS 3-2; 30–11,600 cal. BP) divided by the Last Glacial Maximum (LGM) which occurred during ca. 20,000–18,000 years BP (MIS 2; Clark *et al.*, 2009; Yokoyama *et al.*, 2000) and was followed by strong warming oscillations of the Last Glacial up to 11,600 cal. BP (MIS 2; Lowe *et al.*, 1994) and the subsequent relatively stable warm Holocene Interglacial (MIS 1; from 11,600 cal. BP to recent) (Davis *et al.*, 2003).

The LGM (ca. 20,000–18,000 BP) was the most extreme cooling event during the end of the Late Pleistocene, which was accompanied with the maximum extent of the global ice sheet (Clark *et al.*, 2009). This is true also in Central Europe where the LGM represented the most important transition between the preceding glacial paleoenvironment of the Late Pleistocene and the subsequent post-glacial and Holocene paleoenvironment (Dreslerová *et al.*, 2007;

Skeletal Variation and Adaptation in Europeans: Upper Paleolithic to the Twentieth Century,
First Edition. Edited by Christopher B. Ruff.

Nerudová and Neruda, 2015). The following Last Glacial (ca. 18,000–12,000 BP) is characterized by abrupt and strong warming oscillations with changes in temperature up to 7 °C per 50 years followed by cold fluctuations (Lowe, 1994, 2001; Lowe *et al.*, 1994; Walker, 1995). The Central European paleoenvironment during the Last Glacial is characterized by the termination of pleniglacial processes with appearance of new soils, changes in watercourses, appearance of forest with decline of open landscape but still suitable for long-range movement of large herds (such as reindeer and elk) (Ray and Adams, 2001; Dreslerová *et al.*, 2007; Kuneš *et al.*, 2008; Ehlers *et al.*, 2013).

The Holocene (11,600 cal. BP – recent) represents on a global European geological scale a relatively stable and warm climatic period compared to preceding fluctuations of the Last Glacial. The Central European Holocene only slightly deviates from the global European climatic record (Davis *et al.*, 2003). However, the Central European mean temperature of the warmest moth (MTWM) and the mean temperature of the coldest month (MTCM) reach lower values and are less stable than the overall European global record in the beginning of the Holocene. The Central European Holocene is also characterized by at least a slight mid-Holocene maximum about 7500 BP and a subsequent mid-Holocene reversal (Davis *et al.*, 2003; Kalis *et al.*, 2003). However, the overall oscillation in the Central European mean temperatures is less than 1 °C for the rest of the Holocene.

The absence of strong fluctuations in mean temperature on a geological scale in the Central Europe Holocene does not mean, however, that the Holocene climate was so stable that it had no effect on human occupation. In fact, local climatic conditions may differ significantly from global and overall regional patterns. Thus, short-term climatic oscillations in the range of years may have a substantial impact on demography and human occupation (Le Roy Ladurie, 1971; Piontek, 1992; Brázdil, 1996; Fagan, 2000; Brázdil *et al.*, 2001, 2005; Svoboda *et al.*, 2003; Rotberg and Rabb, 2014; Nerudová and Neruda, 2015), with the magnitude of the impact dependent mostly on the strength of socioeconomic buffers (Kates, 1985; Pfister, 2001; Brown, 2005; Menotti, 2012). Moreover, changes in mean temperature are not the only factor influencing human occupations, as for example indicated by fluctuations in human occupation and subsistence according to the appearance of distinct wet and dry periods during the Central European Holocene (Arbogast *et al.*, 2006; Dreslerová, 2012), and the effect of cold and dry arctic winds on the pattern of human occupation during the LGM (Nerudová and Neruda, 2015). Thus, relatively small fluctuations observed on a global scale may have resulted in substantial risk of death, decrease of conceptions, famine, and disruption of socioeconomic systems, as observed for example from short-term famine during the Little Ice Age in the 14th and 18th centuries (Le Roy Ladurie, 1971; Fagan, 2000; Brázdil *et al.*, 2001). This can also be observed in changes in land occupation patterns in the Early Medieval Great Moravia (Macháček, 2012), and in a shift in land use toward the driest and warmest parts with the best Chernozem soils during the Central European Neolithic and Copper Age (Dreslerová, 2012).

11.1.2 Central Europe: Archaeological Context

Central Europe has provided one of the largest European skeletal samples of Early Upper Paleolithic humans (~32–23 kyr BP; Matiegka, 1934, 1938; Sládek, 2000; Sládek *et al.*, 2000; Teschler-Nicola, 2006; Trinkaus and Svoboda, 2006), but only a limited human sample from the Late Upper Paleolithic (~14–11 kyr BP) and Mesolithic (~8–5.5 kyr BP) (Podborský, 1993; Oliva, 2005; Valoch and Neruda, 2005; Vencl and Fridrich, 2007; Nerudová and Neruda, 2015). The first anatomically modern humans have been associated with hunter-gatherer subsistence specialized on the hunting of large game such as reindeer, which implies long-distance mobility.

The pattern of foraging strategy changed significantly after the Last Glacial Maximum (Binford, 1968, 1984; Flannery, 1969; Shott, 1993; Bogucki and Crabtree, 2004a; Gamble *et al.*, 2004), which resulted in the Late Upper Paleolithic and Mesolithic of an increase in sedentism and some other behavioral adaptations usually associated with the Neolithic (Holt, 2003; Holt and Formicola, 2008).

The first Neolithic farmers appeared in Central Europe between 5600 and 5400 cal BC and formed the distinct Linearbandkeramik (LBK) culture about 5500 cal BC (Pavlů, 2005; Bickle and Whittle, 2013a). The human skeletal samples in the Central European Neolithic are relatively sparse, mainly found in the LBK horizon (Podborský, 1993, 2002; Lenneis *et al.*, 1995; Pavlů and Zápotocká, 2007; Bickle *et al.*, 2013; Whittle *et al.*, 2013). Neolithic groups also employed cremation after the LBK, and therefore skeletal evidence for later Neolithics are limited (Podborský, 1993; Lenneis *et al.*, 1995; Pavlů and Zápotocká, 2007). The LBK humans have been found to be relatively sedentary farmers who practiced intensive year-on-year cereal cultivation employing also intensive tillage and weeding (i.e., intensive garden horticulture) (Bogaard, 2004; Bogaard and Jones, 2007; Bickle and Whittle, 2013b). This implies also integration of domesticated animals (e.g., manuring) in the early Neolithic subsistence and a mixed model of husbandry (Bogucki, 1988; Bogaard, 2004;); however, there is no evidence for nomadic pastoral subsistence except possibly transhumance at least for some of the domestic species (but see discussion in Bickle and Whittle, 2013b).

The Central European farmers experienced from about 4500 BC substantial intensification of subsistence and the introduction of new innovations on a larger scale than has been found in the previous Neolithic occupation (Podborský, 1993; Bogucki and Crabtree, 2004a; Neustupný, 2008; Greenfield, 2010). The intensification of subsistence is described as the Secondary Products Revolution (Sherratt, 1981, 1983; see review in Greenfield, 2010) and the period usually as the Copper Age based on the appearance of the first metallurgy of copper (e.g., Amzallag, 2009) (however, other labels such as the Late Neolithic, Aneolithic, Chalcolithic, Eneolit are also used; see review in Neustupný, 2008). Central European Copper Age agriculture is characterized by an intensification of subsistence activities in a way which persisted during the following Holocene agricultural period, at least up to the Medieval (e.g., the pattern of agricultural field in a landscape; Neustupný, 2008). This major shift in subsistence is associated not directly with the first origins of agriculture but with a new scale in how animals were exploited (Greenfield, 2010) for intensive milking (Bokonyi, 1974), wool production (Flannery, 1965; Ryder, 1987; Good, 2001), traction (Sherratt, 1983), and for wheeled vehicles and plows (Sherratt, 1981, 1983). Central European Copper Age subsistence was based on grain crops with mixed animal breeding (Sjögren *et al.*, 2016), yet independent pastoral subsistence during the Copper Age was rejected, at least at the end of the Central European Copper Age (Sládek *et al.*, 2006a,b, 2007).

Intensification of an agricultural subsistence strategy in Central Europe resulted in economic transformation and social differentiation associated with the first appearance of bronze metallurgy and the first Bronze Age occupational sites in ca. 2200 BC (Podborský, 1993; Shennan, 1993; Neugebauer *et al.*, 1994; Harding, 2000; Kristiansen and Larsson, 2005; Jiráň, 2008). The Central European occupation changed from small farming groups in the Copper Age only weakly linked at a regional level, to the Bronze Age occupation with quasi-political groupings on a larger scale (Harding, 2000). However, the transition between the Central European Copper Age and Early Bronze Age has been reflected to be relatively continual in both the archaeological and skeletal record (Jiráň, 2008; Sládek *et al.*, 2006a,b, 2007; Sosna *et al.*, 2008). Central European Bronze Age societies experienced craft specialization beyond household production, initiated intensive long-distance trading, intensified metal production, and adopted bronze technology (Podborský, 1993; Harding, 2000; Bogucki and Crabtree, 2004b).

During the Bronze Age the first concentration of inhabitants in large settlement units appeared in Central Europe (Harding, 2000; Jiráň, 2008) along with a more complex society differentiated in status, power, and wealth, as reflected for example in burial practices (Sosna, 2007; Sosna *et al.*, 2008). However, the agricultural subsistence compared to the Copper Age remained relatively unchanged (Podborský 1993; Jiráň, 2008).

The Iron Age began in Central Europe about 750 years BC, and is subdivided by archaeological evidence into the earlier Hallstatt period (750 BC–400 BC; Podborský, 1993; Nebelsick *et al.*, 1997; Venclová *et al.*, 2008) and the later La Tène (400 BC–0 BC; Podborský, 1993; Venclová *et al.*, 2008). Subsistence changes between the Central European Bronze Age and Hallstatt period were relatively small (Podborský, 1993; Nebelsick *et al.*, 1997) but the later Iron Age was accompanied by substantial changes in social stratification and inequalities (Kristiansen and Rowlands, 2005), as well as with intensification of long-distance trading such as contacts of Central Europe with ancient Greece and Etruscan territories (Collis, 2003; Venclová *et al.*, 2008). During the Iron Age several technologies intensified, such as ore mining and metallurgy by introduction of iron, pottery production by adoption of the pottery wheel, and transport by horses and wheeled vehicles (see review in Collis, 2003). Iron Age populations also intensified food-grinding technology by the adoption of the rotary quern (Lynch and Rowland, 2005; Wefers, 2011), with an impact on upper limb morphology (Sládek *et al.*, 2016). Iron Age society developed complex social stratification with an elite class and the emergence of warriors as well as slaves (Kristiansen and Rowlands, 2005; Keegan, 2011; Sparacello *et al.*, 2015). Social stratification is well reflected in burial practices by the presence of rich burials in distinct grave mounds along with poor burials in the settlement areas (Venclová *et al.*, 2008). Later La Tène culture is the first archaeological period when the written records are available for Central Europe, which introduced several terms later used in the description of Central European territories (e.g., Boiohaemum for Bohemia; Kruta, 2000; Venclová *et al.*, 2008). The Iron Age also had an influence on the formation of Central European population groupings during the Roman period in the first four centuries AD (Podborský, 1993; Salač *et al.*, 2008).

The end of the Roman period and the collapse of the Western Roman Empire from about AD 476 is characterized by population movements in Europe, and this Migration period also affected Central European regions (Bogucki and Crabtree, 2004b). The end of this period in Central Europe is dated in the sixth century and is accompanied with the appearance of the western branch of Slavs in Central Europe (Třeštík, 1997; Barford, 2001; Měřínský, 2002; Bogucki and Crabtree, 2004b; Beranová and Lutovský, 2009). This migration was caused by the decline of political power and administration of the Roman Empire and socio-political changes in Barbarian societies, as well as by the attack of groups migrating from Eurasian steppes (Bogucki and Crabtree, 2004b). This movement was accompanied by a decrease in population density, changes in population structure, and the destruction of previous settlement areas. The collapse of Roman control of Central European regions was probably not sudden but gradual, and was probably the major factor influencing the living conditions in this region. However, it is also not clear to what degree actual population replacement occurred and how much indigenous populations influenced formation of the Early Medieval societies dominated by migrated Slavic groups.

The emerging Central European Early Medieval societies were directly affected by the impact of Euroasian invasion of the Avars during mid-6th century and the emergence of an Avar Khaganate between the 6th and 9th centuries (Bogucki and Crabtree, 2004b). It is accepted that the Avars were centered in Pannonia (i.e., former Roman province in South-Central Europe situated in present western Hungary, western Slovakia, eastern Austria, and northwest Balkan Peninsula) and represented a highly organized society with a mobile and nomadic subsistence, at least for elites. They also cooperated with Slavic inhabitants and encouraged them to settle in northern boundary regions to buffer outside attacks (Bogucki and Crabtree, 2004b).

However, it remains unclear how strong the effect of nomad subsistence and high mobility was on indigenous Slavic groups in Central Europe. Archaeologists expect that Avars did not transform indigenous, mainly peasant societies into a mobile mode of subsistence, but Avars may have represented in the majority of their Khaganate the political administration, without a direct effect on the daily life of Slavic groups.

The Slavs were the major part of established Central European Early Medieval empires such as Samo's Empire (AD 624–659; Lutovský and Profantová, 1995) and the Great Moravian Empire (AD 833–907; Třeštík, 2001; Měřínský, 2011). The Early Medieval period in Central Europe also represents a transformation from independent, closely related smaller units to the appearance of a complex chiefdom social organization, and replacement and/or transition toward the first state organization from about the 10th–11th century (Třeštík, 1997; Johnson and Earle, 2000; Beliaev *et al.*, 2001; Smith, 2004; Macháček, 2005). Slavs dominated also in the Central European Late Medieval state organizations such as in the Přemyslid state with a center in Bohemia between the 10th and 14th centuries AD (Válka, 1991; Třeštík, 1997). The Early and Late Medieval periods provide in Central Europe voluminous archaeological, anthropological, and historical records that help to detail several aspects of living condition and subsistence changes (see summary in Třeštík, 2001). The subsistence economy in the Early Medieval period in Central Europe was dominated by a mainly rural society based on agriculture, similar to previous Holocene periods, but during the Late Medieval period populations shifted from rural towards more complex urban societies, with economic specialization and complex social stratification (Bogucki and Crabtree, 2004b; Macháček, 2010). Moreover, Central Europe was in close contact mainly with Southern Europe in the Early Medieval and with Western Europe in the Late Medieval (Válka, 1991; Třeštík, 1997, 2001; Měřínský, 2011). Thus, we can expect that the Central European burial record in the medieval period represents populations that experienced more diverse environmental and genetic impacts and subsistence specialization than the relatively more homogeneous groups buried in earlier Holocene cemeteries.

The most recent period of the European Holocene began in the late 18th century with a transition from agrarian to industrial economies during the Industrial Revolution (Ashton, 1950; Stearns, 1993; Cipolla, 1994). The most recent period is represented in Central Europe by an improvement of living conditions after industrialization in the 19th century. The human population also experienced several negative impacts, such as World Wars I and II. The improvement in living conditions is associated mainly with an effort to improve the social system, public health, sanitation, food production and distribution, education, and standard of living. This had a substantial effect on biological variation in Central European populations, as observed for example in the presence of positive secular trend in body size and changes in the growth curve (Vignerová *et al.*, 1998). However, human biologists and ecologists have pointed out that the present population is faced with new problems resulted from the negative consequences of living improvements (Lieberman, 2013), such as a reduction of physical activity and increase of obesity (Bassett, Jr *et al.*, 2008), as well as changes in the structure of families and society (Archer *et al.*, 2013).

11.2 Materials and Methods

11.2.1 Sample

Table 11.1 and Figure 11.1 summarize the Central European sample used in this chapter. The total Central European sample consists of 700 individuals (404 males and 296 females). In addition, as explained below, another 122 individuals from across Europe were included

Table 11.1 Central European sample.

Period/Culture	Abbrev.	N (Sites)	N (Total)	N (M)	N (F)	Years (BP)
Early Upper Paleolithic[a]	EUP	14	23	12 (6)	11 (4)	33,388–26,406
Late Upper Paleolithic[a]	LUP	17	23	16 (3)	8 (1)	21,922–11,615
Mesolithic[1]	Meso	32	93	56 (1)	37 (3)	10,500–5928
Neolithic	Neol	10	29	16	13	7300–5000
Copper Age	CopA	64	102	60	42	4600–4300
Bronze Age	BA	17	180	100	80	4000–2950
Iron Age	IA	4	39	26	13	2200
Avar	Avar	3	98	54	44	1300–1275
Early Medieval	EMed	3	99	60	39	1150–950
Late Medieval[b]	LMed	4	109	59	50	700–650 (260)
Recent	Rec	1	26	19	7	10
Total		**169**	**822**	**478**	**344**	

[a] UP and Mesolithic samples are not limited exclusively to the Central European region but combine all the available European skeletal remains. The maximum number of CE males and females is shown in parentheses, but the number may differ in specific analyses due to preservation.
[b] LMed period includes also Opava-Pivovar sample dated to the Recent (AD 1588–1789).

in analyses of pre-Neolithic periods. The following tables provide sample sizes for each analysis. Because of the very large number of sites (n = 169), we show the locations of sites in Figure 11.1 grouped into four general geographic areas; further details about specific sites are given in Appendix 1. Eight periods were compared in the Holocene (Neol, CopA, BA, IA, Avar, EMed, LMed, Recent; see Table 11.1 for abbreviations). The main criterion for site classification was based on archaeological context and not on direct C^{14} dating; however, the time span overlap between archaeological cultures in the Central European Holocene is minor compared to the broader European analyses.

We also compared Central European Neolithic and later periods with samples from the Early Upper Paleolithic, Late Upper Paleolithic, and Mesolithic to give a broader context for the impact of the adoption of agriculture during the Central European early Holocene. Since postcranial data are available for only a limited number of individuals from the Early and Late Upper Paleolithic and Mesolithic periods in Central Europe, for these analyses we combined data derived from the pooled European sample (see Table 11.1). We are aware of the possible limitations in using such a broad geographic sample in a regionally oriented study and discuss below how this might affect our conclusions regarding early temporal trends observed in the Central European Holocene.

The Central European Holocene sample has a somewhat different time period structure than temporal periods used in some of the other chapters of this volume, mainly because of the specific nature of the regional cultural changes and the availability of appropriate samples. The early agricultural Holocene period is divided into the Neolithic and Copper Age to analyze the impact of one of the major subsistence changes among the early agriculturalists (Greenfield, 2010). We split the Iron Age and Roman periods and carried out Central European analyses only on skeletons from the Iron Age given the very small sample size available for the Roman period (5 males, 7 females). We also distinguished the Avar sample as a specific group with an

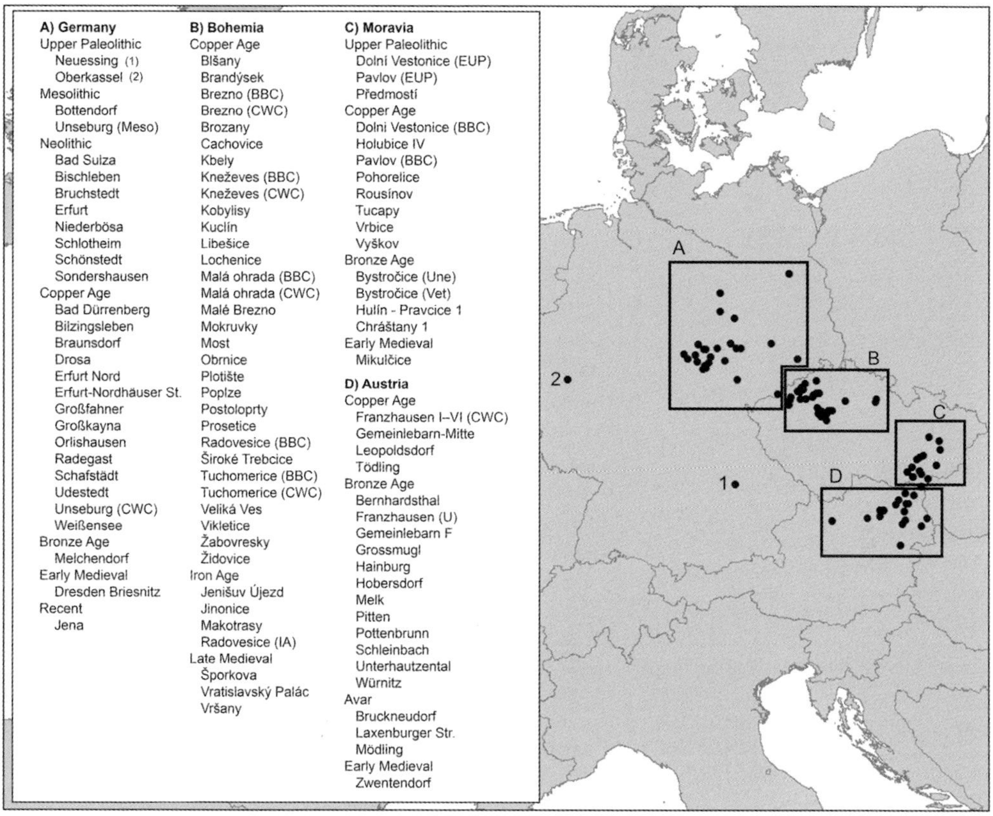

Figure 11.1 Map of the study sites. See Table 11.1 for sample sizes by period.

expected higher level of mobility in the beginning of the Early Medieval. The Recent Central European sample consists only of a very recent autopsy collection from the industrial 20th century; the Late Medieval/Early Modern skeletal sample (Opava-Pivovar; n = 28) is included in the Late Medieval sample.

We use the terms pre-Holocene for EUP and LUP samples, and Holocene for the periods from the Mesolithic to Recent. However, we also distinguish two general periods in the Holocene: an early agricultural Holocene sample (i.e., Neol, CopA, and BA) and a late agricultural sample (i.e., IA–Recent). The Recent sample is included in the late agricultural period to simplify descriptions (details about the non-agricultural origin for the 20th sample is provided in Appendix 1 and the effect on the compared properties in the Discussion).

11.2.2 Measurements and Variables

Table 11.2 summarizes variables used in the Central European analysis as well as the abbreviations, units, and brief descriptions. Standard osteometric measurements followed the measurement protocol from Bräuer (1988). The computation details and biological background for body size and shape measurements are given in Chapters 2 and 4, for cross-sectional properties and their size standardization in Chapter 3, and for bilateral skeletal asymmetry in Chapter 7.

Table 11.2 Summary of variables used for Central European analysis.

Measurement	Abbrev.	Description[a]
Stature	STA	Stature estimated by anatomical and mathematical technique (cm).
Body mass	BM	Body mass estimated either from bi-iliac breadth or femoral head breadth (kg).
Body mass index	BMI	BM/Stature^{-2} (BM in kg, stature in m).
Relative bi-iliac breadth	RBIB	BIB/STA × 100 (BIB and STA in cm; BIB is converted from skeletal bi-iliac breadth using equation (5) in Ruff *et al.*, 1997.
Crural index	CI	Tibial lateral condyle-medial malleolus length (M#1a)/Femoral bicondylar length (M#2) × 100.[b]
Brachial index	BI	Radial maximum length (M#1)/Humeral maximum length (M#1) × 100.[c]
Cortical area	CA	Compressive/tensile strength (adjusted by BM, multiplied by 100; mm^2 kg^{-1}).[d]
Polar section modulus	Z_p	Torsional and average bending strength (adjusted by BM × BML, multiplied by 10^{-4}; [mm^3 (kg × mm)]).[d]
Directional asymmetry	DirA	[(R – L)/(R + L)/2)] × 100; [%]; R is right side, L is left side

[a] See computation details in Chapter 2 and Methods. M#: Bräuer, 1988.
[b] Right side is preferred, if not preserved then left side is used.
[c] Average right and left side is used, if is not preserved then BI is computed either from right side or left side.
[d] CSG properties are taken at 50% of femoral (F) and tibial (T), and 35% of humeral (H) biomechanical length. For femora and tibiae either right or left side was used given the preservation. Body size standardization (std) is used (see Chapter 3), except in paired analyses of directional asymmetry

11.2.3 Statistical Techniques

Linear variables such as stature, body mass, and cross-sectional geometric (CSG) properties were normally distributed; therefore, parametric tests were used in these cases. Similarly to these dimensions, all body shape indices such as the crural and brachial indices did not show deviations from the normal distribution and were treated in the same way. The most important parametric tests we employed were two-way ANOVA with sex, period, and interaction between sex × period. Sex was significant in each of the studied variables; therefore, we used the Fisher LSD post-hoc comparison to further test sexual dimorphism differences. When interactions between sex × period were found to be significant, Fisher LSD post-hoc tests were also used to test for significant variation between periods; however, this was only in the case for humeral CSG properties.

Directional asymmetry values deviate from normality in our Central European sample; therefore, bootstrap estimates and non-parametric statistics were used. The medians were also computed by standard descriptive techniques to allow for direct comparisons in future studies; however, the ±95% CIs were computed using bootstrap replicates. The CIs were computed using a macro written in Visual Basic in Excel 2010 following the approach described in Manly (2006) (see Chapter 7 for details). Several non-parametric tests were further applied to assess the significance of the observed pattern of variation in maximum and directional asymmetry: the multiple group Kruskal–Wallis test to assess significance between periods split up by sex, the Mann–Whitney *U*-test for two independent samples to test sexual dimorphism significance, and the one sample Wilcoxon Signed-Rank test for testing significance of directional asymmetry against $H_0 = 0$ (i.e., against absence of directional asymmetry).

All tables, raw data preparation, and bootstrap calculations were carried out using Excel Microsoft Office Professional 2010 (Microsoft Corporation, 2010). All other tests were computed in Statistica 12 for Windows (64 bits) (StatSoft, Inc. 1984–2013).

11.3 Results

11.3.1 Body Size and Shape Variation

11.3.1.1 Stature, Body Mass, and Body Mass Index

Table 11.3 summarizes means and sample sizes for stature, and Figure 11.2(a) shows temporal distribution by sex and period. Average stature varies significantly between both periods and sex ($p < 0.001$), but the interaction between period and sex is non-significant ($p < 0.21$). Average stature decreases about 14 cm (8%) in males and 12 cm (7%) in females between the Early Upper Paleolithic and Neolithic, whereas from the Neolithic to Recent average stature increases about 10 cm (6%) in males and 7 cm (5%) in females. Increase in average stature after the Neolithic in males and females is not continuous. A relatively abrupt increase is observed between the Neolithic and Copper Age [about 10 cm (6%) in males and 6 cm (4%) in females], and a major decrease occurs between the Copper Age and Avars in males (about 5 cm; 3%) and between the Copper Age and Iron Age in females (about 4 cm; 2.5%). Maxima in stature in the Holocene are reached in the Copper Age, Early Medieval, and Recent in males and Copper Age and Recent in females.

Table 11.3 summarizes statistics for body mass, and Figure 11.2(b) shows temporal distribution by sex and periods. Average body mass also varies significantly between both periods and sex ($p < 0.001$). The overall pattern of variation in body mass is similar to previously described changes in stature. Average body mass decreases between the Early Upper Paleolithic and Neolithic by about 9 kg (13–15%) in both sexes, and increases from Neolithic to Recent by about 4 kg (7%) in males and 6 kg (11%) in females. Average body mass in males is more variable than females during Holocene.

Means and sample sizes for the body mass index (BMI) are shown in Table 11.3, and temporal distribution by sex and period in Figure 11.2(c). Average BMI significantly varies between both periods and sex ($p < 0.01$). In males, average BMI increases between Early Upper Paleolithic and Iron Age by 1.63 $kg\,m^{-2}$ (7%) and decreases from Iron Age to Recent by 2.21 $kg\,m^{-2}$ (9%). In females, the temporal trend is less marked and average BMI remains mostly unchanged in Terminal Pleistocene and during Holocene [two exceptions are observed in females: increase from Early Upper Paleolithic to Mesolithic about 1.23 $kg\,m^{-2}$ (5%) and from Neolithic to Iron Age about 1.41 $kg\,m^{-2}$ (6%)].

11.3.1.2 Changes in Body Size Between Central European Holocene and Living Humans

Holocene variation in stature can be compared with a much large sample of living Central Europeans than are included in our small Recent skeletal sample. The variation in stature among living Central Europeans can be derived from four anthropological surveys taken from 1951 to 2001 in the Czech Republic (Fetter *et al.*, 1963; Prokopec *et al.*, 1986; Lhotská *et al.*, 1993; Vignerová *et al.*, 2006). This living sample consists of children and adolescents aged from 2.5 to 18 years; however, for our purposes we only use stature from the interval from 17.8 to 18 years old. As indicated in Figures 1 and 2 of Vignerová *et al.* (2006), the 17–18 years interval represented completed longitudinal growth at least in surveys from 1981 to 2001 for males, and from

Table 11.3 Body size and shape statistics for Central European humans [Mean (N, ±SE)].[a]

	Males			Females		
	STA	BM	BMI	STA	BM	BMI
EUP	174.3 (14; 1.81)*	71.5 (13; 2.45)*	23.44 (13; 0.51)	163.9 (10; 2.04)*	61.7 (9; 1.73)*	22.99 (9; 0.69)
LUP	165.2 (15; 1.40)*	66.1 (15; 1.79)*	24.3 (15; 0.78)	156.3 (8; 1.56)*	57.0 (7; 2.21)*	22.93 (7; 0.82)
Meso	163.6 (54; 1.18)*	65.4 (42; 1.32)*	23.88 (41; 0.32)	155.3 (35; 0.86)*	58.6 (26; 1.27)*	24.22 (26; 0.47)
Neol	160.6 (16; 1.41)*	62.3 (16; 1.33)*	24.24 (16; 0.63)*	152.3 (13; 1.43)*	52.6 (13; 1.58)*	22.59 (13; 0.44)*
CopA	170.1 (54; 0.78)*	67.8 (59; 0.98)*	23.61 (53; 0.33)	158.1 (38; 0.89)*	57.2 (40; 0.79)*	22.84 (37; 0.26)
BA	167.1 (95; 0.61)*	66.4 (89; 0.79)*	23.75 (87; 0.24)*	157.1 (76; 0.62)*	56.3 (73; 0.50)*	22.85 (71; 0.21)*
IA	165.5 (25; 1.55)*	69.5 (24; 1.96)*	25.07 (23; 0.45)	154.2 (13; 1.54)*	57.2 (12; 1.57)*	24.0 (12; 0.68)
Avar	164.9 (52; 0.86)*	67.1 (53; 0.92)*	24.68 (52; 0.29)*	157.3 (44; 0.71)*	57.2 (44; 0.74)*	23.11 (44; 0.22)*
EMed	169.6 (60; 0.83)*	70.5 (60; 1.03)*	24.45 (60; 0.26)*	157.8 (39; 0.84)*	58.3 (39; 0.79)*	23.39 (39; 0.21)*
LMed	166.9 (54; 0.93)*	66.9 (52; 1.10)*	24.12 (51; 0.34)	157.5 (44; 0.95)*	59.0 (42; 1.12)*	23.6 (40; 0.22)
Rec	170.3 (19; 2.31)*	66.5 (19; 2.45)*	22.86 (19; 0.63)	159.7 (7; 2.50)*	58.3 (7; 2.05)*	22.82 (7; 0.40)

[a] See details about abbreviations and parameters in Tables 11.1 and 11.2. The three earliest period data (EUP, LUP, Meso) are for the pan-European sample.
* ANOVA post-hoc comparison for sexual dimorphism significant at $p < 0.05$.

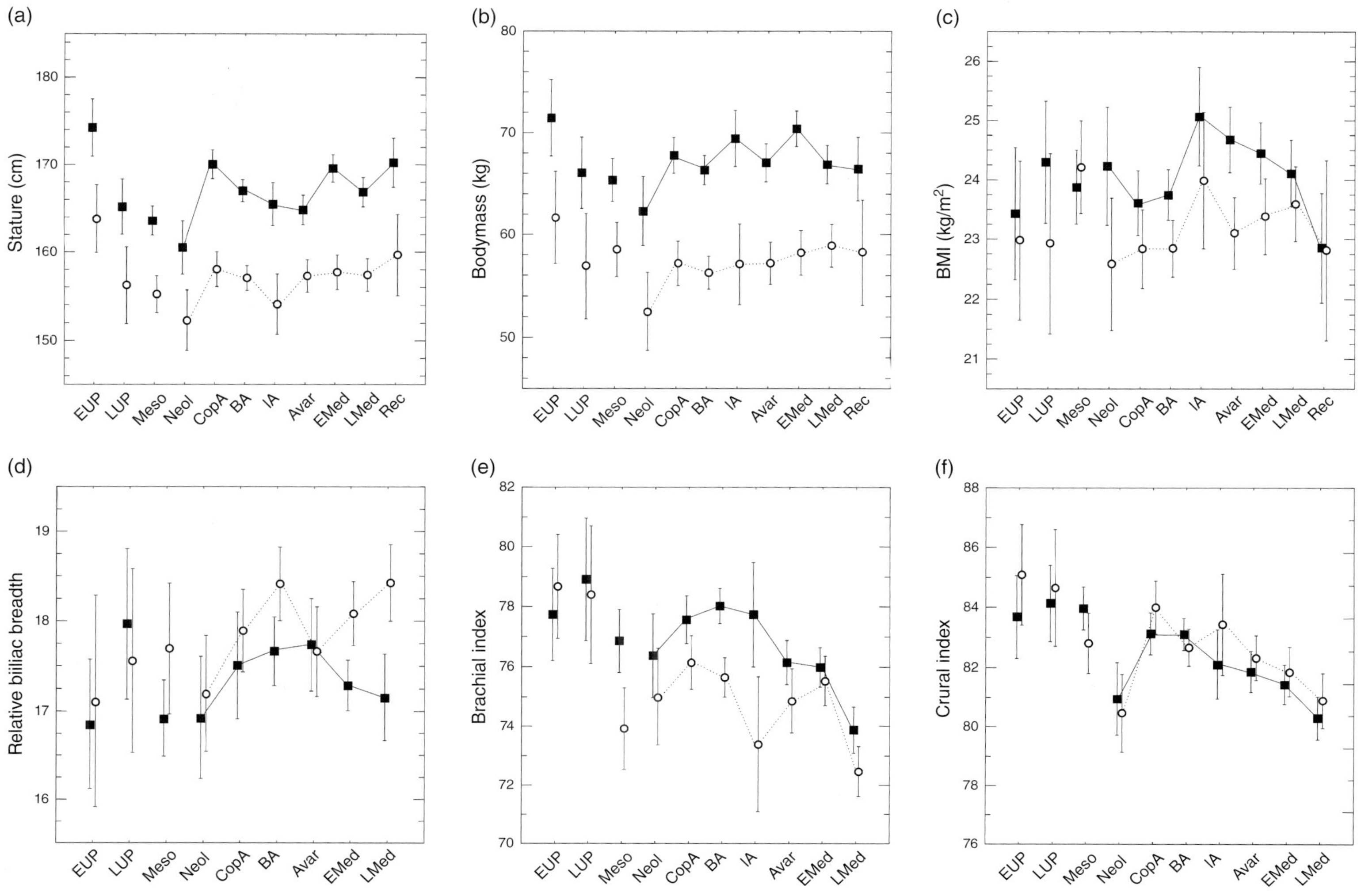

Figure 11.2 Temporal trends in Central European body size and shape by sex and temporal period. Mean ±95% CI. Males: solid symbols; females: open symbols. (a) Stature. (b) Body Mass. (c) Body mass index. (d) Relative bi-iliac breadth. (e) Brachial index. (f) Crural index.

1951 to 2001 for females. Table 11.4 summarizes mean statistics for stature taken from Czech Medieval samples and from living humans measured in Nation-Wide Anthropological surveys between 1951 and 2001. Mean stature significantly increases between periods and between sexes through the last thousand years ($p < 0.001$). Mean stature shows a significant increase from the Medieval to the most recent period, about 10–13 cm in males and 9–10 cm in females. This secular trend slows down in males and females between the two most recent anthropological surveys (i.e., between 1991 and 2001), which may indicate that the increase of mean stature has reached its limits (e.g., Padez, 2003; Larnkjær *et al.*, 2006). Average stature only surpasses that for the pan-European Early Upper Paleolithic sample in the 1981 and subsequent surveys (compare with Table 11.3).

11.3.1.3 Relative Bi-iliac Breadth

Sample statistics for relative bi-iliac breadth are summarized in Table 11.5, and temporal distribution by period and sex are shown in Figure 11.2(d). Relative bi-iliac breadth significantly varies between both periods and sex ($p < 0.02$), but non-significantly in interaction between periods and sex ($p < 0.21$). Central European males and females show a relatively narrow body breadth in the Central European Neolithic, similar to values observed for the European Upper Paleolithic. In males, average relative bi-iliac breadth increases toward larger values in the Holocene between the Neolithic and Avar periods by 0.8 (4.7%) and decreases by 0.6 (3.4%) thereafter. In contrast, females show in the Holocene an overall increase in average relative bi-iliac breadth from the Neolithic to the Late Medieval 1.3 (7.6%) but maximum relative bi-iliac breadths is observed also in the Bronze Age. However, male and female variation in relative bi-iliac breadth is relatively small in the Holocene and largely reflects variation in stature rather than absolute bi-iliac breadth (also see Chapter 4).

11.3.1.4 Crural and Brachial Indices

Sample statistics for crural and brachial indices are summarized in Table 11.5, and temporal distribution by sex and period are shown in Figure 11.2(e,f). Crural index significantly decreases between periods ($p < 0.001$), but is non-significantly different between sex ($p < 0.06$) and in interaction between sex and periods ($p < 0.18$). Brachial index is significantly different between both sex and period ($p < 0.001$) and in the interaction between period and sex ($p < 0.02$). Both indices decrease from more warm-adapted linear body proportions in the Terminal Pleistocene to more cold-adapted linear proportions in the Late Holocene. The changes between Early Upper Paleolithic and Late Medieval are about 4% in males and 5% in females in crural index, and 5% in males and 8% in females in brachial index. Crural and brachial indices strongly decrease from the Terminal Pleistocene toward Neolithic in both sexes; values for crural index in Neolithic are similar to those observed for the Late Holocene periods (LMed). Thus, the Neolithic sample shows more cold-adapted crural and brachial indices similar to linear proportions of Late Holocene groups. Brachial index also decreases among females in the Iron Age.

11.3.1.5 Sexual Dimorphism in Body Size and Shape

ANOVA post-hoc LSD tests indicate that sexual dimorphism (SD) is more marked in body size than in body shape (Tables 11.3 and 11.5). The relative number of significant differences between the sexes in BMI is relatively low compared to those for stature and body mass (Table 11.3). Sexual dimorphism is significant in stature and body mass in all the studied periods, whereas in BMI it is significant only in the Neolithic, Bronze Age, Avar, and Early Medieval. Low SD in body shape is also supported by a non-significant difference in post-hoc LSD test of ANOVA in crural index and a limited number of sexually dimorphic samples in relative bi-iliac breadth (BA, EMed, and LMed; see Table 11.5). The main exception in the studied body shape properties is the brachial index, which is significantly dimorphic (males larger) in five periods (Meso, CopA, BA, IA, and LMed).

Table 11.4 Stature statistics for the Central European Late and Early Medieval skeletal samples and for recent humans from National-Wide Anthropological Surveys in Czech Republic between 1951 and 2001 [Mean (N; ±SD)].

	Stature			
Period	Males	Females	Source[a]	Reference
EMed[b]	170 (33; 6.8)	158 (21; 5.5)	Our study	–
LMed[c]	167 (54; 6.8)	157 (44; 6.3)	Our study	–
1951	172 (243; 7.1)	162 (165; 5.8)	NAS 1951	Fetter *et al.*, 1963
1981	178 (352; 6.9)	165 (423; 6.1)	NAS 1981	Prokopec *et al.*, 1986
1991	179 (300; 6.8)	166 (375; 6.3)	NAS 1991	Lhotská *et al.*, 1993
2001	180 (241; 7.1)	167 (459; 6.6)	NAS 2001	Vignerová *et al.*, 2006

[a] NAS: Nation-Wide Anthropological Survey in the Czech Republic (CAV); data for males and females between 17.8 to 18 years old.
[b] EMed: only samples from Czech Republic included (site: Mikulčice).
[c] LMed: only samples from Czech Republic included (sites: Opava-Pivovar, Šporkova, Vratislavský Palác).

Table 11.5 Body shape statistics for Central European humans [Mean (N, ±SE)].[a]

	Males			Females		
	RBIB	CI	BI	RBIB	CI	BI
EUP	16.9 (8; 0.15)	83.7 (12; 0.66)	77.8 (9; 0.4)	17.2 (3; 0.41)	85.1 (8; 0.83)	78.7 (7; 0.62)
LUP	18.0 (6; 0.51)	84.2 (14; 0.73)	79.0 (5; 1.33)	17.6 (4; 0.79)	84.7 (6; 1.41)	78.5 (4; 1.65)
Meso	17.0 (23; 0.15)	84.0 (44; 0.52)	76.9 (19; 0.44)*	17.7 (8; 0.59)	82.9 (23; 0.56)	74 (11; 0.46)*
Neol	17.0 (9; 0.15)	81.0 (15; 0.66)	76.4 (11; 0.66)	17.2 (10; 0.29)	80.5 (13; 0.8)	75 (8; 0.64)
CopA	17.6 (12; 0.32)	83.2 (47; 0.38)	77.6 (33; 0.41)*	17.9 (20; 0.32)	84 (29; 0.37)	76.2 (27; 0.34)*
BA	17.7 (29; 0.21)*	83.2 (81; 0.25)	78.1 (61; 0.33)*	18.5 (25; 0.21)*	82.7 (61; 0.27)	75.7 (50; 0.34)*
IA	–	82.2 (17; 0.67)	77.8 (7; 1.12)*	–	83.5 (8; 0.75)	73.4 (4; 2.17)*
Avar	17.8 (16; 0.25)	81.9 (47; 0.27)	76.2 (40; 0.38)	17.7 (17; 0.19)	82.4 (41; 0.37)	74.9 (18; 0.47)
EMed	17.3 (54; 0.16)*	81.5 (50; 0.42)	76.0 (50; 0.34)	18.1 (33; 0.18)*	81.9 (33; 0.32)	75.6 (31; 0.47)
LMed	17.2 (18; 0.19)*	80.3 (43; 0.31)	73.9 (34; 0.44)*	18.5 (23; 0.22)*	80.9 (26; 0.51)	72.5 (29; 0.46)*

[a] See details about abbreviations and parameters in Tables 11.1 and 11.2. The three earliest period data (EUP, LUP, Meso) are for the pan-European sample.
* ANOVA post-hoc comparison for sexual dimorphism significant at p <0.05.

11.3.2 Mobility and Sedentism

11.3.2.1 Femoral and Tibial Cortical Area and Bending Strength

Sample statistics for femoral and tibial cortical areas are shown in Table 11.6, and temporal distribution by period and sex in Figure 11.3(a,d). Femoral and tibial cortical areas are significantly different between both periods and sex ($p < 0.001$), and non-significant in interaction between period and sex ($p < 0.23$–0.55). In general, average femoral cortical area is larger in the Upper Paleolithic than in the Recent samples, but the major decline in femoral cortical area is observed between the Early Upper Paleolithic and Bronze Age (about 12% in males and

Table 11.6 Femoral and tibial CSG properties for Central European humans [Mean (N, ±SE)].[a]

	Males			Females		
	CAstd	$Z_p$std	Z_x/Z_y	CAstd	$Z_p$std	Z_x/Z_y
Femora						
EUP	731.7 (10; 41.62)	1254.3 (9; 81.48)	1.27 (11; 0.05)*	736.0 (5; 45.08)	1153.1 (5; 89.66)	1.12 (5; 0.04)*
LUP	721.0 (13; 18.53)	1169.5 (13; 37.08)	1.18 (13; 0.08)*	706.8 (7; 56.68)	1154.5 (7; 88.68)	1.08 (8; 0.05)*
Meso	731.8 (38; 15.11)*	1246.6 (37; 29.40)*	1.04 (51; 0.02)	673.5 (15; 26.24)*	1131.8 (14; 47.58)*	1.03 (23; 0.04)
Neol	652.1 (14; 15.58)	1138.0 (14; 31.57)	0.99 (14; 0.03)	624.5 (12; 25.03)	1056.2 (12; 61.11)	0.94 (12; 0.02)
CopA	643.8 (50; 10.11)*	1082.5 (49; 18.13)	1.01 (51; 0.02)*	607.9 (36; 13.52)*	1035.2 (36; 27.94)	0.92 (36; 0.02)*
BA	640.9 (72; 8.02)*	1082.7 (72; 18.26)*	0.96 (77; 0.02)	567.0 (49; 7.89)*	957.6 (48; 22.75)*	0.94 (51; 0.02)
IA	649.1 (22; 15.35)	1098.8 (20; 35.45)	0.95 (22; 0.03)*	630.0 (10; 18.58)	1104.0 (10; 45.67)	0.86 (10; 0.04)*
Avar	633.5 (50; 9.66)*	1117.0 (50; 22.29)*	0.91 (50; 0.02)	579.0 (43; 9.65)*	982.7 (43; 20.53)*	0.89 (43; 0.02)
EMed	665.9 (59; 10.61)*	1174.1 (59; 21.52)*	0.94 (59; 0.02)	591.4 (38; 12.77)*	1026.5 (38; 23.84)*	0.93 (38; 0.02)
LMed	677.6 (46; 12.81)*	1176.9 (45; 24.60)	0.93 (48; 0.02)	628.4 (34; 13.86)*	1116.3 (34; 26.01)	0.9 (35; 0.02)
Rec	663.7 (19; 18.96)	1101.4 (19; 37.88)	1.02 (19; 0.03)	616.1 (7; 26.22)	1029.3 (7; 59.47)	0.95 (7; 0.05)
Tibiae						
EUP	611.3 (7; 38.36)	1077.2 (7; 76.46)	1.47 (7; 0.1)*	599.5 (3; 118.25)	916.9 (3; 215.23)	1.22 (3; 0.13)
LUP	656.7 (11; 15.07)	1072 (11; 25.09)	1.67 (11; 0.07)*	647.0 (5; 67.68)	1054.1 (5; 78.58)	1.45 (6; 0.07)*
Meso	651.5 (29; 14.16)*	1104.1 (28; 39.11)*	1.51 (42; 0.05)*	540.9 (10; 17.79)*	895.6 (8; 43.64)*	1.38 (15; 0.07)*
Neol	547.3 (14; 28.55)*	1013.5 (14; 67.78)*	1.38 (14; 0.06)	489.5 (13; 20.59)*	860.1 (13; 47.42)*	1.34 (13; 0.04)
CopA	540.7 (50; 10.37)*	942.5 (50; 19.39)*	1.47 (51; 0.03)	486.6 (31; 11.86)*	851.5 (31; 26.38)*	1.48 (32; 0.03)
BA	534.3 (64; 9.62)*	956.4 (64; 20.97)*	1.47 (71; 0.02)*	465.9 (47; 9.02)*	807.5 (47; 20.79)*	1.4 (49; 0.03)*
IA	536.6 (17; 17.58)*	919.7 (17; 37.49)*	1.39 (19; 0.04)	469.2 (8; 21.79)*	795.9 (8; 40.17)*	1.31 (9; 0.03)
Avar	480.3 (49; 8.57)*	818.2 (49; 17.75)*	1.35 (49; 0.03)	413.3 (42; 8.2)*	685.3 (42; 16.76)*	1.38 (42; 0.03)
EMed	498.4 (49; 10.52)*	864.4 (49; 22.02)*	1.27 (49; 0.02)	417.3 (34; 8.94)*	723.0 (34; 18.34)*	1.3 (34; 0.03)
LMed	536.9 (44; 10.02)*	957.8 (44; 18.76)*	1.34 (47; 0.03)	467.7 (30; 11.75)*	834.2 (30; 18.15)*	1.28 (34; 0.03)

[a] See details about abbreviations and parameters in Tables 11.1 and 11.2. The three earliest period data (EUP, LUP, Meso) are for the pan-European sample.
* ANOVA post-hoc comparison for sexual dimorphism significant at $p < 0.05$.

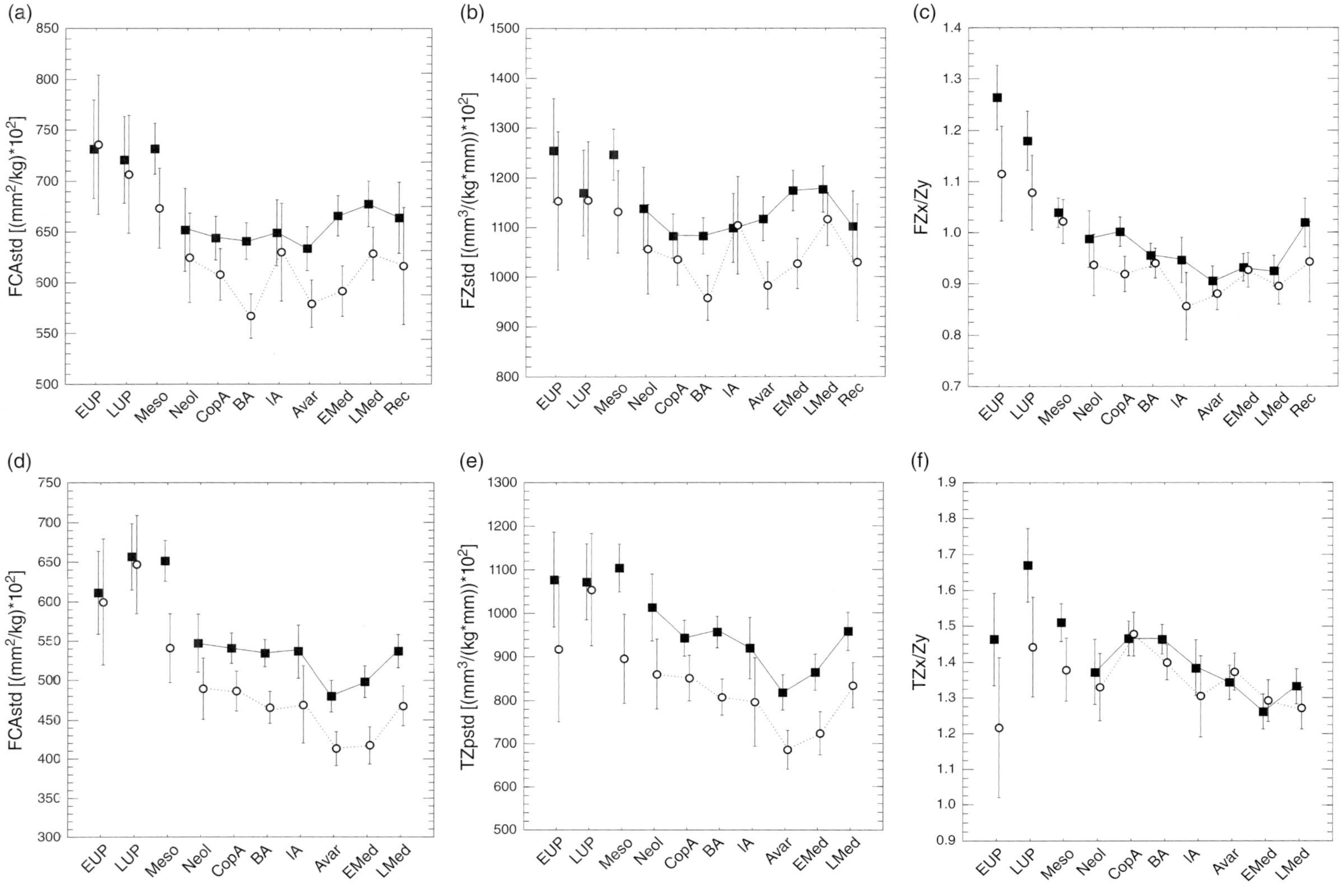

Figure 11.3 Temporal trends in Central European femoral and tibial CSG parameters relative to body size by sex and temporal period. Mean ±95% CI. Males: solid symbols; females: open and dashed symbols. (a,d) Femoral and tibial cortical area. (b,e) Femoral and tibial strength. (c,f) Femoral and tibial shape ratios.

22% in females), with a slight increase from Bronze Age toward Recent sample (about 3.5% in males and 8.5% in females). Similarly, average tibial cortical area is larger in the Upper Paleolithic than in Late Medieval period (total decrease between the LUP and Late Medieval is about 18% in males and 28% in females), but the major decrease of tibial cortical area is from the Late Upper Paleolithic to the Avar period (about 27% in males and 36% in females) followed by an increase to the Late Medieval (12% in males and 13% in females). The overall decrease of average tibial cortical area in the Holocene is also more pronounced than in femoral cortical area, especially in females.

Statistics for femoral and tibial average bending strengths are shown in Table 11.6, and temporal distribution by period and sex in Figure 11.3(b,e). Average femoral and tibial bending strengths are significantly different between periods ($p < 0.001$) and sex ($p < 0.001$), but non-significant in interaction between period and sex (FZ_p: $p < 0.23$; TZ_p: $p < 0.76$). In general, Central European variation in average bending strengths mirrors the trend observed for cortical areas. Average femoral bending strengths decline between Early Upper Paleolithic and Bronze Age by about 8% in males and 11.5% in females, and show no clear overall trend between Bronze Age and LMed/Rec except for a slight increase. In contrast, tibial average bending strength declines from the Terminal Pleistocene (LUP) to Avar by about 24% in males and 35% in females, and increases between the last three periods in the Late Holocene by about 17% in males and 22% in females.

11.3.2.2 Femoral and Tibial Shape Ratio (Z_x/Z_y)

Statistics for femoral and tibial diaphyseal shape ratios are shown in Table 11.6, and temporal distribution by sex and period in Figure 11.3(c,f). Femoral and tibial shape ratios are significantly different between periods ($p < 0.001$) and sex ($p < 0.001$). Interaction between period and sex is significant for tibial shape ratio ($p < 0.01$) but non-significant for femoral shape ratio ($p < 0.12$). Femoral shape ratio shows a decline from relatively higher femoral A-P strength towards a more circular diaphyseal shape between Early Upper Paleolithic and Avar period in males (about 28%) and IA in females (about 23%); however, the major decline to a more circular shape is observed between the Early Upper Paleolithic and Neolithic (22% in males and 16% in females), with less change during the rest of Holocene. Femoral A-P/M-L strength again relatively increases after the Avar period by about 12% in males, and after the Iron Age by about 10.5% in females. Temporal trends in the tibial shape ratio are less consistent, in part because of a relatively low value in the Neolithic; however, in general, the tibial shape ratio also decreases between the Terminal Pleistocene and the end of the Holocene (decrease between LUP and Recent is about 20% in males and 12% in females).

11.3.2.3 Sexual Dimorphism in Lower Limb CSG Properties

Sexual dimorphism (SD) in size-corrected lower limb CSG properties is similar in femora and tibiae (Table 11.6). Both CSG properties have more marked SD (males larger) in cortical areas and section moduli than in diaphyseal shape, although in the tibia the differences are significant in all Holocene periods, whereas in femora the number of significant differences is reduced. Femoral and tibial shapes show more significant SD in the Terminal Pleistocene and Early Holocene than in the Late Holocene, which probably indicates that the degree of mobility was more sexually dimorphic among the earlier Holocene groups than in the later ones.

11.3.3 Manipulative Behavior Changes

11.3.3.1 Humeral Cortical Area

Sample statistics for right and left humeral cortical areas are shown in Table 11.7, and temporal distribution by period and sex in Figure 11.4(a,c). Right and left humeral cortical areas are significantly different between both periods and sex (both $p < 0.001$), and also in the interaction between period and sex ($p < 0.02$). Cortical area on the right side only slightly decreases

Table 11.7 Right and left humeral CSG properties for Central European humans [Mean (N; SE)].[a]

	Males		Females	
	CAstd	$Z_p$std	CAstd	$Z_p$std
Right humeri				
EUP	399.3 (5; 34.29)	643.4 (4; 41.68)*	350.0 (3; 33.8)	494.2 (3; 57.72)
LUP	380.8 (10; 16.4)	621.5 (10; 40.11)	343.7 (5; 34.05)	583.7 (5; 49.56)
Meso	378.8 (18; 18.72)	655.3 (16; 41.63)	357.0 (6; 41.35)	619.3 (6; 87.28)
Neol	380.0 (10; 15.12)*	659.5 (10; 32.75)*	317.2 (10; 8.42)*	527.5 (10; 24.75)*
CopA	358.5 (37; 7.57)*	617.9 (35; 15.06)*	325.3 (26; 7.84)*	563.0 (26; 14.65)*
BA	360.9 (49; 6.13)*	650.8 (46; 12.65)*	301.3 (35; 6.2)*	523.6 (33; 13.05)*
IA	366.4 (8; 14.96)	642.5 (7; 20.02)	321.9 (3; 8.81)	539.5 (3; 13.94)
Avar	336.5 (41; 7.29)*	599.0 (41; 16.11)*	261.7 (24; 10)*	438.8 (24; 16.38)*
EMed	360.7 (48; 8.53)*	661.2 (48; 17.23)*	265.3 (31; 6.8)*	474.2 (31; 10.69)*
LMed	335.7 (28; 9.17)*	587 (27; 21.43)*	287.9 (24; 8.79)*	495.1 (24; 17.34)*
Left humeri				
EUP	282.0 (6; 14.89)	415.1 (6; 23.48)	294.9 (3; 38.27)	440.7 (2; 67.12)
LUP	333.7 (8; 15.85)	536.3 (8; 37.98)	299.2 (6; 18.99)	501.8 (6; 20.6)*
Meso	370.0 (14; 14.94)*	648.5 (13; 32.27)*	301.8 (10; 27.79)*	496.3 (10; 51.09)*
Neol	362.0 (10; 14.18)	619.6 (10; 24.08)	339.6 (11; 13.3)	587.4 (11; 31.26)
CopA	322.7 (36; 7.7)	550.3 (35; 15)	317.9 (25; 10.06)	567.5 (24; 18.76)
BA	329.7 (44; 5.54)*	576.1 (44; 11.84)*	305.4 (35; 6.22)*	526.2 (34; 14.66)*
IA	335.6 (8; 13.31)	546.1 (7; 27.9)	279.1 (4; 30.43)	468.7 (4; 35.81)
Avar	330.8 (38; 8.05)*	577.9 (38; 17.67)*	261.4 (18; 10.97)*	424.1 (18; 19.72)*
EMed	346.7 (41; 8.34)*	600.2 (41; 16.01)*	262.3 (32; 6.81)*	462.2 (32; 12.86)*
LMed	331.4 (29; 8.92)*	571.6 (29; 18.93)*	275.5 (24; 11.03)*	480.3 (24; 19.21)*

[a] See details about abbreviations and parameters in Tables 11.1 and 11.2. The three earliest period data (EUP, LUP, Meso) are for the pan-European sample.
* ANOVA post-hoc comparison for sexual dimorphism significant at $p < 0.05$.

about 5% in males and 9% in females between the Early Upper Paleolithic and Neolithic, whereas the left side shows a strong increase of about 28% in males and 16% in females between the European Early Upper Paleolithic and Central European Neolithic. This indicates that the decline in directional asymmetry in humeral cortical area observed in males and females between the Early Upper Paleolithic and Neolithic (see below) is shaped mainly by an increase in cortical area on the left side, together with some decrease on the right side. Humeral cortical area decreases during the Holocene between the Central European Neolithic and Late Medieval about 12% in males and 9% in females on the right side and about 8% in males and 19% in females on the left side.

11.3.3.2 Humeral Bending Strength

Statistics for right and left humeral strength are shown in Table 11.7, and temporal distribution by period and sex in Figure 11.4(b,d). Temporal and sex differences in right and left humeral strengths are significant, as is the interaction between period and sex ($p < 0.01$). In the right

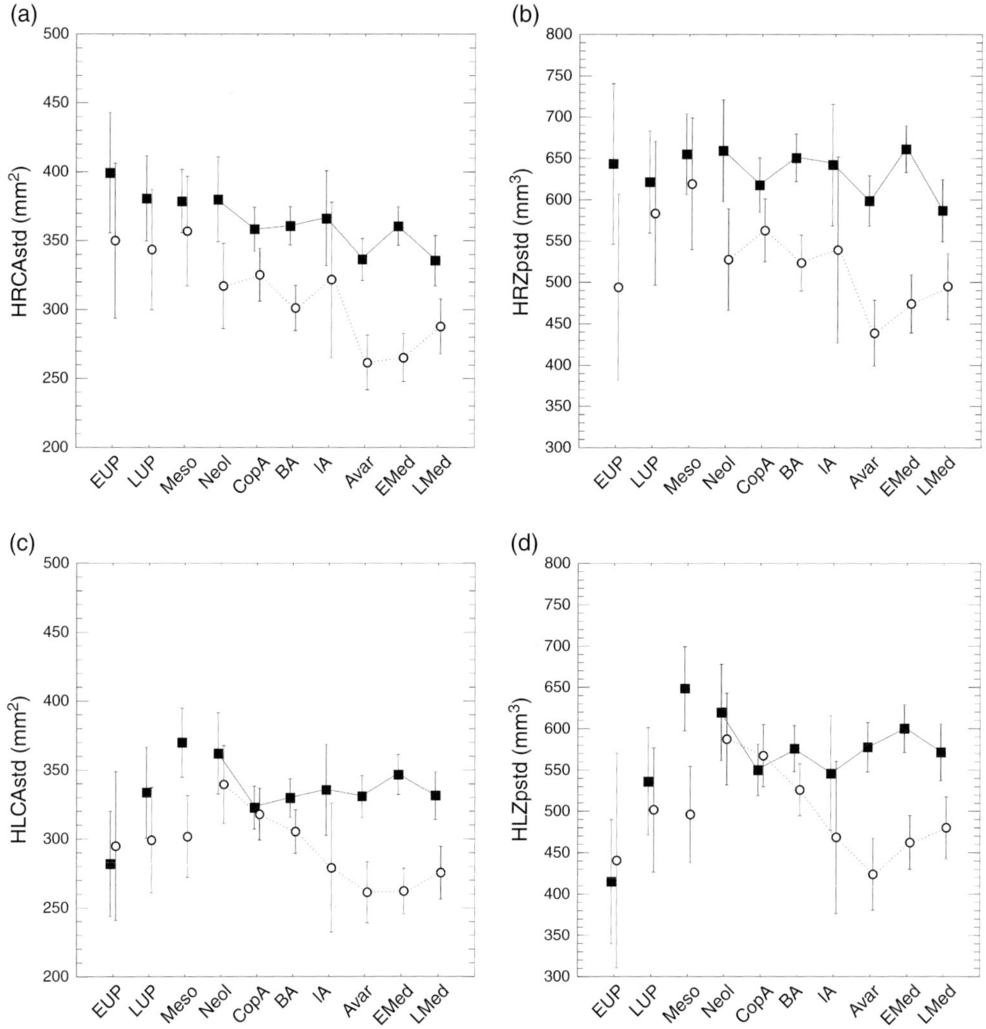

Figure 11.4 Temporal trends in Central European right and left humeral CSG parameters relative to body size by sex and temporal period. Mean ±95% CI. Males: solid symbols; females: open and dashed symbols. (a,c) Right and left humeral cortical area. (b,d) Right and left humeral strength.

humerus, strength shows an overall temporal trend similar to that observed for right humeral cortical area in both males and females, except that right humeral strength in males shows less of a temporal trend (9% decrease between Neol and LMed) and right humeral strength in females increases between Early Upper Paleolithic and Mesolithic (28%). The temporal trend in left humeral strength mirrors the trend observed in left humeral cortical area in both males and females – that is, an increase from EUP to Mesolithic in males (56%) and to Neolithic in females (33%), a minor decline of left humeral strength after the Neolithic in males (8%), and a decrease of left humeral strength after the Neolithic in females (18%).

11.3.3.3 Humeral Directional Asymmetry

Sample statistics for directional asymmetry in humeral cortical area and section modulus are presented in Table 11.8, and temporal distribution by period and sex shown in Figure 11.5(a,b). Since sample sizes for Late Upper Paleolithic and Iron Age periods in females are insufficient

Table 11.8 Directional asymmetry in humeral CSG properties for Central European humans [Median (N; –95% CI– +95%CI)].[a]

	Humeral DirCA		Humeral DirZ_p	
	Males	Females	Males	Females
EUP	**27.85 (6; 3.09–38.68)**	–	**37.57 (6; 13.2–57.26)**	–
LUP	21.01 (8; 0.99–46.35)	6.38 (5; –6.55–29.35)	23.96 (8; 10.91–54.29)	13.42 (5; –3.97–38.34)
Meso	**10.54 (16; 7.86–13.52)**	**9.56 (5; 2.1–21.42)**	**9.99 (16; 4.54–21.88)**	**14.14 (5; 6.12–29.2)**
Neol	**5.17 (7; –5.52–5.74)***	0.46 (9; –5.82–2.34)*	6.24 (7; –2.91–17.12)	–0.69 (9; –6.65–6.77)
CopA	**8.37 (25; 2.63–14.37)***	0.90 (19; –0.37–4.96)*	**12.97 (25; 9.58–23.57)***	–1.12 (19; –7.65–1.72)*
BA	**10.35 (39; 9.85–15.56)***	–1.81 (27; –5.28– –1.45)*	**12.75 (39; 8.2–17.54)***	–0.52 (27; –2.32–2.36)*
IA	9.52 (4; –5.05–23.96)	–	18.53 (4; 7.23–33.38)	–
Avar	3.00 (30; –0.16–6.37)	1.90 (16; –1.71–6.41)	**8.31 (30; 0.32–9.76)**	**8.17 (16; –0.7–14.88)**
EMed	**5.3 (36; 3.43–8.15)***	–0.64 (25; –3.95–1.1)*	**13.73 (36; 9.6–15.82)***	**3.21 (25; 1.46–7.07)***
LMed	3.22 (22; –2.13–7.16)	**3.64 (19; 0.26–6.19)**	**5.46 (22; –4.76–10.6)**	**4.96 (19; 2.15–7.39)**

[a] See details about abbreviations and parameters in Tables 11.1 and 11.2. The three earliest period data (EUP, LUP, Meso) are for the pan-European sample.
* Mann–Whitney *U*-test for sexual dimorphism significant at $p < 0.05$.
Bold text indicates Wilcoxon Matched Pairs Test significantly different from 0 (i.e., different from 0 asymmetry) at $p < 0.05$.

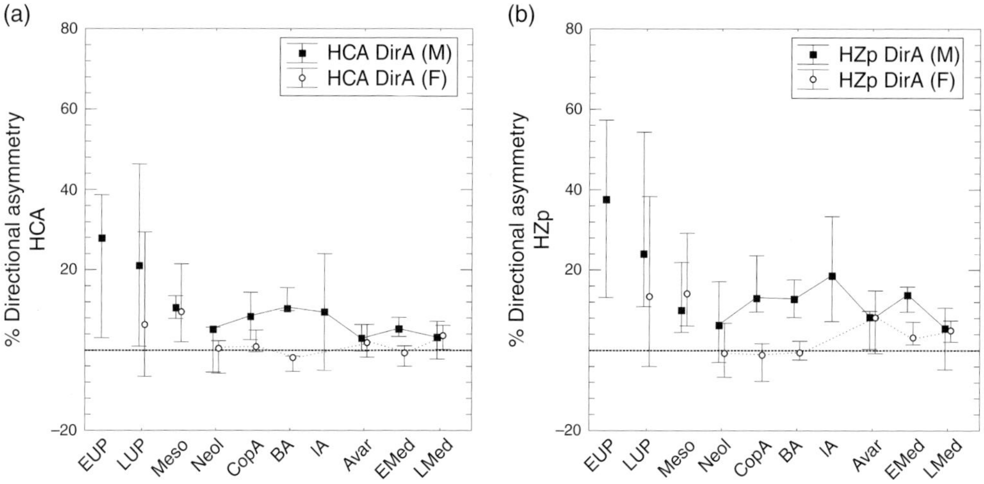

Figure 11.5 Temporal trends in Central European directional asymmetry in cortical area (a) and humeral strength (b). Median ±95% CI. Males: solid symbols; females: open and dashed symbols.

for statistical analysis, non-parametric Kruskal–Wallis multiple group tests were performed using a pooled Early-Late Upper Paleolithic sample and without Iron Age. Directional asymmetries in both cortical area and section modulus show similar patterns of variation, with differences only in the magnitude of the observed directional asymmetry, and this is also supported by similar results of Kruskal–Wallis multiple group tests, which are significant for both parameters between period in each sex at $p < 0.001$.

In males, directional asymmetry decreases greatly between the EUP and Neolithic, by about 23% in cortical area and by 31% in polar section modulus; however, the major decline takes place between the Early Upper Paleolithic and Mesolithic. The changes in directional asymmetry are less marked after the Neolithic in both CSG properties in males. Decrease of directional asymmetry between Terminal Pleistocene and Holocene in males is also supported by multiple significant sample comparisons. Directional asymmetry in humeral cortical area in males is significantly different between Upper Paleolithic and the three late Holocene samples (Avar, EMed, and LMed; $p < 0.001$) and in directional asymmetry in humeral section modulus between Upper Paleolithic and two early Holocene (Meso, Neol; $p < 0.038$–0.004) and late Holocene groups (Avar, LMed; $p < 0.001$).

In females, directional asymmetry declines by about 9% in humeral cortical area and by 15% in section modulus between the Mesolithic and Neolithic toward values that indicate an absence of directional asymmetry in the early agricultural periods (i.e., Neol, CopA, and BA). The absence of directional asymmetry in the early agricultural Holocene groups is also supported by a Wilcoxon matched pairs test, where Neolithic, Copper Age, and Bronze Age females show non-significant differences from 0, relatively narrow 95%CI, low upper value of 95%CI, and shift toward low 95% CI. In females, directional asymmetry again increases in the Late Holocene, but precise temporal change is not possible to indicate with certainty since the sample size for the Iron Age is too small (however, individual values indicate that the increase of asymmetry during the Iron Age is similar in magnitude to that observed for other European regions; see Chapter 7 for further details). Female samples from the Bronze Age and Copper Age are also significantly different from the Upper Paleolithic ($p < 0.02$) and

Mesolithic ($p < 0.05$) and near-significant between the Neolithic and UP ($p < 0.053$) and Mesolithic ($p < 0.064$) in the Kruskal–Wallis multiple group comparisons.

11.3.3.4 Sexual Dimorphism in Humeral CSG Properties

Sexual dimorphism (SD) in humeral CSG properties is frequently present among Central European samples in cortical area and section modulus (Table 11.7) on both right and left sides. Sexual dimorphism in right humeral cortical area and section modulus is significant in all the compared agricultural groups except the Iron Age, whereas left humeri are significant mostly in the late agricultural Holocene (Avar, EMed, and LMed) and less so in the early agricultural Holocene (only BA is significant).

Sexual dimorphism in humeral directional asymmetry is significant by Mann–Whitney *U*-tests with $p < 0.05$ in early agricultural groups (Neol, CopA, and BA in HCA and CopA and BA in HZp) and only in one late agricultural group (LMed) (Table 11.8). This pattern is also supported by bootstrap ±95%CI of medians (Fig. 11.5). Thus, the decrease in directional asymmetry in early agricultural Holocene female samples also increases SD and, by implication, differences in gender-specific manipulative activity.

11.3.4 Comparison of Central European Body Size, Shape, and CSG Properties to European Holocene

Because the European Holocene population as a whole shows a gradient at least in body size properties from southern to northern Europe (see Chapter 4), Central European postcranial variation was compared separately with the southern European sample (SE; sample includes Iberian Peninsula, Italy, and Balkans) and the northern European sample (NE; sample includes England and Scandinavia). Due to incomplete representation of subsamples in each region, comparisons were carried out for a limited number of temporal periods.

In general, average Central European stature, body mass, and BMI are intermediate between shorter and lighter southern Europeans and taller and heavier northern Europeans (Fig. 11.6). This trend is more consistent in BA and EMed than in LMed. The only deviation from this pattern is in the Neolithic for stature, where the Central European Neolithic sample shows minimum average stature compared to average stature observed in southern and north European regions. However, it is important to note that overall ANOVA analysis does not support a significant difference in any of the studied body size properties in interaction between regions, periods, and sex. A similar absence of significant differences in ANOVA analysis in interaction between regions, periods, and sex was obtained also for body shape properties (i.e., relative bi-iliac breadth, brachial index, and crural index; data not shown here). Moreover, the distribution of body shape properties also does not show a marked south–north gradient in a majority of comparisons, and the Central Europeans show similar average values as observed for northern and southern regions. The only exception is variation in brachial and crural indices in the Neolithic, where the Central European Neolithic shows lower brachial and crural indices than the Southern Europeans but similar to the Northern Europeans.

Figures 11.7 and 11.8 plot variation in cross-sectional properties between Central Europe and southern and northern European samples. Cross-sectional properties are non-significant in interaction between regions, periods, and sex. None of the studied Central European periods deviates from the range of variation observed for CSG properties in other regions of Europe. However, where there is apparent geographic variation, as Central Europeans fit well into the south–north gradient. The only exception is the relatively low index of tibial Z_x/Z_y (i.e., measures of mobility) in the Central European Neolithic compared to southern European Neolithic, while similar values were obtained for northern European Neolithic.

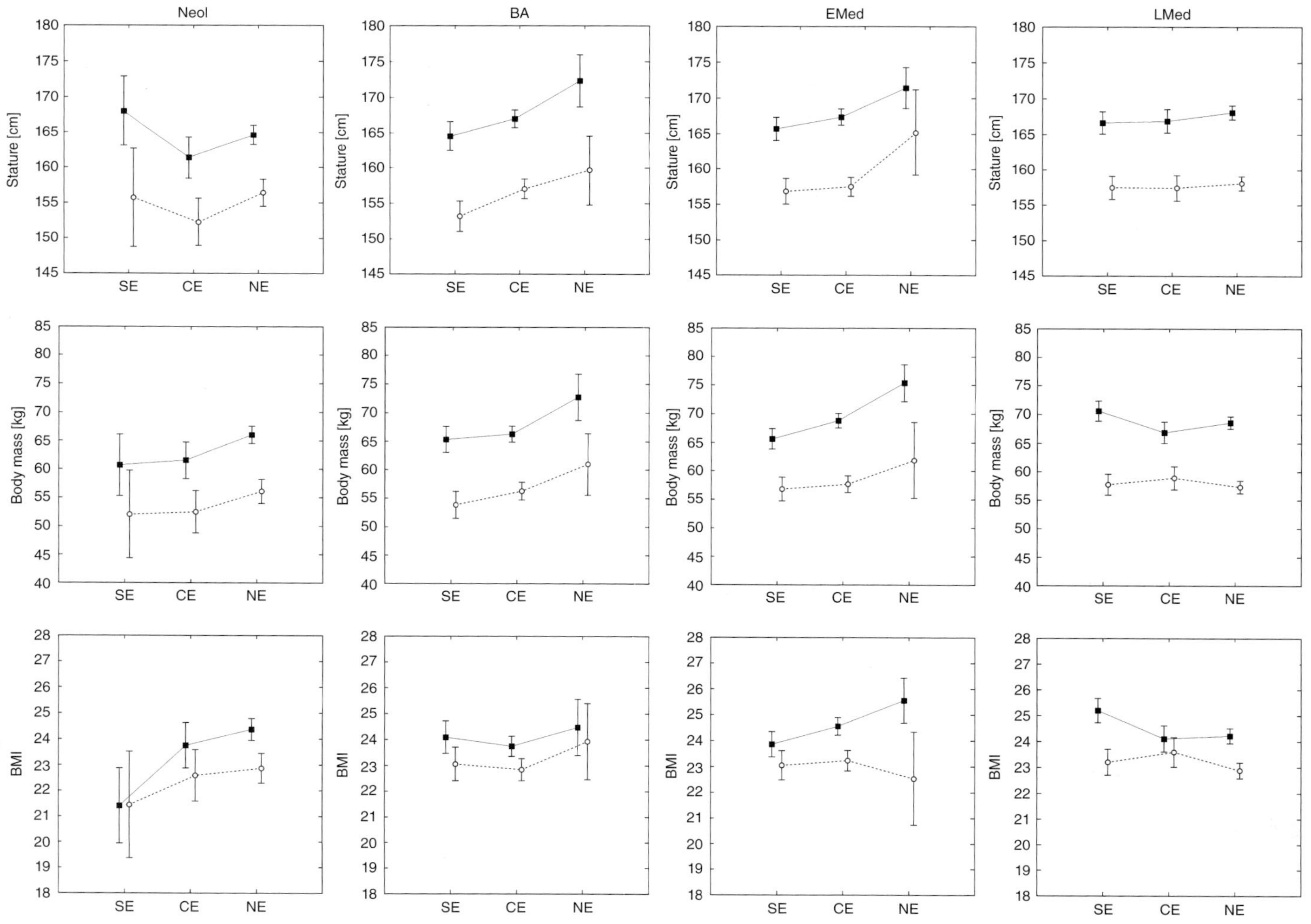

Figure 11.6 Regional comparison of body size and shape. Mean ±95% CI. CE: Central Europe; SE: Southern Europe; NE: Northern Europe. Males: solid symbols; females: open symbols.

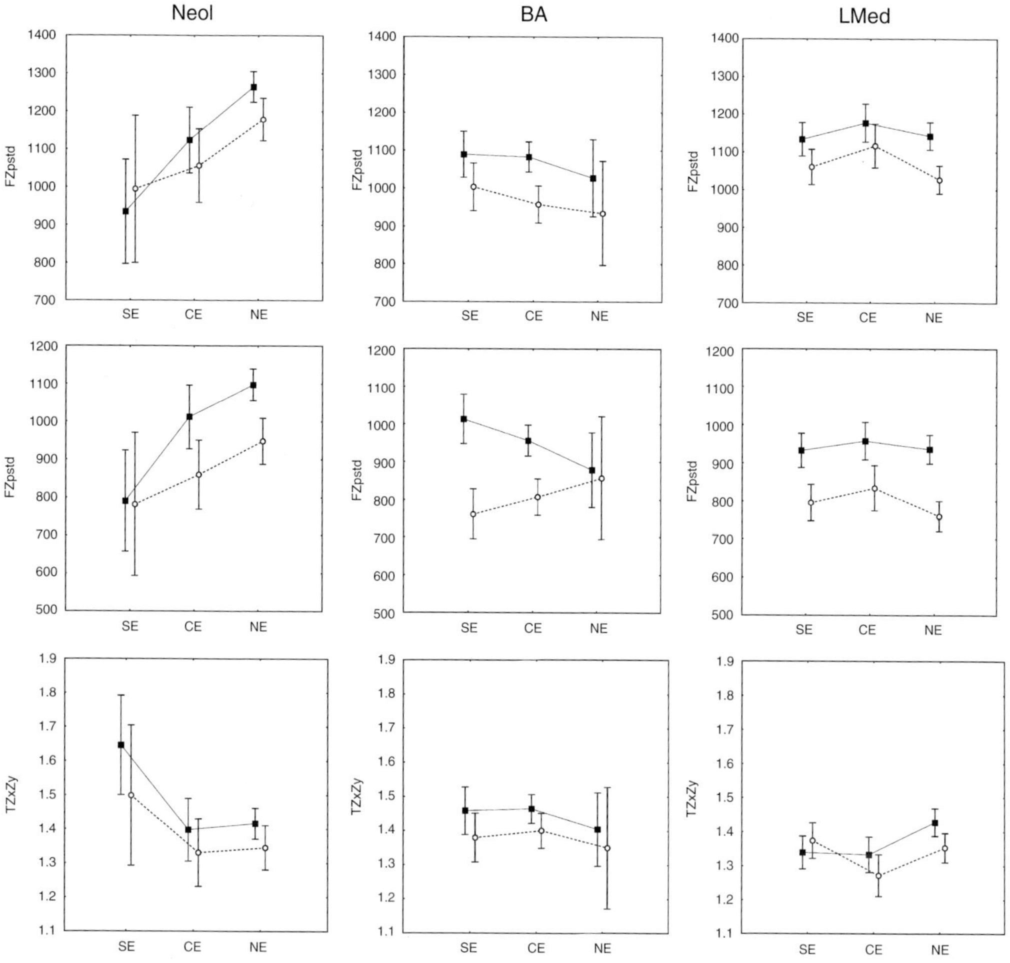

Figure 11.7 Regional comparison of femoral and tibial cross-sectional properties. Mean ±95% CI. CE: Central Europe; SE: Southern Europe; NE: Northern Europe. Males: solid symbols; females: open symbols.

11.4 Discussion

11.4.1 Central European UP and Mesolithic

As indicated in the Materials and Methods section, Central European Upper Paleolithic and Mesolithic temporal periods are represented by only a very limited number of skeletons. This is why the Central European Late Pleistocene sample was pooled together with other European regions to show a broader context for the impact of adoption of agriculture during the early Holocene in Central Europe. We are aware that this approach may be affected by ecogeographic (or other regional) variation within these periods. However, comparisons of some key body size and long bone cross-sectional properties between Central European and other European samples during the Mesolithic and Upper Paleolithic provide little evidence of any significant differences (Figs 11.9 and 11.10). Central European UP and Mesolithic individuals consistently fall within ±2SD of pan-European variation in stature and body mass (Fig. 11.9), although mean UP stature and body mass is lower compared to pan-European means. A general similarity in body

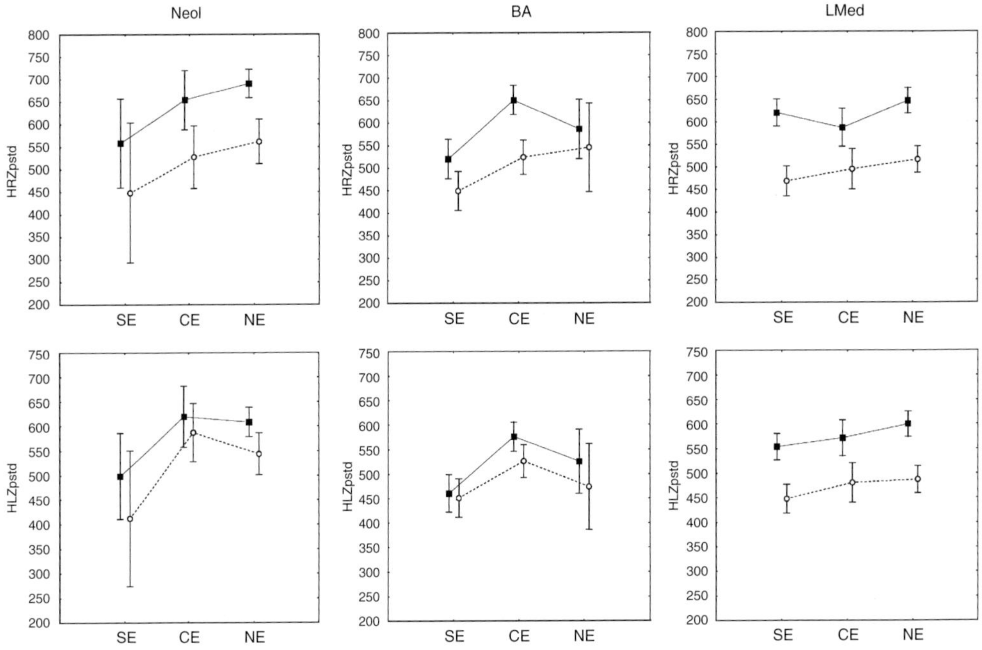

Figure 11.8 Regional comparison of humeral cross-sectional properties. Mean ±95% CI. CE: Central Europe; SE: Southern Europe; NE: Northern Europe. Males: solid symbols; females: open symbols.

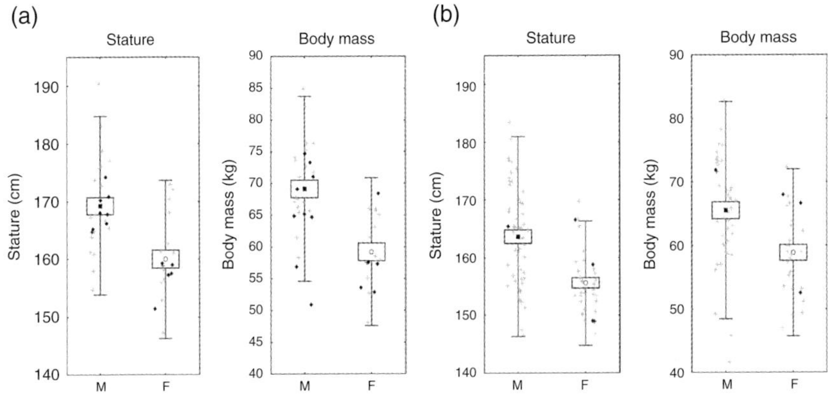

Figure 11.9 Variation in stature and body mass among Central European and other European (a) Upper Paleolithic and (b) Mesolithic. Box plots: Mean ± SE/±2SD; black diamonds: Central European individuals; gray cross: other European individuals.

size between Central European and other European UP samples is also supported by other studies (Trinkaus, 2006). We did find some evidence for somewhat smaller body size (but similar body shape) in Central European compared to other European samples when limited to the EUP (data not shown). Thus, the decline in body size between the EUP and LUP may have been smaller in Central Europe than depicted here; however, overall temporal trends would have been similar.

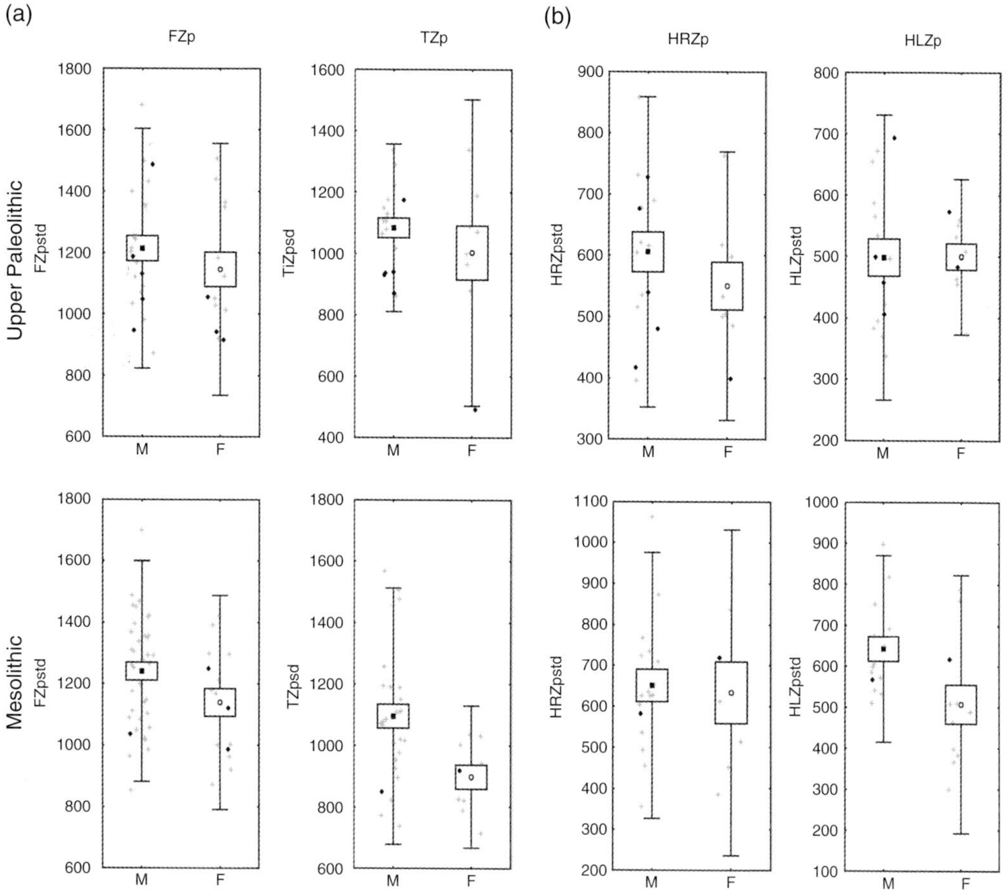

Figure 11.10 Variation in femoral, tibial, and humeral standardized section moduli among Central European and other European (a) Upper Paleolithic and (b) Mesolithic. Box plots: Mean ± SE/±2SD; black diamonds: Central European individuals; gray cross: other European individuals.

Similarly, the available Central European UP and Mesolithic individuals with cross-sectional data fall within pan-European distributions in section moduli, although Central European UP individuals show a slight shift toward lower values for FZ_p and TZ_p.

The only exception here is TZ_p in Upper Paleolithic females. The Central European value is represented by only one tibia from Dolní Věstonice 3 (DV 3), whose TZ_p value is more than 2SD below the pan-European sample mean. However, the size-standardized value of TZ_p for DV 3 may be affected by estimated tibial length because the proximal and distal epiphyses are damaged (Trinkaus and Jelinek, 1997). Moreover, DV 3 also shows a gracile pattern in other postcranial features (see for example its low value for FZp; Sládek *et al.*, 2000; Trinkaus *et al.*, 2000), and is at the lower limit of Gravettian variation in body size indicators (Trinkaus and Jelinek, 1997), so DV3 may be generally unusual even in the Central European UP sample.

In conclusion, we may expect that using pan-European UP and Mesolithic samples in our temporal assessments will not substantially shift our interpretation of the impact of the transition to agriculture in Central Europe, although the decrease between the EUP and Mesolithic as well as between Late Pleistocene and early Holocene may not be quite so profound as indicated when using the pooled European samples.

11.4.2 Adoption of Agriculture

Our data support a view that the adoption of agriculture was one of the major and global events in the human past (Larsen, 1995; Pinhasi and Stock, 2011; Larsen *et al.*, 2015; Ruff *et al.*, 2015) and that the new subsistence also influenced biological characteristics of the first farmers in Central Europe. We observed that Central European Neolithic males and females show about 7–8% lower stature and about 13–15% lower body mass than the European EUP males and females. This is one of the major decreases in body size properties observed in our Terminal Pleistocene and Central European Holocene sample. Since body size within the same physical environment has been found to be sensitive to general health and nutritional status (see Chapter 4 for biological background), decrease of stature in the Central European Neolithic may be another argument to see the Neolithic as a period with substantial deterioration of living conditions due to a new subsistence strategy and sedentary lifestyle (Cohen, 1977; Cohen *et al.*, 1984; Larsen, 1995).

The changes in body shape proportions observed in the Central European Neolithic sample may also be partly related to the impact of migration of the first farmers from the Near East. This may be supported by the observation that the Central European Neolithic also shows a decrease in body breadth, which is less sensitive to short-term (e.g., nutritional) factors (Ruff, 1994 also see Chapter 4). We observed that Neolithic males and females have absolutely and relatively narrow bi-iliac breadths compared to other Central European Holocene groups, but are similar in relative body breadth to preceding pan-European EUP groups. This observation is confirmed also when the Central European Neolithic is compared with the Central European EUP male subsample (Fig. 11.11; only one female bi-iliac breadth is preserved in the Central European EUP subsample and therefore females are not included in this particular comparison). The Neolithic males show significantly lower relative bi-iliac breadth compared to CopA/BA, but non-significantly different relative bi-iliac breadth compared to the Central European EUP subsample by ANOVA post-hoc Fisher LSD test. The relatively narrow bi-iliac breadth observed in the Central European EUP subsample is mainly due to their larger stature, whereas in the CopA and BA the higher ratio is due to a wider bi-iliac breadth. In contrast, the Neolithic shows a proportional decrease in both stature and bi-iliac breadth from the EUP. It has been argued that the pan-European EUP possess 'tropical' (i.e., tall and relatively narrow-bodied) proportions as a result of an African origin and later migration to Europe (Holliday, 1997). Similar migrations from more southerly regions – southeastern Europe and Anatolia – may also have affected the Neolithic population,

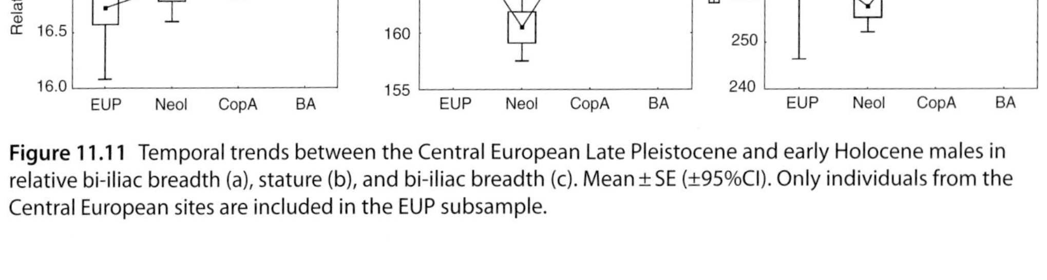

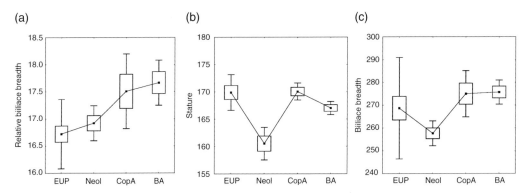

Figure 11.11 Temporal trends between the Central European Late Pleistocene and early Holocene males in relative bi-iliac breadth (a), stature (b), and bi-iliac breadth (c). Mean ± SE (±95%CI). Only individuals from the Central European sites are included in the EUP subsample.

as supported by ancient DNA studies. For example, Mathieson *et al.* (2015) showed in their genome-wide scan analysis that the source of Europe's first farmers is in the Anatolian Neolithic culture. A contribution of Aegean first farmers to the early European Neolithic gene pool was found in other aDNA studies (Bramanti *et al.*, 2009; Haak *et al.*, 2010; Hofmanová *et al.*, 2016). The partial impact of migration on spread of agriculture in Central Europe may be also supported from demographic models. It has been estimated that the admixture contribution to the first farmers from migration was between 28% and 45% (Galeta *et al.*, 2011). Thus, these migrations may have contributed to the decline in both overall body size and especially body breadth in Central European Neolithics.

The Neolithic sample shows about 2% lower brachial and about 4% lower crural indices compared to the pan-European EUP group. The same decrease is observed for both indices between the Neolithic sample and the Central European EUP male subsample. This may indicate adaptation to a colder climate (Trinkaus, 1981; Holliday, 1999); however, this factor seems unlikely as a complete explanation, for two reasons. First, the Neolithic crural index is also about 3–4% lower than in the subsequent Copper Age samples, who lived under similar climatic conditions and had longer to adjust body proportions to those conditions. Second, contributions to the gene pool from populations migrating from more southerly regions would be expected to increase intra-limb indices (Trinkaus, 1981). The relatively low indices (for Late Pleistocene/Early Holocene populations) are thus not consistent with climatic expectations. Rather, they may be related to deprivation in living conditions, which has negative effects on not only body size but also on relative limb length (Boas, 1930; Tanner, 1990; Bogin and Keep, 1999; Bogin and Rios, 2003). A decline in intra-limb indices under such conditions could be related to the fact that distal limb segments grow at a higher rate than proximal segments due to a cephalo-caudal gradient in growth (Schultz, 1926), and therefore distal segments may be more sensitive to environmental condition during postnatal development (Jantz and Jantz, 1999; Wadsworth *et al.*, 2002; Li *et al.*, 2007). Thus, the low brachial and crural indices during the Neolithic might well be in agreement with a view that the adoption of agriculture brought deprivation of living conditions in the Central European population.

11.4.3 Secondary Products Revolution

The Copper Age males and females are about 4–6% taller and 8–9% heavier than the previous Central European Neolithic sample. This increase is one of the largest among studied Holocene groups (see comparison with recent Europeans below). In fact, the Copper Age is the first period with an increase of stature and body mass after the decrease of body size properties observed during the Terminal Pleistocene and during the adoption of agriculture in Central Europe. It may indicate that the Copper Age sample is the first agricultural period that improves living conditions (mainly nutritional intake) due to the Secondary Products Revolution and intensification of agriculture (Greenfield, 2010). This conclusion is also supported by isotope analysis of Corded Ware Culture (CWC) diet and observations that the CWC shifts compared to earlier Neolithic periods toward higher δ15N values, that is, a higher consumption of animal proteins and a greater reliance on milk and milk products (Sjögren *et al.*, 2016).

The Copper Age also shows a slightly higher index of mobility compared to the Neolithic. This may indicate that the Copper Age groups approach pastoral subsistence as expected from the absence of sedentary archaeological features in the Copper Age (Vencl, 1994), but also from aDNA studies. As shown by Haak *et al.* (2015), the Central European Corded Ware groups possess about 75% of the ancestry of steppe migrants of the Yamnaya with a mobile subsistence and probably in part replaced the sedentary Neolithic inhabitants in Europe. This may argue that not only did the Secondary Product Revolution play a role in changes of body size and shape but also that these changes are due in part to the impact of massive migrations approximately 4500 years ago from more eastern regions.

However, archaeologists have pointed out that a nomadic subsistence is not the most plausible explanation for the absence of settlement features in the Late Eneolithic (see review in Sládek *et al.*, 2006a). The Copper Age people may have built their dwellings without the need of elements sunk under the surface (Shennan, 1993; Neustupný, 1997). The similarity between Central European Copper Age and Bronze Age in the index of mobility also advocates for similar mobility patterns, although Central European Bronze Age possess clear archaeological evidence of sedentary subsistence. Moreover, the higher impact of mobility in the Central European Copper Age group was rejected in previous comparison of the Central European Late Copper Age and Early Bronze Age CSG properties (Sládek *et al.*, 2006a,b). Broader comparison in the present study through the Central Holocene shows that TZ_x/Z_y is the only feature that supports higher mobility of the Copper Age sample compared to other Central European groups. Indeed, the index of mobility increased in the Copper Age and subsequent Bronze Age; however, it is accompanied by the absence of differences in femoral Z_x/Z_y index between the Central European Neolithic and Bronze Age and decrease of overall femoral and tibial strength in the Central European Copper Age.

The impact of an eastern steppe migration of the Yamnaya (Haak *et al.* 2015) is also equivocal based on our observations on body shape variation. All compared body shape indices show features for cold adaptation in the Copper Age, as expected from geographic distributions. The Copper Age sample shows a relatively wide bi-iliac breadth accompanied by a large stature, which is in contrast to the relatively narrow body and short stature observed for the Central European Neolithic group (see also Fig. 11.9 and comments above). The Copper Age sample also shows a relatively higher crural and brachial index than the previous Neolithic sample. In fact, apart from intra-limb indices, the observed body shape in the Central European Neolithic and Copper Age is in good agreement with a possible impact of more southern-derived populations during the spread of agriculture into Central Europe in the early Neolithic (Bramanti *et al.*, 2009; Haak *et al.*, 2010; Hofmanová *et al.*, 2016) and later adaptation of body shape to the colder Central European environment in the Copper Age (see Chapter 4 for further biological background about body shape and climate). Moreover, the colder-adapted body shape is also well in agreement with the distribution of the Corded Ware culture, which includes also north Europe and Scandinavia. In contrast, the Yamnaya culture occupied territories from eastern and southern latitudes belonging to Pontic Steppes (i.e., southern Bug, Dniester, and Ural). On the other hand, larger brachial and crural indices in the Copper Age compared to Neolithic may also indicate improvements of living conditions due to improvements of factors influencing postnatal growth (see discussion above). All of this indicates that the Copper Age population is adapted as expected to a colder climate, as well as exhibiting postcranial features for improvement of living conditions as expected from the Secondary Products Revolution.

Haak *et al.* (2015) also argued from their aDNA study that the Late Eneolithic Corded Ware culture shows a contrast in aDNA structure compared to the Late Eneolithic Bell Beakers (BBC). He and his coworkers expect that the Bell Beakers represent a resurgence of an indigenous Neolithic population in Europe whereas the Corded Ware is the result of eastern Yamnaya migration. However, our data do not support this view. The Bell Beakers are non-significantly different from the Corded Ware culture in all of the compared body size and shape features, and do not show similarities in studied postcranial properties with the Neolithic occupation (data not shown here). In fact, values observed for CWC and BBC are mostly identical in the majority of studied postcranial properties. This argues for biological similarity between CWC and BBC and argues against a direct impact of migration into the Central European regions during the Late Eneolithic. Such similarity between CWC and BBC was supported also by other more detailed studies of long bone cross-sectional geometry (Sládek *et al.*, 2006a,b, 2007).

11.4.4 Origin of Metallurgy

Adoption of metallurgy during the Copper Age and intensification of metallurgy in the Iron Age appear to have had little effect on postcranial variation in Central Europe. We observed an increase of directional asymmetry in male humeral CSG properties between the Central European Neolithic and Copper Age, as well as between the Bronze Age and Iron Age (Fig. 11.5). However, this increase is non-significant and is based in the Iron Age on a small sample of only four individuals. Comparing mean Central European Iron Age directional asymmetry in humeral rigidity in a different sample (n = 10), Macintosh *et al.* (2014) also did not find a significant increase of humeral asymmetry in males, which they had predicted from the assumption that metal production is associated with strenuous unilateral tasks with direct impact on the upper limb (see Macintosh *et al.*, 2014 for further details). In fact, their estimation of directional asymmetry in humeral rigidity in males is non-significantly different between the Bronze Age and Iron Age, ranging from 12.4% in the Bronze Age to 10.8% in the Iron Age. Moreover, our assessment of right and left standardized humeral Z_p indicates that the increase of directional asymmetry in Copper Age and Iron Age males is due to a decrease of loading on the left side rather than by increases of loading on the dominant right side (Fig. 11.4). The left humeral strength in males also shows the minimum value for the assessed early metal groups in respect to all Holocene variation, and right humeral strength does not show a significant increase with respect to other Holocene groups (Fig. 11.4). Thus, all of these results do not support a global impact of mining and metallurgy on postcranial properties in Central Europe males, as suggested by Macintosh *et al.* (2014). However, there is still the possibility that metallurgy was a specialized activity among early metallurgy groups and that this specialized group is not included in our samples.

On the other hand, we observed an increase of directional asymmetry in humeral strength among Iron Age females (Fig. 11.4). It is difficult to specify statistical significance because the Iron Age sample is very small (n = 2); however, the increase in directional asymmetry in female humeral Z_p was observed also in other European regions (see Sládek *et al.*, 2016 and Chapter 7). A significant increase in directional asymmetry in humeral rigidity was also observed between the Bronze Age (DirA J = 4.2%) and Iron Age (DirA J = 9%) in Macintosh *et al.* (2014). They explain the increase in asymmetry in the female Iron Age sample as the effect of metallurgy; however, this does not fit our Central European data and also it is unlikely to affect the females sample to any degree. As shown in Figures 11.4 and 11.5, the increase of humeral directional asymmetry in the Iron Age females clearly separates the early agricultural groups with saddle quern grinding, and the late agricultural groups with rotary quern grinding (see also Chapter 7 for further details). It has been shown that rotary quern grinding produces a higher unilateral loading on the upper limb even in the case where bimanual rotation is used, compared to bilateral loading on the upper limb resulting from saddle quern reciprocal grinding (Sládek *et al.*, 2016). In fact, the pattern of asymmetry in muscle activity during grinding well explains the absence of asymmetry in humeral Z_p in the Central European early agricultural females

11.4.5 The Avars and Nomadic Subsistence in Central Europe

Our data do not support an increase in mobility during the Avar period as expected from the high mobility and nomadic subsistence of Avar Khaganate which partly affects Central Europe after the invasion of the Eurasian nomad warriors between the 6th and 9th centuries (Bogucki and Crabtree, 2004b). In fact, femoral and tibial strengths are the lowest in the Avar sample among all Holocene groups, both in males and females. This is partly also true for low femoral Z_x/Z_y. The only exception is higher tibial Z_x/Z_y in the Avar sample compared to EMed; however, the index is similar to the Neolithic and Bronze Age. This supports a view that our Avar sample consists of individuals (probably Slavs?) who approached a sedentary subsistence with mixed agriculture, and is also in agreement with the observation that our Avar sample consists of groups living on the periphery of the Khaganate (Bogucki and Crabtree, 2004b). Thus, we might expect that Avars influenced the Central European Early Medieval indigenous group mainly as political administrators but did not directly transform their sedentary subsistence.

Archaeologists also showed that the Slavic indigenous population may serve in Avar warfare either directly as a part of the Avar's army, or as a more independent unit controlling the northern and western boundaries of Avar's Khaganate (Bogucki and Crabtree, 2004b). Our data do not support the direct impact of warfare manipulation on skeletal properties. As indicated from the humeral directional asymmetry in HCA and HZ_p, the CSG properties show low sexual dimorphism in bilateral asymmetry. The bilateral asymmetry for males from the Avar sample is relatively low compared to, for example, the Early Medieval sample. Moreover, the Avar sample shows relatively low humeral strength both for males and females and low sexual dimorphism. This indicates that males and females approach a similar pattern of bilateral loading, and that males do not exhibit larger bilateral asymmetry compared to other periods, as expected from a high unilateral loading with the use of weapons (Sparacello *et al.*, 2015). On the other hand, it has been shown that individuals serving as temporary warriors had lower humeral rigidity, which reflects an absence of pre-adolescent weapon training compared to elite militia (Sparacello *et al.*, 2015). Thus, we cannot exclude that our Avar males served in a temporary role in the Avar army even though they exhibit low humeral strength and low directional asymmetry in humeral strength.

11.4.6 The Transition to the Archaic State in Central Europe

Our data support a view that the transition between complex chiefdom to state during the Early Medieval and Late Medieval periods had some effect on postcranial variation in Central Europeans. As indicated from body size comparisons, stature and body mass significantly decrease between EMed and LMed periods ($p < 0.021$) in males. This may indicate that living conditions deteriorated during the LMed and after the new state organization, with a possibly greater effect on males. Females may be less responsive to this decline possibly due to female buffering (Greulich, 1951; Greulich *et al.*, 1953; Stini, 1975; Stinson, 1985). However, several other factors besides the change in the organization of society may be influential, such as the effect of the environmental deterioration due to the Little Ice Age cold oscillation (Grove, 1988), demographic pressure due to increase of inhabitants, and deterioration of living conditions due to concentration of inhabitants in the Late Medieval urban centers.

The decrease of body size properties may be also affected by the fact that our Early Medieval cemeteries originate partly from the main center of the Great Moravian Empire with a higher concentration of elites and warriors (Macháček, 2010). This may be related to the observation that the EMed sample exhibits significantly higher sexual dimorphism in humeral directional asymmetry in HZ_p ($p < 0.001$), whereas sexual dimorphism is not significant in subsequent LMed ($p < 0.79$). Moreover, the EMed males also show significantly higher right humeral strength compared to LMed ($p < 0.01$), but this is not observed for females ($p < 0.43$). At the same time, there is significantly lower tibial strength in the EMed male sample compared to LMed males ($p < 0.02$). All this shows that males in the EMed sample exhibit a significant increase of manipulative loading on the right humerus, but have relatively reduced impact of mobility loading on the tibia compared to the LMed males.

11.4.7 Comparison to Living Central Europeans and Impact of Secular Trend

Our data again highlight the effect of the secular trend in stature observed from the late 19th century up to the beginning of the 21st century among children from the Czech Republic (Vignerová *et al.*, 2006). We have shown that the rate of increase in average stature observed during last 50 years in the Czech Republic is not observed during any of the previous time periods of the Holocene. The most rapid previous increase of average stature observed among Central European Holocene populations occurred between the Neolithic and Copper Age over about 1000 years (males increase in mean stature about 9.5 cm and females about 5.8 cm) is in fact only slightly greater than what occurred over 50 years of secular trend in the Czech Republic. Extending the modern data further through use of other available growth data (Matiegka, 1927; Vignerová *et al.*, 2006) indicates that the secular trend in stature between 1895 and 2001 for Czech late adolescents was about 12 cm in boys and 10 cm in girls, even greater than the increase in average stature between the Neolithic and Copper Age, and similar in magnitude to the total decrease in average stature over more than 10,000 years between the EUP and Neolithic (males decreased in mean stature about 13.7 cm and females about 11.6 cm). This indicates that the changes among modern Central Europeans in living conditions and other factors affecting mean stature are exceptional compared to all of Holocene human evolution. It is also apparent that the increase in mean stature observed for the Copper Age groups is not outside the possible range of secular trends related to changes in living conditions, and that the increase during the Copper Age could very well be related to changes in subsistence due to the Secondary Products Revolution. On the other hand, the decrease of stature between the Upper Paleolithic and Neolithic again highlights the potential negative impact related to changes in living conditions and in subsistence changes, accompanied by partial population replacement (Bramanti *et al.*, 2009; Haak *et al.*, 2010; Hofmanová *et al.*, 2016).

Our data also support a view that secular trends in average stature may be dependent on socioeconomic factors (e.g., Boas, 1930; Cole, 2000; Bogin and Rios, 2003). The average living statures taken from the 2001 survey are significantly higher ($p < 0.001$) than our recent sample used in the Terminal Pleistocene and Holocene analysis – about 10 cm for males and 7.5 cm for females. In fact, the Recent sample shows more similarities in body size properties with other Holocene groups than with current populations living in this region. The underestimation of living stature in our Recent sample may be consistent with the origin of the sample. The Recent sample used in our Holocene comparison originates from autopsy bodies obtained in the 1990s in the south of East Germany. Unfortunately, no further information about socioeconomic status is available for this sample; however, it seems probable that the sample consists of individuals from lower social status groups. On the other hand, our observation also indicates that comparison of recent samples may be biased depending on their temporal position relative to the secular trend and the age composition of the selected sample. Our Recent sample consists of individuals with ages between 22 and 59 years, which resulted in individuals born between 1930 and 1970 and subject to different impacts of the recent secular trend. We indeed find a negative correlation of stature with age in the Recent sample where older individuals show shorter statures (male $r = -0.59$, female $r = -0.48$) (note that this is not caused by longitudinal age-related declines in stature, since statures in this sample are based on femoral length). The difference between younger and older individuals may be even more than 10 cm. Thus, the lower stature in the Recent sample argues for sensitivity of stature to variation in living conditions (see Chapter 4 for further details), both within and between temporal periods.

11.5 Conclusions

Central Europeans show a decrease in stature and body mass from the Upper Paleolithic to the Neolithic and an overall increase afterward, but the trend is not continuous and is more variable in males. Central Europeans also show an overall increase in relative bi-iliac breadth from the Neolithic to the Recent. Relative body breadth is more variable among Central European females than males. Brachial and crural index show an overall decrease from warm-adapted proportions observed in the Terminal Pleistocene to cold-adapted proportions in the late Holocene. The only exception is low crural and brachial indices observed in the Neolithic, but we hypothesize that this may be evidence of a biological response to the deprivation of living conditions due to the new subsistence and sedentary lifestyle. The Central Europeans show an overall decrease of femoral and tibial cortical area and average bending strength between the Terminal Pleistocene and early Holocene up to the Bronze Age (femoral CSG) and the Avar (tibial CSG), and a slight increase afterwards. The overall trend in CA and Z_p is more pronounced in tibial than in femoral CSG properties, and is more apparent in females. The femoral and tibial diaphyseal shape ratio (Z_x/Z_y) shows a decline from more A-P elongated shape observed in the Terminal Pleistocene and early Holocene to a more rounded diaphyseal shape found in the late Holocene. The femoral diaphyseal shape ratio shows a larger decline and less variation than what is observed for the tibia. The Central Europeans show a decrease of humeral cortical area and average bending strength on both sides through the Holocene in female samples, whereas only on the right side in male samples. Both humeral CSG properties have no trend between Terminal Pleistocene and Neolithic on the right side but strongly increase on the left side toward the Neolithic. Central Europeans show positive directional asymmetry in humeral CSG properties different from zero through all of the Holocene in males, whereas there is an absence of directional asymmetry in early agricultural females (Neol-BA), followed by an increase of directional asymmetry from the Iron Age. We also showed that Central European variations in postcranial properties fit well either within the south–north gradient or remain within the range of variation observed for the South and North Europe.

Our data support a view that the adoption of agriculture and a sedentary lifestyle has profound and in many ways negative effects on postcranial morphology in the Central European Neolithic, with large declines in body size. Observed patterns of postcranial variation also indicate substantial improvement of living conditions in the subsequent Copper Age, and the ability of humans to adapt to a new sedentary subsistence. We cannot support an impact of massive migration during the Copper Age from East Eurasia into Central Europe since we found similarities in body size and shape between Corded Ware (expected migrant group) and Bell Beakers (expected indigenous group) samples, and the Copper Age shows also cold-adapted body proportions as expected from Central European Holocene variation.

Mobility in Central Europe declined from the Upper Paleolithic through the Iron Age, as reflected in reductions in relative femoral and tibial strength, thus lasting well beyond the initial adoption of agriculture. Changes in humeral strength asymmetry parallel changes in food procurement and processing, and are sex-specific, with males responding primarily to the end of big-game hunting between the Terminal Pleistocene and early Holocene, and females to the introduction of two-handed and then one-handed grinding tools in the early-mid-Holocene. We showed that the adoption of metallurgy had no global impact on postcranial variation in the Central European samples, mainly because no differences have been found between the early agricultural group (Neolithic) and the first metallurgy group (Copper Age) in directional asymmetry in humeral average bending strength. We support a view that individuals from our Avar sample from the Central European periphery practiced a sedentary agricultural subsistence. We also support a view that the transition from complex chiefdom society in the Central

European Early Medieval to a state organization during the Late Medieval has some effect on postcranial variation; however, it can be partly due to the fact that our Early Medieval sample may represent a group belonging to the elite and warriors or include some individuals of this status. We also demonstrate that changes in stature and body mass observed during Holocene variations are well within the range of secular changes observed during last 100 years among Central European living populations. However, we also showed that the Central European secular trend of the last 100 years is unique among all Terminal Pleistocene and Holocene populations because of the exceptional speed of change.

Acknowledgments

The Central European dataset was taken with help of Národní museum in Prague and Naturhistorisches Museum in Vienna. We would like to thank Miluše Dobisíková, Petra Havelková, Anna Pankowská, Maria Teschler-Nicola, Karin Wiltschke-Schrotta, Horst Bruchhaus, and Ronny Bindl for providing access to Central European museum collections. We would also like to thank the students who participated in some data collection. For comments about statistics and archaeological background we would like to thank Patrik Galeta and Daniel Sosna, and for comments about Czech secular trend we would like to thank Jana Vígnerová and Petr Sedlak. The computed tomography (CT) scanning of the majority of the Central European dataset has been done with the support of Sibylle Kneissl, Wolfgang Henninger, and Martin Kolnar (Institute of Radiology, Veterinary University of Vienna). We would also like to thank Horst Bruchhaus for arranging local CT scanning of archaeological material from Jena, and for the support of Jena University Hospital.

References

Amzallag, N. (2009) From metallurgy to bronze age civilizations: The synthetic theory. *Am. J. Archaeol.*, **113**, 497–519.

Arbogast, R.-M., Jacomet, S., Magny, M., and Schibler, J. (2006) The significance of climate fluctuations for lake level changes and shifts in subsistence economy during the late Neolithic (4300–2400 BC) in central Europe. *Veget. Hist. Archaeobot.*, **15** (4), 403–418.

Archer, E., Shook, R.P., Thomas, D.M., Church, T.S., Katzmarzyk, P.T., Hébert, J.R., McIver, K.L., Hand, G.A., Lavie, C.J., and Blair, S.N. (2013) 45-Year trends in women's use of time and household management energy expenditure. *PLoS One*, **8** (2), e56620.

Ashton, T.S. (1950) *The industrial revolution 1760–1830.* CUP Archive.

Barford, P.M. (2001) *The early Slavs: culture and society in early medieval Eastern Europe.* Cornell University Press, New York.

Bassett, D.R., Jr, Pucher, J., Buehler, R., Thompson, D.L., and Crouter, S.E. (2008) Walking, cycling, and obesity rates in Europe, North America, and Australia. *J. Phys. Act. Health*, **5** (6), 795–814.

Beliaev, D.D., Bondarenko, D.M., and Korotayev, A.V. (2001) Origins and evolution of chiefdoms. *Rev. Anthropol.*, **30** (4), 373–395.

Beranová, M. and Lutovský, M. (2009) Slované v Čechách: archeologie 6.–12. století. Libri, Praha.

Bickle, P., Bentley, R.A., Blesl, C., Fibiger, L., Hamilton, J., Lenneis, E., Neugebauer-Maresch, C., Stadler, P., Teschler-Nicola, M., Tiefenböck, B., *et al.* (2013) Austria. In: *The First Farmers of Central Europe: diversity in LBK lifeways* (eds P. Bickle and A. Whittle), Oxbow Books, Oxford.

Bickle, P. and Whittle, A. (eds) (2013a) *The First Farmers of Central Europe: diversity in LBK lifeways*, Oxbow Books, Oxford.
Bickle, P. and Whittle, A. (2013b) LBK lifeways: a search for difference. In: *The First Farmers of Central Europe: diversity in LBK lifeways* (eds P. Bickle and A. Whittle), Oxbow Books, Oxford.
Binford, L.R. (1968) Post-Pleistocene Adaptations. In: *New Perspectives in Archaeology* (eds L.R. Binford and S.R. Binford), Aldine, Chicago, pp. 313–341.
Binford, L.R. (1984) *Faunal remains from Klasies River mouth*. Academic Press, New York.
Boas, F. (1930) Observations on the growth of children. *Science*, **72** (1854), 44–48.
Bogaard, A. (2004) *Neolithic farming in central Europe: an archaeobotanical study of crop husbandry practices*. Routledge, New York.
Bogaard, A. and Jones, G. (2007) Neolithic farming in Britain and central Europe: contrast or continuity? In: *Going over: the Mesolithic-Neolithic transition in north-west Europe* (eds A. Whittle and V. Cummings), Oxford University Press for the British Academy, Oxford, pp. 357–375.
Bogin, B. and Keep, R. (1999) Eight thousand years of economic and political history in Latin America revealed by anthropometry. *Ann. Hum. Biol.*, **26** (4), 333–351.
Bogin, B. and Rios, L. (2003) Rapid morphological change in living humans: implications for modern human origins. *Comp. Biochem. Physiol. Part A, Mol. Integr. Physiol.*, **136** (1), 71–84.
Bogucki, P.I. (1988) *Forest farmers and stockherders: early agriculture and its consequences in north-central Europe*. CUP Archive.
Bogucki, P.I. and Crabtree, P.J. (2004a) *Ancient Europe 8000 B.C.–A.D. 1000: Encyclopedia of the Barbarian world*. Thompson/Gale, New York. Vol. 1
Bogucki, P.I. and Crabtree, P.J. (2004b) *Ancient Europe 8000 B.C.–A.D. 1000: Encyclopedia of the Barbarian world*. Thompson/Gale, New York. Vol. 2
Bokonyi, S. (1974) *History of domestic mammals in Central and Eastern Europe*. Akademiai Kiado, Budapest.
Bramanti, B., Thomas, M., Haak, W., Unterlaender, M., Jores, P., Tambets, K., Antanaitis-Jacobs, I., Haidle, M., Jankauskas, R., and Kind, C.-J. (2009) Genetic discontinuity between local hunter-gatherers and central Europe's first farmers. *Science*, **326** (5949), 137–140.
Bräuer, G. (1988) Osteometrie. In: *Anthropologie: Handbuch der Vergleichenden Biologie des Menschen, Ban 1; Wesen und Methoden der Anthropologie, Teil 1. Wissenschaftstheorie, Geschichte, Morphologische Methoden* (ed. R. Knussman), Gustav Fischer Verlag, Stuttgart, pp. 160–232.
Brázdil, R. (1996) *Reconstructions of past climate from historical sources in the Czech lands. Climatic variations and forcing mechanisms of the last 2000 years*. Springer, pp. 409–431.
Brázdil, R., Pfister, C., Wanner, H., Von Storch, H., and Luterbacher, J. (2005) Historical climatology in Europe – the state of the art. *Clim. Change*, **70** (3), 363–430.
Brázdil, R., Valášek, H., Luterbacher, J., and Macková, J. (2001) *Die Hungerjahre 1770–1772 in den böhmischen Ländern*. Österreichische Zeitschrift für Geschichtswissenschaften, **12**, 44–78.
Brown, N. (2005) *History and climate change: a Eurocentric perspective*. Routledge, New York.
Cipolla, C.M. (1994) *Before the industrial revolution: European society and economy, 1000–1700*. W.W. Norton & Company, New York.
Clark, P.U., Dyke, A.S., Shakun, J.D., Carlson, A.E., Clark, J., Wohlfarth, B., Mitrovica, J.X., Hostetler, S.W., and McCabe, A.M. (2009) The last glacial maximum. *Science*, **325** (5941), 710–714.
Cohen, M.N. (1977) *Food crisis in prehistory: overpopulation and the origins of agriculture*. Yale University Press, New Haven.
Cohen, M.N., Armelagos, G.J., Wenner-Gren Foundation for Anthropological Research, and State University of New York College at Plattsburgh. (1984) *Paleopathology at the origins of agriculture*. Academic Press, New York.

Cole, T.J. (2000) Secular trends in growth. *Proc. Nutr. Soc.*, **59** (02), 317–324.
Collis, J. (2003) *The European Iron Age.* Routledge, Oxford.
Davis, B.A., Brewer, S., Stevenson, A.C., and Guiot, J. (2003) The temperature of Europe during the Holocene reconstructed from pollen data. *Quat. Sci. Rev.*, **22** (15), 1701–1716.
Dreslerová, D., Horáček, I., and Pokorný, P. (2007) Přírodní prostředí Čech a jeho vývoj. In: Archeologie pravěkých Čech (ed M. Kuna), Archeologický ústav AV ČR, Praha, pp. 23–50.
Dreslerová, P. (2012) Human Response to Potential Robust Climate Change around 5500 cal BP in the Territory of Bohemia (the Czech Republic). *Interdisciplinaria Archaeologica*, **3** (1), 43–55.
Ehlers, J., Astakhov, V., Gibbard, P.L., Mangerud, J., and Svendsen, J. (2013) Glaciations Late Pleistocene in Euroasia. In: *Encyclopedia of Quaternary Science* (eds S.A. Elias and C.J. Mock), Elsevier, Amsterdam, pp. 224–235.
Fagan, B.M. (2000) *The Little Ice Age: how climate made history, 1300–1850.* Basic Books, New York.
Fetter, V., Prokopec, M., Suchý, J., and Šobová, A. (1963) Developmental acceleration of the youth by anthropometric surveys from 1951 and 1961. *Cesk. Pediatr.*, **18**, 673–677.
Flannery, K.V. (1965) The Ecology of Early Food Production in Mesopotamia Prehistoric farmers and herders exploited a series of adjacent but contrasting climatic zones. *Science*, **147** (3663), 1247–1256.
Flannery, K.V. (1969) Origins and Ecological Effects of Early Domestication in Iran and the Near East. In: *The Domestication and Exploitation of Plants and Animals* (eds P.J. Ucko and G.W. Dimbleby), Aldine Publishing Co., Chicago, pp. 73–100.
Galeta, P., Sládek, V., Sosna, D., and Brůžek, J. (2011) Modeling Neolithic dispersal in Central Europe: demographic implications. *Am. J. Phys. Anthropol.*, **146**, 104–115.
Gamble, C., Davies, W., Pettitt, P., and Richards, M. (2004) Climate change and evolving human diversity in Europe during the last glacial. *Philos. Trans. Royal Soc. London B: Biol. Sci.*, **359** (1442), 243–254.
Good, I. (2001) Archaeological textiles: a review of current research. *Annu. Rev. Anthropol.*, 209–226.
Greenfield, H.J. (2010) The Secondary Products Revolution: the past, the present and the future. *World Archaeol.*, **42** (1), 29–54.
Greulich, W.W. (1951) The growth and developmental status of Guamanian school children in 1947. *Am. J. Phys. Anthropol.*, **9** (1), 55–70.
Greulich, W.W., Crismon, C.S., Turner, M.L., Greulich, M.L., and Okumoto, Y. (1953) The physical growth and development of children who survived the atomic bombing of Hiroshima or Nagasaki. *J. Pediatr.*, **43** (2), 121–145.
Grove, J.M. (1988) *The Little Ice Age.* Methuen, London, New York.
Haak, W., Balanovsky, O., Sanchez, J.J., Koshel, S., Zaporozhchenko, V., Adler, C.J., Der Sarkissian, C.S., Brandt, G., Schwarz, C., and Nicklisch, N. (2010) Ancient DNA from European early Neolithic farmers reveals their near eastern affinities. *PLoS Biol.*, **8** (11), e1000536.
Haak, W., Lazaridis, I., Patterson, N., Rohland, N., Mallick, S., Llamas, B., Brandt, G., Nordenfelt, S., Harney, E., and Stewardson, K. (2015) Massive migration from the steppe was a source for Indo-European languages in Europe. *Nature*, **522** (7555), 207–211.
Harding, A.F. (2000) *European societies in the Bronze Age.* Cambridge University Press, Cambridge.
Hofmanová, Z., Kreutzer, S., Hellenthal, G., Sell, C., Diekmann, Y., Díez-del-Molino, D., van Dorp, L., López, S., Kousathanas, A., and Link, V. (2016) Early farmers from across Europe directly descended from Neolithic Aegeans. *Proc. Natl Acad. Sci. USA*, **113** (52), 6886–6891.
Holliday, T.W. (1997) Body proportions in Late Pleistocene Europe and modern human origins. *J. Hum. Evol.*, **32** (5), 423–447.

Holliday, T.W. (1999) Brachial and crural indices of European Late Upper Paleolithic and Mesolithic humans. *J. Hum. Evol.*, **36** (5), 549–566.

Holt, B.M. (2003) Mobility in Upper Paleolithic and Mesolithic Europe: Evidence from the lower limb. *Am. J. Phys. Anthropol.*, **122** (3), 200–215.

Holt, B.M. and Formicola, V. (2008) Hunters of the Ice Age: the biology of Upper Paleolithic people. *Am. J. Phys. Anthropol.*, **137** (S47), 70–99.

Jantz, L.M. and Jantz, R.L. (1999) Secular change in long bone length and proportion in the United States, 1800–1970. *Am. J. Phys. Anthropol.*, **110** (1), 57–67.

Jiráň, L. (ed.) (2008) *Archeologie pravěkých Čech 4: Doba bronzová*. Archeologický ústav AV ČR, Praha.

Johnson, A.W. and Earle, T. (2000) *The evolution of human societies: from foraging group to agrarian state*. Stanford University Press, Stanford.

Kalis, A.J., Merkt, J., and Wunderlich, J. (2003) Environmental changes during the Holocene climatic optimum in central Europe - human impact and natural causes. *Quat. Sci. Rev.*, **22** (1), 33–79.

Kates, R.W. (1985) The interaction of climate and society. *SCOPE*, pp. 3–36.

Keegan, J. (2011) *A history of warfare*. Random House, New York.

Kristiansen, K. and Larsson, T.B. (2005) *The rise of Bronze Age society: travels, transmissions and transformations*. Cambridge University Press, New York.

Kristiansen, K. and Rowlands, M. (2005) *Social Transformations in Archaeology: Global and local perspectives*. Routledge, London.

Kruta, V. (2000) *Les Celtes: histoire et dictionnaire*. R. Laffont, Paris.

Kuneš, P., Pelánková, B., Chytrý, M., Jankovská, V., Pokorný, P., and Petr, L. (2008) Interpretation of the last-glacial vegetation of eastern-central Europe using modern analogues from southern Siberia. *J. Biogeogr.*, **35** (12), 2223–2236.

Larnkjær, A., Attrup Schrøder, S., Maria Schmidt, I., Hørby Jørgensen, M., and Fleischer Michaelsen, K. (2006) Secular change in adult stature has come to a halt in northern Europe and Italy. *Acta Paediatr.*, **95** (6), 754–755.

Larsen, C.S. (1995) Biological changes in human populations with agriculture. *Annu. Rev. Anthropol.*, **24**, 185–213.

Larsen, C.S., Hillson, S.W., Boz, B., Pilloud, M.A., Sadvari, J.W., Agarwal, S.C., Glencross, B., Beauchesne, P., Pearson, J., Ruff, C.B., Garofalo, E.M., Hager, L.D., Haddow, S.D., and Knüsel, C.J. (2015) Bioarchaeology of Neolithic Çatalhöyük: Lives and Lifestyles of an Early Farming Society in Transition. *J. World Prehistory*, **28** (1), 27–68.

Le Roy Ladurie, E. (1971) *Times of feast, times of famine. A history of climate since the year 1000*. Doubleday, New York.

Lenneis, E., Neugebauer-Maresch, C., and Ruttkay, E. (1995) *Jungsteinzeit im osten Österreichs*. Verlag Niederösterreichisches Pressehaus, St. Pölten.

Lhotská, L., Bláha, P., Vignerová, J., Roth, Z., and Prokopec, M. (1993) *Vth Nation-wide Anthropological Survey of children and adolescents 1991 (Czech Republic)*. National Institute of Public Health, Prague.

Li, L., Dangour, A.D., and Power, C. (2007) Early life influences on adult leg and trunk length in the 1958 British birth cohort. *Am. J. Hum. Biol.*, **19** (6), 836–843.

Lieberman, D. (2013) *The story of the human body: evolution, health, and disease*. Vintage Books, New York.

Lowe. J. (1994) The Weichselian Late-glacial in southwestern Europe (Iberian Peninsula, Pyrenees, Massif Central, northern Apennines). *J. Quat. Sci.*, **9** (2), 101–107.

Lowe, J., Ammann, B., Birks, H., Björck, S., Coope, G., Cwynar, L., and De Beaulieu, M. (1994) Climatic changes in areas adjacent to the North Atlantic during the last glacial-interglacial transition (14–9 ka BP): a contribution to IGCP-253. *J. Quat. Sci.*, **9** (2), 185–198.

Lowe, J.J. (2001) Abrupt climatic changes in Europe during the last glacial-interglacial transition: the potential for testing hypotheses on the synchroneity of climatic events using tephrochronology. *Glob. Planet Change*, **30** (1), 73–84.

Lutovský, M. and Profantová, N. (1995) *Sámova říše*. Academia, Praha.

Lynch, A.J. and Rowland, C.A. (2005) *The history of grinding*. Society for Mining Metallurgy & Exploration, Littleton, Colorado.

Macintosh, A.A., Pinhasi, R., and Stock, J.T. (2014) Divergence in male and female manipulative behaviors with the intensification of metallurgy in Central Europe. *PLoS ONE*, **9** (11), e112116.

Macháček, J. (2005) Raně středověké Pohansko u Břeclavi: munitio, palatium, nebo emporium moravských panovníků? *Archeol. Rozhl.*, **LVII** (1), 100–138.

Macháček, J. (2010) *The rise of medieval towns and states in East Central Europe: early medieval centres as social and economic systems*. Brill, Leiden.

Macháček, J. (2012) Archeologie údolní nivy aneb Proč možná zanikla Velká Morava. *Vesmír*, **91** (10), 566–569.

Manly, B.F.J. (2006) *Randomization, bootstrap and Monte Carlo methods in biology*. Chapman & Hall/CRC Press, Boca Raton.

Mathieson, I., Lazaridis, I., Rohland, N., Mallick, S., Patterson, N., Roodenberg, S.A., Harney, E., Stewardson, K., Fernandes, D., and Novak, M. (2015) Genome-wide patterns of selection in 230 ancient Eurasians. *Nature*, **528** (7583), 499–503.

Matiegka, J. (1927) *Somatologie školní mládeže*. Česká Akademie, Praha.

Matiegka. J. (1934) *L'Homme fossile de Předmostí en Moravie (Tchécoslovaquie). I. Les crânes*. Académie Tchèque des Sciences et des Arts, Prague.

Matiegka, J. (1938) *Homo předmostensis: fosilní člověk z Předmostí na Moravě. 2. Ostatni části kostrové*. Nákladatelství České Akademie věd a umění, Praha.

Menotti, F. (2012) *Wetland archaeology and beyond: theory and practice*: Oxford University Press, Oxford.

Měřínský, Z. (2002) *České země od příchodu Slovanů po Velkou Moravu I*. Libri, Praha.

Měřínský, Z. (2011) *Morava na úsvitě dějin*. Muzejní a vlastivědná společnost, Brno.

Nebelsick, L.D., Eibner, A., Lauermann, E., and Neugebauer, J.-W. (1997) *Hallstattkultur im osten Österreichs*. Niederösterreichisches Pressehaus, St.Pölten.

Nerudová, Z. and Neruda, P. (2015) Moravia between Gravettian and Magdalenian. In: *Forgotten times and spaces: New perspectives in paleoanthropological, paleoethnological and archaeological studies* (eds S. Sázelová, M. Novák, and A. Mizerová), Institute of Archaelogy of the Czech Academy of Sciences, Masaryk University, Brno, pp. 378–394.

Neugebauer, J.-W., Lochner, M., Neugebauer-Maresch, C., and Teschler-Nicola, M. (1994) *Bronzezeit in Ostösterreich*. Verlag Niederösterreichisches Pressehaus, St.Pölten.

Neustupný, E. (1997) Šňůrová sídliště, kulturní normy a symboly. *Archeol. Rozhl.*, **49**, 304–322.

Neustupný, E. (ed.) (2008) *Archeologie pravěkých Čech 4: Eneolit*. Archeologický ústav AV ČR, Praha.

Oliva. M. (2005) *Civilizace moravského paleolitu a mezolitu*. Moravské Zemské Muzeum, Brno.

Padez, C. (2003) Secular trend in stature in the Portuguese population (1904–2000). *Ann. Hum. Biol.*, **30** (3), 262–278.

Pavlů, I. (2005) Neolitizace střední Evropy. *Archeol. Rozhl.*, **57** (2), 293–302.

Pavlů, I. and Zápotocká, M. (2007) *Archeologie pravěkých Čech 3: Neolit*. Archeologický ústav AV ČR, Praha.

Pfister, C. (2001) Klimawandel in der Geschichte Europas. Zur Entwicklung und zum Potenzial der Historischen Klimatologie OÉZG. *OÉsterr Z. Gesch. Wiss.*, **12** (2), 7–43.

Pinhasi, R. and Stock, J.T. (eds) (2011) *Human bioarchaeology of the transition to agriculture*. Wiley-Blackwell, Chichester.

Piontek. J. (1992) Climatic changes and biological structure of the human populations in Poland in the Middle Ages. In: *European Climate Reconstructed* (ed. F. Burkhard), Gustav Fisher Verlag, Stuttgart, pp. 105–113.
Podborský, V. (ed.) (1993) *Pravěké dějiny Moravy: Vlastivěda Moravská: Země a lid.* Muzejní a vlastivědná společnost v Brně, Brno.
Podborský, V. (2002) Dvě pohřebiště neolitického lidu s lineární keramikou ve Vedrovicích na Moravě. Masarykova univerzita, Brno.
Prokopec, M., Titlbachová, S., Dutková, L., and Zlámalová, H. (1986) Height and weight of Czech children in 1981 based on the results of a nation-wide anthropological survey. *Česk. Pediatr.*, **41** (1), 20–26.
Ray, N. and Adams, J. (2001) A GIS-based vegetation map of the world at the last glacial maximum (25,000–15,000 BP). *Internet Archaeol.*, **11**, 1–44.
Rotberg, R.I. and Rabb, T.K. (2014) *Climate and history: studies in interdisciplinary history.* Princeton University Press, Princeton.
Ruff, C.B. (1994) Morphological adaptation to climate in modern and fossil hominids. *Yrbk. Phys. Anthropol.* **37**, 65–107.
Ruff, C.B., Holt, B., Niskanen, M., Sladek, V., Berner, M., Garofalo, E., Garvin, H., Hora, M., Junno, J.A., Schuplerova, E., *et al.* (2015) Gradual decline in mobility with adoption of food production in Europe. *Proc. Natl Acad. Sci. USA*, **112** (23), 7147–7152.
Ryder, M.L. (1987) The evolution of the fleece. *Sci. Am.*, **256** (1), 112–119.
Salač, V., Droberjar, E., Militký, J., Musil, J., and Urbanová, K. (2008) Archeologie pravěkých Čech 8: Doba římská a stěhování národů. Archeologický ústav AV ČR, Praha.
Shennan, S.J. (1993) Settlement and social change in Central Europe. *J. World Prehistory*, **7** (2), 121–161.
Sherratt, A. (1981) *Plough and pastoralism: aspects of the Secondary Products Revolution.* Cambridge University Press, London.
Sherratt. A. (1983) The secondary exploitation of animals in the Old World. *World Archaeol.*, **15**, 90–104.
Shott, M.J. (1993) Spears, darts, and arrows: Late Woodland hunting techniques in the Upper Ohio Valley. *Am. Antiq.*, **58** (3), 425–443.
Schultz, A.H. (1926) Fetal growth of man and other primates. *Q. Rev. Biol.*, **1**, 465–521.
Sjögren, K.-G., Price, T.D., and Kristiansen, K. (2016) Diet and Mobility in the Corded Ware of Central Europe. *PLoS ONE*, **11** (5), e0155083.
Sládek, V. (2000) Hominid evolution in Central Europe during Upper Pleistocene: Origin of anatomically modern humans. PhD thesis, Université de Bordeaux I, Bordeaux.
Sládek, V., Berner, M., and Sailer, R. (2006a) Mobility in Central European Late Eneolithic and Early Bronze Age: Femoral Cross-sectional Geometry. *Am. J. Phys. Anthropol.*, **130** (3), 320–332.
Sládek, V., Berner, M., and Sailer, R. (2006b) Mobility in Central European Late Eneolithic and Early Bronze Age: Tibial Cross-sectional Geometry. *J. Archaeol. Sci.*, **33** (4), 470–482.
Sládek, V., Berner, M., Sosna, D., and Sailer, R. (2007) Human manipulative behavior in the Central European Late Eneolithic and Early Bronze Age: Humeral bilateral asymmetry. *Am. J. Phys. Anthropol.*, **133** (1), 669–681.
Sládek, V., Hora, M., Farkasova, K., and Rocek, T. (2016) Impact of grinding technology on bilateral asymmetry in muscle activity of the upper limb. *J. Archaeol. Sci.*, **72**, 142–156.
Sládek, V., Trinkaus, E., Holiday, T., and Hillson, S. (2000) *The People of the Pavlovian: Skeletal Catalogue and Osteometrics of the Gravettian fossil hominids from Dolní Věstonice and Pavlov.* Archeologický ústav AVČR, Dolní Věstonice Studies 5, Brno.
Smith, M.E. (2004) The archaeology of ancient state economies. *Annu. Rev. Anthropol.*, 73–102.
Sosna, D. (2007) Social Differentiation in the Late Copper Age and the Early Bronze Age in South Moravia (Czech Republic). Ph. D. Thesis, Florida State University, Florida.

Sosna, D., Galeta, P., and Sládek, V. (2008) A resampling approach to gender relations: the Rebešovice cemetery. *J. Archaeol. Sci.*, **35** (2), 342–354.
Sparacello, V.S., d'Ercole, V., and Coppa, A. (2015) A bioarchaeological approach to the reconstruction of changes in military organization among Iron Age Samnites (Vestini) from Abruzzo, Central Italy. *Am. J. Phys. Anthropol.*, **156** (3), 305–316.
Stearns, P.N. (1993) *The industrial revolution in world history*. Westview Press, Boulder.
Stini, W.A. (1975) Sexual dimorphism and nutrient reserves. In: *Sexual dimorphism in Homo sapiens* (ed. R.L. Hall), Praeger, New York, pp. 391–419.
Stinson, S. (1985) Sex differences in environmental sensitivity during growth and development. *Yrbk. Phys. Anthropol.*, **28**, 123–147.
Svoboda, J., Cílek, V., and Vašků, Z. (2003) *Velká kniha o klimatu zemí Koruny české*. Regia, Praha.
Tanner, J.M. (1990) *Foetus into man: Physical growth from conception to maturity*. Harvard University Press, Cambridge.
Teschler-Nicola, M. (ed.) (2006) *Early Modern Humans at the Moravian Gate: The Mladec Caves and their Remains*. Springer, Vienna.
Trinkaus, E. (1981) Neanderthal limb proportions and cold adaptation. In: *Aspects of human evolution* (ed. C.B. Stringer), Taylor and Francis, London, pp. 187–224.
Trinkaus, E. (2006) Body length and body mass. In: *Early Modern Human Evolution in Central Europe: the People of Dolní Vestŏnice and Pavlov* (eds E. Trinkaus and J. Svoboda), Oxford University Press, Oxford, pp. 233–241.
Trinkaus, E. and Jelinek, J. (1997) Human remains from the Moravian Gravettian: The Dolni Vestonice 3 postcrania. *J. Hum. Evol.*, **33** (1), 33–82.
Trinkaus, E. and Svoboda, J. (eds) (2006) *Early Modern Human Evolution in Central Europe: The People of Dolní Věstonice and Pavlov*. Oxford University Press, Oxford.
Trinkaus, E., Svoboda, J., West, D., Sládek, V., Hillson, S., Drozdová, E., and Fišáková, M. (2000) Human remains from the Moravian Gravettian: Morphology and Taphonomy of isolated elements from the Dolní Věstonice II site. *J. Archaeol. Sci.*, **27**, 1115–1132.
Třeštík, D. (1997) *Počátky Přemyslovců: vstup Čechů do dějin, 530–935*. Nakladatelství Lidové noviny, Praha.
Třeštík, D. (2001) *Vznik Velké Moravy*. Nakladatelství Lidové noviny, Praha.
Válka, J. (1991) *Dějiny Moravy I. Středověká Morava*. Muzejní a vlastivědná společnost, Brno.
Valoch, K. and Neruda, P. (2005) K chronologii moravského magdalénienu. *Archeol. Rozhl.*, **57**, 459–476.
Vencl, S. (1994) K problému sídlišť kultury s keramikou šňůrovou. *Archeol. Rozhl.*, **46**, 3–24.
Vencl, S. and Fridrich, J. (2007) *Paleolit a mezolit*. Archeologický ústav AV ČR, Praha.
Venclová, N., Drda, P., Michálek, J., Militký, J., Salač, V., Sankot, P., and Vokolek, V. (2008) *Archeologie pravěkých Čech 7: Doba laténská*. Archeologický ústav AV ČR, Praha.
Vignerová, J., Bláha, P., and Lhotska, L. (1998) Physical growth of Czech Children and some socio-economic factors. *J. Hum. Ecol.*, **9** (3), 227–231.
Vignerová, J., Brabec, M., and Bláha, P. (2006) Two centuries of growth among Czech children and youth. *Econ. Hum. Biol.*, **4**, 237–252.
Wadsworth, M., Hardy, R., Paul, A., Marshall, S., and Cole, T. (2002) Leg and trunk length at 43 years in relation to childhood health, diet and family circumstances; evidence from the 1946 national birth cohort. *Int. J. Epidemiol.*, **31** (2), 383–390.
Walker, M. (1995) Climatic changes in Europe during the last glacial/interglacial transition. *Quat. Int.*, **28**, 63–76.
Wefers, S. (2011) Still using your saddle quern? A compilation of the oldest known rotary quern in western Europe. In: *Bread for the people: the archaeology of mills and milling* (eds D. Williams and D. Peacock), Archaeopress, Oxford, pp. 67–76.

Whittle, A., Bentley, R.A., Bickle, P., Dočkalová, M., Fibiger, L., Hamilton, J., Hedges, R., Mateiciucová, I., and Pavúk, J. (2013) Moravia and western Slovakia. In: *The first farmers of central Europe: diversity in LBK lifeways* (eds P. Bickle and A. Whittle), Oxbow Books, Oxford.

Yokoyama, Y., Lambeck, K., De Deckker, P., Johnston, P., and Fifield, L.K. (2000) Timing of the Last Glacial Maximum from observed sea-level minima. *Nature*, **406** (6797), 713–716.

12

Scandinavia and Finland

Markku Niskanen[1], *Heli Maijanen*[1], *Juho-Antti Junno*[1], *Sirpa Niinimäki*[1], *Anna-Kaisa Salmi*[1], *Rosa Vilkama*[1], *Tiina Väre*[1], *Kati Salo*[2], *Anna Kjellström*[3], *and Petra Molnar*[3]

[1] *Department of Archaeology, University of Oulu, Oulu, Finland*
[2] *Department of Archaeology, University of Helsinki, Helsinki, Finland*
[3] *Osteological Research Laboratory, Stockholm University, Stockholm, Sweden*

12.1 Introduction to Region, Samples, and Techniques

12.1.1 Region and its Population History

This chapter focuses on Scandinavia, Finland, and adjacent regions of Northwest Russia (Karelia and the Kola Peninsula). This region was the last major region of Europe to be settled by humans due to the late persistence of the Fennoscandian ice sheet. Because ancestors of the earliest inhabitants of this region had dispersed from the Franco-Cantabria in the west and the South Russian Plain in the east after the Last Glacial Maximum (on these locations see Verpoorte, 2009: their Fig. 3) they are here referred to as the 'western' and the 'eastern' colonists, respectively.

The 'western' colonists arrived in the Jutland Peninsula of Denmark ca. 12,848 calBC (Grimm and Weber, 2008) and followed the ice-free Norwegian coast northward arriving in northern Norway ca. 8750 calBC (Nordin and Gustafsson, 2010; Bang-Andersen, 2012). The 'eastern' colonists colonized Northwest Russia and Finland, and encountered their 'western' counterparts in northern Norway ca. 8300–8200 calBC (Rankama and Kankaanpää, 2011; Sørensen *et al.*, 2013). These 'western' and 'eastern' colonists likely represent 'western European hunter-gatherer' (WHG) and 'eastern European hunter-gatherer' (EHG) clusters, respectively. A 'Scandinavian hunter-gatherer' (SHG) cluster emerged as the WHG and EHG intermixed in Scandinavia (on these genetic clusters, see Haak *et al.*, 2015).

The Early Mesolithic people in this region (e.g., the Maglemose culture in South Scandinavia ca. 9000–6000 calBC; the Suomusjärvi culture in Finland ca. 8600–5000 calBC) were largely hunters of big game. Fishing and/or sealing gained importance in a subsistence economy that included pottery-using (e.g., the Ertbølle in South Scandinavia ca. 5400–3900 calBC; Comb Ware in Finland and Karelia ca. 5400–2300 cal BC; Timofeev *et al.*, 2004) and more sedentary Late Mesolithic foragers, at least locally (Zvelebil, 2008; Jessen *et al.*, 2015).

The Late Mesolithic South Scandinavians of the Ertebølle culture (ca. 5400–3900 calBC) were fishers and sealers, who lived in relatively large and sedentary settlements (Price, 2000; Fischer *et al.*, 2007). Their southern neighbors were farmers of the Linear Band Ceramic culture

Skeletal Variation and Adaptation in Europeans: Upper Paleolithic to the Twentieth Century,
First Edition. Edited by Christopher B. Ruff.

(the LBK), which was established in northern Germany ca. 5550–5050 calBC. The northern boundary of this culture formed the longest-lasting forager-farmer boundary in Neolithic Europe (Rowley-Conwy, 2011). About 4400 calBC, the successors of the LBK (the so-called Danubian cultures) with at least some cultural (see Price, 2000; Czekaj-Zastawny *et al.*, 2013a,b) and genetic (see Brand *et al.*, 2013; Haak *et al.*, 2015) contribution from indigenous foragers of the North European Plain gave rise to the Middle Neolithic Funnelbeaker culture (the TRB).

The TRB farmers spread over the southern one-third of Scandinavia between 4000 calBC and 3700 calBC via colonization based on ancient DNA (Skoglund *et al.*, 2012, 2014) and archeological evidence, but foragers survived in some regions for centuries (Sørensen and Karg, 2014). This parallel existence facilitated gene flow in Scandinavia and elsewhere in Europe during Middle Neolithic (Brand *et al.*, 2013; Haak *et al.*, 2015).

The Corded Ware culture (the CWC) replaced the TRB culture in the North European Plain and Scandinavia within ca. 3050–2650 calBC (Wlodarczak, 2009; Iversen, 2013). The CWC people had a largely external origin in Central Europe because ca. 75% of their ancestry is from the Yamnaya steppe people (Haak *et al.*, 2015) and at least a partly external origin in South Scandinavia (Allentoft *et al.*, 2015). Domesticated horses were probably one factor allowing this population expansion at the expense of the local TRB farmers (Anthony and Brown, 2011).

The CWC arrived ca. 2900 calBC in southern Finland, where it replaced foragers of the Comb Ware culture (Bläuer and Kantanen, 2013) and introduced domestic cattle if not farming (Cramp *et al.*, 2014). The CWC people's genetic influence may still be detectable in Southwest Finland (Neuvonen *et al.*, 2015).

The Bell Beaker culture (the BBC) spread also partly via population movements (Brandt *et al.*, 2013). It emerged in the Iberian Peninsula ca. 2800 calBC and spread widely to Western and Central Europe, including Southern Scandinavia, ca. 2500 calBC. It prevailed until ca. 2050 calBC (Vankilde, 2007; Haak *et al.*, 2015, Supplementary Information 3).

The Pitted Ware culture (the PWC) is a special case in Scandinavian prehistory because this coastal foraging culture – dated to 3400–2300 calBC – is contemporary with the late TRB farming culture and the entire date range of the CWC in Scandinavia. The PWC people descended from the Mesolithic Scandinavians (Skoglund *et al.*, 2012, 2014; Haak *et al.*, 2015) and were relatively sedentary foragers. Their sites are located along the eastern middle and southern coastlines of the Scandinavian Peninsula, in Denmark, on the islands of Gotland and Öland of Sweden, and on the Åland Islands of Finland (Molnar, 2008; Iversen, 2013 and references therein).

About 2400–2300 calBC, a fully agricultural Scandinavian Late Neolithic culture (partly contemporary with Early Bronze Age in many other regions of Europe) replaced the CWB and PWC in South Scandinavia (Vankilde, 2007; Iversen, 2013) and a partly agricultural Kiukainen culture replaced the CWB in Southern Finland. Farming did not become fully established in southern Finland until the Bronze Age (1700–500 calBC). The Bronze Age farmers were contemporary with foragers because a foraging way-of-life continued in northern Scandinavia as well as in northern and central Finland to historical times (Bläuer and Kantanen, 2013). The recent Saami descend from these foragers.

Some migration from Sweden to Finland may have occurred during the Bronze and Iron Ages (Edgren, 1998; Tuovinen, 2002; Tarkiainen, 2010; Bläuer and Kantanen, 2013). The Swedes were certainly settling in coastal Finland in the 12th century, and large-scale immigration commenced during the late 13th century when southern and southwestern Finland became part of Sweden (Sawyer and Sawyer, 1993; Tarkiainen, 2010). As a result of these prehistoric and historic migrations, the Western Finnish gene pool is partly Swedish (Salmela *et al.*, 2011), and there are Swedish-speaking communities in coastal southern Finland (Tarkiainen, 2010). A lesser migration from Finland to Sweden during the Medieval and Early Modern periods has affected the Swedish gene pool especially in eastern central Sweden (Salmela *et al.*, 2011; Humphreys *et al.*, 2011). This long history of strong Swedish influence and gene flow in southern Finland gives justification for pooling skeletal samples from Scandinavia and southern Finland. For instance, the Porvoo sample from the Early Modern period probably represents mostly Swedish-speaking people of largely Swedish ancestry (on the Swedish settlement, see Tarkiainen, 2010).

Urbanization commenced relatively late in Scandinavia and Finland. Sigtuna, founded in 975 AD during the late Viking period as a bridgehead for new royal power, is the oldest urban center in Sweden (on Sigtuna see Sawyer and Sawyer, 1993). Turku is the oldest city of Finland, but did not become an actual urban settlement until the very early 14th century (Pihlman, 2010).

12.1.2 Samples

Temporal coverage from southern Scandinavia and southern Finland combined is quite good, but the Bronze Age is represented only by a few individuals from Bröste and Jørgensen (1956) because the prevalent burial custom at that time in Scandinavia was cremation. Table 12.1 provides sample sizes for each period, while Figure 12.1 shows the geographic distribution of these samples and individuals.

Most data used in this study are from a large European dataset collected during an international research collaboration (see Chapter 1 in this volume). Some data (primarily long bone length data) are taken from the literature to fill in poorly represented time periods or regions and to augment sample sizes.

The Mesolithic sample is from Denmark and southern Sweden. Two individuals represent the Early Mesolithic Maglemosian culture and others (N = 27) the Late Mesolithic Ertebølle culture. There are no measurable Mesolithic skeletons from Finland, but early Mesolithic people from Olenii Ostrov in Russian Karelia (dated to ca. 6350 calBC; Price and Jacobs, 1990)

Table 12.1 Scandinavian and Finnish sample sizes by temporal period. Date range of the Very Recent sample is the range of death years. Early Medieval is equivalent to the Viking Age in Scandinavian and Finnish chronology. Date ranges are those of the Scandinavian and Finnish skeletal individuals. They thus differ slightly from those of the total European dataset, but dates are within that of the total date range for each temporal period (see Chapter 1). The Neolithic foragers include the Pitted Ware sample (the PWC) and the Neolithic farmers the Funnel Beaker (the TRB), the Corded Ware (the CWC), and the Scandinavian Late Neolithic skeletons (the SLN) as well as Neolithic individuals with unknown cultural period affiliation ('other'). Site numbers correspond with locations on the map in Figure 12.1.

Period	Period/Association	Date range (cal. yr)	Males	Females	Site number
Very Recent	Autopsy sample	1915–1935 AD	48	22	73
Early Modern	Porvoo, Renko	1600–1850 AD	28	24	71, 72
Late Medieval	Westerhus, Humlegården	1080–1300 AD	95	82	69, 70
Early Medieval	Nunnan Block	800–950 AD	19	4	23, 69
Iron/Roman	Roman Iron Age Danes	0–400 AD	35	30	42–67
Neolithic foragers	PWC from Gotland	3100–2100 BC	46	30	37–41
Neolithic farmers	TRB, CWC, SLN, 'other'	4000–1700 BC	26	6	10–22, 24–36
Mesolithic	Maglemosian, Ertebølle	7100–4000 BC	12	17	1 9, 68
Total		7100 BC–1935 AD	309	215	

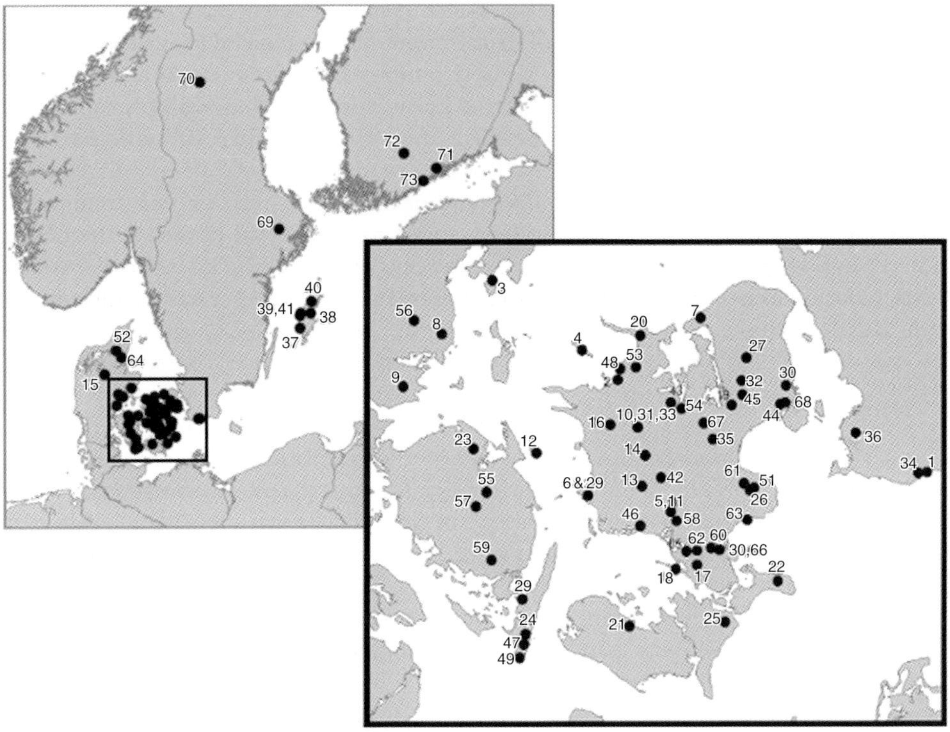

Mesolithic
1 Skateholm
2 Dragsholm
3 Væengesø 2
4 Sejrø
5 Holmegärd
6 Korsør Glaesverk
7 Melby
8 Koelbjerg
9 Tybrind Vig
68 Bloksbjerg

Neolithic
10 Bodal Mose
11 Porsmose
12 Gjemild
13 Kelderød
14 Døjringe
15 Haunø-Lauenkjor
16 Hallebygärd
17 Over Vindinge
18 Langebjerg
19 Marbierg
20 Holmstrup
21 Grydehøj
22 Borre
24 Hjortholm Mose
25 Ästrup Äs
26 Hellested
27 Grose
28 Poregård
29 Korsør Nor
30 St. Tuborg
31 Store Lyng
32 Viksø
33 Østrup Mose
34 Beddinge
35 Viby
36 Frederiksberg
37 Ajvide
38 Västerbjers
39 Fritorp
40 Ire
41 Visby

Iron Age
42 Simonsborg
43 Store Grandløse
44 Landlystvej
45 Lille Vasby
46 Skafterup Mark
47 Nordenbrogård
48 Bøkebjerg
49 Store Keldbjerg
50 Endegårde
51 Varpelev
52 Scheelminde
53 Asnæs
54 Englerup
55 Fraugde
56 Sanderumgård
57 Lumby Grusgrav
58 Næstved Mark
59 Egebjerg
60 Broskov
61 Himlingøje
62 Gammel Lundby
63 Vemmeltofte Skovridergård
64 Smidie
65 Øbjerggård
66 Græsbjerg
67 Lyregård

Medieval
23 Otterup
69 Sigtuna
70 Westerhus

Moderm
71 Porvoo
72 Renko
73 Helsinki

Figure 12.1 Locations of skeletal samples from Scandinavia and Finland.

are used as representatives of their Finnish contemporaries and close neighbors (see also Jacobs, 1993, 1995; Formicola and Giannecchini, 1999).

The Neolithic farmer sample (N = 32) is from Denmark and southern Sweden. Twelve individuals represent the TRB (4000–2800 calBC), seven individuals the CWC (2800–2350 calBC), and six individuals the Late Neolithic (2350–1700 calBC). Five Neolithic individuals cannot be assigned to a particular Neolithic sub-period. Additional osteometric data on the Neolithic Scandinavian is from Bröste and Jørgensen (1956).

The Neolithic PWC foragers include individuals from the sites of Ajvide (N = 36), Västerbjers (N = 23), Fridtorp (N = 6), Ire (N = 4), Visby (N = 7) on the Baltic island of Gotland (on these sites see Molnar, 2008). These enigmatic foragers are treated separately here to perform comparisons with their Mesolithic predecessors and contemporaneous Neolithic farmers.

There is no usable Neolithic skeletal material from Finland. We can only assume that the Finnish representatives of various archaeological cultures (e.g., the PWC and the CWC of southern and southwestern Finland) were similar to representatives of the same cultures in Scandinavia.

The Iron Age sample (N = 65) represents the Roman Iron Age (0–400 AD) of Denmark. Both Early (the first two centuries AD) and Late Roman (the 3rd and 4th centuries AD) Iron Age individuals from many sites are included. Additional Iron Age data are from Sellevold *et al.* (1984).

A Late Viking Period sample (N = 22) from Nunnan Block cemetery of Sigtuna, Sweden (Kjellström, 2005), and one individual from Denmark represent Early Medieval Scandinavians. Late Medieval Scandinavians are represented by an urban sample from Humlegården in Sigtuna (Kjellström, 2005) and a rural sample from Westerhus in Jämtland, Sweden (Gejvall, 1960), which both date to 1100–1300 AD. Additional data on Medieval Scandinavians are from Arcini (1999), Arcini *et al.* (2014), and Kjellström (2005).

The Early Modern sample includes individuals from Porvoo (N = 22) and Renko (N = 30). The Porvoo sample represents the 17th century urban people from Porvoo, which is located on the south coast of Finland and is one of the oldest cities in Finland. Individuals in this sample were probably mostly Swedish-speakers and even descended from the Medieval Period Swedish immigrants. The Renko sample represents the 18th century rural people from Renko in southern Tavastia in the inland of southern Finland. These individuals were more likely Finnish-than Swedish-speakers. Dates of these two samples are based on unpublished radiocarbon dates provided by Markku Oinonen in 2015. Salo (2016) provides archaeological and historical information.

The Recent sample is an autopsy sample of Finns, who were born in 1840–1914 and died in 1915–1935 (on this sample, see Telkkä, 1950). Unfortunately, no vertebral columns were preserved and many individuals have either just upper or lower limb bones. Cadaveric statures are available for 50 individuals. These cadaveric statures converted to living statures (see Chapter 2) represent anatomical statures in this sample, which is biased toward southern Finland, especially Uusimaa Province. Both Swedish- and Finnish-speakers are probably represented.

Statures of the Early Modern and Recent Scandinavians and Finns from published literature are compared with estimated statures of skeletal individuals and to extend the time scale of stature comparisons to living people of this region. Stature and other anthropometric variables of Finns (males N = 25; females N = 25) represent living people of this region in tables and figures of this chapter. This sample is largely composed of individuals included in Ruff *et al.* (2005). Body masses in this sample shown in figures are not true body masses, but are estimated from stature and bi-iliac breadth to make these values more directly comparable with body mass estimates of skeletal samples. The same applies to body mass indices. Estimated and true body masses and body mass indices are provided in the tables. The sample used provides very similar mean values to published mean values computed for larger Scandinavian and

Finnish samples (see Appendix 3(b) and references therein). Raw data (anthropometric measurements of individuals) are also available for inclusion in scatter plots.

12.1.3 Technique Summary

All statistical analyses are performed using IBM SPSS Statistics 22. Body size and shape estimation methods are presented in Chapter 2. Only skeletal reconstructions of anthropometric-equivalent dimensions are used in this chapter to allow comparisons with anthropometric data. Published osteometric data are utilized to augment our osteometric data set, but all statures and body masses are estimated or re-estimated by using equations in Ruff *et al.* (2012). If femoral lengths are not provided in the reference, stature estimates given, for example, by Sjøvold's (1990) equation, are first converted to femoral lengths with a reversal of the regression equation: [(stature estimate – intercept/slope)] used in that particular study. After this, sex-specific femoral equations from Ruff *et al.* (2012) are applied to provide new stature estimates. This procedure accounts for technique differences and ensures that comparable values are being compared. Stature is emphasized in this chapter because it affects many other aspects of body size and shape (see Chapters 2 and 4).

Anthropometric variables are converted to z-scores (to mean of 0 and standard deviation of 1) using sex-specific means and standard deviations for some analyses as in Chapter 4. This transformation allows using pooled-sex samples and thus larger sample sizes because it eliminates male–female differences in size and shape.

Methods to quantify skeletal robusticity are presented in Chapter 3. In this chapter, we concentrate on size-standardized section moduli Z_{xstd} (bending strength in A-P plane), Z_{ystd} (bending strength in M-L plane), and Z_{pstd} (average bending strength), with an emphasis on the first two (Z_{xstd} and Z_{ystd}). These most critical and appropriate measures of mechanical performance are computed from the section moduli values Z_x, Z_y, and Z_p (see equations for converting second moments of area to section moduli in Chapter 3: Table 3.3) by dividing them by the product of body mass (kg) and bone length (mm). We multiplied the resulting value by 10^4 to avoid using too many decimal places. The ratio Z_x/Z_y is used to represent relative A-P/M-L bending strength.

Sexual dimorphism (SD) in anthropometric variables and long-bone shaft cross-sectional variables was calculated as follows:

$$\text{SD} = \left[\left(\text{male value} - \text{female value}\right) / \text{mid-sex mean}\right] \times 100.$$

Humeral bilateral asymmetry describes the magnitude of laterality and was calculated as follows:

$$\text{Bilateral asymmetry} = \left[\left(\text{Right} - \text{Left}\right) / \text{mean of right and left}\right] \times 100.$$

ANOVA is used (with either untransformed variables or z-scores as dependent variables) to determine if differences in body size and shape as well as cross-sectional variables of long bones between temporal groups and between the Fennoscandians and other Europeans are significant. *t*-tests are applied for inter-period comparisons to examine the significance of changes in particular variables between a more recent and an older temporal sample.

As in Chapter 4, many of the untransformed variables and some of the z-score transformed variables are plotted against time periods. Interpolation lines connect sample means in these plots. The PWC foragers are not included in these figures because their much larger sample size than that of the Neolithic farmers would have biased results. These Neolithic period foragers are treated separately due to their large sample size and being an enigma in Scandinavian prehistory.

Temporal trends in body size and shape within Scandinavia and Finland across all time periods are examined first, followed by comparisons with other Europeans as well as other comparisons dealing with the Neolithic period PWC foragers, urban–rural differences, and sexual dimorphism. Next, temporal trends in cross-sectional properties are examined following the same format, that is, temporal trends within Scandinavia and Finland, comparisons with other Europeans, and other comparisons.

12.2 Body Size and Body Shape

12.2.1 Temporal Trends

Mean reconstructed anthropometric variables are provided in Tables 12.2a and 12.2b. Stature estimates based on published long bone length data are presented in Table 12.3. Applying ANOVA using temporal period number as an independent variable indicates that all of these variables vary significantly over time, but stature, lower limb length, trunk length, crural index, and body mass exhibit the most variation (Table 12.4). Mean z-scores of anthropometric variables (sexes combined) are presented in Table 12.5. Significance of difference values between means of earlier and later periods are included between these mean values.

Stature data are plotted by temporal period and sex in Figure 12.2. Stature increased from the Mesolithic to Neolithic, remained high until the Early Medieval Period, but declined thereafter resulting in low statures during the Early Modern and Recent periods. It increased considerably from the Recent Period to the late 20th century (i.e., living sample). There were significant stature changes between Mesolithic and Neolithic, Early Medieval and Late Medieval periods, Late Medieval and Early Modern periods, and between Very Recent and Living periods (see Table 12.5). Temporal trends in stature are discussed in greater detail below.

Both Early Mesolithic Maglemosian individuals in our data set dated to ca. 9000 calBP are taller than Late Mesolithic Scandinavians (males 159.6 cm, N = 11; females 153.7 cm, N = 16). A Holmegården male's estimated stature is 175.5 cm, and a Koelbjerg female's 155.4 cm.

Early Mesolithic people from Olenii Ostrov (dated to around 6350 calBC; Price and Jacobs, 1990) in Russian Karelia were tall based on their mean long bone lengths in Jacobs (1993: their Table 6) and stature estimates (males 173.1 cm, N = 34; females 162.6 cm, N = 17) in Formicola and Giannecchini (1999: their Table 4a), computed using Jacobs's unpublished raw data. Equivalent mean statures estimated using Ruff *et al.* (2012) North European equations are a little higher (males 174.3 cm; females 165.3 cm), whereas estimates provided by South European equations are quite similar (males 172.5 cm; females 163.8 cm). Their contemporaries in Finland represented by the Suomusjärvi Culture (8300–5000 calBC) may have had similar statures due to cultural and environment similarities, as well as likely genetic relatedness.

The Mesolithic–Neolithic stature increase shown in Figure 12.2 is largely due to comparing mainly Late Mesolithic people with our total Neolithic sample. Statures of our Late Mesolithic individuals (see above) are similar to those of Early Neolithic Danes (males 161.4 cm, N = 23; females 151.5 cm, N = 10) in Table 12.3.

Stature increased during the Neolithic. Our TRB males (165.8 cm, N = 10) are much shorter than the CWC (172.5 cm, N = 7) and Late Neolithic (181.9 cm, N = 4) males. Males from unknown sub-periods are also tall (173.3 cm, N = 7).

Based on statures estimated from published mean femoral lengths (see Table 12.3), both male and female statures clearly increased during the Danish Neolithic. The Bronze Age and

Table 12.2(a) Reconstructed anthropometric dimensions of males. True body masses and body mass indices of living Finns are in parentheses.

Period		STA	ASTA	LSHT	LlimbL	LRSHT	BM	BMI	BAB	BIB	RBAB	RBIB	CI	RlegL	BI
Living (Finns)	N	25	25	25	25	25	25	25	25	25	25	25	–	–	–
	Mean	181.8	181.8	95.1	86.7	52.3	76.6 (79.4)	23.1 (24.0)	41.7	29.7	23.0	16.3	–	–	–
	SE	1.4	1.4	0.6	0.9	0.2	1.3 (1.8)	0.2 (0.5)	0.3	0.3	0.2	0.1	–	–	–
Very Recent	N	48	36	34	34	34	46	46	20	35	20	35	45	35	35
	Mean	167.0	166.6	88.2	78.9	52.7	67.0	24.1	40.1	28.6	24.0	17.2	80.8	22.0	74.2
	SE	0.9	1.0	0.6	0.7	0.3	1.0	0.3	0.6	0.3	0.3	0.2	0.4	0.2	0.4
Early Modern	N	28	13	13	13	13	27	27	19	14	19	14	24	13	20
	Mean	164.6	163.8	86.2	77.6	52.6	65.1	24.1	39.0	28.6	23.8	17.5	81.1	22.1	74.7
	SE	1.1	1.4	0.7	1.1	0.4	1.4	0.5	0.4	0.5	0.2	0.3	0.5	0.2	0.5
Late Medieval	N	95	67	67	67	67	95	95	76	80	76	80	88	67	77
	Mean	169.2	169.7	88.9	80.8	52.4	68.6	23.9	39.9	28.8	23.6	17.0	80.3	22.2	75.2
	SE	0.7	0.8	0.4	0.5	0.1	0.8	0.2	0.2	0.2	0.1	0.1	0.2	0.1	0.2
Early Medieval	N	19	7	7	7	7	18	18	12	5	10	5	14	7	9
	Mean	171.4	172.0	90.1	81.8	52.4	74.9	25.4	39.9	30.5	23.3	17.3	80.7	22.1	75.4
	SE	1.8	2.3	0.9	1.7	0.5	2.1	0.5	0.6	1.1	0.3	0.4	0.5	0.2	0.6
Iron Age	N	35	17	17	17	17	34	34	19	23	19	23	29	17	23
	Mean	171.9	170.8	88.3	82.5	51.7	74.5	25.2	39.6	30.7	23.4	17.8	82.2	22.8	76.2
	SE	1.2	1.8	0.9	1.0	0.2	1.5	0.4	0.6	0.5	0.2	0.3	0.4	0.1	0.4
Neolithic foragers (PWC)	N	46	33	33	33	33	46	46	35	28	35	28	44	33	27
	Mean	163.3	163.5	87.7	75.8	53.6	66.2	24.8	39.5	29.7	24.2	18.0	81.3	21.8	75.2
	SE	0.9	1.2	0.6	0.7	0.1	1.1	0.3	0.3	0.3	0.1	0.2	0.3	0.1	0.4
Neolithic farmers	N	26	11	11	11	11	25	25	18	14	18	14	21	11	17
	Mean	171.7	173.0	90.1	82.9	52.1	70.9	24.1	40.6	29.9	24.0	17.3	83.5	22.5	77.9
	SE	1.6	2.5	1.1	1.7	0.4	1.7	0.4	0.6	0.5	0.3	0.2	0.5	0.2	0.4
Mesolithic	N	12	6	6	6	6	11	10	3	3	3	3	9	6	7
	Mean	160.9	162.4	86.4	76.0	53.2	67.0	25.5	40.6	27.4	25.0	16.8	84.1	22.3	77.2
	SE	2.0	1.6	1.0	1.3	0.5	2.7	0.7	0.3	0.2	0.2	0.1	1.1	0.4	0.6

Table 12.2(b) Reconstructed anthropometric dimensions of females. True body masses and body mass indices of living Finns are in parentheses.

Period		STA	ASTA	LSHT	LlimbL	LRSHT	BM	BMI	BAB	BIB	RBAB	RBIB	CI	RlegL	BI
Living (Finns)	N	25	25	25	25	25	25	25	25	25	25	25	–	–	–
	Mean	166.5	166.5	87.8	78.7	52.7	62.3 (60.3)	22.4 (21.7)	36.8	28.2	22.1	17.0	–	–	–
	SE	1.4	1.4	0.6	0.9	0.2	1.3 (1.8)	0.1 (0.5)	0.3	0.3	0.2	0.1	–	–	–
Very Recent	N	22	14	14	14	14	21	21	6	10	6	10	20	14	8
	Mean	155.7	156.8	84.1	72.7	53.6	57.5	23.7	36.2	28.1	23.2	18.1	80.0	21.7	73.4
	SE	1.2	1.5	0.6	0.8	0.3	1.0	0.5	0.5	0.5	0.2	0.3	0.4	0.2	0.8
Early Modern	N	24	11	11	11	11	23	23	14	7	13	7	19	11	12
	Mean	156.5	155.8	83.8	72.0	53.8	56.7	23.0	35.9	27.3	23.2	17.9	82.0	21.9	72.7
	SE	1.3	1.8	0.7	1.3	0.4	1.2	0.4	0.5	1.0	0.4	0.7	0.5	0.2	0.6
Late Medieval	N	82	57	57	57	57	82	82	61	67	61	67	79	57	65
	Mean	158.0	158.2	85.1	73.1	53.8	57.1	22.8	35.6	27.8	22.5	17.6	80.8	21.7	74.0
	SE	0.6	0.7	0.3	0.5	0.2	0.6	0.1	0.2	0.2	0.1	0.1	0.2	0.1	0.3
Early Medieval	N	4	1	1	1	1	4	4	1	2	1	2	1	1	2
	Mean	165.2	160.9	83.3	77.6	51.8	61.9	22.5	34.1	27.3	21.4	16.9	85.3	23.1	73.1
	SE	3.9	–	–	–	–	5.3	0.8	–	0.1	–	0.3	–	–	1.4
Iron Age	N	30	19	19	19	19	29	29	18	22	18	22	28	19	18
	Mean	160.9	159.9	85.2	74.7	53.3	60.9	23.4	36.6	29.1	23.0	18.1	82.1	22.1	74.2
	SE	0.9	0.9	0.5	0.6	0.2	1.1	0.3	0.3	0.5	0.2	0.3	0.5	0.1	0.4
Neolithic foragers (PWC)	N	30	17	17	17	17	29	29	17	21	17	21	29	17	17
	Mean	155.9	156.2	85.6	70.6	54.8	56.3	23.1	35.3	27.7	22.4	17.8	81.5	21.4	74.2
	SE	0.9	1.1	0.7	0.6	0.2	0.8	0.2	0.4	0.3	0.2	0.2	0.4	0.1	0.4
Neolithic farmers	N	6	2	2	2	2	6	6	3	3	3	3	4	2	4
	Mean	157.9	154.8	82.4	72.4	53.2	57.3	23.0	36.3	28.7	23.1	18.2	82.3	22.1	76.0
	SE	1.6	2.5	1.1	1.7	0.4	1.7	0.8	1.0	0.2	0.5	0.3	0.6	0.3	0.7
Mesolithic	N	17	2	2	2	2	15	15	4	5	4	5	12	2	8
	Mean	153.8	152.4	82.8	69.6	54.3	59.2	25.0	35.4	28.5	23.0	18.7	83.1	21.6	73.9
	SE	0.9	3.5	2.0	1.5	0.1	1.3	0.5	0.5	0.9	0.3	0.6	0.8	0.3	0.6

Table 12.3 Statures computed from published data. Statures were estimated from femoral lengths in Neolithic samples and from both femoral and tibial lengths in Iron Age samples. In Medieval samples, statures were estimated directly from femoral length means, when possible, or femoral length means derived by back regressing statures estimated using a regression equation originally used to derive these estimates.

Sample	Date	Males	Females	Source[a]
The 14th and 15th c. Scandinavians	1300–1500 AD	168.2 (398)	155.2 (271)	1, 2
The 12th and 13th c. Scandinavians	1100–1300 AD	169.5 (365)	158.2 (275)	1, 2, 3
The late 11th c. Scandinavians	1050–1100 AD	167.8 (450)	156.4 (316)	1, 2, 3
Viking Period Scandinavians	800–1050 AD	169.2 (94)	156.7 (70)	2, 3, 4
Migration & Germanic Period Danes	400–800 AD	172.0 (7)	164.4 (4)	4
Late Roman Iron Age Danes	200–400 AD	174.3 (25)	162.1 (23)	4
Early Roman Iron Age Danes	0–200 AD	170.7 (50)	160.9 (32)	4
Pre-Roman Iron Age Danes	500–0 BC	172.5 (1)	155.8 & 160.7	4
Bronze Age Danes	1700–500 BC	170.2 (5)	158.0 & 173.1	5
Late Neolithic Danes	2350–1700 BC	173.2 (61)	161.8 (17)	5
Early Neolithic Danes	4000–2350 BC	161.4 (23)	153.7 (16)	5

[a] Data sources:

1. Arcini (1999).
2. Kjellström (2005).
3. Arcini *et al.* (2014).
4. Sellevold *et al.* (1984).
5. Bröste and Jørgensen (1956).

Table 12.4 ANOVA of anthropometric variables (z-scores) using period number as an independent variable. Paleolithic period individuals are excluded. Anthropometric variables are sorted by F ratios.

	F-ratio	Significance
STA	13.898	0.000
CI	10.478	0.000
LlimbL	10.132	0.000
LRSHT	8.437	0.000
BM	8.247	0.000
RlegL	7.970	0.000
BI	7.719	0.000
ASTA	6.939	0.000
BMI	6.477	0.000
BIB	4.918	0.000
RBIB	3.751	0.001
RBAB	2.482	0.017
LSHT	2.313	0.026
BAB	1.117	0.351

Table 12.5 Mean z-scores of anthropometric variables by period. Significance values refer to significance of difference between period and preceding period.

Period		STA	ASTA	LSHT	LlimbL	LRSHT	BM	BMI	BAB	BIB	RBAB	RBIB	CI	RlegL	BI
Living (Finns)	N	50	50	50	50	50	50	50	50	50	50	50	–	–	–
	Mean	1.91	1.95	1.63	1.63	−0.56	1.04	−0.44	0.81	0.25	−0.65	−0.91	–	–	–
	Sig.	0.000	0.000	0.000	0.000	0.011	0.000	0.000	0.015	0.032	0.000	0.000	–	–	–
Very Recent	N	70	50	48	48	48	67	67	29	45	48	45	65	49	43
	Mean	0.01	0.06	0.14	0.06	0.02	0.03	0.10	0.29	−0.11	0.26	−0.12	−0.46	−0.27	−0.55
	Sig.	0.352	0.156	0.107	0.373	0.838	0.318	0.482	0.185	0.612	0.871	0.385	0.045	0.417	0.814
Early Modern	N	52	24	24	24	24	50	50	33	21	32	21	43	24	32
	Mean	−0.14	−0.24	−0.30	−0.15	−0.04	−0.14	−0.04	−0.05	−0.24	0.21	0.13	−0.08	−0.04	−0.50
	Sig.	0.002	0.002	0.001	0.018	0.619	0.097	0.540	0.374	0.480	0.045	0.126	0.012	0.790	0.050
Late Medieval	N	177	124	124	124	124	177	177	137	147	137	147	167	124	142
	Mean	0.32	0.39	0.34	0.33	−0.13	0.12	−0.14	0.12	−0.09	−0.16	−0.31	−0.45	−0.10	−0.20
	Sig.	0.015	0.197	0.274	0.219	0.489	0.000	0.003	0.958	0.106	0.274	0.948	0.400	0.584	0.914
Early Medieval	N	23	8	8	8	8	22	22	13	7	11	7	15	8	11
	Mean	0.83	0.82	0.70	0.73	−0.33	0.93	0.40	0.11	0.48	−0.48	−0.29	−0.26	0.07	−0.23
	Sig.	0.748	0.513	0.250	0.907	0.429	0.643	0.867	0.551	0.608	0.149	0.198	0.101	0.094	0.243
Iron Age	N	65	36	36	36	36	63	63	37	45	37	45	57	36	41
	Mean	0.75	0.61	0.27	0.69	−0.58	0.80	0.36	0.29	0.73	−0.07	0.29	0.16	0.60	0.05
	Sig.	0.580	0.565	0.316	0.874	0.515	0.057	0.054	0.453	0.440	0.276	0.328	0.044	0.280	0.000
Neolithic farmers	N	32	13	13	13	13	31	31	21	17	21	17	25	13	21
	Mean	0.63	0.78	0.60	0.74	−0.42	0.38	−0.04	0.50	0.48	0.19	0.00	0.59	0.34	0.75
	Sig.	0.000	0.008	0.051	0.005	0.033	0.570	0.001	0.395	0.148	0.282	0.553	0.721	0.431	0.010
Mesolithic	N	29	8	8	8	8	26	25	7	8	7	8	21	8	15
	Mean	−0.60	−0.57	−0.35	−0.59	0.41	0.23	0.98	0.15	−0.13	0.63	0.23	0.70	−0.01	0.15

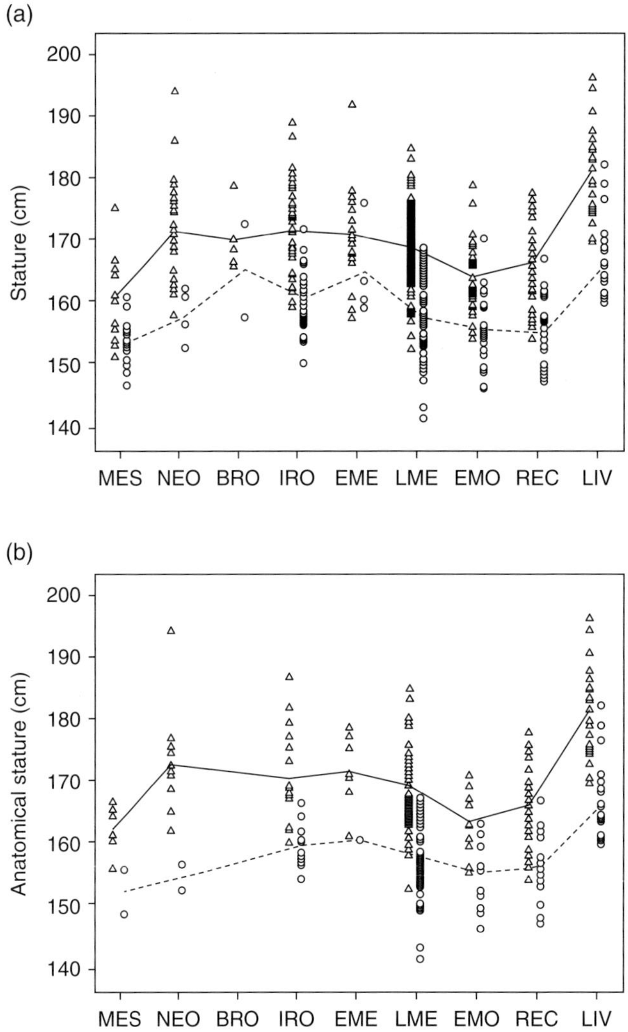

Figure 12.2 Stature. (a) Stature estimated anatomically or from long bone lengths; (b) anatomically estimated stature. Males: triangles and solid lines; females: circles and dashed lines.

the pre-Viking period Iron Age Danes were also tall; stature peaked during the Late Roman Iron Age (200–400 AD). There was some stature decline during the Late Iron Age (includes the Viking Age in Scandinavia) and the late 11th century, followed by a slight increase to the 12th and 13th centuries, and then a decline during the 14th and 15th centuries.

The Late Medieval stature decline continued during the Early Modern Period (Fig. 12.2). Our 17th century Finns from Porvoo were short-statured (males 162.7 cm, N = 11; females 154.7 cm, N = 11). Some recovery occurred by the following century because our 18th century Finns from Renko are clearly taller (males 165.9 cm, N = 17; females 157.9 cm, N = 13). Mean statures of soldiers corrected for a minimum stature requirement or estimated remaining growth (see Fig. 12.3 legend) reveals that a stature decline reoccurred during the later 18th century in Sweden (on this same trend in Finland, see Penttinen *et al.*, 2013), but there has been a considerable increase since the early 19th century to the end of the 20th century (Fig. 12.3).

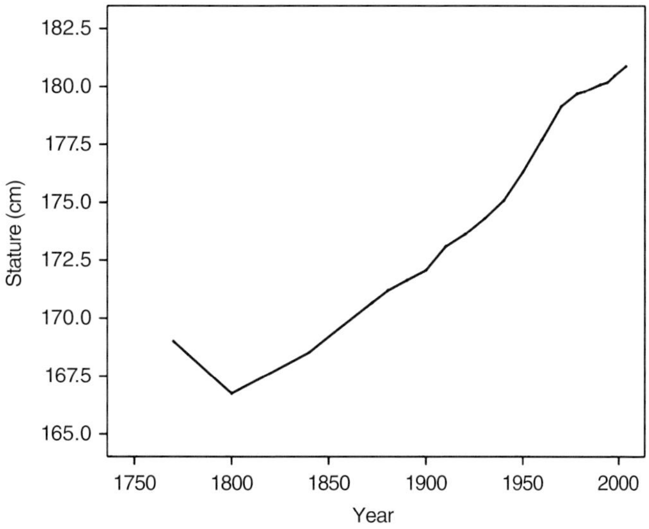

Figure 12.3 Mean statures of Swedish soldiers from the 18th century to current times, corrected for minimum stature requirement or estimated remaining growth. Data from Heintel *et al.* (1998: Figs 1 and 2), Werner (2007: Tables 2 and 4), and Öberg (2014a,b). Remaining growth is estimated using information from Kajava (1926), Hultkrantz (1927:38 referenced in Öberg, 2014b:33), Udjus (1964), Trotter and Gleser (1958:104), and Öberg (2014a,b).

Statures of our Early Modern and Very Recent Period Finns are similar and do not reflect the expected stature increase from the 18th century to the early 20th century (as shown in Fig. 12.3) because our Recent Finns are shorter than expected (males 167.0 cm, N = 48; females 155.7 cm). Measured mean living statures of adult Finnish men and women from the early 20th century are clearly higher (males 171.01 cm, N = 33,252; females 160.11 cm, N = 3181) (Kajava, 1926).

Stature increase commenced in Norway and Sweden apparently earlier than in Denmark and Finland. As a result, in 1921, the Norwegian and Swedish conscripts aged 20 years had taller mean statures (171.56 cm and 172.11 cm, respectively) than their Danish and Finnish counterparts (169.1 cm and 169.0 cm, respectively) (Udjus, 1964: their Table 3; Kajava, 1926).

The above stature differences have been considerably reduced and partly reversed in contemporary populations. Based on recent growth studies, Norwegian males and females average 181.0 cm and 167.2 cm (Júlíusson *et al.*, 2013), Swedish males and females 180.4 cm and 167.7 cm (Werner and Bodin, 2006), Danish males and females 180.4 cm and 169.5 cm (Tinggaard *et al.*, 2014), and Finnish males and females 180.7 cm and 167.2 cm (Saari *et al.*, 2011). The small stature difference between the sexes in Denmark is because males are represented by conscripts and females by a growth study sample. Conscript statures tend to be lower than growth study statures (M. Niskanen, personal observation).

There has been a very clear south-to-north declining stature gradient in Norway and Finland at least since the late 19th century (Kajava, 1926; Udjus, 1964; Dahlström, 1981). Inhabitants of Denmark and southern regions of Sweden and Norway are currently the tallest Scandinavians. The Danes have been the tallest Scandinavian conscripts since ca. 1980 (see Schmidt *et al.*, 1995: their Fig. 1; Larnkjær *et al.*, 2006: their Fig. 1). Late adolescent males (mean age 18.6 years) of Nordic origin from Gothenburg, Sweden, born in 1989–1991, averaged 181.4 cm (Sjöberg *et al.*, 2012), whereas Norwegian conscripts from Aust-Agder and Vest-Agder averaged 181.6 cm and 181.7 cm, respectively, in 2011 (Statistical Yearbook of Norway 2011: see Table 109).

These temporal stature changes affected other aspects of body size and shape. Variation in stature is due mainly to variation in lower limb length rather than trunk length, as indicated by relative sitting height decreasing as stature increases (see Fig. 12.4).

As expected, temporal changes in body mass generally follow stature changes (Fig. 12.5), and significant stature changes are generally accompanied by significant body mass changes (Table 12.5). Some temporal variation in body mass index is noticeable (Fig. 12.6). BMI reduced

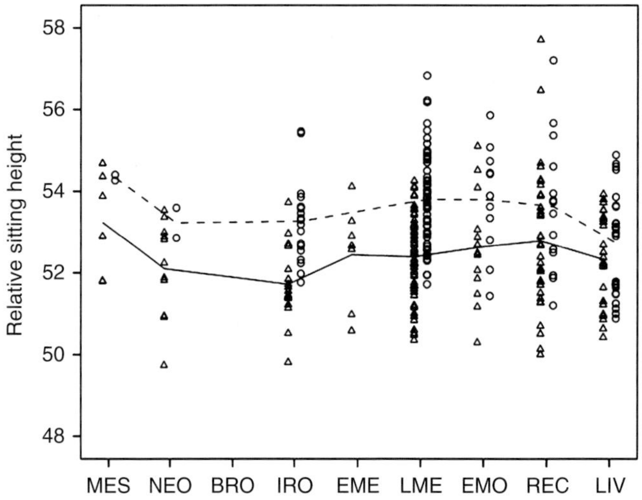

Figure 12.4 Relative sitting height. Males: triangles and solid lines; females: circles and dashed lines.

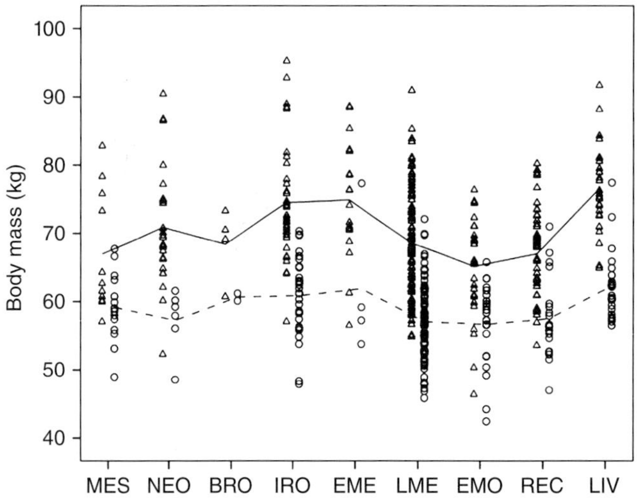

Figure 12.5 Body mass. Males: triangles and solid lines; females: circles and dashed lines.

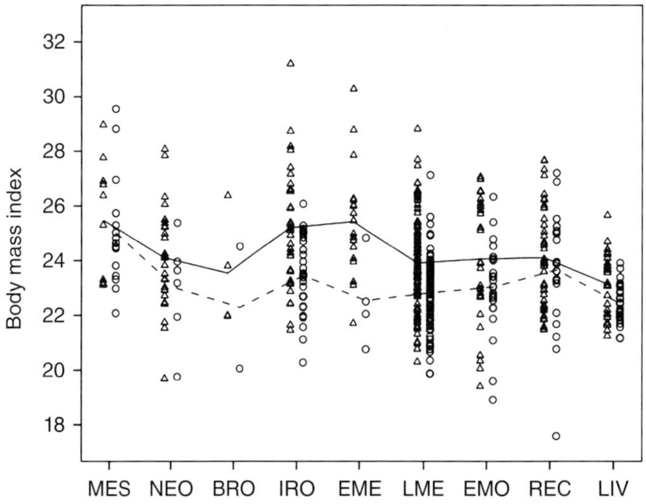

Figure 12.6 Body mass index. Males: triangles and solid lines; females: circles and dashed lines.

significantly across the Mesolithic–Neolithic transition, increased again to the Iron Age, but reduced thereafter. This reduction since the Iron Age was significant between Early and Late Medieval periods and between Very Recent and Living samples (Fig. 12.6; Table 12.5).

Biacromial breadth has been quite stable over time (Fig. 12.7a). Bi-iliac breadth exhibited more temporal variation than expected, but this is mainly due to the very broad bi-iliac breadths of the Roman Iron Age Danes (Fig. 12.7b). Changes in these absolute trunk breadths are significant only between Very Recent and Living samples (Table 12.5).

Relative trunk breadths exhibit a slight tendency for short-statured temporal samples to have relatively broad trunks and tall-statured ones relatively narrow ones, due to less variation in body breadths than in stature. (Fig. 12.8a,b). Relative biacromial breadth increased significantly

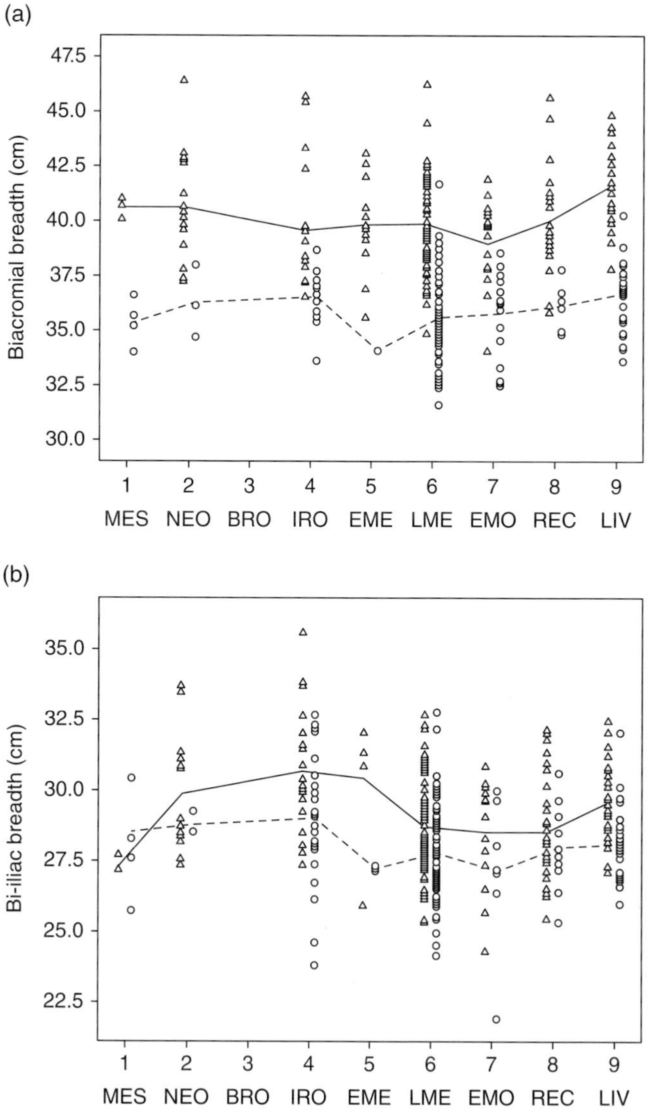

Figure 12.7 Trunk breadths. (a) Bi-acromial shoulder breadth; (b) bi-iliac breadth. Males: triangles and solid lines; females: circles and dashed lines.

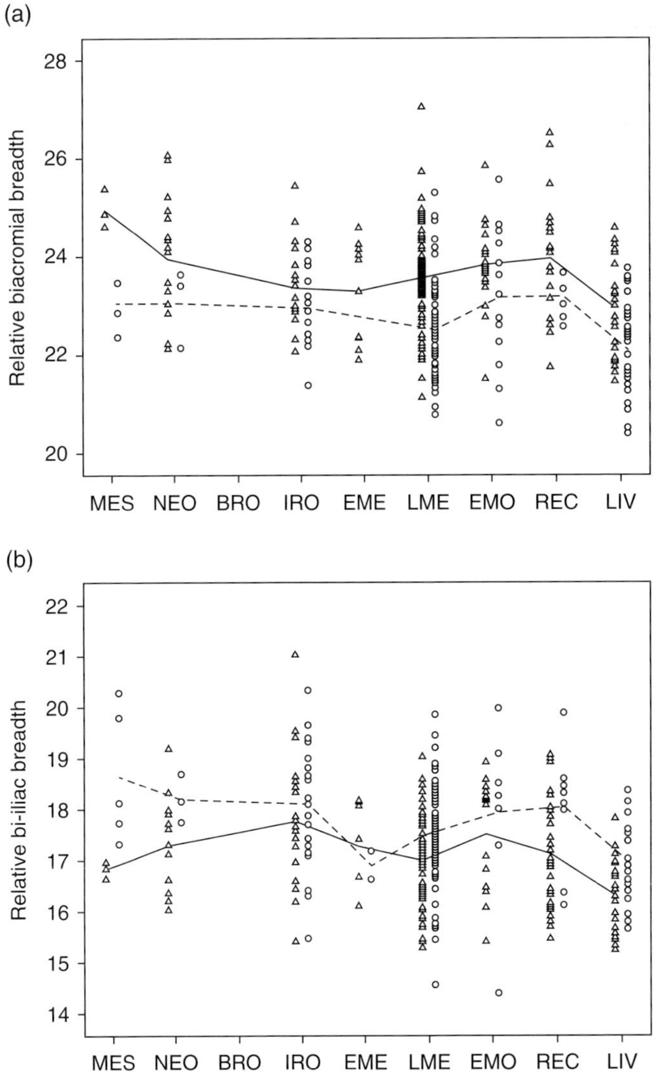

Figure 12.8 Relative trunk breadths. (a) Bi-acromial shoulder breadth relative to stature; (b) bi-iliac breadth relative to stature. Males: triangles and solid lines; females: circles and dashed lines.

between Late Medieval and Early Modern samples and reduced significantly between Very Recent and Living samples. The only significant relative bi-iliac breadth temporal change is its decline between Very Recent and Living samples (Table 12.5).

Tibial length relative to femoral length (crural index) reduced over time, as it did also elsewhere in Europe (Fig. 12.9a). Lower leg length relative to stature (Fig. 12.9b) increased from Mesolithic to Neolithic, remained quite high during the Roman Iron Age, and declined to the Medieval period and remained stable through Early Modern and Recent periods. The Recent period is represented by the Norwegian conscripts measured in the early 1960s (data from Udjus, 1964) and the Finnish recruits measured in the late 1970s (Dahlström, 1981). Interestingly, the Roman Iron Age Danes have relatively longer lower legs relative to stature than the Recent Norwegians Finns, but the Neolithic farmers are similar to living samples.

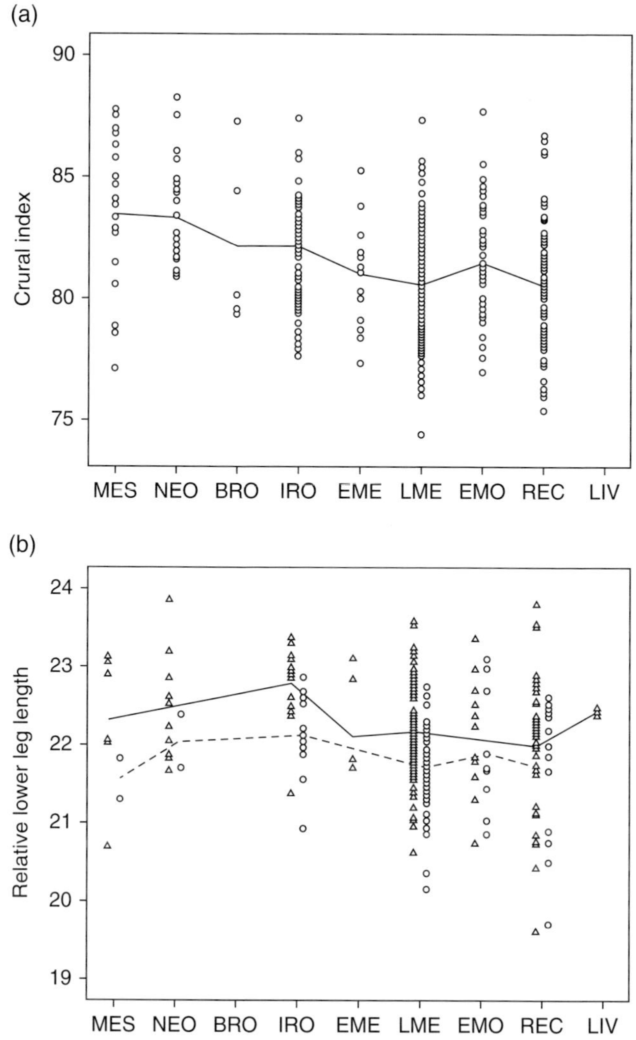

Figure 12.9 Relative tibial lengths. (a) Crural index; (b) relative lower leg length (lower leg length/stature). The Norwegian and Finnish conscript means (Udjus, 1964; Dahlström, 1981) represent relative lower leg length relative to stature of living males.

Temporal changes in crural index are significant between Neolithic and Iron Age periods, between Late Medieval and Early Modern periods, and between Early Modern and Very Recent periods. None of the temporal changes in lower leg length relative to stature is significant (Table 12.5).

Brachial index increased from Mesolithic to Neolithic and decreased thereafter (Fig. 12.10). These changes were significant between Mesolithic and Neolithic periods, Neolithic and Iron Age periods, and Early and Late Medieval periods (Table 12.5). The general temporal pattern is the same as in the pan-European material (Chapter 4).

12.2.2 Comparison with pan-European

ANOVA shows that the Scandinavians and Finns are significantly different from other Europeans by taller statures and longer trunks, heavier body masses, lower crural and brachial indices, absolutely longer lower limbs, broader shoulders, and higher body mass indices.

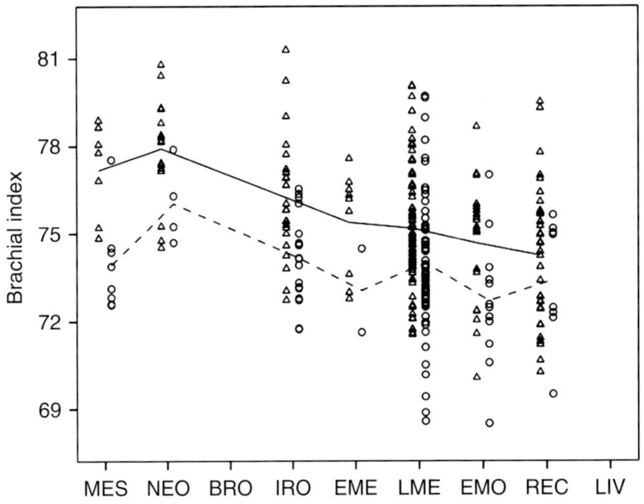

Figure 12.10 Brachial index. Males: triangles and solid lines; females: circles and dashed lines.

Differences in other anthropometric variables are small (Table 12.6(a,b)). There are, however, a few exceptions to this general pattern. For example, the Late Medieval Swedes have significantly narrower bi-iliac breadths (both absolutely and relatively) and lower body mass indices than other Late Medieval Europeans (see Table 12.6(b)); the Iron Age Danes have significantly higher crural indices, relative lower leg lengths, and relative sitting heights than other Iron Age Europeans (data not shown).

In general, stature followed by body mass differentiates the Scandinavians most consistently from other Europeans. Differences in stature, body mass, and body mass index between the Scandinavians and other Europeans are summarized in Figures 12.11, 12.12, and 12.13, respectively. These figures reveal that statures of the Mesolithic Scandinavians are similar to those of non-Scandinavian Mesolithic people, but their heavier body masses result in higher body mass indices. The post-Mesolithic Scandinavians are clearly taller and heavier than other Europeans, but this body size difference was temporarily reduced during the Late Medieval Period. Body mass index does not differentiate post-Mesolithic Scandinavians from other Europeans.

Although the Mesolithic Scandinavians have higher sitting heights than other Mesolithic Europeans, this was reversed during the Neolithic and Iron Age (no data from the Bronze Age), and there were few or no differences thereafter (Fig. 12.14). Lower relative sitting heights of the Neolithic and Iron Age Scandinavians (and presumably those of the Bronze Age Scandinavians) than those of their contemporaries are most probably due to taller average statures in Scandinavia during the periods in question.

Significantly lower crural indices for Scandinavians than for other Europeans based on ANOVA (Table 12.6(a,b)) is somewhat misleading. For instance, the Iron Age Danes had higher crural indices than their non-Scandinavian contemporaries (Fig. 12.15).

The taller stature of the Neolithic Scandinavians farmers in comparison to other Neolithic farmers shown in Figure 12.11 reflects Neolithic period stature increase. The earliest Neolithic farmers in Scandinavia are from the Middle Neolithic TRB period when statures had already increased from very low levels during Early Neolithic (see Chapter 4). The Scandinavian TRB farmer males actually had about the same stature as other Middle Neolithic Europeans (Scandinavian 165.8 cm, N = 10; non-Scandinavian 165.0 cm, N = 26). They were, however, heavier (Scandinavian 68.4 kg, N = 16; non-Scandinavian 61.9 kg, N = 26). We have only two Scandinavian females dated to this period. Statures and body masses estimated from long bone

Table 12.6(a) ANOVA of anthropometric variables (z-scores) of Scandinavians versus other Europeans. Paleolithic period individuals are excluded. Anthropometric variables are sorted by F-ratios.

		Scandinavia & Finland	Other Europe	F-ratio	Sig.
STA	Mean	0.18	−0.08	27.512	0.000
	N	524	1559		
BM	Mean	0.19	−0.07	26.647	0.000
	N	511	1508		
LSHT	Mean	0.23	−0.11	25.692	0.000
	N	311	778		
ASTA	Mean	0.21	−0.11	23.766	0.000
	N	313	778		
CI	Mean	−0.18	0.03	15.864	0.000
	N	466	1317		
LlimbL	Mean	0.16	−0.08	13.610	0.000
	N	311	778		
BAB	Mean	0.13	−0.07	9.973	0.002
	N	326	899		
BI	Mean	−0.17	0.02	9.856	0.002
	N	349	1056		
BMI	Mean	0.09	−0.03	5.531	0.019
	N	510	1488		
BIB	Mean	0.08	−0.03	3.270	0.071
	N	339	844		
RBIB	Mean	−0.06	0.03	1.872	0.171
	N	339	839		
RlegL	Mean	−0.06	−0.01	0.521	0.471
	N	263	770		
LRSHT	Mean	−0.02	0.02	0.420	0.517
	N	311	778		
RBAB	Mean	0.00	0.00	0.013	0.909
	N	323	886		

lengths and femoral head breadths reported in Bröste and Jørgensen (1956: their Table 103) using Ruff *et al.* (2012) equations, however, indicate shorter mean statures and lower body masses for a larger sample of the presumably TRB period Scandinavians. Males and females would have averaged 161.4 cm (N = 23) and 151.5 cm (N = 10), respectively, in height, and 64.6 kg (N = 18) and 56.0 kg (N = 9) in body mass.

The CWC and Late Neolithic Scandinavians were significantly taller and heavier than other Copper Age Europeans. This is not due to sampling because long bone lengths and femoral head breadths in Bröste and Jørgensen (1956: their Table 103) also provide high statures (males 173.2 cm, N = 61; females 161.8 cm, N = 17) and body masses (males 73.0 kg, N = 52; females 58.1 kg, N = 14) for larger numbers of individuals. For comparison, the non-Scandinavian North/Central European Copper Age males' stature and body mass means are 169.3 cm (N = 142) and 67.3 kg (N = 142), respectively.

Current Scandinavians and Finns have similar mean statures as East Baltic people and many Central Europeans. They are taller than most South Europeans but, except for South

Table 12.6(b) ANOVA of anthropometric variables (z-scores) of Late Medieval Scandinavians vs. other Late Medieval Europeans. Anthropometric variables are sorted by F-ratios.

		Scandinavia & Finland	Other Europe	F-ratio	Sig.
RBIB	Mean	−0.31	0.16	22.075	0.000
	N	147	249		
BMI	Mean	−0.14	0.19	16.123	0.000
	N	177	342		
BIB	Mean	−0.09	0.22	9.629	0.002
	N	147	252		
CI	Mean	−0.45	−0.21	8.059	0.005
	N	167	285		
LSHT	Mean	0.34	0.06	6.953	0.009
	N	124	196		
STA	Mean	0.32	0.12	5.052	0.025
	N	177	354		
BI	Mean	−0.20	−0.38	3.803	0.052
	N	142	258		
ASTA	Mean	0.39	0.19	3.671	0.056
	N	124	196		
RlegL	Mean	−0.10	0.08	2.182	0.141
	N	124	195		
BM	Mean	0.12	0.23	1.669	0.197
	N	177	345		
BAB	Mean	0.12	−0.02	1.606	0.206
	N	137	227		
LlimbL	Mean	0.33	0.22	0.890	0.346
	N	124	196		
LRSHT	Mean	−0.13	−0.21	0.347	0.556
	N	124	196		
RBAB	Mean	−0.16	−0.13	0.086	0.770
	N	137	778		

Scandinavians, not as tall as the Dutch and inhabitants of the Northwestern Balkan Peninsula. Body proportions do not differentiate the Scandinavians and Finns from other North/Central Europeans (see Appendix 3(b)).

12.2.3 Other Comparisons

The PWC are a special case because, although foragers, they lived during the Scandinavian Middle Neolithic and were contemporary with Neolithic farmers. They are compared to preceding Mesolithic and contemporary farming Neolithic samples from Scandinavia in Table 12.7.

The PWC foragers differ from Mesolithic Scandinavians by lower body mass indices and shorter tibiae relative to both femora and stature. They differ from their farmer contemporaries in Scandinavia by their shorter statures, absolutely and relatively shorter limbs (especially distal limb segments), absolutely narrower shoulders, and lower body masses (Table 12.7).

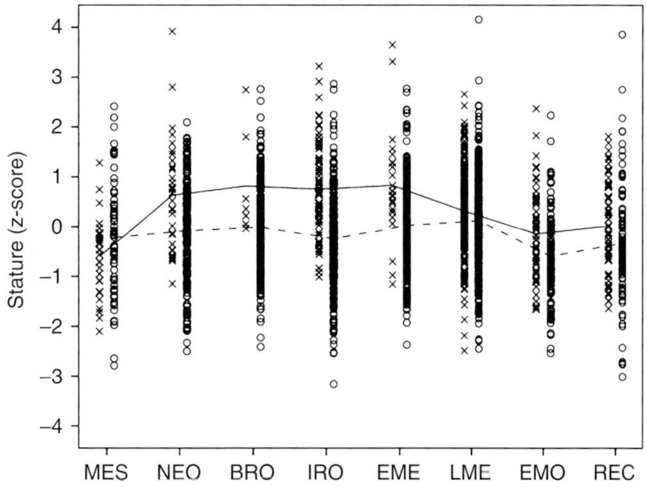

Figure 12.11 Scandinavian (crosses and solid line) versus other European (circles and dashed line) stature (z-score).

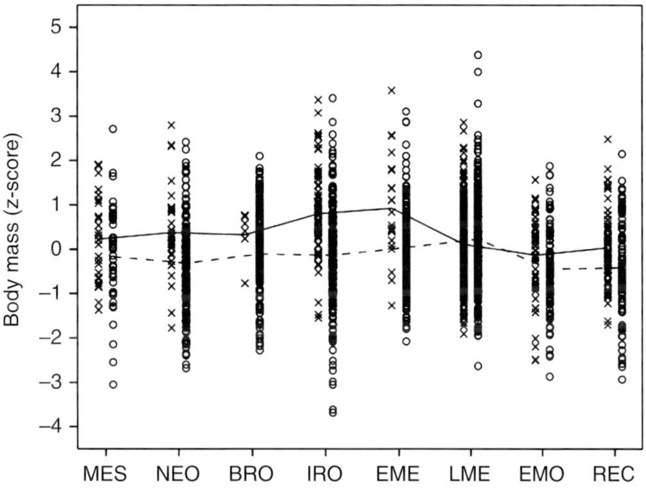

Figure 12.12 Scandinavian (crosses and solid line) versus other European (circles and dashed line) body mass (z-score).

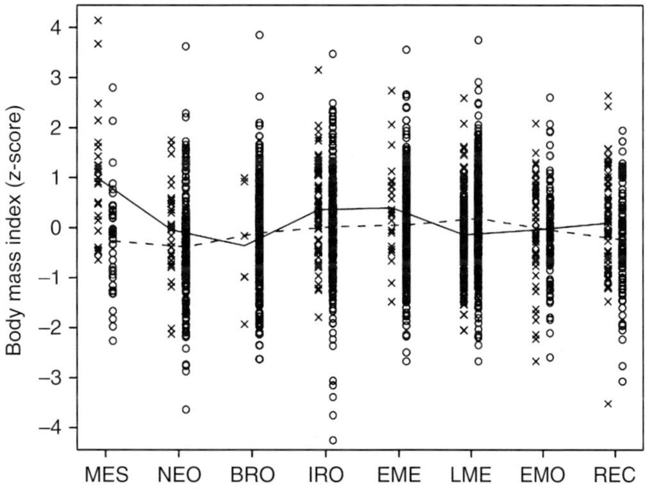

Figure 12.13 Scandinavian (crosses and solid line) versus other European (circles and dashed line) body mass index (z-score).

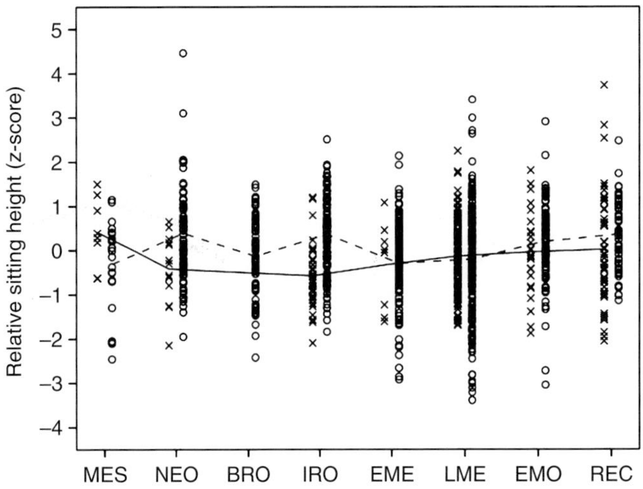

Figure 12.14 Scandinavian (crosses and solid line) versus other European (circles and dashed line) relative sitting height (z-score).

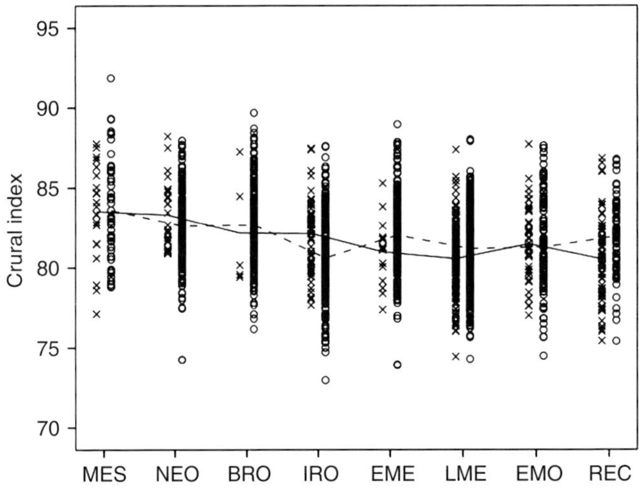

Figure 12.15 Scandinavian (crosses and solid line) versus other European (circles and dashed line) crural index.

There were apparently no urban–rural differences in Medieval Sweden based on comparing skeletal samples from Westerhus (representing rural people) and Sigtuna (representing urban people), but there were in Early Modern Finland. Urban people from Porvoo were shorter (males 162.7 cm, N = 11; females 154.7 cm, N = 11) than rural people from Renko (males 165.9 cm, N = 17; females 157.9 cm, N = 13). However, this finding may be partly due to the fact that the Porvoo sample is from the 17th century and the Renko sample from the 18th century, when statures were higher (see above and Chapter 4).

The pattern of sexual dimorphism in anthropometric dimensions is as expected (Table 12.8), with generally larger absolute dimensions in males, along with relatively wider shoulders and higher brachial indices, and females with relatively larger bi-iliac breadths. Tall-statured temporal samples also tend to exhibit more dimorphism in lower limb length than short-statured ones, possibly due to so-called ‘female buffering’ during growth (see Chapter 6). For example, the PWC and Late Medieval females have similar statures, but the PWC males are much shorter than the Late Medieval males, leading to reduced sexual dimorphism in body size

Table 12.7 Differences in anthropometric dimensions (z-scores) of Mesolithic, Pitted Ware (PWC), and Neolithic farmer samples. Significance values refer to significance of difference between Mesolithic and Pitted Ware and Pitted Ware and Neolithic. Lower figures are sample sizes.

		Mesolithic	Sig.	PWC	Sig.	Neolithic
STA	Mean	−0.60	0.112	−0.31	0.000	0.63
	N	29		76		32
ASTA	Mean	−0.57	0.353	−0.26	0.001	0.78
	N	8		50		13
LSHT	Mean	−0.35	0.151	0.19	0.219	0.60
	N	8		50		13
LlimbL	Mean	−0.59	0.818	−0.53	0.000	0.74
	N	8		50		13
LRSHT	Mean	0.41	0.228	0.72	0.000	−0.42
	N	8		50		13
BM	Mean	0.23	0.098	−0.10	0.013	0.38
	N	26		75		31
BMI	Mean	0.98	0.000	0.17	0.261	−0.04
	N	25		75		31
BAB	Mean	0.15	0.559	−0.05	0.024	0.50
	N	7		52		21
BIB	Mean	−0.13	0.418	0.14	0.182	0.48
	N	8		49		17
RBAB	Mean	0.63	0.190	0.17	0.938	0.19
	N	7		52		21
RBIB	Mean	0.23	0.829	0.31	0.240	0.00
	N	8		49		17
CI	Mean	0.70	0.000	−0.13	0.000	0.59
	N	21		73		25
RlegL	Mean	−0.01	0.043	−0.59	0.000	0.34
	N	8		50		13
BI	Mean	0.15	0.120	−0.17	0.000	0.75
	N	15		44		21

(Table 12.8). A higher frequency of enamel hypoplasias in our PWC sample from Ajvide than in our Late Medieval sample from Humlegården of Sigtuna suggests poorer childhood nutrition and/or health, at least in this sample of PWC foragers (Molnar, 2008: Paper IV and references therein), which apparently had more effect on males than on females.

12.3 Cross-Sectional Properties

12.3.1 Temporal Trends

There was some reduction in lower limb and upper limb bending strengths across the Mesolithic–Neolithic transition, and significant reductions from the Neolithic to the Iron Age. There were significant increases from the Iron Age to the Medieval Period and relatively little change thereafter (Tables 12.9 and 12.10; Figs. 12.16 and 12.17).

Table 12.8 Sexual dimorphism in anthropometric variables. Higher figures are sexual dimorphism (SD) values. Lower figures are sex-specific sample sizes (number of males/number of females).

Period	STA	ASTA	LSHT	LlimbL	RSHT	BM	BMI	BAB	BIB	RBAB	RBIB	CI	RlegL	BI
Living (Finns)	8.8 25/25	8.8 25/25	8.0 25/25	9.7 25/25	−0.8 25/25	20.6 25/25	3.1 25/25	12.7 25/25	4.9 25/25	3.8 25/25	−3.9 25/25	– –	– –	– –
Very Recent	7.0 48/22	6.1 36/14	4.7 34/14	8.2 34/14	−1.6 34/14	15.3 46/21	1.9 46/21	10.2 20/6	1.9 35/10	3.4 20/6	−5.2 35/10	1.0 45/20	1.4 35/14	1.1 35/8
Early Modern	5.1 28/24	5.0 13/11	2.9 13/11	7.5 13/11	−2.2 13/11	13.9 27/23	4.3 27/23	8.5 19/14	4.8 14/7	2.8 19/13	−2.4 14/7	−1.1 24/19	0.0 13/11	2.7 20/12
Late Medieval	6.9 95/82	7.0 67/57	4.4 67/57	10.0 67/57	−2.6 67/57	18.4 95/82	4.7 95/82	11.4 76/61	3.4 80/67	4.5 76/61	−3.4 80/67	−0.6 88/79	2.0 67/57	1.6 77/67
Early Medieval	3.7 19/4	6.6 7/1	7.8 7/1	5.3 7/1	1.3 7/1	19.0 18/4	12.0 18/4	15.5 12/1	10.9 5/2	8.5 10/1	2.2 5/2	−5.6 14/1	−4.3 7/1	3.1 9/2
Iron Age	6.6 35/30	6.6 17/19	3.6 17/19	9.8 17/19	−2.9 17/19	20.2 34/29	7.3 34/29	7.9 19/18	5.6 23/22	1.7 19/18	−2.0 17/19	0.1 29/28	3.0 17/19	2.8 23/18
Neolithic foragers	4.6 46/30	4.6 33/17	2.4 33/17	7.1 33/17	−2.1 33/17	16.3 46/29	7.0 46/29	11.2 35/17	6.8 28/21	7.7 35/17	1.3 28/21	−0.3 44/29	1.9 33/17	1.3 27/17
Neolithic farmers	8.3 26/6	11.1 11/2	8.9 11/2	13.6 11/2	−2.1 11/2	21.3 25/6	4.8 25/6	11.3 18/3	3.6 14/3	3.8 18/3	13.6 11/2	−5.0 14/3	1.5 21/4	1.9 11/2
Mesolithic	4.5 12/17	6.4 6/2	4.3 6/2	8.8 6/2	−2.0 6/2	12.4 11/15	1.8 10/15	13.8 3/4	−4.1 3/5	8.0 3/4	−10.3 3/5	1.1 9/12	3.4 6/2	4.3 7/8

Table 12.9(a) Temporal trends in cross-sectional properties of femur relative to body size in Scandinavia and Finland. Values are size-standardized section moduli. Mean and standard error of mean (SE) are provided.

			FZ_{xstd}		FZ_{ystd}		FZ_{pstd}		Z_x/Z_y	
		N	Mean	SE	Mean	SE	Mean	SE	Mean	SE
Very Recent	Males	44	561.3	12.2	640.8	12.4	1131.2	22.3	87.9	1.4
	Females	20	498.3	14.5	578.2	22.6	1021.8	37.8	87.3	2.3
Early Modern	Males	26	589.9	10.4	659.9	16.0	1167.5	26.3	90.1	1.7
	Females	17	517.7	17.3	607.0	17.4	1049.0	28.0	85.9	2.8
Late Medieval	Males	30	551.1	16.3	622.0	13.0	1098.4	27.2	88.7	2.0
	Females	23	474.6	16.2	564.5	16.1	985.4	29.0	84.2	1.9
Iron Age	Males	33	439.0	15.1	530.0	18.0	913.6	29.7	83.7	2.1
	Females	27	422.1	8.8	549.0	13.8	911.2	20.6	77.5	1.7
Neolithic foragers	Males	44	662.2	16.4	729.4	17.8	1298.9	29.6	91.3	1.6
	Females	26	628.3	15.9	694.4	15.0	1224.0	25.7	90.6	1.4
Neolithic farmers	Males	20	645.2	22.7	670.0	27.6	1218.1	46.8	97.4	2.5
	Females	5	518.2	37.1	635.2	35.3	1099.4	63.6	81.7	4.9
Mesolithic	Males	8	693.1	35.1	704.5	34.4	1304.3	79.8	98.7	3.2
	Females	5	607.7	54.9	625.7	45.5	1153.7	82.0	97.4	6.1

Table 12.9(b) Temporal trends in bending strength of tibia relative to body size in Scandinavia and Finland. Values are size-standardized section moduli. Mean and standard error of mean (SE) are provided.

			TZ_{xstd}		TZ_{ystd}		TZ_{pstd}		Z_x/Z_y	
		N	Mean	SE	Mean	SE	Mean	SE	Mean	SE
Very Recent	Males	45	586.4	12.7	472.4	11.6	920.6	21.0	125.3	2.2
	Females	21	525.0	20.2	416.7	18.5	816.5	34.6	127.4	3.4
Early Modern	Males	24	607.3	17.9	540.9	15.8	965.9	30.2	112.9	2.5
	Females	20	519.5	21.6	449.2	15.2	812.0	26.8	115.9	3.2
Late Medieval	Males	26	621.2	28.8	474.6	23.7	950.7	42.1	132.6	3.7
	Females	21	482.3	18.3	389.4	12.9	754.4	26.8	124.2	3.2
Iron Age	Males	23	480.8	22.4	349.2	15.7	710.0	31.2	138.0	2.6
	Females	25	384.3	10.1	292.8	9.5	503.7	17.0	133.0	3.7
Neolithic foragers	Males	43	790.0	22.0	567.7	18.8	1150.3	29.9	141.5	1.7
	Females	22	636.1	23.6	488.4	16.3	967.0	30.3	131.0	3.7
Neolithic farmers	Males	16	667.0	25.7	467.2	19.5	942.1	30.0	145.2	6.6
	Females	4	588.1	35.4	386.5	17.1	838.6	34.8	153.1	11.7
Mesolithic	Males	4	770.2	70.7	542.1	88.1	1105.9	139.6	150.3	20.1
	Females	2	654.1	119.1	411.7	102.2	875.3	161.3	161.7	11.2

Table 12.9(c) Temporal trends in cross-sectional properties of right humerus relative to body size in Scandinavia and Finland. Values are size-standardized section moduli. Mean and standard error of mean (SE) are provided.

			HRZ_{xstd}		HRZ_{ystd}		HRZ_{pstd}		Z_x/Z_y	
		N	Mean	SE	Mean	SE	Mean	SE	Mean	SE
Very Recent	Males	21	400.6	16.2	358.1	13.7	710.9	26.1	112.2	2.2
	Females	8	290.2	23.8	243.9	18.5	490.1	37.5	119.1	4.0
Early Modern	Males	24	367.3	9.0	317.2	9.8	635.5	16.8	116.6	1.8
	Females	15	292.1	16.8	255.2	11.8	497.8	26.8	113.9	3.7
Late Medieval	Males	23	300.9	12.1	292.0	13.1	555.3	22.8	103.9	2.2
	Females	14	210.6	9.7	195.9	6.6	384.5	16.1	107.2	2.6
Iron Age	Males	17	264.4	12.4	264.0	12.6	495.3	22.8	100.6	1.9
	Females	22	226.7	7.2	226.6	8.9	423.4	14.7	101.1	2.2
Neolithic foragers	Males	31	382.1	14.1	386.5	13.5	717.8	23.5	99.0	1.7
	Females	15	307.2	14.4	277.3	14.1	553.8	26.5	111.6	3.1
Neolithic farmers	Males	17	369.0	20.9	356.4	20.2	674.3	38.7	104.0	2.2
	Females	3	289.0	20.8	258.3	20.6	512.7	34.9	112.3	5.6
Mesolithic	Males	4	392.4	55.6	440.2	56.4	764.7	102.3	88.8	1.7
	Females	3	412.7	82.3	372.4	65.7	734.4	143.1	109.7	3.4

Table 12.9(d) Temporal trends in cross-sectional properties of left humerus relative to body size in Scandinavia and Finland. Values are size-standardized section moduli. Mean and standard error of mean (SE) are provided.

			HLZ_{xstd}		HLZ_{ystd}		HLZ_{pstd}		Z_x/Z_y	
		N	Mean	SE	Mean	SE	Mean	SE	Mean	SE
Very Recent	Males	26	354.7	11.1	319.7	9.4	605.4	17.5	111.1	1.5
	Females	7	279.5	25.0	249.1	15.2	460.5	36.8	111.3	3.8
Early Modern	Males	22	364.5	10.5	328.2	11.3	610.5	18.6	111.7	1.7
	Females	15	293.9	13.4	263.8	11.8	482.5	21.3	111.9	2.9
Late Medieval	Males	22	290.8	9.2	255.9	8.6	521.2	16.5	114.1	2.3
	Females	14	229.1	12.5	204.6	9.0	411.3	18.6	111.8	2.6
Iron Age	Males	18	234.0	11.1	218.1	8.4	425.1	16.5	107.6	3.1
	Females	17	226.7	8.3	212.7	6.0	408.4	12.7	106.4	2.5
Neolithic foragers	Males	27	342.5	13.1	336.6	12.6	624.8	22.6	102.0	1.5
	Females	17	292.6	13.9	295.7	13.5	539.1	21.4	99.4	2.5
Neolithic farmers	Males	16	348.2	19.8	324.9	20.1	622.7	35.7	107.9	2.5
	Females	4	325.2	45.9	327.0	58.9	607.9	105.0	102.1	5.7
Mesolithic	Males	3	334.3	49.8	371.6	64.5	667.2	115.7	90.8	2.9
	Females	4	328.9	68.5	315.2	45.0	610.1	98.3	101.2	8.5

Table 12.10(a) Temporal trends in bending strength of males relative to body size in Scandinavia and Finland. Cross-sectional properties are standardized by body mass and bone length. F: femur; T: tibia; HR: Right humerus; HL: left humerus. Significance values refer to significance of difference between earlier and later periods.

Period		FZ_{xstd}	FZ_{ystd}	TZ_{xstd}	TZ_{ystd}	HRZ_{xstd}	HRZ_{ystd}	HLZ_{xstd}	HLZ_{ystd}
Very Recent	N	44	44	45	45	21	21	26	26
	Mean	561.3	640.8	586.4	472.4	40.6	358.1	354.7	319.7
	Sig.	0.112	0.349	0.342	0.001	0.081	0.017	0.528	0.560
Early Modern	N	26	26	24	24	24	24	22	22
	Mean	589.9	659.9	607.3	540.9	367.3	317.2	364.5	328.2
	Sig.	0.050	0.070	0.690	0.027	0.000	0.129	0.000	0.000
Late Medieval	N	30	30	26	26	23	23	22	22
	Mean	551.1	622.0	621.2	474.6	300.9	292.0	290.8	255.9
	Sig.	0.000	0.001	0.000	0.000	0.046	0.143	0.000	0.004
Iron Age	N	33	33	23	23	17	17	18	18
	Mean	439.0	530.0	480.8	349.2	264.4	264.0	234.0	218.1
	Sig.	0.000	0.000	0.000	0.000	0.000	0.000	0.000	0.000
Neolithic Farmers	N	20	20	16	16	17	17	16	16
	Mean	645.2	670.0	667.0	467.2	369.0	356.4	348.2	324.9
	Sig.	0.267	0.488	0.111	0.463	0.648	0.106	0.786	0.393
Mesolithic	N	8	8	4	4	4	4	3	3
	Mean	693.1	704.5	770.1	542.1	392.4	440.1	334.3	371.6

Table 12.10(b) Temporal trends in bending strength of females relative to body size in Scandinavia and Finland. Cross-sectional properties are standardized by body mass and bone length. F: femur; T: tibia; HR: Right humerus; HL: left humerus. Significance values refer to significance of difference between earlier and later periods.

Period		FZ_{xstd}	FZ_{ystd}	TZ_{xstd}	TZ_{ystd}	HRZ_{xstd}	HRZ_{ystd}	HLZ_{xstd}	HLZ_{ystd}
Very	N	20	20	21	21	8	8	7	7
Recent	Mean	498.3	578.2	525.0	416.7	290.2	243.9	279.5	249.1
	Sig.	0.393	0.334	0.853	0.186	0.947	0.595	0.584	0.474
Early	N	17	17	20	20	15	15	15	15
Modern	Mean	517.7	607.0	519.5	449.2	292.1	255.2	293.9	263.8
	Sig.	0.081	0.084	0.195	0.005	0.000	0.000	0.000	0.000
Late	N	23	23	21	21	13	13	14	14
Medieval	Mean	474.6	564.5	482.3	389.4	210.6	195.9	229.1	204.6
	Sig.	0.007	0.464	0.000	0.000	0.184	0.019	0.847	0.442
Iron Age	N	27	27	25	25	22	22	17	17
	Mean	422.1	549.0	384.3	292.8	226.7	226.6	226.3	212.7
	Sig.	0.001	0.021	0.000	0.001	0.007	0.227	0.002	0.001
Neolithic farmers	N	5	5	4	4	3	3	4	4
	Mean	518.2	635.2	588.1	386.5	289.0	258.3	325.1	327.0
	Sig.	0.214	0.873	0.678	0.847	0.219	0.173	0.966	0.880
Mesolithic	N	5	5	2	2	3	3	4	4
	Mean	607.7	625.7	654.1	411.7	412.7	372.4	328.9	315.3

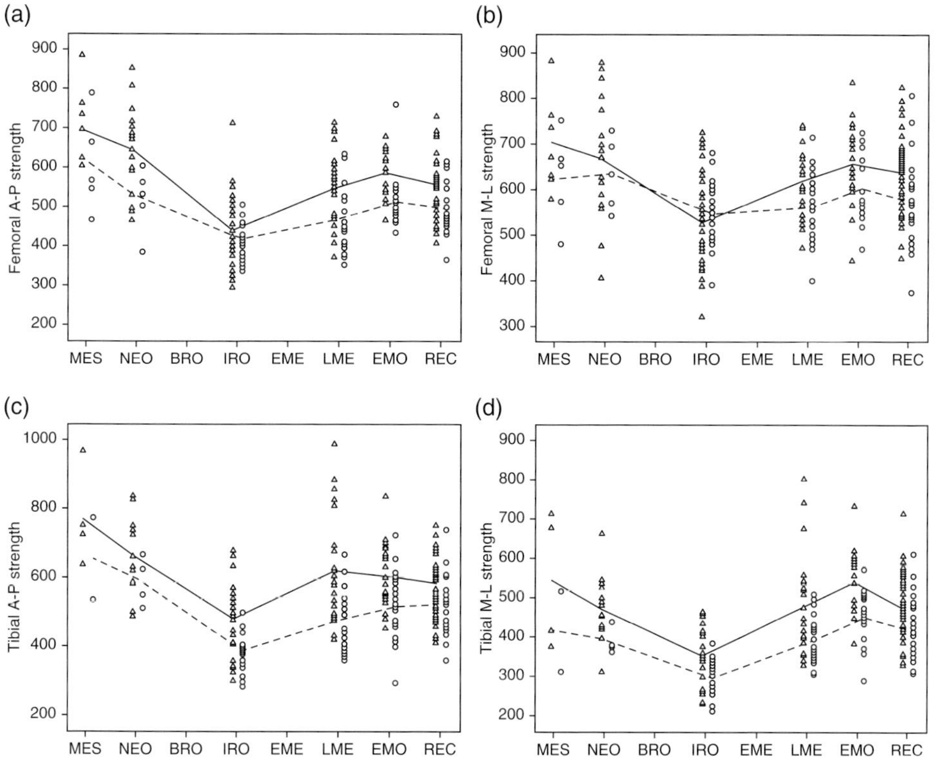

Figure 12.16 Temporal trends in size-standardized bending strengths of lower limb long bones. (a) Femoral A-P strength; (b) femoral M-L strength; (c) tibial A-P strength; (d) tibial M-L strength. Males: triangles and solid lines; females: circles and dashed lines.

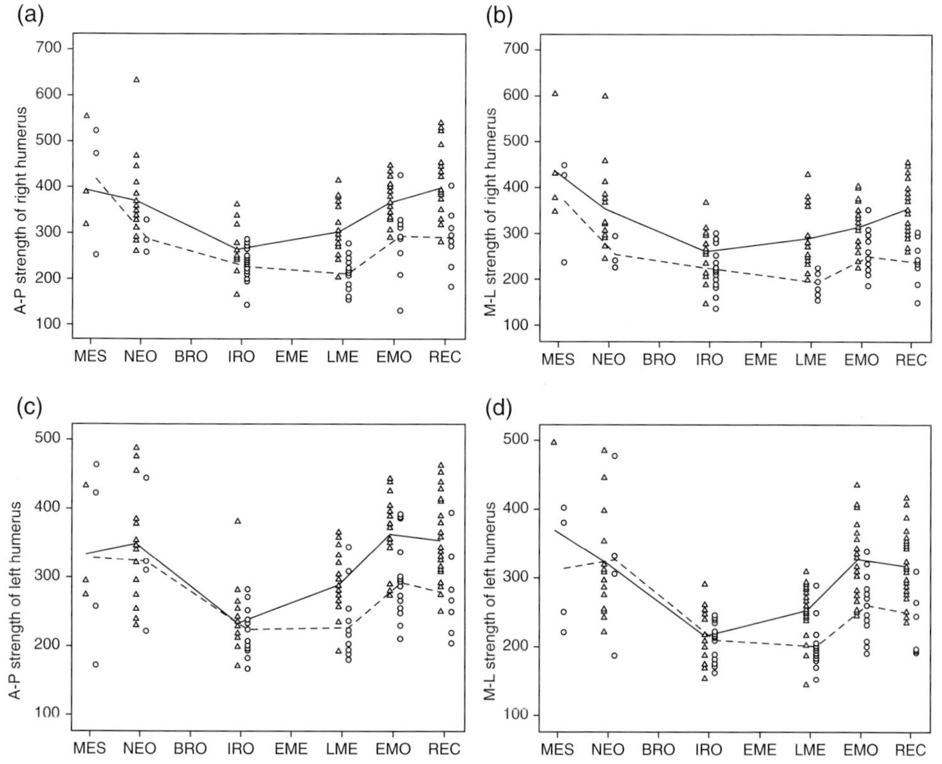

Figure 12.17 Temporal trends in size-standardized bending strengths of right and left humerus. (a) A-P strength of right humerus; (b) M-L strength of right humerus; (c) A-P strength of left humerus; (d) M-L strengths of left humerus. Males: triangles and solid lines; females: circles and dashed lines.

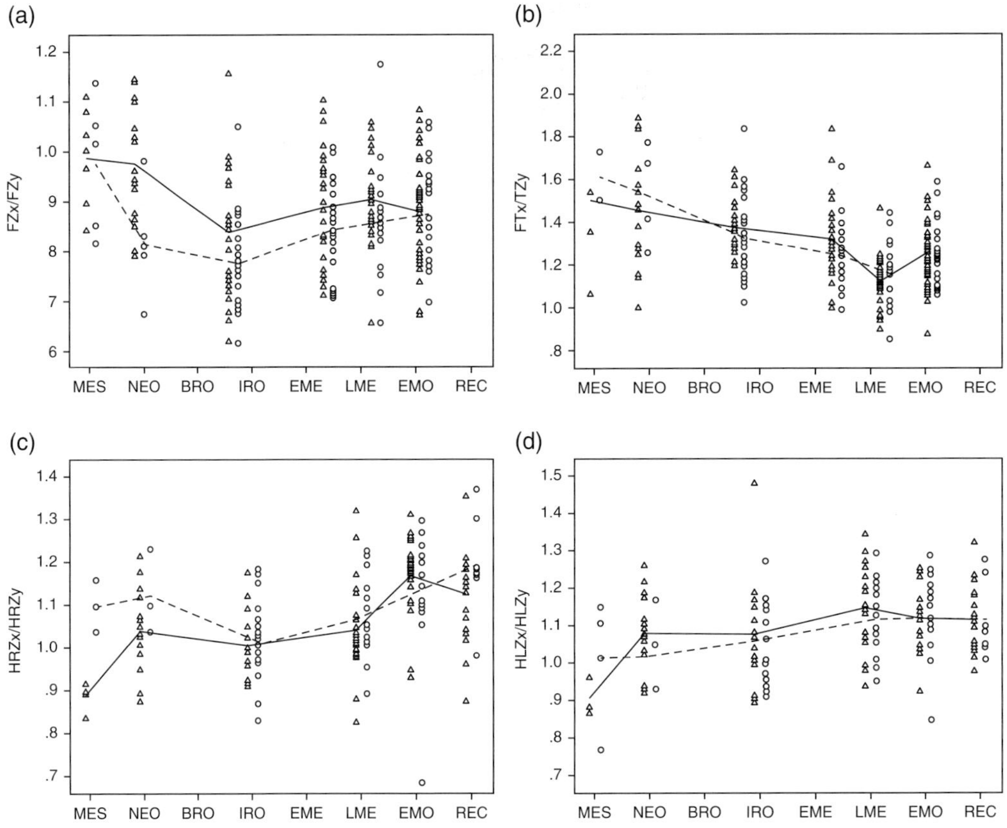

Figure 12.18 Relative size-standardized bending strengths. (a) Femoral A-P/M-L bending strength; (b) tibial A-P/M-L bending strength; (c) right humeral A-P/M-L bending strength; (d) left humeral A-P/M-L bending strength. Males: triangles and solid lines; females: circles and dashed lines.

A-P/M-L bending strength of the femur exhibits some decline across the Mesolithic–Neolithic transition, a noticeable decline to the Iron Age, an increase to the Medieval period, and little change thereafter (Fig. 12.18(a)). A-P/M-L bending strength of the tibia declined gradually from the Mesolithic to the Late Medieval period. It declined more considerably to the Early Modern period, then increased slightly again (Fig. 12.18(b)). The A-P/M-L bending strengths of both right and left humeri exhibit a tendency to increase over time (Fig. 12.18(c,d)). This is due to somewhat greater reductions in M-L than in A-P bending strengths.

The Early Modern Finns exhibited bending strengths similar to that of Medieval people. There was also no decline between Early Modern and Recent periods in Finland.

12.3.2 Comparisons with pan-European

Comparisons between Scandinavians and other Europeans in size-standardized cross-sectional properties are presented for femoral A-P and M-L bending strengths (Fig. 12.19), and torsional/average bending strengths of right and left humeri (Fig. 12.20). The Scandinavians do not differ from other Europeans in femoral bending strengths, with the exception of the Iron Age Danes, who have relatively weak femora for body size (Fig. 12.19). This is more noticeable in males (Fig. 12.19(a,c)) than in females (Fig. 12.19(b,d)), and in A-P bending strength (Fig. 12.19(a,b)) than in M-L bending strength (Fig. 12.19(c,d)). For instance, the Danish females do not differ from other Iron Age females in M-L bending strength (Fig. 12.19(d)).

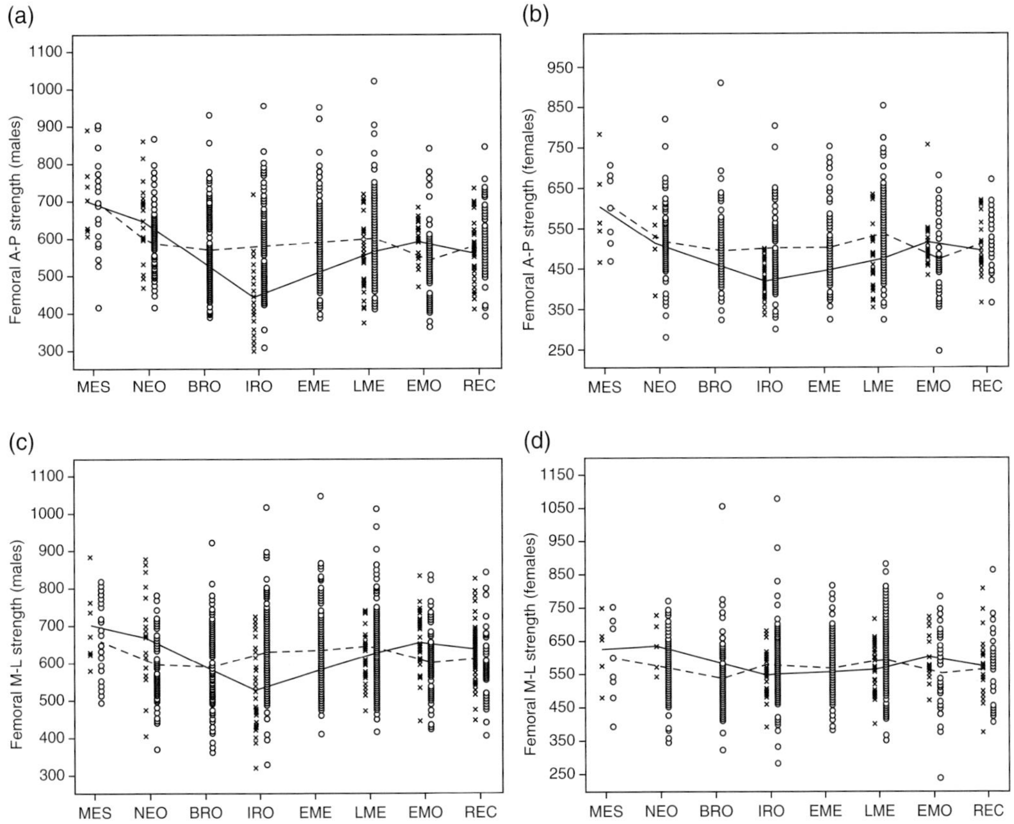

Figure 12.19 Scandinavian (crosses and solid lines) versus other European (circles and dashed lines) femoral strengths. (a) A-P strength, males; (b) A-P strength, females; (c) M-L strength, males; (d) M-L strength, females.

In the case of the right humerus (Fig. 12.20(a,b)), the Mesolithic Scandinavians have stronger right humeri than other Mesolithic Europeans. There are no differences in Neolithic farmers. The Iron Age Danes and Late Medieval Swedes have weaker humeri than their contemporaries. The Early Modern Finnish males differ little from their contemporaries, but females have stronger humeri than other Early Modern females. The Recent Finns, especially males, have stronger humeri than other Recent Europeans.

For the left humerus (Fig. 12.20(c,d)), there are no differences between Scandinavian Mesolithic males and other Mesolithic females, but Scandinavian females may have stronger humeri than other females. Small sample sizes may, however, provide misleading results here. As in case of the right humeri discussed above, the Scandinavian Neolithic farmers do not differ from other Neolithic farmers, and the Iron Age and Late Medieval Scandinavians have weaker humeri than their contemporaries. The Early Modern and Recent Finnish males and females have stronger humeri than their contemporaries. These post-Medieval differences are greater in males than in females.

12.3.3 Other Comparisons

The Neolithic PWC foragers are more similar to the Mesolithic foragers than to the Neolithic farmers in relative long bone bending strength (Table 12.11(a,b)). The PWC males differ from the Mesolithic males only by their significantly greater femoral M-L bending strength, but they

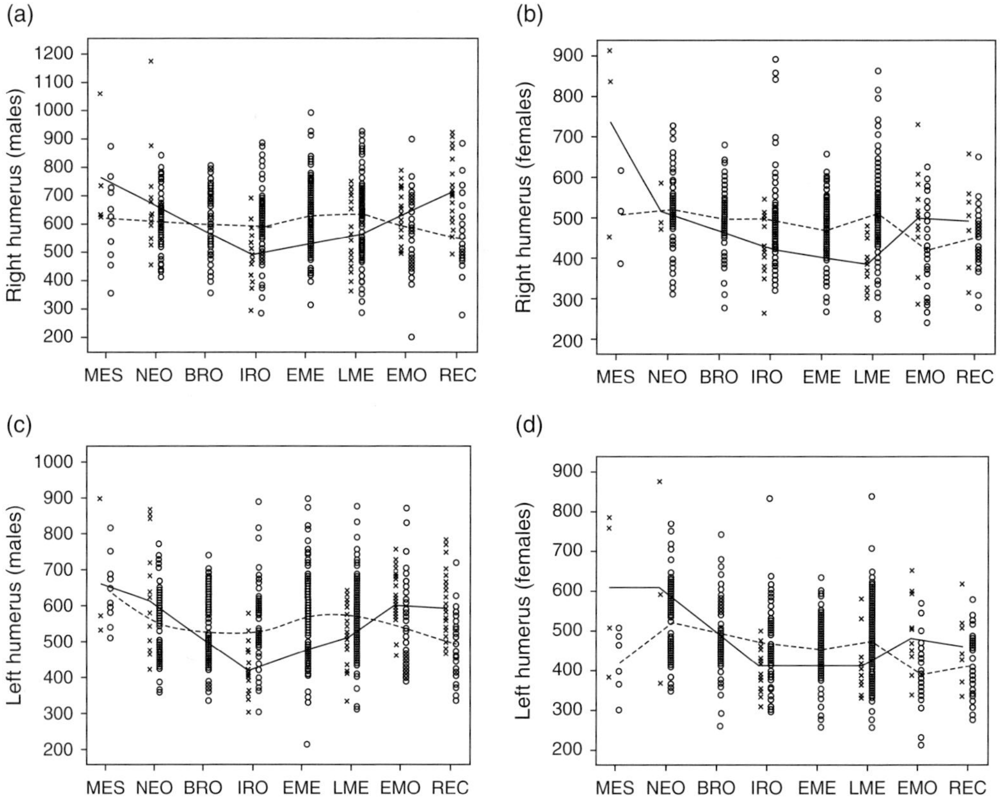

Figure 12.20 Scandinavian (crosses and solid lines) versus other European (circles and dashed lines) humeral torsional/average bending strengths. (a) Right humerus, males; (b) right humerus, females; (c) left humerus, males; (d) left humerus, females.

have significantly higher bending strengths than Neolithic farming males in all bones (Table 12.11(a)). The PWC females differ from the Mesolithic females only by their significantly higher femoral bending strengths. They differ from the Neolithic females by significantly greater bending strengths in the femur and tibia, but not humeri (Table 12.11(b)). The PWC foragers' greater robusticity in the lower limb in comparison to the Neolithic farmers is particularly interesting because it is not preferentially in the A-P direction (see below for discussion).

The Early Modern samples from Porvoo and Renko are our only samples that allow us to examine urban–rural differences in long bone bending strengths. We found no difference in either sex in these bending strengths.

Humeral bilateral asymmetry in bending strengths declined after the Iron Age in males. Females do not exhibit a clear temporal trend in humeral bilateral asymmetry, but this finding is somewhat inconclusive due to small female sample sizes (Table 12.12).

Sexual dimorphism in long bone bending strengths is shown in Table 12.13. Sample sizes of Mesolithic foragers and Neolithic farmers are too small to provide reliable results. The PWC foragers exhibit slightly reduced sexual dimorphism in the femur but closer to average levels in the tibia and humeri.

The Roman/Iron Age Danes have little sexual dimorphism in bending strengths of their femora and humeri largely due to male bending strengths being relatively low. Sexual dimorphism is about average in tibial bending strength.

Table 12.11(a) Size-standardized bending strengths of Mesolithic, Pitted Ware (PWC), and Neolithic males compared. Sample size is given below mean values.

		Mesolithic	**Sig.**	**PWC**	**Sig.**	**Neolithic**
FZ_{xstd}	Mean	694.6	0.191	662.2	0.000	598.6
	N	37		44		112
FZ_{ystd}	Mean	672.9	0.023	729.4	0.000	611.1
	N	37		44		112
TZ_{xstd}	Mean	778.0	0.747	790.0	0.000	679.1
	N	28		43		106
TX_{ystd}	Mean	521.8	0.092	567.7	0.000	470.2
	N	28		43		106
HRZ_{xstd}	Mean	354.8	0.278	382.1	0.006	337.4
	N	16		31		85
HRZ_{ystd}	Mean	356.7	0.248	386.5	0.000	323.9
	N	16		31		85
HLZ_{xstd}	Mean	323.8	0.299	342.5	0.028	312.3
	N	13		27		84
HLZ_{ystd}	Mean	377.1	0.100	336.6	0.004	295.0
	N	13		27		84

Table 12.11(b) Size-standardized bending strengths of Mesolithic, Pitted Ware, and Neolithic females compared. Sample size is given below mean values.

		Mesolithic	**Sig.**	**PWC**	**Sig.**	**Neolithic**
FZ_{xstd}	Mean	608.8	0.499	628.3	0.000	518.1
	N	14		26		82
FZ_{ystd}	Mean	610.1	0.021	694.4	0.000	574.0
	N	14		26		82
TZ_{xstd}	Mean	617.1	0.668	636.1	0.055	584.1
	N	8		22		72
TX_{ystd}	Mean	443.6	0.177	488.4	0.000	399.8
	N	8		22		72
HRZ_{xstd}	Mean	345.4	0.484	307.2	0.160	286.2
	N	6		15		61
HRZ_{ystd}	Mean	314.3	0.460	277.3	0.346	263.4
	N	6		15		61
HLZ_{xstd}	Mean	268.8	0.440	292.6	0.758	287.5
	N	10		17		60
HLZ_{ystd}	Mean	268.0	0.285	295.7	0.167	271.6
	N	10		17		60

We do not have enough data for assessing sexual dimorphism in the Early Medieval period. The Late Medieval individuals with long bone bending strength values are all from Sigtuna, Sweden. There is a considerable amount of sexual dimorphism in right humeri, but about an average amount in other bones. Sexual dimorphism increased from the Iron Age because bending strength values of males increased and female values remained about the same.

Table 12.12 Humeral bilateral asymmetry (HZ_p). Mean and standard error of mean values (SE) are provided.

Period	Sex	N	Mean HZ_p	SE of HZ_p
Very Recent	Male	17	10.4	1.7
	Female	6	7.5	2.3
Early Modern	Male	21	5.9	1.4
	Female	13	5.2	3.5
Late Medieval	Male	20	5.5	2.7
	Female	12	3.0	3.3
Iron Age	Male	15	15.3	3.6
	Female	20	3.9	3.2
Neolithic foragers	Male	25	15.3	2.4
	Female	13	3.1	2.9
Neolithic farmers	Male	16	12.4	3.8
	Female	2	13.8	13.2
Mesolithic	Male	2	17.6	1.2
	Female	1	8.0	–

Table 12.13 Sexual dimorphism in standardized lone bone bending strengths. Lower figures are sex-specific sample sizes (number of males/number of females).

Period	FZ_{xstd}	FZ_{ystd}	TZ_{xstd}	TZ_{ystd}	HRZ_{xstd}	HRZ_{ystd}	HLZ_{xstd}	HLZ_{ystd}
Very Recent	11.9 44/20	10.3 44/20	11.1 45/21	12.5 45/41	32.0 21/8	37.9 21/8	23.7 26/7	24.8 26/7
Early Modern	13.0 26/17	8.4 26/17	15.6 24/20	18.5 24/20	22.8 24/15	21.6 24/15	21.5 22/15	21.8 22/15
Late Medieval	14.9 30/23	9.7 30/23	25.2 26/21	19.7 26/21	35.3 23/14	39.4 23/14	23.7 22/14	22.3 22/14
Iron Age	3.9 33/27	−3.5 33/27	22.3 23/25	17.6 23/25	15.4 17/22	15.3 17/22	3.4 18/17	2.5 18/17
Neolithic foragers	5.3 44/26	4.9 44/26	21.6 43/22	15.0 43/22	21.7 31/15	32.9 31/15	15.7 27/17	13.0 17/17
Neolithic farmers	21.8 20/5	5.3 20/5	12.6 16/4	18.9 16/4	24.3 17/3	31.9 17/3	6.8 16/4	-0.6 16/4
Mesolithic	13.1 8/5	11.8 8/5	16.3 4/2	27.4 4/2	−5.0 4/3	16.7 4/3	1.6 3/4	16.4 3/4

The Early Modern sample exhibits about average levels of sexual dimorphism in bending strengths of all long bones. Sexual dimorphism in the Recent sample is average in femoral bending strengths, reduced in tibial bending strengths, and elevated in both humeri. The elevated male values for humeri are again consistent with heavy manual labor.

12.4 Discussion

12.4.1 Body Size and Body Shape

The finding that both early Mesolithic Maglemosian individuals in our data set are taller than Late Mesolithic Scandinavians is in accord with a stature decline within the Mesolithic.

As discussed in Chapter 4, nutrition and lifestyle changes (e.g., a more sedentary lifestyle) may have resulted in this Mesolithic stature decline. The 18th century Norwegian Saami are a more recent example of relatively short-statured European foragers, because their femoral lengths in Schreiner (1935) indicate very short statures (males 154.9 cm; females 147.0 cm). The 20th century Saami are also shorter-statured than the other Fennoscandians (Näätänen, 1936; Auger *et al.*, 1980). Differences in body proportions (e.g., higher relative sitting height) of the Mesolithic Scandinavians compared to non-Scandinavians may reflect their descent from people who inhabited colder climatic zones than ancestors of most other Mesolithic West Europeans.

Agriculture was not associated with body size reduction because the CWC, and especially the Late Neolithic individuals, averaged much taller than their TRB predecessors. This temporal trend was partly due to nutritional changes brought by the Secondary Product Revolution (see Chapter 4), and partly due to gene flow from people of the Yamnaya steppe culture (Mathieson *et al.*, 2015), whose predisposition to reach tall statures was inherited from their tall-statured eastern European hunter-gatherer ancestors (on Mesolithic stature variation, see Chapter 4). This gene flow from the tall-statured steppe people was mediated to the north via the spread of the CWC people (Haak *et al.*, 2015; Allentoft *et al.*, 2015).

Tall statures of the Late Roman Iron Age people are at least partly due to nutritional factors. Stable isotope studies indicate high protein intake during this period (see Jørkov *et al.*, 2010). Causes of a slight stature decline during the later Iron Age are unknown, but the agricultural revolution starting ca. 1000 AD may explain a slight stature recovery at that time (on this revolution, see Andersen *et al.*, 2016).

The Late Medieval stature decline coincides with that of the Little Ice Age (which commenced ca. 1275–1300 AD according to Miller *et al.*, 2012). Very low statures during the 17th century are expected based on records of famine (see Muroma, 1991) during this possibly worst century of the Little Ice Age. Stature recovery by the early 18th century is probably due to less adverse conditions for growth than during the 17th century.

The late 18th century stature decline is due to a decline in standards of living associated with the Industrial Revolution (see Chapter 4 and references therein). Interestingly, the famine of 1866–1868 (Turpeinen, 1986) is not reflected in the Norwegian and Swedish conscript statures (Udjus, 1964: their Table 3), but may be reflected in the Finnish ones (Nummela, 2000).

Stature recovery commencing during the 19th century was due to improvements of overall nutritional and health statuses, as well as hygiene. As elsewhere in Europe, this stature increase apparently began earlier in urban regions (see Chapter 4 and references therein). For example, in Denmark there was little urban–rural stature difference in 1901, but a greater stature increase made the urban inhabitants taller within ten years after that (see Boldsen, 1993: their Fig. 3). In the early 1920s, Swedish urban conscripts averaged 0.86 cm taller than rural ones (Lundborg and Linders, 1926). In the 1930s, the Finnish urban reservists averaged 0.5–1.8 cm taller than rural ones from the same province (see Nummela, 2000: their Table 8). In 1962, the Norwegian urban conscripts averaged 0.61 cm taller than rural ones (Udjus, 1964: their Table 24).

Socioeconomic status matters in human biology because status differences often affect childhood health and nutrition. Very tall statures of the Late Neolithic Scandinavians from Denmark and southern Sweden may thus partly also reflect a high socioeconomic status of these individuals (on Late Neolithic social elite, see Vankilde, 2007; Iversen, 2013).

Increasing social status differences during the Roman Iron Age (Heather, 2009; Jørkov *et al.*, 2010 and references therein) reflects in the Late Roman Iron Age (200–400 AD) statures. As computed from data in Sellevold *et al.* (1984), 'rich' males and females (at least one piece of jewelry and/or imported item as grave goods) averaged 177.4 cm (N = 10) and 164.0 cm (N = 10), respectively, but 'poor' males and females (no jewelry and/or imports as grave goods) averaged 172.2 cm (N = 15) and 160.6 cm (N = 13), respectively.

Shorter than expected statures of our Recent Finns are not due to underestimation of stature in this sample (see Chapter 2; Table 2.8), but rather to low socioeconomic status. Most individuals in this sample were laborers of different kinds (e.g., 'general' workmen). Socioeconomic status is clearly reflected in stature in early 20th century Finland. For example, 20-year-old males from 'better' schools averaged 174.4 cm in the 1910s (Suomen Tilastollinen Vuosikirja, 1925: their Table 49), but national mean stature of conscripts was still only 169.0 cm about a decade later in the early 1920s (Kajava, 1926).

Students born in 1958 were still 2.0 cm taller than non-students born the same year, based on Finnish conscript data (Dahlström, 1981: their Table 5-15). Stature differences between socioeconomic groups were still noticeable in Sweden among individuals born in the mid-1950s, but have largely disappeared in more recent birth cohorts due to greater stature increase in lower socioeconomic groups (Peck and Vågero, 1987).

It is not yet known if the clear south-to-north stature decline in Scandinavia and Finland is due to differences in growth environment or to gene pool differences. The north-to-south stature difference in Europe as a whole is at least partly due to genetic differences – that is, 'height-increasing alleles' have higher frequencies in taller-statured North Europeans than in shorter-statures South Europeans (Turchin *et al.*, 2012). Future studies may show if height-increasing alleles have higher frequencies in South Scandinavia than in North Fennoscandia.

12.4.2 Cross-Sectional Properties

The finding that the PWC foragers are similar to the Mesolithic foragers in long bone bending strength is expected. Both groups represent maritime foragers and thus probably had largely similar physical activities. The PWC foragers' greater robusticity in the lower limb in comparison to the Neolithic farmers is interesting, because it is not preferentially in the A-P direction. This implies that this greater lower limb robusticity is not so much due to greater mobility than to a heavier overall workload. Reduced upper limb robusticity of the PWC females in comparison to the Mesolithic females indicates that their physical tasks and/or workloads were not identical to those of their Mesolithic predecessors.

A slight reduction in skeletal strength to the Iron Age, followed by a slight increase to the Medieval Period, is intriguing. This reduced skeletal strength of the Iron Age people is apparently more clear in this region than elsewhere in Europe (on pan-European temporal trends see Chapter 5; Ruff *et al.*, 2015). An examination of absolute (not shown here) and size-standardized skeletal strength values reveals that there was little or no reduction in absolute bone strength, but a considerable reduction in bone strength standardized for body size. The Roman Iron Age Danes' large bodies, combined with their average absolute bone shaft strength, resulted in relatively lower skeletal strength. In any event, it is apparent that the South Scandinavian Iron Age people represented by the Roman Iron Age Danes did not need to perform as much heavy physical labor as the Neolithic South Scandinavian farmers and the Medieval urban people from Sigtuna.

There was apparently little or no change in mobility and physical labor between Medieval and Early Modern times, based on samples included in this study, which is similar to broader pan-European comparisons (see Chapter 5; Ruff *et al.*, 2015). Our finding that skeletal strength did not decline from the Early Modern period to the Recent Periods in Finland probably largely reflects sample composition of the Recent Period Finns. Many of these individuals were laborers.

A decline in humeral bilateral asymmetry after the Iron Age likely reflects changes in manipulative behavior especially in males. There were apparently lesser changes in this manipulative behavior in females.

12.5 Conclusions

Temporal patterns in body size and shape, as well as long bone strength changes, were largely similar in this region to those elsewhere in Europe. The general Neolithic stature increase (see Chapter 4) and Iron Age skeletal strength reduction were particularly marked in Scandinavia (see Chapter 5; Ruff *et al.*, 2015). The Recent Finns' relatively robust upper limbs (particularly evident in males) is a slight exception to the overall pattern of robusticity reduction in more recent periods, and is likely due to sample composition, since many individuals in this sample had performed very heavy physical labor during their lives.

Stature increase during the Neolithic was preceded by stature decline during the Mesolithic. This pattern implies that agriculture is not necessarily associated with stature and overall body size decline, but that there are other influences on adult stature than merely subsistence practices and nutrition. For instance, parasitic infestation could explain shorter stature among sedentary foragers (i.e., the Late Mesolithic Scandinavians and the Middle Neolithic PWC foragers), and substantial gene flow from the tall-statured Yamnaya steppe people via the CWC people may have increased genetic predisposition for taller stature during the Late Neolithic. There were also more noticeable shifts in male stature than in female stature with childhood nutrition and health apparently affecting male growth more than female growth. Taller mean stature is associated with greater sexual dimorphism in stature in our temporal samples.

The Little Ice Age (ca. 1300–1870 AD) and Industrial Revolution (since the mid-18th century) probably explain low statures from the 14th century to the late 19th century. Stature increase commencing during the 19th century was due to improvements in nutrition, health, and hygiene, and a reduction of social status differences in overall health and nutrition.

As in the rest of Europe, bone strengths in Fennoscandinavia decline from the Mesolithic through Iron Ages, with preferential declines in A-P bending strengths of the femur and tibia, likely reflecting reductions in mobility associated with adoption and intensification of food production (Chapter 5; Ruff *et al.*, 2015). The PWC foragers are a special case. These Neolithic foragers are more similar to their Mesolithic predecessors and apparent genetic ancestors than to their Neolithic contemporaries and neighbors in body size and shape as well as in skeletal robusticity. However, although they exhibit similar overall skeletal strength levels as the Mesolithic foragers, their pattern of skeletal robusticity may reflect a rigorous workload rather than a high level of mobility, given that their lower limb robusticity is not preferentially in the A-P direction, as in mobile foragers. Also, the PWC females have less robust humeri than the Mesolithic Scandinavian females. Thus, bone structural differences imply significant behavioral variation between Mesolithic, PWC, and Neolithic farming cultures, which need to be clarified through continued archaeological research.

Acknowledgments

We thank the following people who provided access to skeletal collections and/or facilitated data collection in Denmark and Sweden: Jan Storå, Niels Lynnerup, Pia Bennike, Leena Drenzel, Torbjörn Ahlström, Per Karsten, Bernd Gerlach, and Lars Larsson. We also thank the Finnish National Board of Antiquities and the Finnish Museum of Natural History for access to the Finnish skeletal collections.

References

Allentoft, M.E., Sikora, M., Sjögren, K.-G., Rasmussen, S., Rasmussen, M., Stenderup, J., Damgaard, P.B., Schroeder, H., Ahlström, T., Vinner, L., Malaspinas, A.-S., Margaryan, A., Higham, T., Chivall, D., Lynnerup, N., Harvig, L., Baron, J., Casa, P.D., Dabrowski, P., Duffy, P.R., Ebel, A.V., Epimakhov, A., Frei, K., Furmanek, M., Gralak, T., Gromov, A., Gronkiewicz, S., Grupe, G., Hajdu, T., Jarysz, R., Khartanovich, V., Khokhlov, A., Kiss, V., Kolář, J., Kriiska, A., Lasak, I., Longhi, C., McGlynn, G., Merkevicius, A., Merkyte, I., Metspalu, M., Mkrtchyan, R., Moiseyev, V., Paja, L., Pálfi, G., Pokutta, D., Pospiezny, L., Price, T.D., Saag, L., Sablin, M., Shishlina, N., Smrčka, V., Soenov, V.I., Szeverényi, V., Tóth, G., Trifanova, S.V., Varul, L., Vicze, M., Yepiskoposyan, L., Zhitenev, V., Orlando, L., Sicheritz-Pontén, T., Brunak, S., Nielsen, R., Kristiansen, K., and Willerslev, E. (2015) Population genomics of Bronze Age Eurasia. *Nature*, **522**, 167–172.

Andersen, T.M., Jensen, P.S., and Skovsgaard, C.V. (2016) The heavy plow and the agricultural revolution in Medieval Europe. *J. Dev. Econ.*, **118**, 133–149.

Anthony, D.W. and Brown, D.R. (2011) The secondary product revolution, horse-riding, and mounted warfare. *J. World Prehist.*, **24**, 131–160.

Arcini, C. (1999) Health and disease in early Lund. Osteopathologic studies of 3,305 individuals buried in the first cemetery area of Lund 990–1536. Archaeological Lundensia VIII, Lund.

Arcini, C., Ahlström, T., and Tagesson, G. (2014) Variations in diet and stature: Are they linked? Bioarchaeology and paleodietary Bayesian mixing models from Linköping, Sweden. *Int. J. Osteoarchaeol.*, **24**, 543–556.

Auger, F., Jamison, P.L., Balslev-Jorgensen, J., Lewin, T., de Peña, J.F., and Skrobak-Kaczynski, J. (1980) Anthropometry of circumpolar populations. In: *The Human Biology of Circumpolar Populations* (ed. F.A. Milan), Cambridge University Press, Cambridge, pp. 213–225.

Bang-Andersen, S. (2012) Colonizing contrasting landscapes. The pioneer coast settlement and inland utilization in southern Norway 10,000–9500 years before present. *Oxford J. Archaeol.*, **31**, 103–120.

Bläuer, A. and Kantanen, J. (2013) Transition from hunting to animal husbandry in Southern, Western and Eastern Finland: new dated osteological evidence. *J. Archaeol. Sci.*, **40**, 1646–1666.

Boldsen, J.L. (1993) Height variation in Denmark A.D. 1100–1988. In: *Population of the Nordic countries: Human population biology from the present to the Mesolithic* (eds E. Iregren and R. Liljekvist), University of Lund: Institute of Archaeology Report Series No. 46. Bloms Boktryckeri AB, Lund, pp. 52–60.

Brandt, G., Haak, W., Adler, C.J., Roth, C., Szécsényi-Nagy, A., Karimnia, S., Möller-Rieker, S., Meller, H., Ganslmeier, R., Friederich, S., Dresely, V., Nicklisch, N., Pickrell, J.K., Sirocko, F., Reich, D., Cooper, A., Alt, K.W., The Geographic Consortium (2013) Ancient DNA reveals key stages in the formation of Central European mitochondrial genetic diversity. *Science*, **342**, 257–261.

Bröste, K. and Jørgensen, J.B. (1956) *Prehistoric man in Denmark: a study in physical anthropology. Volume I. Stone and Bronze Ages.* Einar Munksgaard Publishers, Copenhagen.

Cramp, L.J.E., Evershed, R.P., Lavento, M., Halinen, P., Mannermaa, K., Oinonen, M., Kettunen, J., Perola, M., Onkamo, P., and Heyd, V. (2014) Neolithic dairy farming at the extreme of agriculture in northern Europe. *Proc. R. Soc. B.*, **281**, 20140819.

Czekaj-Zastawny, A., Kabacinski, J., and Terberger, T. (2013a) The origin of the Funnel Beaker culture from a southern Baltic coast perspective. In: *Environment and subsistence – forty years after Janusz Kruk's settlement studies* (eds S. Kadrow and P. Wlodarczak), Institute of Archaeology, UR & Verlag, Bonn, pp. 409–428.

Czekaj-Zastawny, A., Kabacinski, J., Terberger, T., and Ilkiewicz, J. (2013b) Relations of Mesolithic hunter-gatherers of Pomerania (Poland) with Neolithic cultures of central Europe. *J. Field Archaeol.*, **38**, 195–209.

Dahlström, S. (1981) Suomalaisen nuoren miehen ruumiinrakenne kutsuntamittausten ja 20-vuotiaiden varusmiesten antropometrisen mittauksen perusteella. Academic dissertation, University of Turku, Turku, Finland.

Edgren, T. (1998) *Den förhistoriska tiden*. Finlands Historia 1. Ekenäs.

Fischer, A., Olsen, J., Richards, M., Heinemeier, J., Sveinbjörnsdóttir, A.E., and Bennike, P. (2007) Coast-inland mobility and diet in the Danish Mesolithic and Neolithic: evidence from stable isotope values of humans and dogs. *J. Archaeol. Sci.*, **34**, 2125–2150.

Formicola, V. and Giannecchini, M. (1999) Evolutionary trends of stature in Upper Paleolithic and Mesolithic Europe. *J. Hum. Evol.*, **36**, 319–333.

Gejvall, N.-G. (1960) *Westerhus: Medieval population and church in the light of skeletal remains*. Håkan Ohlssons Boktryckeri, Lund.

Grimm, S.B. and Weber, M.-J. (2008) The chronological framework of the Hamburgian in the light of old and new ^{14}C dates. *Quantär*, **55**, 17–40.

Haak, W., Lozaridis, I., Patterson, N., Rohland, N., Mallick, S., Llamas, B., Brandt, G., Nordenfelt, S., Harney, E., Stewardson, K., Fu, Q., Mittnik, A., Bánffy, E., Economou, C., Francken, M., Friederich, S., Pena, R.G., Hallgren, F., Khartanovich, V., Khokhlov, A., Kunst, M., Kuznetsov, P., Meller, H., Mochalov, O., Moiseyev, V., Nicklisch, N., Pichler, S.L., Risch, R., Guerra, M.A.R., Roth, C., Szécsényi-Nagy, A., Wahl, J., Meyer, M., Krause, J., Brown, D., Anthony, D., Cooper, A., Alt, K.W., and Reich, D. (2015) Massive migration from the steppe was a source for Indo-European languages in Europe. *Nature*, **522**, 207–211.

Heather, P. (2009) *Empires and Barbarians: Migration, Development and the Birth of Europe*. Macmillan, London.

Heintel, M., Sandberg, L., and Steckel, R. (1998) Swedish historical heights revisited: new estimation techniques and results. In: *The biological standards of living in comparative perspectives* (eds J. Komlos and J. Baten), Proceedings of a conference held in Munich January 18–23, 1997. Franz Steiner Verlag, Stuttgart, pp. 449–458.

Hultkrantz, J.V. (1927) *Über Die Zunahme Der Körpergrösse in Schweden in Den Jahren 1840–1926*. Norblads bokh, Uppsala.

Humphreys, K., Grankvist, A., Leu, M., Hall, P., Liu, J., Ripatti, S., Rehnström, K., Groop, L., Klareskog, L., Ding, B., Grönberg, H., Xu, J., Pedersen, N.L., Lichtenstein, P., Mattingsdal, M., Andreassen, O.A., O'Dushlaine, C., Purcell, S.M., Sklar, P., Sullivan, P.F., Hultman, C.M., Palmgren, J., and Magnusson, P.K.E. (2011) The genetic structure of the Swedish population. *PLoS ONE*, **6**, e22547.

Iversen, R. (2013) Beyond the Neolithic transition – the 'de-Neolithisation' of South Scandinavia. In: *NW Europe in transition – The early Neolithic in Britain and South Sweden* (eds M. Larsson and J. Debert), BAR International Series 2475, Chapter 3, pp. 21–17.

Jacobs, K. (1993) Human postcranial variation in the Ukrainian Mesolithic–Neolithic. *Curr. Anthropol.*, **34**, 311–324.

Jacobs, K. (1995) Returning to Oleni'ostrov: social, economic, and skeletal dimensions of a boreal forest Mesolithic cemetery. *J. Anthropol. Archaeol.*, **14**, 359–403.

Jessen, C.A., Pedersen, K.B., Christensen, C., Olsen, J., Mortensen, M.F., and Hansen, K.M. (2015) Early Maglemosian culture in the Preboreal landscape: archaeology and vegetation from the earliest Mesolithic site in Denmark at Lundby Mose, Sjælland. *Quat. Int.*, **378**, 73–87.

Jørkov, M.L., Jørgensen, L., and Lynnerup, N. (2010) Uniform diet in a diverse society. Revealing new dietary evidence of the Danish Roman Iron Age based on stable isotope analysis. *Am. J. Phys. Anthropol.*, **143**, 523–533.

Júlíusson, P.B., Roelants, M., Nordal, E., Furevik, L., Eide, G.E., Moster, D,. Hauspie, R., and Bjerknes, R. (2013) Growth references for 0–19-year-old Norwegian children for length/height, weight, body mass index and head circumference. *Ann. Hum. Biol.*, **40**, 220–227.

Kajava, Y. (1926) Suomalaisten rotuominaisuuksia. In: *Suomen suku* (eds A. Kannisto, E.N. Setälä, U.T. Sirenius, and Y. Wichmann), Kustannusyhtiö Otava, Helsinki, pp. 215–250.

Kjellström, A. (2005) The Urban Farmer: Osteoarchaeological Analysis of Skeletons from Medieval Sigtuna Interpreted in a Socioeconomic Perspective. Theses and Papers in Osteoarchaeology No. 2. Stockholm University, Stockholm.

Larnkjær, A., Schrøder, S.A., Schmidt, I.M., Jørgensen, M.H., and Michaelsen, K.F. (2006) Secular change in adult stature has come to a halt in northern Europe and Italy. *Acta Paediatr.*, **95**, 754–755.

Lundborg, H. and Linders, F.J. (1926) *The racial characters of the Swedish nation.* Almqvist & Wiksell, Uppsala.

Mathieseon, I., Lazaridis, I., Rohland, N., Mallick, S., Patterson, N., Roodenberg, S.A., Harney, E., Stewardson, K., Fernandes, D., Novak, M., Sirak, K., Gamba, C., Jones, E.R., Llamas, B., Dryomov, S., Pickrell, J., Arsuaga, J.L., de Castro, J.M.B., Carbonell, E., Gerritsen, F., Khokhlov, A., Kuznetsov, P., Lozano, M., Meller, H., Mochalov, O., Moiseyev, V., Guerra, M.A.R., Roodenberg, J., Vergès, J.M., Krause, J., Cooper, A., Alt, K.W., Brown, D., Anthony, D., Lalueza-Fox, C., Haak, W., Pinhasi, R., and Reich, D. (2015) Genome-wide patterns of selection in 230 ancient Eurasians. *Nature*, **528**, 499–503.

Miller, G.H., Geirsdóttir, Á., Zhong, Y., Larsen, D.J., Otto-Bliesner, B.L., Holland, M.M., Bailey, D.A., Refsnider, K.A., Lehman, S.J., Southon, J.R., Anderson, C., Björnsson, H., and Thordarson, T. (2012) Abrupt onset of the Little Ice Age triggered by volcanism and sustained by sea-ice/ocean feedbacks. *Geophys. Res. Lett.*, **39**, L02708. doi:10.1029/2011GLO50168.

Molnar, P. (2008) Tracing prehistoric activities: life ways, habitual behaviour and health of hunter-gatherers on Gotland. Theses and Papers in Osteoarchaeology No. 4. Doctoral Thesis. Stockholm University, Stockholm, Sweden.

Muroma, S. (1991) *Suurten kuolovuosien (1696–1697) väestönmenetys Suomessa.* Suomen Historiallinen Seura, Helsinki.

Näätänen, E.K. (1936) Über die Anthropologie der Lappen in Suomi. Ann. Acad. Sci. Fennie. Ser. A, 47/2.

Neuvonen, A.M., Putkonen, M., Översti, S., Sundell, T., Onkamo, P., Sajantila, A., and Palo, J. (2015) Vestiges of an ancient border in the contemporary genetic diversity of North-Eastern Europe. *PLoS ONE*, **10**, e0139331. doi:10.1371/journal.pone.0130331.

Nordin, M. and Gustafsson, P. (2010) Unto a good land: early Mesolithic colonization of eastern central Sweden. In: *Uniting Sea II: Stone Age Societies in the Baltic Sea region* (eds Å.M. Larsson and L. Papmehl-Dufay), OPIA, vol. 51, pp. 81–106.

Nummela, I. (2000) Pätkä tai ei? Suomalaisen pituuskasvun historiaa. In: *'Pane leipään puolet petäjäistä': nälkä- ja pulavuodet Suomen historiassa* (ed. P. Karvonen), Gummeruksen kirjapaino, Jyväskylä, pp. 93–153.

Öberg, S. (2014a) Long-term changes of socioeconomic differences in height among young adult men in Southern Sweden, 1818–1968. *Econ. Hum. Biol.*, **15**, 140–152.

Öberg, S. (2014b) *Social bodies: Family and Community Level Influences on Height and Weight in Southern Sweden 1818–1968.* Department of Economy and Society, School of Business, Economics and Law, University of Gothenburg, Göteborg.

Peck, A.M. and Vågerö, D.H. (1987) Adult body height and childhood socioeconomic group in the Swedish population. *J. Epidemiol. Community Health*, **41**, 333–337.

Penttinen, A., Moltchanovo, E., and Nummela, I. (2013) Bayesian modeling of the evolution of male height in 18th century Finland from incomplete data. *Econ. Hum. Biol.*, **11**, 405–415.

Pihlman, A. (2010) Turun kaupungin muodostuminen ja kaupunkiasutuksen laajeneminen 1300-luvulla. Varhainen Turku. Raportteja 22, 2010, pp. 9–29. Turun museokeskus, Turku. ISBN 978-951-595-146-5.

Price, T.D. (2000) *Europe's First Farmers*. Cambridge University Press, Cambridge.

Price, T.D. and Jacobs, K. (1990) Oleni'ostrov: first radiocarbon dates from a major Mesolithic cemetery in Karelia, USSR. *Antiquity*, **64**, 849–853.

Rankama, T. and Kankaanpää, J. (2011) First evidence of eastern Preboreal pioneers in Arctic Finland and Norway. *Quartät*, **58**, 183–209.

Rowley-Conwy, P. (2011) Westward ho! The spread of agriculture from Central Europe to the Atlantic. *Curr. Anthropol.*, **52**, S431–S451.

Ruff, C.B., Holt, B.M., Sladék, V., Berner, M., Garofalo, E., Garvin, H.M., Hora, M., Maijanen, H., Niinimäki, S., Salo, K., Schuplerová, E., and Tompkins, D. (2012) Stature and body mass estimation from skeletal remains in the European Holocene. *Am. J. Phys. Anthropol.*, **148**, 601–617.

Ruff, C.B., Holt, B., Niskanen, M., Sladek, V., Berner, M., Garofalo, E., Garvin, H.M., Hora, M., Junno, J.-A., Schuplerova, E., Vilkama, R., and Whittey, E. (2015) Gradual decline in mobility with the adoption of food production in Europe. *Proc. Natl Acad. Sci. USA*, **112**, 7147–7152.

Ruff, C.B., Niskanen, M., Junno, J.-A., and Jamison, P. (2005) Body mass prediction from stature and bi-iliac breadth in two high latitude populations, with application to earlier higher latitude humans. *J. Hum. Evol.*, **48**, 381–392.

Saari, A., Sankilampi, U., Hannila, M.-L., Kiviniemi, V., Kesseli, K., and Dunkel, L. (2011) New Finnish growth references for children and adolescents aged 0 to 20 years: Length/height-for-age, weight-for-length/height, and body mass index-for-age. *Ann. Med.*, **43**, 235–248.

Salo, K. (2016) Health in Southern Finland – Bioarchaeological Analysis of 555 Skeletons Excavated from Nine Cemeteries (11th–19th century AD). Ph.D. dissertation. University of Helsinki, Helsinki, Finland.

Salmela, E., Lappalainen, T., Liu, J., Sistonen, P., Andersen, PM., Schreiber, S., Savontaus, M.-L., Lahermo, P., Hall, P., and Kere, J. (2011) Swedish population substructure revealed by genome-wide single nucleotide polymorphism data. *PLoS ONE*, **6**, e16742.

Sawyer, B. and Sawyer, P. (1993) *Medieval Scandinavia: From conversion to reformation circa 800–1500*. University of Minnesota Press, Minneapolis.

Schmidt, I.M., Jørgensen, M.H., and Michaelsen, K.F. (1995) Height of conscripts in Europe: is postneonatal mortality a predictor? *Ann. Hum. Biol.*, **22**, 57–67.

Schreiner, K.E. (1935) *Zur Osteologie der Lappen*. H. Aschehoug & Co., Oslo.

Sellevold, B.J., Hansen, U.L., and Jørgensen, J.B. (1984) *Iron Age man in Denmark. Prehistoric Man in Denmark, Vol. III*. Det Kongelige Nordiske Oldskriftselskab, København.

Sjøvold, T. (1990) Estimation of stature from long bones utilizing the line of organic correlation. *Hum. Evol.*, **5**, 431–447.

Sjöberg, A., Barrenäs, M.-L., Brann, J.E., Chaplin, J.E., Dahlgren, J., Mårild, S., Lissner, L.,and Albertsson-Wikland, K. (2012) Body size and lifestyle in an urban population entering adulthood: The 'Grow up Gothenburg' Study. *Acta Paediatr.*, **101**, 964–972.

Skoglund, P., Malmström, H., Omrak, A., Raghavan, M., Valdiosera, C., Günther, T., Hall, P., Parik, J., Sjögren, K.-G., Apel, J., Willerslev, E., Storå, J., Götherström, A., and Jakobsson, M. (2014) Genomic diversity and admixture differs for Stone-Age Scandinavian foragers and farmers. *Science*, **344**, 747–750.

Skoglund, P., Malmström, H., Raghavan, M., Storå, J., Hall, P., Willerslev, E., Thomas, M., Gilbert, P., Götherström, A., and Jakobsson, M. (2012) Origins and genetic legacy of Neolithic farmers and hunter-gatherers in Europe. *Science*, **27**, 466–469.

Sørensen, M., Rankama, T., Kankaanpää, J., Knutsson, K., Knutsson, H., Melvold, S., Eriksen, B.V., and Glørstad, H. (2013) The first eastern migrations of people and knowledge into Scandinavia: evidence from studies of Mesolithic technology, 9th-8th millennium BC. *Norw. Archaeol. Rev.*, **46**, 19–56.

Sørensen, L. and Karg, S. (2014) The expansion of agrarian societies towards the north e new evidence for agriculture during the Mesolithic/Neolithic transition in Southern Scandinavia. *J. Archaeol. Sci.*, **51**, 98–114.

Statistical Yearbook of Norway 2011. https://www.ssb.no/a/en/histstat/aarbok/2011_en.pdf.

Suomen Tilastollinen Vuosikirja (1925) Tilastollisen Päätoimiston Julkaisema. Valtioneuvoston Kirjapaino, Helsinki. http://www.doria.fi/handle/10024/69243.

Tarkiainen, K. (2010) *Ruotsin Itämää. Esihistoriasta Kustaa Vaasaan. Suomen Ruotsalainen historia 1.* Svenska Litteratursällskapet I Finland, Helsinki.

Telkkä, A. (1950) On the prediction of human stature from the long bones. *Acta Anat.*, **9**, 103–117.

Timofeev, V.I., Zaitseva, G.I., Lavento, M., Dolukhanov, P., and Halinen, P. (2004) The radiocarbon datings of the Stone Age – Early Metal Period on the Karelian Isthmus. *Geochronometria*, **23**, 93–99.

Tinggaard, J., Aksglaede, L., Sørensen, K., Mouritsen, A., Wohlfarth-Veje, C., Hagen, C.P., Mieritz, M.G., Jørgensen, N., Heuck, C., Petersen, J.H., Main, K.M., and Juul, A. (2014) The 2014 Danish references from birth to 20 years for height, weight and body mass index. *Acta Paediatr.*, **103**, 214.224.

Trotter, M. and Gleser, G.C. (1958) A re-evaluation of estimation of stature based on measurements of stature taken during life and of long bones after death. *Am. J. Phys. Anthropol.*, **16**, 79–124.

Tuovinen, T. (2002) The burial cairns and the landscape in the Archipelago of Åboland, SW Finland, in the Bronze Age and the Iron Age. Acta Universitatis Ouluensis Humaniora, B 46.

Turchin, M.C., Chiang, C.W.K., Palmer, C.D., Sankararaman, S., Reich, D., Genetic Investigation of ANthropometric Traits (GIANT) Consortium, Hirschhorn, J.N. (2012) Evidence of widespread selection on standing variation in Europe at height-associated SNPs. *Nat. Genet.*, **44**, 1015–1019.

Turpeinen, O. (1986) *Nälkä vai tauti tappoi? Kauhunvuodet 1866–1868.* Societas Historica Finlandiae, Helsinki.

Udjus, L.G. (1964) *Anthropometric changes in Norwegian men in the twentieth century.* Universitetsforlaget, Oslo.

Vankilde, H. (2007) A review of the Early Late Neolithic Period in Denmark: practice, identity and connectivity. *Offa*, **61/62**, 75–109.

Verpoorte, A. (2009) Limiting factors on early modern human dispersals: the human biogeography of late Pleniglacial Europe. *Quat. Int.*, **201**, 77–85.

Werner, B. (2007) *Growth in Sweden.* Karolinska Institutet, Stockholm.

Werner, B. and Bodin, L. (2006) Growth from birth to age 19 for children in Sweden born in 1981: Descriptive values. *Acta Paediatr.*, **95**, 600–613.

Wlodarczak, P. (2009) Radiocarbon and dendrochronological dates of the Corded Ware culture. *Radiocarbon*, **51**, 737–749.

Zvelebil, M. (2008) Innovating hunter-gatherers: the Mesolithic in the Baltic. In: *Mesolithic Europe* (eds G. Bailey and P. Spikins), Cambridge University Press, Cambridge, pp. 18–59.

13

The Balkans

Christopher B. Ruff[1] *and Brigitte Holt*[2]

[1] *Center for Functional Anatomy and Evolution, Johns Hopkins University School of Medicine, Baltimore, MD, USA*
[2] *Department of Anthropology, University of Massachusetts, Amherst, MA, USA*

13.1 Introduction

Two skeletal samples from the Balkans region were included in this study: a Late Mesolithic (6400–7400 cal BC) sample from Schela Cladovei, in the Iron Gates region along the Danube River in Western Romania (Boroneant *et al.*, 1995; Bonsall, 2008), and a Late Medieval (1400–1500 AD) sample from Mistihalj, in southern Bosnia-Herzegovina near the border with Montenegro (Alexeeva *et al.*, 2003) (Table 13.1; Figure 13.1). Because only two samples from this region were available, with tenuous population continuity at best, temporal trends within the region are de-emphasized. Rather, comparisons focus on differences and similarities between Balkan and other European samples during each time period represented, that is, Mesolithic and Late Medieval.

Seventeen individuals (12 males and 5 females) from the Schela Cladovei site were analyzed, although samples sizes were generally smaller for individual skeletal parameters due to variable preservation. Schela Cladovei belongs to a series of well-known Late Mesolithic and Early Neolithic sites (including Lepenski Vir, Vlasac, and others) along a stretch of the Danube River in and near gorges through the Carpathian-Balkan mountain chain known as the Iron Gates region (Jochim, 2002; Bonsall, 2008). Schela Cladovei lies just east (downstream) from the gorge area, along a terrace of the Danube River (Boroneant *et al.*, 1995; Bonsall, 2008). The date range for the site given above and in Table 13.1 is derived from Bonsall (2008: their Table 10.1), based on ^{14}C ages from samples of human bone, corrected for the Danube freshwater reservoir effect, and then calibrated. All of the material included in this study thus falls before the Neolithic transition in this region, which occurred at ca. 6000 cal BC (Bonsall, 2008; Boric *et al.*, 2014).

Although situated on a floodplain of the Danube River, Schela Cladovei is surrounded by higher-elevation ground and within a few kilometers of the Carpathian-Balkan mountains (Bonsall, 2008); it is thus categorized as 'mountainous' using our terrain coding system (see Chapter 5). Given the large number of deer and wild pig remains at the site, as well as arrowheads (Bonsall *et al.*, 1997; Bonsall, 2008), it is apparent that the inhabitants spent considerable time in the uplands hunting terrestrial fauna. However, the most notable faunal remains are those from fish, particularly large anadromous species such as sturgeons, taken from the Danube River. Fishing was a very important economic activity throughout the Iron Gates

Skeletal Variation and Adaptation in Europeans: Upper Paleolithic to the Twentieth Century, First Edition. Edited by Christopher B. Ruff.

Table 13.1 Balkan samples.

Site	Present country	Period	Date range (cal.)	Male N	Female N	Total N
Schela Cladovei	Romania	Mesolithic	6400–7400 BC	12	5	17
Mistihalj	Bosnia-Herzegovina	Late Medieval	1400–1500 AD	27	27	54

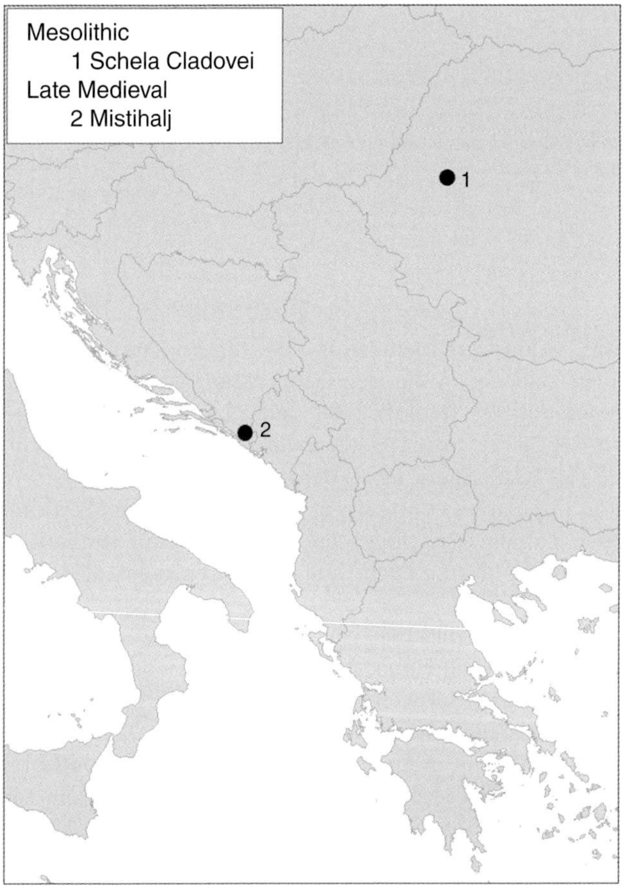

Figure 13.1 Map of Balkan site locations.

Mesolithic sites, with fish constituting perhaps the majority of the protein intake (Voytek and Tringham, 1990; Bonsall *et al.*, 1997; Bonsall, 2008; Boric *et al.*, 2014). Many of the sites, including Schela Cladovei, also have evidence for houses and hearths, as well as possible food storage facilities (Voytek and Tringham, 1990; Boroneant *et al.*, 1995; Bonsall, 2008). Such features, and the relatively large numbers of burials (i.e., 'cemeteries') have led some researchers to consider the occupants to be at least semi-sedentary, if not sedentary (e.g., Voytek and Tringham, 1990; Jochim, 2002), although the actual picture was likely to have been more complex (Bonsall, 2008). This issue is revisited in the Discussion. There is also evidence for intergroup conflict at Schela Cladovei, with a relatively high incidence of arrow wounds and skull and parry fractures (Boroneant *et al.*, 1995; Bonsall, 2008).

The Mistihalj sample includes 27 males and 27 females. The site is in a mountainous region of the Dinaric Alps near the town of Trebinje (Alexeeva *et al.*, 2003). Over 300 burials were originally excavated from a cemetery, including many juveniles as well as more recent (19th century) inhabitants that were not included in this study. Based on grave monuments and historical data, the population represented at Mistihalj was composed of the ethnic group known as Vlachs (or Vlakhs) (Alexeeva *et al.*, 2003). Vlachs were traditionally pastoral herdsmen, or nomadic cattle breeders practicing transhumance (seasonal migration), with a long history as a separate ethnic group in this region, from northern Greece through southern Bulgaria (Malcolm, 1994; Alexeeva *et al.*, 2003). They may be descendants of Illyrians, an ancient group of tribes with an ancestry stretching back into Classical Greek and Roman times (Malcolm, 1994; Alexeeva *et al.*, 2003), although at least one author has proposed that, based on their physical appearance, they may have affinities to the "...massive population [that] could be found in Mesolithic and Early Neolithic populations of Danube basin near Iron Gates" (Alexeeva *et al.*, 2003: 123). This is also further discussed later in this chapter. The traditional diet of Vlachs was apparently milk-based (Alexeeva *et al.*, 2003). In addition to herding, they were known (and sometimes employed) during the Medieval period as fighters, because of their 'hardy mountain-dwell[ing]' physiques and 'military tradition' (Malcolm, 1994: 72–74).

The Mistihalj skeletal sample, housed at the Harvard Peabody Museum, has been used in a number of studies of morphological variation among Late Pleistocene and Holocene humans (Trinkaus, 1981; Holliday, 1997a,b, 1999; Niewoehner, 2001; Cowgill, 2010; Cowgill *et al.*, 2012). Cowgill's 2010 study of ontogenetic changes in long bone robusticity, which included Mistihalj, is particularly relevant to the present analyses and is discussed in more detail later in the chapter.

Body size and shape parameters were calculated as described in Chapter 2. About half of the sample from Schela Cladovei had sufficient skeletal elements to estimate stature from the anatomical method, and body mass using the stature/bi-iliac method, which is a higher than average proportion for Mesolithics in our total European sample (about one-third) and is evidence for the relatively good preservation of the remains (Boroneant *et al.*, 1995). The Mistihalj sample was also relatively well preserved, with anatomical statures possible for about three-fourths of the sample, and stature/bi-iliac body masses for more than 90% (compared with 58% and 73%, respectively, for Late Medieval samples across Europe). Almost all of the Mistihalj individuals had all four long bones (femur, tibia, right and left humeri); the femur and tibia were relatively well represented at Schela Cladovei but humeri less so, with only seven individuals with both humeri. Cross-sectional properties were determined using the molding and biplanar radiography technique for the Schela Cladovei sample (see Chapter 3), and using a small computed tomography (CT; pQCT) scanner for the Mistihalj sample (Ferretti *et al.*, 1996; Ruff *et al.*, 2013).

13.2 Body Size and Shape

13.2.1 Schela Cladovei

Summary statistics for body size and shape parameters of the Schela Cladovei sample, by sex, are given in Table 13.2, along with similar statistics for other pooled European Mesolithic samples. Box plots of stature and body mass of the individual samples are shown in Figure 13.2(a) and (b).

Both males and females of the Schela Cladovei sample are considerably (6–9%) taller than other Mesolithics ($p < 0.01$; t-tests with pooled pan-European sample) (Fig. 13.2(a)). Males are also significantly heavier than other Mesolithics (12%, $p < 0.01$), while females overlap more, although on average they are still the heaviest sample (Fig. 13.2(b)). The increased body mass

Table 13.2 Body size and shape in Schela Cladovei and other Mesolithic samples.

	Schela Cladovei						Other Mesolithic					
	Male			Female			Male			Female		
Property	**N**	**Mean**	**SE**	**N**	**Mean**	**SE**	**N**	**Mean**	**SE**	**N**	**Mean**	**SE**
Stature	12	174.89	1.29	5	163.77	2.27	42	160.37	1.02	30	153.79	0.62
Body Mass	11	70.83	1.70	4	63.02	4.03	31	63.41	1.54	22	57.71	1.27
Bi-iliac Bd.	6	28.88	0.52	2	26.19	0.76	17	28.02	0.19	6	28.04	0.88
Crural	8	85.35	0.97	3	82.37	1.69	36	84.63	0.58	20	83.68	0.58
Rt. Brachial	6	75.90	0.88	2	74.63	0.69	5	77.36	0.93	7	73.99	0.68
Lt. Brachial	6	76.50	0.65				9	77.31	0.56	6	73.42	0.39
Rel. Sit. Ht.	2	37.60	0.42	1	38.04		8	40.24	0.41	3	40.30	0.38
Bi-iliac/Stat.	6	16.31	0.33	2	16.07	0.11	17	17.13	0.12	6	18.23	0.64
Rel. Clav. Len.	6	16.90	0.22	5	17.80	0.20	10	17.59	0.36	6	17.31	0.27

Underlined values are significantly different from other Mesolithic.

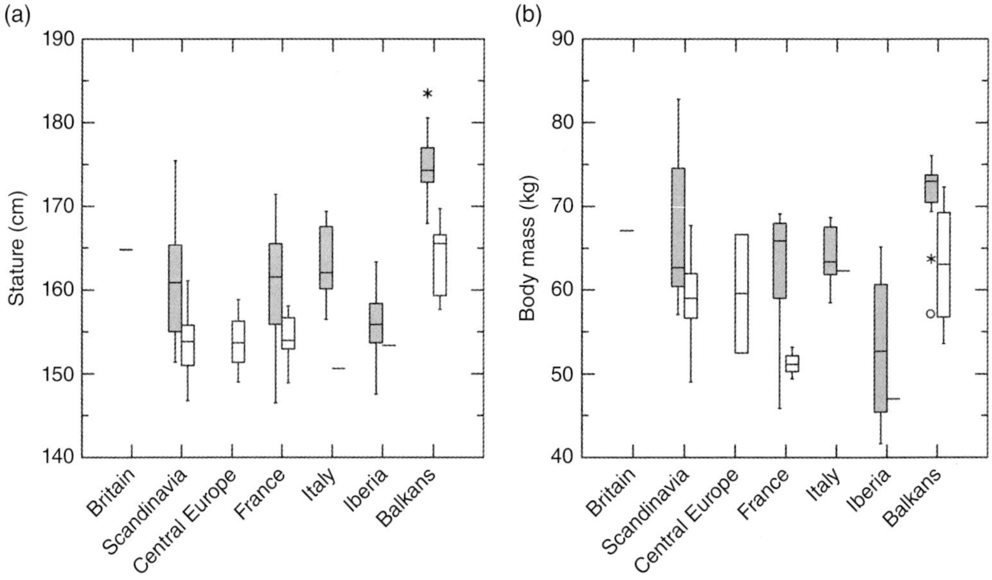

Figure 13.2 Boxplots showing regional variation in (a) stature and (b) body mass in Mesolithic samples. Open boxes: females; gray boxes: males.

at Schela Cladovei appears to be mainly a result of increased stature, not body (bi-iliac) breadth (Table 13.2), although sample sizes for the latter dimension are small. This is also supported by the lower bi-iliac breadth/stature ratios of Schela Cladovei compared to other Europeans (significant ($p < 0.05$) in females and near-significant ($p = 0.06$) in males), although again sample sizes are small, especially for females.

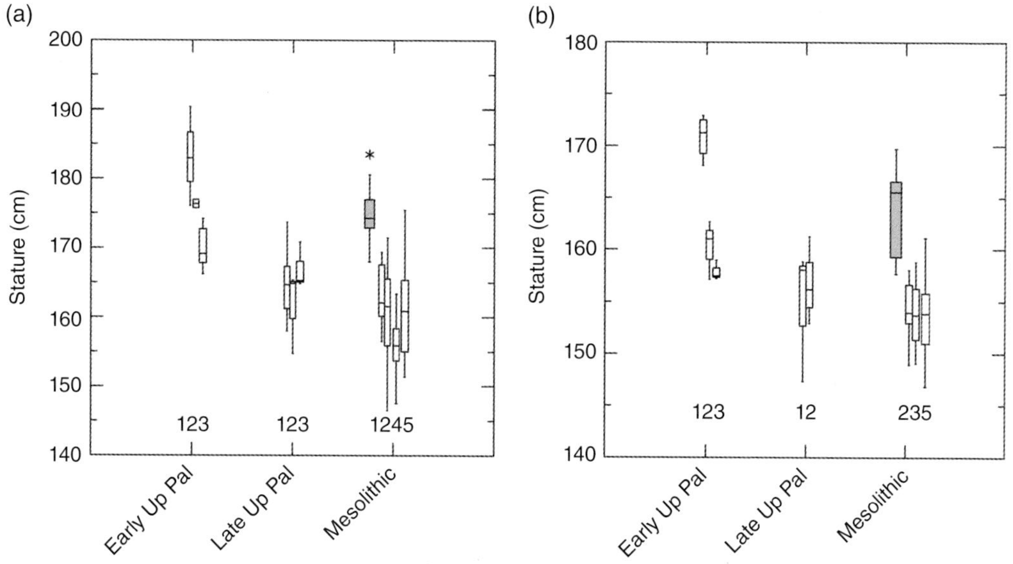

Figure 13.3 Boxplots showing regional variation in stature in (a) males and (b) females in Early Upper Paleolithic, Late Upper Paleolithic, and Mesolithic periods. Gray boxes: Schela Cladovei. Numbers for other boxes: 1) Italy; 2) France; 3) Central Europe; 4) Iberia; 5) Scandinavia. Only regional samples with >1 individual are plotted.

Comparisons of stature were also carried out between Mesolithic and Early and Late Upper Paleolithic samples from across Europe; results are shown in Figure 13.3(a) and (b) for males and females, respectively. The steep decline in stature across Europe between the Early and Late UP (as documented in Chapter 4) is clearly apparent, with a smaller decrease between the Upper Paleolithic and Mesolithic in Italy and France. However, the Schela Cladovei Mesolithic sample is an exception, being as tall, on average, as Early UP samples from other parts of Europe, and taller than other Late UP samples. Thus, the decline in stature during the Terminal Pleistocene and early Holocene characteristic of Western Europe does not appear to apply to this region, although without UP material from the Balkans it is difficult to evaluate temporal trends in this area. This general regional patterning has been noted previously (Formicola and Giannecchini, 1999) and is further discussed below.

The increased stature of Schela Cladovei relative to other Mesolithics appears to be largely attributable to longer lower limbs rather than longer trunks, which is similar to the proportional difference between Early and Late UP (and other Mesolithic) samples across Europe (Chapter 4). Relative sitting height is lower in Schela Cladovei than in other Mesolithics (significantly in males), but small samples sizes (n = 1–2; Table 13.2) limit meaningful comparisons. However, direct comparisons of reconstructed vertebral column length and lower limb length in somewhat larger Schela Cladovei samples (n = 3–8) show greater differences with other Mesolithics in lower limb length (males 14%; females 8%) than in vertebral column length (males 3%; females 5%) (data not shown), supporting this inference. Crural and brachial indices, and relative clavicular length, are similar to those of other Mesolithics (Table 13.2).

Average sexual dimorphism in stature (6.6%) and body mass (11.7%) at Schela Cladovei are similar to, although slightly greater than, that of other Mesolithics (4.2% and 9.4%, respectively) (Table 13.2). Sample sizes for body shape parameters in Schela Cladovei females are generally too small (n = 1–2) for meaningful comparisons of male and female means.

13.2.2 Mistihalj

Body size and shape statistics for Mistihalj and other Late Medieval European samples are given in Table 13.3, with box plots of stature and body mass for each region shown in Figure 13.4(a) and (b). Mistihalj males are slightly (5%) but significantly ($p < 0.05$) heavier than the pooled pan-European sample, although overlap is considerable and some other regional samples have very similar distributions (Fig. 13.4(b)). Their increased body mass is due to wider bodies (i.e., bi-iliac breadth; $p < 0.05$) rather than stature, which is non-significantly different from other Europeans (Table 13.3; Fig. 13.4(a)). Thus, bi-iliac breadth/stature is significantly greater in Mistihalj males ($p < 0.001$). Mistihalj females show a similar, but much less marked pattern (body mass, $p < 0.10$ with other Europeans, non-significant stature and bi-iliac breadth comparisons).

The crural index is significantly higher in both Mistihalj males and females ($p < 0.001$) than in other Late Medieval Europeans (Table 13.3). However, this likely due to a latitudinal effect whereby Southern Europeans have relatively longer tibiae, probably as a result of climatic adaptation (Ruff *et al.*, 2012: also see Chapter 4). If only Southern Europeans are included in the pan-European sample, differences in crural index with Mistihalj become non-significant for both males and females ($p > 0.30$). Latitudinal variation may also influence the significantly lower relative sitting height of Mistihalj males (Table 13.3), but the difference is still significant if only Southern Europeans are included. Brachial indices and relative clavicular length are similar to other Late Medieval Europeans. Average sexual dimorphism in stature (5.6%), body mass (19.6%), and other morphological parameters is similar to that in other Europeans.

13.3 Cross-Sectional Properties

13.3.1 Schela Cladovei

Femoral, tibial, and humeral cross-sectional properties for Schela Cladovei and other Mesolithic Europeans are given in Table 13.4. Standardized tibial and femoral A-P bending strengths (Z_x) for all regions are plotted in Figures 13.5(a) and (b).

Schela Cladovei males tend to have greater than average strengths (both Z_x and Z_p) compared to other Mesolithic males for both the femur and tibia, although differences do not reach significance. However, there is one very low Schela Cladovei outlier for femoral Z_x (Fig. 13.5(b)). This individual is considerably lower for this property than any other Mesolithic male across Europe; he also has a very low (for a male) femoral Z_x/Z_y value (0.84), as well as the lowest values in the sample for humeral strength (no tibia was available for this individual). Although not small in body size by overall male Mesolithic standards, his estimated stature (168 cm) is the shortest among the Schela Cladovei males and overlaps with the five females of the sample (Fig. 13.2(a)). No obvious skeletal pathologies were noted, but his sex designation was noted as being 'probable.' The present results indicate that he may actually have been female, given the smaller body size and lower relative strength of females (Fig. 13.5; also see below). Therefore, bone cross-sectional statistics for males were calculated both with and without this individual (femur and right humerus), and are also shown in Table 13.4. When this individual is not included, size-standardized femoral Z_x for males from Schela Cladovei is significantly (17%; $p < 0.01$) higher than in other Mesolithic males. Tibial A-P bending strength (not affected by this individual) shows almost the same degree of relative difference (15%), but does not reach significance, possibly due in part to the smaller sample sizes available for this bone (Table 13.4).

Table 13.3 Body size and shape in Mistihalj and other Late Medieval samples.

	Mistihalj						Other Late Medieval					
	Male			Female			Male			Female		
	N	Mean	SE	N	Mean	SE	N	Mean	SE	N	Mean	SE
Stature	27	168.32	1.13	27	159.20	0.91	256	167.56	0.41	221	157.63	0.40
Body Mass	27	<u>72.11</u>	1.31	27	59.21	0.89	251	68.61	0.49	217	57.54	0.41
Bi-iliac Bd.	24	<u>29.96</u>	0.33	25	28.77	0.34	184	29.03	0.14	166	28.16	0.15
Crural	27	<u>83.44</u>	0.43	26	<u>83.21</u>	0.32	218	81.57	0.15	180	81.54	0.16
Rt. Brachial	23	75.03	0.42	22	74.14	0.47	153	74.71	0.18	135	73.45	0.21
Lt. Brachial	23	75.02	0.56	23	74.00	0.52	147	74.97	0.19	127	73.70	0.22
Rel. Sit. Ht.	19	<u>38.21</u>	0.37	20	40.00	0.27	62	39.30	0.13	56	40.29	0.16
Bi-iliac/Stat.	24	<u>17.89</u>	0.15	25	18.08	0.20	182	17.31	0.08	165	17.80	0.09
Rel. Clav. Len.	22	17.84	0.17	21	17.47	0.17	171	17.92	0.07	146	17.26	0.07

Underlined values are significantly different from other Medieval.

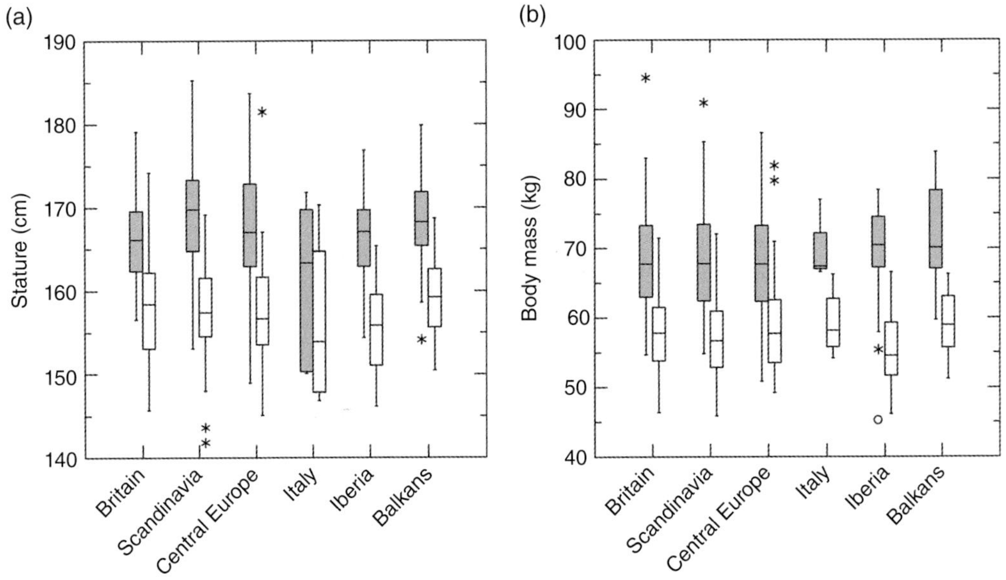

Figure 13.4 Boxplots showing regional variation in (a) stature and (b) body mass in Late Medieval samples. Open boxes: females; gray boxes: males.

In contrast, humeral strength is not increased in Schela Cladovei relative to other Mesolithic males; in fact, it is lower on the right side (Table 13.4), reaching near-significance ($p = 0.054$) with the individual above included, and still about 20% lower on average without this individual. Left humeri are very similar in strength to those of other Mesolithic males. Because of this differential trend on the two sides, average humeral asymmetry is close to 0 among Schela Cladovei males, while it averages about 10% in other Mesolithic males. However, this obscures some significant variation among Schela Cladovei males: as shown in Figure 13.6, three males have asymmetry values of –9% to –14% (i.e., left side larger) and two have values of 11% and 17%. Thus, males were either strongly 'left dominant' or 'right dominant,' with a majority left dominant. This is a very unusual pattern among males, who tend to have a large predominance of right-handers throughout the Terminal Pleistocene and Holocene (see Chapter 7). Only one out of 11 Mesolithic males from other regions (Teviec 11) has strong (>5%) left-dominant humeral strength asymmetry (Fig. 13.6), which is equivalent to the frequency for the entire European male sample across all periods (about 9%). Even given the small number of paired male Schela Cladovei humeri, with a consequent increased chance of sampling bias, this still suggests some unusual use of the upper limb among males (a Fisher exact test comparing the frequency of strong left dominance among Schela Cladovei males and other Mesolithic males is near-significant, $p = 0.063$, and compared to males in the entire sample is highly significant, $p = 0.006$). Possible behavioral implications of this observation are discussed below.

The female sample from Schela Cladovei with cross-sectional data is small (Table 13.4), but shows some interesting differences from the male sample. On the whole, females show no evidence for increased relative bone strength in either the lower or upper limb compared to other Mesolithics. They also show a normal pattern of humeral bilateral asymmetry, with both of the available individuals being 'right dominant' by 15% and 19%, compared to an average of 13% in other Mesolithic females, although sample sizes are quite small in both groups (Fig. 13.6). Because of the relatively increased strength of the lower limb bones among males, but not females, average sexual dimorphism in femoral and tibial strength in Schela Cladovei is generally

Table 13.4 Bone strengths in Schela Cladovei and other Mesolithic samples.

	Schela Cladovei						Other Mesolithic					
	Male			Female			Male			Female		
	N	Mean	SE	N	Mean	SE	N	Mean	SE	N	Mean	SE
Tibia Z_{xstd}	7	864.0	63.3	3	561.8	42.4	21	749.4	34.5	5	650.2	40.5
Femur Z_{xstd}	11 [10]	750.1 [<u>783.6</u>]	41.7 [27.6]	4	632.3	54.3	26	671.0	17.7	10	599.5	29.4
Tibia Z_{pstd}	7	1189.1	68.9	3	871.0	66.3	21	1075.7	46.0	5	910.3	62.2
Femur Z_{pstd}	11 [10]	1271.4 [1313.2]	53.8 [37.4]	4	1186.4	113.3	26	1236.0	35.6	10	1109.9	51.8
Rt. Hum. Z_{pstd}	7 [6]	565.5 [600.5]	56.1 [51.9]	1	613.0		9	725.0	50.5	5	620.5	106.9
Lt. Hum. Z_{pstd}	5	634.8	36.2	4	433.8	32.0	8	657.1	49.0	6	537.9	80.9
Hum. Z_p asym.	5	−1.5	6.4	2	17.0	1.9	11	9.8	3.2	3	12.7	6.0
Femur PCCA	12 [11]	78.8 [79.1]	1.7 [1.8]	5	80.4	2.3	39	77.7	1.0	18	80.1	1.8
Tibia PCCA	8	<u>78.6</u>	1.7	4	75.5	3.7	34	83.5	0.7	11	81.0	1.5
Rt. Hum. PCCA	8 [7]	<u>65.9</u> [68.4]	4.2 [3.9]	2	76.3	6.9	6	78.7	4.0	5	67.9	7.6
Lt. Hum. PCCA	5	67.1	4.7	5	67.2	4.3	6	79.6	3.6	5	73.2	7.2

Bracketed values: without one male outlier (see text).
Underlined values are significantly different from other Mesolithic.

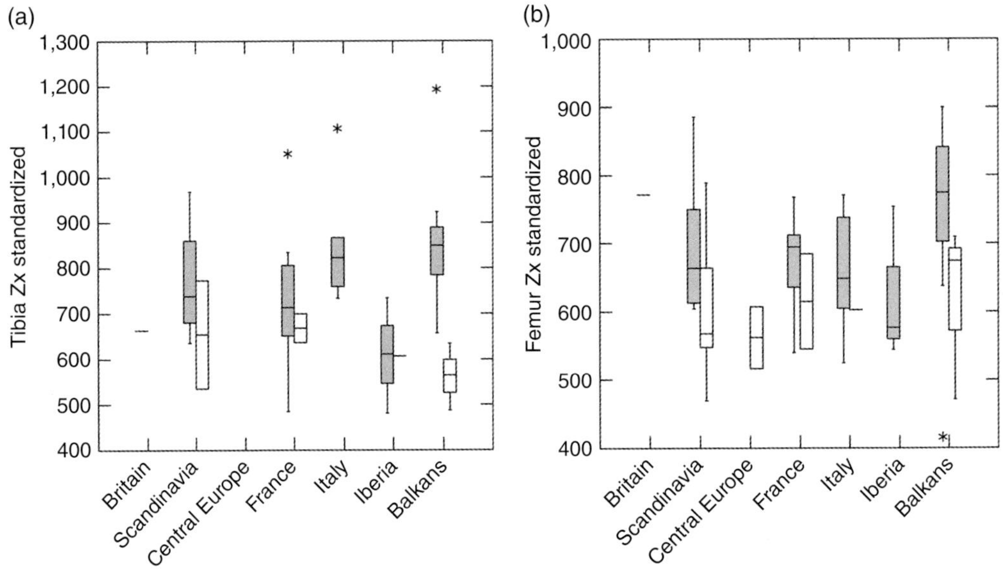

Figure 13.5 Boxplots showing regional variation in (a) tibial and (b) femoral anteroposterior bending strength, standardized for body size, in Mesolithic samples. Open boxes: females; gray boxes: males.

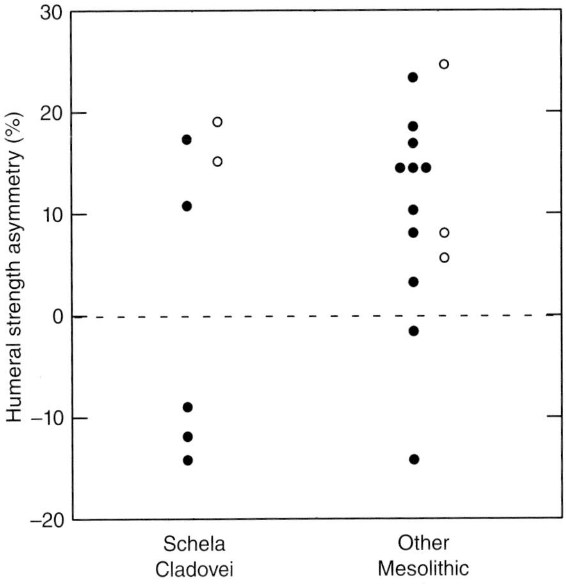

Figure 13.6 Humeral directional asymmetry (R/L) in average bending/torsional strength in Schela Cladovei and other Mesolithic sites. Filled circles: males; open circles: females.

higher (21–42%) than in other Mesolithics (11–17%), except for femoral Z_p (10–11% in both), although this must be interpreted with caution since the pooled non-Schela Cladovei sample combines variable numbers of males and females from different sites (see other chapters in this volume). The ratio of A-P/M-L bending strength (Z_x/Z_y) in the tibia and femur is significantly larger ($p < 0.05$) in Schela Cladovei males (1.665 ± 0.114 and 1.082 ± 0.041 (SE), respectively) than in females (1.314 ± 0.072 and 0.970 ± 0.012, respectively), which is similar to Mesolithic samples as a whole (see Chapter 6). Average sexual dimorphism in left humeral strength is also large in Schela Cladovei (38%, compared to 20% in other Mesolithics). The one value for right humeral strength in Schela Cladovei females is not sufficient for a meaningful comparison with males, although it is interesting that it falls above the male mean.

Percentage cortical areas for the tibia and humeri are significantly ($p < 0.05$) or near-significantly (left humerus, $p = 0.067$) smaller in males from Schela Cladovei than in other Mesolithic males (Table 13.4). This cannot be accounted for by age differences, since age means and distributions are quite similar in the two groups (data not shown). However, no difference in PCCA is found in the bone with the largest samples (femur, $n = 12$ and 39 for Schela Cladovei and other Mesolithic males, respectively), nor are there any differences among females.

13.3.2 Mistihalj

Cross-sectional data for Mistihalj are given in Table 13.5, with regional comparisons for tibial and femoral A-P bending strengths (Z_x) shown in Figures 13.7(a) and (b). Mistihalj males and females fall near the middle of the distributions of Late Medieval Europeans for femoral, tibial, and humeral strengths, which overall show little regional variation. Sexual dimorphism in relative bone strengths is also close to average for Late Medieval samples: 8–17% for the lower limb and 21–27% for humeri in Mistihalj (males greater), which is generally lower than in Schela Cladovei (see above). Sexual dimorphism in A-P/M-L bending strength (Z_x/Z_y) of the tibia and femur is non-significant in Mistihalj. Mistihalj males show significantly ($p < 0.001$) more asymmetry in humeral strength than females, a pattern that mirrors the rest of Europe during this and most other time periods (also see Chapter 7). In stark contrast to Schela Cladovei, all 27 Mistihalj males are 'right dominant,' although two have asymmetry values very close to (within 1% of) 0.

Table 13.5 Bone strengths in Mistihalj and other Late Medieval samples.

	Mistihalj						Other Late Medieval					
	Male			Female			Male			Female		
	N	Mean	SE	N	Mean	SE	N	Mean	SE	N	Mean	SE
Tibia Z_{xstd}	27	650.9	16.1	26	547.0	17.9	155	629.0	10.7	124	515.5	8.6
Femur Z_{xstd}	27	612.7	13.8	27	541.5	17.0	177	592.9	7.5	150	528.5	7.6
Tibia Z_{pstd}	27	949.7	24.0	26	814.4	26.2	155	938.6	14.5	124	783.2	12.2
Femur Z_{pstd}	27	1153.8	24.6	27	1061.2	30.0	177	1152.4	13.0	150	1051.8	14.2
Rt. Hum. Z_{pstd}	27	628.4	20.6	25	478.7	18.3	121	621.3	11.9	103	500.4	11.3
Lt. Hum. Z_{pstd}	27	569.0	16.4	27	461.7	17.0	120	578.4	10.8	97	473.1	9.8
Hum. Z_p asym.	27	11.4	1.5	25	4.6	1.2	96	8.7	1.3	78	5.0	1.1
Femur PCCA	27	72.8	1.5	27	<u>66.7</u>	1.9	182	73.5	0.5	153	72.1	0.6
Tibia PCCA	27	73.7	1.5	26	66.3	2.2	162	73.2	0.6	130	69.7	0.7
Rt. Hum. PCCA	27	74.0	2.2	25	<u>64.2</u>	2.5	130	73.2	0.8	112	69.9	0.9
Lt. Hum. PCCA	27	73.2	1.8	27	<u>63.1</u>	2.4	129	73.7	0.8	104	69.1	1.0

Underlined values are significantly different from other Late Medieval.

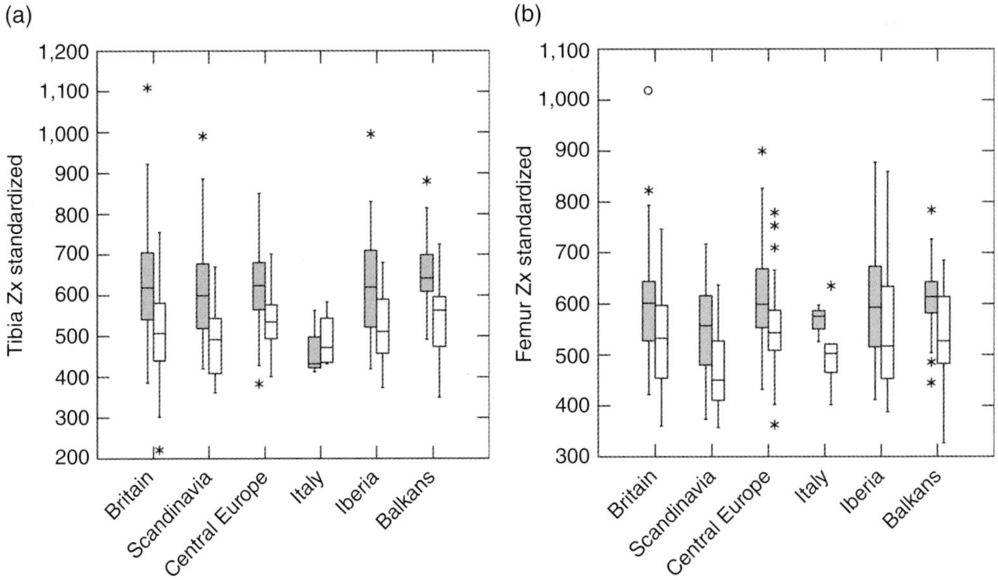

Figure 13.7 Boxplots showing regional variation in (a) tibial and (b) femoral anteroposterior bending strength, standardized for body size, in Late Medieval samples. Open boxes: females; gray boxes: males.

Percentage cortical areas (%CA) of Mistihalj males are similar to those of other Mesolithic males, but Mistihalj females have significantly lower %CA than other Mesolithics in the femur and both humeri (Table 13.5).

13.4 Discussion

13.4.1 Body Size and Shape

The large body size and 'robust' skeletons of the Mesolithic inhabitants of the Iron Gates region have been noted by multiple authors (Prinz, 1987; Boroneant *et al.*, 1995; Formicola and Giannecchini, 1999). As shown in Figure 13.3, they are also clear outliers in terms of the general stature reduction observed across Europe between the Early Upper Paleolithic and Mesolithic (Frayer, 1980, 1981, 1984; Angel, 1984; Meiklejohn *et al.*, 1984; Jacobs, 1985b; Holt and Formicola, 2008; Meiklejohn and Babb, 2011; also see Chapter 4). In fact, some of the variation in previously reported results regarding whether stature continued to decline between the Late UP and Mesolithic were likely influenced by differences in sample composition, that is, whether Eastern European Mesolithic samples were included or not (in some cases results were also affected by pooling of the entire UP).

The Schela Cladovei sample also shows greater similarity in body shape to Early UP European specimens than to Western European Mesolithic samples. Relative sitting height and bi-iliac breadth/stature were both reduced in Schela Cladovei, similar to the Early UP (Chapter 4) (mean values are actually slightly lower in Schela Cladovei than in the pooled Early UP sample). Both features are related to relatively increased lower limb length.

As shown most clearly by Formicola and Giannecchini (1999), tall statures were generally characteristic of all available Eastern European (Balkan, Russian, Ukranian) Mesolithic samples, in comparison to Western European samples (extending through Germany and Sweden).

The various factors – climatic, demographic, nutritional, technological – that may have contributed to the decline of stature following the Early UP in Western Europe have been discussed at length in the above references, and are summarized in Chapter 4 (also see below). As pointed out by Formicola and Giannecchini (1999), similar temporal analyses in Eastern Europe are made more tenuous by the paucity of UP specimens from this region. However, the available data from Sunghir, Russia, and the Moravian sites of Pavlov, Dolni Vestonice, and Premosti (all Early UP, as defined here, i.e., ≥26,000 cal BP) are consistent with the relatively tall statures of Western European Early Upper Paleolithic specimens (see Fig. 13.3; Sunghir 1, not included in that figure, has an estimated stature of 178.8 cm). Thus, it is most parsimonious to conclude that Eastern European UP (at least Early UP) populations were also tall, and that stature did not decline through the Mesolithic in this region to the same extent as in Western Europe. The same is true, to a somewhat lesser extent, of body mass (Table 13.2; also see Chapter 4). It is, of course, possible that body size declined in the Late UP in Eastern Europe and then increased again in the Mesolithic, but this seems less likely.

Why should body size have remained large in the Eastern European Mesolithic, and in the Iron Gates region in particular? One possible explanation is nutritional variation between regions, given the correlation between nutrition and body size among modern populations (e.g., Grasgruber *et al.*, 2014). The nutritional and general health status of the Mesolithic Iron Gates populations has been debated. Meiklejohn and Zvelebil (1991), based largely on y'Ednak's (1989) analysis of dental enamel hypoplasia in a sample from Vlasac, concluded that Mesolithic populations in the Iron Gates region (and other 'hunter-fisher groups in northern and eastern Europe') were characterized by poorer health than those from the Western Mediterranean, which they attributed to "...increased residential permanence and dependence on a fish diet [that] resulted in parasitic infestations and infections which prejudiced the health of these hunter-fisher communities" (Meikeljohn and Zvelebil, 1991: 136). However, in another study, y'Ednak and Fleisch (1983) note that enamel hypoplasia was more prevalent in a Neolithic sample from Lepenski-Vir III (also from the Iron Gates region) than in Vlasac, and that "In either case, enamel hypoplasia is never severe. Only small pits or fine lines manifest the disorder" (p. 291). With respect to the remains from Schela Cladovei, Boroneant *et al.* (1995) noted that the "...population was well nourished and generally in good health ... There were no carious lesions on teeth, and no obvious signs of malnutrition" (p. 389), although they did find evidence of periodontal disease and calculus, which they attributed to 'poor dental hygiene.' Bonsall *et al.* (1997) similarly noted that for the region as a whole "There is little unequivocal evidence for nutritional stress or diet-related diseases among the Mesolithic population" (p. 85). As discussed earlier, the diet of these people was based largely on freshwater fish, with additional contributions from terrestrial mammals (Bonsall *et al.*, 1997). Such a high-protein diet, specifically from fish and animal products, has been shown to be positively correlated with height among male samples from Europe and European-derived living populations (Grasgruber *et al.*, 2014). Thus, it is possible that the specific diet made possible by the particular environment of the Iron Gates region contributed to their large body size. Whether this would also apply to other Eastern European Mesolithic sites (Formicola and Giannecchini, 1999) is unclear. Also, other European Mesolithic regions that also relied heavily on fish in their diets, such as Scandinavia (e.g., Anderson, 2004; Bogucki, 2004), do not show the same large body size (Formicola and Giannecchini, 1999; also see Chapter 12). More detailed analyses of specific dietary components across Mesolithic sites are needed to more fully evaluate this hypothesis.

Climatic adaptation (i.e., Bergmann's and Allen's Rules) has also been considered as a possible factor contributing to the reduction in relative limb length and consequently stature among most Late UP and Mesolithic populations (Trinkaus, 1981; Ruff, 1994; Holliday, 1997a, 1999).

However, this would not explain why Eastern European Mesolithic populations specifically maintained taller statures and relatively longer limbs than Western European Mesolithics, since temporal variation in climate should (broadly) have been similar in both regions. Jacobs (1985a) reported a correlation between intra-limb (crural and brachial) indices and temperature and latitude among European Mesolithic specimens (colder/higher latitude associated with smaller indices), suggesting a climatic explanation for geographic variation and a possible explanation why the Iron Gates samples might have retained longer limbs. However, relatively warmer conditions also characterized southwestern Europe during the Mesolithic, and our Iberian Mesolithic samples are relatively short, with low crural indices (see Chapter 10). Furthermore, North-Central Europeans do not have higher relative sitting heights (i.e., shorter lower limbs) than Southern Europeans as a whole during the Mesolithic (even eliminating the Schela Cladovei sample), although they do in most of the rest of the Holocene (see Chapter 4).

Changes in technology and food acquisition techniques, specifically, a reduction in big-game hunting and close-quarter killing of prey, and consequent reduced selection for large body size, has also been proposed as a driving force behind body size trends between the UP and Mesolithic (Frayer, 1980, 1981, 1984). However, these same general trends characterized both Eastern and Western Europe, so it is difficult to see how they could explain the overall geographic variation observed. Behavioral implications of Schela Cladovei long bone strength characteristics are discussed below.

Paleodemographic factors involving decreased mobility, increased density and regionalization of populations, and reduced gene flow between populations, may also have contributed to reductions in body size between the UP and Mesolithic (Formicola and Giannecchini, 1999; Holt and Formicola, 2008). The Iron Gates region, however, was certainly densely settled during the Mesolithic, and the evidence for extensive inter-group conflict suggests increased local differentiation of populations (Boroneant *et al.*, 1995; Bonsall, 2008), which should have had the same effects.

In additional to possible environmental effects, genetic factors may also play a role in explaining the large body size of Mesolithic Iron Gates populations. Recent DNA analyses of both living (Turchin *et al.*, 2012) and early Holocene (Mathieson *et al.*, 2015) European samples have demonstrated the presence of genetic variants associated with shorter stature in Southern Europeans and taller stature in Northern Europeans, as a result of population history and migration patterns. 'Southern Europeans' in these studies included Iberian Peninsula and Italian samples, but not southeastern European populations. Interestingly, modern inhabitants of Bosnia-Herzegovina, Serbia, and Montenegro, particularly males, are a striking exception to this general North-South trend: they are among the tallest, if not the tallest populations in Europe (Coon, 1939; Pineau *et al.*, 2005; Grasgruber *et al.*, 2014; NCD, 2016). They also constitute an exception to general trends relating socioeconomic factors (GDP, per capita health expenditure) and stature among European populations, since they have among the lowest scores for the former and yet surpass almost all populations in the latter (Grasgruber *et al.*, 2014). On the whole, 'Eastern' Europeans (which in the above study included Czechs, Slovaks, and Hungarians) fell above 'Western' Europeans in stature predicted from these variables, with the Balkan samples particularly high outliers. The authors identified several genetic variants that appear to be correlated with this pattern, some of which may have their roots as far back as the Upper Paleolithic. Overall patterns of variation in height across Europe were best explained as a combined result of environmental (dietary) and genetic factors. Thus, the tall statures of the Iron Gates as well as other Eastern European Mesolithic samples may be in part due to long-standing genetic variation, based on the particular population history of this region.

Unlike the Schela Cladovei sample, the Mistihalj sample does not stand out from other contemporaneous European samples in stature, although Mistihalj males are slightly greater in body mass due to their relatively wider bodies. Both Mistihalj males and females have relatively wider bodies than those from Schela Cladovei ($p < 0.001$), due to a combination of shorter stature and absolutely wider bodies. Trunk shape in the Mistihalj sample is thus relatively stocky; however, in keeping with Southern Europeans in general (see Chapter 4), lower limbs are relatively long.

Alexeeva *et al.* (2003) considered the Mistihalj sample to represent a 'massive population' of 'high stature,' with postcranial bones that were of 'very large size and robusticity,' similar to Mesolithic Iron Gates samples, and also pointed out similarities to modern high-statured peoples of the region. However, as shown here, Mistihalj, while certainly not small-bodied, was not particularly large by European Late Medieval standards, and in this respect does fit into the pattern represented by earlier and later populations. It is possible that the average stature and relatively stocky bodies of the Mistihalj sample may be due to their association with a specific subpopulation, that is, the Vlach ethnic group. No physical description of the Vlachs other than that they were 'hardy mountain dwellers' (Malcolm, 1994: 74) seems to exist. Similarly, no physical size data for other contemporaneous non-Vlach samples from the Balkans are available, so comparisons are not possible. As discussed at length by Malcolm (1994: 70–81), while formerly the Vlachs may have been a relatively distinct group, more recently there has been more assimilation into the populations of Bosnia-Herzegovina, Serbia, and Montenegro. Thus, some Vlach heritage may be represented in the modern anthropometric surveys of this region (Pineau *et al.*, 2005; Grasgruber *et al.*, 2014; NCD, 2016). However, given their former minority status, it seems very unlikely that they would have contributed more than a small percentage to the current gene pool. Contra Alexeeva *et al.* (2003), no special relationship between the Mistihalj sample and earlier Iron Gates samples is supported by the present data.

In terms of environmental variables, the diet of the Vlachs is said to have been based on 'milk products' (Alexeeva *et al.*, 2003), in keeping with their pastoralist lifestyle (Malcolm, 1994; Alexeeva *et al.*, 2003). Dependence on milk products (e.g., cheese) is strongly associated today with increased stature, both in Europe and world-wide (Grasgruber *et al.*, 2014, 2016). Thus, at least in this sense, the Mistihalj inhabitants were likely relatively well-nourished. Based on their observations of high percentages of certain pathologies at Mistihalj, in particular cribra orbitalia in infants, Alexeeva *et al.* (2003) felt that there was "...stress influence of negative condition" on juveniles, but that the "...adult population demonstrated adaptive patterns to local environment" (p. 121), which implies recovery among older individuals.

13.4.2 Bone Structure

As noted earlier, the inhabitants of the Iron Gates region are often cited as examples of at least semi-sedentary, if not sedentary Mesolithic populations, with this lifestyle largely made possible by the abundant aquatic resources of the region (Voytek and Tringham, 1990; Jochim, 2002; Bonsall, 2008). Certainly, the archaeological evidence is consistent with a degree of residential sedentism. However, the extent to which these sites were actually permanent, year-round settlements has been debated (Bonsall, 2004, 2008). The abundance of mammalian faunal remains at the sites, as well as more recent isotopic evidence for exploitation of terrestrial as well as aquatic resources in the Late Mesolithic (Nehlich *et al.*, 2010), indicates that there was likely considerable logistic mobility, at least. Seasonal residential mobility, that is, relocation to more inland (relative to the Danube) sites, is also possible, especially given the limited seasonal availability of some important aquatic resources such as sturgeon (Dinu, 2010). As noted by a number of researchers (e.g., Kelly, 1992), it is often difficult to identify the types and levels of mobility and sedentism practiced by past populations from archaeological remains alone. This is one area where biomechanical analyses of the human skeletal remains themselves may provide some additional insights.

Schela Cladovei males have high femoral and tibial A-P bending strengths compared to other Mesolithic males, especially when the one low male (possibly female) outlier is removed. This in itself argues against this sample being more sedentary than average for European Mesolithics. While some of the other Mesolithic samples, such as some of those from coastal Portugal and Scandinavia, were likely fairly sedentary (see other chapters in this volume), others were not, and Mesolithics as a whole had higher relative A-P bending strengths of the femur and tibia than later, more sedentary populations (Ruff *et al.*, 2015). Interestingly, Schela Cladovei females are average or even somewhat below average in relative strengths of the femur and tibia compared to other Mesolithics, creating more sexual dimorphism in strength, especially A-P bending strength. This implies more sexual dimorphism in mobility than average for Mesolithic samples. This observation would be consistent with greater logistical mobility in males, perhaps in association with inland hunting of game, with a more sedentary lifestyle among females. Stable isotope analyses of Iron Gates samples are mixed in terms of sexual dimorphism in diet, and the number of individuals analyzed to date from the Schela Cladovei sample is too small (only two females) to assess possible sex differences (Bonsall *et al.*, 1997). Given observed variation in stable isotope results for different Mesolithic Iron Gates samples (Bonsall *et al.*, 1997, 2015; Nehlich *et al.*, 2010), it would be quite interesting to compare bone structural properties across sites as well, to further explore possible differences in mobility and subsistence strategies.

Another factor that may have influenced lower limb loading in the Schela Cladovei sample is terrain. As noted previously, although the site is located on a riverine terrace east of the main gorge area, it is within a few kilometers of much more rugged terrain, and it likely that hunting, at least, involved forays into this upland country. As shown in Chapter 5, mountainous and hilly terrain is associated with increases in lower limb strength, particularly A-P bending strength. When Schela Cladovei males are compared to only similar, 'mountainous' terrain Mesolithic males ($n = 8$), they are still significantly greater ($p < 0.01$) in femoral A-P bending strength if the one low outlier is removed ($p = 0.08$ with the outlier). Unfortunately, there are insufficient data from other regions to carry out the same comparison among females, although given their position relative to Mesolithics as a whole, it seems unlikely that Schela Cladovei females would have greater lower limb strength than mountainous terrain females. At the least, then, these results do not support a sedentary lifestyle for Schela Cladovei males, while Schela Cladovei females may have been relatively sedentary.

The contrast between lower and upper limb bone strength in Schela Cladovei males, with males showing reduced right humeral strength compared to other Mesolithic males, reinforces this conclusion, that is, increased lower limb bone strength in Schela Cladovei males was not simply a result of generally increased strength throughout the skeleton. The relatively large number of upper limb left-dominant males (but not females) is also striking. Symmetric use of the upper limbs is characteristic of early agricultural females in Europe (see Chapter 7), but is obviously different in pattern (many females with very little asymmetry, versus either strong right or left asymmetry in Schela Cladovei males) and underlying causes (two-handed grain grinding in early agricultural females). What behavioral explanation could account for the high proportion of left-dominant males at Schela Cladovei?

One of the most remarkable aspects of the subsistence economy at Schela Cladovei was the large number of very large-bodied anadromous fish remains, including sturgeons that may have weighed as much as 150 kg (Bartosiewicz *et al.*, 2008). Even larger specimens are known from this region in later historic times (Bartosiewicz *et al.*, 2008; Dinu, 2010). Methods for catching such large fish (and other fish) along the Danube in later periods consisted of fencing and 'basketing' or weirs (ibid.). No definitive evidence for use of such techniques has been recovered in the Iron Gates Mesolithic (although 'sling balls' have been found which may have been used as net weights; Prinz, 1987: 97), but they were commonly used throughout Northern and Eastern (Russian) European sites during the Mesolithic (Anderson, 2004; Bogucki, 2004; Zhilin, 2014), so it is very likely that they were also used along the Danube. This is also suggested by the specific placement of Iron Gates sites near the entrance of smaller tributaries or islands, both of which are favorable for use of this technique, as well as by the gradient of the Danube itself near sites where sturgeon remains have been found (Bartosiewicz *et al.*, 2008; Dinu, 2010). A somewhat similar fishing technique involving the use of nets (although in a marine environment) was practiced by a Ligurian (Italian) Medieval population sample studied by Sparacello and Marchi (2008). They also found reduced directional asymmetry in humeral strength among males from the sample, compared to an earlier sample from the same region that relied more on agriculture, which they attributed to bimanual activities associated with net casting and dragging. Underwater hand-netting of fish has also been suggested as an alternative capture technique at Vlasac, based on the incidence and form of auditory exostoses observed in the sample (Frayer, 1988); this would have involved swimming, which also leads to reduced directional asymmetry in upper limb bone strength (Shaw and Stock, 2009).

Interestingly, left clavicles have also been noted to be particularly 'robust' in the Vlasac sample, although this was reported to be characteristic of both sexes (Prinz, 1987: 83). 'Robust' was not defined, but probably refers to clavicular shaft breadth/length; clavicular breadths were not measured in the present study so this cannot be assessed in our sample. Clavicular shaft breadth is normally right-biased and is correlated with humeral shaft breadth asymmetry, so it appears to reflect the same biomechanical factors, that is, a greater mechanical loading of the 'dominant' upper limb (Auerbach and Raxter, 2008). The apparent left bias of the Schela Cladovei clavicles is thus consistent with our results for humeri. The author suggested that this unusual pattern may have resulted from left-handed activities associated with fishing at Vlasac, although she left open the possibility of other behavioral explanations (Prinz, 1987).

The Mistihalj sample is similar in relative bone strength to other Late Medieval samples across Europe. As shown in Chapter 5 (also see Ruff *et al.*, 2015), the Late Medieval period represents a late (post-Bronze Age) peak in relative bone strength, which may be attributed to the intense manual labor associated with the agrarian economies typical of most of these samples. The cattle-breeding and herding in rough terrain carried out by the Vlachs buried at Mistihalj would be expected to similarly create high mechanical loadings on the limbs. Sexual dimorphism in limb bone strength and humeral bilateral asymmetry is also normal for Late Medieval (and Holocene populations in general), unlike the unusual patterns observed at Schela Cladovei. Despite the Vlachs' reputation and exploitation as warriors (Malcolm, 1994), there is no evidence for unusual levels of humeral right dominance in strength among males, as has been reported for some other 'warrior' samples (Sparacello *et al.*, 2011). In terms of direct comparisons across periods within the Balkans, Mesolithic males are significantly ($p < 0.01$) stronger in the femur and tibia, but not humerus, than Medieval males, and females do not differ significantly between periods. This again highlights the probable high mobility of Schela Cladovei males.

As noted earlier, The Mistihalj sample has been included in a number of previous studies of Holocene postcranial morphological variation, with Cowgill's (2010) ontogenetic study of limb bone strength being particularly relevant to the present study. In comparisons to several other Holocene ontogenetic samples, Mistihalj, along with Point Hope Inupiats, exhibited consistently greater relative bone strength of the right femur and humerus throughout development. The other samples included a range of subsistence strategies, but in general (except for Point Hope Inupiats) represented more sedentary populations, so the increased strength observed in Mistihalj was not surprising. However, the presence of increased strength from a very early age (<1 year) called into question a purely environmental explanation (i.e., increased mechanical loading) and suggested possible genetic contributions to inter-populational variation. While there are certainly genetic influences on long bone cross-sectional geometry, in part mediated by genetic effects on body size (e.g., Karasik *et al.*, 2008), it is also possible that pre-natal environmental factors played a role in these results. Pre-natal muscle contractions and fetal motility are well known to influence skeletal morphology, including cross-sectional diaphyseal geometry (Muller, 2003; Sharir *et al.*, 2011). Dietary and metabolic variation in the mother can also affect the bone structural properties of the offspring (Tatara *et al.*, 2007). No study has investigated the effects of maternal activity *per se* on long bone structure of newborns, although there is some experimental evidence that animals bred for a running predisposition give birth to offspring with stronger bones (Wallace *et al.*, 2010; the authors interpreted their results as most likely due to genetic or metabolic effects). It is possible, therefore, that the maternal environment at Mistihalj (and Point Hope) influenced pre-natal bone development, either through increased activity or more general (e.g., dietary) effects. Overall, though, as shown here, despite previous depictions of Mistihalj as a very skeletally robust sample (Niewoehner, 2001; Alexeeva *et al.*, 2003), it is in fact not unusually so for Late Medieval European populations. Given the varying population histories of different European regions (e.g., Lazaridis *et al.*, 2014), it seems unlikely that specific genetic factors would be responsible for the generally increased relative bone strength observed across the continent during the Medieval period.

Percentage cortical areas generally decline between the Schela Cladovei and Mistihalj samples (significantly or near-significantly for the femur and tibia), mirroring temporal trends across Europe as a whole (also see Chapter 5 and other chapters in this volume). The only exception is %CA in Schela Cladovei male humeri, which is relatively low compared to Mistihalj males, or Schela Cladovei females. The reason for this is not clear, but it is possibly related to the unusual loading of the upper limbs in this subsample.

13.5 Conclusions

The two available skeletal samples for the Balkans region represent two very different time periods, subsistence strategies, and likely genetic heritages. They share some similarities in that both lived in a rugged physical environment and neither practiced food production, that is, agriculture. The Schela Cladovei inhabitants were very large-bodied by European Mesolithic standards, which may be a result of both dietary (high-protein food sources) and genetic factors (tall statures characteristic of both Eastern European Mesolithics as well as modern populations from the Balkans). Lower limb bone strength in males was higher than average for Mesolithics, indicating that although the population as a whole may have been semi-sedentary (or seasonally sedentary), the males at least were logistically mobile, probably spending significant time during the year engaged in upland hunting. Schela Cladovei males also exhibit an unusual frequency of left-biased upper limb strength, which may be related to the use of bimanual fishing techniques involving nets and weirs. The Mistihalj sample is close to average in both body size and relative bone strength to other Late Medieval samples. The average (rather than larger) body size may be due to their ethnic status as Vlachs. European Medieval long bones are relatively strong overall compared to other later Holocene periods, and Mistihalj fits this pattern.

Acknowledgments

For assistance in data collection and processing, we thank Trang Diem Vu and Heather Garvin. For allowing access to collections, we thank Michèle Morgan, Clive Bonsall, and Adina Boroneant.

References

Alexeeva, T., Bogatenkov, D., and Lebedinskaya, G. (2003) Anthropology of Medieval Vlakhs in comparative study: On date of Mistikhaly Burial Site (in Russian, summary in English). Scientific World, Moscow.

Anderson, S.H. (2004) Tybrind Vig. In: *Ancient Europe 8000 BC–AD 1000 Vol. 1: The Mesolithic to Copper Age (c 8000–2000 BC)* (eds P. Bogucki and P.J. Crabtree), Charles Scribner's Sons, New York, pp. 141–143.

Angel, J.L. (1984) From hunting to farming in the Eastern Mediterranean. In: *Paleopathology at the Origins of Agriculture* (eds M.N. Cohen and G.J. Armelagos), Academic Press, New York, pp. 51–73.

Auerbach, B.M. and Raxter, M.H. (2008) Patterns of clavicular bilateral asymmetry in relation to the humerus: variation among humans. *J. Hum. Evol.*, **54**, 663–674.

Bartosiewicz, L., Bonsall, C., and Sisu, V. (2008) Sturgeon fishing in the Middle and Lower Danube region. In: *The Iron Gates in Prehistory* (eds C. Bonsall, V. Boroneant, and I. Radovanovic), Archeopress, Oxford, pp. 39–54.

Bogucki, P. (2004). The Mesolithic of Northern Europe. In: *Ancient Europe 8000 BC–AD 1000 Vol. 1: The Mesolithic to Copper Age (c 8000–2000 BC)* (eds P. Bogucki and P.J. Crabtree), Charles Scribner's Sons, New York, pp. 132–140.

Bonsall, C. (2004) Iron Gates Mesolithic In: *Ancient Europe 8000B.C.–A.D. 1000. Vol. 1: The Mesolithic to Copper Age (c. 8000–2000B.C.)* (eds P. Bogucki and P.J. Crabtree), Charles Scribner's Sons, New York, pp. 175–179.

Bonsall, C. (2008) The Mesolithic of the Iron Gates. In: *Mesolithic Europe* (eds G. Bailey and P. Spikins), Cambridge University Press, Cambridge, pp. 238–279.

Bonsall, C., Cook, G., Pickard, C., McSweeney, K., Sayle, K., Bartosiewicz, L., Radovanovic, I., Higham, T., Soficaru, A., and Boroneant, A. (2015) Food for thought: Re-assessing Mesolithic diets in the Iron Gates. *Radiocarbon*, **57**, 689–699.

Bonsall, C., Lennon, R., McSweeney, K., Harkness, D., Boroneant, V., Bartosiewicz, L., Payton, R., and Chapman, J. (1997) Mesolithic and Early Neolithic in the Iron Gates: a palaeodietary perspective. *J. Eur. Archaeol.*, **5** (1), 50–92.

Boric, D., French, C.A.I., Stefanovic, S., Dimitrijevic, V., Cristiani, E., Gurova, M., Antonovic, D., Allue, E., and Filipovic, D. (2014) Late Mesolithic lifeways and deathways at Vlasac (Serbia). *J. Field Archaeol.*, **39**, 4–31.

Boroneant, V., Bonsall, C., McSweeney, K., Payton, R., and Macklin, M. (1995) A Mesolithic burial area at Schela Cladovei, Romania. In: *L'Europe des derniers chasseurs, Epipaleolithique et Mesolithique* (ed. A. Thevenin), Editions du CTHS, Paris, pp. 385–390.

Coon, C.S. (1939) *The Races of Europe.* MacMillan, New York.

Cowgill, L.W. (2010) The ontogeny of Holocene and Late Pleistocene human postcranial strength. *Am. J. Phys. Anthropol.*, **141**, 16–37.

Cowgill, L.W., Eleazer, C.D., Auerbach, B.M., Temple, D.H., and Okazaki, K. (2012) Developmental variation in ecogeographic body proportions. *Am. J. Phys. Anthropol.*, **148** (4), 557–570.

Dinu, A. (2010) Mesolithic fish and fishermen of the Lower Danube (Iron Gates). *Documenta Praehistorica*, **37**, 299–310.

Ferretti, J.L., Capozza, R.F., and Zanchetta, J.R. (1996) Mechanical validation of a tomographic (pQCT) index for noninvasive estimation of rat femur bending strength. *Bone*, **18**, 97–102.
Formicola, V. and Giannecchini, M. (1999) Evolutionary trends of stature in Upper Paleolithic and Mesolithic Europe. *J. Hum. Evol.*, **36**, 319–333.
Frayer, D.W. (1980) Sexual dimorphism and cultural evolution in the late Pleistocene and Holocene of Europe. *J. Hum. Evol.*, **9**, 399–415.
Frayer, D.W. (1981) Body size, weapon Use, and natural selection in the European Upper Paleolithic and Mesolithic. *Am. Anthropol.*, **83**, 57–73.
Frayer, D.W. (1984) Biological and cultural change in the European Late Pleistocene and Early Holocene. In: *The Origins of Modern Humans: A World Survey of the Fossil Evidence* (eds F.H. Smith and F. Spencer), Wiley-Liss, New York, pp. 211–250.
Frayer, D.W. (1988) Auditory exostoses and evidence for fishing at Vlasac. *Curr. Anthropol.*, **29**, 346–349.
Grasgruber, P., Cacek, J., Kalina, T., and Sebera, M. (2014) The role of nutrition and genetics as key determinants of the positive height trend. *Econ. Hum. Biol.*, **15**, 81–100.
Grasgruber, P., Sebera, M., Hrazdira, E., Cacek, J., and Kalina, T. (2016) Major correlates of male height: A study of 105 countries. *Econ. Hum. Biol.*, **21**, 172–195.
Holliday, T.W. (1997a) Body proportions in Late Pleistocene Europe and modern human origins. *J. Hum. Evol.*, **32**, 423–447.
Holliday, T.W. (1997b) Postcranial evidence of cold adaptation in European Neandertals. *Am. J. Phys. Anthropol.*, **104**, 245–258.
Holliday, T.W. (1999) Brachial and crural indices of European Late Upper Paleolithic and Mesolithic humans. *J. Hum. Evol.*, **36**, 549–566.
Holt, B.M. and Formicola, V. (2008) Hunters of the Ice Age: The Biology of Upper Paleolithic People. *Yrbk Phys. Anthropol.*, **51**, 70–99.
Jacobs, K.H. (1985a) Climate and the hominid postcranial skeleton in Würm and early Holocene Europe. *Curr. Anthropol.*, **26**, 512–514.
Jacobs, K.H. (1985b) Evolution in the postcranial skeleton of Late Glacial and early Postglacial European hominids. *Z. Morphol. Anthropol.*, **75**, 307–326.
Jochim, M. (2002) The Mesolithic. In: *European Prehistory: A Survey* (ed. S. Milisauskas), Kluwer Academic/Plenum, Pp. 115–141.
Karasik, D., Shimabuku, N.A., Zhou, Y., Zhang, Y., Cupples, L.A., Kiel, D.P., and Demissie, S. (2008) A genome wide linkage scan of metacarpal size and geometry in the Framingham Study. *Am. J. Hum. Biol.*, **20**, 663–670.
Kelly, R.L. (1992) Mobility/sedentism – Concepts, archaeological measures, and effects. *Annu. Rev. Anthropol.*, **21**, 43–66.
Lazaridis, I., Patterson, N., Mittnik, A., Renaud, G., Mallick, S., Kirsanow, K., Sudmant, P.H., Schraiber, J.G., Castellano, S., Lipson, M., Berger, B., Economou, C., Bollongino, R., Fu, Q.M., Bos, K.I., Nordenfelt, S., Li, H., de Filippo, C., Prufer, K., Sawyer, S., Posth, C., Haak, W., Hallgren, F., Fornander, E., Rohland, N., Delsate, D., Francken, M., Guinet, J.M., Wahl, J., Ayodo, G., Babiker, H.A., Bailliet, G., Balanovska, E., Balanovsky, O., Barrantes, R., Bedoya, G., Ben-Ami, H., Bene, J., Berrada, F., Bravi, C.M., Brisighelli, F., Busby, G.B.J., Cali, F., Churnosov, M., Cole, D.E.C., Corach, D., Damba, L., van Driem, G., Dryomov, S., Dugoujon, J.M., Fedorova, S.A., Romero, I.G., Gubina, M., Hammer, M., Henn, B.M., Hervig, T., Hodoglugil, U., Jha, A.R., Karachanak-Yankova, S., Khusainova, R., Khusnutdinova, E., Kittles, R., Kivisild, T., Klitz, W., Kucinskas, V., Kushniarevich, A., Laredj, L., Litvinov, S., Loukidis, T., Mahley, R.W., Melegh, B., Metspalu, E., Molina, J., Mountain, J., Nakkalajarvi, K., Nesheva, D., Nyambo, T., Osipova, L., Parik, J., Platonov, F., Posukh, O., Romano, V., Rothhammer, F., Rudan, I., Ruizbakiev, R., Sahakyan, H., Sajantila, A., Salas, A., Starikovskaya, E.B., Tarekegn, A.,

Toncheva, D., Turdikulova, S., Uktveryte, I., Utevska, O., Vasquez, R., Villena, M., Voevoda, M., Winkler, C.A., Yepiskoposyan, L., Zalloua, P., Zemunik, T., Cooper, A., Capelli, C., Thomas, M.G., Ruiz-Linares, A., Tishkoff, S.A., Singh, L., Thangaraj, K., Villems, R., Comas, D., Sukernik, R., Metspalu, M., Meyer, M., Eichler, E.E., Burger, J., Slatkin, M., Paabo, S., Kelso, J., Reich, D., and Krause, J. (2014) Ancient human genomes suggest three ancestral populations for present-day Europeans. *Nature*, **513**, 409–413.

Malcolm, N. (1994) *Bosnia: a Short History*. MacMillan, London.

Mathieson, I., Lazaridis, I., Rohland, N., Mallick, S., Patterson, N., Roodenberg, S.A., Harney, E., Stewardson, K., Fernandes, D., Novak, M., Sirak, K., Gamba, C., Jones, E.R., Llamas, B., Dryomov, S., Pickrell, J., Arsuaga, J.L., de Castro, J.M., Carbonell, E., Gerritsen, F., Khokhlov, A., Kuznetsov, P., Lozano, M., Meller, H., Mochalov, O., Moiseyev, V., Guerra, M.A., Roodenberg, J., Verges, J.M., Krause, J., Cooper, A., Alt, K.W., Brown, D., Anthony, D., Lalueza-Fox, C., Haak, W., Pinhasi, R., and Reich, D. (2015) Genome-wide patterns of selection in 230 ancient Eurasians. *Nature*, **528**, 499–503.

Meikeljohn, C. and Zvelebil, M. (1991) Health status of European populations at the agricultural transition and the implications for the adoption of farming. BAR International Series, British Archaeological Reports 567.

Meiklejohn, C. and Babb, J. (2011) Long bone length, stature and time in the European Late Pleistocene and Early Holocene. In: *Human Bioarchaeology of the Transition to Agriculture* (eds R. Pinhasi and J.T. Stock), Wiley-Blackwell, Chichester, pp. 153–175.

Meiklejohn, C., Schentag, C., Venema, A., and Key, P. (1984) Socioeconomic change and patterns of pathology and variation in the Mesolithic and Neolithic of Western Europe: Some suggestions. In: *Paleopathology at the Origins of Agriculture* (eds M.N. Cohen and G.J. Armelagos), Academic Press, Orlando, pp. 75–100.

Muller, G.B. (2003) Embryonic motility: environmental influences and evolutionary innovation. *Evol. Dev.*, **5**, 56–60.

NCD RFC (2016) A century of trends in adult human height. Elife **5**:e13410.

Nehlich, O., Boric, D., Stefanovic, S., and Richards, M.P. (2010) Sulphur isotope evidence for freshwater fish consumption: a case study from the Danube Gorges, SE Europe. *J. Archaeol. Sci.*, **37**, 1131–1139.

Niewoehner, W.A. (2001) Behavioral inferences from the Skhul/Qafzeh early modern human hand remains. *Proc. Natl Acad. Sci. USA*, **98**, 2979–2984.

Pineau, J.C., Delamarche, P., and Bozinovic, S. (2005) Average height of adolescents in the Dinaric Alps. *Cr. Biol.*, **328**, 841–846.

Prinz, B. (1987) Mesolithic adaptations on the Lower Danube: Vlasac and the Iron Gates Gorge. BAR International Series, British Archaeological Reports 330.

Ruff, C.B. (1994) Morphological adaptation to climate in modern and fossil hominids. *Yrbk Phys. Anthropol.*, **37**, 65–107.

Ruff, C.B., Burgess, M.L., Bromage, T.G., Mudakikwa, A., and McFarlin, S.C. (2013) Ontogenetic changes in limb bone structural proportions in mountain gorillas (*Gorilla beringei beringei*). *J. Hum. Evol.*, **65**, 693–703.

Ruff, C.B., Holt, B.M., Niskanen, M., Sladek, V., Berner, M., Garofalo, E., Garvin, H.M., Hora, M., Junno, J.-A., Schuplerova, E., Vilkama, R., and Whittey, E. (2015) Gradual decline in mobility with the adoption of food production in Europe. *Proc. Natl Acad. Sci. USA*, **112**, 7147–7152.

Ruff, C.B., Holt, B.M., Niskanen, M., Sladek, V., Berner, M., Garofalo, E., Garvin, H.M., Hora, M., Maijanen, H., Niinimaki, S., Salo, K., Schuplerova, E., and Tompkins, D. (2012) Stature and body mass estimation from skeletal remains in the European Holocene. *Am. J. Phys. Anthropol.*, **148**, 601–617.

Sharir, A., Stern, T., Rot, C., Shahar, R., and Zelzer, E. (2011) Muscle force regulates bone shaping for optimal load-bearing capacity during embryogenesis. *Development*, **138**, 3247–3259.

Shaw, C.N. and Stock, J.T. (2009) Habitual throwing and swimming correspond with upper limb diaphyseal strength and shape in modern human athletes. *Am. J. Phys. Anthropol.*, **140**, 160–172.

Sparacello, V. and Marchi, D. (2008) Mobility and subsistence economy: a diachronic comparison between two groups settled in the same geographical area (Liguria, Italy). *Am. J. Phys. Anthropol.*, **136**, 485–495.

Sparacello, V.S., Pearson, O.M., Coppa, A., and Marchi, D. (2011) Changes in skeletal robusticity in an iron age agropastoral group: the Samnites from the Alfedena necropolis (Abruzzo, Central Italy). *Am. J. Phys. Anthropol.*, **144**, 119–130.

Tatara, M.R., Sliwa, E., and Krupski, W. (2007) Prenatal programming of skeletal development in the offspring: effects of maternal treatment with beta-hydroxy-beta-methylbutyrate (HMB) on femur properties in pigs at slaughter age. *Bone*, **40**, 1615–1622.

Trinkaus, E. (1981) Neanderthal limb proportions and cold adaptation. In: *Aspects of Human Evolution* (ed. C.B. Stringer), Taylor and Francis, London, pp. 187–224.

Turchin, M.C., Chiang, C.W., Palmer, C.D., Sankararaman, S., Reich, D., Genetic Investigation of ATC, and Hirschhorn, J.N. (2012) Evidence of widespread selection on standing variation in Europe at height-associated SNPs. *Nat. Genet.*, **44**, 1015–1019.

Voytek, B.A. and Tringham, R. (1990) Rethinking the Mesolithic: the case of South-East Europe. In: *The Mesolithic in Europe. Proceedings of the Third International Symposium, Edinburgh 1985* (ed. C. Bonsall), John Donald Ltd, Edinburgh, pp. 492–499.

Wallace, I.J., Middleton, K.M., Lublinsky, S., Kelly, S.A., Judex, S., Garland, T., Jr, and Demes, B. (2010) Functional significance of genetic variation underlying limb bone diaphyseal structure. *Am. J. Phys. Anthropol.*, **143**, 21–30.

y'Edynak, G. (1989) Yugoslav Mesolithic dental reduction. *Am. J. Phys. Anthropol.*, **78**, 17–36.

y'Edynak, G. and Fleisch, S. (1983) Microevolution and biological adaptability in the transition from food-collecting to food-producing in the Iron Gates of Yugoslavia. *J. Hum. Evol.*, **12**, 279–296.

Zhilin, M.G. (2014) Early Mesolithic hunting and fishing activities in Central Russia: A review of the faunal and artefactual evidence from wetland sites. *J. Wetland Archaeol.*, **14**, 91–105.

14

Conclusions

Christopher B. Ruff[1], Brigitte Holt[2], Markku Niskanen[3], Vladimir Sládek[4], and Margit Berner[5]

[1] *Center for Functional Anatomy and Evolution, Johns Hopkins University School of Medicine, Baltimore, MD, USA*
[2] *Department of Anthropology, University of Massachusetts, Amherst, MA, USA*
[3] *Department of Archaeology, University of Oulu, Oulu, Finland*
[4] *Department of Anthropology and Human Genetics, Faculty of Science, Charles University, Prague, Czech Republic*
[5] *Department of Anthropology, Natural History Museum, Vienna, Austria*

The approximately 2000 individuals included in this project represent the largest study carried out to date of variation in skeletal structure among modern humans within a defined geographic region. As such, it allows analysis of both broader temporal trends and geographic variability within this region, as well as more specific localized trends. The time span of the sample encompasses early hunter-gatherers, transitional Mesolithic and sedentary 'Neolithic foragers,' early agriculturalists, later more intensive agriculturalists, and Medieval and more recent populations. A variety of terrains, from flat to mountainous, as well as both rural and urban settlements, are represented in the sample. Thus, it was possible to examine the potential impact of a number of environmental factors on modern human skeletal form.

14.1 Body Size and Shape

The two most dramatic changes in body size over the last 30,000 years in Europe as a whole were a marked decrease from the Early through the Late Upper Paleolithic, continuing at a slower pace through the succeeding Mesolithic, and a marked increase over the past 150 years. As a result of these two contrasting trends, European populations from the mid-20th century are about as large-bodied as those from the Early Upper Paleolithic. The earlier trend has been attributed to various factors, including dietary, climatic, technological, and demographic (Frayer, 1984; Holliday, 1997; Formicola and Giannecchini, 1999; Formicola and Holt, 2007). The later trend is associated with recent improvements in health and living conditions (Tanner, 1992). Another more modest but general decline in body size occurred during the end of the Late Medieval ('Little Ice Age') and early industrial period (1300–1800 AD), when environmental conditions deteriorated (Chapter 4).

Changes in stature account for most of the temporal variation in body mass in our sample, with body breadth – at least pelvic breadth – remaining much more constant. This is consistent with other observations relating morphological adaptation to environmental change, in which stature appears to be much more responsive in the short term than body breadth

Skeletal Variation and Adaptation in Europeans: Upper Paleolithic to the Twentieth Century,
First Edition. Edited by Christopher B. Ruff.

(Ruff, 1994). Stature variation in our sample is in turn primarily due to variation in lower limb length, with less dramatic temporal changes in trunk length. This suggests nutritional explanations for these trends, since poorer nutrition leads to a reduction in relative limb length, and vice versa (Bogin *et al.*, 2002). However, climatic influences may also play a role, since relative limb length affects thermoregulation and shows ecogeographic patterning (Trinkaus, 1981; Ruff, 1994; Holliday, 1997). It is possible that both factors were in part responsible for the changes in body form occurring during the terminal Pleistocene and early Holocene, although it seems most likely that variation in nutritional and general health levels explain more recent changes. These major temporal trends are apparent in all regions of Europe (where the available data permit evaluation), implying widespread adaptation to common environmental factors.

Another temporal trend observed in Britain and central Europe is an increase in body size between the Neolithic and early Bronze, or Copper Age (Eneolithic). This may be attributable to the so-called 'Secondary Products Revolution' (Greenfield, 2010), characterized by increased consumption of milk and other dairy products. The Neolithic period is characterized by relatively small body size in some, although not all, regions of Europe, partially supporting the concept of generally declining health levels with the advent of food production (Cohen, 1989; Larsen, 2006; Cohen and Crane-Kramer, 2007). However, overall, stature is as low in late Mesolithic as in early Neolithic samples (Chapter 4), suggesting that food production *per se* was not the only factor involved. This also argues against a purely genetic (i.e., population replacement) explanation for reduced body size in Europe in the early Holocene. Large-scale movements of people into Europe at the beginning of the Neolithic are well documented through DNA analyses (Bramanti *et al.*, 2009; Haak *et al.*, 2010; Bollongino *et al.*, 2013; Brandt *et al.*, 2013). However, source populations from the Near East were not necessarily small-bodied compared to late hunter-gatherers in Europe (Hillson *et al.*, 2013; Mathieson *et al.*, 2015). Body size and shape were likely affected by population movements throughout European prehistory (Haak *et al.*, 2015; Mathieson *et al.*, 2015), but the major effects of subsistence changes and other environmental factors are still discernible. More localized changes are also apparent within specific regions, and can often be related to environmental variables, as described in the individual chapters of this volume.

Genetic effects have also been implicated in geographic clines in body size (stature) within both ancient and modern European populations (Turchin *et al.*, 2012; Grasgruber *et al.*, 2014; Mathieson *et al.*, 2015). One of the most consistent geographic patterns that we observed in our data set was a tendency for larger body size – both stature and body mass – in northern compared to southern Europeans throughout the Holocene, which is concordant with genetic evidence (Grasgruber *et al.*, 2014; Mathieson *et al.*, 2015). We corroborated earlier results showing larger body size in Eastern relative to Western Europe during the Mesolithic (Formicola and Giannecchini, 1999), which may also in part have a genetic basis (Chapter 13).

We also found a north–south cline in relative limb length, with southern populations (after the Mesolithic) exhibiting relatively longer lower limbs, particularly in the tibia. This is similar to geographic clines observed worldwide, and may be related to climatic adaptation, that is, Allen's Rule (Ruff, 1994). Given the evidence summarized above for smaller body size in southern Europeans, it seems unlikely to be related to nutritional effects. Systematic variation in linear proportions has implications for estimating stature from long bone lengths, with different equations being more appropriate for northern and southern Europe when using the tibia (Ruff *et al.*, 2012). We also observed a general temporal decline through the Holocene in intralimb (crural and brachial) indices, which is currently unexplained.

Some of the temporal trends in stature documented here have been noted by previous authors for particular time ranges or regions within Europe (Frayer, 1984; Holliday, 1997; Jaeger *et al.*, 1998; Roberts and Cox, 2003; Steckel, 2004). However, the present study is the first attempt to assess such trends in large samples from throughout Europe and for all time periods from the Upper Paleolithic through the present. The development of region-specific stature formulae for this study (Ruff *et al.*, 2012) should also have resulted in more accurate stature estimates. In addition, by including a number of other skeletal dimensions and reconstruction techniques, we were able to assess trends in other aspects of body form – that is, body mass and body proportions – which give further insights into possible underlying factors, as discussed above and in other chapters of this volume.

We found some evidence for 'female buffering,' the concept that females are more protected from environmental effects during growth than males (Stini, 1976; Gray and Wolfe, 1980; Stinson, 1985), although this was mainly limited to later temporal periods. Thus, stature and body mass decline more in males than in females in the Late Medieval and early Modern periods, and then increase more in males during the 20th century, paralleling overall health trends. In the modern Lisbon sample, there is more sexual dimorphism in body size among high-SES individuals than among low-SES individuals. The large-bodied early Bronze Age sample from Britain exhibits higher sexual dimorphism in stature than any other temporal period from this region, until the late 20th century.

In matched regional comparisons from the Iron Age to the 19th century, rural samples were taller, heavier, and had relatively longer limbs than urban samples, all indications of better health conditions. This is consistent with observations based on only more recent (18th and 19th century) populations (see references in Martinez-Carrion and Moreno-Lazaro, 2007), and suggests that the same generally negative effects of urbanization on health, prior to the 20th century, were also characteristic of much earlier populations.

14.2 Long Bone Strength

The most pronounced change in relative long bone strength observed in our sample was a marked decrease occurring between the Upper Paleolithic and early agricultural periods (Neolithic and Bronze Age). This decline in inferred mechanical loadings of the skeleton can be attributed to increased sedentism and changes in methods of food procurement and processing associated with the shift in subsistence strategy (Ruff *et al.*, 2015; Sládek *et al.*, 2016). Relative bone strength (of the tibia and humerus) continued to decline through the Iron/Roman period, probably reflecting the intensification of agriculture and sedentism, as well as the introduction and increasing use of labor-saving devices such as the horse, wheeled vehicles, and the iron plow (Milisauskas, 2002). Relative bone strength then increased somewhat in the Medieval period, although variably by skeletal location and sex (femur in both sexes, humerus in males), suggesting an increasing workload possibly associated with intensive labor among at least some segments of the population (Chapter 5). A slight decline then occurred in most bones in the Early Modern and Very Recent periods, likely associated with increasing mechanization and urbanization.

Cross-sectional shape – that is, relative anteroposterior (A-P) to mediolateral (M-L) bending strength – also shows systematic temporal changes in the lower limb bones, with the largest change (decline) again occurring during the early Holocene. The preferential decrease in A-P bending strength of the femur and tibia is also indicative of increasing sedentism (Ruff *et al.*, 2015). In the tibia, the trend continues through more recent periods, which is likely a result of generally decreasing mobility (at least, on foot) through the Holocene.

Changes in humeral strength are tied to manipulative behaviors. In addition to assessments of overall strength, bilateral asymmetry in strength of the upper limb bones can also provide important insights into behaviors associated with changes in subsistence strategy and food processing. Bilateral asymmetry in humeral strength declines precipitously in males from the Early Upper Paleolithic to the Mesolithic, reflecting the transition from more unimanual to bimanual hunting techniques (e.g., use of the bow and arrow). Females show a large decline in humeral strength asymmetry from the Mesolithic to early food production time periods, which is associated with the adoption of bimanual food-grinding technology (saddle querns). Females then increase in asymmetry as one-handed (rotary quern) techniques (as well as more mechanized methods) of food processing are introduced, while males continue to decline slowly. Both sexes reach similar and moderately low levels of asymmetry in the most recent periods, consistent with a relative lack of sex-specific activities involving unimanual versus bimanual tasks. However, interestingly, sexual dimorphism in overall humeral strength actually increases during the later Holocene, which could be the result of shifting patterns of workloads, with heavier tasks stressing the upper limb being preferentially performed by males (Chapter 6). Sexual dimorphism in lower limb bone strength remains similar throughout the Holocene.

The major temporal patterns in bone strength that we observed, such as the decline in lower limb bone strength, especially A-P strength, with increasing sedentism, and the changing patterns of upper limb bone asymmetry paralleling food procurement and processing techniques, are characteristic of European populations in general, reflecting their associations with continent-wide changing subsistence strategies and technology. However, specific regional patterns could also be discriminated, in part on the basis of comparisons with broader samples. For example, in Iberia, long bones from the Mesolithic are relatively weaker than those from the later Bronze Age, reversing the usual temporal trend. This is likely a result of both the relatively sedentary lifestyle of these Mesolithic people as well as their coastal or near-coastal environment, which is in sharp contrast to the very rugged environments of the interior Bronze Age samples (also see below on terrain effects in general). Lower limb bones of Neolithic Period Pitted Ware samples from southern Scandinavia are relatively stronger than those of contemporaneous farmers and more similar to Mesolithic samples from the region, as would be expected for a foraging population. However, they do not show the preferential increase in A-P bending strength characteristic of most European (including Scandinavian) Mesolithic samples, suggesting less mobility, which is again consistent with their relatively more sedentary subsistence economy, centered around fishing, sealing, and waterfowl hunting. An unusual subsistence economy among the Mesolithic inhabitants of the Iron Gates region involving the bimanual capturing of large anadromous river fish may explain the lack of directional asymmetry in humeral strength among males, which is unlike the right-biased pattern observed more generally across Europe. However, the same males also exhibited increased A-P bending strength of their lower limb bones relative to other Mesolithic samples, suggesting that while the population may have been semi-sedentary in terms of residency, there was considerable logistical mobility in connection with hunting in the nearby rugged uplands (as also evidenced by the numerous pig and deer remains at the site). Such examples show how consideration of multiple bone structural traits, within a broader context, can highlight specific behavioral strategies to address local environmental conditions and opportunities.

As noted above, terrain likely played a major role in mechanical loadings of the lower limb bones in our samples, depending in part on how mobile the populations were. Over all of our samples, those from 'mountainous' and 'hilly' terrains had relatively stronger femora and tibiae, especially for A-P bending, than those from 'flat' terrains. As would be expected, there was no effect of terrain on humeral strength. The effects in the lower limb were more marked when Upper Paleolithic and Mesolithic samples were included, which suggests, not surprisingly, that their generally greater mobility – that is, long-distance travel over different types of terrain – was also a factor. We did not find consistent differences in relative bone strengths between urban and rural populations, which may reflect in part the rather loose (and temporally changing) definitions of these categories, as well as the probability that many 'urban' dwellers, in pre-modern times, actually worked outside of the towns or cities that they resided in. Class differences also likely played a role. The Late Medieval urban sample from Leiria in Portugal contained a large proportion of middle and higher-status individuals, and males from this sample were significantly weaker, particularly in the humerus, than males from a rural Late Medieval Iberian site, or even than low-SES (manual laboring) 20th century males from Lisbon. Within-sample variability in bone strength across Europe increased between the Neolithic and Iron Age, particularly among males, possibly reflecting increased social complexity and class structure.

14.3 Other Bone Structural Observations

Another cross-sectional diaphyseal parameter that we assessed in our sample was percentage cortical area (%CA), or the percentage of the total subperiosteal area of the section filled by cortical bone (CA/TA). Variation in %CA reflects differences in relative cortical thickness, with reductions linked to dietary deficiencies (Garn *et al.*, 1969). Average %CA declines virtually continuously throughout the temporal range of our sample, with larger declines in the lower limb bones. This pattern differs in some ways from that for relative bone strength, which also declines from the Terminal Pleistocene through the early agricultural periods, but then increases during the Medieval and shows only a small decline through the most recent periods (especially in the lower limb). The difference in temporal variation between bone strength and %CA is well illustrated in Britain (Chapter 8), which shows large declines in lower limb bone %CA throughout the Holocene, despite little change in relative strength after the Bronze Age. Physiological 'stress indicators' in British skeletal samples generally increase through time (Roberts and Cox, 2003, 2007), which is consistent with this pattern. Percentage cortical area is also particularly low across Europe for several bone/sex comparisons in the Early Modern period, which as noted earlier represents a time period characterized by a general deterioration in living conditions. Although cortical thinning is also associated with aging (Garn, 1970), this is unlikely to explain the temporal trend in %CA, since average age varies relatively little between periods (35–43 years).

Therefore, it seems likely that at least part of the reduction in relative cortical thickness observed in our sample through the Holocene may be attributable to generally declining health levels, due to nutritional deficits and negative effects of increasing urbanization (also see above on body size variation). Mechanical loading can also affect the distribution of cortical bone in long bone diaphyses, through either subperiosteal expansion and/or endosteal contraction under conditions of increased loading (Ruff *et al.*, 1994), and it is possible that this also contributed to declines in %CA in early periods – that is, between the Upper Paleolithic and early agricultural periods. However, this seems less likely of an explanation for changes in later periods, where there is a disjunction between bone strength and %CA temporal patterns. As with all bone structural parameters, consideration of the total physiological context is necessary to interpret observed patterns of variation.

The highly unusual lower limb bilateral asymmetry of the 'Amesbury Archer' from the early Bronze Age in Britain (Fitzpatrick, 2011) provided an opportunity to examine the effects of extreme mechanical loading situations on bone structure, as well as more general principles of bone adaptation. As described in Chapter 8, the asymmetry in this individual is associated with the lack of a left patella, which was almost certainly congenital: sections from the left side exhibit both endosteal expansion and periosteal contraction relative to those from the right side, indicative of reduced mechanical loading beginning in childhood (Ruff *et al.*, 1994). The resulting strength values on the left side are below those of any other British Bronze Age male, while the right side shows a compensatory increase in strength to among the highest values for other males. The disability also apparently affected the upper limbs: unlike any other British Bronze Age male, the Archer has slight left dominance in humeral strength, which may have resulted from changes in balance and bracing during use of the bow and arrow, due to the only partially functioning left lower limb. Interestingly, long bone length and articular dimensions, including of the left lower limb, are not unusual in the Amesbury Archer, illustrating the variable effects of mechanical loading on different skeletal features (Trinkaus *et al.*, 1994). This demonstrates the developmental plasticity of long bone diaphyseal cross-sections and their responsiveness to specifically mechanical factors, in addition to any general systemic factors. In the case of the Archer, general growth was apparently not affected by his disability (he is average in size for a British Bronze Age male). This also indirectly supports our use of articular and length dimensions to estimate body size in mechanical analyses, that is, to standardize mechanical properties, while avoiding potential circular reasoning. Finally, the analysis of the Amesbury Archer again shows the value of temporal and geographic context in this type of study: without the other British Bronze Age males, it would have been difficult to determine whether his very large lower limb strength asymmetry was due simply to compensatory changes on the right side, or reduced strength on the left side, which has consequences for interpreting the etiology of the condition.

This example illustrates one of the primary aims of this project which, as stated in Chapter 1, was to provide a broad comparative database that could be used in future studies of other European (and world-wide) samples (see Appendix 1 and the on-line data file: http://www.hopkinsmedicine.org/fae/CBR.html). While the approximately 2000 individuals included here allowed assessment of a number of interesting temporal and regional trends in morphology and behavior, additional samples are needed to further test and refine these findings, particularly on a regional and local level, as well as in contexts outside of Europe. Our hope is that the methods described here will be applied to many other such samples in the future, providing further insights into the skeletal adaptations of modern humans to their various environmental challenges.

References

Bogin, B., Smith, P., Orden, A.B., Varela Silva, M.I., and Loucky, J. (2002) Rapid change in height and body proportions of Maya American children. *Am. J. Hum. Biol.*, **14**, 753–761.

Bollongino, R., Nehlich, O., Richards, M.P., Orschiedt, J., Thomas, M.G., Sell, C., Fajkosova, Z., Powell, A., and Burger, J. (2013) 2000 years of parallel societies in Stone Age Central Europe. *Science*, **342**, 479–481.

Bramanti, B., Thomas, M.G., Haak, W., Unterlaender, M., Jores, P., Tambets, K., Antanaitis-Jacobs, I., Haidle, M.N., Jankauskas, R., Kind, C.J., Lueth, F., Terberger, T., Hiller, J., Matsumura, S., Forster, P., and Burger, J. (2009) Genetic discontinuity between local hunter-gatherers and central Europe's first farmers. *Science*, **326**, 137–140.

Brandt, G., Haak, W., Adler, C.J., Roth, C., Szecsenyi-Nagy, A., Karimnia, S., Moller-Rieker, S., Meller, H., Ganslmeier, R., Friederich, S., Dresely, V., Nicklisch, N., Pickrell, J.K., Sirocko, F., Reich, D., Cooper, A., and Alt, K.W. (2013) Ancient DNA reveals key stages in the formation of central European mitochondrial genetic diversity. *Science*, **342**, 257–261.

Cohen, M.N. (1989) *Health and the Rise of Civilization*. Yale University Press, New Haven.

Cohe, M.N. and Crane-Kramer, G.M.M. (eds) (2007) *Ancient Health: Skeletal Indicators of Agricultural and Economic Intensification*. University Press of Florida, Gainesville.

Fitzpatrick, A.P. (2011) The Amesbury Archer and Boscombe Bowmen. Bell Beaker Burials on Boscombe Down, Amesbury, Wiltshire. Wessex Archaeology Report 27. Wessex Archaeology Ltd, Salisbury, UK.

Formicola, V. and Giannecchini, M. (1999) Evolutionary trends of stature in Upper Paleolithic and Mesolithic Europe. *J. Hum. Evol.*, **36**, 319–333.

Formicola, V. and Holt, B.M. (2007) Resource availability and stature decrease in Upper Paleolithic Europe. *J. Anthropol. Sci.*, **85**, 147–155.

Frayer, D.W. (1984) Biological and cultural change in the European Late Pleistocene and Early Holocene. In: *The Origins of Modern Humans: A World Survey of the Fossil Evidence* (eds F.H. Smith and F. Spencer), Wiley-Liss, New York, pp. 211–250.

Garn, S.M. (1970) *The Earlier Gain and the Later Loss of Cortical Bone*. Charles C. Thomas, Springfield.

Garn, S.M., Guzman, M.A., and Wagner, B. (1969) Subperiosteal gain and endosteal loss in protein-calorie malnutrition. *Am. J. Phys. Anthropol.*, **30**, 153–155.

Grasgruber, P., Cacek, J., Kalina, T., and Sebera, M. (2014) The role of nutrition and genetics as key determinants of the positive height trend. *Econ. Hum. Biol.*, **15**, 81–100.

Gray, J.P. and Wolfe, L.D. (1980) Height and sexual dimorphism of stature among human societies. *Am. J. Phys. Anthropol.*, **53**, 441–456.

Greenfield, H.J. (2010) The Secondary Products Revolution: the past, the present and the future. *World Archaeol.*, **42**, 29–54.

Haak, W., Balanovsky, O., Sanchez, J.J., Koshel, S., Zaporozhchenko, V., Adler, C.J., Der Sarkissian, C.S., Brandt, G., Schwarz, C., Nicklisch, N., Dresely, V., Fritsch, B., Balanovska, E., Villems, R., Meller, H., Alt, K.W., Cooper, A., and Members of the Genographic C. (2010) Ancient DNA from European early Neolithic farmers reveals their near eastern affinities. *PLoS Biol.*, **8**, e1000536.

Haak, W., Lazaridis, I., Patterson, N., Rohland, N., Mallick, S., Llamas, B., Brandt, G., Nordenfelt, S., Harney, E., Stewardson, K., Fu, Q., Mittnik, A., Banffy, E., Economou, C., Francken, M., Friederich, S., Pena, R.G., Hallgren, F., Khartanovich, V., Khokhlov, A., Kunst, M., Kuznetsov, P., Meller, H., Mochalov, O., Moiseyev, V., Nicklisch, N., Pichler, S.L., Risch, R., Rojo Guerra, M.A., Roth, C., Szecsenyi-Nagy, A., Wahl, J., Meyer, M., Krause, J., Brown, D., Anthony, D., Cooper, A., Alt, K.W., and Reich, D. (2015) Massive migration from the steppe was a source for Indo-European languages in Europe. *Nature*, **522**, 207–211.

Hillson, S.W., Larsen, C.S., Boz, B., Pilloud, M.A., Sadvari, J.W., Agarwal, S.C., Glencross, B., Beauchesne, P., Pearson, J., Ruff, C.B., Garofalo, E.M., Hager, L., and Haddow, S.D. (2013) The Human Remains I: Interpreting Community Structure, Health, and Diet in Neolithic Çatalhöyük. In: *Humans and Landscapes of Çatalhöyük* (ed. I. Hodder), Cotsen Institute of Archaeology Press, Los Angeles, pp. 335–389.

Holliday, T.W. (1997) Body proportions in Late Pleistocene Europe and modern human origins. *J. Hum. Evol.*, **32**, 423–447.

Jaeger, U., Bruchhaus, H., Finke, L., Kromeyer-Hauschild, K., and Zellner, K. (1998) Säkularer trend bei der körperhöhe seit dem Neolithikum. *Anthropol. Anz.*, **56**, 117–130.

Larsen, C.S. (2006) The agricultural revolution as environmental catastrophe: Implications for health and lifestyle in the Holocene. *Quat. Int.*, **150**, 12–20.

Martinez-Carrion, J.M. and Moreno-Lazaro, J. (2007) Was there an urban height penalty in Spain, 1840–1913? *Econ. Hum. Biol.*, **5**, 144–164.

Mathieson, I., Lazaridis, I., Rohland, N., Mallick, S., Patterson, N., Roodenberg, S.A., Harney, E., Stewardson, K., Fernandes, D., Novak, M., Sirak, K., Gamba, C., Jones, E.R., Llamas, B.,

Dryomov, S., Pickrell, J., Arsuaga, J.L., de Castro, J.M., Carbonell, E., Gerritsen, F., Khokhlov, A., Kuznetsov, P., Lozano, M., Meller, H., Mochalov, O., Moiseyev, V., Guerra, M.A., Roodenberg, J., Verges, J.M., Krause, J., Cooper, A., Alt, K.W., Brown, D., Anthony, D., Lalueza-Fox, C., Haak, W., Pinhasi, R., and Reich, D. (2015) Genome-wide patterns of selection in 230 ancient Eurasians. *Nature*, **528**, 499–503.

Milisauskas, S. (ed.) (2002) *European Prehistory: A Survey*. Kluwer Academic/Plenum, New York.

Roberts, C. and Cox, M. (2003) *Health and Disease in Britain: From Prehistory to the Present Day*. Sutton Publishing, Stroud, UK.

Roberts, C. and Cox, M. (2007) The impact of economic intensification and social complexity on human health in Britain from 6000 BP (Neolithic) and the introduction of farming to the mid-nineteenth century AD. In: *Ancient Health: Skeletal Indicators of Agricultural and Economic Intensification* (eds M.N. Cohen and G.M.M. Crane-Kramer), University Press of Florida, Gainesville, pp. 149–163.

Ruff, C.B. (1994) Morphological adaptation to climate in modern and fossil hominids. *Yrbk Phys. Anthropol.*, **37**, 65–107.

Ruff, C.B., Holt, B.M., Niskanen, M., Sladek, V., Berner, M., Garofalo, E., Garvin, H.M., Hora, M., Junno, J.-A., Schuplerova, E., Vilkama, R., and Whittey, E. (2015) Gradual decline in mobility with the adoption of food production in Europe. *Proc. Natl Acad. Sci. USA*, **112**, 7147–7152.

Ruff, C.B., Holt, B.M., Niskanen, M., Sladek, V., Berner, M., Garofalo, E., Garvin, H.M., Hora, M., Maijanen, H., Niinimaki, S., Salo, K., Schuplerova, E., and Tompkins, D. (2012) Stature and body mass estimation from skeletal remains in the European Holocene. *Am. J. Phys. Anthropol.*, **148**, 601–617.

Ruff, C.B., Walker, A., and Trinkaus, E. (1994) Postcranial robusticity in *Homo*, III: Ontogeny. *Am. J. Phys. Anthropol.*, **93**, 35–54.

Sládek, V., Ruff, C.B., Berner, M., Holt, B., Niskanen, M., Schuplerová, E., and Hora, M. (2016) The impact of subsistence changes on humeral bilateral asymmetry in Terminal Pleistocene and Holocene Europe. *J. Hum. Evol.*, **92**, 37–49.

Steckel, R.H. (2004) New light on the 'dark ages'. The remarkably tall stature of northern European men during the Medieval period. *Soc. Sci. Hist.*, **28**, 211–229.

Stini, W.A. (1976) Adaptive strategies of human populations under nutritional stress. In: *Biosocial Interrelations in Population Adaptation* (eds E.S. Watts, F.E. Johnston, and G.W. Lasker), Moutan, The Hague, pp. 19–40.

Stinson, S. (1985) Sex differences in environmental sensitivity during growth and development. *Yrbk Phys. Anthropol.*, **28**, 123–147.

Tanner, J.M. (1992) Growth as a measure of the nutritional and hygienic status of a population. *Horm. Res.*, **38** (Suppl. 1), 106–115.

Trinkaus, E. (1981) Neanderthal limb proportions and cold adaptation. In: *Aspects of Human Evolution* (ed. C.B. Stringer), Taylor and Francis, London, pp. 187–224.

Trinkaus, E., Churchill, S.E., and Ruff, C.B. (1994) Postcranial robusticity in *Homo*, II: Humeral bilateral asymmetry and bone plasticity. *Am. J. Phys. Anthropol.*, **93**, 1–34.

Turchin, M.C., Chiang, C.W., Palmer, C.D., Sankararaman, S., Reich, D., Genetic Investigation of ANthropometric Traits (GIANT) Consortium, and Hirschhorn, J.N. (2012) Evidence of widespread selection on standing variation in Europe at height-associated SNPs. *Nat. Genet.*, **44**, 1015–1019.

Appendix 1

Study Samples

Skeletal Variation and Adaptation in Europeans: Upper Paleolithic to the Twentieth Century,
First Edition. Edited by Christopher B. Ruff.

Region	Period	Site	Culture	Date range (cal.)	M/F N	Reference
Britain	Early Modern	Spitalfields		1700–1850 AD	20/20	Molleson and Cox, 1993
	Late Medieval	York (Fishergate)		1000–1350 AD	30/30	Burt, 2013
		Blackgate		1000–1150 AD	30/31	Nolan, 2010
	Iron/Roman	Poundbury	Romano/British	300–400 AD	20/20	Farwell and Molleson, 1993
		Poundbury Farm	Romano/British	0–400 AD	3/8	Dinwiddy and Bradley, 2011
		Battlesbury Bowl	Iron	200 BC–0 AD	2/1	Ellis and Powell, 2008
		Cockey Down	Iron	400 BC–50 AD	2/1	Lovell *et al.*, 1999
		Wetwang Slack	Iron	100–400 BC	30/30	Jay and Richards, 2006
	Bronze	Cliff's End	Late Bronze	920–1190 BC	0/2	McKinley *et al.*, 2015
		Wilsford S. Lake	Bell Beaker	1750–2000 BC	0/2	Smith, 1991
		Crichel Down	Bell Beaker	1500–2400 BC	2/0	Piggott and Piggott, 1944
		Shrewton	Bell Beaker	1900–2250 BC	3/1	Green and Rollo-Smith, 1984
		Staxton	Bell Beaker	1800–2600 BC	2/1	Stead, 1957
		A303	Bell Beaker	2000–2500 BC	1/0	Fitzpatrick, 2011
		Stonehenge	Bell Beaker	2300–2400 BC	1/0	Fitzpatrick, 2011
		Boscombe	Bell Beaker	2200–2500 BC	1/0	Fitzpatrick, 2011
		Amesbury	Bell Beaker	2300–2500 BC	1/1	Fitzpatrick, 2011
	Neolithic	Lanhill		2500–3500 BC	2/2	Keiller and Piggott, 1938
		Skendleby (Giant's Hill 1)		2800–3300 BC	1/2	Phillips, 1932
		Hazelton North		3300–3700 BC	1/1	Saville, A.
		Wayland's Smithy I		3500–3600 BC	5/2	Whittle, 1991
		West Kennet		3600–3700 BC	2/1	Piggott, 1962
	Mesolithic	Gough's Cave		8000–8450 BC	1/0	Seligman and Parsons, 1914
	Early Up. Pal.	Paviland		33,400–34,000 BP	1/0	Sollas, 1913
France	Early Modern	Moirans		1600-mid 1800 AD	10/5	Diverrez *et al.*, 2012
		L'Observance		1722 AD	14/6	Signoli *et al.*, 1997
	Early Medieval	Jau Dignac		600–700 AD	10/7	Cartron and Castex, 2006
	Iron Age/Roman	Rue Jacque Brel	Gallo-Roman	100–250 AD	8/11	Baigl *et al.*, 1997
	Neolithic	Pontcharraud		4330–4060 BC	14/20	Civetta *et al.*, 2009
	Mesolithic	Le Rastel		6575 ± 55	1/0	Constandse-Westermann *et al.*,1982
		Hoedic		6390–6940 BP	5/3	Vallois and de Felice, 1977
		Cuzoul de Gramat		<8400 BP	1/0	Newell *et al.*, 1979
		Culoz		10,220–9260 BP	2/0	Combier and Genet-Varcin, 1959
		Loschbour		8047 ± 65	1/0	Delsate *et al.*, 2009
		Bonifacio		>8520 BP	0/1	Duday,1975

		Teviec		10,267 ± 63	5/4	Vallois and de Felice, 1977
	Late Up. Pal.	Rochereil		Late Tardiglacial	1/0	Ferembach, 1974
		Chancelade		Tardiglacial	1/0	Vallois, 1941-1946
		Cap Blanc		Tardiglacial	0/1	von Bonin, 1935
		Le Peyrat		13,330 ± 185	1/0	Patte, 1968
		Bruniquel		18,394 ± 303	0/1	Gambier *et al.*, 2000
		St Germain la Riviere		19,013 ± 264	0/1	Gambier *et al.*, 2000
	Early Up. Pal.	Cro-Magnon		32,285 ± 319	2/1	Henry-Gambier, 2002
		La Rochette		28,563 ± 390	0/1	Oakley *et al.*, 1971
Italy	Very Recent	Sassari		late 1800-early 1900 AD	5/5	Becastro *et al.*, 2008
	Early Modern	Siracusani		1800 AD	17/6	Parenti, 1952
	Late Medieval	Piazza della Signoria		800–1300 AD	6/8	Boccone *et al.*, 2011
	Early Medieval	Roselle		500–800 AD	12/11	Boccone *et al.*, 2011
		Vicenne Campo Chiaro		600–700 AD	11/7	Belcastro and Facchini, 2001
	Iron Age/Roman	Lucus Feroniae	Roman	100–300 AD	11/15	Manzi *et al.*, 1999
		Quadrella	Roman	100–400 AD	12/7	Belcastro *et al.*, 2007
	Bronze Age	Olmo di Nogarra		1600–1100 BC	17/ 17	Capitanio and Corrain, 1998
	Neolithic	Fontenoce	Eneolithic	5438 ± 113 BP	6/3	Silvestrini and Pignocchi, 1997
	Mesolithic	Mondeval		8138 ± 74 BP	1//0	Alciati *et al.*, 2005
		Vatte di Zambana		8612 ± 185 BP	0/1	Alciati *et al.*, 2005
		Molara		9632 ± 108 BP	1/0	Alciati *et al.*, 2005
		Uzzo		10,455 ± 131 BP	3/0	Alciati *et al.*, 2005
	Late Up. Pal.	Romanelli		Tardiglacial	1/0	Alciati *et al.*, 2005
		San Teodoro		Tardiglacial	0/1	Alciati *et al.*, 2005
		Arene Candide		11,367 ± 90–12,709 ± 46	5/0	Alciati *et al.*, 2005
		Riparo Continenza		12,095 ± 283 BP	1/0	Alciati *et al.*, 2005
		Grotte des Enfants		13,034 ± 147	0/1	Alciati *et al.*, 2005
		Romito		13,051 ± 170	1/2	Alciati *et al.*, 2005
		Villabruna		14,171 ± 243	1/0	Alciati *et al.*, 2005
		Riparo Tagliente		16,332 ± 469 BP	1/0	Alciati *et al.*, 2005
	Early Up. Pal.	Parabita (Veneri)		c. 22,000 BP	1/1	Alciati *et al.*, 2005
		Grotte des Enfants		28,304 ± 308	1/1	Alciati *et al.*, 2005
		Caviglione		28,304 ± 308	0/1	Alciati *et al.*, 2005
		Paglicci		28,396 ± 565	0/1	Alciati *et al.*, 2005
		Ostuni		29,190 ± 544	0/1	Alciati *et al.*, 2005
		Barma Grande		29,576 ± 902	1/0	Alciati *et al.*, 2005

(*Continued*)

Region	Period	Site	Culture	Date range (cal.)	M/F N	Reference
Iberia	Very Recent	Luis Lopez Collection		1933–1973 AD	26/25	Cardoso, 2006
	Late Medieval	Leiria		1200–1550 AD	16/10	Cardoso and Garcia, 2009
		San Baudelio de Berlanga		1100–1200 AD	12/10	al-Oumaoui *et al.*, 2004
	Early Medieval	Villanueva de Soportiva		850–1100 AD	13/10	al-Oumaoui *et al.*, 2004
		Santa Maria de Hito		500–1200 AD	19/21	Galera and Garralda, 1992
	Bronze	Castellon Alto		1710–1200 BC	15/15	al-Oumaoui *et al.*, 2004
		Terrera del Reloj		1710–1200 BC	2/2	Jiménez-Brobeil *et al.*, 2010
	Mesolithic	Los Canes		5210–5830 BC	2/1	Arias and Garralda, 1996
		Muge Arruda		5370–5840 BC	1/0	Jackes and Lubell, 1999
		Moita de Sebastiao		5710–5910 BC	5/2	Jackes and Lubell, 1999
Austria	Neolithic	Franzhausen I (Bad)	Baden	2900–2300 BC	3/0	Neugebauer and Neugebauer, 1997
		Ratzersdorf	Baden	3320–3080 BC	2/2	Krumpel, 2008; Wiltschke-Schrotta *et al.*, 2008
	Copper Age	Leopoldsdorf	Bell Beaker	2600–2000 BC	1/0	Neugebauer-Maresch, 1994
		Tödling	Bell Beaker	2600–2000 BC	3/2	Pertlwieser, 2002
		Franzhausen (CWC)	Corded Ware	2020–1770 BC	1/0	Neugebauer-Maresch, 1994; Kern, 2012
		Franzhausen I (CWC)	Corded Ware	3320–3080 BC	1/0	Neugebauer and Neugebauer, 1997; Kern, 2012
		Franzhausen II (CWC)	Corded Ware	2900–2300 BC	2/4	Neugebauer-Maresch, 1994; Kern, 2012
		Franzhausen III (CWC)	Corded Ware	2900–2300 BC	3/2	Neugebauer-Maresch, 1994; Kern, 2012
		Franzhausen IV (CWC)	Corded Ware	2900–2300 BC	5/2	Neugebauer-Maresch, 1994; Kern, 2012
		Franzhausen V (CWC)	Corded Ware	2900–2300 BC	1/0	Neugebauer-Maresch, 1994; Kern, 2012
		Franzhausen VI CWC)	Corded Ware	2900–2300 BC	0/2	Neugebauer-Maresch, 1994; Kern, 2012
		Gemeinlebarn-Mitte	Corded Ware	2900–2300 BC	2/2	Neugebauer-Maresch, 1994; Kern, 2012
	Bronze Age	Pitten	Middle Bronze Age	1600–1300 BC	5/0	Hampl *et al.*, 1985; Teschler-Nicola, 1985
		Bernhardsthal	Únětice	2200–1820 BC	4/2	Teschler-Nicola, 1992
		Grossmugl	Únětice	2200–1820 BC	1/0	Neugebauer, 1994; Teschler-Nicola, 1992

		Hobersdorf	Únětice	2200–1820 BC	0/2	Neugebauer, 1994; Teschler-Nicola, 1992
		Schleinbach	Únětice	2200–1820 BC	0/2	Teschler-Nicola, 1992
		Unterhautzental	Únětice	2200–1820 BC	9/14	Lauermann, 1995
		Würnitz	Únětice	2200–1820 BC	3/0	Teschler-Nicola, 1992
		Franzhausen (U)	Unterwölbling	2900–2300 BC	0/2	Neugebauer, 1994
		Franzhausen I (U)	Unterwölbling	2020–1770 BC	7/20	Neugebauer and Neugebauer, 1997; Sprenger, 1999
		Gemeinlebarn F	Unterwölbling	2020–1770 BC	9/8	Neugebauer, 1991; Heinrich and Teschler-Nicola, 1991
		Melk	Unterwölbling	2020–1770 BC	4/8	Teschler-Nicola, 1992
		Pottenbrunn	Unterwölbling	2020–1770 BC	8/4	Blesl, 2006
		Hainburg	Wieselburger	2000–1700 BC	24/74	Teschler-Nicola, 1992
	Roman	Halbturn	Roman	100–500 AD	6/14	Doneus, 2014; Berner, 2014
	Early Medieval	Bruckneudorf	Avar	600–800 AD	13/16	Sauer, 2013
		Laxenburger Str.	Avar	600–800 AD	20/34	Sauer, 2007; Pany-Kucera and Wiltschke-Schrotta, 2017
		Mödling	Avar	630–820 AD	21/38	Marhold, 1977
		Zwentendorf	Slavic	1000–1100 AD	13/20	Heinrich, 2001
Czech Republic	Early Up. Pal.	Dolní Věstonice (EUP)		25,570 ± 280 BP	2/2	Sládek *et al.*, 2000; Trinkaus and Svoboda, 2006
		Pavlov (EUP)		26,170 ± 450 BP	1/0	Sládek *et al.*, 2000; Trinkaus and Svoboda, 2006
		Předmostí		26,000 BP	3/4	Matiegka, 1938
	Copper Age	Brandýsek	Bell Beaker	2600–2000 BC	3/2	Kytlicová, 1960
		Březno (BBC)	Bell Beaker	2600–2000 BC	1/0	Neustupný, 2008
		Dolní Věstonice (BBC)	Bell Beaker	2600–2000 BC	0/2	Podborský, 1993
		Holubice IV	Bell Beaker	2600–2000 BC	0/2	Čižmář and Geisler, 1998
		Kbely	Bell Beaker	2600–2000 BC	1/0	Neustupný, 2008
		Kněževes (BBC)	Bell Beaker	2600–2000 BC	1/2	Kytlicová, 1956
		Lochenice	Bell Beaker	2600–2000 BC	1/0	Buchvaldek, 1990
		Malá Ohrada (BBC)	Bell Beaker	2600–2000 BC	0/2	Neustupný, 2008
		Mokrůvky	Bell Beaker	2600–2000 BC	1/0	Neustupný, 2008
		Pavlov (BBC)	Bell Beaker	2600–2000 BC	1/2	Podborský, 1993
		Plotiště	Bell Beaker	2600–2000 BC	1/0	Neustupný, 2008
		Radovesice (BBC)	Bell Beaker	2600–2000 BC	3/0	Neustupný, 2008
		Tuchoměřice (BBC)	Bell Beaker	2600–2000 BC	1/0	Neustupný, 2008
		Žabovřesky	Bell Beaker	2600–2000 BC	1/0	Neustupný, 2008

(*Continued*)

Region	Period	Site	Culture	Date range (cal.)	M/F N	Reference
		Židovice	Bell Beaker	2600–2000 BC	0/2	Neustupný, 2008
		Blšany	Corded Ware	2900–2300 BC	0/2	Buchvaldek, 1986, Neustupný, 2008
		Březno (CWC)	Corded Ware	2900–2300 BC	0/2	Dobeš, 1993
		Brozany	Corded Ware	2900–2300 BC	1/0	Buchvaldek, 1986, Neustupný, 2008
		Čachovice	Corded Ware	2900–2300 BC	1/0	Neustupný and Smrž, 1989
		Kněževes (CWC)	Corded Ware	2900–2300 BC	0/2	Neustupný, 2008
		Kobylisy	Corded Ware	2900–2300 BC	1/0	Buchvaldek, 1986, Neustupný, 2008
		Kučlín	Corded Ware	2900–2300 BC	1/0	Buchvaldek, 1986, Neustupný, 2008
		Libešice	Corded Ware	2900–2300 BC	0/2	Dobeš *et al*, 1991
		Malá Ohrada (CWC)	Corded Ware	2900–2300 BC	1/0	Buchvaldek, 1986, Neustupný, 2008
		Malé Březno	Corded Ware	2900–2300 BC	0/2	Dobeš, 1993
		Most	Corded Ware	2900–2300 BC	1/0	Dobeš and Buchvaldek, 1993
		Obrnice	Corded Ware	2900–2300 BC	1/0	Dobeš and Buchvaldek, 1993
		Pohořelice	Corded Ware	2900–2300 BC	2/0	Šebela, 1999
		Poplze	Corded Ware	2900–2300 BC	1/0	Buchvaldek, 1986, Neustupný, 2008
		Postoloprty	Corded Ware	2900–2300 BC	1/2	Buchvaldek, 1986, Neustupný, 2008
		Prosetice	Corded Ware	2900–2300 BC	0/2	Buchvaldek *et al.*, 1987
		Rousínov	Corded Ware	2900–2300 BC	2/0	Šebela, 1999
		Široké Třebčice	Corded Ware	2900–2300 BC	0/4	Buchvaldek, 1986, Neustupný, 2008
		Tučapy	Corded Ware	2900–2300 BC	0/2	Šebela, 1999
		Tuchoměrice (CWC)	Corded Ware	2900–2300 BC	0/4	Buchvaldek, 1986, Neustupný, 2008
		Veliká Ves	Corded Ware	2900–2300 BC	0/2	Buchvaldek, 1986, Neustupný, 2008
		Vikletice	Corded Ware	2900–2300 BC	2/2	Buchvaldek and Koutecký, 1970
		Vrbice	Corded Ware	2900–2300 BC	0/2	Šebela, 1999
		Vyškov	Corded Ware	2900–2300 BC	0/2	Šebela, 1999
	Bronze Age	Bystročice (Une)	Únětice	1700–1500 BC	2/0	Pankowská, 2008
		Chrášťany 1	Únětice/Věteřov	2200–1500 BC	1/6	Daňhel and Pankowská, 2010
		Bystročice (Vet)	Věteřov	2200–1500 BC	1/0	Pankowská, 2008
		Hulín-Pravčice 1	Věteřov	1700–1500 BC	2/8	Daňhel and Pankowská, 2010
	Iron Age	Jenišův Újezd	La Tene	400–0 BC	6/2	Waldhauser and Dehn, 1978
		Jinonice	La Tene	400–0 BC	9/14	Velemínský and Dobisíková, 1998
		Makotřasy	La Tene	400–0 BC	5/8	Čižmář, 1978, Chochol, 1978
		Radovesice (IA)	La Tene	400–0 BC	6/2	Kuželka *et al.*, 2004, Velemínský *et al.*, 2004
	Early Medieval	Mikulčice	Slavic	800–900 AD	33/42	Poláček, 2008
	Late Medieval	Šporkova ulice	Medieval	1100–1500 AD	24/34	Stloukal, 2013

		Vratislavský Palác	Medieval	1100–1500 AD	2/6	Dobisíková *et al.*, 1996
		Vršany	Medieval	1100–1500 AD	16/38	Velemínský, 1997
		Opava-Pivovar	New Age	1588–1789 AD	17/22	Kozák *et al.*, 2010
Germany	Late Up. Pal.	Neuessing		18,200 ± 0.2 BP	1/0	Giesler, 1953
		Oberkassel		12,180 ± 110 BP	1/2	Henke, 1984
	Mesolithic	Bottendorf		4850 calBC	0/2	Vlček, 1967
		Unseburg (Meso)		2400–2800 BC	1/2	Bach and Bruchhaus, 1988
	Neolithic	Bad Sulza	Bandkeramik	5500–4900 BC	1/0	Galagher *et al.*, 2008; Preuß, 1998a,b
		Bischleben	Bandkeramik	5500–5200 BC	1/0	Galagher *et al.*, 2008; Preuß, 1998a,b
		Bruchstedt	Bandkeramik	5200–4900 BC	1/8	Kahlke, 2004
		Erfurt	Bandkeramik	5500–4900 BC	0/2	Galagher *et al.*, 2008; Preuß, 1998a,b
		Schlotheim	Bandkeramik	5500–4900 BC	1/0	Preuß, 1998a,b
		Sondershausen	Bandkeramik	5500–5200 BC	4/6	Kahlke, 2004
		Niederbösa	Walternieburger	2800–3200 BC	2/2	Preuß, 1998a,b
		Schönstedt	Walternieburger	2800–3200 BC	3/6	Preuß, 1998a,b
	Copper Age	Bad Dürrenberg	Corded Ware	2400–2800 BC	1/0	Buchvaldek and Strahm, 1992
		Bilzingsleben	Corded Ware	2400–2800 BC	0/6	Buchvaldek and Strahm, 1992
		Braunsdorf	Corded Ware	2400–2800 BC	2/0	Matthias, 1982
		Drosa	Corded Ware	2400–2800 BC	0/2	Buchvaldek and Strahm, 1992
		Erfurt Nord	Corded Ware	2400–2800 BC	0/2	Buchvaldek and Strahm, 1992
		Erfurt-Nordhäuser St.	Corded Ware	2400–2800 BC	0/2	Noll, 2001
		Großfahner	Corded Ware	2400–2800 BC	1/0	Buchvaldek and Strahm, 1992
		Großkayna	Corded Ware	2400–2800 BC	0/2	Matthias, 1982
		Orlishausen	Corded Ware	2400–2800 BC	1/0	Buchvaldek and Strahm, 1992
		Radegast	Corded Ware	2400–2800 BC	1/0	Buchvaldek and Strahm, 1992
		Schafstädt	Corded Ware	2400–2800 BC	2/4	Buchvaldek and Strahm, 1992
		Udestedt (2)	Corded Ware	2400–2800 BC	0/2	Buchvaldek and Strahm, 1992
		Unseburg (CWC)	Corded Ware	6550 calBC	0/2	Buchvaldek and Strahm, 1992
		Weißensee	Corded Ware	2400–2800 BC	1/0	Buchvaldek and Strahm, 1992
	Bronze Age	Melchendorf		1200–700 BC	19/12	Dusek, 1999
	Early Medieval	Dresden Briesnitz		900–1200 AD	14/16	Beberhold, 2003
	Recent	Jena	Modern autopsy	1990 AD	19/14	Scherf *et al.*, 2016
Scandinavia	Late Medieval	Westerhus		1080–1300 AD	48/56	Geijvall, 1960
		Humlegården of Sigtuna	Urban farmer	1080–1300 AD	47/26	Kjellström, 2005
	Early Medieval	Nunnan Block of Sigtuna	Urban farmer	800–950 AD	19/4	Kjellström, 2005
		Otterup		700–1000 AD	1/0	Niels Lynnerup, pers. comm.

(*Continued*)

Region	Period	Site	Culture	Date range (cal.)	M/F N	Reference
	Iron/Roman	Simonsborg	Roman Iron Age	0–200 AD	9/13	Sellevold *et al.*, 1984
		Store Grandløse	Roman Iron Age	0–200 AD	0/1	Sellevold *et al.*, 1984
		Landlystvej	Roman Iron Age	0–200 AD	1/0	Sellevold *et al.*, 1984
		Lille Vasby	Roman Iron Age	0–200 AD	1/5	Sellevold *et al.*, 1984
		Skafterup Mark	Roman Iron Age	0–200 AD	0/1	Sellevold *et al.*, 1984
		Nordensbrogård	Roman Iron Age	0–200 AD	2/0	Sellevold *et al.*, 1984
		Bøgebjerg	Roman Iron Age	0–200 AD	1/0	Sellevold *et al.*, 1984
		Store Keldbjerg	Roman Iron Age	0–200 AD	1/0	Sellevold *et al.*, 1984
		Endegårde	Roman Iron Age	200–400 AD	1/0	Sellevold *et al.*, 1984
		Varpelev	Roman Iron Age	200–400 AD	3/0	Sellevold *et al.*, 1984
		Scheelminde	Roman Iron Age	0–200 AD	1/0	Sellevold *et al.*, 1984
		Asnæs	Roman Iron Age	0–200 AD	1/1	Sellevold *et al.*, 1984
		Englerup	Roman Iron Age	200–400 AD	1/0	Sellevold *et al.*, 1984
		Fraugde	Roman Iron Age	200–400 AD	1/1	Sellevold *et al.*, 1984
		Sanderumgård	Roman Iron Age	200–400 AD	1/1	Sellevold *et al.*, 1984
		Lumby Grusgrav	Roman Iron Age	200–400 AD	1/0	Sellevold *et al.*, 1984
		Næstved Mark	Roman Iron Age	200–400 AD	1/0	Sellevold *et al.*, 1984
		Egebjerg	Roman Iron Age	200–400 AD	0/1	Sellevold *et al.*, 1984
		Broskov	Roman Iron Age	200–400 AD	1/0	Sellevold *et al.*, 1984
		Himlingøje	Roman Iron Age	200–400 AD	1/0	Sellevold *et al.*, 1984
		Gammel Lundby	Roman Iron Age	200–400 AD	1/0	Sellevold *et al.*, 1984
		Vemmeltofte Skovridegård	Roman Iron Age	200–400 AD	0/1	Sellevold *et al.*, 1984
		Smidie	Roman Iron Age	0–400 AD	1/0	Sellevold *et al.*, 1984
		Øbjerggård	Roman Iron Age	0–400 AD	3/1	Sellevold *et al.*, 1984
		Græsbjerg	Roman Iron Age	0–400 AD	1/3	Sellevold *et al.*, 1984
		Lyregård	Roman Iron Age	0–400 AD	1/0	Sellevold *et al.*, 1984
	Neolithic	Bodal Mose	TRB	ca. 3800 BC	1/0	Fischer *et al.*, 2007
		Porsmose	TRB	ca. 3500 BC	1/0	Fischer *et al.*, 2007
		Gjemild (Gjerrild)	Corded Ware	2850–2350 BC	2/1	Petersen, 1993
		Kelderød	TRB	4000–3200 BC	1/0	Bennike, 1985
		Døjringe	TRB	4000–3200 BC	2/0	Bennike, 1985
		Haunø-Lauenkjor	Neolithic	4000–1700 BC	1/0	Panum Institute
		Hallebygård	TRB	ca. 3500 BC	1/0	Fischer *et al.*, 2007
		Over Vindinge	Late Neolithic	2350–1700 BC	1/0	Panum Institute
		Langebjerg	Late Neolithic?	ca. 1700 BC	1/0	Panum Institute
		Marbjerg	Late Neolithic	2191–1972 BC	1/0	Allentoft *et al.*, 2015
		Holmstrup	TRB	3500–3300 BC	1/0	Panum Institute

		Grydehøj	Neolithic	4000–1700 BC	1/0	Bennike, 1985
		Borre	Late Neolithic	2350–1700 BC	1/0	Panum Institute
		Hjortholm Mose	Neolithic	4000–1700 BC	1/0	Panum Institute
		Åstrup Ås	TRB	4000–1700 BC	1/0	Sørensen, 2014
		Hellested	TRB	3300–2850 BC	1/0	Panum Institute
		Grose	Late Neolithic	2350–1700 BC	0/1	Panum Institute
		Poregård	TRB	3300–2850 BC	0/1	Panum Institute
		Korsør Nor	Neolithic	4000–1700 BC	1/0	Fischer *et al.*, 2007
		St. Tuborg	Late Neolithic	2350–1700 BC	0/1	Panum Institute
		Store Lyng	TRB	ca. 3200 BC	1/0	Fischer *et al.*, 2007
		Viksø	TRB	ca. 3800 BC	0/1	Fischer *et al.*, 2007
		Østrup Mose	TRB	ca. 3200 BC	1/0	Fischer *et al.*, 2007
		Beddinge	Corded Ware	2275–2032 BC	4/1	Allentoft *et al.*, 2015
		Viby	Corded Ware	2621–2472 BC	1/0	Allentoft *et al.*, 2015
		Frederiskberg	Late Neolithic	2025–1885 BC	0/1	Allentoft *et al.*, 2015
		Ajvide	Pitted Ware	3100–2100 AD	22/14	Molnar, 2008
		Västerbjers	Pitted Ware	3100–2100 AD	13/10	Molnar, 2008
		Fritorp	Pitted Ware	3100–2100 AD	3/3	Molnar, 2008
		Ire	Pitted Ware	3100–2100 AD	3/1	Molnar, 2008
		Visby	Pitted Ware	3100–2100 AD	5/2	Molnar, 2008
	Mesolithic	Skateholm	Ertebølle	5200–5000 BC	7/12	Larsson, 1988
		Dragsholm	Ertebølle	ca. 3900 BC	0/2	Fischer *et al.*, 2007
		Vængesø 2	Ertebølle	ca. 4350 BC	0/1	Fischer *et al.*, 2007
		Sejrø	Ertebølle	ca. 4000 BC	1/0	Fischer *et al.*, 2007
		Holmegård	Maglemose	7000–5000 BC	2/0	Fischer *et al.*, 2007
		Korsor Glaesverk	Ertebølle	ca. 5100 BC	1/0	Fischer *et al.*, 2007
		Melby	Ertebølle	ca. 5200 BC	1/0	Fischer *et al.*, 2007
		Koelbjerg	Maglemose	ca. 7000 BC	0/1	Fischer *et al.*, 2007
		Tybrind Vig	Ertebølle	ca. 5700 BC	0/1	Fischer *et al.*, 2007
		Bloksbjerg	Ertebølle	ca. 4200 BC	1/0	Fischer *et al.*, 2007
Finland	Very Recent	Helsinki		1915–1935 AD	48/22	Telkkä, 1950
	Early Modern	Porvoo		1660–1720 AD	11/11	Markku Oinonen, pers. comm.; Salo, 2016
		Renko		1680–1850 AD	17/14	Markku Oinonen, pers. comm.; Salo, 2016
Balkans	Late Medieval	Mistihalj	Vlach	1400–1500 AD	27/27	Alexeeva *et al.*, 2003
	Mesolithic	Schela Cladovei		6400–7400 BC	12/5	Boroneant *et al.*, 1995

References

al-Oumaoui, I., Jimenez-Brobeil, S., and du Souich, P. (2004) Markers of activity patterns in some populations of the Iberian Peninsula. *Int. J. Osteoarchaeol.*, **14** (5), 343–359.

Alciati, G., Pesce Delfino, V., and Vacca E. (2005) Catalogue of Italian fossil human remains from the Palaeolithic to the Mesolithic. *J. Anthropol. Sci.* (*Riv. Antrop.*), suppl. 84.

Alexeeva, T., Bogatenkov, D., and Lebedinskaya, G. (2003) Anthropology of Medieval Vlakhs in comparative study: On date of Mistikhaly Burial Site (in Russian, summary in English). Scientific World, Moscow.

Allentoft, M.E., Sikora, M., Sjögren, K.-G., Rasmussen, S., Rasmussen, M., Stenderup, J., Damgaard, P.B., Schroeder, H., Ahlström, T., Vinner, L., Malaspinas, A.-S., Margaryan, A., Higham, T., Chivall, D., Lynnerup, N., Harvig, L., Baron, J., Casa, P.D., Dabrowski, P., Duffy, P.R., Ebel, A.V., Epimakhov, A., Frei, K., Furmanek, M., Gralak, T., Gromov, A., Gronkiewicz, S., Grupe, G., Hajdu, T., Jarysz, R., Khartanovich, V., Khokhlov, A., Kiss, V., Kolář, J., Kriiska, A., Lasak, I., Longhi, C., McGlynn, G., Merkevicius, A., Merkyte, I., Metspalu, M., Mkrtchyan, R., Moiseyev, V., Paja, L., Pálfi, G., Pokutta, D., Pospiezny, L., Price, T.D., Saag, L., Sablin, M., Shishlina, N., Smrčka, V., Soenov, V.I., Szeverényi, V., Tóth, G., Trifanova, S.V., Varul, L., Vicze, M., Yepiskoposyan, L., Zhitenev, V., Orlando, L., Sicheritz-Pontén, T., Brunak, S., Nielsen, R., Kristiansen, K., and Willerslev, E. (2015) Population genomics of Bronze Age Eurasia. *Nature*, **522**, 167–172.

Arias, P. and Garralda, M.D. (1996) Mesolithic burials in Los Canes cave (Asturias, Spain). *Hum. Evol.*, **11**, 129–138.

Bach, A. and Bruchhaus, H. (1988) Das mesolitische Skelett von Unseburg, Kr. Stafrest. *Jahresschrift für mitteldeutsche Vorgeschichte*, **71** (21–36).

Baigl, J.P., Farago-Szekeres, B., and Roger, J. (1997) *Saintes: La Nécropole de la Rue Jacques Brel.* SRA Poitou-Charentes, Poitiers.

Beberhold, O. (2003) Zu Möglichkeiten und Grenzen der Geschlechsbestimmung und Körperhöhenschätzung durch Femurmaße am Beispiel eines mittelalterlichen Gräberfeldes (Dresden-Briesnitz), Jena.

Belcastro, G., Rastelli, E., Mariotti, V., Consiglio, C., Facchini, F., and Bonfiglioli, B. (2007) Continuity or discontinuity of the life-style in Central Italy during the Roman Imperial Age-Early Middle Ages transition: diet, health, and behavior. *Am. J. Phys. Anthropol.*, **132**, 381–394.

Belcastro, M.G., Rastelli, E., and Mariotti, V. (2008) Variation of the degree of sacral vertebral body fusion in adulthood in two European modern skeletal collections. *Am. J. Phys. Anthropol.*, **135**, 149–160.

Belcastro, M.G. and Facchini, F. (2001) Anthropological and Cultural Features of a Skeletal Sample of Horsemen from the Medieval Necropolis of Vicenne-Campochiaro (Molise, Italy). *Coll. Antropol.*, **25**, 387–401.

Bennike, P. (1985) *Paleopathology of Danish Skeletons*. Akademisk Forlag, Copenhagen.

Berner, M. (2014) Demographische und paläopathologische Untersuchungen der Skelette aus dem römerzeitlichen Gräberfeld Halbturn I. In: *Das kaiserzeitliche Gräberfeld von Halbturn, Burgenland* (ed. N. Doneus), Monographien des Römisch-Germanischen Zentralmuseums, Mainz, 122, 2, pp. 309–485.

Blesl, C. (2005) *Das frühbronzezeitliche Gräberfeld von Pottenbrunn*. Fundberichte aus Österreich, Materialhefte A, Wien.

Boccone, S., Chilleri, F., Pacciani, E., Moggi-Cecchi, J., and Salvini, M. (2011) The skeleton of a medieval male with multiple traumatic fractures from piazza della signoria, Florence, Italy. *Int. J. Osteoarchaeol.*, **21**, 602–612.

Borgonini-Tarli, S., Canci, A., Piperno, M., and Repetto, E. (1993) Dental archeologici e antroplogici sulle sepolture mesolithiche della Grotta dell'Uzzo (Trapani). *Bull. Paleontol. Ital.*, **84**, 85–179.
Boroneant, V., Bonsall, C., McSweeney, K., Payton, R., and Macklin, M. (1995) A Mesolithic burial area at Schela Cladovei, Romania. In: *L'Europe des derniers chasseurs, Epipaleolithique et Mesolithique* (ed. A. Thevenin), Editions du CTHS, Paris, pp. 385–390.
Buchvaldek, M. (1986) Kultura se šňůrovou keramikou ve střední Evropě. *Praehistorica*, **12**, 1–160.
Buchvaldek, M. (1990) Pohřebiště lidu se zvoncovitými poháry. *Praehistorica*, **16**, 29–50.
Buchvaldek, M. and Koutecký, D. (1970) *Vikletice, ein schnurkeramisches Gräberfeld.* Karlova Univerzita, Praha.
Buchvaldek, M. and Strahm, C. (1992) Die kontinentaleuropäischen Gruppen der Kultur mit Schnurkeramik: Schnurkeramik-Symposium 1990. *Praehistorica*, **XIX**, 1–363.
Buchvaldek, M., Cvrčková, M., and Budinský, T. (1987) Katalog šňůrové keramiky v Čechách III. Ústecko a Teplicko. *Praehistorica*, **13**, 123–146.
Burt, N.M. (2013) Stable isotope ratio analysis of breastfeeding and weaning practices of children from Fishergate House York, UK. *Am. J. Phys. Anthropol.*, **152** (3), 407–416.
Capitanio, M.A. and Corrain, C. (1998) Resti scheletrici umani di Olmo di Nogara (Verona) dell' eta del Bronzo. *Archivio per l'antropologia e l'etnologia*, **126-127**, 155–188.
Cardoso, H.F. (2006) Brief communication: the collection of identified human skeletons housed at the Bocage Museum (National Museum of Natural History), Lisbon, Portugal. *Am. J. Phys. Anthropol.*, **129** (2), 173–176.
Cardoso, H.F. and Garcia, S. (2009) The Not-so-Dark Ages: ecology for human growth in medieval and early twentieth century Portugal as inferred from skeletal growth profiles. *Am. J. Phys. Anthropol.*, **138** (2), 136–147.
Cartron, I. and Castex, D. (2006) L'occupation d'un ancien îlot de l'estuaire de la Gironde: Du temple antique à la chapelle Saint-Siméon (Jau-Dignac-et-Loirac). *Aquitania*, **22**, 253–282.
Chochol, J. (1978) Antropologická charakteristika laténské skupiny z Makotřas. *Památky archeologické*, **69**, 145–170.
Civetta, A., Schmitt, A., Saliba-Serre, B., Gisclon, J.-L., and Loison, G. (2009) Comparaison de deux populations de la deuxieme moitié du ve millenaire avant notre ere: Approche anthropométrique. *Bull. Mém. Soc. Anthropol. Paris*, **21** (3-4), 141–158.
Čižmář, M. (1978) Keltské pohřebiště v Makotřasích, okres Kladno. *Památky archeologické*, **69**, 117–144.
Čižmář, M. and Geisler, M. (1998) Hroby se šňůrovou keramikou z prostoru dálnice Brno-Vyškov (Pravěk (nová řada, Ústav archeologické památkové péče, Brno). *Supplementum* **1**, 1–102.
Combier, J. and Genet-Varcin, E. (1959) L'homme mésolithique de Culoz et son gisement. *Ann. Paléont.*, **45**, 141–174.
Constandse-Westermann, T.S., Meiklejohn, C., and Newell, R.R. (1982) A reconsideration of the mesolithic skeleton from Rastel, commune de Peillon, Alpes-Maritimes, *France. Bull. du Musée d'Anthropol. Préhistorique de Monaco*, **26**, 75–89.
Daňhel, M. and Pankowská, A. (2010) Pohřby na sídlištích ze starší doby bronzové z Hulínska. Hroby, pohřby a lidské pozůstatky na pravěkých a středověkých sídlištích, 125–136.
Delsate, D., Guinet, J.-M., and Saverwyns, S. (2009) De l'ocre sur le crâne mésolithique (haplogroupe U5a) de Reuland-Loschbour (Grand-Duché de Luxembourg)? *Bull. Soc. Préhist. Luxembourgeoise*, **31**, 7–30.
Dinwiddy, K.E. and Bradley, P. (2011) *Prehistoric Activity and a Romano-British Settlement at Poundbury Farm, Dorchester, Dorset.* Wessex Archaeology Reports. Trust for Wessex Archaeology Ltd.

Diverrez, F., Poulmarc'h, M., and Schmitt, A. (2012) Nouvelles données sur les inhumations ad sanctos à l'époque moderne en milieu rural: le cas de l'église Saint-Pierre de Moirans (Isère). *Bull. Mém. Soc. Anthropol. Paris*, pp. 1–12.

Dobeš, M. (1993) Katalog šňůrové keramiky v SZ Čechách VII. Chomutovsko. *Praehistorica*, **XX**, 175–196.

Dobeš, M. and Buchvaldek, M. (1993) Katalog šňůrové keramiky v Čechách VIII. Mostecko. *Praehistorica*, **XX**, 197–258.

Dobeš, M., Budínský, P., Buchvaldek, M., and Muška, J. (1991) Katalog šňůrové keramiky v Čechách V. Bílinsko. *Praehistorica*, **17**, 75–145.

Dobisíková, M., Velemínský, P., and Kuželka, P. (1996) Obyvatelé Malé Strany v raném středověku. Středověké pohřebiště v areálu Vratislavského paláce z pohledu antropologa. Muzejní a Vlastivědná práce/ Časopis společnosti přátel starožitností 34/104, 201–212.

Doneus, N. (2014) *Das kaiserzeitliche Gräberfeld von Halbturn, Burgenland*. Römisch-Germanischen Zentralmuseums, Mainz, vol. **122**, 1.

Duday, H. (1975) Le squelette du sujet feminin de la sépulture pré-néolithique de Bonifacio (Corse): étude anthropologique. Essai d'interprétation paléoethnographique. Cahiers d'Anthropologie, Mémoires du Laboratoire d'Anatomie de la Faculté de Médecine de Paris, 24.

Dusek, S. (1999) Ph. D. Thesis, Weimar: Ur- und Frühgeschichte Thüringens.

Ellis, C. and Powell, A.B. (2008) An Iron Age settlement outside Balttlesbury Hillfort, Warminster, and sites along the Southern Range Road. Wessex Archaeology Report 22: Wessex Archaeology and Defense Estates.

Europe. Trustees of the British Museum (Natural History), London.

Farwell, D.E. and Molleson, T.I. (1993) *Excavations at Poundbury, Dorset 1966–1982. Volume II: The Cemeteries*. Monograph Series Number 11. Dorset Natural History and Archaeological Society, Dorchester.

Ferembach, D. (1974) Le squelette humain azilien de Rochereil (Dordogne). *Bull. Mém. Soc. Anthropol. Paris Sér.*, **13** (2), 271–291.

Fischer, A., Olsen, J., Richards, M., Heinemeier, J., Sveinbjörnsdóttir, A.E., and Bennike, P. (2007) Coast-inland mobility and diet in the Danish Mesolithic and Neolithic: evidence from stable isotope values of humans and dogs. *J. Archaeol. Sci.*, **34**, 2125–2150.

Fitzpatrick, A.P. (2011) The Amesbury Archer and Boscombe Bowmen. Bell Beaker Burials on Boscombe Down, Amesbury, Wiltshire. Wessex Archaeology Report 27. Wessex Archaeology Ltd, Salisbury, UK.

Galera, V. and Garralda, M.D. (1993) Enthesopathies in a Spanish Medieval population: anthropological, epidemiological, and ethnohistorical aspects. *Int. J. Anthropol.*, **8**, 247–258.

Gallagher, A., Gunther, M., and Bruchhaus, H. (2009) Population continuity, demic diffusion and Neolithic origins in central-southern Germany: The evidence from body proportions. *HOMO J. Comp. Hum. Biol.*, **60** (2), 95–126.

Gambier, D., Valladas, H., Tisnérat-Laborde, N., Arnold, M., and Bresson, F. (2000) Datation de vestiges humains présumés du Paléolithique supérieur par la méthode du carbone 14 en spectrométrie de masse par accélerateur. *Paleo*, **12**, 201–212.

Gejvall, N.-G. (1960) *Westerhus: Medieval population and church in the light of skeletal remains*. Håkan Ohlssons Boktryckeri, Lund.

Gieseler, W. (1953) Das jungpaläolithische Skelett von Neuessing. *Natur. Mschr. Aus der Heimat.*, **61**, 161–174.

Green, C. and Rollo-Smith, S. (1984) The excavation of eighteen round barrows near Shrewton, *Wiltshire. Proc. Prehist. Soc.*, **50**, 255–318.

Grifoni Cremonesi, R., Borgognini Tarli, S.M., Formicola, V., and Paoli, G. (1995) La sepoltura epigravetttianascoperta nel 1993 nella Grotta Continenza di Trasacco (l'Aquila). *Rivista di Antropologia (Roma)*, **73**, 225–36.

Hampl, F. and Benkovsky-Pivovarová, Z. (1985) Das mittelbronzezeitliche Gräberfeld von Pitten in Niederösterreich. Mitteilungen der Prähistorischen Kommission, 21–22.
Heinrich, W. and Teschler-Nicola, M. (1991) Zur Anthropologie des Gräberfeldes F von Gemeinlebarn, Niederösterreich. *Röm. German Forschung.*, **49**, 222–262.
Heinrich, W. (2001) *Zwentendorf–Ein Gräberfeld aus dem 10–11 Jahrhundert.* Prähistorischen Kommission, Akademie der Wissenschaften, Wien.
Henke, W. (1984) Vergleichend-morphologische Kennzeichnung der Jungpaläolithiker von Oberkassel bei Bonn. *Z. Morphol. Anthropol.*, **75**, 27–44.
Henry-Gambier, D. (2002) Les fossiles de Cro-Magnon (Les Eyzies-de-Tayac, Dordogne): nouvelle données sur leur position chronologique et leur attribution culturelle. *Bull. Mém. Soc. Anthropol. Paris*, **14**, 89–112.
Jackes, M. and Lubell, D. (1999) Human biological variability in the Portuguese Mesolithic. *Arqueologia*, **24**, 25–42.
Jay, M. and Richards, M.P. (2006) Diet in the iron age cemetery population at Wetwang Slack, East Yorkshire, UK: carbon and nitrogen stable isotope evidence. *J. Archaeol. Sci.*, **33** (5), 653–662.
Jiménez-Brobeil, S.A., Al Oumaoui, I., and Du Souich, P. (2010) Some types of vertebral pathologies in the Argar Culture (Bronze Age, SE Spain). *Int. J. Osteoarchaeol.*, **20** (1), 36–46.
Kahlke, H.D. (2004) *Sondershausen und Bruchstedt: zwei Gräberfelder mit älterer Linienbandkeramik in Thüringen.* Weimarer Monographien zur Ur- und Frühgeschichte. Beier & Beran, Weimar.
Keiller, A. and Piggott, S. (1938) Excavation of an untouched chamber in the Lanhill Long Barrow. *Proc. Prehist. Soc.*, **4**, 122–150.
Kern, D. (2012) Migration and mobility in the latest Neolithic of the Traisen valley, Lower Austria: Archaeology. In: *Population Dynamics in Prehistory and Early History: New Approaches Using Stable Isotopes and Genetics* (eds E. Kaiser, J. Burger, and W. Schier W), de Gruyter, Berlin, pp. 213–224.
Kjellström, A. (2005) *The Urban Farmer: Osteoarchaeological Analysis of Skeletons from Medieval Sigtuna Interpreted in a Socioeconomic Perspective.* Theses and Papers in Osteoarchaeology No. 2, Stockholm University, Stockholm.
Kozák, P., Pankowská, A., Plaštiaková, M., and Zezula, M. (2010) Michal: Opava (k. ú. Opava-Město, okr. Opava). Areál pivovaru (tzv. horní dvůr), p.č. 128/7, 128/8. Středověk, novověk. Město. Zjišťovací výzkum. Přehled výzkumů AU AV ČR v Brně:458–462.
Krumpel, J. (2008) Vier Gräber der Badener Kultur aus Ratzersdorf, Niederösterreich. *Eine Neubewertung der Bestattungssitten der Badener Kultur in ihrer österreichischen Verbreitung, Fundberichte aus Österreich*, **47**, 99–150.
Kuželka, P., Velemínský, P., and Hanáková, H. (2004) Antropologická expertiza kosterních pozůstatků z Radovesic. In: *Druhé keltské pohřebiště z Radovesic (okres Teplice) v severozápadních Čechách* (eds P. Budinský and J. Waldhauser), Regionální muzeum v Teplicích Archeologické výzkumy v severních Čechách 31, pp. 37–42.
Kytlicová, O. (1956) Pohřebiště kultury zvoncovitých pohárů v Kněževsi. *Archeologické rozhledy*, **8**, 328–356.
Kytlicová, O. (1960) Eneolitické pohřebiště v Brandýsku. *Památky archeologické*, **61**, 442–474.
Larsson, L. (1988) *The Skateholm Project I. Man and Environment.* Almqvist & Wiksell International, Stockholm.
Lauermann, E. (1995) *Ein frühbronzezeitliches Gräberfeld aus Unterhautzenthal.* Österreichische Gesellschaft für Ur-und Frühgeschichte, Stockerau.
Lovell, J., Hamilton-Dyer, S.A., Loader, E., and McKinley, J.I. (1999) Further investigation of an Iron Age and Romano-British farmstead on Cockey Down, near Salisbury. *Wiltshire Arch. Nat. Hist. Mag.*, **92**, 33–38.

Manzi, G., Salvadei, L., Vienna, A., and Passarello, P. (1999) Discontinuity of life conditions at the transition from the Roman Imperial Age to Early Middle Ages: example from Central Italy evaluated by pathological dento-alveolar lesions. *Am. J. Hum. Biol.*, **11**, 327–341.

Marhold, F.-J. (1977) Anthropologische Untersuchung der Skelette des awarenzeitlichen Gräberfeldes in Mödling 'Goldene Stiege'. Dissertation, Universität Wien, Wien.

Matiegka, J. (1938) Homo předmostensis: fosilní člověk z Předmostí na Moravě. 2. Ostatni části kostrové: Nákladatelství České Akademie Věd a Umění.

Matthias, W. (1982) *Kataloge zur mitteldeutschen Schnurkeramik*. Niemeyer, Halle.

McKinley, J.I., Leivers, M., Schuster, J., Marshall, P., Barclay, A.J., and Stoodley, N.A. (2015) Cliffs End Farm, Isle of Thanet, Kent: a mortuary and ritual site of the Bronze Age, Iron Age and Anglo-Saxon period with evidence for long-distance maritime mobility. Wessex Archaeology Monograph 31, Wessex Archaeology Ltd.

Molleson, T. and Cox, M. (1993) *The Spitalfields Project, Vol. 2*. CBA Research report 86. Council for British Archaeology, York.

Molnar. P. (2008) Tracing prehistoric activities: lefe ways, habitual behaviour and health of hunter-gatherers on Gotland. Theses and Papers in Osteoarchaeology No. 4. Doctoral Thesis. Stockholm University, Stockholm, Sweden.

Neugebauer, J.-W. (1991) Die Nekropole F von Gemeinlebarn, Niederösterreich: Untersuchungen zu den Bestattungssitten und zum Grabraub in der ausgehenden Frühbronzezeit in Niederösterreich südlich der Donau zwischen Enns und Wienerwald. Römisch Germanische Forschungen 49.

Neugebauer, J.-W. (1994) *Bronzezeit in Ostösterreich*. Wissenschaftliche Schriftenreihe Niederösterreichs.

Neugebauer-Maresch, C. (1994) Endneolithikum. Die Lokalgruppe der Schnurkeramik des Unteren Traisentales. In: *Bronzezeit in Ostösterreich St Polten* (ed. J.-W. Neugebauer), Wissenschaftliche Schriftenreihe Niederösterreichs, Vienna, 98/101, pp. 23–48.

Neugebauer-Maresch, C. and Neugebauer, J.-W. (1997) Franzhausen. Das frühbronzezeitliche Gräberfeld I. Fundberichte aus Österreich Materialhefte A 5/1.

Neustupný, E. (ed.) (2008) Archeologie pravěkých Čech. 4: Eneolit. Archeologický ústav AV ČR, Prague.

Neustupný, E. and Smrž, Z. (1989) Čachovice-pohřebiště kultury se šňůrovou keramikou a zvoncovitých pohárů. *Památky archeologické*, **80**, 282–383.

Newell, R.R., Constandse-Westermann, T.S., and Meiklejohn, C. (1979) The skeletal remains of mesolithic man in western Europe: an evaluative catalogue. *J. Hum. Evol.*, **8**, 1–228.

Nolan, J. (2010) The Early Medieval cemetery at the Castle, Newcastle-upon-Tyne. *Archaeologia Aeliana*, **39**, 147–287.

Noll, G. (2001) Schnurkeramische Bestattungen auf dem Gelände der Nordhäuser Straße. Archäologie und Bauforschung in Erfurt, 17–35.

Oakley, K.P., Campbell, B.G., and Molleson, T.I. (1971) *Catalogue of Fossil Hominids. Part II: Europe*. British Museum, London.

Pany-Kucera, D. and Wiltschke-Schrotta, K. (2017) Die awarische Bevölkerung von Vösendorf/S1. *Annalen des Naturhistorischen Museums in Wien Serie A*, **119**, 5–31.

Parenti, R. (1952) Caratteristiche angolari del cranio umano nel piano sagittale. Studio biometrico di una serie di crani siracusani. *Archivio per l'antropologia e la etnologia*, **82**, 33–82.

Patte, E. (1968) L'homme et la femme de l'Azilien de Saint Rabier. *Mém. Mus. Hist. Nat. C*, **19**, 1–56.

Pertlwieser, T., Gemering, K.G., St Florian, M.G., and Linz-Land, V.B. (2001) Grabungsbericht Jungsteinzeit bis Latènezeit, Siedlung bei Tödling. *Fundberichte aus Österreich*, **40**, 579–581.

Petersen, H.C. (1993) An anthropological investigation of the Single Grave Culture in Denmark. In: *Population of the Nordic countries: Human population biology from the present to the Mesolithic* (eds E. Iregren and R. Liljekvist), Institute of Archaeology Report Series No. 46, University of Lund. Bloms Boktryckeri AB, Lund, pp.178–188.

Phillips, C.W. (1932) The excavations of the Giant's Hill Long Barrow, Skendleby, Lincolnshire. *Archaeologia*, **85**, 174–202.
Piggott, S. and Piggott, C.M. (1944) Excavation of barrows on Crhichel and Launceston Downs. *Archaeologia*, **90**, 47–80.
Piggott, S. (1962) *The West Kennet Long Barrow*. Ministry of Works Arch. Rep. 4. Her Majesty's Stationery Office, London
Podborský, V. (ed.) (1993) *Pravěké dějiny Moravy: Vlastivěda Moravská: Země a lid*. Muzejní a vlastivědná společnost v Brně, Brno.
Poláček, L. (2008) Great Moravia, the power centre at Mikulčice and the issue of the socioeconomic structure. In: *Studien zum Burgwall von Mikulčice VIII* (eds P. Velemínský and L. Poláček), Archeologický ústav, Brno, pp. 11–44.
Preuß, J. (1998a) *Das Neolithikum in Mitteleuropa. Teil A*. Beier and Beran, Weissbach.
Preuß, J. (1998b) *Das Neolithikum in Mitteleuropa. Teil B*. Beier and Beran, Weissbach.
Salo, K. (2016) Health in Southern Finland – Bioarchaeological Analysis of 555 Skeletons Excavated from Nine Cemeteries (11th–19th century AD). Ph.D. dissertation. University of Helsinki, Helsinki, Finland.
Sauer, F. (ed.) (2007) *Die archäologischen Grabungen auf der Trasse der A6: Fundstelle Vösendorf/ Laxenburgerstrasse*. Bundesdenkmalamt und Asfinag, Wien.
Saville, A. (1990) Hazleton North, Gloucestershire, 1979–1982: The excavation of a Neolithic long cairn of the Cotswold-Severn group. English Heritage Arch. Rep. 13: Historic Buildings and Monuments Commission for England.
Scherf, H., Wahl, J., Hublin, J.J., and Harvati, K. (2016) Patterns of activity adaptation in humeral trabecular bone in Neolithic humans and present-day people. *Am. J. Phys. Anthropol.*, **159** (1), 106–115.
Šebela, L. (1999) *The Corded Ware Culture in Moravia and in the adjacent part of Silesia* (catalogue). Archeologický ústav Akademie věd České republiky, Brno.
Seligman, C.G. and Parsons, F.G. (1914) The Cheddar Man: a skeleton of Late Palaeolithic date. *J. Roy. Anth. Inst.*, **44**, 241–263.
Signoli, M., Leonetti, G., and Dutour, O. (1997) The great plague of Marseille (1720–1722): New anthropological data. *Acta Biol.*, **42**, 123–133.
Silvestrini, M. and Pignocchi, G. (1997) La necropolis Eneolitica di Fontenoce di Recanati: Lo scavo 1992. *Rivista di Scienze Preistoriche*, **48**, 309–366.
Sládek, V., Trinkaus, E., Holiday, T., and Hillson, S. (2000) The People of the Pavlovian: Skeletal Catalogue and Osteometrics of the Gravettian fossil hominids from Dolní Věstonice and Pavlov. Archeologický ústav AVČR, Dolní Věstonice Studies 5, Brno.
Smith, I.F. (1991) Round Barrows Wilsford cum Lake G51-G54: Excavations by Ernest Greenfield in 1958. *Wiltshire Arch. Nat. Hist. Mag.*, **84**, 11–39.
Sollas, W.J. (1913) Paviland Cave: An Auragnacian Station in Wales. *JRAI*, **43**, 325–374.
Sørensen, L. (2014) From Hunter to Farmer in Northern Europe: Migration and Adaptation During the Neolithic and Bronze Age. Ph.D. dissertation. University of Copenhagen, Copenhagen. Published by ACTA ARCHAEOLOGICA & Centre of World Archaeology.
Sprenger, S. (1999) Zur Bedeutung des Grabraubes für sozialanthropologische Gräberfeldanalysen. Fundberichte aus Österreich Materialhefte (Wien) A 7.
Stead, I.M. (1969) The excavation of beaker burials at Staxton, East Riding, 1957. *Yorkshire Arch. J.*, **15**, 129–144.
Stloukal, M. (2013) Základní určení koster z pohřebiště ve Šporkově ul. čp. 322/III, výzkum č. 30/0. Archiv Antropologického oddělení Národního Muzea v Praze (unpublished manuscript).
Telkkä, A. (1950) On the prediction of human stature from the long bones. *Acta Anat.*, **9**, 103–117.

Teschler-Nicola, M. (1985) Die Körper- und Brandbestattungen des mittel-bronzezeitlichen Gräberfeldes von Pitten, Niederösterreich. Demographische und anthropologische Analyse. *Mitteilungen der prähistorischen Kommission der österreichischen Akademie der Wissenschaften*, **21-22**, 127–272.

Teschler-Nicola, M. (1992) Untersuchungen zur Bevölkerungsbiologie der Bronzezeit in Ostösterreich. Phänetische Analyse kontinuierlicher und nichtkontinuierlicher Skelettmerkmale. Habilitationsschrift zur Erlangung der venia legendi. Formal- und Naturwissenschaftliche Fakultät, Wien.

Trinkaus, E. and Svoboda, J. (eds) (2006) *Early Modern Human Evolution in Central Europe: The People of Dolní Věstonice and Pavlov*. Oxford University Press.

Trňáčková. Z. (1970) Věteřovské sídliště u Bystročic (okr. Olomouc). Přehledy výzkumů 1968 (AVČR v Brně), pp. 20–21.

Vallois, H.V. (1941–1946) Nouvelles recherches sur le squelette de Chancelade. *L'Anthropol.*, **50**, 165–202.

Vallois, H.V. and de Felice, S. (1977) Les mésolithiques de France. *Archives de l''Institut de Paléontologie humaine*, **37**.

Velemínský, P. (1997) Antropologický posudek pohřebiště Vršany. Archiv Antropologického oddělení Národního Muzea v Praze (unpublished manuscript).

Velemínský, P. and Dobisíková, M. (1998) Demografie a základní antropologická charakteristika pravěkých pohřebišť v Praze 5-Jinonicích (Eneolit, kultura únětická, laténské období). *Archaeologica Pragensia*, **14**, 229–271.

Velemínský, P., Kuželka, P., and Hanáková, H. (2004) Demografická a antropologická charakteristika druhého keltského pohřebiště z Radovesic. In: *Druhé keltské pohřebiště z Radovesic (okres Teplice) v severozápadních Čechách* (eds P. Budinský and J. Waldhauser), Regionální muzeum v Teplicích Archeologické výzkumy v severních Čechách 31, pp. 57–66.

Vlček, E. (1967) Die Anthropologie der mittelsteinzeitlichen Graber von Bettendorf Kr. *Artern. Jahresschrift. Kitteldeutsche Vorgeschichte*, **51**, 53–64.

von Bonin, G. (1935) European races of the Upper Paleolithic. *Hum. Biol.*, **7**, 196–221.

Waldhauser, J. and Dehn, W. (1978) *Das keltische Gräberfeld bei Jenišův Újezd in Böhmen*. Krajské Muzeum Teplice, Teplice.

Whittle, A. (1991) Wayland's Smithy, Oxfordshire: Excavations at the Neolithic Tomb in 1962–63 by R.J.C. Atkinson and S. Piggott. *Proc. Prehist. Soc.*, **57**, 61–101.

Wiltschke-Schrotta, K., Cemper-Kiesslich, J., and Höger, A.-M. (2008) Die badenzeitlichen Skelette von Ratzersdorf/Traisen, Niederösterreich. *Niederösterreich Fundberrichte aus Österreich*, **47**, 151–166.

Appendix 2(a)

Skeletal Variation and Adaptation in Europeans: Upper Paleolithic to the Twentieth Century,
First Edition. Edited by Christopher B. Ruff.

Appendix 2(a) Total (C2–L5) vertebral column length estimations from different combinations of vertebral heights (i.e., anterior midline heights, posterior heights, and/or maximum heights anterior to pedicles) with sex-specific regression equations based on a sample of 36 males and 26 females with complete presacral columns (M. Niskanen, unpublished data). Notes and sources provide, in addition to literature and other sources used, a list of available vertebrae for each individual. Heights of these vertebrae were summed to provide the sum given in the table. The regression equation provided in the table was then applied to this sum to derive regression-predicted C2–L5 (presacral column) length.

Specimen	Sex	Period	No. of vertebra	Sum present	Percent present	Slope	Intercept	r	Predicted C2–L5 presacral column length	95% C.I.	Notes and sources
Barma Grande 2	M	EUP	16	419.7	71.68	1.289	38.975	0.944	579.84	559.20–600.49	1
Cro-Magnon 4327 & 4332	M	EUP	6	155.4	30.65	2.656	95.710	0.904	508.53	484.18–532.87	2
Cro-Magnon 4324 & 4330	F	EUP	9	206.3	43.87	1.351	194.536	0.891	473.23	453.38–493.09	3
Dolni Vestonice 3	F	EUP	12	216.9	47.94	1.670	95.170	0.892	457.41	437.25–477.57	4
Dolni Vestonice 13	M	EUP	15	327.0	65.43	1.348	60.310	0.941	501.10	481.76–520.43	5
Grotte des Enfants 4	M	EUP	17	393.4	78.42	1.168	42.925	0.925	502.42	480.74–524.10	6
Predmost 3	M	EUP	15	302.1	62.99	1.528	19.144	0.967	480.75	466.03–495.47	7
Predmost 4	F	EUP	17	324.1	73.38	1.161	70.876	0.931	447.00	430.31–463.68	8
Predmost 9	F	EUP	7	125.2	25.21	2.048	231.021	0.613	487.40	452.50–522.30	9
Predmost 10	F	EUP	7	151.8	32.03	2.250	133.429	0.893	474.96	455.33–494.59	10
Predmost 14	M	EUP	16	316.6	65.86	1.456	20.975	0.958	481.96	465.39–498.52	11
Parabita 1	M	EUP	3	79.2	15.56	4.843	125.710	0.861	509.32	480.33–538.29	12
Parabita 2	F	EUP	3	86.8	16.34	3.774	183.049	0.871	510.60	487.78–533.41	13
Arena Candide 2	M	LUP	10	236.8	45.97	1.972	47.831	0.925	514.74	493.11–536.36	14
Arena Candide 4	M	LUP	11	232.6	47.96	1.863	54.494	0.894	487.76	461.89–513.62	15
Arena Candide 10	M	LUP	6	148.6	31.66	2.528	102.012	0.906	477.60	452.83–502.37	16
Arena Candide 12	M	LUP	7	157.5	32.78	2.709	57.170	0.882	483.82	456.49–511.14	17
Bichon 1	F	LUP	15	338.1	68.61	1.122	109.887	0.932	489.26	473.31–505.22	18
Bruniquel 1	F	LUP	4	113.2	21.54	2.296	241.373	0.744	501.29	470.81–531.78	19
Cap Blanc 1	F	LUP	11	240.7	53.88	1.539	81.437	0.932	451.95	435.61–468.29	20
Chancelade 1	M	LUP	18	367.7	75.80	1.329	−3.854	0.967	484.88	470.05–499.70	21
Continenza 1	M	LUP	8	160.0	34.42	2.340	+99.908	0.852	474.27	443.40–505.13	22

Cough's Cave	M	LUP	13	264.5	57.43	1.628	+33.095	0.937	463.82	442.95–484.69	23
Grotte des Enfants 3	F	LUP	3	78.8	16.85	3.872	+166.027	0.735	471.17	441.46–500.87	24
Oberkassel 1	M	LUP	11	259.3	50.65	1.891	+21.685	0.935	502.68	482.46–522.90	25
Oberkassel 2	F	LUP	8	194.4	41.07	1.990	+87.167	0.909	474.05	455.83–492.28	26
Le Peyrat 5	M	LUP	12	291.0	58.54	1.501	+61.944	0.920	498.74	484.12–515.55	27
Le Peyrat 6	F	LUP	9	210.7	46.20	1.360	+177.498	0.891	464.06	444.06–484.05	28
Romito 4	F	LUP	12	259.7	57.28	1.275	+128.657	0.924	459.89	442.93–476.85	29
St. Germain la Rivière	F	LUP	9	227.0	46.08	1.528	+141.250	0.906	488.14	469.60–506.69	30
Veyrier 1	M	NEO	9	229.6	42.20	2.013	+76.942	0.911	539.01	515.13–562.89	31
Hoëdic 8	F	MES	5	112.7	23.71	3.188	+116.501	0.903	475.80	457.01–494.58	32
Hoëdic 9	M	MES	10	208.3	48.14	1.828	+61.173	0.916	442.01	416.87–467.15	33
Rastel 1	M	MES	16	377.1	73.48	1.279	+30.860	0.932	513.07	494.34–533.80	34
Villabruna 1	M	LUP	17	384.5	78.42	1.168	+42.925	0.925	492.03	470.21–513.84	35
Skateholm 24	F	MES	5	156.7	30.58	2.779	+71.778	0.774	507.10	477.65-536.56	36
Skateholm 37	F	MES	9	169.1	36.84	2.140	+101.016	0.923	462.99	445-97-480.01	37
Skateholm II	M	MES	10	210.47	43.60	1.913	+84.668	0.935	487.34	466-92-507-76	38
Skateholm Xa	M	MES	7	115.97	25.87	2.965	+119.014	0.855	462.90	431.66-494.15	39
Skateholm XVI	F	MES	6	156.86	30.77	2.568	+100.220	0.755	503.06	472.97-533.16	40
Skateholm XVII	M	MES	18	388.8	80.41	1.293	−20.127	0.989	473.47	473.74-491.20	41
Frederiksberg 1	F	NEO	19	352.45	76.10	1.161	+55.807	0.965	464.83	453.35-476.31	42
Ajvide 14	M	NEO	23	437.61	98.68	1.016	−1.117	0.993	443.31	436.22-450.41	43
Västerbjers 5	M	NEO	21	438.99	88.65	1.126	+0.837	0.978	495.21	483.20-507.33	44
Västerbjers 9	M	NEO	17	351.67	67.11	1.525	−11.901	0.978	524.30	512.38-536.22	45
Västerbjers 12	M	NEO	19	407.49	82.21	1.202	+5.843	0.983	495.83	485.22-506.43	46
Västerbjers 42	M	NEO	20	448.47	86.14	1.180	−8.289	0.982	520.80	509.91-531.69	47
Västerbjers 61	M	NEO	20	437.40	83.75	1.161	+14.320	0.971	521.93	508.19-535.68	48
Västerbjers 80	F	NEO	23	465.65	98.28	0.912	+49.616	0.984	474.19	466.35-482.02	49
Västerbjers 84	F	NEO	15	284.63	56.68	1.552	+57.316	0.937	499.17	483.52-514.82	50
Hjortholm Mose	M	NEO	14	325.06	65.19	1.446	+29.244	0.985	499.35	489.36-509.33	51
Grose	F	NEO	11	217.92	45.13	1.590	+134.998	0.936	481.50	466.15-496.85	52
St. Tuborg	F	NEO	9	156.44	36.15	2.712	+9.184	0.917	433.44	414.21-452.67	53

Notes and Sources

1 Anterior midline heights of T1–L5 (Holliday, 1995).
2 Posterior midline heights of T2 and L1–L5 (Holliday, 1995).
3 Posterior midline heights of T6-T12, L2, and L5 (Holliday, 1995).
4 Posterior midline heights of C5–C7, T2–T4, anterior midline heights of T9–T12, posterior heights of L4 and L5 (Holliday, 1995; Sládek *et al.*, 2000).
5 Anterior midline heights C6–T1 and T6–L5 (Sládek *et al.*, 2000).
6 Posterior midline height of T1–L5 (Holliday, 1995).
7 Anterior midline heights of C2–T1, T4, T9–T11, L4 and L5; posterior midline heights of T7, T11, and T12 (Matiegka, 1938).
8 Anterior midline heights of C2–T1 and T10, T11, L4 and L5; posterior midline heights of T2, T6, T7, T12, L2, and L3 (Matiegka, 1938).
9 Anterior heights of C2, and C5–C7; posterior midline heights of C4, T1, and T2 (Matiegka, 1938).
10 Anterior midline heights of C3, T11, L1, and L2; posterior midline heights of T9, T10, and L5 (Matiegka, 1938).
11 Anterior midline heights of C2–C6, T1, T10, and T12; posterior midline heights of C7, T2, T7–T9, T11, L4, and L5 (Matiegka, 1938).
12 Posterior midline heights of T11, L1, and L5 (Holliday, 1995).
13 Posterior midline heights of T11–L1 (Holliday, 1995).
14 Posterior midline heights of T1, T2, T4, T5, T7, T12, and L2–L5 (Holliday, 1995).
15 Posterior midline heights of T1–T5, T7-T9, T12, L1, and L5 (Holliday, 1995).
16 Anterior midline height of T9; posterior midline heights of T12–L4 (Holliday, 1995).
17 Posterior midline heights of T1, T5, T9–T12, and L2 (Holliday, 1995).
18 Posterior midline heights of T1–T8, T10–T12, and L2–L5 (Holliday, 1995).
19 Posterior midline heights of L2–L5 (Holliday, 1995).
20 Anterior midline heights of T1, T3, T8, and L2–L4; posterior midline heights of T10–L1, and L5 (Holliday, 1995).
21 Anterior midline heights of C2, C3, C5-T3, T6–T10, L1, and L2; posterior midline heights of T11, L3, and L4 (Billy, 1969, 1975; Holliday, 1995).
22 Anterior midline heights of T5–T8, and T10–L1 (Holliday, 1995).
23 Anterior midline heights of T1, T4, T6, T7, and T9–L2; posterior midline heights of T5, T8, and L3–L5 (Holliday, 1995).
24 Anterior midline heights of L1, L4, and L5 (Holliday, 1995).
25 Anterior midline heights of T2, T9, L1, and L4; posterior midline heights of T1, T3, T8, T10–T12, L3, and L5 (Holliday, 1995).
26 Anterior height of C2; posterior heights of C3, T5, T6, T12–L2, and L4 (Billy, 1975; Holliday, 1995).
27 Posterior midline heights of T5–T10 and T12–L5 (Holliday, 1995).
28 Posterior midline heights of T8–L4 (Holliday, 1995).
29 Anterior midline height of T8; posterior midline heights of T1–T3 and T9–L4 (Holliday, 1995).
30 Posterior midline heights of T4, T7, and T11–L5 (Holliday, 1995).
31 Anterior midline heights of T7–T11, L1, and L3; posterior midline heights of T12 and L3 (Holliday, 1995).
32 Posterior midline heights of T1, T3, T11, L2, and L3 (Holliday, 1995).
33 Posterior midline heights of T2, T6, T8, T9, T10, T12, L1, and L3–L5 (Holliday, 1995).
34 Anterior midline height of L3; posterior midline heights of T1–L2 and L4 (Holliday, 1995).

35 Posterior midline heights of T1–L5 derived by subtracting anterior sacral length (100 mm) from skeletal trunk height (484.5 mm) in Vercellotti *et al.* (2008), derived as in Holliday (1995).
36 Posterior midline height of L1; maximum heights anterior to pedicles C2 and L2–L5 (Niskanen).
37 Posterior midline heights of T1–T3; maximum heights anterior to pedicles of T4–T8 and L2 (Niskanen).
38 Anterior midline heights of T11–L5; maximum heights anterior to pedicles of C5–C7 and T9 (Niskanen).
39 Maximum heights anterior C2–C6, T1 and T2 (Niskanen).
40 Posterior midline height of C5; maximum heights anterior to pedicles of C2, C6, L3, L4, and L5 (Niskanen).
41 Maximum heights anterior to pedicles C2, C3, and C7-L3 (Niskanen).
42 Anterior midline heights of C4–T11 and L2–L5 (Niskanen).
43 Maximum heights anterior to pedicles C3–T5 and T7–L3 (Niskanen).
44 Posterior midline heights of T2, T4, L1–L3, and L5; maximum heights anterior to pedicles C3–T1, T3, and T5–T11 (Niskanen).
45 Maximum heights anterior to pedicles of C2–T11 (Niskanen).
46 Posterior midline heights of L1, L2, and L4; maximum heights anterior to pedicles of C2–T3, T5, T6, T10–T12, L3, and L5 (Niskanen).
47 Maximum heights anterior to pedicles of C2, C3, C5–L3 (Niskanen).
48 Posterior midline heights of L1–L5; maximum heights anterior to pedicles of C4–T9 (Niskanen).
49 Posterior midline heights of L2–L5; maximum heights anterior to pedicles of C2–L1 (Niskanen).
50 Maximum heights anterior to pedicles of C2–T9 (Niskanen).
51 Maximum heights anterior to pedicles of C2, C5, C6, T5–T12, and L2–L4 (Niskanen).
52 Maximum heights anterior to pedicles of T1–T11 (Niskanen).
53 Maximum heights anterior to pedicles of C5–T1, T4, T5, L1, L2, and L5 (Niskanen).

References

Billy, G. (1969) Le squelette post-cranien de l'homme de Chancelade. *L'Anthropologie*, **73**, 207–246.

Billy, G. (1975) Etude anthropologique des restes humains d L'Abri Pataud. In: *The Excavations of the Abri Pataud* (ed. H.L. Movius), Peabody Museum, Cambridge, pp. 201–261.

Holliday, T. (1995) Body Size and Proportions in the Late Pleistocene Western Old World and the Origins of Modern Humans. Ph.D. dissertation. The University of New Mexico, Albuquerque, New Mexico.

Matiegka, J. (1938) *Homo Předmostensis. Folsilni Člověk z Předmostí na Moravě II.* Česká Akademie Věd a Uměni, Prague.

Sládek, V., Trinkaus, E., Hillson, S.W., and Holliday, T.W. (2000) The People of the Pavlovian: Skeletal Catalogue and Osteometrics of the Gravettian Fossil Hominids from Dolni Věstonice and Pavlov. The Dolni Věstonice Studies, Volume 5/2000. The Academy of Sciences of the Czech Republic, Brno.

Vercellotti, G., Alciati, G., Richards, M.P., and Formicola, V. (2008) The Late Upper Paleolithic skeleton Villabruna 1 (Italy): a source of data on biology and behavior of 1 14,000 year-old hunter. *J. Anthropol. Sci.*, **86**, 143–163.

Appendix 2(b)

Appendix 2(b) Column length estimation in Formicola sample (Arena Candide 5, Bonifacio 1, Gramat 1, Lochbour 1, Mondeval 1, Romanelli 1, Teviec 1, Teviec 6, and Teviec 16). Correction factors for converting maximum midline vertebral heights to corresponding maximum heights anterior to pedicles (maximum height anterior to pedicles = maximum midline height × correction factor).

Vertebra	Correction factor	Vertebra	Correction factor
C2	1.000	T8	0.968
C3	0.970	T9	0.973
C4	0.974	T10	0.965
C5	0.960	T11	0.927
C6	0.979	T12	0.927
C7	0.901	L1	0.935
T1	0.968	L2	0.977
T2	0.998	L3	1.001
T3	0.986	L4	1.020
T4	0.979	L5	1.017
T5	0.976	L6	1.007
T6	0.970	S1	1.031
T7	0.975		

Skeletal Variation and Adaptation in Europeans: Upper Paleolithic to the Twentieth Century, First Edition. Edited by Christopher B. Ruff.

Appendix 3(a)

Appendix 3(a) Statures of late adolescent and young adult Europeans. Statures are adjusted to maximum adult statures if necessary based on information from longitudinal studies (Fredriks *et al.*, 2000; Cacciari *et al.*, 2002; Roelants *et al.*, 2009; Schönbeck *et al.*, 2013). These adjustments are 1.0 cm if age 18, 0.36 cm if age 19, and 0.12 cm if age 20 in males. In females, 0.1 cm is added if age is 18. The resultant estimates of maximum adult stature are further adjusted by averaging male and female means and adding 6.96 cm to the result to derive male stature and subtracting the same amount to derive female stature. In case of a missing (*Swiss females and Italian females*) or highly dubious sex-specific stature value (*Latvian males*), 13.92 cm is subtracted from male stature or 13.92 cm is added to female stature (values in italics).

	Birth year	Sex	Age (years)	Observed stature	Max. stature	Adjusted max. stature	Source
England	1978–1987	M	25–34	177.80	177.80	177.51	Health Survey for
		F	25–34	163.30	163.30	163.59	England, 2012
Scotland	1974–1983	M	25–34	178.20	178.20	177.81	The Scottish Health
		F	25–34	163.50	163.50	163.89	Survey, 2008
Netherlands	1988	M	21	183.80	183.80	184.21	Schönbeck *et al.*, 2013
		F	21	170.70	170.70	170.29	
Belgium	1981–1983	M	21	181.10	181.10	180.81	Roelants *et al.*, 2009
(Flanders)		F	21	166.60	166.60	166.89	
Denmark	1987	M	20	180.40	180.40	182.02	Tinggaard *et al.*, 2014
		F	18	169.50	169.60	168.10	
Norway	1984–1987	M	19	181.00	181.36	181.24	Júlíusson *et al.*, 2013
(Bergen)		F	19	167.20	167.20	167.32	
Sweden	1981	M	19	180.40	180.76	181.19	Werner and
		F	19	167.70	167.70	167.27	Bodin, 2006
Finland	1983–	M	20	180.70	180.82	180.97	Saari *et al.*, 2011
(Espoo)		F	20	167.20	167.20	167.05	
Estonia	1979–1984	M	20–24	180.94	180.94	181.15	Kaarma *et al.*, 2008
		F	20–24	167.44	167.44	167.23	
Latvia	?	M	Unknown	177.63	177.63	*181.04*	Gerhards, 2005
		F	Unknown	167.12	167.12	167.12	
Lithuania	1990	M	18	179.72	180.72	181.32	Suchomlinov and
(Vilnius)		F	18	167.90	168.00	167.40	Tutkuviene, 2013

(*Continued*)

Skeletal Variation and Adaptation in Europeans: Upper Paleolithic to the Twentieth Century, First Edition. Edited by Christopher B. Ruff.

Appendix 3(a) (Continued)

	Birth year	Sex	Age (years)	Observed stature	Max. stature	Adjusted max. stature	Source
France	1977–1988	M	18–29	177.80	177.80	178.02	Etude nationale nutrition santé, ENNS, 2006–2007
		F	18–29	164.20	164.20	164.10	
Germany	1986–1989	M	18	179.10	180.10	179.96	Rosario *et al.*, 2011
		F	18	165.80	165.90	166.04	
Austria	1990	M	19	179.60	179.96	179.64	Gleiss *et al.*, 2013
		F	19	165.40	165.40	165.72	
Switzerland	1990	M	18.5	178.30	178.98	178.98	Staub *et al.*, 2011
		F				*165.06*	
Czech Republic	1984	M	Early 20s	180.60	180.60	181.06	Hermanussen *et al.*, 2010
		F	Early 20s	167.60	167.60	167.14	
Slovakia	1983	M	18	179.35	180.35	180.94	Ševčiková *et al.*, 2004
		F	18	165.61	165.71	167.02	
Poland	1989	M	18	178.68	179.68	179.38	Kulaga *et al.*, 2011
		F	18	165.06	165.16	165.46	
Hungary	1987	M	18	177.31	178.31	178.31	Mészáros *et al.*, 2008
						164.39	
Spain	1980–1988	M	18.1–24	177.33	177.54	177.71	Lezcano *et al.*, 2008
		F	18.1–24	163.96	163.96	163.79	
Portugal	1982	M	18	172.13	173.13	173.13	Padez, 2007
		F				159.21	
Italy	1980	M	18	174.58	175.58	175.58	Arcaleni, 2006
		F				*161.66*	
Slovenia (Ljubljana)	1992	M	19	180.20	180.56	180.94	Starc and Strel, 2011
		F	19	167.40	167.40	167.02	
Croatia	1988	M	18	180.50	181.50	180.91	Jureša *et al.*, 2012
		F	18	166.30	166.40	166.99	
Serbia	?	M	20.13	181.96	181.96	181.41	Popovic *et al.* (2013)
		F	19.59	166.82	166.82	167.49	
Montenegro	?	M	20.97	183.21	183.21	182.81	Bjelica *et al.*, 2012
		F	20.86	168.37	168.37	168.89	
Greece	1980–1988	M	18–26	178.06	178.06	178.22	Papadimitriou *et al.*, 2008
		F				164.30	
Russia	1976–1980	M	21–25?	177.20	177.20	177.61	Brainerd, 2008,
		F	21–25?	164.10	164.10	163.69	
Turkey (Istanbul)	1974–1984	M	18	176.00	177.00	177.06	Neyzi *et al.*, 2006
		F	18	163.10	163.20	163.14	

References

Arcaleni, E. (2006) Secular trend and regional differences in the stature of Italians, 1954–1980. *Econ. Hum. Biol.*, **4** 24–38.

Bjelica, D., Popovic, S., Kezunovic, M., Petkovic, J., Jurak, G., and Grasgruber, P. (2012) Body height and its estimation utilizing arm span measurements in Montenegrin adults. *Anthropological Notebooks*, **18**, 69–83.

Brainerd, E. (2008) Reassessing the standard of living in the Soviet Union: An analysis using archival and anthropometric data. University of Michigan, Ann Arbor, MI. web.williams.edu/Economics/brainerd/.../ussr_july08.p...

Cacciari, E., Milani, S., Balsamo, A., Demmacco, F., De Luca, F., Chiarelli, F., Pasquino, A.M., Tonini, G., and Vanelli, M. (2002) Italian cross-sectional growth charts for height, weight and BMI (6–20 y). *Eur. J. Clin. Nutr.*, **56**, 171–180.

Etude national nutrition santé, (ENNS) (2006) Mesures anthropométriques – Adultes 18–74 and. Taleaux de distribution – Eude national nutrition santé (ENNS). Tableau 1.5.1. http:www.invs.sante.fr/content/download/12842/77060/version/3/file/etat_nutritionnel_adultes_anthropometrie+(1).pdf

Fredriks, A.M., Van Bureen, S., Burgmeijer, R.J.F., Meulmeester, J.F., Beuker, R.J., Brugman, E., Roede, M.J., Verloove-Vanhorick, S.P., and Wit, J.-M. (2000) Continuing positive secular growth change in the Netherlands 1955–1997. *Pediatr. Res.*, **47**, 316–323.

Gerhards, G. (2005) Secular variations in the body stature of the inhabitants of Latvia (7th millennium BC–20th c. AD). *Acta Med. Litu.*, **12**, 33–39.

Gleiss, A., Lassi, M., Blümel, P., Borkenstein, M., Kapelari, K., Mayer, M., Schemper, M., and Häusler, G. (2013) Austrian height and body proportion references for children aged 4 to under 19 years. *Ann. Hum. Biol.*, **40**, 324–332.

Hermanussen, M., Godina, E., Rühli, F.J., Blaha, P., Boldsen, J.L., van Buuren, S., MacIntyre, M., Aßmann, C., Ghost, A., de Stefano, G.F., Sonkin, V.D., Tresguerres, J.A.F., Meigen, C., Scheffler, C., Geiger, C., and Lieberman, L.S. (2010) Growth variation, final height and secular trend. Proceedings of the 17th Aschauer Soiree, 7th November 2009. *Homo*, **61**, 277–284.

Health Survey for England – 2012, Trend Tables http://www.hscic.gov.uk/searchcatalogue?productid=13888&q=height&sort=Relevance&size=10&page=1#top.

Júlíusson, P.B., Roelants, M., Nordal, E., Furevik, L., Eide, G.E., Moster, D., Hauspie, R., and Bjerknes, R. (2013) Growth references for 0–19 year-old Norwegian children for length/height, weight, body mass index and head circumference. *Ann. Hum. Biol.*, **40**, 220–227.

Jureša, V., Musil, V., and Tiljak, M.K. (2012) Growth charts for Croatian school children and secular trends in past twenty years. *Coll. Anthropol.*, **36** (Suppl. 1), 47–57.

Kaarma, H., Saluste, L., Lintsi, M., Kasmel, J., Veldre, G., Tiit, E.-M., Koskel, S., and Arend, A. (2008) Height and weight norms for adult Estonian men and women (aged 20–70 years) and ways of somatotyping using height-weight classification. *Papers in Anthropology*, **XVII**, 113–130.

Kulaga, Z., Litwin, M., Tkaczyk, M., Palczewska, I., Zajaczkowska, M., Zwolinska, D., Krynicki, T., Wasilwska, A., Moczulska, A., Morawiec-Knysak, A., Barwicka, K., Grajda, A., Gurzkowska, B., Napieralska, E., and Pan, H. (2011) Polish 2010 growth references for school-aged children and adolescents. *Eur. K. Pediatr.*, **170**, 599–609.

Lezcano, A.C., García, J.M.F., Ramos, C.F., Longás, A.F., López-Siguero, J.P., Gonzáöez, E.S., Ruiz, B.S., Fernandez, D.Y., y Grupo Colaborador Español (2008) Estudio transversal español de crecimiento 2008. Parte II: valores de talla, peso e índice de masa corporal desde el nacimiento a la talla adulta. *Ann. Pediatr. (Barc.)*, **68**, 552–569.

Mészáros, Z., Mészáros, J., Völgyi, E., Sziva, Á., Pampakas, P., Prókai, A., and Szmodis, M. (2008) Body mass and body fat in Hungarian schoolboys: Differences between 1980–2005. *J. Physiol. Anthropol.*, **27**, 241–245.

Mironov, B. (2007) Birth weight and physical stature in St. Petersburg: Living standards of women in Russia, 1980–2005. *Econ. Hum. Biol.*, **5**, 123–143.

Neyzi, O., Furman, A., Bundak, R., Gunoz, H., Darendeliler, F., and Bas, F. (2006) Growth references for Turkish children aged 6 to 18 years. *Acta Paediatr.*, **95**, 1635–1641.

Padez, C. (2007) Secular trend in Portugal. *J. Hum. Ecol.*, **22**, 15–22.

Papadimitriou, A., Fytanidis, G., Douros, K., Papadimitriou, D.T., Nicolaidou, P., and Fretzayas, A. (2008) Greek young men grow taller. *Acta Paediatr.*, **97**, 1105–1107.

Popovic, S., Bjelica, D., Monlar, S., Jaksic, D., and Akpinar, S. (2013) Body height and its estimation utilizing arm span measurements in Serbian adults. *Int. J. Morphol.*, **31**, 271–279.
Roelants, M., Hauspie, R., and Hoppenbrouwers, K. (2009) References for growth and pubertal development from birth to 21 years in Flanders, Belgium. *Ann. Hum. Biol.*, **36**, 680–694.
Rosario, A.S., Schienkiewitz, A., and Neuhauser, H. (2011) German height references for children aged 0 to under 18 years compared to WHO and CDC growth charts. *Ann. Hum. Biol.*, **38**, 121–130.
Saari, A., Sankilampi, U., Hannila, M.-L., Kiviniemi, V., Kesseli, K., and Dunkel, L. (2011) New Finnish growth references for children and adolescents aged 0 to 20 years: Length/height-for-age, weight-for-length/height, and body mass index-for-age. *Ann. Med.*, **43**, 235–248.
Schönbeck, Y., Talma, H., van Dommelen, P., Buitendijk, S.E., HiraSing, R.A., and van Buuren, S. (2013) The world's tallest nation has stopped growing taller: the height of Dutch children from 1955 to 2009. *Pediatr. Res.*, **73**, 371–377.
Ševčiková, L., Nováková, J., Hamade, J., and Tatara, M. (2004) Rast a vývojové trendy slovenských detí a mládeže za posledných 10 rokov. In: *Životné podmienku a zdravie* (ed. L. Ághová), Bratislava. pp. 192–208.
Staub, K., Rühli, F.J., Woitek, U., and Pfister, C. (2011) The average height of 18- and 19-year-old conscripts (N = 458,322) in Switzerland from 1992 to 2009, and the secular height trend since 1878. *Swiss Med. Wkly*, **141**, x13238.
Statistical Yearbook of Norway 2013. Table 108: Conscripts, by height and weight. http://www.ssb.no/a/english/aarbok/tab/tab-108.html.
Starc, G. and Strel, J. (2011) Is there a rationale for establishing Slovenian body mass index references of school-aged children and adolescents? *Anthropological Notebooks*, **17**, 89–100.
Suchomlinov, A. and Tutkuviene, J. (2013) Growth tendencies of the 'generation of independence': the relation between socioeconomic factors and growth indices. *Acta Med. Litu.*, **20**, 19–26.
Tinggaard, J., Aksglaede, L., Sørensen, K., Mouritsen, A., Wohlfarth-Veje, C., Hagen, C.P., Mieritz, M.G., Jørgensen, N., Heuck, C., Petersen, J.H., Main, K.M., and Juul, A. (2014) The 2014 Danish references from birth to 20 years for height, weight and body mass index. *Acta Paediatr.*, **103**, 214–224.
The Scottish Health Survey 2008. Table 7.2 Height, by age and sex. http://www.scotland.gov.uk/Publications/2009/09/28102003/79.
Werner, B. and Bodin, L. (2006) Growth from birth to age 19 for children in Sweden born in 1981: Descriptive values. *Acta Paediatr.*, **95**, 600–613.

Appendix 3(b)

Appendix 3(b(i)) Anthropometric dimensions of recent and current European males published in 1960s or more recently.

Sample	Age (years)	Birth year	Stature	Sitting height	Lower limb length	Relative sitting height	Biacromial breadth	Bi-iliac breadth	Weight (kg)	Source
UK (London)	18	1947	174.7	92.0	82.7	52.7	39.6	27.6	63.0	Eveleth and Tanner, 1976
UK	45	1958	176.0	92.1	84.0	52.3				Le *et al.*, 2007
Finland (Saami)	19 and 20 (mean 19.4)	1947–1955	166.73				37.04	27.78	58.18	Auger *et al.*, 1980
	20–25	1942–1954	166.67	89.57	77.10	53.7	38.15	27.55	61.20	
	20–30	1937–1954	167.0	89.2	77.8	53.4	37.9	27.7	62.3	
	31–40	1927–1943	162.7	86.2	76.5	53.0	37.2	27.2	63.3	
	20–40	1927–1954	165.12	87.92	77.2	53.2	37.61	27.49	62.74	
Finland (Finns)	20	1958	179.1	94.1	85.1	52.5	40.3	28.5	72.1	Dahlström, 1981
Finland (Finns)	27.28 22–39	1961–1978	181.82	95.12	86.70	52.3	41.70	29.67	79.4	Based on Ruff *et al.*, 2005 sample
Sweden (South & Middle Sweden)	18? Inductees	Early or mid 1950s	178.7	92.4	86.3	51.7	38.9	27.9		Lewin, 1973, referenced in Dahlström, 1981
Norway	20	1942	177.49	92.65	84.85	52.2				Udjus, 1964
Norway	20–29	1970–1979	179.3	93.2	86.1	52.0	40.9		76.2	Bolstad *et al.*, 2001
	30–39	1960–1969	179.9	93.3	86.6	51.9	40.6		80.1	
	30.0 20–39	1960–1980?	179.6	93.3	86.3	51.9	40.8		78.2	
Netherlands	21	1975 & 1976	183.798	94.654	89.144	51.5				Fredriks *et al.*, 2007
Netherlands (Oosterwolde)	17	1962–1963	181.2	94.4	86.8	52.1	39.6	27.8	65.6	Eveleth and Tanner, 1990
Germany	20	1948–1949	175.8	92.1	83.7	52.4	39.4	28.4	68.7	Jürgens, 1970
Germany	25–40	1925–1946	174.5	91.9	82.6	52.7	40.0	29.2	76.3	Jürgens, 1973
Belgium (Brussels; students)	21–25	1930s	175.5	91.3	84.2	52.0	39.0	28.1		Eveleth and Tanner, 1976
Belgium (Brussels)	21–25	1935–1940	174.5				39.9	28.6	66.9	Eveleth and Tanner, 1976

Belgium (Brussels)	18	1942–1943	173.9				39.2	28.0	63.5	Eveleth and Tanner, 1976
Belgium (National)	18	1949–1956/ 1962	176.5	91.3	85.2	51.7	39.4		66.7	Eveleth and Tanner, 1990
France (Paris)	18–42	1918–1942?	172.0	90.2	81.8	52.4		28.9		Eveleth and Tanner, 1976
Czech	25–45	1916–1946	172.6	90.6	82.0	52.5	38.9	28.9	75.2	Rokopec, 1977
Czech	18–29	1982–1994	181.0	95.2	85.8	52.6				Grasgruber and Hrazdíra, 2013
Poland (Warsaw)	18–20	Early 1940s	173.0				39.6	28.4?	64.0	Eveleth and Tanner, 1976
Poland (Cracow)	21–25	1950s	173.2	90.0	83.2	52.0			69.0	Eveleth and Tanner, 1976
Poland (Warsaw)	18	1953	174.4	91.3	83.1	52.4	38.9	27.8	66.1	Eveleth and Tanner, 1976
Poland (Rural)	18	1942–1944	165.2				38.0	28.0	60.3	Eveleth and Tanner, 1976
Poland (Warsaw)	18	1958–1962	176.8				38.7	27.8	66.1	Eveleth and Tanner, 1990
Hungary (railway workers)	?	1930s & 1940s?	170.9	89.0	81.9	52.1	39.2		74.6	Eveleth and Tanner, 1976
Hungary (Budapest)	18	1950–1951	175.9	91.1	84.8	51.8	39.8	28.6	66.8	Eveleth and Tanner, 1976
Hungary (Szegeb)	18	1940–1941	172.1	90.0	82.1	52.3			63.2	Eveleth and Tanner, 1976
Hungary (National)	18	1964–1967	175.3	91.7	83.6	52.3	40.1	27.6	67.0	Eveleth and Tanner, 1990
Hungary (Kormend)	18	1960	172.8	89.8	83.0	52.0	40.0	28.6	64.4	Eveleth and Tanner, 1990
Croatia (Zagreb)	18	1981	180.4	92.9	87.5	51.5	39.37	28.38	–	Zivicnjak *et al.*, 2003, 2008
Romania (Students)	21?	Early 1940s?	171.2				38.1	27.9	60.7	Eveleth and Tanner, 1976
Bulgaria (Sofia)	24	Ca. 1940?	171.3	89.6	81.7	52.3	37.7	28.1	68.8	Eveleth and Tanner, 1976
Bulgaria (National)	24	Ca. 1940?	169.8	88.8	81.0	52.3	38.2	28.2	67.0	Eveleth and Tanner, 1976
Bulgaria (Sofia)	18	1945	171.9	90.2	81.7	52.5	39.1	28.2		Eveleth and Tanner, 1976
Basque (France)	?	Ca. 1910–1940?	169.2	88.7	80.5	52.4	39.5	29.3		Eveleth and Tanner, 1976
Basque (Spain)	?	Ca. 1910–1940?	170.0	89.6	80.4	52.7	39.1	29.6		Eveleth and Tanner, 1976
Basque (Spain)	21.48 20–25	1976–81	175.72	91.46	84.26	52.0	39.2	29.03	73.15	Martini *et al.*, 2005
Sardinian	22.01 20–25	1970–78	169.89	87.96	81.93	51.8	39.42	27.31	67.22	Martini *et al.*, 2005

(*Continued*)

Appendix 3(b(i)) (Continued)

Sample	Age (years)	Birth year	Stature	Sitting height	Lower limb length	Relative sitting height	Biacromial breadth	Bi-iliac breadth	Weight (kg)	Source
Greece	22.5 17–43	1916–1942?	170.51	90.25	80.26	52.9	38.87		67.03	Hertzberg *et al.*, 1963
Italy	23.58 18–59	1900–1941?	170.60	89.66	80.94	52.6	39.84		70.26	Hertzberg *et al.*, 1963
Italy (Genoa)	18	1944 & 1947	172.0					27.7	65.7	Eveleth and Tanner, 1976
Greece (Students)	20–24	1946–1954	173.7	91.4	82.3	52.6				Manolis *et al.*, 1995
	20–24	1951–1959	175.9	92.1	83.8	52.4				
	20–24	1956–1964	177.0	92.1	84.9	52.0				
	20–24	1961–1970	177.6	91.8	85.8	51.7				
Russia	17	1965–1967	175.8				38.6	28.0	68.1	Eveleth and Tanner, 1990

Appendix 3(b(ii)) Anthropometric dimensions of recent and current European females published in 1960s or more recently.

Sample	Age	Birth year	Stature	Sitting height	Lower limb length	Relative sitting height	Biacromial breadth	Bi-iliac breadth	Weight (kg)	Source
Ireland	Mean age 28	1940s or earlier?	159.8	86.0	73.8	53.8	–	29.2	61.5	Eveleth and Tanner, 1976
UK (London)	18	1947	162.2	87.0	75.2	53.6	36.6	27.9	56.6	Eveleth and Tanner, 1976
UK	45	1958	162.4	86.2	76.4	53.1				Li *et al.*, 2007
Finland (Saami)	19	1948–1955	153.6				34.1	26.6	53.2	Auger *et al.*, 1980
	20–25	1942–1954	153.92	83.42	70.50	54.2	34.60	27.32	53.60	
	20–30	1937–1954	154.7	83.7	71.0	54.1	34.9	27.3	55.2	
	31–40	1927–1943	151.5	82.0	69.5	54.1	34.1	27.4	57.2	
	20–40	1927–1954	153.21	82.88	70.33	54.1	34.52	27.35	56.14	
Finland (Finns)	24.40 20–35	1965–1980	166.53	87.81	78.72	52.7	36.78	28.96	60.29	Based on Ruff *et al.*, 2005 sample

Estonia	17–23	1973–1980	167.19				35.8	26.78	60.395	Kaarma *et al.*, 2009
Norway	20–29	1970–1980?	166.3	87.4	78.9	52.6	36.8		63.0	Bolstad *et al.*, 2001
	30–39	1960–69	166.0	87.0	79.0	52.4	36.1		63.9	
	30.0 20–39	1960–1980?	166.1	87.2	78.9	52.5	36.4		63.4	
Netherlands	21	1975 & 1976	170.684	89.627	81.057	52.5				Fredriks *et al.*, 2007
Netherlands (Oosterwolde)	17	1962–1963	169.8	90.3	79.5	53.2	36.2	27.1	59.1	Eveleth and Tanner, 1990
Belgium (Brussels; students)	21–25	1930s	163.1	86.4	76.7	53.0	35.3	27.2		Eveleth and Tanner, 1976
Belgium (students and workers)	21–25	1935–1940	162.0				35.8	27.4	56.2	Eveleth and Tanner, 1976
Belgium (Brussels)	18	1942–1943	161.0				35.2	27.1	55.5	Eveleth and Tanner, 1976
Belgium	18	1949–1956/1962	164.2	87.2	77.0	53.1	35.4	27.2	57.0	Eveleth and Tanner, 1990
France (Paris)	18–42	1921–1942?	157.8	84.3	73.5	53.4		27.7		Eveleth and Tanner, 1976
France (National)	20–34	1920s & 1930s?	160.4	85.1	75.3	53.1	35.8	29.8	55.5	Eveleth and Tanner, 1976
Czech	25–45	1916–1946	158.9	84.6	74.3	53.2	35.9	28.6	66.9	Prokopec, 1977
Czech	18–29	1982–1994	168.8	90.1	78.7	53.4				Grasgruber and Hrazdíra, 2013
Poland (Warsaw)	18–20	Early 1940s	158.9				35.2	28.3?	56.0	Eveleth and Tanner, 1976
Poland (Cracow)	21–25	1950s	160.2	84.0	76.2	52.4			59.0	Eveleth and Tanner, 1976
Poland (Warsaw)	18	1953	161.0	85.7	75.3	53.2	34.6	26.8	56.0	Eveleth and Tanner, 1976
Poland (Rural)	18	1942–1944	157.9				35.9	28.1	52.7	Eveleth and Tanner, 1976
Poland (Warsaw)	18	1958–1962	163.1				35.5	27.4	55.0	Eveleth and Tanner, 1990
Hungary (Students)	18–25	1940s?	159.8	84.5	75.3	52.9	36.6		56.2	Eveleth and Tanner, 1976
Hungary (Budapest)	18	1950–1951	160.8	84.8	76.0	52.7	34.7	26.7	55.8	Eveleth and Tanner, 1976
Hungary (Szegeb)	17	1941–1942	160.1	85.2	74.9	53.2			54.6	Eveleth and Tanner, 1976
Hungary (National)	18	1964–1967	162.3	86.5	75.8	53.3	36.2	27.1	55.7	Eveleth and Tanner, 1990

(*Continued*)

Appendix 3(b(ii) (Continued)

Sample	Age	Birth year	Stature	Sitting height	Lower limb length	Relative sitting height	Biacromial breadth	Bi-iliac breadth	Weight (kg)	Source
Hungary (Kormend)	18	1960	161.2	84.8	76.4	52.6	35.8	27.5	52.4	Eveleth and Tanner, 1990
Croatia (Zagreb)	18	1981	166.5	87.4	79.1	52.5	34.95	27.64	–	Zivicnjak *et al.*, 2003, 2008
Romania (Students)	ca. 20?	Early 1940s?	157.1				35.4	27.9	53.9	Eveleth and Tanner, 1976
Bulgaria (Sofia)	24	ca. 1940?	160.2	84.6	75.6	52.8	34.4	27.5	60.2	Eveleth and Tanner, 1976
Bulgaria (National)	24	Ca. 1940?	157.7	83.1	74.6	52.7	35.2	27.8	58.7	Eveleth and Tanner, 1976
Bulgaria (Sofia)	18	1945	159.0	81.9	77.1	51.5	35.4	27.2		Eveleth and Tanner, 1976
France (Basque)	?	1910–1940?	156.4	84.1	72.3	53.8	35.9	29.4		Eveleth and Tanner, 1976
Spain (Basque)	?	1910–1940?	157.3	83.8	73.5	53.3	35.6	29.4		Eveleth and Tanner, 1976
Spain (Basque)	21.93 20–25	1976–1981	162.10	85.86	76.24	53.0	35.60	27.05	58.96	Martini *et al.*, 2005
Italy (Genoa)	17	1945 & 1948	159.2					27.6	55.7	Eveleth and Tanner, 1976
Italy (Sardinia)	22.05 20–25	1970–78	156.70	83.17	73.53	53.1	35.14	26.37	53.71	Martini *et al.*, 2005
Italy (Sassari)	?		155.4				35.4	27.5	51.4	Eveleth and Tanner, 1976
Greece (students)	20–24	1946–1954	161.4	86.3	75.2	53.5				Manolis *et al.*, 1995
	20–24	1951–1959	162.7	86.3	76.5	53.0				
	20–24	1956–1964	162.7	84.6	77.9	52.0				
	20–24	1961–1970	163.6	86.7	76.8	53.0				
Russia	17	1965–1967	164.0				35.4	29.9	57.4	Eveleth and Tanner, 1990

References

Auger, F., Jamison, P.L., Balslev-Jorgensen, J., Lewin, T., de Peña, J.F., and Skrobak-Kaczynski, J. (1980) Anthropometry of circumpolar populations. In: The Human Biology of Circumpolar Populations (ed. F.A. Milan), Cambridge University Press, Cambridge. pp. 213–225.

Bolstad, G., Benum, B., and Rokne, A. (2001) Anthropometry of Norwegian light industry and office workers. *Appl. Ergon.*, **32**, 239–246.

Dahström, S. (1891) Suomalaisen nuoren miehen ruumiinrakenne kutsuntamittausten ja 20-vuotiaiden varusmiesten antropometristen mittausten perusteella. Ph.D. Dissertation. University of Turku. Turku, Finland.

Eveleth, P.B. and Tanner, J.M. (1976) *Wordwide Variation in Human Growth.* Cambridge University Press, Cambridge.

Eveleth, P.B. and Tanner, J.M. (1990) *Wordwide Variation in Human Growth.* Cambridge University Press, Cambridge.

Fredriks, A.M., van Buuren, S., van Heel, W.J.M., Dijkman-Neerincx, R.H.M., Verloove-Vanhorick, S.P., and Wit, J.M. (2005) Nationwide age references for sitting height, leg length, and sitting height/height ratio, and their diagnostic value for disproportionate growth disorders. *Arch. Dis. Child.*, **90**, 807–812.

Grasgruber, P. and Hrazdíra, E. (2013) Anthropometric characteristics of the young Czech population and their relationship to the national sports potential. *J. Hum. Sport Exercise*, **8**, S120–S134.

Jürgens, H.W. (1970) Körpermasse 20-jährigen Männer als Grundöage für die Gestaltung von Arbeitsgerät, Ausrustung und Arbeitsplatz. Forschungsbericht aus der Wehrmedizin, BWVg InSan Nr 2/69.

Jürgens, H.W. (1973) Körpermasse 25-40 jährigen Männer zur Prüfung der anthropometrisch-ergonomischen Bedeutung altersbedingter Veränderungen der Körperform. Forschungsberich aus der Wehrmedizin, BWVg-FBWM 73-1.

Kaarma, H., Peterson, J., Kasmel, J., Lintsi, M., Saluste, L., Koskel, S., and Arend, A. (2009) The role of body height, weight and BMI in body build classification. *Papers on Anthropology*, **XVIII**, 155–173.

Lewin, T. (1973) Body build of young males. Part I. Preliminary results of a study of Swedish inductees with special reference to secular changes. Report FMI-73-1, Institute of Aviation Medicine. Malmslätt.

Li, L., Dangour, A.D., and Power, C. (2007) Early life influences on adult leg and trunk length in the 1958 British birth cohort. *Am. J. Hum. Biol.*, **19**, 836–843.

Manolis, S., Neroutsos, A., Zafeiratos, C., and Pentzou-Daponte, A. (1995) Secular changes in body formation of Greek students. *Hum. Evol.*, **10**, 199–204.

Martini, E., Rabato, E., Racugno, W., Buffa, R., Salces, I., and Tarli, S.M.B. (2005) Dispersion dimorphism in human populations. *Am. J. Phys. Anthropol.*, **127**, 342–350.

Prokopec, M. (1977) An anthropometric study of the Rembarranga: Comparisons with other populations. *J. Hum. Evol.*, **6**, 371–391.

Ruff, C., Niskanen, M., Junno, J.-A., and Jamison, P. (2005) Body mass prediction from stature and bi-iliac breadth in two high latitude populations, with application to earlier higher latitude humans. *J. Hum. Evol.*, **48**, 381–392.

Udjus, L.G. (1964) *Anthropometric changes in Norwegian men in the twentieth century.* Univ. forl., Oslo.

Živičnjak, M., Narančić, N.M., Szirovicza, L., Franke, D., Hrenović, J., and Bišof, V. (2003) Gender-specific growth pattern for stature, sitting height and limbs length in Croatian children and youth (3 to 18 years of Age). *Coll. Anthropol.*, **27**, 321–334.

Živičnjak, M., Narančić, N.M., Szirovicza, L., Franke, D., Hrenović, J., Bišof, V., Tomas, Ž., and Škarić-Jurić, T. (2008) Gender-specific growth pattern of transversal body dimensions in Croatian children and youth (2 to 18 years of age). *Coll. Anthropol.*, **32**, 419–431.

Appendix 4

Appendix 4(a) Untransformed measurements of males for each period and both main geographic areas (i.e., latitude groups 'North' and 'South'). Sample sizes, means, and standard errors of means (SE) are included. Variable abbreviations are as in Table 2.2 in Chapter 2.

			STA	ASTA	LSHT	LlimbL	LRSHT	BM	BMI
Very Recent	North	N	67					65	65
		Mean	167.6					66.8	23.8
		SE	0.9					1.0	0.3
	South	N	31	31	31	31	31	31	31
		Mean	162.8	162.8	87.0	75.9	53.4	61.8	23.2
		SE	1.2	1.2	0.6	0.7	0.2	1.6	0.4
Early Modern	North	N	48	30	30	30	30	47	47
		Mean	164.2	163.7	86.7	77.0	53.0	63.6	23.6
		SE	0.9	1.1	0.5	0.8	0.2	1.2	0.3
	South	N	41	33	33	33	33	40	40
		Mean	160.8	161.0	84.9	76.2	52.7	63.0	24.2
		SE	1.0	1.2	0.6	0.8	0.2	1.2	0.3
Late Medieval	North	N	223	135	135	135	135	220	219
		Mean	167.9	168.1	87.9	80.2	52.3	68.5	24.3
		SE	0.4	0.5	0.3	0.4	0.1	0.5	0.1
	South	N	60	41	41	41	41	58	58
		Mean	166.7	167.2	86.9	80.4	52.0	70.6	25.2
		SE	0.8	1.0	0.3	0.8	0.2	0.9	0.2
Early Medieval	North	N	116	43	43	43	43	116	115
		Mean	167.7	169.2	88.8	80.3	52.5	69.2	24.6
		SE	0.6	1.1	0.5	0.7	0.2	0.7	0.2
	South	N	65	35	35	35	35	64	63
		Mean	165.4	165.3	86.3	79.0	52.2	66.2	24.2
		SE	0.7	1.1	0.5	0.8	0.2	0.8	0.3
Iron/ Roman	North	N	123	65	65	65	65	121	120
		Mean	166.5	165.8	87.8	78.0	53.0	69.7	25.0
		SE	0.7	0.9	0.4	0.6	0.1	0.8	0.2
	South	N	31	16	16	16	16	27	27
		Mean	161.6	161.1	85.9	75.1	53.4	59.0	22.6
		SE	1.0	1.4	0.7	1.0	0.3	2.2	0.7

(*Continued*)

Skeletal Variation and Adaptation in Europeans: Upper Paleolithic to the Twentieth Century, First Edition. Edited by Christopher B. Ruff.

Appendix 4(a) (Continued)

			STA	ASTA	LSHT	LlimbL	LRSHT	BM	BMI
Bronze	North	N	106	32	32	32	32	100	98
		Mean	167.6	167.7	87.9	79.8	52.4	67.0	23.8
		SE	0.6	1.0	0.5	0.7	0.2	0.8	0.2
	South	N	34	19	19	19	19	33	33
		Mean	164.6	165.5	86.9	78.7	52.5	65.4	24.1
		SE	0.9	1.1	0.6	0.8	0.2	1.3	0.4
Neolithic	North	N	155	73	73	73	73	159	153
		Mean	166.9	165.1	87.8	77.2	53.2	66.7	24.0
		SE	0.6	0.9	0.4	0.6	0.1	0.6	0.2
	South	N	20	15	15	15	15	17	17
		Mean	162.2	162.0	85.4	76.6	52.8	60.0	22.6
		SE	1.6	1.7	0.6	1.3	0.3	2.0	0.6
Mesolithic	North	N	39	17	17	17	17	31	30
		Mean	165.1	167.7	87.7	80.0	52.4	67.1	24.2
		SE	1.4	2.0	0.7	1.5	0.4	1.4	0.4
	South	N	14	3	3	3	3	10	10
		Mean	159.6	166.2	88.4	77.8	53.2	60.1	23.1
		SE	1.8	3.7	2.6	1.2	0.5	2.8	0.6
Late Upper Paleolithic	North	N	4	3	3	3	3	4	4
		Mean	166.9	167.4	87.7	79.7	52.4	65.7	23.6
		SE	1.4	1.8	1.1	1.2	0.4	3.1	0.9
	South	N	12	8	8	8	8	12	12
		Mean	163.7	164.5	87.5	77.0	53.2	67.4	25.2
		SE	1.4	1.7	0.6	1.2	0.3	1.4	0.5
Early Upper Paleolithic	North	N	9	4	4	4	4	8	8
		Mean	172.1	169.1	87.2	81.8	51.6	68.6	23.1
		SE	1.5	1.8	0.7	2.1	0.8	3.0	0.8
	South	N	6	4	4	4	4	6	6
		Mean	178.0	181.3	91.9	89.4	50.7	74.0	23.1
		SE	3.3	3.5	2.2	1.9	0.6	3.7	0.5
Total	North & South	N	1204	607	607	607	607	1169	1156
		Mean	166.3	166.0	87.4	78.7	52.6	67.1	24.2
		SE	0.2	0.3	0.1	0.2	0.1	0.2	0.1

Appendix 4(a) Male data continued.

			BAB	BIB	RBAB	RBIB	CI	RLegL	BI
Very Recent	North	N	20	35	20	35	45		35
		Mean	40.1	28.6	24.1	17.2	80.6		74.2
		SE	0.6	0.3	0.3	0.2	0.4		0.4
	South	N	28	29	28	29	31	31	31
		Mean	39.0	27.8	24.0	17.0	82.0	22.0	74.5
		SE	0.4	0.4	0.2	0.2	0.5	0.1	0.5
Early Modern	North	N	30	34	30	34	42	30	38
		Mean	38.8	27.9	23.7	17.1	81.2	22.1	73.4
		SE	0.4	0.3	0.2	0.2	0.4	0.1	0.4
	South	N	34	23	34	23	41	33	36
		Mean	39.1	27.7	24.3	17.2	81.2	22.2	75.3
		SE	0.3	0.4	0.2	0.2	0.5	0.1	0.4
Late Medieval	North	N	151	162	149	160	191	135	156
		Mean	39.6	29.0	23.6	17.3	80.5	22.2	74.8
		SE	0.2	0.1	0.1	0.1	0.2	0.1	0.2
	South	N	45	46	44	46	54	41	53
		Mean	40.0	29.8	23.8	17.8	82.1	22.6	75.0
		SE	0.4	0.3	0.2	0.1	0.3	0.1	0.3
Early Medieval	North	N	81	62	78	62	100	43	88
		Mean	39.6	29.1	23.6	17.3	81.6	22.2	76.1
		SE	0.2	0.3	0.1	0.1	0.2	0.1	0.2
	South	N	43	31	42	30	56	35	39
		Mean	38.5	28.7	23.3	17.5	81.7	22.5	75.8
		SE	0.3	0.3	0.1	0.2	0.3	0.1	0.4
Iron/ Roman	North	N	63	65	63	65	105	65	83
		Mean	39.4	29.6	23.7	17.7	80.9	22.0	76.0
		SE	0.3	0.3	0.1	0.1	0.3	0.1	0.3
	South	N	19	5	19	5	25	16	11
		Mean	39.2	29.2	24.4	17.9	80.8	21.8	73.9
		SE	0.6	0.8	0.4	0.4	0.6	0.2	1.1

(Continued)

Appendix 4(a) (Continued)

			BAB	BIB	RBAB	RBIB	CI	RLegL	BI
Bronze	North	N	57	37	56	37	92	32	72
		Mean	39.9	29.3	23.9	17.6	83.0	22.5	78.0
		SE	0.3	0.3	0.2	0.2	0.2	0.1	0.3
	South	N	20	16	20	16	27	19	28
		Mean	39.0	29.4456	23.7	17.8	82.4	22.5	77.3
		SE	0.5	0.39166	0.2	0.2	0.6	0.1	0.3
Neolithic	North	N	93	71	90	71	137	73	96
		Mean	39.7	29.1	23.9	17.5	82.3	22.0	76.7
		SE	0.2	0.2	0.1	0.1	0.2	0.1	0.2
	South	N	11	12	11	12	19	15	16
		Mean	38.6	27.6	24.1	17.0	82.9	22.5	78.9
		SE	0.5	0.5	0.3	0.3	0.4	0.1	0.5
Mesolithic	North	N	13	15	13	15	32	17	16
		Mean	39.0	28.4	23.2	16.9	84.0	22.5	76.7
		SE	0.6	0.3	0.3	0.2	0.5	0.2	0.5
	South	N	2	7	2	7	11	3	2
		Mean	38.5	27.9	22.7	17.1	84.0	22.0	78.1
		SE	0.1	0.3	0.1	0.1	1.4	0.2	2.5
Late Upper Paleolithic	North	N	4	2	4	2	4	3	4
		Mean	38.8	27.0	23.3	16.3	84.8	23.0	80.3
		SE	1.0	1.3	0.5	0.6	1.8	0.3	2.2
	South	N	9	6	9	6	11	8	4
		Mean	38.2	29.7	23.4	18.3	83.7	22.3	78.2
		SE	0.7	0.4	0.5	0.2	0.7	0.1	0.2
Early Upper Paleolithic	North	N	7	5	7	5	8	4	8
		Mean	42.9	28.6	24.9	16.6	83.6	23.0	77.7
		SE	1.4	0.6	0.7	0.4	0.5	0.4	0.5
	South	N	2	4	2	4	5	4	2
		Mean	43.0	29.6	22.9	16.5	83.7	23.4	76.7
		SE	0.4	1.0	0.7	0.2	1.5	0.5	0.4
Total	North & South	N	732	667	721	664	1036	607	818
		Mean	39.5	28.9	23.8	17.4	81.7	22.2	75.8
		SE	0.1	0.1	0.0	0.0	0.1	0.0	0.1

Appendix 4(b) Untransformed measurements of females for each period and both main geographic areas (i.e., latitude groups 'North' and 'South'). Sample sizes, means, and standard errors of means (SE) are included. Variable abbreviations are as in Table 2.2 in Chapter 2.

			STA	ASTA	LSHT	LlimbL	LRSHT	BM	BMI
Very Recent	North	N	29					28	28
		Mean	157.0					57.7	23.3
		SE	1.1					1.1	0.3
	South	N	30	27	27	27	27	30	30
		Mean	151.2	152.0	82.2	69.9	54.1	53.5	23.4
		SE	1.0	0.9	0.5	0.6	0.2	1.1	0.3
Early Modern	North	N	44	31	31	31	31	43	43
		Mean	155.2	154.4	83.6	70.7	54.2	55.5	23.0
		SE	0.9	1.0	0.4	0.7	0.2	0.8	0.2
	South	N	17	11	11	11	11	14	14
		Mean	153.8	152.3	81.8	70.5	53.7	56.8	24.0
		SE	1.6	1.6	0.9	1.2	0.5	1.8	0.5
Late Medieval	North	N	195	108	108	108	108	192	190
		Mean	157.9	158.4	85.6	72.9	54.0	57.7	23.1
		SE	0.4	0.5	0.3	0.4	0.1	0.4	0.1
	South	N	53	35	35	35	35	52	52
		Mean	157.4	158.6	84.8	73.9	53.5	57.7	23.2
		SE	0.8	1.0	0.5	0.7	0.2	0.7	0.2
Early Medieval	North	N	79	25	25	25	25	79	79
		Mean	157.9	157.7	84.2	73.5	53.4	57.8	23.2
		SE	0.6	1.1	0.5	0.8	0.2	0.6	0.2
	South	N	52	31	31	31	31	48	48
		Mean	156.4	156.8	83.4	73.4	53.2	56.5	23.1
		SE	0.7	0.9	0.5	0.7	0.2	0.7	0.2
Iron/ Roman	North	N	111	75	75	75	75	109	109
		Mean	157.6	157.5	85.1	72.4	54.0	57.9	23.3
		SE	0.5	0.6	0.3	0.4	0.1	0.5	0.1
	South	N	33	11	11	11	11	30	30
		Mean	155.4	159.5	84.9	74.6	53.2	54.3	22.4
		SE	1.3	2.3	1.2	1.3	0.3	1.3	0.3

(Continued)

Appendix 4(b) (Continued)

			STA	ASTA	LSHT	LlimbL	LRSHT	BM	BMI
Bronze	North	N	82	11	11	11	11	79	77
		Mean	157.2	158.1	85.3	72.9	53.9	56.6	22.9
		SE	0.6	1.6	0.8	1.0	0.3	0.5	0.2
	South	N	32	21	21	21	21	32	31
		Mean	153.2	152.6	81.9	70.8	53.6	53.8	23.1
		SE	1.0	1.4	0.7	0.9	0.2	0.9	0.3
Neolithic	North	N	95	38	38	38	38	96	93
		Mean	156.6	155.6	85.2	70.4	54.8	56.1	22.8
		SE	0.6	0.8	0.4	0.6	0.2	0.5	0.1
	South	N	23	12	12	12	12	22	22
		Mean	153.20	152.8	82.6	70.2	54.1	51.1	21.8
		SE	1.3	1.9	1.0	1.0	0.2	1.1	0.3
Mesolithic	North	N	33	8	8	8	8	25	25
		Mean	155.4	158.0	84.8	73.2	53.7	59.0	24.4
		SE	0.9	2.1	1.0	1.4	0.4	1.2	0.5
	South	N	2	1	1	1	1	1	1
		Mean	152.9	152.5	83.6	68.9	54.8	47.0	20.0
		SE	0.5	–	–	–	–	–	–
Late Upper Paleolithic	North	N	1					1	1
		Mean	156.9					56.6	23.0
		SE	–					–	–
	South	N	7	5	5	5	5	6	6
		Mean	155.8	157.1	85.3	71.8	54.3	56.8	23.0
		SE	1.7	1.4	0.9	0.9	0.4	2.6	0.9
Early Upper Paleolithic	North	N	3	3	3	3	3	3	3
		Mean	157.9	157.9	84.0	73.9	53.2	59.5	23.8
		SE	0.6	0.6	1.1	0.5	0.5	4.6	1.7
	South	N	7	3	3	3	3	6	6
		Mean	166.3	168.7	88.6	80.1	52.5	62.7	22.6
		SE	2.3	3.1	1.3	2.0	0.4	1.5	0.7
Total	North & South	N	928	456	456	456	456	896	888
		Mean	156.7	156.7	84.5	72.3	53.9	56.8	23.1
		SE	0.2	0.3	0.1	0.2	0.1	0.2	0.1

Appendix 4(b) Female data continued.

			BAB	BIB	RBAB	RBIB	CI	RLegL	BI
Very Recent	North	N	6	10	6	10	20		8
		Mean	36.2	28.1	23.0	18.1	80.0		73.4
		SE	0.5	0.5	0.2	0.3	0.4		0.8
	South	N	26	28	26	28	30	27	30
		Mean	34.7	27.7	23.0	18.3	81.9	21.8	72.3
		SE	0.3	0.4	0.2	0.2	0.4	0.1	0.4
Early Modern	North	N	32	27	31	27	39	31	31
		Mean	35.0	27.3	22.8	17.9	81.6	21.7	72.0
		SE	0.3	0.4	0.2	0.2	0.4	0.1	0.4
	South	N	11	7	11	7	15	11	12
		Mean	35.3	29.1	23.2	18.9	81.3	22.0	73.1
		SE	0.3	1.0	0.2	0.4	0.8	0.2	0.7
Late Medieval	North	N	129	152	129	152	159	108	145
		Mean	35.5	28.1	22.5	17.8	80.7	21.6	73.7
		SE	0.2	0.2	0.1	0.1	0.2	0.1	0.2
	South	N	39	39	38	38	47	35	46
		Mean	35.7	28.8	22.7	18.1	82.3	22.1	73.6
		SE	0.3	0.3	0.1	0.2	0.3	0.1	0.4
Early Medieval	North	N	49	45	49	45	69	25	47
		Mean	35.9	28.2	22.7	17.8	82.3	22.1	75.3
		SE	0.2	0.2	0.1	0.1	0.3	0.1	0.3
	South	N	31	28	31	28	45	31	35
		Mean	35.4	27.9	22.8	17.8	82.6	22.3	75.6
		SE	0.3	0.4	0.1	0.3	0.4	0.1	0.5
Iron/ Roman	North	N	52	73	52	73	100	75	79
		Mean	36.5	28.2	23.1	17.9	81.1	21.7	74.1
		SE	0.3	0.2	0.2	0.1	0.2	0.1	0.3
	South	N	16	2	16	2	25	11	12
		Mean	34.4	28.6	22.0	18.1	80.4884	22.0326	71.8411
		SE	0.5	1.2	0.4	0.1	0.56868	0.27378	1.00170

(Continued)

Appendix 4(b) (Continued)

			BAB	BIB	RBAB	RBIB	CI	RLegL	BI
Bronze	North	N	34	29	33	29	66	11	54
		Mean	35.7	28.8	22.7	18.4	82.6	21.8	75.7
		SE	0.3	0.3	0.2	0.2	0.3	0.2	0.3
	South	N	22	11	21	11	27	21	24
		Mean	35.4	28.0	23.1	18.0	82.1	22.2	76.2
		SE	0.3	0.5	0.2	0.3	0.5	0.2	0.5
Neolithic	North	N	42	59	41	58	80	38	56
		Mean	35.4	27.6	22.6	17.7	82.4	21.4	75.4
		SE	0.3	0.2	0.1	0.2	0.3	0.1	0.2
	South	N	15	15	15	15	18	12	15
		Mean	35.3	26.0	23.0	17.0	81.2	21.8	77.8
		SE	0.6	0.5	0.2	0.3	0.4	0.1	0.5
Mesolithic	North	N	10	8	10	8	22	8	10
		Mean	36.5	27.6	22.9	17.7	82.8	21.9	74.1
		SE	0.5	0.7	0.2	0.6	0.6	0.2	0.5
	South	N	1		1		1	1	1
		Mean	33.1		21.6		82.9	21.5	72.4
		SE	–		–		–	–	–
Late Upper Paleolithic	North	N		1		1			
		Mean		27.7		17.6			
		SE		–		–			
	South	N	3	3	3	3	6	5	4
		Mean	35.2	27.7	22.4	17.6	84.6	21.9	78.4
		SE	1.3	1.6	0.9	1.1	1.4	0.3	1.7
Early Upper Paleolithic	North	N	2	1	2	1	3	3	3
		Mean	39.4	28.0	24.9	17.8	84.6	22.7	77.9
		SE	0.7	–	0.6	–	0.7	0.2	0.7
	South	N	4	2	4	2	5	3	4
		Mean	37.7	28.8	22.3	16.8	85.4	23.0	79.3
		SE	1.2	0.4	0.5	0.4	1.3	0.3	0.9
Total	North & South	N	524	540	519	538	777	456	616
		Mean	35.6	28.0	22.7	17.9	81.7	21.8	74.4
		SE	0.1	0.1	0.0	0.0	0.1	0.0	0.1

Appendix 5

Appendix 5 Z-scores of variables computed using sex-specific mean and standard deviation values. Variable abbreviations are as in Table 2.2 in Chapter 2.

			STA	ASTA	LSHT	LlimbL	LRSHT	BM	BMI
Very Recent	North	Mean	0.1373					0.0237	−0.0686
		N	96					93	93
	South	Mean	−0.7029	−0.6111	−0.4148	−0.5880	0.2809	−0.6019	−0.1313
		N	61	58	58	58	58	61	61
Early Modern	North	Mean	−0.2820	−0.3638	−0.2361	−0.3640	0.4135	−0.3250	−0.1690
		N	92	61	61	61	61	90	90
	South	Mean	−0.6877	−0.7174	−0.8028	−0.4944	−0.1335	−0.3697	0.1659
		N	58	44	44	44	44	54	54
Late Medieval	North	Mean	0.2114	0.2855	0.2446	0.2399	−0.0381	0.1635	0.0213
		N	418	243	243	243	243	412	409
	South	Mean	0.0831	0.2361	−0.0413	0.3686	−0.4728	0.2982	0.2823
		N	113	76	76	76	76	110	110
Early Medieval	North	Mean	0.1960	0.3407	0.2584	0.3284	−0.3699	0.2176	0.1306
		N	195	68	68	68	68	195	194
	South	Mean	−0.0951	−0.0536	−0.3337	0.1584	−0.4244	−0.0824	−0.0065
		N	117	66	66	66	66	112	111
Iron/ Roman	North	Mean	0.0841	0.0508	0.1654	−0.0435	0.1936	0.2541	0.2657
		N	234	140	140	140	140	230	229
	South	Mean	−0.4372	−0.2279	−0.2099	−0.1893	−0.1796	−0.6872	−0.5773
		N	64	27	27	27	27	57	57
Bronze	North	Mean	0.1380	0.2361	0.1906	0.2112	−0.0594	−0.0200	−0.1298
		N	188	43	43	43	43	179	175
	South	Mean	−0.4116	−0.3955	−0.5186	−0.1985	−0.1073	−0.3599	−0.0249
		N	66	40	40	40	40	65	64
Neolithic	North	Mean	0.0388	−0.1590	0.1707	−0.3523	0.5358	−0.0738	−0.1062
		N	250	111	111	111	111	255	246
	South	Mean	−0.5818	−0.6085	−0.6059	−0.4648	−0.0098	−0.9287	−0.8082
		N	43	27	27	27	27	39	39
Mesolithic	North	Mean	−0.1985	0.2274	0.1094	0.2553	−0.2127	0.1693	0.3675
		N	72	25	25	25	25	56	55

	South	Mean	−0.9046	−0.1673	0.1633	−0.3478	0.3544	−0.9181	−0.6090
		N	16	4	4	4	4	11	11
Late Upper Paleolithic	North	Mean	0.1468	0.1961	0.1143	0.2080	−0.1155	−0.1115	−0.2711
		N	5	3	3	3	3	5	5
	South	Mean	−0.2900	−0.1135	0.1289	−0.2582	0.3990	0.0208	0.2743
		N	19	13	13	13	13	18	18
Early Upper Paleolithic	North	Mean	0.6585	0.3273	−0.0886	0.5405	−0.6592	0.2535	−0.2111
		N	12	7	7	7	7	11	11
	South	Mean	1.6290	2.0894	1.3744	2.0889	−1.1771	0.9288	−0.3752
		N	13	7	7	7	7	12	12
Total	North and South	Mean	0.0000	0.0000	0.0000	0.0000	0.0000	0.0000	0.0000
		N	2132	1063	1063	1063	1063	2065	2044

Appendix 5 (continued) Z-scores of variables computed using sex-specific mean and standard deviation values. Variable abbreviations are as in Table 2.2 in Chapter 2.

			BAB	BIB	RBAB	RLBIB	CI	RlegL	BI
Very Recent	North	Mean	0.2871	−0.1089	0.2795	−0.1062	−0.4557		−0.5501
		N	26	45	26	45	65		43
	South	Mean	−0.3391	−0.3861	0.2256	0.0201	0.0938	−0.1412	−0.6312
		N	54	57	54	57	61	58	61
Early Modern	North	Mean	−0.2910	−0.4573	−0.0225	−0.1770	−0.1218	−0.2095	−0.8789
		N	62	61	61	61	81	61	69
	South	Mean	−0.1730	−0.3360	0.4634	0.0639	−0.1678	0.0336	−0.2580
		N	45	30	45	30	56	44	48
Late Medieval	North	Mean	−0.0044	0.0298	−0.1897	−0.1013	−0.4265	−0.1351	−0.3194
		N	280	314	278	312	350	243	301
	South	Mean	0.1650	0.4060	0.0244	0.3276	0.1810	0.4805	−0.2916
		N	84	85	82	84	101	76	99
Early Medieval	North	Mean	0.0961	0.0938	−0.0861	−0.0634	0.0758	0.1527	0.1885
		N	130	107	127	107	169	68	135
	South	Mean	−0.2722	−0.0826	−0.1779	0.0537	0.1610	0.4616	0.2105
		N	74	59	73	58	101	66	74
Iron/ Roman	North	Mean	0.1967	0.2193	0.1343	0.1686	−0.2645	−0.2537	−0.0191
		N	115	138	115	138	205	140	162
	South	Mean	−0.3509	0.1772	−0.0021	0.4175	−0.4074	−0.2572	−0.8315
		N	35	7	35	7	50	27	23
Bronze	North	Mean	0.1584	0.2955	0.0871	0.3401	0.4126	0.2173	0.6794
		N	91	66	89	66	158	43	126
	South	Mean	−0.1593	0.1534	0.1894	0.2768	0.2243	0.4584	0.6027
		N	42	27	41	27	54	40	52
Neolithic	North	Mean	0.0397	−0.0554	0.0368	−0.0175	0.2349	−0.3945	0.3403
		N	135	130	131	129	217	111	152
	South	Mean	−0.2383	−0.8791	0.2367	−0.5727	0.1364	0.1704	1.2075
		N	26	27	26	27	37	27	31
Mesolithic	North	Mean	0.0834	−0.2578	−0.2324	−0.4057	0.6787	0.2906	0.1502
		N	23	23	23	23	54	25	26
	South	Mean	−0.7354	−0.4969	−0.9973	−0.2979	0.8240	−0.3637	0.3275
		N	3	7	3	7	12	4	3

Late Upper Paleolithic	North	Mean	−0.2935	−0.7330	−0.5256	−0.8676	1.1427	1.0188	1.1641
		N	3	3	3	3	4	3	3
	South	Mean	−0.4657	0.1983	−0.3184	0.5183	0.8911	0.0947	1.1877
		N	12	9	12	9	17	13	8
Early Upper Paleolithic	North	Mean	1.5144	−0.1197	1.1464	−0.6056	0.8263	1.1239	0.8715
		N	9	6	9	6	11	7	11
	South	Mean	1.2654	0.3637	−0.4492	−0.8588	1.1027	1.5637	1.3043
		N	6	6	6	6	10	7	6
Total	North and South	Mean	0.0000	0.0000	0.0000	0.0000	0.0000	0.0000	0.0000
		N	1255	1207	1239	1202	1813	1063	1433

Index

Skeletal Variation and Adaptation in Europeans: Upper Paleolithic to the Twentieth Century, First Edition. Edited by Christopher B. Ruff.

d

u